Studies in Computational Intelligence

Volume 608

Series editor

Janusz Kacprzyk, Polish Academy of Sciences, Warsaw, Poland
e-mail: kacprzyk@ibspan.waw.pl

About this Series

The series "Studies in Computational Intelligence" (SCI) publishes new developments and advances in the various areas of computational intelligence—quickly and with a high quality. The intent is to cover the theory, applications, and design methods of computational intelligence, as embedded in the fields of engineering, computer science, physics and life sciences, as well as the methodologies behind them. The series contains monographs, lecture notes and edited volumes in computational intelligence spanning the areas of neural networks, connectionist systems, genetic algorithms, evolutionary computation, artificial intelligence, cellular automata, self-organizing systems, soft computing, fuzzy systems, and hybrid intelligent systems. Of particular value to both the contributors and the readership are the short publication timeframe and the worldwide distribution, which enable both wide and rapid dissemination of research output.

More information about this series at http://www.springer.com/series/7092

George A. Anastassiou

Intelligent Systems II: Complete Approximation by Neural Network Operators

Springer

George A. Anastassiou
Department of Mathematical Sciences
University of Memphis
Memphis, TN
USA

ISSN 1860-949X ISSN 1860-9503 (electronic)
Studies in Computational Intelligence
ISBN 978-3-319-37053-8 ISBN 978-3-319-20505-2 (eBook)
DOI 10.1007/978-3-319-20505-2

Springer Cham Heidelberg New York Dordrecht London

Softcover reprint of the hardcover 1st edition 2016

Printed on acid-free paper

Springer International Publishing AG Switzerland is part of Springer Science+Business Media
(www.springer.com)

To my wife Koula and my daughters
Angela and Peggy

Preface

This monograph is the continuation and completion of the author's earlier 2011 monograph, "Intelligent Systems: Approximation by Artificial Neural Networks", Springer, Intelligent Systems Reference Library, Volume 19.

In this monograph we present the complete recent work of the last four years of the author in approximation by neural networks. It is the natural outgrowth of his related publications. Chapters are self-contained and can be read independently and several advanced courses can be taught out of this book. An extensive list of references is given per chapter.

The topics covered are from A to Z of this research area, all studied for the first time by the author. We list these topics:

Rate of convergence of basic neural network operators to the unit-univariate case.

Rate of convergence of basic multivariate neural network operators to the unit.

Fractional neural network approximation.

Fractional approximation by Cardaliaguet-Euvrard and squashing neural network operators.

Fractional Voronovskaya type asymptotic expansions for quasi-interpolation neural network operators.

Voronovskaya type asymptotic expansions for multivariate quasi-interpolation neural network operators.

Fractional approximation by normalized bell and squashing type neural network operators.

Fractional Voronovskaya type asymptotic expansions for bell and squashing type neural network operators.

Multivariate Voronovskaya type asymptotic expansions for normalized bell and squashing type neural network operators.

Multivariate fuzzy-random normalized neural network approximation operators.

Fuzzy fractional approximations by fuzzy normalized bell and squashing type neural network operators.

Fuzzy fractional neural network approximation by fuzzy quasi-interpolation operators.

Higher order multivariate fuzzy approximation by basic neural network operators.

High degree multivariate fuzzy approximation by quasi-interpolation neural network operators.

Multivariate fuzzy-random quasi-interpolation neural network approximation operators.

Approximation by Kantorovich and quadrature type quasi-interpolation neural network operators.

Univariate error function-based neural network approximation.

Multivariate error function-based neural network approximations.

Voronovskaya type asymptotic expansions for error function-based quasi-interpolation neural network operators.

Fuzzy fractional error function-based neural network approximation.

High order multivariate fuzzy approximation by neural network operators based on the error function.

Multivariate fuzzy-random error function-based neural network approximation.

Approximation by perturbed neural network operators.

Approximation by multivariate perturbed neural network operators.

Voronovskaya type asymptotic expansions for perturbed neural network operators.

Approximation by fuzzy perturbed neural network operators.

Multivariate fuzzy perturbed neural network operators approximation.

Multivariate fuzzy-random perturbed neural network approximation.

The book's results are expected to find applications in many areas of applied mathematics, computer science and engineering. As such this monograph is suitable for researchers, graduate students, and seminars of the above subjects, also to be in all science and engineering libraries.

The preparation of book took place during 2014–2015 in Memphis, Tennessee, USA.

I would like to thank Prof. Alina Alb Lupas of University of Oradea, Romania, for checking and reading the manuscript.

Memphis, USA
March 2015

George A. Anastassiou

Contents

About the Author

George A. Anastassiou was born in Athens, Greece in 1952. He received his B.Sc. degree in Mathematics from Athens University, Greece in 1975. He received his Diploma in Operations Research from Southampton University, UK in 1976. He also received his M.A. in Mathematics from University of Rochester, USA in 1981. He was awarded his Ph.D. in Mathematics from University of Rochester, USA in 1984. During 1984–1986 he served as a visiting assistant professor at the University of Rhode Island, USA. Since 1986 till now 2015, he is a faculty member at the University of Memphis, USA. He is currently a full Professor of Mathematics since 1994. His research area is "Computational Analysis" in the very broad sense. He has published over 400 research articles in international mathematical journals and over 27 monographs, proceedings and textbooks in well-known publishing houses. Several awards have been awarded to George Anastassiou. In 2007 he received the Honorary Doctoral Degree from University of Oradea, Romania. He is associate editor in over 60 international mathematical journals and editor-in-chief in three journals, most notably in the well-known "Journal of Computational Analysis and Applications".

About the Author

[illegible]

Chapter 1
Rate of Convergence of Basic Neural Network Operators to the Unit-Univariate Case

This chapter deals with the determination of the rate of convergence to the unit of some neural network operators, namely, "the normalized bell and squashing type operators". This is given through the modulus of continuity of the involved function or its derivative and that appears in the right-hand side of the associated Jackson type inequalities. It follows [2].

1.1 Introduction

The Cardaliaguet-Euvrard operators were first introduced and studied extensively in [3], where the authors among many other things proved that these operators converge uniformly on compacta, to the unit over continuous and bounded functions. Our "normalized bell and squashing type operators" (1.1) and (1.18) were motivated and inspired by [3]. The work in [3] is qualitative where the used bell-shaped function is general. However, our work, though greatly motivated by [3], is quantitative and the used bell-shaped and "squashing" functions are of compact support. We produce a series of inequalities giving close upper bounds to the errors in approximating the unit operator by the above neural network induced operators. All involved constants there are well determined. These are mainly pointwise estimates involving the first modulus of continuity of the engaged continuous function or some of its derivatives. This work is a continuation and simplification of our earlier work in [1].

1.2 Convergence with Rates of the Normalized Bell Type Neural Network Operators

We need the following (see [3]).

G.A. Anastassiou, *Intelligent Systems II: Complete Approximation by Neural Network Operators*, Studies in Computational Intelligence 608,
DOI 10.1007/978-3-319-20505-2_1

Definition 1.1 A function $b : \mathbb{R} \rightarrow \mathbb{R}$ is said to be bell-shaped if b belongs to L^1 and its integral is nonzero, if it is nondecreasing on $(-\infty, a)$ and nonincreasing on $[a, +\infty)$, where a belongs to $\mathbb{R}$. In particular $b(x)$ is a nonnegative number and at a b takes a global maximum; it is the center of the bell-shaped function. A bell-shaped function is said to be centered if its center is zero. The function $b(x)$ may have jump discontinuities. In this work we consider only centered bell-shaped functions of compact support $[-T, T]$, $T > 0$.

Example 1.2 (1) $b(x)$ can be the characteristic function over $[-1, 1]$.

(2) $b(x)$ can be the hat function over $[-1, 1]$, i.e.,

$$b(x) = \begin{cases} 1 + x, & -1 \leq x \leq 0, \\ 1 - x, & 0 < x \leq 1 \\ 0, & \text{elsewhere.} \end{cases}$$

Here we consider functions $f : \mathbb{R} \rightarrow \mathbb{R}$ that are either continuous and bounded, or uniformly continuous.

In this chapter we study the pointwise convergence with rates over the real line, to the unit operator, of the "normalized bell type neural network operators",

$$(H_n(f))(x) := \frac{\sum_{k=-n^2}^{n^2} f\left(\frac{k}{n}\right) b\left(n^{1-\alpha}\left(x - \frac{k}{n}\right)\right)}{\sum_{k=-n^2}^{n^2} b\left(n^{1-\alpha}\left(x - \frac{k}{n}\right)\right)}, \tag{1.1}$$

where $0 < \alpha < 1$ and $x \in \mathbb{R}$, $n \in \mathbb{N}$. The terms in the ratio of sums (1.1) can be nonzero iff

$$\left| n^{1-\alpha}\left(x - \frac{k}{n}\right)\right| \leq T, \text{ i.e. } \left| x - \frac{k}{n} \right| \leq \frac{T}{n^{1-\alpha}}$$

iff

$$nx - Tn^{\alpha} \leq k \leq nx + Tn^{\alpha}. \tag{1.2}$$

In order to have the desired order of numbers

$$-n^2 \leq nx - Tn^{\alpha} \leq nx + Tn^{\alpha} \leq n^2, \tag{1.3}$$

it is sufficient enough to assume that

$$n \geq T + |x|. \tag{1.4}$$

When $x \in [-T, T]$ it is enough to assume $n \geq 2T$ which implies (1.3).

Proposition 1.3 (see [1]) *Let $a \leq b$, $a, b \in \mathbb{R}$. Let $card(k)$ (≥ 0) be the maximum number of integers contained in $[a, b]$. Then*

$$\max(0, (b - a) - 1) \leq card(k) \leq (b - a) + 1.$$

Note 1.4 We would like to establish a lower bound on $card\,(k)$ over the interval $[nx - Tn^{\alpha}, nx + Tn^{\alpha}]$. From Proposition 1.3 we get that

$$card\,(k) \geq \max\left(2Tn^{\alpha} - 1, 0\right).$$

We obtain $card\,(k) \geq 1$, if

$$2Tn^{\alpha} - 1 \geq 1 \text{ iff } n \geq T^{-\frac{1}{\alpha}}.$$

So to have the desired order (1.3) and $card\,(k) \geq 1$ over $[nx - Tn^{\alpha}, nx + Tn^{\alpha}]$, we need to consider

$$n \geq \max\left(T + |x|, T^{-\frac{1}{\alpha}}\right). \tag{1.5}$$

Also notice that $card\,(k) \to +\infty$, as $n \to +\infty$.

Denote by $[\cdot]$ the integral part of a number and by $\lceil \cdot \rceil$ its ceiling. Here comes our first main result.

Theorem 1.5 *Let $x \in \mathbb{R}$, $T > 0$ and $n \in \mathbb{N}$ such that $n \geq \max\left(T + |x|, T^{-\frac{1}{\alpha}}\right)$. Then*

$$|(H_n(f))(x) - f(x)| \leq \omega_1\left(f, \frac{T}{n^{1-\alpha}}\right), \tag{1.6}$$

where ω_1 is the first modulus of continuity of f.

Proof Call

$$V(x) := \sum_{k=\lceil nx - Tn^{\alpha}\rceil}^{[nx+Tn^{\alpha}]} b\left(n^{1-\alpha}\left(x - \frac{k}{n}\right)\right).$$

Clearly we obtain

$$\Delta := |(H_n(f))(x) - f(x)|$$

$$= \left|\frac{\sum_{k=\lceil nx-Tn^{\alpha}\rceil}^{[nx+Tn^{\alpha}]} f\left(\frac{k}{n}\right) b\left(n^{1-\alpha}\left(x - \frac{k}{n}\right)\right)}{V(x)} - f(x)\right|.$$

The last comes by the compact support $[-T, T]$ of b and (1.2).

Hence it holds

$$\Delta = \left|\sum_{k=\lceil nx-Tn^{\alpha}\rceil}^{[nx+Tn^{\alpha}]} \frac{\left(f\left(\frac{k}{n}\right) - f(x)\right)}{V(x)} b\left(n^{1-\alpha}\left(x - \frac{k}{n}\right)\right)\right|$$

$$\leq \sum_{k=\lceil nx-Tn^{\alpha}\rceil}^{[nx+Tn^{\alpha}]} \frac{\omega_1\left(f, \left|\frac{k}{n} - x\right|\right)}{V(x)} b\left(n^{1-\alpha}\left(x - \frac{k}{n}\right)\right).$$

Thus

$$|(H_n(f))(x) - f(x)| \leq$$

$$\omega_1\left(f, \frac{T}{n^{1-\alpha}}\right) \frac{\left(\sum_{k=\lceil nx-Tn^{\alpha}\rceil}^{[nx+Tn^{\alpha}]} b\left(n^{1-\alpha}\left(x-\frac{k}{n}\right)\right)\right)}{V(x)} = \omega_1\left(f, \frac{T}{n^{1-\alpha}}\right),$$

proving the claim. ■

Our second main result follows.

Theorem 1.6 *Let $x \in \mathbb{R}$, $T > 0$ and $n \in \mathbb{N}$ such that $n \geq \max\left(T + |x|, T^{-\frac{1}{\alpha}}\right)$. Let $f \in C^N(\mathbb{R})$, $N \in \mathbb{N}$, such that $f^{(N)}$ is a uniformly continuous function or $f^{(N)}$ is continuous and bounded. Then*

$$|(H_n(f))(x) - f(x)| \leq \left(\sum_{j=1}^{N} \frac{\left|f^{(j)}(x)\right| T^j}{n^{j(1-\alpha)} j!}\right) + \tag{1.7}$$

$$\omega_1\left(f^{(N)}, \frac{T}{n^{1-\alpha}}\right) \cdot \frac{T^N}{N! n^{N(1-\alpha)}}.$$

Notice that as $n \to \infty$ we have that R.H.S.(1.7)$\to 0$, therefore L.H.S.(1.7)$\to 0$, i.e., (1.7) gives us with rates the pointwise convergence of $(H_n(f))(x) \to f(x)$, as $n \to +\infty$, $x \in \mathbb{R}$.

Proof Note that b here is of compact support $[-T, T]$ and all assumptions are as earlier. By Taylor's formula we have that

$$f\left(\frac{k}{n}\right) = \sum_{j=0}^{N-1} \frac{f^{(j)}(x)}{j!}\left(\frac{k}{n} - x\right)^j +$$

$$\int_x^{\frac{k}{n}} \left(f^{(N)}(t) - f^{(N)}(x)\right) \frac{\left(\frac{k}{n} - t\right)^{N-1}}{(N-1)!} dt.$$

Call

$$V(x) := \sum_{k=\lceil nx-Tn^{\alpha}\rceil}^{[nx+Tn^{\alpha}]} b\left(n^{1-\alpha}\left(x - \frac{k}{n}\right)\right).$$

Hence

$$\frac{f\left(\frac{k}{n}\right) b\left(n^{1-\alpha}\left(x-\frac{k}{n}\right)\right)}{V(x)} = \sum_{j=0}^{N} \frac{f^{(j)}(x)}{j!}\left(\frac{k}{n} - x\right)^j \frac{b\left(n^{1-\alpha}\left(x-\frac{k}{n}\right)\right)}{V(x)} +$$

$$\frac{b\left(n^{1-\alpha}\left(x-\frac{k}{n}\right)\right)}{V(x)}\int_x^{\frac{k}{n}}\left(f^{(N)}(t)-f^{(N)}(x)\right)\frac{\left(\frac{k}{n}-t\right)^{N-1}}{(N-1)!}dt.$$

Thus

$$(H_n(f))(x)-f(x)=\sum_{k=\lceil nx-Tn^{\alpha}\rceil}^{[nx+Tn^{\alpha}]}\frac{f\left(\frac{k}{n}\right)b\left(n^{1-\alpha}\left(x-\frac{k}{n}\right)\right)}{V(x)}-f(x)$$

$$=\sum_{j=1}^{N}\frac{f^{(j)}(x)}{j!}\left(\sum_{k=\lceil nx-Tn^{\alpha}\rceil}^{[nx+Tn^{\alpha}]}\left(\frac{k}{n}-x\right)^{j}\frac{b\left(n^{1-\alpha}\left(x-\frac{k}{n}\right)\right)}{V(x)}\right)+R,$$

where

$$R:=\sum_{k=\lceil nx-Tn^{\alpha}\rceil}^{[nx+Tn^{\alpha}]}\frac{b\left(n^{1-\alpha}\left(x-\frac{k}{n}\right)\right)}{V(x)}\int_x^{\frac{k}{n}}\left(f^{(N)}(t)-f^{(N)}(x)\right)\frac{\left(\frac{k}{n}-t\right)^{N-1}}{(N-1)!}dt. \tag{1.8}$$

So that

$$|(H_n(f))(x)-f(x)|\leq$$

$$\sum_{j=1}^{N}\frac{\left|f^{(j)}(x)\right|}{j!}\left(\sum_{k=\lceil nx-Tn^{\alpha}\rceil}^{[nx+Tn^{\alpha}]}\frac{T^{j}}{n^{j(1-\alpha)}}\frac{b\left(n^{1-\alpha}\left(x-\frac{k}{n}\right)\right)}{V(x)}\right)+|R|.$$

And hence

$$|(H_n(f))(x)-f(x)|\leq\left(\sum_{j=1}^{N}\frac{T^{j}\left|f^{(j)}(x)\right|}{n^{j(1-\alpha)}j!}\right)+|R|. \tag{1.9}$$

Next we estimate

$$|R|=\left|\sum_{k=\lceil nx-Tn^{\alpha}\rceil}^{[nx+Tn^{\alpha}]}\frac{b\left(n^{1-\alpha}\left(x-\frac{k}{n}\right)\right)}{V(x)}\int_x^{\frac{k}{n}}\left(f^{(N)}(t)-f^{(N)}(x)\right)\frac{\left(\frac{k}{n}-t\right)^{N-1}}{(N-1)!}dt\right|$$

$$\leq\sum_{k=\lceil nx-Tn^{\alpha}\rceil}^{[nx+Tn^{\alpha}]}\frac{b\left(n^{1-\alpha}\left(x-\frac{k}{n}\right)\right)}{V(x)}\left|\int_x^{\frac{k}{n}}\left(f^{(N)}(t)-f^{(N)}(x)\right)\frac{\left(\frac{k}{n}-t\right)^{N-1}}{(N-1)!}dt\right|$$

$$\leq\sum_{k=\lceil nx-Tn^{\alpha}\rceil}^{[nx+Tn^{\alpha}]}\frac{b\left(n^{1-\alpha}\left(x-\frac{k}{n}\right)\right)}{V(x)}\cdot\gamma\leq(*),$$

where

$$\gamma := \left| \int_x^{\frac{k}{n}} | f^{(N)}(t) - f^{(N)}(x) | \frac{|\frac{k}{n} - t|^{N-1}}{(N-1)!} dt \right|, \tag{1.10}$$

and

$$(*) \leq \sum_{k=\lceil nx - Tn^{\alpha} \rceil}^{[nx+Tn^{\alpha}]} \frac{b\left(n^{1-\alpha}\left(x - \frac{k}{n}\right)\right)}{V(x)} \cdot \varphi = \varphi, \tag{1.11}$$

where

$$\varphi := \omega_1\left(f^{(N)}, \frac{T}{n^{1-\alpha}}\right) \frac{T^N}{n! n^{N(1-\alpha)}}. \tag{1.12}$$

The last part of inequality (1.11) comes from the following:

(i) Let $x \leq \frac{k}{n}$, then

$$\gamma = \int_x^{\frac{k}{n}} \left| f^{(N)}(t) - f^{(N)}(x) \right| \frac{\left|\frac{k}{n} - t\right|^{N-1}}{(N-1)!} dt$$

$$\leq \int_x^{\frac{k}{n}} \omega_1\left(f^{(N)}, |t - x|\right) \frac{\left|\frac{k}{n} - t\right|^{N-1}}{(N-1)!} dt$$

$$\leq \omega_1\left(f^{(N)}, \left|x - \frac{k}{n}\right|\right) \int_x^{\frac{k}{n}} \frac{\left(\frac{k}{n} - t\right)^{N-1}}{(N-1)!} dt$$

$$\leq \omega_1\left(f^{(N)}, \frac{T}{n^{1-\alpha}}\right) \frac{\left(\frac{k}{n} - x\right)^N}{N!} \leq \omega_1\left(f^{(N)}, \frac{T}{n^{1-\alpha}}\right) \frac{T^N}{N! n^{N(1-\alpha)}};$$

i.e., when $x \leq \frac{k}{n}$ we get

$$\gamma \leq \omega_1\left(f^{(N)}, \frac{T}{n^{1-\alpha}}\right) \frac{T^N}{N! n^{N(1-\alpha)}}. \tag{1.13}$$

(ii) Let $x \geq \frac{k}{n}$, then

$$\gamma = \left| \int_{\frac{k}{n}}^{x} \left| f^{(N)}(t) - f^{(N)}(x) \right| \frac{\left|t - \frac{k}{n}\right|^{N-1}}{(N-1)!} dt \right| =$$

$$\int_{\frac{k}{n}}^{x} \left| f^{(N)}(t) - f^{(N)}(x) \right| \frac{\left(t - \frac{k}{n}\right)^{N-1}}{(N-1)!} dt \leq$$

$$\int_{\frac{k}{n}}^{x} \omega_1 \left(f^{(N)}, |t - x| \right) \frac{\left(t - \frac{k}{n}\right)^{N-1}}{(N-1)!} dt \leq$$

$$\omega_1 \left(f^{(N)}, \left| x - \frac{k}{n} \right| \right) \int_{\frac{k}{n}}^{x} \frac{\left(t - \frac{k}{n}\right)^{N-1}}{(N-1)!} dt =$$

$$\omega_1 \left(f^{(N)}, \left| x - \frac{k}{n} \right| \right) \frac{\left(x - \frac{k}{n}\right)^{N}}{N!} \leq \omega_1 \left(f^{(N)}, \frac{T}{n^{1-\alpha}} \right) \frac{T^N}{N! n^{N(1-\alpha)}}.$$

Thus in both cases we have

$$\gamma \leq \omega_1 \left(f^{(N)}, \frac{T}{n^{1-\alpha}} \right) \frac{T^N}{N! n^{N(1-\alpha)}}. \tag{1.14}$$

Consequently from (1.11), (1.12) and (1.14) we obtain

$$|R| \leq \omega_1 \left(f^{(N)}, \frac{T}{n^{1-\alpha}} \right) \frac{T^N}{N! n^{N(1-\alpha)}}. \tag{1.15}$$

Finally from (1.15) and (1.9) we conclude inequality (1.7). ■

Corollary 1.7 *Let $b(x)$ be a centered bell-shaped continuous function on $\mathbb{R}$ of compact support $[-T, T]$. Let $x \in \left[-T^*, T^*\right]$, $T^* > 0$, and $n \in \mathbb{N}$ be such that $n \geq \max\left(T + T^*, T^{-\frac{1}{\alpha}}\right)$, $0 < \alpha < 1$. Consider $p \geq 1$. Then*

$$\|H_n(f) - f\|_{p,[-T^*,T^*]} \leq \omega_1 \left(f, \frac{T}{n^{1-\alpha}} \right) \cdot 2^{\frac{1}{p}} \cdot T^{*\frac{1}{p}}. \tag{1.16}$$

From (1.16) we get the L_p convergence of $H_n(f)$ to f with rates.

Corollary 1.8 *Let $b(x)$ be a centered bell-shaped continuous function on $\mathbb{R}$ of compact support $[-T, T]$. Let $x \in \left[-T^*, T^*\right]$, $T^* > 0$, and $n \in \mathbb{N}$ be such that $n \geq \max\left(T + T^*, T^{-\frac{1}{\alpha}}\right)$, $0 < \alpha < 1$. Consider $p \geq 1$. Then*

$$\|H_n(f) - f\|_{p,[-T^*,T^*]} \leq \tag{1.17}$$

$$\left(\sum_{j=1}^{N} \frac{T^j \cdot \left\| f^{(j)} \right\|_{p,[-T^*,T^*]}}{n^{j(1-\alpha)} j!} \right) + \omega_1 \left(f^{(N)}, \frac{T}{n^{1-\alpha}} \right) \frac{2^{\frac{1}{p}} T^N T^{*\frac{1}{p}}}{N! n^{N(1-\alpha)}},$$

where $N \geq 1$.

Here from (1.17) we get again the L_p convergence of $H_n(f)$ to f with rates.

Proof Inequality (1.17) now comes by integration of (1.7) and the properties of the L_p-norm. ∎

1.3 The "Normalized Squashing Type Operators" and Their Convergence to the Unit with Rates

We need

Definition 1.9 Let the nonnegative function $S : \mathbb{R} \to \mathbb{R}$, S has compact support $[-T, T]$, $T > 0$, and is nondecreasing there and it can be continuous only on either $(-\infty, T]$ or $[-T, T]$. S can have jump discontinuities. We call S the "squashing function" (see also [3]).

Let $f : \mathbb{R} \to \mathbb{R}$ be either uniformly continuous or continuous and bounded. For $x \in \mathbb{R}$ we define the "normalized squashing type operator"

$$(K_n(f))(x) := \frac{\sum_{k=-n^2}^{n^2} f\left(\frac{k}{n}\right) \cdot S\left(n^{1-\alpha} \cdot \left(x - \frac{k}{n}\right)\right)}{\sum_{k=-n^2}^{n^2} S\left(n^{1-\alpha} \cdot \left(x - \frac{k}{n}\right)\right)}, \tag{1.18}$$

$0 < \alpha < 1$ and $n \in \mathbb{N} : n \geq \max\left(T + |x|, T^{-\frac{1}{\alpha}}\right)$. It is clear that

$$(K_n(f))(x) = \frac{\sum_{k=\lceil nx - Tn^{\alpha}\rceil}^{[nx + Tn^{\alpha}]} f\left(\frac{k}{n}\right) \cdot S\left(n^{1-\alpha} \cdot \left(x - \frac{k}{n}\right)\right)}{W(x)}, \tag{1.19}$$

where

$$W(x) := \sum_{k=\lceil nx - Tn^{\alpha}\rceil}^{[nx + Tn^{\alpha}]} S\left(n^{1-\alpha} \cdot \left(x - \frac{k}{n}\right)\right).$$

Here we give the pointwise convergence with rates of $(K_n f)(x) \to f(x)$, as $n \to +\infty$, $x \in \mathbb{R}$.

Theorem 1.10 *Under the above terms and assumptions we obtain*

$$|K_n(f)(x) - f(x)| \leq \omega_1\left(f, \frac{T}{n^{1-\alpha}}\right). \tag{1.20}$$

Proof As in Theorem 1.5. ∎

We also give

Theorem 1.11 *Let $x \in \mathbb{R}$, $T > 0$ and $n \in \mathbb{N}$ such that $n \geq \max\left(T + |x|, T^{-\frac{1}{\alpha}}\right)$. Let $f \in C^N(\mathbb{R})$, $N \in \mathbb{N}$, such that $f^{(N)}$ is a uniformly continuous function or $f^{(N)}$ is continuous and bounded. Then*

$$|(K_n(f))(x) - f(x)| \leq \left(\sum_{j=1}^{N} \frac{\left|f^{(j)}(x)\right| T^j}{j! n^{j(1-\alpha)}}\right) + \tag{1.21}$$

$$\omega_1\left(f^{(N)}, \frac{T}{n^{1-\alpha}}\right) \cdot \frac{T^N}{N! n^{N(1-\alpha)}}.$$

So we obtain the pointwise convergence of $K_n(f)$ to f with rates.

Proof As similar to Theorem 1.6 is omitted. ■

Note 1.12 The maps H_n, K_n are positive linear operators reproducing constants, in particular

$$H_n(1) = K_n(1) = 1. \tag{1.22}$$

References

1. G.A. Anastassiou, Rate of convergence of some neural network operators to the unit-univariate case. J. Math. Anal. Appl. **212**, 237–262 (1997)
2. G.A. Anastassiou, Rate of convergence of some neural network operators to the unit-univariate case, revisited. Mat. Vesnik **65**(4), 511–518 (2013)
3. P. Cardaliaguet, G. Euvrard, Approximation of a function and its derivative with a neural network. Neural Netw. **5**, 207–220 (1992)

Chapter 2
Rate of Convergence of Basic Multivariate Neural Network Operators to the Unit

This chapter deals with the determination of the rate of convergence to the unit of some multivariate neural network operators, namely the normalized "bell" and "squashing" type operators. This is given through the multidimensional modulus of continuity of the involved multivariate function or its partial derivatives of specific order that appear in the right-hand side of the associated multivariate Jackson type inequality. It follows [3].

2.1 Introduction

The multivariate Cardaliaguet-Euvrard operators were first introduced and studied thoroughly in [4], where the authors among many other interesting things proved that these multivariate operators converge uniformly on compacta, to the unit over continuous and bounded multivariate functions. Our multivariate normalized "bell" and "squashing" type operators (2.1) and (2.16) were motivated and inspired by the "bell" and "squashing" functions of [4].

The work in [4] is qualitative where the used multivariate bell-shaped function is general. However, though our work is greatly motivated by [4], it is quantitative and the used multivariate "bell-shaped" and "squashing" functions are of compact support.

This paper is the continuation and simplification of [1, 2], in the multidimensional case. We produce a set of multivariate inequalities giving close upper bounds to the errors in approximating the unit operator by the above multidimensional neural network induced operators. All appearing constants there are well determined. These are mainly pointwise estimates involving the first multivariate modulus of continuity of the engaged multivariate continuous function or its partial derivatives of some fixed order.

G.A. Anastassiou, *Intelligent Systems II: Complete Approximation by Neural Network Operators*, Studies in Computational Intelligence 608,
DOI 10.1007/978-3-319-20505-2_2

2.2 Convergence with Rates of Multivariate Neural Network Operators

We need the following (see [4]) definitions.

Definition 2.1 A function $b : \mathbb{R} \to \mathbb{R}$ is said to be bell-shaped if b belongs to L^1 and its integral is nonzero, if it is nondecreasing on $(-\infty, a)$ and nonincreasing on $[a, +\infty)$, where a belongs to $\mathbb{R}$. In particular $b(x)$ is a nonnegative number and at a, b takes a global maximum; it is the center of the bell-shaped function. A bell-shaped function is said to be centered if its center is zero.

Definition 2.2 (*see* [4]) A function $b : \mathbb{R}^d \to \mathbb{R}$ $(d \geq 1)$ is said to be a d-dimensional bell-shaped function if it is integrable and its integral is not zero, and for all $i = 1, \ldots, d$,

$$t \to b(x_1, \ldots, t, \ldots, x_d)$$

is a centered bell-shaped function, where $\overrightarrow{x} := (x_1, \ldots, x_d) \in \mathbb{R}^d$ arbitrary.

Example 2.3 (from [4]) Let b be a centered bell-shaped function over $\mathbb{R}$, then $(x_1, \ldots, x_d) \to b(x_1) \ldots b(x_d)$ is a d-dimensional bell-shaped function.

Assumption 2.4 Here $b(\overrightarrow{x})$ is of compact support $\mathcal{B} := \prod_{i=1}^{d} [-T_i, T_i]$, $T_i > 0$ and it may have jump discontinuities there. Let $f : \mathbb{R}^d \to \mathbb{R}$ be a continuous and bounded function or a uniformly continuous function.

In this chapter, we study the pointwise convergence with rates over $\mathbb{R}^d$, to the unit operator, of the "normalized bell" multivariate neural network operators

$$M_n(f)(\overrightarrow{x}) :=$$

$$\frac{\sum_{k_1=-n^2}^{n^2} \cdots \sum_{k_d=-n^2}^{n^2} f\left(\frac{k_1}{n}, \ldots, \frac{k_d}{n}\right) b\left(n^{1-\alpha}\left(x_1 - \frac{k_1}{n}\right), \ldots, n^{1-\alpha}\left(x_d - \frac{k_d}{n}\right)\right)}{\sum_{k_1=-n^2}^{n^2} \cdots \sum_{k_d=-n^2}^{n^2} b\left(n^{1-\alpha}\left(x_1 - \frac{k_1}{n}\right), \ldots, n^{1-\alpha}\left(x_d - \frac{k_d}{n}\right)\right)}, \tag{2.1}$$

where $0 < \alpha < 1$ and $\overrightarrow{x} := (x_1, \ldots, x_d) \in \mathbb{R}^d$, $n \in \mathbb{N}$. Clearly M_n is a positive linear operator.

The terms in the ratio of multiple sums (2.1) can be nonzero iff simultaneously

$$\left| n^{1-\alpha}\left(x_i - \frac{k_i}{n}\right)\right| \leq T_i, \text{ all } i = 1, \ldots, d,$$

i.e., $\left|x_i - \frac{k_i}{n}\right| \leq \frac{T_i}{n^{1-\alpha}}$, all $i = 1, \ldots, d$, iff

$$nx_i - T_i n^{\alpha} \leq k_i \leq nx_i + T_i n^{\alpha}, \text{ all } i = 1, \ldots, d. \tag{2.2}$$

To have the order

$$-n^2 \leq nx_i - T_i n^\alpha \leq k_i \leq nx_i + T_i n^\alpha \leq n^2, \tag{2.3}$$

we need $n \geq T_i + |x_i|$, all $i = 1, \ldots, d$. So (2.3) is true when we take

$$n \geq \max_{i \in \{1,\ldots,d\}} \left(T_i + |x_i|\right). \tag{2.4}$$

When $\overrightarrow{x} \in \mathcal{B}$ in order to have (2.3) it is enough to assume that $n \geq 2T^*$, where $T^* := \max\{T_1, \ldots, T_d\} > 0$. Consider

$$\widetilde{I}_i := \left[nx_i - T_i n^\alpha, nx_i + T_i n^\alpha\right], i = 1, \ldots, d, \ n \in \mathbb{N}.$$

The length of $\widetilde{I}_i$ is $2T_i n^\alpha$. By Proposition 1 of [1], we get that the cardinality of $k_i \in \mathbb{Z}$ that belong to $\widetilde{I}_i := card\,(k_i) \geq \max\left(2T_i n^\alpha - 1, 0\right)$, any $i \in \{1, \ldots, d\}$. In order to have $card\,(k_i) \geq 1$, we need $2T_i n^\alpha - 1 \geq 1$ iff $n \geq T_i^{-\frac{1}{\alpha}}$, any $i \in \{1, \ldots, d\}$.

Therefore, a sufficient condition in order to obtain the order (2.3) along with the interval $\widetilde{I}_i$ to contain at least one integer for all $i = 1, \ldots, d$ is that

$$n \geq \max_{i \in \{1,\ldots,d\}} \left\{T_i + |x_i|, T_i^{-\frac{1}{\alpha}}\right\}. \tag{2.5}$$

Clearly as $n \to +\infty$ we get that $card\,(k_i) \to +\infty$, all $i = 1, \ldots, d$. Also notice that $card\,(k_i)$ equals to the cardinality of integers in $[\lceil nx_i - T_i n^\alpha \rceil, [nx_i + T_i n^\alpha]]$ for all $i = 1, \ldots, d$. Here, $[\cdot]$ denotes the integral part of the number while $\lceil \cdot \rceil$ denotes its ceiling.

From now on, in this chapter we will assume (2.5). Furthermore it holds

$$\left(M_n(f)\right)\left(\overrightarrow{x}\right) = \frac{\sum_{k_1=\lceil nx_1-T_1n^\alpha\rceil}^{[nx_1+T_1n^\alpha]} \cdots \sum_{k_d=\lceil nx_d-T_dn^\alpha\rceil}^{[nx_d+T_dn^\alpha]} f\left(\frac{k_1}{n}, \ldots, \frac{k_d}{n}\right)}{V\left(\overrightarrow{x}\right)}. \tag{2.6}$$

$$b\left(n^{1-\alpha}\left(x_1 - \frac{k_1}{n}\right), \ldots, n^{1-\alpha}\left(x_d - \frac{k_d}{n}\right)\right)$$

all $\overrightarrow{x} := (x_1, \ldots, x_d) \in \mathbb{R}^d$, where

$$V\left(\overrightarrow{x}\right) :=$$

$$\sum_{k_1=\lceil nx_1-T_1n^\alpha\rceil}^{[nx_1+T_1n^\alpha]} \cdots \sum_{k_d=\lceil nx_d-T_dn^\alpha\rceil}^{[nx_d+T_dn^\alpha]} b\left(n^{1-\alpha}\left(x_1 - \frac{k_1}{n}\right), \ldots, n^{1-\alpha}\left(x_d - \frac{k_d}{n}\right)\right).$$

Denote by $\|\cdot\|_\infty$ the maximum norm on $\mathbb{R}^d$, $d \geq 1$. So if $\left|n^{1-\alpha}\left(x_i - \frac{k_i}{n}\right)\right| \leq T_i$, all $i = 1, \ldots, d$, we get that

$$\left\| \overrightarrow{x} - \frac{\overrightarrow{k}}{n} \right\|_\infty \leq \frac{T^*}{n^{1-\alpha}},$$

where $\overrightarrow{k} := (k_1, \ldots, k_d)$.

Definition 2.5 Let $f : \mathbb{R}^d \to \mathbb{R}$. We call

$$\omega_1(f, h) := \sup_{\substack{\text{all } \overrightarrow{x}, \overrightarrow{y}: \\ \|\overrightarrow{x} - \overrightarrow{y}\|_\infty \leq h}} \left| f\left(\overrightarrow{x}\right) - f\left(\overrightarrow{y}\right) \right|, \tag{2.7}$$

where $h > 0$, the first modulus of continuity of f.

Here is our first main result.

Theorem 2.6 *Let* $\overrightarrow{x} \in \mathbb{R}^d$; *then*

$$\left| (M_n(f))\left(\overrightarrow{x}\right) - f\left(\overrightarrow{x}\right) \right| \leq \omega_1\left(f, \frac{T^*}{n^{1-\alpha}}\right). \tag{2.8}$$

Inequality (2.8) is attained by constant functions.

Inequality (2.8) gives $M_n(f)\left(\overrightarrow{x}\right) \to f\left(\overrightarrow{x}\right)$*, pointwise with rates, as* $n \to +\infty$*, where* $\overrightarrow{x} \in \mathbb{R}^d$, $d \geq 1$.

Proof Next, we estimate

$$\left| (M_n(f))\left(\overrightarrow{x}\right) - f\left(\overrightarrow{x}\right) \right| \overset{(2.6)}{=}$$

$$\left| \sum_{k_1 = \lceil nx_1 - T_1 n^\alpha \rceil}^{[nx_1 + T_1 n^\alpha]} \cdots \sum_{k_d = \lceil nx_d - T_d n^\alpha \rceil}^{[nx_d + T_d n^\alpha]} f\left(\frac{k_1}{n}, \ldots, \frac{k_d}{n}\right) \cdot \right.$$

$$\left. \frac{b\left(n^{1-\alpha}\left(x_1 - \frac{k_1}{n}\right), \ldots, n^{1-\alpha}\left(x_d - \frac{k_d}{n}\right)\right)}{V\left(\overrightarrow{x}\right)} - f\left(\overrightarrow{x}\right) \right| =$$

$$\left| \frac{\sum_{\overrightarrow{k} = \lceil n\overrightarrow{x} - \overrightarrow{T} n^\alpha \rceil}^{[n\overrightarrow{x} + \overrightarrow{T} n^\alpha]} \left(f\left(\frac{\overrightarrow{k}}{n}\right) - f\left(\overrightarrow{x}\right)\right) b\left(n^{1-\alpha}\left(\overrightarrow{x} - \frac{\overrightarrow{k}}{n}\right)\right)}{V\left(\overrightarrow{x}\right)} \right| \leq$$

$$\sum_{\overrightarrow{k}=\lceil n\overrightarrow{x}-\overrightarrow{T}n^{\alpha}\rceil}^{[n\overrightarrow{x}+\overrightarrow{T}n^{\alpha}]} \frac{\left|f\left(\frac{\overrightarrow{k}}{n}\right)-f\left(\overrightarrow{x}\right)\right|}{V\left(\overrightarrow{x}\right)} b\left(n^{1-\alpha}\left(\overrightarrow{x}-\frac{\overrightarrow{k}}{n}\right)\right) \leq$$

$$\sum_{\overrightarrow{k}=\lceil n\overrightarrow{x}-\overrightarrow{T}n^{\alpha}\rceil}^{[n\overrightarrow{x}+\overrightarrow{T}n^{\alpha}]} \frac{\omega_1\left(f,\left\|\overrightarrow{x}-\frac{\overrightarrow{k}}{n}\right\|_{\infty}\right)}{V\left(\overrightarrow{x}\right)} b\left(n^{1-\alpha}\left(\overrightarrow{x}-\frac{\overrightarrow{k}}{n}\right)\right).$$

That is

$$\left|\left(M_n(f)\right)\left(\overrightarrow{x}\right)-f\left(\overrightarrow{x}\right)\right| \leq \frac{\omega_1\left(f,\frac{T^*}{n^{1-\alpha}}\right)}{V\left(\overrightarrow{x}\right)}\cdot$$

$$\sum_{k_1=\lceil nx_1-T_1n^{\alpha}\rceil}^{[nx_1+T_1n^{\alpha}]} \cdots \sum_{k_d=\lceil nx_d-T_dn^{\alpha}\rceil}^{[nx_d+T_dn^{\alpha}]} b\left(n^{1-\alpha}\left(x_1-\frac{k_1}{n}\right),\ldots,n^{1-\alpha}\left(x_d-\frac{k_d}{n}\right)\right)$$

$$=\omega_1\left(f,\frac{T^*}{n^{1-\alpha}}\right), \tag{2.9}$$

proving the claim. ∎

Our second main result follows.

Theorem 2.7 *Let* $\overrightarrow{x} \in \mathbb{R}^d$, $f \in C^N\left(\mathbb{R}^d\right)$, $N \in \mathbb{N}$, *such that all of its partial derivatives* $f_{\widetilde{\alpha}}$ *of order* N, $\widetilde{\alpha} : |\widetilde{\alpha}| = N$, *are uniformly continuous or continuous are bounded. Then,*

$$\left|\left(M_n(f)\right)\left(\overrightarrow{x}\right)-f\left(\overrightarrow{x}\right)\right| \leq \tag{2.10}$$

$$\left\{\sum_{j=1}^{N}\frac{(T^*)^j}{j!n^{j(1-\alpha)}}\left(\left(\sum_{i=1}^{d}\left|\frac{\partial}{\partial x_i}\right|\right)^j f\left(\overrightarrow{x}\right)\right)\right\}+$$

$$\frac{(T^*)^N d^N}{N!n^{N(1-\alpha)}}\cdot \max_{\widetilde{\alpha}:|\widetilde{\alpha}|=N}\omega_1\left(f_{\widetilde{\alpha}},\frac{T^*}{n^{1-\alpha}}\right).$$

Inequality (2.10) is attained by constant functions. Also, (2.10) gives us with rates the pointwise convergence of $M_n(f) \to f$ *over* $\mathbb{R}^d$, *as* $n \to +\infty$.

Proof Set

$$g_{\frac{\overrightarrow{k}}{n}}(t) := f\left(\overrightarrow{x}+t\left(\frac{\overrightarrow{k}}{n}-\overrightarrow{x}\right)\right),\ 0 \leq t \leq 1.$$

Then

$$g_{\frac{\overrightarrow{k}}{n}}^{(j)}(t)=\left[\left(\sum_{i=1}^{d}\left(\frac{k_i}{n}-x_i\right)\frac{\partial}{\partial x_i}\right)^{j}f\right]\left(x_1+t\left(\frac{k_1}{n}-x_1\right),\ldots,x_d+t\left(\frac{k_d}{n}-x_d\right)\right)$$

and $g_{\frac{\overrightarrow{k}}{n}}(0)=f\left(\overrightarrow{x}\right)$. By Taylor's formula, we get

$$f\left(\frac{k_1}{n},\ldots,\frac{k_d}{n}\right)=g_{\frac{\overrightarrow{k}}{n}}(1)=\sum_{j=0}^{N}\frac{g_{\frac{\overrightarrow{k}}{n}}^{(j)}(0)}{j!}+R_N\left(\frac{\overrightarrow{k}}{n},0\right),$$

where

$$R_N\left(\frac{\overrightarrow{k}}{n},0\right)=\int_0^1\left(\int_0^{t_1}\ldots\left(\int_0^{t_{N-1}}\left(g_{\frac{\overrightarrow{k}}{n}}^{(N)}(t_N)-g_{\frac{\overrightarrow{k}}{n}}^{(N)}(0)\right)dt_N\right)\ldots\right)dt_1.$$

Here we denote by

$$f_{\widetilde{\alpha}}:=\frac{\partial^{\widetilde{\alpha}}f}{\partial x^{\widetilde{\alpha}}},\ \widetilde{\alpha}:=(\alpha_1,\ldots,\alpha_d),\alpha_i\in\mathbb{Z}^+,$$

$i=1,\ldots,d$, such that $|\widetilde{\alpha}|:=\sum_{i=1}^{d}\alpha_i=N$. Thus,

$$\frac{f\left(\frac{\overrightarrow{k}}{n}\right)b\left(n^{1-\alpha}\left(\overrightarrow{x}-\frac{\overrightarrow{k}}{n}\right)\right)}{V\left(\overrightarrow{x}\right)}=$$

$$\sum_{j=0}^{N}\frac{g_{\frac{\overrightarrow{k}}{n}}^{(j)}(0)}{j!}\frac{b\left(n^{1-\alpha}\left(\overrightarrow{x}-\frac{\overrightarrow{k}}{n}\right)\right)}{V\left(\overrightarrow{x}\right)}+\frac{b\left(n^{1-\alpha}\left(\overrightarrow{x}-\frac{\overrightarrow{k}}{n}\right)\right)}{V\left(\overrightarrow{x}\right)}\cdot R_N\left(\frac{\overrightarrow{k}}{n},0\right).$$

Therefore

$$\left(M_n(f)\right)\left(\overrightarrow{x}\right)-f\left(\overrightarrow{x}\right)=$$

$$\sum_{\overrightarrow{k}=\left\lceil n\overrightarrow{x}-\overrightarrow{T}n^{\alpha}\right\rceil}^{\left[n\overrightarrow{x}+\overrightarrow{T}n^{\alpha}\right]}\frac{f\left(\frac{\overrightarrow{k}}{n}\right)}{V\left(\overrightarrow{x}\right)}b\left(n^{1-\alpha}\left(\overrightarrow{x}-\frac{\overrightarrow{k}}{n}\right)\right)-f\left(\overrightarrow{x}\right)=$$

$$\sum_{j=1}^{N}\frac{1}{j!}\left(\sum_{\overrightarrow{k}=\left\lceil n\overrightarrow{x}-\overrightarrow{T}n^{\alpha}\right\rceil}^{\left[n\overrightarrow{x}+\overrightarrow{T}n^{\alpha}\right]}g_{\frac{\overrightarrow{k}}{n}}^{(j)}(0)\frac{b\left(n^{1-\alpha}\left(\overrightarrow{x}-\frac{\overrightarrow{k}}{n}\right)\right)}{V\left(\overrightarrow{x}\right)}\right)+R^*,$$

where

$$R^* := \sum_{\overrightarrow{k}=\lceil n\overrightarrow{x}-\overrightarrow{T}n^{\alpha}\rceil}^{[n\overrightarrow{x}+\overrightarrow{T}n^{\alpha}]} \frac{b\left(n^{1-\alpha}\left(\overrightarrow{x}-\frac{\overrightarrow{k}}{n}\right)\right)}{V\left(\overrightarrow{x}\right)} \cdot R_N\left(\frac{\overrightarrow{k}}{n},0\right).$$

Consequently, we obtain

$$\left|\left(M_n\left(f\right)\right)\left(\overrightarrow{x}\right)-f\left(\overrightarrow{x}\right)\right| \leq$$

$$\sum_{j=1}^{N}\frac{1}{j!}\left(\sum_{\overrightarrow{k}=\lceil n\overrightarrow{x}-\overrightarrow{T}n^{\alpha}\rceil}^{[n\overrightarrow{x}+\overrightarrow{T}n^{\alpha}]} \frac{\left|g_{\frac{\overrightarrow{k}}{n}}^{(j)}(0)\right| b\left(n^{1-\alpha}\left(\overrightarrow{x}-\frac{\overrightarrow{k}}{n}\right)\right)}{V\left(\overrightarrow{x}\right)}\right)+\left|R^*\right| =: \Theta.$$

Notice that

$$\left|g_{\frac{\overrightarrow{k}}{n}}^{(j)}(0)\right| \leq \left(\frac{T^*}{n^{1-\alpha}}\right)^j\left(\left(\sum_{i=1}^{d}\left|\frac{\partial}{\partial x_i}\right|\right)^j f\left(\overrightarrow{x}\right)\right)$$

and

$$\Theta \leq \left\{\sum_{j=1}^{N}\frac{1}{j!}\left(\frac{T^*}{n^{1-\alpha}}\right)^j\left(\left(\sum_{i=1}^{d}\left|\frac{\partial}{\partial x_i}\right|\right)^j f\left(\overrightarrow{x}\right)\right)\right\}+\left|R^*\right|. \tag{2.11}$$

That is, by (2.11), we get

$$\left|\left(M_n\left(f\right)\right)\left(\overrightarrow{x}\right)-f\left(\overrightarrow{x}\right)\right| \leq$$

$$\left\{\sum_{j=1}^{N}\frac{(T^*)^j}{j!n^{j(1-\alpha)}}\left(\left(\sum_{i=1}^{d}\left|\frac{\partial}{\partial x_i}\right|\right)^j f\left(\overrightarrow{x}\right)\right)\right\}+\left|R^*\right|. \tag{2.12}$$

Next, we need to estimate $|R^*|$. For that, we observe $(0 \leq t_N \leq 1)$

$$\left|g_{\frac{\overrightarrow{k}}{n}}^{(N)}(t_N)-g_{\frac{\overrightarrow{k}}{n}}^{(N)}(0)\right| =$$

$$\left|\left(\sum_{i=1}^{d}\left(\frac{k_i}{n}-x_i\right)\frac{\partial}{\partial x_i}\right)^N f\left(\overrightarrow{x}+t_N\left(\frac{\overrightarrow{k}}{n}-\overrightarrow{x}\right)\right)-\right.$$

$$\left(\sum_{i=1}^{d}\left(\frac{k_i}{n}-x_i\right)\frac{\partial}{\partial x_i}\right)^N f\left(\overrightarrow{x}\right)\Bigg|$$

$$\leq \frac{(T^*)^N d^N}{n^{N(1-\alpha)}}\cdot \max_{\widetilde{\alpha}:|\widetilde{\alpha}|=N}\omega_1\left(f_{\widetilde{\alpha}},\frac{T^*}{n^{1-\alpha}}\right).$$

Thus,

$$\left|R_N\left(\frac{\overrightarrow{k}}{n},0\right)\right| \leq \int_0^1\left(\int_0^{t_1}\ldots\left(\int_0^{t_{N-1}}\left|g_{\frac{\overrightarrow{k}}{n}}^{(N)}(t_N)-g_{\frac{\overrightarrow{k}}{n}}^{(N)}(0)\right|dt_N\right)\ldots\right)dt_1$$

$$\leq \frac{(T^*)^N d^N}{N!n^{N(1-\alpha)}}\cdot \max_{\widetilde{\alpha}:|\widetilde{\alpha}|=N}\omega_1\left(f_{\widetilde{\alpha}},\frac{T^*}{n^{1-\alpha}}\right).$$

Therefore,

$$|R^*| \leq \sum_{\overrightarrow{k}=\lceil n\overrightarrow{x}-\overrightarrow{T}n^{\alpha}\rceil}^{[n\overrightarrow{x}+\overrightarrow{T}n^{\alpha}]}\frac{b\left(n^{1-\alpha}\left(\overrightarrow{x}-\frac{\overrightarrow{k}}{n}\right)\right)}{V\left(\overrightarrow{x}\right)}\left|R_N\left(\frac{\overrightarrow{k}}{n},0\right)\right|$$

$$\leq \frac{(T^*)^N d^N}{N!n^{N(1-\alpha)}}\cdot \max_{\widetilde{\alpha}:|\widetilde{\alpha}|=N}\omega_1\left(f_{\widetilde{\alpha}},\frac{T^*}{n^{1-\alpha}}\right). \quad (2.13)$$

By (2.12) and (2.13) we get (2.10). ■

Corollary 2.8 *Here, additionally assume that b is continuous on* $\mathbb{R}^d$. *Let*

$$\Gamma := \prod_{i=1}^{d}[-\gamma_i,\gamma_i] \subset \mathbb{R}^d,\ \gamma_i > 0,$$

and take

$$n \geq \max_{i\in\{1,\ldots,d\}}\left(T_i+\gamma_i,T_i^{-\frac{1}{\alpha}}\right).$$

Consider $p \geq 1$. *Then,*

$$\|M_n f - f\|_{p,\Gamma} \leq \omega_1\left(f,\frac{T^*}{n^{1-\alpha}}\right)2^{\frac{d}{p}}\prod_{i=1}^{d}\gamma_i^{\frac{1}{p}}, \quad (2.14)$$

attained by constant functions. From (2.14), we get the L_p *convergence of* $M_n f$ *to* f *with rates.*

Proof By (2.8). ■

Corollary 2.9 *Same assumptions as in Corollary 2.8. Then*

$$\|M_n f - f\|_{p,\Gamma} \leq \left\{ \sum_{j=1}^{N} \frac{(T^*)^j}{j! n^{j(1-\alpha)}} \left\| \left(\sum_{i=1}^{d} \left| \frac{\partial}{\partial x_i} \right| \right)^j f \right\|_{p,\Gamma} \right\} +$$

$$\frac{(T^*)^N d^N}{N! n^{N(1-\alpha)}} \cdot \max_{\widetilde{\alpha}:|\widetilde{\alpha}|=N} \omega_1 \left(f_{\widetilde{\alpha}}, \frac{T^*}{n^{1-\alpha}} \right) 2^{\frac{d}{p}} \prod_{i=1}^{d} \gamma_i^{\frac{1}{p}}, \tag{2.15}$$

attained by constants. Here, from (2.15), we get again the L_p convergence of $M_n(f)$ to f with rates.

Proof By the use of (2.10). ■

2.3 The Multivariate "Normalized Squashing Type Operators" and Their Convergence to the Unit with Rates

We give the following definition

Definition 2.10 Let the nonnegative function $S : \mathbb{R}^d \to \mathbb{R}$, $d \geq 1$, S has compact support $\mathcal{B} := \prod_{i=1}^{d} [-T_i, T_i]$, $T_i > 0$ and is nondecreasing there for each coordinate. S can be continuous only on either $\prod_{i=1}^{d} (-\infty, T_i]$ or $\mathcal{B}$ and can have jump discontinuities. We call S the multivariate "squashing function" (see also [4]).

Example 2.11 Let $\widehat{S}$ as above when $d = 1$. Then,

$$S\left(\overrightarrow{x}\right) := \widehat{S}(x_1) \ldots \widehat{S}(x_d), \ \overrightarrow{x} := (x_1, \ldots, x_d) \in \mathbb{R}^d,$$

is a multivariate "squashing function".

Let $f : \mathbb{R}^d \to \mathbb{R}$ be either uniformly continuous or continuous and bounded function.

For $\overrightarrow{x} \in \mathbb{R}^d$, we define the multivariate "normalized squashing type operator",

$$L_n(f)\left(\overrightarrow{x}\right) :=$$

$$\frac{\sum_{k_1=-n^2}^{n^2} \cdots \sum_{k_d=-n^2}^{n^2} f\left(\frac{k_1}{n}, \ldots, \frac{k_d}{n}\right) S\left(n^{1-\alpha}\left(x_1 - \frac{k_1}{n}\right), \ldots, n^{1-\alpha}\left(x_d - \frac{k_d}{n}\right)\right)}{W\left(\overrightarrow{x}\right)}, \tag{2.16}$$

where $0 < \alpha < 1$ and $n \in \mathbb{N}$:

$$n \geq \max_{i \in \{1,\ldots,d\}} \left\{ T_i + |x_i|, T_i^{-\frac{1}{\alpha}} \right\}, \tag{2.17}$$

and

$$W\left(\overrightarrow{x}\right) := \sum_{k_1=-n^2}^{n^2} \ldots \sum_{k_d=-n^2}^{n^2} S\left(n^{1-\alpha}\left(x_1 - \frac{k_1}{n}\right), \ldots, n^{1-\alpha}\left(x_d - \frac{k_d}{n}\right)\right). \tag{2.18}$$

Obviously L_n is a positive linear operator. It is clear that

$$\left(L_n(f)\right)\left(\overrightarrow{x}\right) = \sum_{\overrightarrow{k}=\left\lceil n\overrightarrow{x}-\overrightarrow{T}n^{\alpha}\right\rceil}^{\left[n\overrightarrow{x}+\overrightarrow{T}n^{\alpha}\right]} \frac{f\left(\frac{\overrightarrow{k}}{n}\right)}{\Phi\left(\overrightarrow{x}\right)} S\left(n^{1-\alpha}\left(\overrightarrow{x} - \frac{\overrightarrow{k}}{n}\right)\right), \tag{2.19}$$

where

$$\Phi\left(\overrightarrow{x}\right) := \sum_{\overrightarrow{k}=\left\lceil n\overrightarrow{x}-\overrightarrow{T}n^{\alpha}\right\rceil}^{\left[n\overrightarrow{x}+\overrightarrow{T}n^{\alpha}\right]} S\left(n^{1-\alpha}\left(\overrightarrow{x} - \frac{\overrightarrow{k}}{n}\right)\right). \tag{2.20}$$

Here, we study the pointwise convergence with rates of $\left(L_n(f)\right)\left(\overrightarrow{x}\right) \to f\left(\overrightarrow{x}\right)$, as $n \to +\infty$, $\overrightarrow{x} \in \mathbb{R}^d$.

This is given by the next result.

Theorem 2.12 *Under the above terms and assumptions, we find that*

$$\left|\left(L_n(f)\right)\left(\overrightarrow{x}\right) - f\left(\overrightarrow{x}\right)\right| \leq \omega_1\left(f, \frac{T^*}{n^{1-\alpha}}\right). \tag{2.21}$$

Inequality (2.21) is attained by constant functions.

Proof Similar to (2.8). ■

We also give

Theorem 2.13 *Let $\overrightarrow{x} \in \mathbb{R}^d$, $f \in C^N\left(\mathbb{R}^d\right)$, $N \in \mathbb{N}$, such that all of its partial derivatives $f_{\widetilde{\alpha}}$ of order N, $\widetilde{\alpha} : |\widetilde{\alpha}| = N$, are uniformly continuous or continuous are bounded. Then,*

$$\left|\left(L_n(f)\right)\left(\overrightarrow{x}\right) - f\left(\overrightarrow{x}\right)\right| \leq \tag{2.22}$$

$$\left\{ \sum_{j=1}^{N} \frac{(T^*)^j}{j! n^{j(1-\alpha)}} \left(\left(\sum_{i=1}^{d} \left| \frac{\partial}{\partial x_i} \right| \right)^j f\left(\overrightarrow{x}\right) \right) \right\} +$$

$$\frac{(T^*)^N d^N}{N! n^{N(1-\alpha)}} \cdot \max_{\widetilde{\alpha}:|\widetilde{\alpha}|=N} \omega_1 \left(f_{\widetilde{\alpha}}, \frac{T^*}{n^{1-\alpha}} \right).$$

Inequality (2.22) is attained by constant functions. Also, (2.22) gives us with rates the pointwise convergence of $L_n(f) \to f$ *over* $\mathbb{R}^d$*, as* $n \to +\infty$.

Proof Similar to (2.10). ■

Note 2.14 We see that

$$M_n(1) = L_n(1) = 1.$$

References

1. G.A. Anastassiou, Rate of convergence of some neural network operators to the unit-univariate case. J. Math. Anal. Appl. **212**, 237–262 (1997)
2. G.A. Anastassiou, Rate of convergence of some multivariate neural network operators to the unit. Comput. Math. Appl. **40**(1), 1–19 (2000)
3. G.A. Anastassiou, Rate of convergence of some multivariate neural network operators to the unit, revisited. J. Comput. Anal. Appl. **15**(7), 1300–1309 (2013)
4. P. Cardaliaguet, G. Euvrard, Approximation of a function and its derivative with a neural network. Neural Netw. **5**, 207–220 (1992)

Chapter 3
Fractional Neural Network Operators Approximation

Here we study the univariate fractional quantitative approximation of real valued functions on a compact interval by quasi-interpolation sigmoidal and hyperbolic tangent neural network operators. These approximations are derived by establishing Jackson type inequalities involving the moduli of continuity of the right and left Caputo fractional derivatives of the engaged function. The approximations are pointwise and with respect to the uniform norm. The related feed-forward neural networks are with one hidden layer. Our fractional approximation results into higher order converges better than the ordinary ones. It follows [12].

3.1 Introduction

The author in [1, 2], see Chaps. 2–5, was the first to establish neural network approximations to continuous functions with rates by very specifically defined neural network operators of Cardaliaguet-Euvrard and "Squashing" types, by employing the modulus of continuity of the engaged function or its high order derivative, and producing very tight Jackson type inequalities. He treats there both the univariate and multivariate cases. The defining these operators "bell-shaped" and "squashing" function are assumed to be of compact support. Also in [2] he gives the Nth order asymptotic expansion for the error of weak approximation of these two operators to a special natural class of smooth functions, see Chaps. 4 and 5 there.

The author inspired by [13], continued his studies on neural networks approximation by introducing and using the proper quasi-interpolation operators of sigmoidal and hyperbolic tangent type which resulted into [6, 8–11], by treating both the univariate and multivariate cases.

Continuation of the author's last work is this chapter where neural network approximation is taken at the fractional level resulting into higher rates of approximation. We involve the right and left Caputo fractional derivatives of the function under

G.A. Anastassiou, *Intelligent Systems II: Complete Approximation by Neural Network Operators*, Studies in Computational Intelligence 608,
DOI 10.1007/978-3-319-20505-2_3

approximation and we establish tight Jackson type inequalities. An extensive background is given on fractional calculus and neural networks, all needed to expose our work. Applications are presented at the end.

Feed-forward neural networks (FNNs) with one hidden layer, the only type of networks we deal with in this chapter, are mathematically expressed as

$$N_n(x) = \sum_{j=0}^{n} c_j \sigma\left(\left\langle a_j \cdot x\right\rangle + b_j\right), \quad x \in \mathbb{R}^s, \quad s \in \mathbb{N},$$

where for $0 \leq j \leq n$, $b_j \in \mathbb{R}$ are the thresholds, $a_j \in \mathbb{R}^s$ are the connection weights, $c_j \in \mathbb{R}$ are the coefficients, $\left\langle a_j \cdot x\right\rangle$ is the inner product of a_j and x, and σ is the activation function of the network. About neural networks in general, see [17–19].

3.2 Background

We need

Definition 3.1 Let $f \in C([a, b])$ and $0 \leq h \leq b - a$. The first modulus of continuity of f at h is given by

$$\omega_1(f, h) = \sup\{|f(x) - f(y); x, y \in [a, b], |x - y| \leq h|\} \tag{3.1}$$

If $h > b - a$, then we define

$$\omega_1(f, h) = \omega_1(f, b - a). \tag{3.2}$$

Notice here that

$$\lim_{h \to 0} \omega_1(f, h) = 0.$$

We also need

Definition 3.2 Let $\nu \geq 0$, $n = \lceil \nu \rceil$ ($\lceil \cdot \rceil$ is the ceiling of the number), $f \in AC^n([a, b])$ (space of functions f with $f^{(n-1)} \in AC([a, b])$, absolutely continuous functions). We call left Caputo fractional derivative (see [14], pp. 49–52, [16, 20]) the function

$$D_{*a}^{\nu} f(x) = \frac{1}{\Gamma(n - \nu)} \int_a^x (x - t)^{n-\nu-1} f^{(n)}(t)\, dt, \tag{3.3}$$

$\forall\, x \in [a, b]$, where Γ is the gamma function $\Gamma(\nu) = \int_0^{\infty} e^{-t} t^{\nu-1} dt$, $\nu > 0$.

Notice $D_{*a}^{\nu} f \in L_1([a, b])$ and $D_{*a}^{\nu} f$ exists a.e. on $[a, b]$.

We set $D_{*a}^{0} f(x) = f(x)$, $\forall\, x \in [a, b]$.

Lemma 3.3 ([5]) *Let* $\nu > 0$, $\nu \notin \mathbb{N}$, $n = \lceil \nu \rceil$, $f \in C^{n-1}([a,b])$ *and* $f^{(n)} \in L_{\infty}([a,b])$. *Then* $D_{*a}^{\nu} f(a) = 0$.

Definition 3.4 (*see also* [3, 15, 16]) Let $f \in AC^m([a,b])$, $m = \lceil \alpha \rceil$, $\alpha > 0$. The right Caputo fractional derivative of order $\alpha > 0$ is given by

$$D_{b-}^{\alpha} f(x) = \frac{(-1)^m}{\Gamma(m-\alpha)} \int_x^b (\zeta - x)^{m-\alpha-1} f^{(m)}(\zeta) \, d\zeta, \tag{3.4}$$

$\forall\, x \in [a,b]$. We set $D_{b-}^{0} f(x) = f(x)$. Notice $D_{b-}^{\alpha} f \in L_1([a,b])$ and $D_{b-}^{\alpha} f$ exists a.e. on $[a,b]$.

Lemma 3.5 ([5]) *Let* $f \in C^{m-1}([a,b])$, $f^{(m)} \in L_{\infty}([a,b])$, $m = \lceil \alpha \rceil$, $\alpha > 0$. *Then* $D_{b-}^{\alpha} f(b) = 0$.

Convention 3.6 *We assume that*

$$D_{*x_0}^{\alpha} f(x) = 0, \text{ for } x < x_0, \tag{3.5}$$

and

$$D_{x_0-}^{\alpha} f(x) = 0, \text{ for } x > x_0, \tag{3.6}$$

for all $x, x_0 \in (a,b]$.

We mention

Proposition 3.7 ([5]) *Let* $f \in C^n([a,b])$, $n = \lceil \nu \rceil$, $\nu > 0$. *Then* $D_{*a}^{\nu} f(x)$ *is continuous in* $x \in [a,b]$.

Also we have

Proposition 3.8 ([5]) *Let* $f \in C^m([a,b])$, $m = \lceil \alpha \rceil$, $\alpha > 0$. *Then* $D_{b-}^{\alpha} f(x)$ *is continuous in* $x \in [a,b]$.

We further mention

Proposition 3.9 ([5]) *Let* $f \in C^{m-1}([a,b])$, $f^{(m)} \in L_{\infty}([a,b])$, $m = \lceil \alpha \rceil$, $\alpha > 0$ *and*

$$D_{*x_0}^{\alpha} f(x) = \frac{1}{\Gamma(m-\alpha)} \int_{x_0}^{x} (x-t)^{m-\alpha-1} f^{(m)}(t) \, dt, \tag{3.7}$$

for all $x, x_0 \in [a,b] : x \geq x_0$.

Then $D_{*x_0}^{\alpha} f(x)$ *is continuous in* x_0.

Proposition 3.10 ([5]) *Let* $f \in C^{m-1}([a,b])$, $f^{(m)} \in L_{\infty}([a,b])$, $m = \lceil \alpha \rceil$, $\alpha > 0$ *and*

$$D_{x_0-}^{\alpha} f(x) = \frac{(-1)^m}{\Gamma(m-\alpha)} \int_x^{x_0} (\zeta - x)^{m-\alpha-1} f^{(m)}(\zeta)\, d\zeta, \tag{3.8}$$

for all $x, x_0 \in [a, b] : x \leq x_0$.

Then $D_{x_0-}^{\alpha} f(x)$ *is continuous in* x_0.

We need

Proposition 3.11 ([5]) *Let* $g \in C([a, b])$, $0 < c < 1$, $x, x_0 \in [a, b]$. *Define*

$$L(x, x_0) = \int_{x_0}^{x} (x - t)^{c-1} g(t)\, dt, \text{ for } x \geq x_0, \tag{3.9}$$

and $L(x, x_0) = 0$, *for* $x < x_0$.

Then L *is jointly continuous in* (x, x_0) *on* $[a, b]^2$.

We mention

Proposition 3.12 ([5]) *Let* $g \in C([a, b])$, $0 < c < 1$, $x, x_0\, [a, b]$. *Define*

$$K(x, x_0) = \int_{x}^{x_0} (\zeta - x)^{c-1} g(\zeta)\, d\zeta, \text{ for } x \leq x_0, \tag{3.10}$$

and $K(x, x_0) = 0$, *for* $x > x_0$.

Then $K(x, x_0)$ *is jointly continuous from* $[a, b]^2$ *into* $\mathbb{R}$.

Based on Propositions 3.11, 3.12 we derive

Corollary 3.13 ([5]) *Let* $f \in C^m([a, b])$, $m = \lceil \alpha \rceil$, $\alpha > 0$, $x, x_0 \in [a, b]$. *Then* $D_{*x_0}^{\alpha} f(x)$, $D_{x_0-}^{\alpha} f(x)$ *are jointly continuous functions in* (x, x_0) *from* $[a, b]^2$ *into* $\mathbb{R}$.

We need

Theorem 3.14 ([5]) *Let* $f : [a, b]^2 \to \mathbb{R}$ *be jointly continuous. Consider*

$$G(x) = \omega_1(f(\cdot, x), \delta, [x, b]), \tag{3.11}$$

$\delta > 0$, $x \in [a, b]$.

Then G *is continuous in* $x \in [a, b]$.

Also it holds

Theorem 3.15 ([5]) *Let* $f : [a, b]^2 \to \mathbb{R}$ *be jointly continuous. Then*

$$H(x) = \omega_1(f(\cdot, x), \delta, [a, x]), \tag{3.12}$$

$x \in [a, b]$, *is continuous in* $x \in [a, b]$, $\delta > 0$.

We make

Remark 3.16 ([5]) Let $f \in C^{n-1}([a,b])$, $f^{(n)} \in L_{\infty}([a,b])$, $n = \lceil \nu \rceil$, $\nu > 0$, $\nu \notin \mathbb{N}$. Then we have

$$\left| D_{*a}^{\nu} f(x) \right| \leq \frac{\left\| f^{(n)} \right\|_{\infty}}{\Gamma(n-\nu+1)} (x-a)^{n-\nu}, \ \forall x \in [a,b]. \tag{3.13}$$

Thus we observe

$$\omega_1\left(D_{*a}^{\nu} f, \delta\right) = \sup_{\substack{x,y \in [a,b] \\ |x-y| \leq \delta}} \left| D_{*a}^{\nu} f(x) - D_{*a}^{\nu} f(y) \right|$$

$$\leq \sup_{\substack{x,y \in [a,b] \\ |x-y| \leq \delta}} \left(\frac{\left\| f^{(n)} \right\|_{\infty}}{\Gamma(n-\nu+1)} (x-a)^{n-\nu} + \frac{\left\| f^{(n)} \right\|_{\infty}}{\Gamma(n-\nu+1)} (y-a)^{n-\nu} \right)$$

$$\leq \frac{2 \left\| f^{(n)} \right\|_{\infty}}{\Gamma(n-\nu+1)} (b-a)^{n-\nu}.$$

Consequently

$$\omega_1\left(D_{*a}^{\nu} f, \delta\right) \leq \frac{2 \left\| f^{(n)} \right\|_{\infty}}{\Gamma(n-\nu+1)} (b-a)^{n-\nu}. \tag{3.14}$$

Similarly, let $f \in C^{m-1}([a,b])$, $f^{(m)} \in L_{\infty}([a,b])$, $m = \lceil \alpha \rceil$, $\alpha > 0$, $\alpha \notin \mathbb{N}$, then

$$\omega_1\left(D_{b-}^{\alpha} f, \delta\right) \leq \frac{2 \left\| f^{(m)} \right\|_{\infty}}{\Gamma(m-\alpha+1)} (b-a)^{m-\alpha}. \tag{3.15}$$

So for $f \in C^{m-1}([a,b])$, $f^{(m)} \in L_{\infty}([a,b])$, $m = \lceil \alpha \rceil$, $\alpha > 0$, $\alpha \notin \mathbb{N}$, we find

$$s_1(\delta) := \sup_{x_0 \in [a,b]} \omega_1\left(D_{*x_0}^{\alpha} f, \delta\right)_{[x_0,b]} \leq \frac{2 \left\| f^{(m)} \right\|_{\infty}}{\Gamma(m-\alpha+1)} (b-a)^{m-\alpha}, \tag{3.16}$$

and

$$s_2(\delta) := \sup_{x_0 \in [a,b]} \omega_1\left(D_{x_0-}^{\alpha} f, \delta\right)_{[a,x_0]} \leq \frac{2 \left\| f^{(m)} \right\|_{\infty}}{\Gamma(m-\alpha+1)} (b-a)^{m-\alpha}. \tag{3.17}$$

By Proposition 15.114, p. 388 of [4], we get here that $D_{*x_0}^{\alpha} f \in C([x_0,b])$, and by [7] we obtain that $D_{x_0-}^{\alpha} f \in C([a,x_0])$.

We consider here the sigmoidal function of logarithmic type

$$s(x) = \frac{1}{1+e^{-x}}, \quad x \in \mathbb{R}.$$

It has the properties $\lim_{x \to +\infty} s(x) = 1$ and $\lim_{x \to -\infty} s(x) = 0$.

This function plays the role of an activation function in the hidden layer of neural networks.

As in [13], we consider

$$\Phi(x) := \frac{1}{2}(s(x+1) - s(x-1)), \quad x \in \mathbb{R}. \tag{3.18}$$

We notice the following properties:

(i) $\Phi(x) > 0, \ \forall\, x \in \mathbb{R}$,
(ii) $\sum_{k=-\infty}^{\infty} \Phi(x-k) = 1, \ \forall\, x \in \mathbb{R}$,
(iii) $\sum_{k=-\infty}^{\infty} \Phi(nx-k) = 1, \ \forall\, x \in \mathbb{R}; n \in \mathbb{N}$,
(iv) $\int_{-\infty}^{\infty} \Phi(x)\, dx = 1$,
(v) Φ is a density function,
(vi) Φ is even: $\Phi(-x) = \Phi(x)$, $x \geq 0$.

We see that ([13])

$$\Phi(x) = \left(\frac{e^2-1}{2e}\right) \frac{e^{-x}}{\left(1+e^{-x-1}\right)\left(1+e^{-x+1}\right)} = \tag{3.19}$$
$$\left(\frac{e^2-1}{2e^2}\right) \frac{1}{\left(1+e^{x-1}\right)\left(1+e^{-x-1}\right)}.$$

(vii) By [13] Φ is decreasing on $\mathbb{R}_+$, and increasing on $\mathbb{R}_-$.
(viii) By [11] for $n \in \mathbb{N}$, $0 < \beta < 1$, we get

$$\sum_{\substack{k=-\infty \\ : |nx-k| > n^{1-\beta}}}^{\infty} \Phi(nx-k) < \left(\frac{e^2-1}{2}\right) e^{-n^{(1-\beta)}} = 3.1992 e^{-n^{(1-\beta)}}. \tag{3.20}$$

Denote by $\lceil \cdot \rceil$ the ceiling of a number, and by $\lfloor \cdot \rfloor$ the integral part of a number. Consider $x \in [a, b] \subset \mathbb{R}$ and $n \in \mathbb{N}$ such that $\lceil na \rceil \leq \lfloor nb \rfloor$.

(ix) By [11] it holds

$$\frac{1}{\sum_{k=\lceil na \rceil}^{\lfloor nb \rfloor} \Phi(nx-k)} < \frac{1}{\Phi(1)} = 5.250312578, \ \forall x \in [a, b]. \tag{3.21}$$

(x) By [11] it holds $\lim_{n\to\infty} \sum_{k=\lceil na\rceil}^{\lfloor nb\rfloor} \Phi(nx-k) \neq 1$, for at least some $x \in [a,b]$.

Let $f \in C([a,b])$ and $n \in \mathbb{N}$ such that $\lceil na\rceil \leq \lfloor nb\rfloor$.

We study further (see also [11]) the positive linear neural network operator

$$G_n(f,x) := \frac{\sum_{k=\lceil na\rceil}^{\lfloor nb\rfloor} f\left(\frac{k}{n}\right)\Phi(nx-k)}{\sum_{k=\lceil na\rceil}^{\lfloor nb\rfloor} \Phi(nx-k)}, \quad x \in [a,b]. \tag{3.22}$$

For large enough n we always obtain $\lceil na\rceil \leq \lfloor nb\rfloor$. Also $a \leq \frac{k}{n} \leq b$, iff $\lceil na\rceil \leq k \leq \lfloor nb\rfloor$.

We study here further at fractional level the pointwise and uniform convergence of $G_n(f,x)$ to $f(x)$ with rates.

For convenience we call

$$G_n^*(f,x) := \sum_{k=\lceil na\rceil}^{\lfloor nb\rfloor} f\left(\frac{k}{n}\right)\Phi(nx-k), \tag{3.23}$$

that is

$$G_n(f,x) := \frac{G_n^*(f,x)}{\sum_{k=\lceil na\rceil}^{\lfloor nb\rfloor} \Phi(nx-k)}. \tag{3.24}$$

Thus,

$$G_n(f,x) - f(x) = \frac{G_n^*(f,x)}{\sum_{k=\lceil na\rceil}^{\lfloor nb\rfloor} \Phi(nx-k)} - f(x)$$

$$= \frac{G_n^*(f,x) - f(x)\sum_{k=\lceil na\rceil}^{\lfloor nb\rfloor} \Phi(nx-k)}{\sum_{k=\lceil na\rceil}^{\lfloor nb\rfloor} \Phi(nx-k)}. \tag{3.25}$$

Consequently we derive

$$|G_n(f,x) - f(x)| \leq \frac{1}{\Phi(1)} \left| G_n^*(f,x) - f(x) \sum_{k=\lceil na\rceil}^{\lfloor nb\rfloor} \Phi(nx-k) \right|. \tag{3.26}$$

That is

$$|G_n(f,x) - f(x)| \leq (5.250312578) \left| \sum_{k=\lceil na\rceil}^{\lfloor nb\rfloor} \left(f\left(\frac{k}{n}\right) - f(x) \right) \Phi(nx-k) \right|. \tag{3.27}$$

We will estimate the right hand side of (3.27) involving the right and left Caputo fractional derivatives of f.

We also consider here the hyperbolic tangent function $\tanh x$, $x \in \mathbb{R}$:

$$\tanh x := \frac{e^x - e^{-x}}{e^x + e^{-x}} = \frac{e^{2x} - 1}{e^{2x} + 1}.$$

It has the properties $\tanh 0 = 0$, $-1 < \tanh x < 1$, $\forall\, x \in \mathbb{R}$, and $\tanh(-x) = -\tanh x$. Furthermore $\tanh x \to 1$ as $x \to \infty$, and $\tanh x \to -1$, as $x \to -\infty$, and it is strictly increasing on $\mathbb{R}$. Furthermore it holds $\frac{d}{dx} \tanh x = \frac{1}{\cosh^2 x} > 0$.

This function plays also the role of an activation function in the hidden layer of neural networks.

We further consider

$$\Psi(x) := \frac{1}{4} \left(\tanh(x+1) - \tanh(x-1) \right) > 0, \;\; \forall x \in \mathbb{R}. \tag{3.28}$$

We easily see that $\Psi(-x) = \Psi(x)$, that is Ψ is even on $\mathbb{R}$. Obviously Ψ is differentiable, thus continuous.

Here we follow [8]

Proposition 3.17 *$\Psi(x)$ for $x \geq 0$ is strictly decreasing.*

Obviously $\Psi(x)$ is strictly increasing for $x \leq 0$. Also it holds $\lim_{x \to -\infty} \Psi(x) = 0 = \lim_{x \to \infty} \Psi(x)$.

Infact Ψ has the bell shape with horizontal asymptote the x-axis. So the maximum of Ψ is at zero, $\Psi(0) = 0.3809297$.

Theorem 3.18 *We have that $\sum_{i=-\infty}^{\infty} \Psi(x-i) = 1$, $\forall\, x \in \mathbb{R}$.*

Thus

$$\sum_{i=-\infty}^{\infty} \Psi(nx - i) = 1, \quad \forall n \in \mathbb{N},\; \forall x \in \mathbb{R}.$$

Furthermore we get:

Since Ψ is even it holds $\sum_{i=-\infty}^{\infty} \Psi(i - x) = 1$, $\forall x \in \mathbb{R}$.

Hence $\sum_{i=-\infty}^{\infty} \Psi(i + x) = 1$, $\forall\, x \in \mathbb{R}$, and $\sum_{i=-\infty}^{\infty} \Psi(x + i) = 1$, $\forall\, x \in \mathbb{R}$.

Theorem 3.19 *It holds $\int_{-\infty}^{\infty} \Psi(x)\, dx = 1$.*

So $\Psi(x)$ is a density function on $\mathbb{R}$.

Theorem 3.20 *Let $0 < \beta < 1$ and $n \in \mathbb{N}$. It holds*

$$\sum_{\substack{k=-\infty \\ : |nx - k| \geq n^{1-\beta}}}^{\infty} \Psi(nx - k) \leq e^4 \cdot e^{-2n^{(1-\beta)}}. \tag{3.29}$$

Theorem 3.21 *Let $x \in [a, b] \subset \mathbb{R}$ and $n \in \mathbb{N}$ so that $\lceil na \rceil \leq \lfloor nb \rfloor$. It holds*

$$\frac{1}{\sum_{k=\lceil na \rceil}^{\lfloor nb \rfloor} \Psi(nx-k)} < 4.1488766 = \frac{1}{\Psi(1)}. \tag{3.30}$$

Also by [8], we obtain

$$\lim_{n\to\infty} \sum_{k=\lceil na \rceil}^{\lfloor nb \rfloor} \Psi(nx-k) \neq 1,$$

for at least some $x \in [a, b]$.

Definition 3.22 Let $f \in C([a, b])$ and $n \in \mathbb{N}$ such that $\lceil na \rceil \leq \lfloor nb \rfloor$.

We further study, as in [8], the positive linear neural network operator

$$F_n(f, x) := \frac{\sum_{k=\lceil na \rceil}^{\lfloor nb \rfloor} f\left(\frac{k}{n}\right) \Psi(nx-k)}{\sum_{k=\lceil na \rceil}^{\lfloor nb \rfloor} \Psi(nx-k)}, \quad x \in [a, b]. \tag{3.31}$$

We study F_n similarly to G_n.

For convenience we call

$$F_n^*(f, x) := \sum_{k=\lceil na \rceil}^{\lfloor nb \rfloor} f\left(\frac{k}{n}\right) \Psi(nx-k), \tag{3.32}$$

that is

$$F_n(f, x) := \frac{F_n^*(f, x)}{\sum_{k=\lceil na \rceil}^{\lfloor nb \rfloor} \Psi(nx-k)}. \tag{3.33}$$

Thus,

$$F_n(f, x) - f(x) = \frac{F_n^*(f, x)}{\sum_{k=\lceil na \rceil}^{\lfloor nb \rfloor} \Psi(nx-k)} - f(x)$$

$$= \frac{F_n^*(f, x) - f(x) \sum_{k=\lceil na \rceil}^{\lfloor nb \rfloor} \Psi(nx-k)}{\sum_{k=\lceil na \rceil}^{\lfloor nb \rfloor} \Psi(nx-k)}. \tag{3.34}$$

Consequently we derive

$$|F_n(f, x) - f(x)| \leq \frac{1}{\Psi(1)} \left| F_n^*(f, x) - f(x) \sum_{k=\lceil na \rceil}^{\lfloor nb \rfloor} \Psi(nx-k) \right|. \tag{3.35}$$

That is

$$|F_n(f,x) - f(x)| \leq (4.1488766) \left| \sum_{k=\lceil na \rceil}^{\lfloor nb \rfloor} \left(f\left(\frac{k}{n}\right) - f(x) \right) \Psi(nx-k) \right|. \tag{3.36}$$

We will estimate the right hand side of (3.36).

3.3 Main Results

We present our first main result

Theorem 3.23 *Let $\alpha > 0$, $N = \lceil \alpha \rceil$, $\alpha \notin \mathbb{N}$, $f \in AC^N([a,b])$, with $f^{(N)} \in L_\infty([a,b])$, $0 < \beta < 1$, $x \in [a,b]$, $n \in \mathbb{N}$. Then*

(i)

$$\left| G_n(f,x) - \sum_{j=1}^{N-1} \frac{f^{(j)}(x)}{j!} G_n\left((\cdot - x)^j\right)(x) - f(x) \right| \leq \tag{3.37}$$

$$\frac{(5.250312578)}{\Gamma(\alpha+1)} \cdot \left\{ \frac{\left(\omega_1\left(D_{x-}^{\alpha} f, \frac{1}{n^\beta}\right)_{[a,x]} + \omega_1\left(D_{*x}^{\alpha} f, \frac{1}{n^\beta}\right)_{[x,b]} \right)}{n^{\alpha\beta}} + \right.$$

$$\left. 3.1992 e^{-n^{(1-\beta)}} \left(\left\| D_{x-}^{\alpha} f \right\|_{\infty,[a,x]} (x-a)^\alpha + \left\| D_{*x}^{\alpha} f \right\|_{\infty,[x,b]} (b-x)^\alpha \right) \right\},$$

(ii) if $f^{(j)}(x) = 0$, for $j = 1, \ldots, N-1$, we have

$$|G_n(f,x) - f(x)| \leq \frac{(5.250312578)}{\Gamma(\alpha+1)} \cdot \tag{3.38}$$

$$\left\{ \frac{\left(\omega_1\left(D_{x-}^{\alpha} f, \frac{1}{n^\beta}\right)_{[a,x]} + \omega_1\left(D_{*x}^{\alpha} f, \frac{1}{n^\beta}\right)_{[x,b]} \right)}{n^{\alpha\beta}} + \right.$$

$$\left. 3.1992 e^{-n^{(1-\beta)}} \left(\left\| D_{x-}^{\alpha} f \right\|_{\infty,[a,x]} (x-a)^\alpha + \left\| D_{*x}^{\alpha} f \right\|_{\infty,[x,b]} (b-x)^\alpha \right) \right\},$$

when $\alpha > 1$ notice here the extremely high rate of convergence at $n^{-(\alpha+1)\beta}$,

(iii)

$$|G_n(f,x) - f(x)| \leq (5.250312578) \cdot \tag{3.39}$$

$$\left\{ \sum_{j=1}^{N-1} \frac{|f^{(j)}(x)|}{j!} \left\{ \frac{1}{n^{\beta j}} + (b-a)^j (3.1992) e^{-n^{(1-\beta)}} \right\} + \right.$$

$$\frac{1}{\Gamma(\alpha+1)} \left\{ \frac{\left(\omega_1 \left(D_{x-}^{\alpha} f, \frac{1}{n^{\beta}} \right)_{[a,x]} + \omega_1 \left(D_{*x}^{\alpha} f, \frac{1}{n^{\beta}} \right)_{[x,b]} \right)}{n^{\alpha\beta}} + \right.$$

$$\left. \left. 3.1992 e^{-n^{(1-\beta)}} \left(\left\| D_{x-}^{\alpha} f \right\|_{\infty,[a,x]} (x-a)^{\alpha} + \left\| D_{*x}^{\alpha} f \right\|_{\infty,[x,b]} (b-x)^{\alpha} \right) \right\} \right\},$$

$\forall\, x \in [a,b]$,

and

(iv)

$$\|G_n f - f\|_{\infty} \leq (5.250312578) \cdot \tag{3.40}$$

$$\left\{ \sum_{j=1}^{N-1} \frac{\|f^{(j)}\|_{\infty}}{j!} \left\{ \frac{1}{n^{\beta j}} + (b-a)^j (3.1992) e^{-n^{(1-\beta)}} \right\} + \right.$$

$$\frac{1}{\Gamma(\alpha+1)} \left\{ \frac{\left(\sup_{x\in[a,b]} \omega_1 \left(D_{x-}^{\alpha} f, \frac{1}{n^{\beta}} \right)_{[a,x]} + \sup_{x\in[a,b]} \omega_1 \left(D_{*x}^{\alpha} f, \frac{1}{n^{\beta}} \right)_{[x,b]} \right)}{n^{\alpha\beta}} + \right.$$

$$\left. \left. 3.1992 e^{-n^{(1-\beta)}} (b-a)^{\alpha} \left(\sup_{x\in[a,b]} \left\| D_{x-}^{\alpha} f \right\|_{\infty,[a,x]} + \sup_{x\in[a,b]} \left\| D_{*x}^{\alpha} f \right\|_{\infty,[x,b]} \right) \right\} \right\}.$$

Above, when $N = 1$ the sum $\sum_{j=1}^{N-1} \cdot = 0$.

As we see here we obtain fractionally type pointwise and uniform convergence with rates of $G_n \to I$ the unit operator, as $n \to \infty$.

Proof Let $x \in [a, b]$. We have that $D_{x-}^{\alpha} f(x) = D_{*x}^{\alpha} f(x) = 0$.

From [14], p. 54, we get by the left Caputo fractional Taylor formula that

$$f\left(\frac{k}{n}\right) = \sum_{j=0}^{N-1} \frac{f^{(j)}(x)}{j!} \left(\frac{k}{n} - x\right)^j + \tag{3.41}$$

$$\frac{1}{\Gamma(\alpha)} \int_x^{\frac{k}{n}} \left(\frac{k}{n} - J\right)^{\alpha-1} \left(D_{*x}^{\alpha} f(J) - D_{*x}^{\alpha} f(x)\right) dJ,$$

for all $x \leq \frac{k}{n} \leq b$.

Also from [3], using the right Caputo fractional Taylor formula we get

$$f\left(\frac{k}{n}\right) = \sum_{j=0}^{N-1} \frac{f^{(j)}(x)}{j!} \left(\frac{k}{n} - x\right)^j + \tag{3.42}$$

$$\frac{1}{\Gamma(\alpha)} \int_{\frac{k}{n}}^{x} \left(J - \frac{k}{n}\right)^{\alpha-1} \left(D_{x-}^{\alpha} f(J) - D_{x-}^{\alpha} f(x)\right) dJ,$$

for all $a \leq \frac{k}{n} \leq x$.

Hence we have

$$f\left(\frac{k}{n}\right) \Phi(nx-k) = \sum_{j=0}^{N-1} \frac{f^{(j)}(x)}{j!} \Phi(nx-k) \left(\frac{k}{n} - x\right)^j + \tag{3.43}$$

$$\frac{\Phi(nx-k)}{\Gamma(\alpha)} \int_x^{\frac{k}{n}} \left(\frac{k}{n} - J\right)^{\alpha-1} \left(D_{*x}^{\alpha} f(J) - D_{*x}^{\alpha} f(x)\right) dJ,$$

for all $x \leq \frac{k}{n} \leq b$, iff $\lceil nx \rceil \leq k \leq \lfloor nb \rfloor$, and

$$f\left(\frac{k}{n}\right) \Phi(nx-k) = \sum_{j=0}^{N-1} \frac{f^{(j)}(x)}{j!} \Phi(nx-k) \left(\frac{k}{n} - x\right)^j + \tag{3.44}$$

$$\frac{\Phi(nx-k)}{\Gamma(\alpha)} \int_{\frac{k}{n}}^{x} \left(J - \frac{k}{n}\right)^{\alpha-1} \left(D_{x-}^{\alpha} f(J) - D_{x-}^{\alpha} f(x)\right) dJ,$$

for all $a \leq \frac{k}{n} \leq x$, iff $\lceil na \rceil \leq k \leq \lfloor nx \rfloor$.

We have that $\lceil nx \rceil \leq \lfloor nx \rfloor + 1$.

Therefore it holds

$$\sum_{k=\lfloor nx\rfloor+1}^{\lfloor nb\rfloor} f\left(\frac{k}{n}\right)\Phi(nx-k)=\sum_{j=0}^{N-1}\frac{f^{(j)}(x)}{j!}\sum_{k=\lfloor nx\rfloor+1}^{\lfloor nb\rfloor}\Phi(nx-k)\left(\frac{k}{n}-x\right)^{j}+$$
$$\frac{1}{\Gamma(\alpha)}\sum_{k=\lfloor nx\rfloor+1}^{\lfloor nb\rfloor}\Phi(nx-k)\int_{x}^{\frac{k}{n}}\left(\frac{k}{n}-J\right)^{\alpha-1}\left(D_{*x}^{\alpha}f(J)-D_{*x}^{\alpha}f(x)\right)dJ, \tag{3.45}$$

and

$$\sum_{k=\lceil na\rceil}^{\lfloor nx\rfloor} f\left(\frac{k}{n}\right)\Phi(nx-k)=\sum_{j=0}^{N-1}\frac{f^{(j)}(x)}{j!}\sum_{k=\lceil na\rceil}^{\lfloor nx\rfloor}\Phi(nx-k)\left(\frac{k}{n}-x\right)^{j}+$$
$$\frac{1}{\Gamma(\alpha)}\sum_{k=\lceil na\rceil}^{\lfloor nx\rfloor}\Phi(nx-k)\int_{\frac{k}{n}}^{x}\left(J-\frac{k}{n}\right)^{\alpha-1}\left(D_{x-}^{\alpha}f(J)-D_{x-}^{\alpha}f(x)\right)dJ. \tag{3.46}$$

Adding the last two equalities (3.45) and (3.46) we obtain

$$G_n^*(f,x)=\sum_{k=\lceil na\rceil}^{\lfloor nb\rfloor} f\left(\frac{k}{n}\right)\Phi(nx-k)= \tag{3.47}$$

$$\sum_{j=0}^{N-1}\frac{f^{(j)}(x)}{j!}\sum_{k=\lceil na\rceil}^{\lfloor nb\rfloor}\Phi(nx-k)\left(\frac{k}{n}-x\right)^{j}+$$

$$\frac{1}{\Gamma(\alpha)}\left\{\sum_{k=\lceil na\rceil}^{\lfloor nx\rfloor}\Phi(nx-k)\int_{\frac{k}{n}}^{x}\left(J-\frac{k}{n}\right)^{\alpha-1}\left(D_{x-}^{\alpha}f(J)-D_{x-}^{\alpha}f(x)\right)dJ+\right.$$
$$\left.\sum_{k=\lfloor nx\rfloor+1}^{\lfloor nb\rfloor}\Phi(nx-k)\int_{x}^{\frac{k}{n}}\left(\frac{k}{n}-J\right)^{\alpha-1}\left(D_{*x}^{\alpha}f(J)-D_{*x}^{\alpha}f(x)\right)dJ\right\}.$$

So we have derived

$$G_n^*(f,x)-f(x)\left(\sum_{k=\lceil na\rceil}^{\lfloor nb\rfloor}\Phi(nx-k)\right)= \tag{3.48}$$

$$\sum_{j=1}^{N-1}\frac{f^{(j)}(x)}{j!}G_n^*\left((\cdot-x)^{j}\right)(x)+\theta_n(x),$$

where

$$\theta_n(x) := \frac{1}{\Gamma(\alpha)} \left\{ \sum_{k=\lceil na \rceil}^{\lfloor nx \rfloor} \Phi(nx-k) \int_{\frac{k}{n}}^{x} \left(J - \frac{k}{n}\right)^{\alpha-1} \left(D_{x-}^{\alpha} f(J) - D_{x-}^{\alpha} f(x)\right) dJ \right.$$

$$\left. + \sum_{k=\lfloor nx \rfloor+1}^{\lfloor nb \rfloor} \Phi(nx-k) \int_{x}^{\frac{k}{n}} \left(\frac{k}{n} - J\right)^{\alpha-1} \left(D_{*x}^{\alpha} f(J) - D_{*x}^{\alpha} f(x)\right) dJ \right\}. \quad (3.49)$$

We set

$$\theta_{1n}(x) := \frac{1}{\Gamma(\alpha)} \sum_{k=\lceil na \rceil}^{\lfloor nx \rfloor} \Phi(nx-k) \int_{\frac{k}{n}}^{x} \left(J - \frac{k}{n}\right)^{\alpha-1} \left(D_{x-}^{\alpha} f(J) - D_{x-}^{\alpha} f(x)\right) dJ, \quad (3.50)$$

and

$$\theta_{2n} := \frac{1}{\Gamma(\alpha)} \sum_{k=\lfloor nx \rfloor+1}^{\lfloor nb \rfloor} \Phi(nx-k) \int_{x}^{\frac{k}{n}} \left(\frac{k}{n} - J\right)^{\alpha-1} \left(D_{*x}^{\alpha} f(J) - D_{*x}^{\alpha} f(x)\right) dJ, \quad (3.51)$$

i.e.

$$\theta_n(x) = \theta_{1n}(x) + \theta_{2n}(x). \quad (3.52)$$

We assume $b - a > \frac{1}{n^{\beta}}$, $0 < \beta < 1$, which is always the case for large enough $n \in \mathbb{N}$, that is when $n > \left\lceil (b-a)^{-\frac{1}{\beta}} \right\rceil$. It is always true that either $\left|\frac{k}{n} - x\right| \leq \frac{1}{n^{\beta}}$ or $\left|\frac{k}{n} - x\right| > \frac{1}{n^{\beta}}$.

For $k = \lceil na \rceil, \ldots, \lfloor nx \rfloor$, we consider

$$\gamma_{1k} := \left| \int_{\frac{k}{n}}^{x} \left(J - \frac{k}{n}\right)^{\alpha-1} \left(D_{x-}^{\alpha} f(J) - D_{x-}^{\alpha} f(x)\right) dJ \right| \quad (3.53)$$

$$= \left| \int_{\frac{k}{n}}^{x} \left(J - \frac{k}{n}\right)^{\alpha-1} D_{x-}^{\alpha} f(J)\, dJ \right| \leq \int_{\frac{k}{n}}^{x} \left(J - \frac{k}{n}\right)^{\alpha-1} \left|D_{x-}^{\alpha} f(J)\right| dJ$$

$$\leq \left\| D_{x-}^{\alpha} f \right\|_{\infty,[a,x]} \frac{\left(x - \frac{\kappa}{n}\right)^{\alpha}}{\alpha} \leq \left\| D_{x-}^{\alpha} f \right\|_{\infty,[a,x]} \frac{(x-a)^{\alpha}}{\alpha}. \quad (3.54)$$

That is

$$\gamma_{1k} \leq \left\| D_{x-}^{\alpha} f \right\|_{\infty,[a,x]} \frac{(x-a)^{\alpha}}{\alpha}, \quad (3.55)$$

for $k = \lceil na \rceil, \ldots, \lfloor nx \rfloor$.

Also we have in case of $\left|\frac{k}{n}-x\right| \le \frac{1}{n^{\beta}}$ that

$$\gamma_{1k} \le \int_{\frac{k}{n}}^{x} \left(J-\frac{k}{n}\right)^{\alpha-1} \left|D_{x-}^{\alpha} f(J) - D_{x-}^{\alpha} f(x)\right| dJ \tag{3.56}$$

$$\le \int_{\frac{k}{n}}^{x} \left(J-\frac{k}{n}\right)^{\alpha-1} \omega_1 \left(D_{x-}^{\alpha} f, |J-x|\right)_{[a,x]} dJ$$

$$\le \omega_1 \left(D_{x-}^{\alpha} f, \left|x-\frac{k}{n}\right|\right)_{[a,x]} \int_{\frac{k}{n}}^{x} \left(J-\frac{k}{n}\right)^{\alpha-1} dJ$$

$$\le \omega_1 \left(D_{x-}^{\alpha} f, \frac{1}{n^{\beta}}\right)_{[a,x]} \frac{\left(x-\frac{k}{n}\right)^{\alpha}}{\alpha} \le \omega_1 \left(D_{x-}^{\alpha} f, \frac{1}{n^{\beta}}\right)_{[a,x]} \frac{1}{\alpha n^{a\beta}}. \tag{3.57}$$

That is when $\left|\frac{k}{n}-x\right| \le \frac{1}{n^{\beta}}$, then

$$\gamma_{1k} \le \frac{\omega_1 \left(D_{x-}^{\alpha} f, \frac{1}{n^{\beta}}\right)_{[a,x]}}{\alpha n^{a\beta}}. \tag{3.58}$$

Consequently we obtain

$$|\theta_{1n}(x)| \le \frac{1}{\Gamma(\alpha)} \sum_{k=\lceil na \rceil}^{\lfloor nx \rfloor} \Phi(nx-k) \gamma_{1k} = \tag{3.59}$$

$$\frac{1}{\Gamma(\alpha)} \left\{ \sum_{\substack{k=\lceil na \rceil \\ : \left|\frac{k}{n}-x\right| \le \frac{1}{n^{\beta}}}}^{\lfloor nx \rfloor} \Phi(nx-k) \gamma_{1k} + \sum_{\substack{k=\lceil na \rceil \\ : \left|\frac{k}{n}-x\right| > \frac{1}{n^{\beta}}}}^{\lfloor nx \rfloor} \Phi(nx-k) \gamma_{1k} \right\} \le$$

$$\frac{1}{\Gamma(\alpha)} \left\{ \left(\sum_{\substack{k=\lceil na \rceil \\ : \left|\frac{k}{n}-x\right| \le \frac{1}{n^{\beta}}}}^{\lfloor nx \rfloor} \Phi(nx-k) \right) \frac{\omega_1 \left(D_{x-}^{\alpha} f, \frac{1}{n^{\beta}}\right)_{[a,x]}}{\alpha n^{\alpha\beta}} + \right.$$

$$\left(\sum_{\left\{ \substack{k = \lceil na \rceil \\ : \left| \frac{k}{n} - x \right| > \frac{1}{n^{\beta}}} \right.}^{\lfloor nx \rfloor} \Phi(nx - k) \right) \left\| D_{x-}^{\alpha} f \right\|_{\infty,[a,x]} \frac{(x-a)^{\alpha}}{\alpha} \right\} \leq \tag{3.60}$$

$$\frac{1}{\Gamma(\alpha+1)} \left\{ \frac{\omega_1 \left(D_{x-}^{\alpha} f, \frac{1}{n^{\beta}} \right)_{[a,x]}}{n^{\alpha\beta}} + \right.$$

$$\left(\sum_{\left\{ \substack{k = -\infty \\ : |nx - k| > n^{1-\beta}} \right.}^{\infty} \Phi(nx - k) \right) \left\| D_{x-}^{\alpha} f \right\|_{\infty,[a,x]} (x-a)^{\alpha} \right\} \leq \tag{3.61}$$

$$\frac{1}{\Gamma(\alpha+1)} \left\{ \frac{\omega_1 \left(D_{x-}^{\alpha} f, \frac{1}{n^{\beta}} \right)_{[a,x]}}{n^{\alpha\beta}} + 3.1992 e^{-n^{(1-\beta)}} \left\| D_{x-}^{\alpha} f \right\|_{\infty,[a,x]} (x-a)^{\alpha} \right\}.$$

So we have proved that

$$|\theta_{1n}(x)| \leq \frac{1}{\Gamma(\alpha+1)} \left\{ \frac{\omega_1 \left(D_{x-}^{\alpha} f, \frac{1}{n^{\beta}} \right)_{[a,x]}}{n^{\alpha\beta}} + \right. \tag{3.62}$$

$$\left. 3.1992 e^{-n^{(1-\beta)}} \left\| D_{x-}^{\alpha} f \right\|_{\infty,[a,x]} (x-a)^{\alpha} \right\}.$$

Next when $k = \lfloor nx \rfloor + 1, \ldots, \lfloor nb \rfloor$ we consider

$$\gamma_{2k} := \left| \int_x^{\frac{k}{n}} \left(\frac{k}{n} - J \right)^{\alpha-1} \left(D_{*x}^{\alpha} f(J) - D_{*x}^{\alpha} f(x) \right) dJ \right| \leq$$

$$\int_x^{\frac{k}{n}} \left(\frac{k}{n} - J \right)^{\alpha-1} \left| D_{*x}^{\alpha} f(J) - D_{*x}^{\alpha} f(x) \right| dJ =$$

$$\int_x^{\frac{k}{n}} \left(\frac{k}{n} - J\right)^{\alpha-1} \left|D_{*x}^{\alpha} f(J)\right| dJ \leq \left\|D_{*x}^{\alpha} f\right\|_{\infty,[x,b]} \frac{\left(\frac{k}{n} - x\right)^{\alpha}}{\alpha} \leq \tag{3.63}$$

$$\left\|D_{*x}^{\alpha} f\right\|_{\infty,[x,b]} \frac{(b-x)^{\alpha}}{\alpha}. \tag{3.64}$$

Therefore when $k = \lfloor nx \rfloor + 1, \ldots, \lfloor nb \rfloor$ we get that

$$\gamma_{2k} \leq \left\|D_{*x}^{\alpha} f\right\|_{\infty,[x,b]} \frac{(b-x)^{\alpha}}{\alpha}. \tag{3.65}$$

In case of $\left|\frac{k}{n} - x\right| \leq \frac{1}{n^{\beta}}$, we get

$$\gamma_{2k} \leq \int_x^{\frac{k}{n}} \left(\frac{k}{n} - J\right)^{\alpha-1} \omega_1\left(D_{*x}^{\alpha} f, |J-x|\right)_{[x,b]} dJ \leq \tag{3.66}$$

$$\omega_1\left(D_{*x}^{\alpha} f, \left|\frac{k}{n} - x\right|\right)_{[x,b]} \int_x^{\frac{k}{n}} \left(\frac{k}{n} - J\right)^{\alpha-1} dJ \leq$$

$$\omega_1\left(D_{*x}^{\alpha} f, \frac{1}{n^{\beta}}\right)_{[x,b]} \frac{\left(\frac{k}{n} - x\right)^{\alpha}}{\alpha} \leq \omega_1\left(D_{*x}^{\alpha} f, \frac{1}{n^{\beta}}\right)_{[x,b]} \frac{1}{\alpha n^{\alpha\beta}}. \tag{3.67}$$

So when $\left|\frac{k}{n} - x\right| \leq \frac{1}{n^{\beta}}$ we derived that

$$\gamma_{2k} \leq \frac{\omega_1\left(D_{*x}^{\alpha} f, \frac{1}{n^{\beta}}\right)_{[x,b]}}{\alpha n^{\alpha\beta}}. \tag{3.68}$$

Similarly we have that

$$|\theta_{2n}(x)| \leq \frac{1}{\Gamma(\alpha)} \left(\sum_{k=\lfloor nx \rfloor + 1}^{\lfloor nb \rfloor} \Phi(nx-k)\, \gamma_{2k}\right) = \tag{3.69}$$

$$\frac{1}{\Gamma(\alpha)} \left\{ \sum_{\substack{k = \lfloor nx \rfloor + 1 \\ : \left|\frac{k}{n} - x\right| \leq \frac{1}{n^{\beta}}}}^{\lfloor nb \rfloor} \Phi(nx-k)\, \gamma_{2k} + \sum_{\substack{k = \lfloor nx \rfloor + 1 \\ : \left|\frac{k}{n} - x\right| > \frac{1}{n^{\beta}}}}^{\lfloor nb \rfloor} \Phi(nx-k)\, \gamma_{2k} \right\} \leq$$

$$\frac{1}{\Gamma(\alpha)}\left\{\left(\sum_{\substack{k=\lfloor nx\rfloor+1 \\ :\left|\frac{k}{n}-x\right|\leq\frac{1}{n^{\beta}}}}^{\lfloor nb\rfloor}\Phi(nx-k)\right)\frac{\omega_1\left(D_{*x}^{\alpha}f,\frac{1}{n^{\beta}}\right)_{[x,b]}}{\alpha n^{\alpha\beta}}+\right.$$

$$\left.\left(\sum_{\substack{k=\lfloor nx\rfloor+1 \\ :\left|\frac{k}{n}-x\right|>\frac{1}{n^{\beta}}}}^{\lfloor nb\rfloor}\Phi(nx-k)\right)\left\|D_{*x}^{\alpha}f\right\|_{\infty,[x,b]}\frac{(b-x)^{\alpha}}{\alpha}\right\}\leq \tag{3.70}$$

$$\frac{1}{\Gamma(\alpha+1)}\left\{\frac{\omega_1\left(D_{*x}^{\alpha}f,\frac{1}{n^{\beta}}\right)_{[x,b]}}{n^{\alpha\beta}}+\right.$$

$$\left.\left(\sum_{\substack{k=-\infty \\ :\left|\frac{k}{n}-x\right|>\frac{1}{n^{\beta}}}}^{\infty}\Phi(nx-k)\right)\left\|D_{*x}^{\alpha}f\right\|_{\infty,[x,b]}(b-x)^{\alpha}\right\}\leq \tag{3.71}$$

$$\frac{1}{\Gamma(\alpha+1)}\left\{\frac{\omega_1\left(D_{*x}^{\alpha}f,\frac{1}{n^{\beta}}\right)_{[x,b]}}{n^{\alpha\beta}}+3.1992e^{-n^{(1-\beta)}}\left\|D_{*x}^{\alpha}f\right\|_{\infty,[x,b]}(b-x)^{\alpha}\right\}.$$

So we have proved that

$$|\theta_{2n}(x)|\leq\frac{1}{\Gamma(\alpha+1)}\left\{\frac{\omega_1\left(D_{*x}^{\alpha}f,\frac{1}{n^{\beta}}\right)_{[x,b]}}{n^{\alpha\beta}}+\right. \tag{3.72}$$

$$\left.3.1992e^{-n^{(1-\beta)}}\left\|D_{*x}^{\alpha}f\right\|_{\infty,[x,b]}(b-x)^{\alpha}\right\}.$$

Therefore

$$|\theta_n(x)| \le |\theta_{1n}(x)| + |\theta_{2n}(x)| \le \tag{3.73}$$

$$\frac{1}{\Gamma(\alpha+1)} \left\{ \frac{\omega_1\left(D_{x-}^{\alpha} f, \frac{1}{n^{\beta}}\right)_{[a,x]} + \omega_1\left(D_{*x}^{\alpha} f, \frac{1}{n^{\beta}}\right)_{[x,b]}}{n^{\alpha\beta}} + \right. \tag{3.74}$$

$$\left. 3.1992 e^{-n^{(1-\beta)}} \left(\left\| D_{x-}^{\alpha} f \right\|_{\infty,[a,x]} (x-a)^{\alpha} + \left\| D_{*x}^{\alpha} f \right\|_{\infty,[x,b]} (b-x)^{\alpha} \right) \right\}.$$

From [6], p. 15 we get that

$$\left| G_n^*\left((\cdot - x)^j \right)(x) \right| \le \frac{1}{n^{\beta j}} + (b-a)^j (3.1992) e^{-n^{(1-\beta)}}, \tag{3.75}$$

for $j = 1, \ldots, N-1$, $\forall\, x \in [a, b]$.

Putting things together, we have established

$$\left| G_n^*(f, x) - f(x) \left(\sum_{k=\lceil na \rceil}^{\lfloor nb \rfloor} \Phi(nx - k) \right) \right| \le \tag{3.76}$$

$$\sum_{j=1}^{N-1} \frac{\left| f^{(j)}(x) \right|}{j!} \left[\frac{1}{n^{\beta j}} + (b-a)^j (3.1992) e^{-n^{(1-\beta)}} \right] +$$

$$\frac{1}{\Gamma(\alpha+1)} \left\{ \frac{\left(\omega_1\left(D_{x-}^{\alpha} f, \frac{1}{n^{\beta}}\right)_{[a,x]} + \omega_1\left(D_{*x}^{\alpha} f, \frac{1}{n^{\beta}}\right)_{[x,b]} \right)}{n^{\alpha\beta}} + \right.$$

$$\left. 3.1992 e^{-n^{(1-\beta)}} \left(\left\| D_{x-}^{\alpha} f \right\|_{\infty,[a,x]} (x-a)^{\alpha} + \left\| D_{*x}^{\alpha} f \right\|_{\infty,[x,b]} (b-x)^{\alpha} \right) \right\} =: A_n(x). \tag{3.77}$$

As a result we derive

$$|G_n(f, x) - f(x)| \le (5.250312578)\, A_n(x), \tag{3.78}$$

$\forall\, x \in [a, b]$.

We further have that

$$\|A_n\|_\infty \leq \sum_{j=1}^{N-1} \frac{\|f^{(j)}\|_\infty}{j!} \left[\frac{1}{n^{\beta j}} + (b-a)^j (3.1992) e^{-n^{(1-\beta)}} \right] + \tag{3.79}$$

$$\frac{1}{\Gamma(\alpha+1)} \left\{ \frac{\left\{ \sup_{x\in[a,b]} \left(\omega_1 \left(D_{x-}^{\alpha} f, \frac{1}{n^\beta} \right)_{[a,x]} \right) + \sup_{x\in[a,b]} \left(\omega_1 \left(D_{*x}^{\alpha} f, \frac{1}{n^\beta} \right)_{[x,b]} \right) \right\}}{n^{\alpha\beta}} \right.$$

$$+ 3.1992 e^{-n^{(1-\beta)}} (b-a)^\alpha \cdot$$

$$\left. \left\{ \left(\sup_{x\in[a,b]} \left(\| D_{x-}^{\alpha} f \|_{\infty,[a,x]} \right) + \sup_{x\in[a,b]} \left(\| D_{*x}^{\alpha} f \|_{\infty,[x,b]} \right) \right) \right\} \right\} =: B_n.$$

Hence it holds

$$\|G_n f - f\|_\infty \leq (5.250312578)\, B_n. \tag{3.80}$$

Since $f \in AC^N([a,b])$, $N = \lceil \alpha \rceil$, $\alpha > 0$, $\alpha \notin \mathbb{N}$, $f^{(N)} \in L_\infty([a,b])$, $x \in [a,b]$, then we get that $f \in AC^N([a,x])$, $f^{(N)} \in L_\infty([a,x])$ and $f \in AC^N([x,b])$, $f^{(N)} \in L_\infty([x,b])$.

We have

$$\left(D_{x-}^{\alpha} f\right)(y) = \frac{(-1)^N}{\Gamma(N-\alpha)} \int_y^x (J-y)^{N-\alpha-1} f^{(N)}(J)\, dJ, \tag{3.81}$$

$\forall\, y \in [a,x]$ and

$$\left| \left(D_{x-}^{\alpha} f\right)(y) \right| \leq \frac{1}{\Gamma(N-\alpha)} \left(\int_y^x (J-y)^{N-\alpha-1} dJ \right) \left\| f^{(N)} \right\|_\infty \tag{3.82}$$

$$= \frac{1}{\Gamma(N-\alpha)} \frac{(x-y)^{N-\alpha}}{(N-\alpha)} \left\| f^{(N)} \right\|_\infty =$$

$$\frac{(x-y)^{N-\alpha}}{\Gamma(N-\alpha+1)} \left\| f^{(N)} \right\|_\infty \leq \frac{(b-a)^{N-\alpha}}{\Gamma(N-\alpha+1)} \left\| f^{(N)} \right\|_\infty.$$

That is

$$\left\| D_{x-}^{\alpha} f \right\|_{\infty,[a,x]} \leq \frac{(b-a)^{N-\alpha}}{\Gamma(N-\alpha+1)} \left\| f^{(N)} \right\|_{\infty}, \tag{3.83}$$

and

$$\sup_{x\in[a,b]} \left\| D_{x-}^{\alpha} f \right\|_{\infty,[a,x]} \leq \frac{(b-a)^{N-\alpha}}{\Gamma(N-\alpha+1)} \left\| f^{(N)} \right\|_{\infty}. \tag{3.84}$$

Similarly we have

$$\left(D_{*x}^{\alpha} f \right)(y) = \frac{1}{\Gamma(N-\alpha)} \int_{x}^{y} (y-t)^{N-\alpha-1} f^{(N)}(t)\, dt, \tag{3.85}$$

$\forall\, y \in [x, b]$.

Thus we get

$$\left| \left(D_{*x}^{\alpha} f \right)(y) \right| \leq \frac{1}{\Gamma(N-\alpha)} \left(\int_{x}^{y} (y-t)^{N-\alpha-1} dt \right) \left\| f^{(N)} \right\|_{\infty} \leq$$

$$\frac{1}{\Gamma(N-\alpha)} \frac{(y-x)^{N-\alpha}}{(N-\alpha)} \left\| f^{(N)} \right\|_{\infty} \leq \frac{(b-a)^{N-\alpha}}{\Gamma(N-\alpha+1)} \left\| f^{(N)} \right\|_{\infty}.$$

Hence

$$\left\| D_{*x}^{\alpha} f \right\|_{\infty,[x,b]} \leq \frac{(b-a)^{N-\alpha}}{\Gamma(N-\alpha+1)} \left\| f^{(N)} \right\|_{\infty}, \tag{3.86}$$

and

$$\sup_{x\in[a,b]} \left\| D_{*x}^{\alpha} f \right\|_{\infty,[x,b]} \leq \frac{(b-a)^{N-\alpha}}{\Gamma(N-\alpha+1)} \left\| f^{(N)} \right\|_{\infty}. \tag{3.87}$$

From (3.17) and (3.18) we get

$$\sup_{x\in[a,b]} \omega_1 \left(D_{x-}^{\alpha} f, \frac{1}{n^{\beta}} \right)_{[a,x]} \leq \frac{2 \left\| f^{(N)} \right\|_{\infty}}{\Gamma(N-\alpha+1)} (b-a)^{N-\alpha}, \tag{3.88}$$

and

$$\sup_{x\in[a,b]} \omega_1 \left(D_{*x}^{\alpha} f, \frac{1}{n^{\beta}} \right)_{[x,b]} \leq \frac{2 \left\| f^{(N)} \right\|_{\infty}}{\Gamma(N-\alpha+1)} (b-a)^{N-\alpha}. \tag{3.89}$$

So that $B_n < \infty$.

We finally notice that

$$G_n(f,x) - \sum_{j=1}^{N-1} \frac{f^{(j)}(x)}{j!} G_n\left((\cdot - x)^j\right)(x) - f(x) = \frac{G_n^*(f,x)}{\left(\sum_{k=\lceil na \rceil}^{\lfloor nb \rfloor} \Phi(nx-k)\right)}$$

$$- \frac{1}{\left(\sum_{k=\lceil na \rceil}^{\lfloor nb \rfloor} \Phi(nx-k)\right)} \left(\sum_{j=1}^{N-1} \frac{f^{(j)}(x)}{j!} G_n^*\left((\cdot - x)^j\right)(x)\right) - f(x)$$

$$= \frac{1}{\left(\sum_{k=\lceil na \rceil}^{\lfloor nb \rfloor} \Phi(nx-k)\right)} \cdot \tag{3.90}$$

$$\left[G_n^*(f,x) - \left(\sum_{j=1}^{N-1} \frac{f^{(j)}(x)}{j!} G_n^*\left((\cdot - x)^j\right)(x)\right) - \left(\sum_{k=\lceil na \rceil}^{\lfloor nb \rfloor} \Phi(nx-k)\right) f(x)\right].$$

Therefore we get

$$\left|G_n(f,x) - \sum_{j=1}^{N-1} \frac{f^{(j)}(x)}{j!} G_n\left((\cdot - x)^j\right)(x) - f(x)\right| \leq (5.250312578) \cdot$$

$$\left|G_n^*(f,x) - \left(\sum_{j=1}^{N-1} \frac{f^{(j)}(x)}{j!} G_n^*\left((\cdot - x)^j\right)(x)\right) - \left(\sum_{k=\lceil na \rceil}^{\lfloor nb \rfloor} \Phi(nx-k)\right) f(x)\right|, \tag{3.91}$$

$\forall\, x \in [a,b]$.

The proof of the theorem is now complete. ∎

We give our second main result

Theorem 3.24 *Let $\alpha > 0$, $N = \lceil \alpha \rceil$, $\alpha \notin \mathbb{N}$, $f \in AC^N([a,b])$, with $f^{(N)} \in L_\infty([a,b])$, $0 < \beta < 1$, $x \in [a,b]$, $n \in \mathbb{N}$. Then*

(i)

$$\left|F_n(f,x) - \sum_{j=1}^{N-1} \frac{f^{(j)}(x)}{j!} F_n\left((\cdot - x)^j\right)(x) - f(x)\right| \leq$$

$$\frac{(4.1488766)}{\Gamma(\alpha+1)} \cdot \left\{ \frac{\left(\omega_1\left(D_{x-}^{\alpha} f, \frac{1}{n^{\beta}}\right)_{[a,x]} + \omega_1\left(D_{*x}^{\alpha} f, \frac{1}{n^{\beta}}\right)_{[x,b]}\right)}{n^{\alpha\beta}} + \right.$$

$$e^4 e^{-2n^{(1-\beta)}} \left(\left\| D_{x-}^{\alpha} f \right\|_{\infty,[a,x]} (x-a)^{\alpha} + \left\| D_{*x}^{\alpha} f \right\|_{\infty,[x,b]} (b-x)^{\alpha} \right) \Bigg\}, \quad (3.92)$$

(ii) if $f^{(j)}(x) = 0$, for $j = 1, \ldots, N-1$, we have

$$|F_n(f,x) - f(x)| \leq \frac{(4.1488766)}{\Gamma(\alpha+1)} \cdot$$

$$\left\{ \frac{\left(\omega_1 \left(D_{x-}^{\alpha} f, \frac{1}{n^{\beta}} \right)_{[a,x]} + \omega_1 \left(D_{*x}^{\alpha} f, \frac{1}{n^{\beta}} \right)_{[x,b]} \right)}{n^{\alpha\beta}} + \right.$$

$$e^4 e^{-2n^{(1-\beta)}} \left(\left\| D_{x-}^{\alpha} f \right\|_{\infty,[a,x]} (x-a)^{\alpha} + \left\| D_{*x}^{\alpha} f \right\|_{\infty,[x,b]} (b-x)^{\alpha} \right) \Bigg\}, \quad (3.93)$$

when $\alpha > 1$ notice here the extremely high rate of convergence of $n^{-(\alpha+1)\beta}$,
(iii)

$$|F_n(f,x) - f(x)| \leq (4.1488766) \cdot$$

$$\left\{ \sum_{j=1}^{N-1} \frac{\left| f^{(j)}(x) \right|}{j!} \left\{ \frac{1}{n^{\beta j}} + (b-a)^j e^4 e^{-2n^{(1-\beta)}} \right\} + \right.$$

$$\frac{1}{\Gamma(\alpha+1)} \left\{ \frac{\left(\omega_1 \left(D_{x-}^{\alpha} f, \frac{1}{n^{\beta}} \right)_{[a,x]} + \omega_1 \left(D_{*x}^{\alpha} f, \frac{1}{n^{\beta}} \right)_{[x,b]} \right)}{n^{\alpha\beta}} + \right.$$

$$e^4 e^{-2n^{(1-\beta)}} \left(\left\| D_{x-}^{\alpha} f \right\|_{\infty,[a,x]} (x-a)^{\alpha} + \left\| D_{*x}^{\alpha} f \right\|_{\infty,[x,b]} (b-x)^{\alpha} \right) \Bigg\} \Bigg\}, \quad (3.94)$$

$\forall\, x \in [a,b]$,
and
(iv)

$$\|F_n f - f\|_{\infty} \leq (4.1488766) \cdot$$

$$\left\{ \sum_{j=1}^{N-1} \frac{\|f^{(j)}\|_\infty}{j!} \left\{ \frac{1}{n^{\beta j}} + (b-a)^j e^4 e^{-2n^{(1-\beta)}} \right\} + \right.$$

$$\frac{1}{\Gamma(\alpha+1)} \left\{ \frac{\left(\sup_{x\in[a,b]} \omega_1 \left(D_{x-}^\alpha f, \frac{1}{n^\beta} \right)_{[a,x]} + \sup_{x\in[a,b]} \omega_1 \left(D_{*x}^\alpha f, \frac{1}{n^\beta} \right)_{[x,b]} \right)}{n^{\alpha\beta}} + \right.$$

$$\left. \left. e^4 e^{-2n^{(1-\beta)}} (b-a)^\alpha \left(\sup_{x\in[a,b]} \left\| D_{x-}^\alpha f \right\|_{\infty,[a,x]} + \sup_{x\in[a,b]} \left\| D_{*x}^\alpha f \right\|_{\infty,[x,b]} \right) \right\} \right\}. \tag{3.95}$$

Above, when $N = 1$ *the sum* $\sum_{j=1}^{N-1} \cdot = 0$.

As we see here we obtain fractionally type pointwise and uniform convergence with rates of $F_n \to I$ *the unit operator, as* $n \to \infty$.

Proof Let $x \in [a,b]$. We have that $D_{x-}^\alpha f(x) = D_{*x}^\alpha f(x) = 0$.

From [14], p. 54, we get by the left Caputo fractional Taylor formula that

$$f\left(\frac{k}{n}\right) = \sum_{j=0}^{N-1} \frac{f^{(j)}(x)}{j!} \left(\frac{k}{n} - x\right)^j + \tag{3.96}$$

$$\frac{1}{\Gamma(\alpha)} \int_x^{\frac{k}{n}} \left(\frac{k}{n} - J\right)^{\alpha-1} \left(D_{*x}^\alpha f(J) - D_{*x}^\alpha f(x) \right) dJ,$$

for all $x \le \frac{k}{n} \le b$.

Also from [3], using the right Caputo fractional Taylor formula we get

$$f\left(\frac{k}{n}\right) = \sum_{j=0}^{N-1} \frac{f^{(j)}(x)}{j!} \left(\frac{k}{n} - x\right)^j + \tag{3.97}$$

$$\frac{1}{\Gamma(\alpha)} \int_{\frac{k}{n}}^x \left(J - \frac{k}{n}\right)^{\alpha-1} \left(D_{x-}^\alpha f(J) - D_{x-}^\alpha f(x) \right) dJ,$$

for all $a \le \frac{k}{n} \le x$.

Hence we have

$$f\left(\frac{k}{n}\right)\Psi(nx-k)=\sum_{j=0}^{N-1}\frac{f^{(j)}(x)}{j!}\Psi(nx-k)\left(\frac{k}{n}-x\right)^{j}+ \tag{3.98}$$

$$\frac{\Psi(nx-k)}{\Gamma(\alpha)}\int_{x}^{\frac{k}{n}}\left(\frac{k}{n}-J\right)^{\alpha-1}\left(D_{*x}^{\alpha}f(J)-D_{*x}^{\alpha}f(x)\right)dJ,$$

for all $x\leq\frac{k}{n}\leq b$, iff $\lceil nx\rceil\leq k\leq\lfloor nb\rfloor$, and

$$f\left(\frac{k}{n}\right)\Psi(nx-k)=\sum_{j=0}^{N-1}\frac{f^{(j)}(x)}{j!}\Psi(nx-k)\left(\frac{k}{n}-x\right)^{j}+ \tag{3.99}$$

$$\frac{\Psi(nx-k)}{\Gamma(\alpha)}\int_{\frac{k}{n}}^{x}\left(J-\frac{k}{n}\right)^{\alpha-1}\left(D_{x-}^{\alpha}f(J)-D_{x-}^{\alpha}f(x)\right)dJ,$$

for all $a\leq\frac{k}{n}\leq x$, iff $\lceil na\rceil\leq k\leq\lfloor nx\rfloor$.

Therefore it holds

$$\sum_{k=\lfloor nx\rfloor+1}^{\lfloor nb\rfloor}f\left(\frac{k}{n}\right)\Psi(nx-k)=\sum_{j=0}^{N-1}\frac{f^{(j)}(x)}{j!}\sum_{k=\lfloor nx\rfloor+1}^{\lfloor nb\rfloor}\Psi(nx-k)\left(\frac{k}{n}-x\right)^{j}+ \tag{3.100}$$

$$\frac{1}{\Gamma(\alpha)}\sum_{k=\lfloor nx\rfloor+1}^{\lfloor nb\rfloor}\Psi(nx-k)\int_{x}^{\frac{k}{n}}\left(\frac{k}{n}-J\right)^{\alpha-1}\left(D_{*x}^{\alpha}f(J)-D_{*x}^{\alpha}f(x)\right)dJ,$$

and

$$\sum_{k=\lceil na\rceil}^{\lfloor nx\rfloor}f\left(\frac{k}{n}\right)\Psi(nx-k)=\sum_{j=0}^{N-1}\frac{f^{(j)}(x)}{j!}\sum_{k=\lceil na\rceil}^{\lfloor nx\rfloor}\Psi(nx-k)\left(\frac{k}{n}-x\right)^{j}+ \tag{3.101}$$

$$\frac{1}{\Gamma(\alpha)}\sum_{k=\lceil na\rceil}^{\lfloor nx\rfloor}\Psi(nx-k)\int_{\frac{k}{n}}^{x}\left(J-\frac{k}{n}\right)^{\alpha-1}\left(D_{x-}^{\alpha}f(J)-D_{x-}^{\alpha}f(x)\right)dJ.$$

Adding the last two equalities (3.100) and (3.101) obtain

$$F_{n}^{*}(f,x)=\sum_{k=\lceil na\rceil}^{\lfloor nb\rfloor}f\left(\frac{k}{n}\right)\Psi(nx-k)= \tag{3.102}$$

$$\sum_{j=0}^{N-1} \frac{f^{(j)}(x)}{j!} \sum_{k=\lceil na \rceil}^{\lfloor nb \rfloor} \Psi(nx-k)\left(\frac{k}{n}-x\right)^j +$$

$$\frac{1}{\Gamma(\alpha)}\left\{\sum_{k=\lceil na \rceil}^{\lfloor nx \rfloor} \Psi(nx-k) \int_{\frac{k}{n}}^{x}\left(J-\frac{k}{n}\right)^{\alpha-1}\left(D_{x-}^{\alpha} f(J)-D_{x-}^{\alpha} f(x)\right) dJ+\right.$$

$$\left.\sum_{k=\lfloor nx \rfloor+1}^{\lfloor nb \rfloor} \Psi(nx-k) \int_{x}^{\frac{k}{n}}\left(\frac{k}{n}-J\right)^{\alpha-1}\left(D_{*x}^{\alpha} f(J)-D_{*x}^{\alpha} f(x)\right) dJ\right\}.$$

So we have derived

$$F_n^*(f, x)-f(x)\left(\sum_{k=\lceil na \rceil}^{\lfloor nb \rfloor} \Psi(nx-k)\right)= \tag{3.103}$$

$$\sum_{j=1}^{N-1} \frac{f^{(j)}(x)}{j!} F_n^*\left((\cdot-x)^j\right)(x)+u_n(x),$$

where

$$u_n(x):=\frac{1}{\Gamma(\alpha)}\left\{\sum_{k=\lceil na \rceil}^{\lfloor nx \rfloor} \Psi(nx-k) \int_{\frac{k}{n}}^{x}\left(J-\frac{k}{n}\right)^{\alpha-1}\left(D_{x-}^{\alpha} f(J)-D_{x-}^{\alpha} f(x)\right) dJ\right.$$

$$\left.+\sum_{k=\lfloor nx \rfloor+1}^{\lfloor nb \rfloor} \Psi(nx-k) \int_{x}^{\frac{k}{n}}\left(\frac{k}{n}-J\right)^{\alpha-1}\left(D_{*x}^{\alpha} f(J)-D_{*x}^{\alpha} f(x)\right) dJ\right\}. \tag{3.104}$$

We set

$$u_{1n}(x):=\frac{1}{\Gamma(\alpha)} \sum_{k=\lceil na \rceil}^{\lfloor nx \rfloor} \Psi(nx-k) \int_{\frac{k}{n}}^{x}\left(J-\frac{k}{n}\right)^{\alpha-1}\left(D_{x-}^{\alpha} f(J)-D_{x-}^{\alpha} f(x)\right) dJ, \tag{3.105}$$

and

$$u_{2n}:=\frac{1}{\Gamma(\alpha)} \sum_{k=\lfloor nx \rfloor+1}^{\lfloor nb \rfloor} \Psi(nx-k) \int_{x}^{\frac{k}{n}}\left(\frac{k}{n}-J\right)^{\alpha-1}\left(D_{*x}^{\alpha} f(J)-D_{*x}^{\alpha} f(x)\right) dJ, \tag{3.106}$$

i.e.

$$u_n(x)=u_{1n}(x)+u_{2n}(x). \tag{3.107}$$

We assume $b - a > \frac{1}{n^\beta}$, $0 < \beta < 1$, which is always the case for large enough $n \in \mathbb{N}$, that is when $n > \left\lceil (b-a)^{-\frac{1}{\beta}} \right\rceil$. It is always true that either $\left|\frac{k}{n} - x\right| \leq \frac{1}{n^\beta}$ or $\left|\frac{k}{n} - x\right| > \frac{1}{n^\beta}$.

For $k = \lceil na \rceil, \ldots, \lfloor nx \rfloor$, we consider

$$\gamma_{1k} := \left| \int_{\frac{k}{n}}^{x} \left(J - \frac{k}{n} \right)^{\alpha - 1} \left(D_{x-}^{\alpha} f(J) - D_{x-}^{\alpha} f(x) \right) dJ \right|. \tag{3.108}$$

As in the proof of Theorem 3.23 we get

$$\gamma_{1k} \leq \left\| D_{x-}^{\alpha} f \right\|_{\infty,[a,x]} \frac{(x-a)^{\alpha}}{\alpha}, \tag{3.109}$$

for $k = \lceil na \rceil, \ldots, \lfloor nx \rfloor$, and when $\left|x - \frac{k}{n}\right| \leq \frac{1}{n^\beta}$ then

$$\gamma_{1k} \leq \frac{\omega_1 \left(D_{x-}^{\alpha} f, \frac{1}{n^\beta} \right)_{[a,x]}}{\alpha n^{a\beta}}. \tag{3.110}$$

Consequently we obtain

$$|u_{1n}(x)| \leq \frac{1}{\Gamma(\alpha)} \sum_{k=\lceil na \rceil}^{\lfloor nx \rfloor} \Psi(nx - k) \gamma_{1k} = \tag{3.111}$$

$$\frac{1}{\Gamma(\alpha)} \left\{ \sum_{\substack{k = \lceil na \rceil \\ : \left|\frac{k}{n} - x\right| \leq \frac{1}{n^\beta}}}^{\lfloor nx \rfloor} \Psi(nx - k) \gamma_{1k} + \sum_{\substack{k = \lceil na \rceil \\ : \left|\frac{k}{n} - x\right| > \frac{1}{n^\beta}}}^{\lfloor nx \rfloor} \Psi(nx - k) \gamma_{1k} \right\} \leq$$

$$\frac{1}{\Gamma(\alpha)} \left\{ \left(\sum_{\substack{k = \lceil na \rceil \\ : \left|\frac{k}{n} - x\right| \leq \frac{1}{n^\beta}}}^{\lfloor nx \rfloor} \Psi(nx - k) \right) \frac{\omega_1 \left(D_{x-}^{\alpha} f, \frac{1}{n^\beta} \right)_{[a,x]}}{\alpha n^{\alpha\beta}} + \right.$$

$$\left(\sum_{\begin{cases} k = \lceil na \rceil \\ : \left| \frac{k}{n} - x \right| > \frac{1}{n^\beta} \end{cases}}^{\lfloor nx \rfloor} \Psi(nx - k) \right) \left\| D_{x-}^\alpha f \right\|_{\infty,[a,x]} \frac{(x-a)^\alpha}{\alpha} \Bigg\} \leq \tag{3.112}$$

$$\frac{1}{\Gamma(\alpha+1)} \left\{ \frac{\omega_1\left(D_{x-}^\alpha f, \frac{1}{n^\beta}\right)_{[a,x]}}{n^{\alpha\beta}} + \right.$$

$$\left(\sum_{\begin{cases} k = -\infty \\ : |nx - k| > n^{1-\beta} \end{cases}}^{\infty} \Psi(nx - k) \right) \left\| D_{x-}^\alpha f \right\|_{\infty,[a,x]} (x-a)^\alpha \Bigg\} \leq$$

$$\frac{1}{\Gamma(\alpha+1)} \left\{ \frac{\omega_1\left(D_{x-}^\alpha f, \frac{1}{n^\beta}\right)_{[a,x]}}{n^{\alpha\beta}} + e^4 e^{-2n^{(1-\beta)}} \left\| D_{x-}^\alpha f \right\|_{\infty,[a,x]} (x-a)^\alpha \right\}.$$

So we have proved that

$$|u_{1n}(x)| \leq \frac{1}{\Gamma(\alpha+1)} \left\{ \frac{\omega_1\left(D_{x-}^\alpha f, \frac{1}{n^\beta}\right)_{[a,x]}}{n^{\alpha\beta}} + \right. \tag{3.113}$$

$$\left. e^4 e^{-2n^{(1-\beta)}} \left\| D_{x-}^\alpha f \right\|_{\infty,[a,x]} (x-a)^\alpha \right\}.$$

Next when $k = \lfloor nx \rfloor + 1, \dots, \lfloor nb \rfloor$ we consider

$$\gamma_{2k} := \left| \int_x^{\frac{k}{n}} \left(\frac{k}{n} - J \right)^{\alpha-1} \left(D_{*x}^\alpha f(J) - D_{*x}^\alpha f(x) \right) dJ \right|. \tag{3.114}$$

As in the proof of Theorem 3.23, when $k = \lfloor nx \rfloor + 1, \ldots, \lfloor nb \rfloor$ we get that

$$\gamma_{2k} \leq \left\| D_{*x}^{\alpha} f \right\|_{\infty,[x,b]} \frac{(b-x)^{\alpha}}{\alpha}, \tag{3.115}$$

and when $\left|\frac{k}{n} - x\right| \leq \frac{1}{n^{\beta}}$, we derive

$$\gamma_{2k} \leq \frac{\omega_1 \left(D_{*x}^{\alpha} f, \frac{1}{n^{\beta}} \right)_{[x,b]}}{\alpha n^{\alpha\beta}}. \tag{3.116}$$

Similarly we have that

$$|u_{2n}(x)| \leq \frac{1}{\Gamma(\alpha)} \left(\sum_{k=\lfloor nx \rfloor+1}^{\lfloor nb \rfloor} \Psi(nx-k) \gamma_{2k} \right) =$$

$$\frac{1}{\Gamma(\alpha)} \left\{ \sum_{\substack{k = \lfloor nx \rfloor + 1 \\ : \left|\frac{k}{n} - x\right| \leq \frac{1}{n^{\beta}}}}^{\lfloor nb \rfloor} \Psi(nx-k) \gamma_{2k} + \sum_{\substack{k = \lfloor nx \rfloor + 1 \\ : \left|\frac{k}{n} - x\right| > \frac{1}{n^{\beta}}}}^{\lfloor nb \rfloor} \Psi(nx-k) \gamma_{2k} \right\} \leq \tag{3.117}$$

$$\frac{1}{\Gamma(\alpha)} \left\{ \left(\sum_{\substack{k = \lfloor nx \rfloor + 1 \\ : \left|\frac{k}{n} - x\right| \leq \frac{1}{n^{\beta}}}}^{\lfloor nb \rfloor} \Psi(nx-k) \right) \frac{\omega_1 \left(D_{*x}^{\alpha} f, \frac{1}{n^{\beta}} \right)_{[x,b]}}{\alpha n^{\alpha\beta}} + \right.$$

$$\left. \left(\sum_{\substack{k = \lfloor nx \rfloor + 1 \\ : \left|\frac{k}{n} - x\right| > \frac{1}{n^{\beta}}}}^{\lfloor nb \rfloor} \Psi(nx-k) \right) \left\| D_{*x}^{\alpha} f \right\|_{\infty,[x,b]} \frac{(b-x)^{\alpha}}{\alpha} \right\} \leq$$

$$\frac{1}{\Gamma(\alpha+1)} \left\{ \frac{\omega_1 \left(D_{*x}^{\alpha} f, \frac{1}{n^{\beta}} \right)_{[x,b]}}{n^{\alpha\beta}} + \right.$$

$$\left(\sum_{\substack{k=-\infty \\ \left\{ : \left|\frac{k}{n}-x\right| > \frac{1}{n^{\beta}} \right.}}^{\infty} \Psi(nx-k) \right) \left\| D_{*x}^{\alpha} f \right\|_{\infty,[x,b]} (b-x)^{\alpha} \Bigg\} \leq \tag{3.118}$$

$$\frac{1}{\Gamma(\alpha+1)} \left\{ \frac{\omega_1\left(D_{*x}^{\alpha} f, \frac{1}{n^{\beta}}\right)_{[x,b]}}{n^{\alpha\beta}} + e^4 e^{-2n^{(1-\beta)}} \left\| D_{*x}^{\alpha} f \right\|_{\infty,[x,b]} (b-x)^{\alpha} \right\}.$$

So we have proved that

$$|u_{2n}(x)| \leq \frac{1}{\Gamma(\alpha+1)} \left\{ \frac{\omega_1\left(D_{*x}^{\alpha} f, \frac{1}{n^{\beta}}\right)_{[x,b]}}{n^{\alpha\beta}} + \right. \tag{3.119}$$

$$\left. e^4 e^{-2n^{(1-\beta)}} \left\| D_{*x}^{\alpha} f \right\|_{\infty,[x,b]} (b-x)^{\alpha} \right\}.$$

Therefore

$$|u_n(x)| \leq |u_{1n}(x)| + |u_{2n}(x)| \leq$$

$$\frac{1}{\Gamma(\alpha+1)} \left\{ \frac{\omega_1\left(D_{x-}^{\alpha} f, \frac{1}{n^{\beta}}\right)_{[a,x]} + \omega_1\left(D_{*x}^{\alpha} f, \frac{1}{n^{\beta}}\right)_{[x,b]}}{n^{\alpha\beta}} + \right. \tag{3.120}$$

$$\left. e^4 e^{-2n^{(1-\beta)}} \left(\left\| D_{x-}^{\alpha} f \right\|_{\infty,[a,x]} (x-a)^{\alpha} + \left\| D_{*x}^{\alpha} f \right\|_{\infty,[x,b]} (b-x)^{\alpha} \right) \right\}.$$

From [8] we get that

$$\left| F_n^*\left((\cdot - x)^j \right)(x) \right| \leq \frac{1}{n^{\beta j}} + (b-a)^j e^4 e^{-2n^{(1-\beta)}}, \tag{3.121}$$

for $j = 1, \ldots, N-1$, $\forall\, x \in [a, b]$.

Putting things together, we have established

$$\left| F_n^* (f, x) - f (x) \left(\sum_{k=\lceil na \rceil}^{\lfloor nb \rfloor} \Psi (nx - k) \right) \right| \leq \tag{3.122}$$

$$\sum_{j=1}^{N-1} \frac{\left| f^{(j)} (x) \right|}{j!} \left[\frac{1}{n^{\beta j}} + (b - a)^j e^4 e^{-2n^{(1-\beta)}} \right] +$$

$$\frac{1}{\Gamma (\alpha + 1)} \left\{ \frac{\left(\omega_1 \left(D_{x-}^{\alpha} f, \frac{1}{n^{\beta}} \right)_{[a,x]} + \omega_1 \left(D_{*x}^{\alpha} f, \frac{1}{n^{\beta}} \right)_{[x,b]} \right)}{n^{\alpha \beta}} + \right.$$

$$\left. e^4 e^{-2n^{(1-\beta)}} \left(\left\| D_{x-}^{\alpha} f \right\|_{\infty,[a,x]} (x - a)^{\alpha} + \left\| D_{*x}^{\alpha} f \right\|_{\infty,[x,b]} (b - x)^{\alpha} \right) \right\} =: \overline{A}_n (x) . \tag{3.123}$$

As a result we derive

$$|F_n (f, x) - f (x)| \leq (4.1488766) \overline{A}_n (x) , \tag{3.124}$$

$\forall \, x \in [a, b]$.

We further have that

$$\left\| \overline{A}_n \right\|_{\infty} \leq \sum_{j=1}^{N-1} \frac{\left\| f^{(j)} \right\|_{\infty}}{j!} \left[\frac{1}{n^{\beta j}} + (b - a)^j e^4 e^{-2n^{(1-\beta)}} \right] + \tag{3.125}$$

$$\frac{1}{\Gamma (\alpha + 1)} \left\{ \frac{\left\{ \sup_{x \in [a,b]} \left(\omega_1 \left(D_{x-}^{\alpha} f, \frac{1}{n^{\beta}} \right)_{[a,x]} \right) + \sup_{x \in [a,b]} \left(\omega_1 \left(D_{*x}^{\alpha} f, \frac{1}{n^{\beta}} \right)_{[x,b]} \right) \right\}}{n^{\alpha \beta}} \right.$$

$$+ e^4 e^{-2n^{(1-\beta)}} (b - a)^{\alpha} \cdot$$

$$\left. \left\{ \left(\sup_{x \in [a,b]} \left(\left\| D_{x-}^{\alpha} f \right\|_{\infty,[a,x]} \right) + \sup_{x \in [a,b]} \left(\left\| D_{*x}^{\alpha} f \right\|_{\infty,[x,b]} \right) \right) \right\} \right\} =: \overline{B}_n .$$

Hence it holds

$$\|F_n f - f\|_\infty \leq (4.1488766)\,\overline{B}_n. \tag{3.126}$$

Similarly, as in the proof of Theorem 3.23, we can prove that $\overline{B}_n < \infty$.

We finally notice that

$$F_n(f,x) - \sum_{j=1}^{N-1} \frac{f^{(j)}(x)}{j!} F_n\left((\cdot - x)^j\right)(x) - f(x) = \frac{F_n^*(f,x)}{\left(\sum_{k=\lceil na \rceil}^{\lfloor nb \rfloor} \Psi(nx-k)\right)}$$

$$- \frac{1}{\left(\sum_{k=\lceil na \rceil}^{\lfloor nb \rfloor} \Psi(nx-k)\right)} \left(\sum_{j=1}^{N-1} \frac{f^{(j)}(x)}{j!} F_n^*\left((\cdot - x)^j\right)(x)\right) - f(x)$$

$$= \frac{1}{\left(\sum_{k=\lceil na \rceil}^{\lfloor nb \rfloor} \Psi(nx-k)\right)}. \tag{3.127}$$

$$\left[F_n^*(f,x) - \left(\sum_{j=1}^{N-1} \frac{f^{(j)}(x)}{j!} F_n^*\left((\cdot - x)^j\right)(x)\right) - \left(\sum_{k=\lceil na \rceil}^{\lfloor nb \rfloor} \Psi(nx-k)\right) f(x)\right].$$

Therefore we get

$$\left|F_n(f,x) - \sum_{j=1}^{N-1} \frac{f^{(j)}(x)}{j!} F_n\left((\cdot - x)^j\right)(x) - f(x)\right| \leq (4.1488766)\cdot$$

$$\left|F_n^*(f,x) - \left(\sum_{j=1}^{N-1} \frac{f^{(j)}(x)}{j!} F_n^*\left((\cdot - x)^j\right)(x)\right) - \left(\sum_{k=\lceil na \rceil}^{\lfloor nb \rfloor} \Psi(nx-k)\right) f(x)\right|, \tag{3.128}$$

$\forall\, x \in [a,b]$.

The proof of the theorem is now finished. ■

Next we apply Theorem 3.23 for $N = 1$.

Corollary 3.25 *Let* $0 < \alpha, \beta < 1$, $f \in AC([a,b])$, $f' \in L_\infty([a,b])$, $n \in \mathbb{N}$. *Then*

$$\|G_n f - f\|_\infty \leq \frac{(5.250312578)}{\Gamma(\alpha+1)}. \tag{3.129}$$

$$\left\{ \frac{\left(\sup_{x\in[a,b]} \omega_1 \left(D_{x-}^{\alpha} f, \frac{1}{n^{\beta}} \right)_{[a,x]} + \sup_{x\in[a,b]} \omega_1 \left(D_{*x}^{\alpha} f, \frac{1}{n^{\beta}} \right)_{[x,b]} \right)}{n^{\alpha\beta}} + \right.$$

$$\left. 3.1992 e^{-n^{(1-\beta)}} (b-a)^{\alpha} \left(\sup_{x\in[a,b]} \left\| D_{x-}^{\alpha} f \right\|_{\infty,[a,x]} + \sup_{x\in[a,b]} \left\| D_{*x}^{\alpha} f \right\|_{\infty,[x,b]} \right) \right\}.$$

Also we apply Theorem 3.24 for $N = 1$.

Corollary 3.26 *Let* $0 < \alpha, \beta < 1$, $f \in AC([a,b])$, $f' \in L_{\infty}([a,b])$, $n \in \mathbb{N}$. *Then*

$$\|F_n f - f\|_{\infty} \leq \frac{(4.1488766)}{\Gamma(\alpha+1)} \cdot \tag{3.130}$$

$$\left\{ \frac{\left(\sup_{x\in[a,b]} \omega_1 \left(D_{x-}^{\alpha} f, \frac{1}{n^{\beta}} \right)_{[a,x]} + \sup_{x\in[a,b]} \omega_1 \left(D_{*x}^{\alpha} f, \frac{1}{n^{\beta}} \right)_{[x,b]} \right)}{n^{\alpha\beta}} + \right.$$

$$\left. e^4 e^{-2n^{(1-\beta)}} (b-a)^{\alpha} \left(\sup_{x\in[a,b]} \left\| D_{x-}^{\alpha} f \right\|_{\infty,[a,x]} + \sup_{x\in[a,b]} \left\| D_{*x}^{\alpha} f \right\|_{\infty,[x,b]} \right) \right\}.$$

We make

Remark 3.27 Let $0 < \beta < 1$, $\alpha > 0$, $N = \lceil \alpha \rceil$, $\alpha \notin \mathbb{N}$, $f \in C^N([a,b])$, $n \in \mathbb{N}$. Then, by Corollary 3.13 and Theorems 3.14, 3.15, there exist $x_1, x_2 \in [a,b]$ depended on n, such that

$$\sup_{x\in[a,b]} \omega_1 \left(D_{x-}^{\alpha} f, \frac{1}{n^{\beta}} \right)_{[a,x]} = \omega_1 \left(D_{x_1-}^{\alpha} f, \frac{1}{n^{\beta}} \right)_{[a,x_1]}, \tag{3.131}$$

and

$$\sup_{x\in[a,b]} \omega_1 \left(D_{*x}^{\alpha} f, \frac{1}{n^{\beta}} \right)_{[x,b]} = \omega_1 \left(D_{*x_2}^{\alpha} f, \frac{1}{n^{\beta}} \right)_{[x_2,b]}. \tag{3.132}$$

Clearly here we have in particular that

$$\omega_1\left(D^{\alpha}_{x_1-}f,\frac{1}{n^{\beta}}\right)_{[a,x_1]} \rightarrow 0, \tag{3.133}$$
$$\omega_1\left(D^{\alpha}_{*x_2}f,\frac{1}{n^{\beta}}\right)_{[x_2,b]} \rightarrow 0,$$

as $n\rightarrow\infty$. Notice that to each n may correspond different $x_1, x_2 \in [a,b]$.

Remark 3.28 Let $0<\alpha<1$, then by (3.84), we get

$$\sup_{x\in[a,b]}\left\|D^{\alpha}_{x-}f\right\|_{\infty,[a,x]} \leq \frac{(b-a)^{1-\alpha}}{\Gamma(2-\alpha)}\left\|f'\right\|_{\infty}, \tag{3.134}$$

and by (3.87), we obtain

$$\sup_{x\in[a,b]}\left\|D^{\alpha}_{*x}f\right\|_{\infty,[x,b]} \leq \frac{(b-a)^{1-\alpha}}{\Gamma(2-\alpha)}\left\|f'\right\|_{\infty}, \tag{3.135}$$

given that $f\in AC([a,b])$ and $f'\in L_{\infty}([a,b])$.

Next we specialize to $\alpha=\frac{1}{2}$.

Corollary 3.29 *Let* $0<\beta<1$, $f\in AC([a,b])$, $f'\in L_{\infty}([a,b])$, $n\in\mathbb{N}$. *Then*
(i)

$$\|G_n f - f\|_{\infty} \leq \frac{(10.50062516)}{\sqrt{\pi}}\cdot$$
$$\left\{\frac{\left(\sup_{x\in[a,b]}\omega_1\left(D^{\frac{1}{2}}_{x-}f,\frac{1}{n^{\beta}}\right)_{[a,x]}+\sup_{x\in[a,b]}\omega_1\left(D^{\frac{1}{2}}_{*x}f,\frac{1}{n^{\beta}}\right)_{[x,b]}\right)}{n^{\frac{\beta}{2}}}+\right.$$
$$\left.3.1992e^{-n^{(1-\beta)}}\sqrt{b-a}\left(\sup_{x\in[a,b]}\left\|D^{\frac{1}{2}}_{x-}f\right\|_{\infty,[a,x]}+\sup_{x\in[a,b]}\left\|D^{\frac{1}{2}}_{*x}f\right\|_{\infty,[x,b]}\right)\right\}, \tag{3.136}$$

and
(ii)

$$\|F_n f - f\|_{\infty} \leq \frac{(8.2977532)}{\sqrt{\pi}}\cdot$$

$$\left\{ \frac{\left(\sup_{x\in[a,b]} \omega_1 \left(D_{x-}^{\frac{1}{2}} f, \frac{1}{n^\beta} \right)_{[a,x]} + \sup_{x\in[a,b]} \omega_1 \left(D_{*x}^{\frac{1}{2}} f, \frac{1}{n^\beta} \right)_{[x,b]} \right)}{n^{\frac{\beta}{2}}} + \right.$$

$$\left. e^4 e^{-2n^{(1-\beta)}} \sqrt{b-a} \left(\sup_{x\in[a,b]} \left\| D_{x-}^{\frac{1}{2}} f \right\|_{\infty,[a,x]} + \sup_{x\in[a,b]} \left\| D_{*x}^{\frac{1}{2}} f \right\|_{\infty,[x,b]} \right) \right\}. \tag{3.137}$$

We finish with

Remark 3.30 (to Corollary 3.29) Assume that

$$\omega_1 \left(D_{x-}^{\frac{1}{2}} f, \frac{1}{n^\beta} \right)_{[a,x]} \leq \frac{K_1}{n^\beta}, \tag{3.138}$$

and

$$\omega_1 \left(D_{*x}^{\frac{1}{2}} f, \frac{1}{n^\beta} \right)_{[x,b]} \leq \frac{K_2}{n^\beta}, \tag{3.139}$$

$\forall\, x \in [a, b]$, $\forall\, n \in \mathbb{N}$, where $K_1, K_2 > 0$.

Then for large enough $n \in \mathbb{N}$, by (3.136) and (3.137), we obtain

$$\| G_n f - f \|_\infty, \| F_n f - f \|_\infty \leq \frac{T}{n^{\frac{3}{2}\beta}}, \tag{3.140}$$

for some $T > 0$.

The speed of convergence in (3.140) is much higher than the corresponding speeds achieved in [8, 11], which were there $\frac{1}{n^\beta}$.

References

1. G.A. Anastassiou, Rate of convergence of some neural network operators to the unit-univariate case. J. Math. Anal. Appl. **212**, 237–262 (1997)
2. G.A. Anastassiou, *Quantitative Approximations* (Chapman & Hall/CRC, Boca Raton, 2001)
3. G.A. Anastassiou, On right fractional calculus. Chaos, Solitons Fractals **42**, 365–376 (2009)
4. G.A. Anastassiou, *Fractional Differentiation Inequalities* (Springer, New York, 2009)
5. G. Anastassiou, Fractional Korovkin theory. Chaos, Solitons Fractals **42**(4), 2080–2094 (2009)
6. G.A. Anastassiou, *Inteligent Systems: Approximation by Artificial Neural Networks*, Intelligent Systems Reference Library, vol. 19 (Springer, Heidelberg, 2011)

7. G.A. Anastassiou, Fractional representation formulae and right fractional inequalities. Math. Comput. Model. **54**(11–12), 3098–3115 (2011)
8. G.A. Anastassiou, Univariate hyperbolic tangent neural network approximation. Math. Comput. Model. **53**, 1111–1132 (2011)
9. G.A. Anastassiou, Multivariate hyperbolic tangent neural network approximation. Comput. Math. **61**, 809–821 (2011)
10. G.A. Anastassiou, Multivariate sigmoidal neural network approximation. Neural Netw. **24**, 378–386 (2011)
11. G.A. Anastassiou, Univariate sigmoidal neural network approximation. J. Comput. Anal. Appl. **14**(4), 659–690 (2012)
12. G.A. Anastassiou, Fractional neural network approximation. Comput. Math. **64**, 1655–1676 (2012)
13. Z. Chen, F. Cao, The approximation operators with sigmoidal functions. Comput. Math. Appl. **58**, 758–765 (2009)
14. K. Diethelm, *The Analysis of Fractional Differential Equations*, Lecture Notes in Mathematics, vol. 2004 (Springer, Berlin, 2010)
15. A.M.A. El-Sayed, M. Gaber, On the finite Caputo and finite Riesz derivatives. Electron. J. Theor. Phys. **3**(12), 81–95 (2006)
16. G.S. Frederico, D.F.M. Torres, Fractional optimal control in the sense of Caputo and the fractional Noether's theorem. Int. Math. Forum **3**(10), 479–493 (2008)
17. S. Haykin, *Neural Networks: A Comprehensive Foundation*, 2nd edn. (Prentice Hall, New York, 1998)
18. W. McCulloch, W. Pitts, A logical calculus of the ideas immanent in nervous activity. Bull. Math. Biophys. **7**, 115–133 (1943)
19. T.M. Mitchell, *Machine Learning* (WCB-McGraw-Hill, New York, 1997)
20. S.G. Samko, A.A. Kilbas, O.I. Marichev, *Fractional Integrals and Derivatives, Theory and Applications* (Gordon and Breach, Amsterdam, 1993) [English translation from the Russian, Integrals and Derivatives of Fractional Order and Some of Their Applications (Nauka i Tekhnika, Minsk, 1987)]

Chapter 4
Fractional Approximation Using Cardaliaguet-Euvrard and Squashing Neural Networks

This chapter deals with the determination of the fractional rate of convergence to the unit of some neural network operators, namely, the Cardaliaguet-Euvrard and "squashing" operators. This is given through the moduli of continuity of the involved right and left Caputo fractional derivatives of the approximated function and they appear in the right-hand side of the associated Jackson type inequalities. It follows [7].

4.1 Introduction

The Cardaliaguet-Euvrard (4.22) operators were first introduced and studied extensively in [8], where the authors among many other things proved that these operators converge uniformly on compacta, to the unit over continuous and bounded functions. Our "squashing operator" (see [1]) (4.74) was motivated and inspired by the "squashing functions" and related Theorem 6 of [8]. The work in [8] is qualitative where the used bell-shaped function is general. However, our work, though greatly motivated by [8], is quantitative and the used bell-shaped and "squashing" functions are of compact support. We produce a series of Jackson type inequalities giving close upper bounds to the errors in approximating the unit operator by the above neural network induced operators. All involved constants there are well determined. These are pointwise, uniform and L_p, $p \geq 1$, estimates involving the first moduli of continuity of the engaged right and left Caputo fractional derivatives of the function under approximation. We give all necessary background of fractional calculus.

Initial work of the subject was done in [1], where we involved only ordinary derivatives. Article [1] motivated the current work.

G.A. Anastassiou, *Intelligent Systems II: Complete Approximation by Neural Network Operators*, Studies in Computational Intelligence 608,
DOI 10.1007/978-3-319-20505-2_4

4.2 Background

We need

Definition 4.1 Let $f \in C(\mathbb{R})$ which is bounded or uniformly continuous, $h > 0$. We define the first modulus of continuity of f at h as follows

$$\omega_1(f, h) = \sup\{|f(x) - f(y)|; x, y \in \mathbb{R}, \ |x - y| \le h\} \tag{4.1}$$

Notice that $\omega_1(f, h)$ is finite for any $h > 0$, and

$$\lim_{h \to 0} \omega_1(f, h) = 0.$$

We also need

Definition 4.2 Let $f : \mathbb{R} \to \mathbb{R}$, $\nu \ge 0$, $n = \lceil \nu \rceil$ ($\lceil \cdot \rceil$ is the ceiling of the number), $f \in AC^n([a, b])$ (space of functions f with $f^{(n-1)} \in AC([a, b])$, absolutely continuous functions), $\forall [a, b] \subset \mathbb{R}$. We call left Caputo fractional derivative (see [9], pp. 49–52) the function

$$D_{*a}^{\nu} f(x) = \frac{1}{\Gamma(n - \nu)} \int_a^x (x - t)^{n-\nu-1} f^{(n)}(t) \, dt, \tag{4.2}$$

$\forall x \ge a$, where Γ is the gamma function $\Gamma(\nu) = \int_0^{\infty} e^{-t} t^{\nu-1} dt$, $\nu > 0$. Notice $D_{*a}^{\nu} f \in L_1([a, b])$ and $D_{*a}^{\nu} f$ exists a.e. on $[a, b]$, $\forall b > a$. We set $D_{*a}^0 f(x) = f(x)$, $\forall x \in [a, \infty)$.

Lemma 4.3 ([5]) *Let* $\nu > 0$, $\nu \notin \mathbb{N}$, $n = \lceil \nu \rceil$, $f \in C^{n-1}(\mathbb{R})$ *and* $f^{(n)} \in L_{\infty}(\mathbb{R})$. *Then* $D_{*a}^{\nu} f(a) = 0$, $\forall a \in \mathbb{R}$.

Definition 4.4 (*see also* [2, 10, 11]) Let $f : \mathbb{R} \to \mathbb{R}$, such that $f \in AC^m([a, b])$, $\forall [a, b] \subset \mathbb{R}$, $m = \lceil \alpha \rceil$, $\alpha > 0$. The right Caputo fractional derivative of order $\alpha > 0$ is given by

$$D_{b-}^{\alpha} f(x) = \frac{(-1)^m}{\Gamma(m - \alpha)} \int_x^b (J - x)^{m-\alpha-1} f^{(m)}(J) \, dJ, \tag{4.3}$$

$\forall x \le b$. We set $D_{b-}^0 f(x) = f(x)$, $\forall x \in (-\infty, b]$. Notice that $D_{b-}^{\alpha} f \in L_1([a, b])$ and $D_{b-}^{\alpha} f$ exists a.e. on $[a, b]$, $\forall a < b$.

Lemma 4.5 ([5]) *Let* $f \in C^{m-1}(\mathbb{R})$, $f^{(m)} \in L_{\infty}(\mathbb{R})$, $m = \lceil \alpha \rceil$, $\alpha > 0$. *Then* $D_{b-}^{\alpha} f(b) = 0$, $\forall b \in \mathbb{R}$.

Convention 4.6 *We assume that*

$$D_{*x_0}^{\alpha} f(x) = 0, \ \textit{for } x < x_0, \tag{4.4}$$

and

$$D^{\alpha}_{x_0-} f(x) = 0, \text{ for } x > x_0, \tag{4.5}$$

for all $x, x_0 \in \mathbb{R}$.

We mention

Proposition 4.7 (by [3]) *Let* $f \in C^n(\mathbb{R})$, *where* $n = \lceil \nu \rceil$, $\nu > 0$. *Then* $D^{\nu}_{*a} f(x)$ *is continuous in* $x \in [a, \infty)$.

Also we have

Proposition 4.8 (by [3]) *Let* $f \in C^m(\mathbb{R})$, $m = \lceil \alpha \rceil$, $\alpha > 0$. *Then* $D^{\alpha}_{b-} f(x)$ *is continuous in* $x \in (-\infty, b]$.

We further mention

Proposition 4.9 (by [3]) *Let* $f \in C^{m-1}(\mathbb{R})$, $f^{(m)} \in L_{\infty}(\mathbb{R})$, $m = \lceil \alpha \rceil$, $\alpha > 0$ *and*

$$D^{\alpha}_{*x_0} f(x) = \frac{1}{\Gamma(m-\alpha)} \int_{x_0}^{x} (x-t)^{m-\alpha-1} f^{(m)}(t)\, dt, \tag{4.6}$$

for all $x, x_0 \in \mathbb{R} : x \geq x_0$.

Then $D^{\alpha}_{*x_0} f(x)$ *is continuous in* x_0.

Proposition 4.10 (by [3]) *Let* $f \in C^{m-1}(\mathbb{R})$, $f^{(m)} \in L_{\infty}(\mathbb{R})$, $m = \lceil \alpha \rceil$, $\alpha > 0$ *and*

$$D^{\alpha}_{x_0-} f(x) = \frac{(-1)^m}{\Gamma(m-\alpha)} \int_{x}^{x_0} (J-x)^{m-\alpha-1} f^{(m)}(J)\, dJ, \tag{4.7}$$

for all $x, x_0 \in \mathbb{R} : x_0 \geq x$.

Then $D^{\alpha}_{x_0-} f(x)$ *is continuous in* x_0.

Proposition 4.11 ([5]) *Let* $g \in C_b(\mathbb{R})$ *(continuous and bounded)*, $0 < c < 1$, $x, x_0 \in \mathbb{R}$. *Define*

$$L(x, x_0) = \int_{x_0}^{x} (x-t)^{c-1} g(t)\, dt, \text{ for } x \geq x_0, \tag{4.8}$$

and $L(x, x_0) = 0$, *for* $x < x_0$.

Then L *is jointly continuous in* $(x, x_0) \in \mathbb{R}^2$.

We mention

Proposition 4.12 ([5]) *Let* $g \in C_b(\mathbb{R})$, $0 < c < 1$, $x, x_0 \in \mathbb{R}$. *Define*

$$K(x, x_0) = \int_{x}^{x_0} (J-x)^{c-1} g(J)\, dJ, \text{ for } x \leq x_0, \tag{4.9}$$

and $K(x, x_0) = 0$, *for* $x > x_0$.

Then $K(x, x_0)$ *is jointly continuous in* $(x, x_0) \in \mathbb{R}^2$.

Based on Propositions 4.11, 4.12 we derive

Corollary 4.13 ([5]) *Let* $f \in C^m(\mathbb{R})$, $f^{(m)} \in L_\infty(\mathbb{R})$, $m = \lceil \alpha \rceil$, $\alpha > 0$, $\alpha \notin \mathbb{N}$, $x, x_0 \in \mathbb{R}$. *Then* $D^\alpha_{*x_0} f(x)$, $D^\alpha_{x_0-} f(x)$ *are jointly continuous functions in* (x, x_0) *from* $\mathbb{R}^2$ *into* $\mathbb{R}$.

We need

Proposition 4.14 ([5]) *Let* $f : \mathbb{R}^2 \to \mathbb{R}$ *be jointly continuous. Consider*

$$G(x) = \omega_1 (f(\cdot, x), \delta)_{[x,+\infty)}, \ \delta > 0, x \in \mathbb{R}. \tag{4.10}$$

(Here ω_1 *is defined over* $[x, +\infty)$ *instead of* $\mathbb{R}$*.)*

Then G *is continuous on* $\mathbb{R}$.

Proposition 4.15 ([5]) *Let* $f : \mathbb{R}^2 \to \mathbb{R}$ *be jointly continuous. Consider*

$$H(x) = \omega_1 (f(\cdot, x), \delta)_{(-\infty,x]}, \ \delta > 0, x \in \mathbb{R}. \tag{4.11}$$

(Here ω_1 *is defined over* $(-\infty, x]$ *instead of* $\mathbb{R}$*.)*

Then H *is continuous on* $\mathbb{R}$.

By Propositions 4.14, 4.15 and Corollary 4.13 we derive

Proposition 4.16 ([5]) *Let* $f \in C^m(\mathbb{R})$, $\left\| f^{(m)} \right\|_\infty < \infty$, $m = \lceil \alpha \rceil$, $\alpha \notin \mathbb{N}$, $\alpha > 0$, $x \in \mathbb{R}$. *Then* $\omega_1 \left(D^\alpha_{*x} f, h \right)_{[x,+\infty)}$, $\omega_1 \left(D^\alpha_{x-} f, h \right)_{(-\infty,x]}$ *are continuous functions of* $x \in \mathbb{R}$, $h > 0$ *fixed.*

We make

Remark 4.17 Let g be continuous and bounded from $\mathbb{R}$ to $\mathbb{R}$. Then

$$\omega_1 (g, t) \leq 2 \|g\|_\infty < \infty. \tag{4.12}$$

Assuming that $\left(D^\alpha_{*x} f \right)(t)$, $\left(D^\alpha_{x-} f \right)(t)$, are both continuous and bounded in $(x, t) \in \mathbb{R}^2$, i.e.

$$\left\| D^\alpha_{*x} f \right\|_\infty \leq K_1, \forall \, x \in \mathbb{R}; \tag{4.13}$$

$$\left\| D^\alpha_{x-} f \right\|_\infty \leq K_2, \forall \, x \in \mathbb{R}, \tag{4.14}$$

where $K_1, K_2 > 0$, we get

$$\begin{aligned} &\omega_1 \left(D^\alpha_{*x} f, \xi \right)_{[x,+\infty)} \leq 2K_1; \\ &\omega_1 \left(D^\alpha_{x-} f, \xi \right)_{(-\infty,x]} \leq 2K_2, \forall \, \xi \geq 0, \end{aligned} \tag{4.15}$$

for each $x \in \mathbb{R}$.

Therefore, for any $\xi \geq 0$,

$$\sup_{x\in\mathbb{R}} \left[\max \left(\omega_1 \left(D_{*x}^{\alpha} f, \xi \right)_{[x,+\infty)}, \omega_1 \left(D_{x-}^{\alpha} f, \xi \right)_{(-\infty,x]} \right) \right] \leq 2 \max (K_1, K_2) < \infty. \tag{4.16}$$

So in our setting for $f \in C^m(\mathbb{R})$, $\left\| f^{(m)} \right\|_{\infty} < \infty$, $m = \lceil \alpha \rceil$, $\alpha \notin \mathbb{N}$, $\alpha > 0$, by Corollary 4.13 both $\left(D_{*x}^{\alpha} f \right)(t)$, $\left(D_{x-}^{\alpha} f \right)(t)$ are jointly continuous in (t, x) on $\mathbb{R}^2$. Assuming further that they are both bounded on $\mathbb{R}^2$ we get (4.16) valid. In particular, each of $\omega_1 \left(D_{*x}^{\alpha} f, \xi \right)_{[x,+\infty)}$, $\omega_1 \left(D_{x-}^{\alpha} f, \xi \right)_{(-\infty,x]}$ is finite for any $\xi \geq 0$.

Let us now assume only that $f \in C^{m-1}(\mathbb{R})$, $f^{(m)} \in L_{\infty}(\mathbb{R})$, $m = \lceil \alpha \rceil$, $\alpha > 0$, $\alpha \notin \mathbb{N}$, $x \in \mathbb{R}$. Then, by Proposition 15.114, p. 388 of [4], we find that $D_{*x}^{\alpha} f \in C([x, +\infty))$, and by [6] we obtain that $D_{x-}^{\alpha} f \in C((-\infty, x])$.

We make

Remark 4.18 Again let $f \in C^m(\mathbb{R})$, $m = \lceil \alpha \rceil$, $\alpha \notin \mathbb{N}$, $\alpha > 0$; $f^{(m)}(x) = 1$, $\forall x \in \mathbb{R}$; $x_0 \in \mathbb{R}$. Notice $0 < m - \alpha < 1$. Then

$$D_{*x_0}^{\alpha} f(x) = \frac{(x - x_0)^{m-\alpha}}{\Gamma(m - \alpha + 1)}, \ \forall\, x \geq x_0. \tag{4.17}$$

Let us consider $x, y \geq x_0$, then

$$\left| D_{*x_0}^{\alpha} f(x) - D_{*x_0}^{\alpha} f(y) \right| = \frac{1}{\Gamma(m - \alpha + 1)} \left| (x - x_0)^{m-\alpha} - (y - x_0)^{m-\alpha} \right|$$

$$\leq \frac{|x - y|^{m-\alpha}}{\Gamma(m - \alpha + 1)}. \tag{4.18}$$

So it is not strange to assume that

$$\left| D_{*x_0}^{\alpha} f(x_1) - D_{*x_0}^{\alpha} f(x_2) \right| \leq K |x_1 - x_2|^{\beta}, \tag{4.19}$$

$K > 0$, $0 < \beta \leq 1$, $\forall\, x_1, x_2 \in \mathbb{R}$, $x_1, x_2 \geq x_0 \in \mathbb{R}$, where more generally it is $\left\| f^{(m)} \right\|_{\infty} < \infty$. Thus, one may assume

$$\omega_1 \left(D_{x-}^{\alpha} f, \xi \right)_{(-\infty,x]} \leq M_1 \xi^{\beta_1}, \text{ and} \tag{4.20}$$

$$\omega_1 \left(D_{*x}^{\alpha} f, \xi \right)_{[x,+\infty)} \leq M_2 \xi^{\beta_2},$$

where $0 < \beta_1, \beta_2 \leq 1$, $\forall\, \xi > 0$, $M_1, M_2 > 0$; any $x \in \mathbb{R}$.

Setting $\beta = \min(\beta_1, \beta_2)$ and $M = \max(M_1, M_2)$, in that case we obtain

$$\sup_{x\in\mathbb{R}} \left\{ \max \left(\omega_1 \left(D_{x-}^{\alpha} f, \xi \right)_{(-\infty,x]}, \omega_1 \left(D_{*x}^{\alpha} f, \xi \right)_{[x,+\infty)} \right) \right\} \leq M \xi^{\beta} \to 0, \text{ as } \xi \to 0+. \tag{4.21}$$

4.3 Results

4.3.1 Fractional Convergence with Rates of the Cardaliaguet-Euvrard Neural Network Operators

We need the following (see [8]).

Definition 4.19 A function $b : \mathbb{R} \to \mathbb{R}$ is said to be bell-shaped if b belongs to L^1 and its integral is nonzero, if it is nondecreasing on $(-\infty, a)$ and nonincreasing on $[a, +\infty)$, where a belongs to $\mathbb{R}$. In particular $b(x)$ is a nonnegative number and at a b takes a global maximum; it is the center of the bell-shaped function. A bell-shaped function is said to be centered if its center is zero. The function $b(x)$ may have jump discontinuities. In this work we consider only centered bell-shaped functions of compact support $[-T, T]$, $T > 0$. Call $I := \int_{-T}^{T} b(t)\,dt$. Note that $I > 0$.

We follow [1, 8].

Example 4.20 (1) $b(x)$ can be the characteristic function over $[-1, 1]$.

(2) $b(x)$ can be the hat function over $[-1, 1]$, i.e.,

$$b(x) = \begin{cases} 1 + x, & -1 \le x \le 0, \\ 1 - x, & 0 < x \le 1 \\ 0, & \text{elsewhere.} \end{cases}$$

These are centered bell-shaped functions of compact support.

Here we consider functions $f : \mathbb{R} \to \mathbb{R}$ that are continuous.

In this chapter we study the fractional convergence with rates over the real line, to the unit operator, of the Cardaliaguet-Euvrard neural network operators (see [8]),

$$(F_n(f))(x) := \sum_{k=-n^2}^{n^2} \frac{f\left(\frac{k}{n}\right)}{I \cdot n^{\alpha}} \cdot b\left(n^{1-\alpha} \cdot \left(x - \frac{k}{n}\right)\right), \tag{4.22}$$

where $0 < \alpha < 1$ and $x \in \mathbb{R}$, $n \in \mathbb{N}$. The terms in the sum (4.22) can be nonzero iff

$$\left| n^{1-\alpha} \left(x - \frac{k}{n}\right) \right| \le T, \text{ i.e. } \left| x - \frac{k}{n} \right| \le \frac{T}{n^{1-\alpha}}$$

iff

$$nx - Tn^{\alpha} \le k \le nx + Tn^{\alpha}. \tag{4.23}$$

In order to have the desired order of numbers

$$-n^2 \le nx - Tn^{\alpha} \le nx + Tn^{\alpha} \le n^2, \tag{4.24}$$

it is sufficient enough to assume that

$$n \geq T + |x|. \tag{4.25}$$

When $x \in [-T, T]$ it is enough to assume $n \geq 2T$ which implies (4.24).

Proposition 4.21 *Let $a \leq b$, $a, b \in \mathbb{R}$. Let $card(k)$ (≥ 0) be the maximum number of integers contained in $[a, b]$. Then*

$$\max(0, (b-a) - 1) \leq card(k) \leq (b-a) + 1. \tag{4.26}$$

Remark 4.22 We would like to establish a lower bound on $card(k)$ over the interval $[nx - Tn^{\alpha}, nx + Tn^{\alpha}]$. From Proposition 4.21 we get that

$$card(k) \geq \max\left(2Tn^{\alpha} - 1, 0\right).$$

We obtain $card(k) \geq 1$, if

$$2Tn^{\alpha} - 1 \geq 1 \text{ iff } n \geq T^{-\frac{1}{\alpha}}.$$

So to have the desired order (4.24) and $card(k) \geq 1$ over $[nx - Tn^{\alpha}, nx + Tn^{\alpha}]$, we need to consider

$$n \geq \max\left(T + |x|, T^{-\frac{1}{\alpha}}\right). \tag{4.27}$$

Also notice that $card(k) \to +\infty$, as $n \to +\infty$. We call $b^* := b(0)$ the maximum of $b(x)$.

Denote by $[\cdot]$ the integral part of a number.

Following [1] we have

$$\sum_{k=\lceil nx - Tn^{\alpha} \rceil}^{[nx + Tn^{\alpha}]} \frac{1}{I \cdot n^{\alpha}} \cdot b\left(n^{1-\alpha} \cdot \left(x - \frac{k}{n}\right)\right)$$

$$\leq \frac{b^*}{I \cdot n^{\alpha}} \cdot \sum_{k=\lceil nx - Tn^{\alpha} \rceil}^{[nx + Tn^{\alpha}]} 1$$

$$\leq \frac{b^*}{I \cdot n^{\alpha}} \cdot \left(2Tn^{\alpha} + 1\right) = \frac{b^*}{I} \cdot \left(2T + \frac{1}{n^{\alpha}}\right). \tag{4.28}$$

We will use

Lemma 4.23 *It holds that*

$$S_n(x) := \sum_{k=\lceil nx-Tn^{\alpha}\rceil}^{[nx+Tn^{\alpha}]} \frac{1}{I \cdot n^{\alpha}} \cdot b\left(n^{1-\alpha}\left(x-\frac{k}{n}\right)\right) \to 1, \tag{4.29}$$

pointwise, as $n \to +\infty$, *where* $x \in \mathbb{R}$.

Remark 4.24 Clearly we have that

$$nx - Tn^{\alpha} \leq nx \leq nx + Tn^{\alpha}. \tag{4.30}$$

We prove in general that

$$nx - Tn^{\alpha} \leq [nx] \leq nx \leq \lceil nx \rceil \leq nx + Tn^{\alpha}. \tag{4.31}$$

Indeed we have that, if $[nx] < nx - Tn^{\alpha}$, then $[nx] + Tn^{\alpha} < nx$, and $[nx] + [Tn^{\alpha}] \leq [nx]$, resulting into $[Tn^{\alpha}] = 0$, which for large enough n is not true. Therefore $nx - Tn^{\alpha} \leq [nx]$. Similarly, if $\lceil nx \rceil > nx + Tn^{\alpha}$, then $nx + Tn^{\alpha} \geq nx + [Tn^{\alpha}]$, and $\lceil nx \rceil - [Tn^{\alpha}] > nx$, thus $\lceil nx \rceil - [Tn^{\alpha}] \geq \lceil nx \rceil$, resulting into $[Tn^{\alpha}] = 0$, which again for large enough n is not true.

Therefore without loss of generality we may assume that

$$nx - Tn^{\alpha} \leq [nx] \leq nx \leq \lceil nx \rceil \leq nx + Tn^{\alpha}. \tag{4.32}$$

Hence $\lceil nx - Tn^{\alpha} \rceil \leq [nx]$ and $\lceil nx \rceil \leq [nx + Tn^{\alpha}]$. Also if $[nx] \neq \lceil nx \rceil$, then $\lceil nx \rceil = [nx] + 1$. If $[nx] = \lceil nx \rceil$, then $nx \in \mathbb{Z}$; and by assuming $n \geq T^{-\frac{1}{\alpha}}$, we get $Tn^{\alpha} \geq 1$ and $nx + Tn^{\alpha} \geq nx + 1$, so that $[nx + Tn^{\alpha}] \geq nx + 1 = [nx] + 1$.

We present our first main result

Theorem 4.25 *We consider* $f : \mathbb{R} \to \mathbb{R}$. *Let* $\beta > 0$, $N = \lceil \beta \rceil$, $\beta \notin \mathbb{N}$, $f \in AC^N([a,b])$, $\forall [a,b] \subset \mathbb{R}$, *with* $f^{(N)} \in L_{\infty}(\mathbb{R})$. *Let also* $x \in \mathbb{R}$, $T > 0$, $n \in \mathbb{N} : n \geq \max\left(T + |x|, T^{-\frac{1}{\alpha}}\right)$. *We further assume that* $D_{*x}^{\beta} f$, $D_{x-}^{\beta} f$ *are uniformly continuous functions or continuous and bounded on* $[x, +\infty)$, $(-\infty, x]$, *respectively.*

Then

(1)

$$|F_n(f)(x) - f(x)| \leq |f(x)| \cdot \tag{4.33}$$

$$\left| \sum_{k=\lceil nx-Tn^{\alpha}\rceil}^{[nx+Tn^{\alpha}]} \frac{1}{In^{\alpha}} b\left(n^{1-\alpha}\left(x-\frac{k}{n}\right)\right) - 1 \right| +$$

$$\frac{b^*}{I}\left(2T + \frac{1}{n^{\alpha}}\right)\left(\sum_{j=1}^{N-1} \frac{\left|f^{(j)}(x)\right| T^j}{j! n^{(1-\alpha)j}}\right)$$

$$+\frac{b^*}{I}\left(2T+\frac{1}{n^{\alpha}}\right)\frac{T^{\beta}}{\Gamma(\beta+1)\,n^{(1-\alpha)\beta}}\cdot$$

$$\left\{\omega_1\left(D_{*x}^{\beta}f,\frac{T}{n^{1-\alpha}}\right)_{[x,+\infty)}+\omega_1\left(D_{x-}^{\beta}f,\frac{T}{n^{1-\alpha}}\right)_{(-\infty,x]}\right\},$$

above $\sum_{j=1}^{0}\cdot=0$,
(2)

$$\left|(F_n(f))(x)-\sum_{j=0}^{N-1}\frac{f^{(j)}(x)}{j!}\left(F_n\left((\cdot-x)^j\right)\right)(x)\right|\leq \tag{4.34}$$

$$\frac{b^*}{I}\left(2T+\frac{1}{n^{\alpha}}\right)\frac{T^{\beta}}{\Gamma(\beta+1)\,n^{(1-\alpha)\beta}}\cdot$$

$$\left\{\omega_1\left(D_{*x}^{\beta}f,\frac{T}{n^{1-\alpha}}\right)_{[x,+\infty)}+\omega_1\left(D_{x-}^{\beta}f,\frac{T}{n^{1-\alpha}}\right)_{(-\infty,x]}\right\}=:\lambda_n(x),$$

(3) assume further that $f^{(j)}(x)=0$, *for* $j=0,1,\ldots,N-1$, *we get*

$$|F_n(f)(x)|\leq\lambda_n(x), \tag{4.35}$$

(4) in case of $N=1$, *we obtain*

$$|F_n(f)(x)-f(x)|\leq|f(x)|\cdot \tag{4.36}$$

$$\left|\sum_{k=\lceil nx-Tn^{\alpha}\rceil}^{[nx+Tn^{\alpha}]}\frac{1}{In^{\alpha}}b\left(n^{1-\alpha}\left(x-\frac{k}{n}\right)\right)-1\right|+$$

$$\frac{b^*}{I}\left(2T+\frac{1}{n^{\alpha}}\right)\frac{T^{\beta}}{\Gamma(\beta+1)\,n^{(1-\alpha)\beta}}\cdot$$

$$\left\{\omega_1\left(D_{*x}^{\beta}f,\frac{T}{n^{1-\alpha}}\right)_{[x,+\infty)}+\omega_1\left(D_{x-}^{\beta}f,\frac{T}{n^{1-\alpha}}\right)_{(-\infty,x]}\right\}.$$

Here we get fractionally with rates the pointwise convergence of $(F_n(f))(x)\to f(x)$, *as* $n\to\infty$, $x\in\mathbb{R}$.

Proof Let $x\in\mathbb{R}$. We have that

$$D_{x-}^{\beta}f(x)=D_{*x}^{\beta}f(x)=0\,. \tag{4.37}$$

From [9], p. 54, we get by the left Caputo fractional Taylor formula that

$$f\left(\frac{k}{n}\right)=\sum_{j=0}^{N-1}\frac{f^{(j)}(x)}{j!}\left(\frac{k}{n}-x\right)^{j}+ \tag{4.38}$$

$$\frac{1}{\Gamma(\beta)}\int_{x}^{\frac{k}{n}}\left(\frac{k}{n}-J\right)^{\beta-1}\left(D_{*x}^{\beta}f(J)-D_{*x}^{\beta}f(x)\right)dJ,$$

for all $x\leq\frac{k}{n}\leq x+Tn^{\alpha-1}$, iff $\lceil nx\rceil\leq k\leq[nx+Tn^{\alpha}]$, where $k\in\mathbb{Z}$.

Also from [2], using the right Caputo fractional Taylor formula we get

$$f\left(\frac{k}{n}\right)=\sum_{j=0}^{N-1}\frac{f^{(j)}(x)}{j!}\left(\frac{k}{n}-x\right)^{j}+ \tag{4.39}$$

$$\frac{1}{\Gamma(\beta)}\int_{\frac{k}{n}}^{x}\left(J-\frac{k}{n}\right)^{\beta-1}\left(D_{x-}^{\beta}f(J)-D_{x-}^{\beta}f(x)\right)dJ,$$

for all $x-Tn^{\alpha-1}\leq\frac{k}{n}\leq x$, iff $\lceil nx-Tn^{\alpha}\rceil\leq k\leq[nx]$, where $k\in\mathbb{Z}$.

Notice that $\lceil nx\rceil\leq[nx]+1$.

Hence we have

$$\frac{f\left(\frac{k}{n}\right)b\left(n^{1-\alpha}\left(x-\frac{k}{n}\right)\right)}{In^{\alpha}}=\sum_{j=0}^{N-1}\frac{f^{(j)}(x)}{j!}\left(\frac{k}{n}-x\right)^{j}\frac{b\left(n^{1-\alpha}\left(x-\frac{k}{n}\right)\right)}{In^{\alpha}}+ \tag{4.40}$$

$$\frac{b\left(n^{1-\alpha}\left(x-\frac{k}{n}\right)\right)}{In^{\alpha}\Gamma(\beta)}\int_{x}^{\frac{k}{n}}\left(\frac{k}{n}-J\right)^{\beta-1}\left(D_{*x}^{\beta}f(J)-D_{*x}^{\beta}f(x)\right)dJ,$$

and

$$\frac{f\left(\frac{k}{n}\right)b\left(n^{1-\alpha}\left(x-\frac{k}{n}\right)\right)}{In^{\alpha}}=\sum_{j=0}^{N-1}\frac{f^{(j)}(x)}{j!}\left(\frac{k}{n}-x\right)^{j}\frac{b\left(n^{1-\alpha}\left(x-\frac{k}{n}\right)\right)}{In^{\alpha}}+ \tag{4.41}$$

$$\frac{b\left(n^{1-\alpha}\left(x-\frac{k}{n}\right)\right)}{In^{\alpha}\Gamma(\beta)}\int_{\frac{k}{n}}^{x}\left(J-\frac{k}{n}\right)^{\beta-1}\left(D_{x-}^{\beta}f(J)-D_{x-}^{\beta}f(x)\right)dJ.$$

Therefore we obtain

$$\frac{\sum_{k=[nx]+1}^{[nx+Tn^{\alpha}]}f\left(\frac{k}{n}\right)b\left(n^{1-\alpha}\left(x-\frac{k}{n}\right)\right)}{In^{\alpha}}= \tag{4.42}$$

$$\sum_{j=0}^{N-1} \frac{f^{(j)}(x)}{j!} \left(\frac{\sum_{k=[nx]+1}^{[nx+Tn^{\alpha}]} \left(\frac{k}{n} - x\right)^j b\left(n^{1-\alpha}\left(x - \frac{k}{n}\right)\right)}{In^{\alpha}} \right) +$$

$$\sum_{k=[nx]+1}^{[nx+Tn^{\alpha}]} \frac{b\left(n^{1-\alpha}\left(x - \frac{k}{n}\right)\right)}{In^{\alpha}\Gamma(\beta)} \int_x^{\frac{k}{n}} \left(\frac{k}{n} - J\right)^{\beta-1} \left(D_{*x}^{\beta} f(J) - D_{*x}^{\beta} f(x)\right) dJ,$$

and

$$\frac{\sum_{k=\lceil nx-Tn^{\alpha}\rceil}^{[nx]} f\left(\frac{k}{n}\right) b\left(n^{1-\alpha}\left(x - \frac{k}{n}\right)\right)}{In^{\alpha}} = \tag{4.43}$$

$$\sum_{j=0}^{N-1} \frac{f^{(j)}(x)}{j!} \frac{\sum_{k=\lceil nx-Tn^{\alpha}\rceil}^{[nx]} \left(\frac{k}{n} - x\right)^j b\left(n^{1-\alpha}\left(x - \frac{k}{n}\right)\right)}{In^{\alpha}} +$$

$$\frac{\sum_{k=\lceil nx-Tn^{\alpha}\rceil}^{[nx]} b\left(n^{1-\alpha}\left(x - \frac{k}{n}\right)\right)}{In^{\alpha}\Gamma(\beta)} \int_{\frac{k}{n}}^{x} \left(J - \frac{k}{n}\right)^{\beta-1} \left(D_{x-}^{\beta} f(J) - D_{x-}^{\beta} f(x)\right) dJ.$$

We notice here that

$$(F_n(f))(x) := \sum_{k=-n^2}^{n^2} \frac{f\left(\frac{k}{n}\right)}{In^{\alpha}} b\left(n^{1-\alpha}\left(x - \frac{k}{n}\right)\right) = \tag{4.44}$$

$$\sum_{k=\lceil nx-Tn^{\alpha}\rceil}^{[nx+Tn^{\alpha}]} \frac{f\left(\frac{k}{n}\right)}{In^{\alpha}} b\left(n^{1-\alpha}\left(x - \frac{k}{n}\right)\right).$$

Adding the two equalities (4.42) and (4.43) we obtain

$$(F_n(f))(x) =$$

$$\sum_{j=0}^{N-1} \frac{f^{(j)}(x)}{j!} \left(\frac{\sum_{k=\lceil nx-Tn^{\alpha}\rceil}^{[nx+Tn^{\alpha}]} \left(\frac{k}{n} - x\right)^j b\left(n^{1-\alpha}\left(x - \frac{k}{n}\right)\right)}{In^{\alpha}} \right) + \theta_n(x), \tag{4.45}$$

where

$$\theta_n(x) := \frac{\sum_{k=\lceil nx-Tn^{\alpha}\rceil}^{[nx]} b\left(n^{1-\alpha}\left(x - \frac{k}{n}\right)\right)}{In^{\alpha}\Gamma(\beta)} \cdot$$

$$\int_{\frac{k}{n}}^{x} \left(J - \frac{k}{n}\right)^{\beta-1} \left(D_{x-}^{\beta} f(J) - D_{x-}^{\beta} f(x)\right) dJ +$$

$$\sum_{k=[nx]+1}^{[nx+Tn^{\alpha}]} \frac{b\left(n^{1-\alpha}\left(x-\frac{k}{n}\right)\right)}{In^{\alpha}\Gamma(\beta)} \int_{x}^{\frac{k}{n}} \left(\frac{k}{n}-J\right)^{\beta-1} \left(D_{*x}^{\beta} f(J) - D_{*x}^{\beta} f(x)\right) dJ. \tag{4.46}$$

We call

$$\theta_{1n}(x) := \frac{\sum_{k=\lceil nx-Tn^{\alpha}\rceil}^{[nx]} b\left(n^{1-\alpha}\left(x-\frac{k}{n}\right)\right)}{In^{\alpha}\Gamma(\beta)} \cdot$$

$$\int_{\frac{k}{n}}^{x} \left(J-\frac{k}{n}\right)^{\beta-1} \left(D_{x-}^{\beta} f(J) - D_{x-}^{\beta} f(x)\right) dJ, \tag{4.47}$$

and

$$\theta_{2n}(x) := \sum_{k=[nx]+1}^{[nx+Tn^{\alpha}]} \frac{b\left(n^{1-\alpha}\left(x-\frac{k}{n}\right)\right)}{In^{\alpha}\Gamma(\beta)} \cdot$$

$$\int_{x}^{\frac{k}{n}} \left(\frac{k}{n}-J\right)^{\beta-1} \left(D_{*x}^{\beta} f(J) - D_{*x}^{\beta} f(x)\right) dJ. \tag{4.48}$$

I.e.

$$\theta_n(x) = \theta_{1n}(x) + \theta_{2n}(x). \tag{4.49}$$

We further have

$$(F_n(f))(x) - f(x) = f(x)\left(\frac{\sum_{k=\lceil nx-Tn^{\alpha}\rceil}^{[nx+Tn^{\alpha}]} b\left(n^{1-\alpha}\left(x-\frac{k}{n}\right)\right)}{In^{\alpha}} - 1\right) + \tag{4.50}$$

$$\sum_{j=0}^{N-1} \frac{f^{(j)}(x)}{j!} \left(\frac{\sum_{k=\lceil nx-Tn^{\alpha}\rceil}^{[nx+Tn^{\alpha}]} \left(\frac{k}{n}-x\right)^{j} b\left(n^{1-\alpha}\left(x-\frac{k}{n}\right)\right)}{In^{\alpha}}\right) + \theta_n(x),$$

and

$$|(F_n(f))(x) - f(x)| \le |f(x)| \left| \sum_{k=\lceil nx-Tn^{\alpha}\rceil}^{[nx+Tn^{\alpha}]} \frac{1}{In^{\alpha}} b\left(n^{1-\alpha}\left(x-\frac{k}{n}\right)\right) - 1 \right| +$$

$$\sum_{j=1}^{N-1} \frac{\left|f^{(j)}(x)\right|}{j!} \left(\frac{\sum_{k=\lceil nx-Tn^{\alpha}\rceil}^{[nx+Tn^{\alpha}]} \left|x-\frac{k}{n}\right|^{j} b\left(n^{1-\alpha}\left(x-\frac{k}{n}\right)\right)}{In^{\alpha}}\right) + |\theta_n(x)| \le$$

$$|f(x)| \left| \sum_{k=\lceil nx-Tn^{\alpha}\rceil}^{[nx+Tn^{\alpha}]} \frac{1}{In^{\alpha}} b\left(n^{1-\alpha}\left(x-\frac{k}{n}\right)\right) - 1 \right| + \tag{4.51}$$

$$\sum_{j=1}^{N-1} \frac{\left|f^{(j)}(x)\right|}{j!} \frac{T^j}{n^{(1-\alpha)j}} \left(\frac{\sum_{k=\lceil nx-Tn^{\alpha}\rceil}^{[nx+Tn^{\alpha}]} b\left(n^{1-\alpha}\left(x-\frac{k}{n}\right)\right)}{In^{\alpha}} \right) + |\theta_n(x)| =: (*) .$$

But we have

$$\sum_{k=\lceil nx-Tn^{\alpha}\rceil}^{[nx+Tn^{\alpha}]} \frac{1}{In^{\alpha}} b\left(n^{1-\alpha}\left(x-\frac{k}{n}\right)\right) \leq \frac{b^*}{I}\left(2T+\frac{1}{n^{\alpha}}\right), \tag{4.52}$$

by (4.28).

Therefore we obtain

$$|(F_n(f))(x) - f(x)| \leq |f(x)| \left| \sum_{k=\lceil nx-Tn^{\alpha}\rceil}^{[nx+Tn^{\alpha}]} \frac{1}{In^{\alpha}} b\left(n^{1-\alpha}\left(x-\frac{k}{n}\right)\right) - 1 \right| +$$

$$\frac{b^*}{I}\left(2T+\frac{1}{n^{\alpha}}\right) \left(\sum_{j=1}^{N-1} \frac{\left|f^{(j)}(x)\right| T^j}{j! n^{(1-\alpha)j}} \right) + |\theta_n(x)| . \tag{4.53}$$

Next we see that

$$\gamma_{1n} := \frac{1}{\Gamma(\beta)} \left| \int_{\frac{k}{n}}^{x} \left(J-\frac{k}{n}\right)^{\beta-1} \left(D_{x-}^{\beta} f(J) - D_{x-}^{\beta} f(x)\right) dJ \right| \leq \tag{4.54}$$

$$\frac{1}{\Gamma(\beta)} \int_{\frac{k}{n}}^{x} \left(J-\frac{k}{n}\right)^{\beta-1} \left|D_{x-}^{\beta} f(J) - D_{x-}^{\beta} f(x)\, dJ\right| \leq$$

$$\frac{1}{\Gamma(\beta)} \int_{\frac{k}{n}}^{x} \left(J-\frac{k}{n}\right)^{\beta-1} \omega_1\left(D_{x-}^{\beta} f, |J-x|\right)_{(-\infty,x]} dJ \leq$$

$$\frac{1}{\Gamma(\beta)} \omega_1\left(D_{x-}^{\beta} f, \left|x-\frac{k}{n}\right|\right)_{(-\infty,x]} \int_{\frac{k}{n}}^{x} \left(J-\frac{k}{n}\right)^{\beta-1} dJ \leq$$

$$\frac{1}{\Gamma(\beta)} \omega_1\left(D_{x-}^{\beta} f, \frac{T}{n^{1-\alpha}}\right)_{(-\infty,x]} \frac{\left(x-\frac{k}{n}\right)^{\beta}}{\beta} \leq$$

$$\frac{1}{\Gamma(\beta+1)}\omega_1\left(D_{x-}^{\beta}f,\frac{T}{n^{1-\alpha}}\right)_{(-\infty,x]}\frac{T^{\beta}}{n^{(1-\alpha)\beta}}.$$

That is

$$\gamma_{1n} \leq \frac{T^{\beta}}{\Gamma(\beta+1)\,n^{(1-\alpha)\beta}}\omega_1\left(D_{x-}^{\beta}f,\frac{T}{n^{1-\alpha}}\right)_{(-\infty,x]}. \tag{4.55}$$

Furthermore

$$|\theta_{1n}(x)| \leq \sum_{k=\lceil nx-Tn^{\alpha}\rceil}^{[nx]} \frac{b\left(n^{1-\alpha}\left(x-\frac{k}{n}\right)\right)}{In^{\alpha}}\gamma_{1n} \leq \tag{4.56}$$

$$\left(\sum_{k=\lceil nx-Tn^{\alpha}\rceil}^{[nx]} \frac{b\left(n^{1-\alpha}\left(x-\frac{k}{n}\right)\right)}{In^{\alpha}}\right)\frac{T^{\beta}}{\Gamma(\beta+1)\,n^{(1-\alpha)\beta}}\omega_1\left(D_{x-}^{\beta}f,\frac{T}{n^{1-\alpha}}\right)_{(-\infty,x]} \leq$$

$$\left(\sum_{k=\lceil nx-Tn^{\alpha}\rceil}^{[nx+Tn^{\alpha}]} \frac{b\left(n^{1-\alpha}\left(x-\frac{k}{n}\right)\right)}{In^{\alpha}}\right)\frac{T^{\beta}}{\Gamma(\beta+1)\,n^{(1-\alpha)\beta}}\omega_1\left(D_{x-}^{\beta}f,\frac{T}{n^{1-\alpha}}\right)_{(-\infty,x]} \leq$$

$$\frac{b^*}{I}\left(2T+\frac{1}{n^{\alpha}}\right)\frac{T^{\beta}}{\Gamma(\beta+1)\,n^{(1-\alpha)\beta}}\omega_1\left(D_{x-}^{\beta}f,\frac{T}{n^{1-\alpha}}\right)_{(-\infty,x]}.$$

So that

$$|\theta_{1n}(x)| \leq \frac{b^*}{I}\left(2T+\frac{1}{n^{\alpha}}\right)\frac{T^{\beta}}{\Gamma(\beta+1)\,n^{(1-\alpha)\beta}}\omega_1\left(D_{x-}^{\beta}f,\frac{T}{n^{1-\alpha}}\right)_{(-\infty,x]}. \tag{4.57}$$

Similarly we derive

$$\gamma_{2n} := \frac{1}{\Gamma(\beta)}\left|\int_x^{\frac{k}{n}}\left(\frac{k}{n}-J\right)^{\beta-1}\left(D_{*x}^{\beta}f(J)-D_{*x}^{\beta}f(x)\right)dJ\right| \leq \tag{4.58}$$

$$\frac{1}{\Gamma(\beta)}\int_x^{\frac{k}{n}}\left(\frac{k}{n}-J\right)^{\beta-1}\left|D_{*x}^{\beta}f(J)-D_{*x}^{\beta}f(x)\right|dJ \leq$$

$$\frac{\omega_1\left(D_{*x}^{\beta}f,\frac{T}{n^{1-\alpha}}\right)_{[x,+\infty)}}{\Gamma(\beta+1)}\left(\frac{k}{n}-x\right)^{\beta} \leq$$

$$\frac{\omega_1\left(D_{*x}^{\beta}f,\frac{T}{n^{1-\alpha}}\right)_{[x,+\infty)}}{\Gamma(\beta+1)}\frac{T^{\beta}}{n^{(1-\alpha)\beta}}.$$

That is

$$\gamma_{2n} \leq \frac{T^{\beta}}{\Gamma(\beta+1)\, n^{(1-\alpha)\beta}} \omega_1 \left(D_{*x}^{\beta} f, \frac{T}{n^{1-\alpha}} \right)_{[x,+\infty)}. \tag{4.59}$$

Consequently we find

$$|\theta_{2n}(x)| \leq \left(\sum_{k=[nx]+1}^{[nx+Tn^{\alpha}]} \frac{b\left(n^{1-\alpha}\left(x-\frac{k}{n}\right)\right)}{In^{\alpha}} \right) \cdot$$

$$\frac{T^{\beta}}{\Gamma(\beta+1)\, n^{(1-\alpha)\beta}} \omega_1 \left(D_{*x}^{\beta} f, \frac{T}{n^{1-\alpha}} \right)_{[x,+\infty)} \leq \tag{4.60}$$

$$\frac{b^*}{I}\left(2T+\frac{1}{n^{\alpha}}\right) \frac{T^{\beta}}{\Gamma(\beta+1)\, n^{(1-\alpha)\beta}} \omega_1 \left(D_{*x}^{\beta} f, \frac{T}{n^{1-\alpha}} \right)_{[x,+\infty)}.$$

So we have proved that

$$|\theta_n(x)| \leq \frac{b^*}{I}\left(2T+\frac{1}{n^{\alpha}}\right) \frac{T^{\beta}}{\Gamma(\beta+1)\, n^{(1-\alpha)\beta}} \cdot \tag{4.61}$$

$$\left\{ \omega_1 \left(D_{*x}^{\beta} f, \frac{T}{n^{1-\alpha}} \right)_{[x,+\infty)} + \omega_1 \left(D_{x-}^{\beta} f, \frac{T}{n^{1-\alpha}} \right)_{(-\infty,x]} \right\}.$$

Combining (4.53) and (4.61) we have (4.33). ■

As an application of Theorem 4.25 we give

Theorem 4.26 *Let $\beta > 0$, $N = \lceil \beta \rceil$, $\beta \notin \mathbb{N}$, $f \in C^N(\mathbb{R})$, with $f^{(N)} \in L_{\infty}(\mathbb{R})$. Let also $T > 0$, $n \in \mathbb{N} : n \geq \max\left(2T, T^{-\frac{1}{\alpha}}\right)$. We further assume that $D_{*x}^{\beta} f(t)$, $D_{x-}^{\beta} f(t)$ are both bounded in $(x,t) \in \mathbb{R}^2$. Then*
(1)

$$\|F_n(f) - f\|_{\infty,[-T,T]} \leq \|f\|_{\infty,[-T,T]} \cdot \tag{4.62}$$

$$\left\| \sum_{k=\lceil nx-Tn^{\alpha} \rceil}^{[nx+Tn^{\alpha}]} \frac{1}{In^{\alpha}} b\left(n^{1-\alpha}\left(x-\frac{k}{n}\right)\right) - 1 \right\|_{\infty,[-T,T]} +$$

$$\frac{b^*}{I}\left(2T+\frac{1}{n^{\alpha}}\right) \left(\sum_{j=1}^{N-1} \frac{\|f^{(j)}\|_{\infty,[-T,T]} T^j}{j!\, n^{(1-\alpha)j}} \right) + \frac{b^*}{I}\left(2T+\frac{1}{n^{\alpha}}\right) \frac{T^{\beta}}{\Gamma(\beta+1)\, n^{(1-\alpha)\beta}} \cdot$$

$$\left\{ \sup_{x \in [-T,T]} \omega_1 \left(D_{*x}^{\beta} f, \frac{T}{n^{1-\alpha}} \right)_{[x,+\infty)} + \sup_{x \in [-T,T]} \omega_1 \left(D_{x-}^{\beta} f, \frac{T}{n^{1-\alpha}} \right)_{(-\infty,x]} \right\},$$

(2) in case of $N = 1$, we obtain

$$\| F_n(f) - f \|_{\infty,[-T,T]} \le \| f \|_{\infty,[-T,T]} \cdot \tag{4.63}$$

$$\left\| \sum_{k=\lceil nx - Tn^{\alpha} \rceil}^{[nx+Tn^{\alpha}]} \frac{1}{In^{\alpha}} b \left(n^{1-\alpha} \left(x - \frac{k}{n} \right) \right) - 1 \right\|_{\infty,[-T,T]} +$$

$$\frac{b^*}{I} \left(2T + \frac{1}{n^{\alpha}} \right) \frac{T^{\beta}}{\Gamma(\beta+1) n^{(1-\alpha)\beta}} \cdot$$

$$\left\{ \sup_{x \in [-T,T]} \omega_1 \left(D_{*x}^{\beta} f, \frac{T}{n^{1-\alpha}} \right)_{[x,+\infty)} + \sup_{x \in [-T,T]} \omega_1 \left(D_{x-}^{\beta} f, \frac{T}{n^{1-\alpha}} \right)_{(-\infty,x]} \right\}.$$

An interesting case is when $\beta = \frac{1}{2}$.

Assuming further that $\left\| \sum_{k=\lceil nx-Tn^{\alpha} \rceil}^{[nx+Tn^{\alpha}]} \frac{1}{In^{\alpha}} b \left(n^{1-\alpha} \left(x - \frac{k}{n} \right) \right) - 1 \right\|_{\infty,[-T,T]} \to 0$, as $n \to \infty$, we get fractionally with rates the uniform convergence of $F_n(f) \to f$, as $n \to \infty$.

Proof From (4.33), (4.36) of Theorem 4.25, and by Remark 4.17.

Also by

$$\sum_{k=\lceil nx - Tn^{\alpha} \rceil}^{[nx+Tn^{\alpha}]} \frac{1}{In^{\alpha}} b \left(n^{1-\alpha} \left(x - \frac{k}{n} \right) \right) \le \frac{b^*}{I} (2T+1), \tag{4.64}$$

we get that

$$\left\| \sum_{k=\lceil nx - Tn^{\alpha} \rceil}^{[nx+Tn^{\alpha}]} \frac{1}{In^{\alpha}} b \left(n^{1-\alpha} \left(x - \frac{k}{n} \right) \right) - 1 \right\|_{\infty,[-T,T]} \le \left(\frac{b^*}{I} (2T+1) + 1 \right). \tag{4.65}$$

■

One can also apply Remark 4.18 to the last Theorem 4.26, to get interesting and simplified results.

We make

Remark 4.27 Let $b(x)$ be a centered bell-shaped continuous function on $\mathbb{R}$ of compact support $[-T, T]$, $T > 0$. Let $x \in [-T^*, T^*]$, $T^* > 0$, and $n \in \mathbb{N} : n \ge \max\left(T + T^*, T^{-\frac{1}{\alpha}}\right)$, $0 < \alpha < 1$. Consider $p \ge 1$.

Clearly we get here that

$$\left| \sum_{k=\lceil nx-Tn^{\alpha} \rceil}^{[nx+Tn^{\alpha}]} \frac{1}{In^{\alpha}} b\left(n^{1-\alpha}\left(x-\frac{k}{n}\right)\right) - 1 \right|^{p} \leq \left(\frac{b^{*}}{I}(2T+1)+1\right)^{p}, \tag{4.66}$$

for all $x \in [-T^{*}, T^{*}]$, for any $n \geq \max\left(T+T^{*}, T^{-\frac{1}{\alpha}}\right)$.

By Lemma 4.23, we obtain that

$$\lim_{n\to\infty} \left| \sum_{k=\lceil nx-Tn^{\alpha} \rceil}^{[nx+Tn^{\alpha}]} \frac{1}{In^{\alpha}} b\left(n^{1-\alpha}\left(x-\frac{k}{n}\right)\right) - 1 \right|^{p} = 0, \tag{4.67}$$

all $x \in [-T^{*}, T^{*}]$.

Now it is clear, by the bounded convergence theorem, that

$$\lim_{n\to\infty} \left\| \sum_{k=\lceil nx-Tn^{\alpha} \rceil}^{[nx+Tn^{\alpha}]} \frac{1}{In^{\alpha}} b\left(n^{1-\alpha}\left(x-\frac{k}{n}\right)\right) - 1 \right\|_{p,[-T^{*},T^{*}]} = 0. \tag{4.68}$$

Let $\beta > 0$, $N = \lceil \beta \rceil$, $\beta \notin \mathbb{N}$, $f \in C^{N}(\mathbb{R})$, $f^{(N)} \in L_{\infty}(\mathbb{R})$. Here both $D_{*x}^{\alpha} f(t)$, $D_{x-}^{\alpha} f(t)$ are bounded in $(x,t) \in \mathbb{R}^{2}$.

By Theorem 4.25 we have

$$|F_n(f)(x) - f(x)| \leq \|f\|_{\infty,[-T^{*},T^{*}]} \cdot \tag{4.69}$$

$$\left| \sum_{k=\lceil nx-Tn^{\alpha} \rceil}^{[nx+Tn^{\alpha}]} \frac{1}{In^{\alpha}} b\left(n^{1-\alpha}\left(x-\frac{k}{n}\right)\right) - 1 \right| +$$

$$\frac{b^{*}}{I}\left(2T+\frac{1}{n^{\alpha}}\right)\left(\sum_{j=1}^{N-1} \frac{\left|f^{(j)}(x)\right| T^{j}}{j! n^{(1-\alpha)j}}\right)$$

$$+\frac{b^{*}}{I}\left(2T+\frac{1}{n^{\alpha}}\right)\frac{T^{\beta}}{\Gamma(\beta+1) n^{(1-\alpha)\beta}} \cdot$$

$$\left\{ \sup_{x\in[-T^{*},T^{*}]} \omega_1\left(D_{*x}^{\beta} f, \frac{T}{n^{1-\alpha}}\right)_{[x,+\infty)} + \sup_{x\in[-T^{*},T^{*}]} \omega_1\left(D_{x-}^{\beta} f, \frac{T}{n^{1-\alpha}}\right)_{(-\infty,x]} \right\}.$$

Applying to the last inequality (4.69) the monotonicity and subadditive property of $\|\cdot\|_{p}$, we derive the following L_p, $p \geq 1$, interesting result.

Theorem 4.28 *Let $b(x)$ be a centered bell-shaped continuous function on $\mathbb{R}$ of compact support $[-T, T]$, $T > 0$. Let $x \in [-T^*, T^*]$, $T^* > 0$, and $n \in \mathbb{N}$: $n \geq \max\left(T + T^*, T^{-\frac{1}{\alpha}}\right)$, $0 < \alpha < 1$, $p \geq 1$. Let $\beta > 0$, $N = \lceil \beta \rceil$, $\beta \notin \mathbb{N}$, $f \in C^N(\mathbb{R})$, with $f^{(N)} \in L_\infty(\mathbb{R})$. Here both $D^\beta_{*x} f(t)$, $D^\beta_{x-} f(t)$ are bounded in $(x, t) \in \mathbb{R}^2$. Then*

(1)

$$\|F_n f - f\|_{p,[-T^*,T^*]} \leq \|f\|_{\infty,[-T^*,T^*]} \cdot \tag{4.70}$$

$$\left\| \sum_{k=\lceil nx - Tn^\alpha \rceil}^{[nx+Tn^\alpha]} \frac{1}{In^\alpha} b\left(n^{1-\alpha}\left(x - \frac{k}{n}\right)\right) - 1 \right\|_{p,[-T^*,T^*]} +$$

$$\frac{b^*}{I}\left(2T + \frac{1}{n^\alpha}\right)\left(\sum_{j=1}^{N-1} \frac{\|f^{(j)}\|_{p,[-T^*,T^*]} T^j}{j! n^{(1-\alpha)j}}\right) +$$

$$\frac{2^{\frac{1}{p}} T^{*\frac{1}{p}} b^*}{I}\left(2T + \frac{1}{n^\alpha}\right) \frac{T^\beta}{\Gamma(\beta+1) n^{(1-\alpha)\beta}} \cdot$$

$$\left\{ \sup_{x \in [-T^*,T^*]} \omega_1\left(D^\beta_{*x} f, \frac{T}{n^{1-\alpha}}\right)_{[x,+\infty)} + \sup_{x \in [-T^*,T^*]} \omega_1\left(D^\beta_{x-} f, \frac{T}{n^{1-\alpha}}\right)_{(-\infty,x]} \right\},$$

(2) When $N = 1$, we derive

$$\|F_n f - f\|_{p,[-T^*,T^*]} \leq \|f\|_{\infty,[-T^*,T^*]} \cdot \tag{4.71}$$

$$\left\| \sum_{k=\lceil nx - Tn^\alpha \rceil}^{[nx+Tn^\alpha]} \frac{1}{In^\alpha} b\left(n^{1-\alpha}\left(x - \frac{k}{n}\right)\right) - 1 \right\|_{p,[-T^*,T^*]} +$$

$$\frac{2^{\frac{1}{p}} T^{*\frac{1}{p}} b^*}{I}\left(2T + \frac{1}{n^\alpha}\right) \frac{T^\beta}{\Gamma(\beta+1) n^{(1-\alpha)\beta}} \cdot$$

$$\left\{ \sup_{x \in [-T^*,T^*]} \omega_1\left(D^\beta_{*x} f, \frac{T}{n^{1-\alpha}}\right)_{[x,+\infty)} + \sup_{x \in [-T^*,T^*]} \omega_1\left(D^\beta_{x-} f, \frac{T}{n^{1-\alpha}}\right)_{(-\infty,x]} \right\}.$$

By (4.70), (4.71) we derive the fractional L_p, $p \geq 1$, convergence with rates of $F_n f$ to f.

4.3.2 The "Squashing Operators" and Their Fractional Convergence to the Unit with Rates

We need (see also [1, 8]).

Definition 4.29 Let the nonnegative function $S : \mathbb{R} \to \mathbb{R}$, S has compact support $[-T, T]$, $T > 0$, and is nondecreasing there and it can be continuous only on either $(-\infty, T]$ or $[-T, T]$. S can have jump discontinuities. We call S the "squashing function".

Let $f : \mathbb{R} \to \mathbb{R}$ be continuous. Assume that

$$I^* := \int_{-T}^{T} S(t)\, dt > 0. \tag{4.72}$$

Obviously

$$\max_{x \in [-T,T]} S(x) = S(T). \tag{4.73}$$

For $x \in \mathbb{R}$ we define the "squashing operator" [1]

$$(G_n(f))(x) := \sum_{k=-n^2}^{n^2} \frac{f\left(\frac{k}{n}\right)}{I^* \cdot n^{\alpha}} \cdot S\left(n^{1-\alpha} \cdot \left(x - \frac{k}{n}\right)\right), \tag{4.74}$$

$0 < \alpha < 1$ and $n \in \mathbb{N} : n \geq \max\left(T + |x|, T^{-\frac{1}{\alpha}}\right)$. It is clear that

$$(G_n(f))(x) = \sum_{k=\lceil nx - Tn^{\alpha}\rceil}^{[nx+Tn^{\alpha}]} \frac{f\left(\frac{k}{n}\right)}{I^* \cdot n^{\alpha}} \cdot S\left(n^{1-\alpha} \cdot \left(x - \frac{k}{n}\right)\right). \tag{4.75}$$

Here we study the fractional convergence with rates of $(G_n f)(x) \to f(x)$, as $n \to +\infty$, $x \in \mathbb{R}$.

Notice that

$$\sum_{k=\lceil nx - Tn^{\alpha}\rceil}^{[nx+Tn^{\alpha}]} 1 \leq \left(2Tn^{\alpha} + 1\right). \tag{4.76}$$

From [1] we need

Lemma 4.30 *It holds that*

$$D_n(x) := \sum_{k=\lceil nx - Tn^{\alpha}\rceil}^{[nx+Tn^{\alpha}]} \frac{1}{I^* \cdot n^{\alpha}} \cdot S\left(n^{1-\alpha} \cdot \left(x - \frac{k}{n}\right)\right) \to 1, \tag{4.77}$$

pointwise, as $n \to +\infty$, where $x \in \mathbb{R}$.

We present our second main result

Theorem 4.31 *We consider $f : \mathbb{R} \to \mathbb{R}$. Let $\beta > 0$, $N = \lceil \beta \rceil$, $\beta \notin \mathbb{N}$, $f \in AC^N([a,b])$, $\forall [a,b] \subset \mathbb{R}$, with $f^{(N)} \in L_\infty(\mathbb{R})$. Let also $x \in \mathbb{R}$, $T > 0$, $n \in \mathbb{N} : n \geq \max\left(T + |x|, T^{-\frac{1}{\alpha}}\right)$. We further assume that $D^\beta_{*x} f$, $D^\beta_{x-} f$ are uniformly continuous functions or continuous and bounded on $[x, +\infty)$, $(-\infty, x]$, respectively.*

Then

(1)

$$|G_n(f)(x) - f(x)| \leq |f(x)| \cdot \tag{4.78}$$

$$\left| \sum_{k=\lceil nx - Tn^\alpha \rceil}^{[nx+Tn^\alpha]} \frac{1}{I^* n^\alpha} S\left(n^{1-\alpha}\left(x - \frac{k}{n}\right)\right) - 1 \right| +$$

$$\frac{S(T)}{I^*}\left(2T + \frac{1}{n^\alpha}\right)\left(\sum_{j=1}^{N-1} \frac{\left|f^{(j)}(x)\right| T^j}{j! n^{(1-\alpha)j}}\right)$$

$$+ \frac{S(T)}{I^*}\left(2T + \frac{1}{n^\alpha}\right)\frac{T^\beta}{\Gamma(\beta+1) n^{(1-\alpha)\beta}} \cdot$$

$$\left\{\omega_1\left(D^\beta_{*x} f, \frac{T}{n^{1-\alpha}}\right)_{[x,+\infty)} + \omega_1\left(D^\beta_{x-} f, \frac{T}{n^{1-\alpha}}\right)_{(-\infty,x]}\right\},$$

above $\sum_{j=1}^{0} \cdot = 0$,

(2)

$$\left|(G_n(f))(x) - \sum_{j=0}^{N-1} \frac{f^{(j)}(x)}{j!}\left(G_n\left((\cdot - x)^j\right)\right)(x)\right| \leq \tag{4.79}$$

$$\frac{S(T)}{I^*}\left(2T + \frac{1}{n^\alpha}\right)\frac{T^\beta}{\Gamma(\beta+1) n^{(1-\alpha)\beta}} \cdot$$

$$\left\{\omega_1\left(D^\beta_{*x} f, \frac{T}{n^{1-\alpha}}\right)_{[x,+\infty)} + \omega_1\left(D^\beta_{x-} f, \frac{T}{n^{1-\alpha}}\right)_{(-\infty,x]}\right\} =: \lambda_n^*(x),$$

(3) assume further that $f^{(j)}(x) = 0$, for $j = 0, 1, \ldots, N-1$, we get

$$|G_n(f)(x)| \leq \lambda_n^*(x), \tag{4.80}$$

(4) in case of $N = 1$*, we obtain*

$$|G_n(f)(x) - f(x)| \leq |f(x)| \cdot \tag{4.81}$$

$$\left| \sum_{k=\lceil nx - Tn^{\alpha} \rceil}^{[nx+Tn^{\alpha}]} \frac{1}{I^* n^{\alpha}} S\left(n^{1-\alpha}\left(x - \frac{k}{n}\right)\right) - 1 \right| +$$

$$\frac{S(T)}{I^*}\left(2T + \frac{1}{n^{\alpha}}\right) \frac{T^{\beta}}{\Gamma(\beta+1)\, n^{(1-\alpha)\beta}} \cdot$$

$$\left\{ \omega_1\left(D_{*x}^{\beta} f, \frac{T}{n^{1-\alpha}}\right)_{[x,+\infty)} + \omega_1\left(D_{x-}^{\beta} f, \frac{T}{n^{1-\alpha}}\right)_{(-\infty,x]} \right\}.$$

Here we get fractionally with rates the pointwise convergence of $(G_n(f))(x) \to f(x)$*, as* $n \to \infty$*,* $x \in \mathbb{R}$*.*

Proof Let $x \in \mathbb{R}$. We have that

$$D_{x-}^{\beta} f(x) = D_{*x}^{\beta} f(x) = 0\,.$$

From [9], p. 54, we get by the left Caputo fractional Taylor formula that

$$f\left(\frac{k}{n}\right) = \sum_{j=0}^{N-1} \frac{f^{(j)}(x)}{j!}\left(\frac{k}{n} - x\right)^j + \tag{4.82}$$

$$\frac{1}{\Gamma(\beta)} \int_x^{\frac{k}{n}} \left(\frac{k}{n} - J\right)^{\beta-1} \left(D_{*x}^{\beta} f(J) - D_{*x}^{\beta} f(x)\right) dJ,$$

for all $x \leq \frac{k}{n} \leq x + Tn^{\alpha-1}$, iff $\lceil nx \rceil \leq k \leq [nx + Tn^{\alpha}]$, where $k \in \mathbb{Z}$.

Also from [2], using the right Caputo fractional Taylor formula we get

$$f\left(\frac{k}{n}\right) = \sum_{j=0}^{N-1} \frac{f^{(j)}(x)}{j!}\left(\frac{k}{n} - x\right)^j + \tag{4.83}$$

$$\frac{1}{\Gamma(\beta)} \int_{\frac{k}{n}}^{x} \left(J - \frac{k}{n}\right)^{\beta-1} \left(D_{x-}^{\beta} f(J) - D_{x-}^{\beta} f(x)\right) dJ,$$

for all $x - Tn^{\alpha-1} \leq \frac{k}{n} \leq x$, iff $\lceil nx - Tn^{\alpha} \rceil \leq k \leq [nx]$, where $k \in \mathbb{Z}$.

Hence we have

$$\frac{f\left(\frac{k}{n}\right)S\left(n^{1-\alpha}\left(x-\frac{k}{n}\right)\right)}{I^{*}n^{\alpha}}=\sum_{j=0}^{N-1}\frac{f^{(j)}(x)}{j!}\left(\frac{k}{n}-x\right)^{j}\frac{S\left(n^{1-\alpha}\left(x-\frac{k}{n}\right)\right)}{I^{*}n^{\alpha}}+ \tag{4.84}$$

$$\frac{S\left(n^{1-\alpha}\left(x-\frac{k}{n}\right)\right)}{I^{*}n^{\alpha}\Gamma(\beta)}\int_{x}^{\frac{k}{n}}\left(\frac{k}{n}-J\right)^{\beta-1}\left(D_{*x}^{\beta}f(J)-D_{*x}^{\beta}f(x)\right)dJ,$$

and

$$\frac{f\left(\frac{k}{n}\right)S\left(n^{1-\alpha}\left(x-\frac{k}{n}\right)\right)}{I^{*}n^{\alpha}}=\sum_{j=0}^{N-1}\frac{f^{(j)}(x)}{j!}\left(\frac{k}{n}-x\right)^{j}\frac{S\left(n^{1-\alpha}\left(x-\frac{k}{n}\right)\right)}{I^{*}n^{\alpha}}+ \tag{4.85}$$

$$\frac{S\left(n^{1-\alpha}\left(x-\frac{k}{n}\right)\right)}{I^{*}n^{\alpha}\Gamma(\beta)}\int_{\frac{k}{n}}^{x}\left(J-\frac{k}{n}\right)^{\beta-1}\left(D_{x-}^{\beta}f(J)-D_{x-}^{\beta}f(x)\right)dJ.$$

Therefore we obtain

$$\frac{\sum_{k=[nx]+1}^{[nx+Tn^{\alpha}]}f\left(\frac{k}{n}\right)S\left(n^{1-\alpha}\left(x-\frac{k}{n}\right)\right)}{I^{*}n^{\alpha}}= \tag{4.86}$$

$$\sum_{j=0}^{N-1}\frac{f^{(j)}(x)}{j!}\left(\frac{\sum_{k=[nx]+1}^{[nx+Tn^{\alpha}]}\left(\frac{k}{n}-x\right)^{j}S\left(n^{1-\alpha}\left(x-\frac{k}{n}\right)\right)}{I^{*}n^{\alpha}}\right)+$$

$$\sum_{k=[nx]+1}^{[nx+Tn^{\alpha}]}\frac{S\left(n^{1-\alpha}\left(x-\frac{k}{n}\right)\right)}{I^{*}n^{\alpha}\Gamma(\beta)}\int_{x}^{\frac{k}{n}}\left(\frac{k}{n}-J\right)^{\beta-1}\left(D_{*x}^{\beta}f(J)-D_{*x}^{\beta}f(x)\right)dJ,$$

and

$$\frac{\sum_{k=\lceil nx-Tn^{\alpha}\rceil}^{[nx]}f\left(\frac{k}{n}\right)S\left(n^{1-\alpha}\left(x-\frac{k}{n}\right)\right)}{I^{*}n^{\alpha}}= \tag{4.87}$$

$$\sum_{j=0}^{N-1}\frac{f^{(j)}(x)}{j!}\frac{\sum_{k=\lceil nx-Tn^{\alpha}\rceil}^{[nx]}\left(\frac{k}{n}-x\right)^{j}S\left(n^{1-\alpha}\left(x-\frac{k}{n}\right)\right)}{I^{*}n^{\alpha}}+$$

$$\frac{\sum_{k=\lceil nx-Tn^{\alpha}\rceil}^{[nx]}S\left(n^{1-\alpha}\left(x-\frac{k}{n}\right)\right)}{I^{*}n^{\alpha}\Gamma(\beta)}\int_{\frac{k}{n}}^{x}\left(J-\frac{k}{n}\right)^{\beta-1}\left(D_{x-}^{\beta}f(J)-D_{x-}^{\beta}f(x)\right)dJ.$$

Adding the two equalities (4.86) and (4.87) we obtain

$$(G_n(f))(x) =$$

$$\sum_{j=0}^{N-1} \frac{f^{(j)}(x)}{j!} \left(\frac{\sum_{k=\lceil nx-Tn^{\alpha} \rceil}^{[nx+Tn^{\alpha}]} \left(\frac{k}{n}-x\right)^j S\left(n^{1-\alpha}\left(x-\frac{k}{n}\right)\right)}{I^* n^{\alpha}} \right) + M_n(x), \quad (4.88)$$

where

$$M_n(x) := \frac{\sum_{k=\lceil nx-Tn^{\alpha} \rceil}^{[nx]} S\left(n^{1-\alpha}\left(x-\frac{k}{n}\right)\right)}{I^* n^{\alpha} \Gamma(\beta)}.$$

$$\int_{\frac{k}{n}}^{x} \left(J-\frac{k}{n}\right)^{\beta-1} \left(D_{x-}^{\beta} f(J) - D_{x-}^{\beta} f(x)\right) dJ +$$

$$\sum_{k=[nx]+1}^{[nx+Tn^{\alpha}]} \frac{S\left(n^{1-\alpha}\left(x-\frac{k}{n}\right)\right)}{I^* n^{\alpha} \Gamma(\beta)} \int_{x}^{\frac{k}{n}} \left(\frac{k}{n}-J\right)^{\beta-1} \left(D_{*x}^{\beta} f(J) - D_{*x}^{\beta} f(x)\right) dJ. \quad (4.89)$$

We call

$$M_{1n}(x) := \sum_{k=\lceil nx-Tn^{\alpha} \rceil}^{[nx]} \frac{S\left(n^{1-\alpha}\left(x-\frac{k}{n}\right)\right)}{I^* n^{\alpha} \Gamma(\beta)}.$$

$$\int_{\frac{k}{n}}^{x} \left(J-\frac{k}{n}\right)^{\beta-1} \left(D_{x-}^{\beta} f(J) - D_{x-}^{\beta} f(x)\right) dJ, \quad (4.90)$$

and

$$M_{2n}(x) := \sum_{k=[nx]+1}^{[nx+Tn^{\alpha}]} \frac{S\left(n^{1-\alpha}\left(x-\frac{k}{n}\right)\right)}{I^* n^{\alpha} \Gamma(\beta)}.$$

$$\int_{x}^{\frac{k}{n}} \left(\frac{k}{n}-J\right)^{\beta-1} \left(D_{*x}^{\beta} f(J) - D_{*x}^{\beta} f(x)\right) dJ. \quad (4.91)$$

I.e.

$$M_n(x) = M_{1n}(x) + M_{2n}(x). \quad (4.92)$$

We further have

$$(G_n(f))(x) - f(x) = f(x) \left(\frac{\sum_{k=\lceil nx-Tn^{\alpha} \rceil}^{[nx+Tn^{\alpha}]} S\left(n^{1-\alpha}\left(x-\frac{k}{n}\right)\right)}{I^* n^{\alpha}} - 1 \right) + \quad (4.93)$$

$$\sum_{j=0}^{N-1} \frac{f^{(j)}(x)}{j!} \left(\frac{\sum_{k=\lceil nx-Tn^{\alpha} \rceil}^{[nx+Tn^{\alpha}]} \left(\frac{k}{n} - x\right)^j S\left(n^{1-\alpha}\left(x - \frac{k}{n}\right)\right)}{I^* n^{\alpha}} \right) + M_n(x),$$

and

$$|(G_n(f))(x) - f(x)| \leq |f(x)| \left| \sum_{k=\lceil nx-Tn^{\alpha} \rceil}^{[nx+Tn^{\alpha}]} \frac{1}{I^* n^{\alpha}} S\left(n^{1-\alpha}\left(x - \frac{k}{n}\right)\right) - 1 \right| + \tag{4.94}$$

$$\sum_{j=1}^{N-1} \frac{\left|f^{(j)}(x)\right|}{j!} \left(\frac{\sum_{k=\lceil nx-Tn^{\alpha} \rceil}^{[nx+Tn^{\alpha}]} \left|x - \frac{k}{n}\right|^j S\left(n^{1-\alpha}\left(x - \frac{k}{n}\right)\right)}{I^* n^{\alpha}} \right) + |M_n(x)| \leq$$

$$|f(x)| \left| \sum_{k=\lceil nx-Tn^{\alpha} \rceil}^{[nx+Tn^{\alpha}]} \frac{1}{I^* n^{\alpha}} S\left(n^{1-\alpha}\left(x - \frac{k}{n}\right)\right) - 1 \right| +$$

$$\sum_{j=1}^{N-1} \frac{\left|f^{(j)}(x)\right|}{j!} \frac{T^j}{n^{(1-\alpha)j}} \left(\frac{\sum_{k=\lceil nx-Tn^{\alpha} \rceil}^{[nx+Tn^{\alpha}]} S\left(n^{1-\alpha}\left(x - \frac{k}{n}\right)\right)}{I^* n^{\alpha}} \right) + |M_n(x)| =: (*). \tag{4.95}$$

Therefore we obtain

$$|(G_n(f))(x) - f(x)| \leq |f(x)| \left| \sum_{k=\lceil nx-Tn^{\alpha} \rceil}^{[nx+Tn^{\alpha}]} \frac{1}{I^* n^{\alpha}} S\left(n^{1-\alpha}\left(x - \frac{k}{n}\right)\right) - 1 \right| +$$

$$\frac{S(T)}{I^*} \left(2T + \frac{1}{n^{\alpha}}\right) \left(\sum_{j=1}^{N-1} \frac{\left|f^{(j)}(x)\right| T^j}{j! n^{(1-\alpha)j}} \right) + |M_n(x)|. \tag{4.96}$$

We call

$$\gamma_{1n} := \frac{1}{\Gamma(\beta)} \left| \int_{\frac{k}{n}}^{x} \left(J - \frac{k}{n}\right)^{\beta-1} \left(D_{x-}^{\beta} f(J) - D_{x-}^{\beta} f(x)\right) dJ \right|. \tag{4.97}$$

As in the proof of Theorem 4.25 we have

$$\gamma_{1n} \leq \frac{T^{\beta}}{\Gamma(\beta+1) n^{(1-\alpha)\beta}} \omega_1 \left(D_{x-}^{\beta} f, \frac{T}{n^{1-\alpha}} \right)_{(-\infty, x]}. \tag{4.98}$$

Furthermore

$$|M_{1n}(x)| \leq \sum_{k=\lceil nx-Tn^{\alpha}\rceil}^{[nx]} \frac{S\left(n^{1-\alpha}\left(x-\frac{k}{n}\right)\right)}{I^{*}n^{\alpha}} \gamma_{1n} \leq \tag{4.99}$$

$$\left(\sum_{k=\lceil nx-Tn^{\alpha}\rceil}^{[nx]} \frac{S\left(n^{1-\alpha}\left(x-\frac{k}{n}\right)\right)}{I^{*}n^{\alpha}}\right) \frac{T^{\beta}}{\Gamma(\beta+1)\,n^{(1-\alpha)\beta}} \omega_{1}\left(D_{x-}^{\beta}f, \frac{T}{n^{1-\alpha}}\right)_{(-\infty,x]} \leq$$

$$\left(\sum_{k=\lceil nx-Tn^{\alpha}\rceil}^{[nx+Tn^{\alpha}]} \frac{S\left(n^{1-\alpha}\left(x-\frac{k}{n}\right)\right)}{I^{*}n^{\alpha}}\right) \frac{T^{\beta}}{\Gamma(\beta+1)\,n^{(1-\alpha)\beta}} \omega_{1}\left(D_{x-}^{\beta}f, \frac{T}{n^{1-\alpha}}\right)_{(-\infty,x]} \leq$$

$$\frac{S(T)}{I^{*}}\left(2T+\frac{1}{n^{\alpha}}\right) \frac{T^{\beta}}{\Gamma(\beta+1)\,n^{(1-\alpha)\beta}} \omega_{1}\left(D_{x-}^{\beta}f, \frac{T}{n^{1-\alpha}}\right)_{(-\infty,x]}.$$

So that

$$|M_{1n}(x)| \leq \frac{S(T)}{I^{*}}\left(2T+\frac{1}{n^{\alpha}}\right) \frac{T^{\beta}}{\Gamma(\beta+1)\,n^{(1-\alpha)\beta}} \omega_{1}\left(D_{x-}^{\beta}f, \frac{T}{n^{1-\alpha}}\right)_{(-\infty,x]}. \tag{4.100}$$

We also call

$$\gamma_{2n} := \frac{1}{\Gamma(\beta)}\left|\int_{x}^{\frac{k}{n}}\left(\frac{k}{n}-J\right)^{\beta-1}\left(D_{*x}^{\beta}f(J)-D_{*x}^{\beta}f(x)\right)dJ\right|. \tag{4.101}$$

As in the proof of Theorem 4.25 we get

$$\gamma_{2n} \leq \frac{T^{\beta}}{\Gamma(\beta+1)\,n^{(1-\alpha)\beta}} \omega_{1}\left(D_{*x}^{\beta}f, \frac{T}{n^{1-\alpha}}\right)_{[x,+\infty)}. \tag{4.102}$$

Consequently we find

$$|M_{2n}(x)| \leq \left(\sum_{k=[nx]+1}^{[nx+Tn^{\alpha}]} \frac{S\left(n^{1-\alpha}\left(x-\frac{k}{n}\right)\right)}{I^{*}n^{\alpha}}\right)\cdot$$

$$\frac{T^{\beta}}{\Gamma(\beta+1)\,n^{(1-\alpha)\beta}} \omega_{1}\left(D_{*x}^{\beta}f, \frac{T}{n^{1-\alpha}}\right)_{[x,+\infty)} \leq \tag{4.103}$$

$$\frac{S(T)}{I^{*}}\left(2T+\frac{1}{n^{\alpha}}\right) \frac{T^{\beta}}{\Gamma(\beta+1)\,n^{(1-\alpha)\beta}} \omega_{1}\left(D_{*x}^{\beta}f, \frac{T}{n^{1-\alpha}}\right)_{[x,+\infty)}.$$

So we have proved that

$$|M_n(x)| \leq \frac{S(T)}{I^*}\left(2T + \frac{1}{n^\alpha}\right)\frac{T^\beta}{\Gamma(\beta+1)\,n^{(1-\alpha)\beta}} \cdot \tag{4.104}$$

$$\left\{\omega_1\left(D_{*x}^\beta f, \frac{T}{n^{1-\alpha}}\right)_{[x,+\infty)} + \omega_1\left(D_{x-}^\beta f, \frac{T}{n^{1-\alpha}}\right)_{(-\infty,x]}\right\}.$$

Combining (4.96) and (4.104) we have (4.78). ■

As an application of Theorem 4.31 we give

Theorem 4.32 *Let* $\beta > 0$, $N = \lceil \beta \rceil$, $\beta \notin \mathbb{N}$, $f \in C^N(\mathbb{R})$, *with* $f^{(N)} \in L_\infty(\mathbb{R})$. *Let also* $T > 0$, $n \in \mathbb{N} : n \geq \max\left(2T, T^{-\frac{1}{\alpha}}\right)$. *We further assume that* $D_{*x}^\beta f(t)$, $D_{x-}^\beta f(t)$ *are both bounded in* $(x,t) \in \mathbb{R}^2$. *Then*

(1)

$$\|G_n(f) - f\|_{\infty,[-T,T]} \leq \|f\|_{\infty,[-T,T]} \cdot \tag{4.105}$$

$$\left\| \sum_{k=\lceil nx - Tn^\alpha \rceil}^{[nx+Tn^\alpha]} \frac{1}{I^* n^\alpha} S\left(n^{1-\alpha}\left(x - \frac{k}{n}\right)\right) - 1 \right\|_{\infty,[-T,T]} +$$

$$\frac{S(T)}{I^*}\left(2T + \frac{1}{n^\alpha}\right)\left(\sum_{j=1}^{N-1} \frac{\|f^{(j)}\|_{\infty,[-T,T]}\, T^j}{j!\,n^{(1-\alpha)j}}\right) +$$

$$\frac{S(T)}{I^*}\left(2T + \frac{1}{n^\alpha}\right)\frac{T^\beta}{\Gamma(\beta+1)\,n^{(1-\alpha)\beta}} \cdot$$

$$\left\{\sup_{x\in[-T,T]} \omega_1\left(D_{*x}^\beta f, \frac{T}{n^{1-\alpha}}\right)_{[x,+\infty)} + \sup_{x\in[-T,T]} \omega_1\left(D_{x-}^\beta f, \frac{T}{n^{1-\alpha}}\right)_{(-\infty,x]}\right\},$$

(2) in case of $N = 1$, *we obtain*

$$\|G_n(f) - f\|_{\infty,[-T,T]} \leq \|f\|_{\infty,[-T,T]} \cdot \tag{4.106}$$

$$\left\| \sum_{k=\lceil nx - Tn^\alpha \rceil}^{[nx+Tn^\alpha]} \frac{1}{I^* n^\alpha} S\left(n^{1-\alpha}\left(x - \frac{k}{n}\right)\right) - 1 \right\|_{\infty,[-T,T]} +$$

$$\frac{S(T)}{I^*}\left(2T + \frac{1}{n^\alpha}\right)\frac{T^\beta}{\Gamma(\beta+1)\,n^{(1-\alpha)\beta}} \cdot$$

$$\left\{ \sup_{x \in [-T,T]} \omega_1 \left(D^{\beta}_{*x} f, \frac{T}{n^{1-\alpha}} \right)_{[x,+\infty)} + \sup_{x \in [-T,T]} \omega_1 \left(D^{\beta}_{x-} f, \frac{T}{n^{1-\alpha}} \right)_{(-\infty,x]} \right\}.$$

An interesting case is when $\beta = \frac{1}{2}$.

Assuming further that $\left\| \sum_{k=\lceil nx-Tn^{\alpha}\rceil}^{[nx+Tn^{\alpha}]} \frac{1}{I^* n^{\alpha}} S\left(n^{1-\alpha}\left(x - \frac{k}{n}\right)\right) - 1 \right\|_{\infty,[-T,T]} \to 0$, *as* $n \to \infty$, *we get fractionally with rates the uniform convergence of* $G_n(f) \to f$, *as* $n \to \infty$.

Proof From (4.78), (4.81) of Theorem 4.31, and by Remark 4.17.

Also by

$$\sum_{k=\lceil nx-Tn^{\alpha}\rceil}^{[nx+Tn^{\alpha}]} \frac{1}{I^* n^{\alpha}} S\left(n^{1-\alpha}\left(x - \frac{k}{n}\right)\right) \le \frac{S(T)}{I^*}(2T+1), \tag{4.107}$$

we get that

$$\left\| \sum_{k=\lceil nx-Tn^{\alpha}\rceil}^{[nx+Tn^{\alpha}]} \frac{1}{I^* n^{\alpha}} S\left(n^{1-\alpha}\left(x - \frac{k}{n}\right)\right) - 1 \right\|_{\infty,[-T,T]} \le \left(\frac{S(T)}{I^*}(2T+1) + 1 \right). \tag{4.108}$$

■

One can also apply Remark 4.18 to the last Theorem 4.32, to get interesting and simplified results.

Note 4.33 The maps F_n, G_n, $n \in \mathbb{N}$, are positive linear operators.

We finish with

Remark 4.34 The condition of Theorem 4.26 that

$$\left\| \sum_{k=\lceil nx-Tn^{\alpha}\rceil}^{[nx+Tn^{\alpha}]} \frac{1}{In^{\alpha}} b\left(n^{1-\alpha}\left(x - \frac{k}{n}\right)\right) - 1 \right\|_{\infty,[-T,T]} \to 0, \tag{4.109}$$

as $n \to \infty$, is not uncommon.

We give an example related to that.

We take as $b(x)$ the characteristic function over $[-1, 1]$, that is $\chi_{[-1,1]}(x)$. Here $T = 1$ and $I = 2$, $n \ge 2$, $x \in [-1, 1]$.

We get that

$$\sum_{k=\lceil nx-n^{\alpha}\rceil}^{[nx+n^{\alpha}]} \frac{1}{2n^{\alpha}} \chi_{[-1,1]}\left(n^{1-\alpha}\left(x - \frac{k}{n}\right)\right) \overset{(4.23)}{=} \sum_{k=\lceil nx-n^{\alpha}\rceil}^{[nx+n^{\alpha}]} \frac{1}{2n^{\alpha}} =$$

$$\frac{1}{2n^{\alpha}}\left(\sum_{k=\lceil nx-n^{\alpha}\rceil}^{[nx+n^{\alpha}]} 1\right) = \frac{([nx+n^{\alpha}] - \lceil nx-n^{\alpha}\rceil + 1)}{2n^{\alpha}}. \tag{4.110}$$

But we have

$$\left[nx+n^{\alpha}\right] - \left\lceil nx-n^{\alpha}\right\rceil + 1 \leq 2n^{\alpha} + 1,$$

hence

$$\frac{([nx+n^{\alpha}] - \lceil nx-n^{\alpha}\rceil + 1)}{2n^{\alpha}} \leq 1 + \frac{1}{2n^{\alpha}}. \tag{4.111}$$

Also it holds

$$\left[nx+n^{\alpha}\right] - \left\lceil nx-n^{\alpha}\right\rceil + 1 \geq 2n^{\alpha} - 2 + 1 = 2n^{\alpha} - 1,$$

and

$$\frac{([nx+n^{\alpha}] - \lceil nx-n^{\alpha}\rceil + 1)}{2n^{\alpha}} \geq 1 - \frac{1}{2n^{\alpha}}. \tag{4.112}$$

Consequently we derive that

$$-\frac{1}{2n^{\alpha}} \leq \left(\sum_{k=\lceil nx-n^{\alpha}\rceil}^{[nx+n^{\alpha}]} \frac{1}{2n^{\alpha}}\chi_{[-1,1]}\left(n^{1-\alpha}\left(x-\frac{k}{n}\right)\right) - 1\right) \leq \frac{1}{2n^{\alpha}}, \tag{4.113}$$

for any $x \in [-1, 1]$ and for any $n \geq 2$.

Hence we get

$$\left\| \sum_{k=\lceil nx-n^{\alpha}\rceil}^{[nx+n^{\alpha}]} \frac{1}{2n^{\alpha}}\chi_{[-1,1]}\left(n^{1-\alpha}\left(x-\frac{k}{n}\right)\right) - 1 \right\|_{\infty,[-1,1]} \to 0, \text{ as } n \to \infty. \tag{4.114}$$

References

1. G.A. Anastassiou, Rate of convergence of some neural network operators to the unit-univariate case. J. Math. Anal. Appl. **212**, 237–262 (1997)
2. G.A. Anastassiou, On right fractional calculus. Chaos, Solitons Fractals **42**, 365–376 (2009)
3. G.A. Anastassiou, Fractional Korovkin theory. Chaos, Solitons Fractals **42**(4), 2080–2094 (2009)
4. G.A. Anastassiou, *Fractional Differentiation Inequalities* (Springer, New York, 2009)
5. G.A. Anastassiou, Quantitative approximation by fractional smooth picard singular operators. Math. Eng. Sci. Aerosp. **2**(1), 71–87 (2011)
6. G.A. Anastassiou, Fractional representation formulae and right fractional inequalities. Math. Comput. Model. **54**(11–12), 3098–3115 (2011)

7. G.A. Anastassiou, Fractional approximation by Cardaliaguet-Euvrard and squashing neural network operators. Studia Math. Babes Bolyai **57**(3), 331–354 (2012)
8. P. Cardaliaguet, G. Euvrard, Approximation of a function and its derivative with a neural network. Neural Netw. **5**, 207–220 (1992)
9. K. Diethelm, *The Analysis of Fractional Differential Equations*, Lecture Notes in Mathematics, vol. 2004 (Springer, Berlin, 2010)
10. A.M.A. El-Sayed, M. Gaber, On the finite Caputo and finite Riesz derivatives. Electron. J. Theor. Phys. **3**(12), 81–95 (2006)
11. G.S. Frederico, D.F.M. Torres, Fractional optimal control in the sense of Caputo and the fractional Noether's theorem. Int. Math. Forum **3**(10), 479–493 (2008)

Chapter 5
Fractional Voronovskaya Type Asymptotic Expansions for Quasi-interpolation Neural Networks

Here we study further the quasi-interpolation of sigmoidal and hyperbolic tangent types neural network operators of one hidden layer. Based on fractional calculus theory we derive fractional Voronovskaya type asymptotic expansions for the error of approximation of these operators to the unit operator. It follows [12].

5.1 Background

We need

Definition 5.1 Let $\nu > 0$, $n = \lceil \nu \rceil$ ($\lceil \cdot \rceil$ is the ceiling of the number), $f \in AC^n([a, b])$ (space of functions f with $f^{(n-1)} \in AC([a, b])$, absolutely continuous functions). We call left Caputo fractional derivative (see [14], pp. 49–52) the function

$$D_{*a}^{\nu} f(x) = \frac{1}{\Gamma(n-\nu)} \int_a^x (x-t)^{n-\nu-1} f^{(n)}(t)\, dt, \tag{5.1}$$

$\forall\, x \in [a, b]$, where Γ is the gamma function $\Gamma(\nu) = \int_0^{\infty} e^{-t} t^{\nu-1} dt$, $\nu > 0$. Notice $D_{*a}^{\nu} f \in L_1([a, b])$ and $D_{*a}^{\nu} f$ exists a.e. on $[a, b]$.

We set $D_{*a}^{0} f(x) = f(x)$, $\forall\, x \in [a, b]$.

Definition 5.2 (*see also* [3, 15, 16]) Let $f \in AC^m([a, b])$, $m = \lceil \alpha \rceil$, $\alpha > 0$. The right Caputo fractional derivative of order $\alpha > 0$ is given by

$$D_{b-}^{\alpha} f(x) = \frac{(-1)^m}{\Gamma(m-\alpha)} \int_x^b (\zeta - x)^{m-\alpha-1} f^{(m)}(\zeta)\, d\zeta, \tag{5.2}$$

$\forall\, x \in [a, b]$. We set $D_{b-}^{0} f(x) = f(x)$. Notice $D_{b-}^{\alpha} f \in L_1([a, b])$ and $D_{b-}^{\alpha} f$ exists a.e. on $[a, b]$.

G.A. Anastassiou, *Intelligent Systems II: Complete Approximation by Neural Network Operators*, Studies in Computational Intelligence 608,
DOI 10.1007/978-3-319-20505-2_5

Convention 5.3 *We assume that*

$$D_{*x_0}^{\alpha} f(x) = 0, \text{ for } x < x_0, \tag{5.3}$$

and

$$D_{x_0-}^{\alpha} f(x) = 0, \text{ for } x > x_0, \tag{5.4}$$

for all $x, x_0 \in [a, b]$.

We mention

Proposition 5.4 (by [5]) *Let* $f \in C^n([a, b])$, $n = \lceil \nu \rceil$, $\nu > 0$. *Then* $D_{*a}^{\nu} f(x)$ *is continuous in* $x \in [a, b]$.

Also we have

Proposition 5.5 (by [5]) *Let* $f \in C^m([a, b])$, $m = \lceil \alpha \rceil$, $\alpha > 0$. *Then* $D_{b-}^{\alpha} f(x)$ *is continuous in* $x \in [a, b]$.

Theorem 5.6 ([5]) *Let* $f \in C^m([a, b])$, $m = \lceil \alpha \rceil$, $\alpha > 0$, $x, x_0 \in [a, b]$. *Then* $D_{*x_0}^{\alpha} f(x)$, $D_{x_0-}^{\alpha} f(x)$ *are jointly continuous functions in* (x, x_0) *from* $[a, b]^2$ *into* $\mathbb{R}$.

We mention the left Caputo fractional Taylor formula with integral remainder.

Theorem 5.7 ([14], p. 54) *Let* $f \in AC^m([a, b])$, $[a, b] \subset \mathbb{R}$, $m = \lceil \alpha \rceil$, $\alpha > 0$. *Then*

$$f(x) = \sum_{k=0}^{m-1} \frac{f^{(k)}(x_0)}{k!} (x - x_0)^k + \frac{1}{\Gamma(\alpha)} \int_{x_0}^{x} (x - J)^{\alpha-1} D_{*x_0}^{\alpha} f(J)\, dJ, \tag{5.5}$$

$\forall\, x \geq x_0;\ x, x_0 \in [a, b]$.

Also we mention the right Caputo fractional Taylor formula.

Theorem 5.8 ([3]) *Let* $f \in AC^m([a, b])$, $[a, b] \subset \mathbb{R}$, $m = \lceil \alpha \rceil$, $\alpha > 0$. *Then*

$$f(x) = \sum_{j=0}^{m-1} \frac{f^{(k)}(x_0)}{k!} (x - x_0)^k + \frac{1}{\Gamma(\alpha)} \int_{x}^{x_0} (J - x)^{\alpha-1} D_{x_0-}^{\alpha} f(J)\, dJ, \tag{5.6}$$

$\forall\, x \leq x_0;\ x, x_0 \in [a, b]$.

For more on fractional calculus related to this work see [2, 4, 7].

We consider here the sigmoidal function of logarithmic type

$$s(x) = \frac{1}{1 + e^{-x}}, \quad x \in \mathbb{R}.$$

It has the properties $\lim_{x \to +\infty} s(x) = 1$ and $\lim_{x \to -\infty} s(x) = 0$.

This function plays the role of an activation function in the hidden layer of neural networks.

As in [13], we consider

$$\Phi(x) := \frac{1}{2}(s(x+1) - s(x-1)), \quad x \in \mathbb{R}. \tag{5.7}$$

We notice the following properties:

(i) $\Phi(x) > 0, \ \forall\, x \in \mathbb{R}$,
(ii) $\sum_{k=-\infty}^{\infty} \Phi(x-k) = 1, \ \forall\, x \in \mathbb{R}$,
(iii) $\sum_{k=-\infty}^{\infty} \Phi(nx-k) = 1, \ \forall\, x \in \mathbb{R}; n \in \mathbb{N}$,
(iv) $\int_{-\infty}^{\infty} \Phi(x)\, dx = 1$,
(v) Φ is a density function,
(vi) Φ is even: $\Phi(-x) = \Phi(x), \ x \geq 0$.

We see that [13]

$$\Phi(x) = \left(\frac{e^2-1}{2e}\right) \frac{e^{-x}}{\left(1+e^{-x-1}\right)\left(1+e^{-x+1}\right)} = \tag{5.8}$$

$$\left(\frac{e^2-1}{2e^2}\right) \frac{1}{\left(1+e^{x-1}\right)\left(1+e^{-x-1}\right)}.$$

(vii) By [13] Φ is decreasing on $\mathbb{R}_+$, and increasing on $\mathbb{R}_-$.
(viii) By [11] for $n \in \mathbb{N}, 0 < \beta < 1$, we get

$$\sum_{\substack{k=-\infty \\ : |nx-k| > n^{1-\beta}}}^{\infty} \Phi(nx-k) < \left(\frac{e^2-1}{2}\right) e^{-n^{(1-\beta)}} = 3.1992 e^{-n^{(1-\beta)}}. \tag{5.9}$$

Denote by $\lfloor \cdot \rfloor$ the integral part of a number. Consider $x \in [a,b] \subset \mathbb{R}$ and $n \in \mathbb{N}$ such that $\lceil na \rceil \leq \lfloor nb \rfloor$.

(ix) By [11] it holds

$$\frac{1}{\sum_{k=\lceil na \rceil}^{\lfloor nb \rfloor} \Phi(nx-k)} < \frac{1}{\Phi(1)} = 5.250312578, \ \forall\, x \in [a,b]. \tag{5.10}$$

(x) By [11] it holds $\lim_{n \to \infty} \sum_{k=\lceil na \rceil}^{\lfloor nb \rfloor} \Phi(nx-k) \neq 1$, for at least some $x \in [a,b]$.

Let $f \in C([a,b])$ and $n \in \mathbb{N}$ such that $\lceil na \rceil \leq \lfloor nb \rfloor$.

We study further (see also [11]) the quasi-interpolation positive linear neural network operator

$$G_n(f,x) := \frac{\sum_{k=\lceil na \rceil}^{\lfloor nb \rfloor} f\left(\frac{k}{n}\right) \Phi(nx-k)}{\sum_{k=\lceil na \rceil}^{\lfloor nb \rfloor} \Phi(nx-k)}, \quad x \in [a,b]. \tag{5.11}$$

For large enough n we always obtain $\lceil na \rceil \leq \lfloor nb \rfloor$. Also $a \leq \frac{k}{n} \leq b$, iff $\lceil na \rceil \leq k \leq \lfloor nb \rfloor$.

We also consider here the hyperbolic tangent function $\tanh x$, $x \in \mathbb{R}$:

$$\tanh x := \frac{e^x - e^{-x}}{e^x + e^{-x}} = \frac{e^{2x} - 1}{e^{2x} + 1}.$$

It has the properties $\tanh 0 = 0$, $-1 < \tanh x < 1$, $\forall\, x \in \mathbb{R}$, and $\tanh(-x) = -\tanh x$. Furthermore $\tanh x \to 1$ as $x \to \infty$, and $\tanh x \to -1$, as $x \to -\infty$, and it is strictly increasing on $\mathbb{R}$. Furthermore it holds $\frac{d}{dx} \tanh x = \frac{1}{\cosh^2 x} > 0$.

This function plays also the role of an activation function in the hidden layer of neural networks.

We further consider

$$\Psi(x) := \frac{1}{4}(\tanh(x+1) - \tanh(x-1)) > 0, \quad \forall\, x \in \mathbb{R}. \tag{5.12}$$

We easily see that$\Psi(-x) = \Psi(x)$, that is Ψ is even on $\mathbb{R}$. Obviously Ψ is differentiable, thus continuous.

Here we follow [8]

Proposition 5.9 *$\Psi(x)$ for $x \geq 0$ is strictly decreasing.*

Obviously $\Psi(x)$ is strictly increasing for $x \leq 0$. Also it holds $\lim_{x \to -\infty} \Psi(x) = 0 = \lim_{x \to \infty} \Psi(x)$.

Infact Ψ has the bell shape with horizontal asymptote the x-axis. So the maximum of Ψ is at zero, $\Psi(0) = 0.3809297$.

Theorem 5.10 *We have that $\sum_{i=-\infty}^{\infty} \Psi(x-i) = 1$, $\forall\, x \in \mathbb{R}$.*

Thus

$$\sum_{i=-\infty}^{\infty} \Psi(nx-i) = 1, \quad \forall\, n \in \mathbb{N}, \ \forall\, x \in \mathbb{R}.$$

Furthermore we get:

Since Ψ is even it holds $\sum_{i=-\infty}^{\infty} \Psi(i-x) = 1$, $\forall x \in \mathbb{R}$.

Hence $\sum_{i=-\infty}^{\infty} \Psi(i+x) = 1$, $\forall\, x \in \mathbb{R}$, and $\sum_{i=-\infty}^{\infty} \Psi(x+i) = 1$, $\forall\, x \in \mathbb{R}$.

Theorem 5.11 *It holds* $\int_{-\infty}^{\infty} \Psi(x)\,dx = 1$.

So $\Psi(x)$ is a density function on $\mathbb{R}$.

Theorem 5.12 *Let* $0 < \beta < 1$ *and* $n \in \mathbb{N}$. *It holds*

$$\sum_{\substack{k=-\infty \\ : |nx-k| \geq n^{1-\beta}}}^{\infty} \Psi(nx-k) \leq e^4 \cdot e^{-2n^{(1-\beta)}}. \tag{5.13}$$

Theorem 5.13 *Let* $x \in [a,b] \subset \mathbb{R}$ *and* $n \in \mathbb{N}$ *so that* $\lceil na \rceil \leq \lfloor nb \rfloor$. *It holds*

$$\frac{1}{\sum_{k=\lceil na \rceil}^{\lfloor nb \rfloor} \Psi(nx-k)} < 4.1488766 = \frac{1}{\Psi(1)}. \tag{5.14}$$

Also by [8], we obtain

$$\lim_{n \to \infty} \sum_{k=\lceil na \rceil}^{\lfloor nb \rfloor} \Psi(nx-k) \neq 1, \tag{5.15}$$

for at least some $x \in [a,b]$.

Definition 5.14 Let $f \in C([a,b])$ and $n \in \mathbb{N}$ such that $\lceil na \rceil \leq \lfloor nb \rfloor$.

We further study, as in [8], the quasi-interpolation positive linear neural network operator

$$F_n(f,x) := \frac{\sum_{k=\lceil na \rceil}^{\lfloor nb \rfloor} f\left(\frac{k}{n}\right) \Psi(nx-k)}{\sum_{k=\lceil na \rceil}^{\lfloor nb \rfloor} \Psi(nx-k)}, \quad x \in [a,b]. \tag{5.16}$$

We find here fractional Voronovskaya type asymptotic expansions for $G_n(f,x)$ and $F_n(f,x)$, $x \in [a,b]$.

For related work on neural networks also see [1, 6, 9, 10]. For neural networks in general see [17–19].

5.2 Main Results

We present our first main result

Theorem 5.15 *Let* $\alpha > 0$, $N \in \mathbb{N}$, $N = \lceil \alpha \rceil$, $f \in AC^N([a,b])$, $0 < \beta < 1$, $x \in [a,b]$, $n \in \mathbb{N}$ *large enough. Assume that* $\left\| D_{x-}^{\alpha} f \right\|_{\infty,[a,x]}$, $\left\| D_{*x}^{\alpha} f \right\|_{\infty,[x,b]} \leq M$, $M > 0$. *Then*

$$G_n(f,x) - f(x) = \sum_{j=1}^{N-1} \frac{f^{(j)}(x)}{j!} G_n\left((\cdot - x)^j\right)(x) + o\left(\frac{1}{n^{\beta(\alpha-\varepsilon)}}\right), \quad (5.17)$$

where $0 < \varepsilon \leq \alpha$.

If $N = 1$, *the sum in (5.17) collapses.*

The last (5.17) implies that

$$n^{\beta(\alpha-\varepsilon)} \left[G_n(f,x) - f(x) - \sum_{j=1}^{N-1} \frac{f^{(j)}(x)}{j!} G_n\left((\cdot - x)^j\right)(x) \right] \to 0, \quad (5.18)$$

as $n \to \infty$, $0 < \varepsilon \leq \alpha$.

When $N = 1$, *or* $f^{(j)}(x) = 0$, $j = 1, \ldots, N-1$, *then we derive that*

$$n^{\beta(\alpha-\varepsilon)} \left[G_n(f,x) - f(x) \right] \to 0$$

as $n \to \infty$, $0 < \varepsilon \leq \alpha$. *Of great interest is the case of* $\alpha = \frac{1}{2}$.

Proof From [14], p. 54; (5), we get by the left Caputo fractional Taylor formula that

$$f\left(\frac{k}{n}\right) = \sum_{j=0}^{N-1} \frac{f^{(j)}(x)}{j!} \left(\frac{k}{n} - x\right)^j + \frac{1}{\Gamma(\alpha)} \int_x^{\frac{k}{n}} \left(\frac{k}{n} - J\right)^{\alpha-1} D_{*x}^{\alpha} f(J)\, dJ, \quad (5.19)$$

for all $x \leq \frac{k}{n} \leq b$.

Also from [3]; (6), using the right Caputo fractional Taylor formula we get

$$f\left(\frac{k}{n}\right) = \sum_{j=0}^{N-1} \frac{f^{(j)}(x)}{j!} \left(\frac{k}{n} - x\right)^j + \frac{1}{\Gamma(\alpha)} \int_{\frac{k}{n}}^{x} \left(J - \frac{k}{n}\right)^{\alpha-1} D_{x-}^{\alpha} f(J)\, dJ, \quad (5.20)$$

for all $a \leq \frac{k}{n} \leq x$.

We call

$$V(x) := \sum_{k=\lceil na \rceil}^{\lfloor nb \rfloor} \Phi(nx - k). \quad (5.21)$$

Hence we have

$$\frac{f\left(\frac{k}{n}\right)\Phi(nx-k)}{V(x)} = \sum_{j=0}^{N-1} \frac{f^{(j)}(x)}{j!} \frac{\Phi(nx-k)}{V(x)} \left(\frac{k}{n} - x\right)^j + \quad (5.22)$$

$$\frac{\Phi(nx-k)}{V(x)\Gamma(\alpha)} \int_x^{\frac{k}{n}} \left(\frac{k}{n} - J\right)^{\alpha-1} D_{*x}^{\alpha} f(J)\, dJ,$$

all $x \leq \frac{k}{n} \leq b$, iff $\lceil nx \rceil \leq k \leq \lfloor nb \rfloor$, and

$$\frac{f\left(\frac{k}{n}\right)\Phi(nx-k)}{V(x)} = \sum_{j=0}^{N-1} \frac{f^{(j)}(x)}{j!} \frac{\Phi(nx-k)}{V(x)} \left(\frac{k}{n}-x\right)^j + \tag{5.23}$$

$$\frac{\Phi(nx-k)}{V(x)\Gamma(\alpha)} \int_{\frac{k}{n}}^{x} \left(J-\frac{k}{n}\right)^{\alpha-1} D_{x-}^{\alpha} f(J)\,dJ,$$

for all $a \leq \frac{k}{n} \leq x$, iff $\lceil na \rceil \leq k \leq \lfloor nx \rfloor$.

We have that $\lceil nx \rceil \leq \lfloor nx \rfloor + 1$.

Therefore it holds

$$\sum_{k=\lfloor nx \rfloor+1}^{\lfloor nb \rfloor} \frac{f\left(\frac{k}{n}\right)\Phi(nx-k)}{V(x)} = \sum_{j=0}^{N-1} \frac{f^{(j)}(x)}{j!} \sum_{k=\lfloor nx \rfloor+1}^{\lfloor nb \rfloor} \frac{\Phi(nx-k)\left(\frac{k}{n}-x\right)^j}{V(x)} + \tag{5.24}$$

$$\frac{1}{\Gamma(\alpha)}\left(\frac{\sum_{k=\lfloor nx \rfloor+1}^{\lfloor nb \rfloor}\Phi(nx-k)}{V(x)} \int_{x}^{\frac{k}{n}} \left(\frac{k}{n}-J\right)^{\alpha-1} D_{*x}^{\alpha} f(J)\,dJ\right),$$

and

$$\sum_{k=\lceil na \rceil}^{\lfloor nx \rfloor} f\left(\frac{k}{n}\right)\frac{\Phi(nx-k)}{V(x)} = \sum_{j=0}^{N-1} \frac{f^{(j)}(x)}{j!} \sum_{k=\lceil na \rceil}^{\lfloor nx \rfloor} \frac{\Phi(nx-k)}{V(x)}\left(\frac{k}{n}-x\right)^j + \tag{5.25}$$

$$\frac{1}{\Gamma(\alpha)}\left(\sum_{k=\lceil na \rceil}^{\lfloor nx \rfloor} \frac{\Phi(nx-k)}{V(x)} \int_{\frac{k}{n}}^{x} \left(J-\frac{k}{n}\right)^{\alpha-1} D_{x-}^{\alpha} f(J)\,dJ\right).$$

Adding the last two equalities (5.24) and (5.25) we obtain

$$G_n(f,x) = \sum_{k=\lceil na \rceil}^{\lfloor nb \rfloor} f\left(\frac{k}{n}\right)\frac{\Phi(nx-k)}{V(x)} = \tag{5.26}$$

$$\sum_{j=0}^{N-1} \frac{f^{(j)}(x)}{j!} \sum_{k=\lceil na \rceil}^{\lfloor nb \rfloor} \frac{\Phi(nx-k)}{V(x)}\left(\frac{k}{n}-x\right)^j +$$

$$\frac{1}{\Gamma(\alpha)V(x)}\left\{\sum_{k=\lceil na \rceil}^{\lfloor nx \rfloor} \Phi(nx-k) \int_{\frac{k}{n}}^{x} \left(J-\frac{k}{n}\right)^{\alpha-1} D_{x-}^{\alpha} f(J)\,dJ +\right.$$

$$\left. \sum_{k=\lfloor nx \rfloor+1}^{\lfloor nb \rfloor} \Phi(nx-k) \int_x^{\frac{k}{n}} \left(\frac{k}{n} - J\right)^{\alpha-1} \left(D_{*x}^{\alpha} f(J)\right) dJ \right\}.$$

So we have derived

$$T(x) := G_n(f,x) - f(x) - \sum_{j=1}^{N-1} \frac{f^{(j)}(x)}{j!} G_n\left((\cdot - x)^j\right)(x) = \theta_n^*(x), \quad (5.27)$$

where

$$\theta_n^*(x) := \frac{1}{\Gamma(\alpha) V(x)} \left\{ \sum_{k=\lceil na \rceil}^{\lfloor nx \rfloor} \Phi(nx-k) \int_{\frac{k}{n}}^{x} \left(J - \frac{k}{n}\right)^{\alpha-1} D_{x-}^{\alpha} f(J)\, dJ \right.$$

$$\left. + \sum_{k=\lfloor nx \rfloor+1}^{\lfloor nb \rfloor} \Phi(nx-k) \int_x^{\frac{k}{n}} \left(\frac{k}{n} - J\right)^{\alpha-1} D_{*x}^{\alpha} f(J)\, dJ \right\}. \quad (5.28)$$

We set

$$\theta_{1n}^*(x) := \frac{1}{\Gamma(\alpha)} \left(\frac{\sum_{k=\lceil na \rceil}^{\lfloor nx \rfloor} \Phi(nx-k)}{V(x)} \int_{\frac{k}{n}}^{x} \left(J - \frac{k}{n}\right)^{\alpha-1} D_{x-}^{\alpha} f(J)\, dJ \right), \quad (5.29)$$

and

$$\theta_{2n}^* := \frac{1}{\Gamma(\alpha)} \left(\frac{\sum_{k=\lfloor nx \rfloor+1}^{\lfloor nb \rfloor} \Phi(nx-k)}{V(x)} \int_x^{\frac{k}{n}} \left(\frac{k}{n} - J\right)^{\alpha-1} D_{*x}^{\alpha} f(J)\, dJ \right), \quad (5.30)$$

i.e.

$$\theta_n^*(x) = \theta_{1n}^*(x) + \theta_{2n}^*(x). \quad (5.31)$$

We assume $b - a > \frac{1}{n^{\beta}}$, $0 < \beta < 1$, which is always the case for large enough $n \in \mathbb{N}$, that is when $n > \left\lceil (b-a)^{-\frac{1}{\beta}} \right\rceil$. It is always true that either $\left|\frac{k}{n} - x\right| \leq \frac{1}{n^{\beta}}$ or $\left|\frac{k}{n} - x\right| > \frac{1}{n^{\beta}}$.

For $k = \lceil na \rceil, \ldots, \lfloor nx \rfloor$, we consider

$$\gamma_{1k} := \left| \int_{\frac{k}{n}}^{x} \left(J - \frac{k}{n}\right)^{\alpha-1} D_{x-}^{\alpha} f(J)\, dJ \right| \leq \quad (5.32)$$

$$\int_{\frac{k}{n}}^{x} \left(J - \frac{k}{n}\right)^{\alpha-1} \left|D_{x-}^{\alpha} f(J)\right| dJ$$

$$\leq \left\| D_{x-}^{\alpha} f \right\|_{\infty,[a,x]} \frac{\left(x - \frac{\kappa}{n}\right)^{\alpha}}{\alpha} \leq \left\| D_{x-}^{\alpha} f \right\|_{\infty,[a,x]} \frac{(x-a)^{\alpha}}{\alpha}. \tag{5.33}$$

That is

$$\gamma_{1k} \leq \left\| D_{x-}^{\alpha} f \right\|_{\infty,[a,x]} \frac{(x-a)^{\alpha}}{\alpha}, \tag{5.34}$$

for $k = \lceil na \rceil, \ldots, \lfloor nx \rfloor$.

Also we have in case of $\left| \frac{k}{n} - x \right| \leq \frac{1}{n^{\beta}}$ that

$$\gamma_{1k} \leq \int_{\frac{k}{n}}^{x} \left(J - \frac{k}{n} \right)^{\alpha-1} \left| D_{x-}^{\alpha} f(J) \right| dJ \tag{5.35}$$

$$\leq \left\| D_{x-}^{\alpha} f \right\|_{\infty,[a,x]} \frac{\left(x - \frac{\kappa}{n}\right)^{\alpha}}{\alpha} \leq \left\| D_{x-}^{\alpha} f \right\|_{\infty,[a,x]} \frac{1}{n^{\alpha\beta}\alpha}.$$

So that, when $\left(x - \frac{k}{n}\right) \leq \frac{1}{n^{\beta}}$, we get

$$\gamma_{1k} \leq \left\| D_{x-}^{\alpha} f \right\|_{\infty,[a,x]} \frac{1}{\alpha n^{\alpha\beta}}. \tag{5.36}$$

Therefore

$$\left| \theta_{1n}^{*}(x) \right| \leq \frac{1}{\Gamma(\alpha)} \left(\frac{\sum_{k=\lceil na \rceil}^{\lfloor nx \rfloor} \Phi(nx-k)}{V(x)} \gamma_{1k} \right) = \frac{1}{\Gamma(\alpha)} \cdot$$

$$\left\{ \frac{\sum_{\substack{k=\lceil na \rceil \\ : \left| \frac{k}{n} - x \right| \leq \frac{1}{n^{\beta}}}}^{\lfloor nx \rfloor} \Phi(nx-k)}{V(x)} \gamma_{1k} + \frac{\sum_{\substack{k=\lceil na \rceil \\ : \left| \frac{k}{n} - x \right| > \frac{1}{n^{\beta}}}}^{\lfloor nx \rfloor} \Phi(nx-k)}{V(x)} \gamma_{1k} \right\}$$

$$\leq \frac{1}{\Gamma(\alpha)} \left\{ \left(\frac{\sum_{\substack{k=\lceil na \rceil \\ : \left| \frac{k}{n} - x \right| \leq \frac{1}{n^{\beta}}}}^{\lfloor nx \rfloor} \Phi(nx-k)}{V(x)} \right) \left\| D_{x-}^{\alpha} f \right\|_{\infty,[a,x]} \frac{1}{\alpha n^{\alpha\beta}} + \right.$$

$$\frac{1}{V(x)}\left\{\left(\sum_{\substack{k=\lceil na\rceil \\ :\left|\frac{k}{n}-x\right|>\frac{1}{n^{\beta}}}}^{\lfloor nx\rfloor}\Phi(nx-k)\right)\left\|D_{x-}^{\alpha}f\right\|_{\infty,[a,x]}\frac{(x-a)^{\alpha}}{\alpha}\right\}\overset{\text{(by (5.9), (5.10))}}{\leq}$$

$$\frac{\left\|D_{x-}^{\alpha}f\right\|_{\infty,[a,x]}}{\Gamma(\alpha+1)}\left\{\frac{1}{n^{\alpha\beta}}+(5.250312578)(3.1992)\,e^{-n^{(1-\beta)}}(x-a)^{\alpha}\right\}. \tag{5.37}$$

Therefore we proved

$$\left|\theta_{1n}^{*}(x)\right|\leq\frac{\left\|D_{x-}^{\alpha}f\right\|_{\infty,[a,x]}}{\Gamma(\alpha+1)}\left\{\frac{1}{n^{\alpha\beta}}+(16.7968)\,e^{-n^{(1-\beta)}}(x-a)^{\alpha}\right\}. \tag{5.38}$$

But for large enough $n\in\mathbb{N}$ we get

$$\left|\theta_{1n}^{*}(x)\right|\leq\frac{2\left\|D_{x-}^{\alpha}f\right\|_{\infty,[a,x]}}{\Gamma(\alpha+1)\,n^{\alpha\beta}}. \tag{5.39}$$

Similarly we have

$$\gamma_{2k}:=\left|\int_{x}^{\frac{k}{n}}\left(\frac{k}{n}-J\right)^{\alpha-1}D_{*x}^{\alpha}f(J)\,dJ\right|\leq$$

$$\int_{x}^{\frac{k}{n}}\left(\frac{k}{n}-J\right)^{\alpha-1}\left|D_{*x}^{\alpha}f(J)\right|dJ\leq$$

$$\left\|D_{*x}^{\alpha}f\right\|_{\infty,[x,b]}\frac{\left(\frac{k}{n}-x\right)^{\alpha}}{\alpha}\leq\left\|D_{*x}^{\alpha}f\right\|_{\infty,[x,b]}\frac{(b-x)^{\alpha}}{\alpha}. \tag{5.40}$$

That is

$$\gamma_{2k}\leq\left\|D_{*x}^{\alpha}f\right\|_{\infty,[x,b]}\frac{(b-x)^{\alpha}}{\alpha}, \tag{5.41}$$

for $k=\lfloor nx\rfloor+1,\ldots,\lfloor nb\rfloor$.

Also we have in case of $\left|\frac{k}{n}-x\right|\leq\frac{1}{n^{\beta}}$ that

$$\gamma_{2k}\leq\frac{\left\|D_{*x}^{\alpha}f\right\|_{\infty,[x,b]}}{\alpha n^{\alpha\beta}}. \tag{5.42}$$

Consequently it holds

$$\left|\theta_{2n}^{*}(x)\right| \leq \frac{1}{\Gamma(\alpha)}\left(\frac{\sum_{k=\lfloor nx\rfloor+1}^{\lfloor nb\rfloor}\Phi(nx-k)}{V(x)}\gamma_{2k}\right) =$$

$$\frac{1}{\Gamma(\alpha)}\left\{\left(\frac{\sum_{\substack{k=\lfloor nx\rfloor+1 \\ :\left|\frac{k}{n}-x\right| \leq \frac{1}{n^{\beta}}}}^{\lfloor nb\rfloor}\Phi(nx-k)}{V(x)}\right)\frac{\left\|D_{*x}^{\alpha}f\right\|_{\infty,[x,b]}}{\alpha n^{\alpha\beta}} +\right.$$

$$\left.\frac{1}{V(x)}\left(\sum_{\substack{k=\lfloor nx\rfloor+1 \\ :\left|\frac{k}{n}-x\right| > \frac{1}{n^{\beta}}}}^{\lfloor nb\rfloor}\Phi(nx-k)\right)\left\|D_{*x}^{\alpha}f\right\|_{\infty,[x,b]}\frac{(b-x)^{\alpha}}{\alpha}\right\} \leq$$

$$\frac{\left\|D_{*x}^{\alpha}f\right\|_{\infty,[x,b]}}{\Gamma(\alpha+1)}\left\{\frac{1}{n^{\alpha\beta}}+(16.7968)\,e^{-n^{(1-\beta)}}(b-x)^{\alpha}\right\}. \tag{5.43}$$

That is

$$\left|\theta_{2n}^{*}(x)\right| \leq \frac{\left\|D_{*x}^{\alpha}f\right\|_{\infty,[x,b]}}{\Gamma(\alpha+1)}\left\{\frac{1}{n^{\alpha\beta}}+(16.7968)\,e^{-n^{(1-\beta)}}(b-x)^{\alpha}\right\}. \tag{5.44}$$

But for large enough $n \in \mathbb{N}$ we get

$$\left|\theta_{2n}^{*}(x)\right| \leq \frac{2\left\|D_{*x}^{\alpha}f\right\|_{\infty,[x,b]}}{\Gamma(\alpha+1)\,n^{\alpha\beta}}. \tag{5.45}$$

Since $\left\|D_{x-}^{\alpha}f\right\|_{\infty,[a,x]}, \left\|D_{*x}^{\alpha}f\right\|_{\infty,[x,b]} \leq M$, $M > 0$, we derive

$$\left|\theta_{n}^{*}(x)\right| \leq \left|\theta_{1n}^{*}(x)\right| + \left|\theta_{2n}^{*}(x)\right| \overset{\text{(by (5.39), (5.45))}}{\leq} \frac{4M}{\Gamma(\alpha+1)\,n^{\alpha\beta}}. \tag{5.46}$$

That is for large enough $n \in \mathbb{N}$ we get

$$|T(x)| = \left|\theta_{n}^{*}(x)\right| \leq \left(\frac{4M}{\Gamma(\alpha+1)}\right)\left(\frac{1}{n^{\alpha\beta}}\right), \tag{5.47}$$

resulting to

$$|T(x)| = O\left(\frac{1}{n^{\alpha\beta}}\right), \tag{5.48}$$

and

$$|T(x)| = o(1). \tag{5.49}$$

And, letting $0 < \varepsilon \leq \alpha$, we derive

$$\frac{|T(x)|}{\left(\frac{1}{n^{\beta(\alpha-\varepsilon)}}\right)} \leq \left(\frac{4M}{\Gamma(\alpha+1)}\right)\left(\frac{1}{n^{\beta\varepsilon}}\right) \to 0, \tag{5.50}$$

as $n \to \infty$.

I.e.

$$|T(x)| = o\left(\frac{1}{n^{\beta(\alpha-\varepsilon)}}\right), \tag{5.51}$$

proving the claim. ■

We present our second main result

Theorem 5.16 *Let $\alpha > 0$, $N \in \mathbb{N}$, $N = \lceil \alpha \rceil$, $f \in AC^N([a,b])$, $0 < \beta < 1$, $x \in [a,b]$, $n \in \mathbb{N}$ large enough. Assume that $\left\| D_{x-}^{\alpha} f \right\|_{\infty,[a,x]}$, $\left\| D_{*x}^{\alpha} f \right\|_{\infty,[x,b]} \leq M$, $M > 0$. Then*

$$F_n(f,x) - f(x) = \sum_{j=1}^{N-1} \frac{f^{(j)}(x)}{j!} F_n\left((\cdot - x)^j\right)(x) + o\left(\frac{1}{n^{\beta(\alpha-\varepsilon)}}\right), \tag{5.52}$$

where $0 < \varepsilon \leq \alpha$.

If $N = 1$, the sum in (5.52) collapses.

The last (5.52) implies that

$$n^{\beta(\alpha-\varepsilon)}\left[F_n(f,x) - f(x) - \sum_{j=1}^{N-1} \frac{f^{(j)}(x)}{j!} F_n\left((\cdot - x)^j\right)(x)\right] \to 0, \tag{5.53}$$

as $n \to \infty$, $0 < \varepsilon \leq \alpha$.

When $N = 1$, or $f^{(j)}(x) = 0$, $j = 1, \ldots, N-1$, then we derive that

$$n^{\beta(\alpha-\varepsilon)}\left[F_n(f,x) - f(x)\right] \to 0$$

as $n \to \infty$, $0 < \varepsilon \leq \alpha$. Of great interest is the case of $\alpha = \frac{1}{2}$.

Proof Similar to Theorem 5.15, using (5.13) and (5.14). ■

References

1. G.A. Anastassiou, Rate of convergence of some neural network operators to the unit-univariate case. J. Math. Anal. Appl. **212**, 237–262 (1997)
2. G.A. Anastassiou, *Quantitative Approximations* (Chapman & Hall/CRC, Boca Raton, 2001)
3. G.A. Anastassiou, On right fractional calculus. Chaos, Solitons Fractals **42**, 365–376 (2009)
4. G.A. Anastassiou, *Fractional Differentiation Inequalities* (Springer, New York, 2009)
5. G. Anastassiou, Fractional Korovkin theory. Chaos, Solitons Fractals **42**(4), 2080–2094 (2009)
6. G.A. Anastassiou, *Intelligent Systems: Approximation by Artificial Neural Networks*, Intelligent Systems Reference Library, vol. 19 (Springer, Heidelberg, 2011)
7. G.A. Anastassiou, Fractional representation formulae and right fractional inequalities. Math. Comput. Model. **54**(11–12), 3098–3115 (2011)
8. G.A. Anastassiou, Univariate hyperbolic tangent neural network approximation. Math. Comput. Model. **53**, 1111–1132 (2011)
9. G.A. Anastassiou, Multivariate hyperbolic tangent neural network approximation. Comput. Math. **61**, 809–821 (2011)
10. G.A. Anastassiou, Multivariate sigmoidal neural network approximation. Neural Netw. **24**, 378–386 (2011)
11. G.A. Anastassiou, Univariate sigmoidal neural network approximation, submitted for publication. J. Comput. Anal. Appl. **14**(4), 659–690 (2012)
12. G.A. Anastassiou, Fractional Voronovskaya type asymptotic expansions for quasi-interpolation neural network operators. Cubo **14**(03), 71–83 (2012)
13. Z. Chen, F. Cao, The approximation operators with sigmoidal functions. Comput. Math. Appl. **58**, 758–765 (2009)
14. K. Diethelm, *The Analysis of Fractional Differential Equations*, Lecture Notes in Mathematics, vol. 2004 (Springer, Berlin, 2010)
15. A.M.A. El-Sayed, M. Gaber, On the finite Caputo and finite Riesz derivatives. Electron. J. Theor. Phys. **3**(12), 81–95 (2006)
16. G.S. Frederico, D.F.M. Torres, Fractional optimal Control in the sense of Caputo and the fractional Noether's theorem. Int. Math. Forum **3**(10), 479–493 (2008)
17. S. Haykin, *Neural Networks: A Comprehensive Foundation*, 2nd edn. (Prentice Hall, New York, 1998)
18. W. McCulloch, W. Pitts, A logical calculus of the ideas immanent in nervous activity. Bull. Math. Biophys. **7**, 115–133 (1943)
19. T.M. Mitchell, *Machine Learning* (WCB-McGraw-Hill, New York, 1997)

Chapter 6
Voronovskaya Type Asymptotic Expansions for Multivariate Quasi-interpolation Neural Networks

Here we study further the multivariate quasi-interpolation of sigmoidal and hyperbolic tangent types neural network operators of one hidden layer. We derive multivariate Voronovskaya type asymptotic expansions for the error of approximation of these operators to the unit operator. It follows [7].

6.1 Background

Here we follow [5, 6].

We consider here the sigmoidal function of logarithmic type

$$s_i(x_i) = \frac{1}{1+e^{-x_i}}, \quad x_i \in \mathbb{R}, i = 1, \ldots, N; \ x := (x_1, \ldots, x_N) \in \mathbb{R}^N,$$

each has the properties $\lim_{x_i \to +\infty} s_i(x_i) = 1$ and $\lim_{x_i \to -\infty} s_i(x_i) = 0$, $i = 1, \ldots, N$.

These functions play the role of activation functions in the hidden layer of neural networks.

As in [8], we consider

$$\Phi_i(x_i) := \frac{1}{2}\left(s_i(x_i+1) - s_i(x_i-1)\right), \quad x_i \in \mathbb{R}, i = 1, \ldots, N.$$

We notice the following properties:

(i) $\Phi_i(x_i) > 0, \ \forall\, x_i \in \mathbb{R}$,
(ii) $\sum_{k_i=-\infty}^{\infty} \Phi_i(x_i - k_i) = 1, \ \forall\, x_i \in \mathbb{R}$,
(iii) $\sum_{k_i=-\infty}^{\infty} \Phi_i(nx_i - k_i) = 1, \ \forall\, x_i \in \mathbb{R}; n \in \mathbb{N}$,
(iv) $\int_{-\infty}^{\infty} \Phi_i(x_i)\, dx_i = 1$,

G.A. Anastassiou, *Intelligent Systems II: Complete Approximation by Neural Network Operators*, Studies in Computational Intelligence 608,
DOI 10.1007/978-3-319-20505-2_6

(v) Φ_i is a density function,
(vi) Φ_i is even: $\Phi_i(-x_i) = \Phi_i(x_i)$, $x_i \geq 0$, for $i = 1, \ldots, N$.

We see that [8]

$$\Phi_i(x_i) = \left(\frac{e^2-1}{2e^2}\right) \frac{1}{\left(1+e^{x_i-1}\right)\left(1+e^{-x_i-1}\right)}, \quad i = 1, \ldots, N.$$

(vii) Φ_i is decreasing on $\mathbb{R}_+$, and increasing on $\mathbb{R}_-$, $i = 1, \ldots, N$.

Let $0 < \beta < 1$, $n \in \mathbb{N}$. Then as in [6] we get

(viii)

$$\sum_{\substack{k_i = -\infty \\ : |nx_i - k_i| > n^{1-\beta}}}^{\infty} \Phi_i(nx_i - k_i) \leq 3.1992 e^{-n^{(1-\beta)}}, \quad i = 1, \ldots, N.$$

Denote by $\lceil \cdot \rceil$ the ceiling of a number, and by $\lfloor \cdot \rfloor$ the integral part of a number. Consider here $x \in \left(\prod_{i=1}^N [a_i, b_i]\right) \subset \mathbb{R}^N$, $N \in \mathbb{N}$ such that $\lceil na_i \rceil \leq \lfloor nb_i \rfloor$, $i = 1, \ldots, N$; $a := (a_1, \ldots, a_N)$, $b := (b_1, \ldots, b_N)$.

As in [6] we obtain

(ix)

$$0 < \frac{1}{\sum_{k_i=\lceil na_i \rceil}^{\lfloor nb_i \rfloor} \Phi_i(nx_i - k_i)} < \frac{1}{\Phi_i(1)} = 5.250312578,$$

$\forall\, x_i \in [a_i, b_i]$, $i = 1, \ldots, N$.

(x) As in [6], we see that

$$\lim_{n\to\infty} \sum_{k_i=\lceil na_i \rceil}^{\lfloor nb_i \rfloor} \Phi_i(nx_i - k_i) \neq 1,$$

for at least some $x_i \in [a_i, b_i]$, $i = 1, \ldots, N$.

We will use here

$$\Phi(x_1, \ldots, x_N) := \Phi(x) := \prod_{i=1}^N \Phi_i(x_i), \quad x \in \mathbb{R}^N. \tag{6.1}$$

It has the properties:

(i)' $\Phi(x) > 0, \ \forall\, x \in \mathbb{R}^N$,
We see that

$$\sum_{k_1=-\infty}^{\infty} \sum_{k_2=-\infty}^{\infty} \dots \sum_{k_N=-\infty}^{\infty} \Phi(x_1 - k_1, x_2 - k_2, \dots, x_N - k_N) =$$

$$\sum_{k_1=-\infty}^{\infty} \sum_{k_2=-\infty}^{\infty} \dots \sum_{k_N=-\infty}^{\infty} \prod_{i=1}^{N} \Phi_i(x_i - k_i) = \prod_{i=1}^{N} \left(\sum_{k_i=-\infty}^{\infty} \Phi_i(x_i - k_i) \right) = 1.$$

That is

(ii)'

$$\sum_{k=-\infty}^{\infty} \Phi(x - k) := \sum_{k_1=-\infty}^{\infty} \sum_{k_2=-\infty}^{\infty} \dots \sum_{k_N=-\infty}^{\infty} \Phi(x_1 - k_1, \dots, x_N - k_N) = 1,$$

$k := (k_1, \dots, k_N)$, $\forall\, x \in \mathbb{R}^N$.

(iii)'

$$\sum_{k=-\infty}^{\infty} \Phi(nx - k) :=$$

$$\sum_{k_1=-\infty}^{\infty} \sum_{k_2=-\infty}^{\infty} \dots \sum_{k_N=-\infty}^{\infty} \Phi(nx_1 - k_1, \dots, nx_N - k_N) = 1,$$

$\forall\, x \in \mathbb{R}^N$; $n \in \mathbb{N}$.

(iv)'

$$\int_{\mathbb{R}^N} \Phi(x)\, dx = 1,$$

that is Φ is a multivariate density function.

Here $\|x\|_\infty := \max\{|x_1|, \dots, |x_N|\}$, $x \in \mathbb{R}^N$, also set $\infty := (\infty, \dots, \infty)$, $-\infty := (-\infty, \dots, -\infty)$ upon the multivariate context, and

$$\lceil na \rceil := (\lceil na_1 \rceil, \dots, \lceil na_N \rceil),$$
$$\lfloor nb \rfloor := (\lfloor nb_1 \rfloor, \dots, \lfloor nb_N \rfloor).$$

For $0 < \beta < 1$ and $n \in \mathbb{N}$, fixed $x \in \mathbb{R}^N$, have that

$$\sum_{k=\lceil na \rceil}^{\lfloor nb \rfloor} \Phi(nx - k) =$$

$$\sum_{\begin{cases} k = \lceil na \rceil \\ \left\| \frac{k}{n} - x \right\|_{\infty} \leq \frac{1}{n^{\beta}} \end{cases}}^{\lfloor nb \rfloor} \Phi(nx - k) + \sum_{\begin{cases} k = \lceil na \rceil \\ \left\| \frac{k}{n} - x \right\|_{\infty} > \frac{1}{n^{\beta}} \end{cases}}^{\lfloor nb \rfloor} \Phi(nx - k).$$

In the last two sums the counting is over disjoint vector of k's, because the condition $\left\| \frac{k}{n} - x \right\|_{\infty} > \frac{1}{n^{\beta}}$ implies that there exists at least one $\left| \frac{k_r}{n} - x_r \right| > \frac{1}{n^{\beta}}$, $r \in \{1, \ldots, N\}$.

It holds

(v)'

$$\sum_{\begin{cases} k = \lceil na \rceil \\ \left\| \frac{k}{n} - x \right\|_{\infty} > \frac{1}{n^{\beta}} \end{cases}}^{\lfloor nb \rfloor} \Phi(nx - k) \leq 3.1992 e^{-n^{(1-\beta)}},$$

$0 < \beta < 1, n \in \mathbb{N}, x \in \left(\prod_{i=1}^{N} [a_i, b_i] \right)$.

Furthermore it holds

(vi)'

$$0 < \frac{1}{\sum_{k=\lceil na \rceil}^{\lfloor nb \rfloor} \Phi(nx - k)} < (5.250312578)^N,$$

$\forall\, x \in \left(\prod_{i=1}^{N} [a_i, b_i] \right), \; n \in \mathbb{N}$.

It is clear also that

(vii)'

$$\sum_{\begin{cases} k = -\infty \\ \left\| \frac{k}{n} - x \right\|_{\infty} > \frac{1}{n^{\beta}} \end{cases}}^{\infty} \Phi(nx - k) \leq 3.1992 e^{-n^{(1-\beta)}},$$

$0 < \beta < 1, n \in \mathbb{N}, x \in \mathbb{R}^N$.

By (x) we obviously see that

(viii)'

$$\lim_{n \to \infty} \sum_{k=\lceil na \rceil}^{\lfloor nb \rfloor} \Phi(nx - k) \neq 1$$

for at least some $x \in \left(\prod_{i=1}^{N} [a_i, b_i] \right)$.

Let $f \in C\left(\prod_{i=1}^{N} [a_i, b_i] \right)$ and $n \in \mathbb{N}$ such that $\lceil na_i \rceil \leq \lfloor nb_i \rfloor$, $i = 1, \ldots, N$.

We define the multivariate positive linear neural network operator ($x := (x_1, \ldots, x_N) \in \left(\prod_{i=1}^{N}[a_i, b_i]\right)$)

$$G_n(f, x_1, \ldots, x_N) := G_n(f, x) := \frac{\sum_{k=\lceil na \rceil}^{\lfloor nb \rfloor} f\left(\frac{k}{n}\right) \Phi(nx - k)}{\sum_{k=\lceil na \rceil}^{\lfloor nb \rfloor} \Phi(nx - k)} \tag{6.2}$$

$$:= \frac{\sum_{k_1=\lceil na_1 \rceil}^{\lfloor nb_1 \rfloor} \sum_{k_2=\lceil na_2 \rceil}^{\lfloor nb_2 \rfloor} \cdots \sum_{k_N=\lceil na_N \rceil}^{\lfloor nb_N \rfloor} f\left(\frac{k_1}{n}, \ldots, \frac{k_N}{n}\right) \left(\prod_{i=1}^{N} \Phi_i(nx_i - k_i)\right)}{\prod_{i=1}^{N} \left(\sum_{k_i=\lceil na_i \rceil}^{\lfloor nb_i \rfloor} \Phi_i(nx_i - k_i)\right)}.$$

For large enough n we always obtain $\lceil na_i \rceil \leq \lfloor nb_i \rfloor$, $i = 1, \ldots, N$. Also $a_i \leq \frac{k_i}{n} \leq b_i$, iff $\lceil na_i \rceil \leq k_i \leq \lfloor nb_i \rfloor$, $i = 1, \ldots, N$.

Notice here that for large enough $n \in \mathbb{N}$ we get that

$$e^{-n^{(1-\beta)}} < n^{-\beta j}, \; j = 1, \ldots, m \in \mathbb{N}, 0 < \beta < 1.$$

Thus be given fixed $A, B > 0$, for the linear combination $\left(An^{-\beta j} + Be^{-n^{(1-\beta)}}\right)$ the (dominant) rate of convergence to zero is $n^{-\beta j}$. The closer β is to 1 we get faster and better rate of convergence to zero.

By $AC^m\left(\prod_{i=1}^{N}[a_i, b_i]\right)$, $m, N \in \mathbb{N}$, we denote the space of functions such that all partial derivatives of order $(m-1)$ are coordinatewise absolutely continuous functions, also $f \in C^{m-1}\left(\prod_{i=1}^{N}[a_i, b_i]\right)$.

Let $f \in AC^m\left(\prod_{i=1}^{N}[a_i, b_i]\right)$, $m, N \in \mathbb{N}$. Here f_α denotes a partial derivative of f, $\alpha := (\alpha_1, \ldots, \alpha_N)$, $\alpha_i \in \mathbb{Z}^+$, $i = 1, \ldots, N$, and $|\alpha| := \sum_{i=1}^{N} \alpha_i = l$, where $l = 0, 1, \ldots, m$. We write also $f_\alpha := \frac{\partial^\alpha f}{\partial x^\alpha}$ and we say it is order l.

We denote

$$\|f_\alpha\|_{\infty,m}^{\max} := \max_{|\alpha|=m} \{\|f_\alpha\|_\infty\}, \tag{6.3}$$

where $\|\cdot\|_\infty$ is the supremum norm.

We assume here that $\|f_\alpha\|_{\infty,m}^{\max} < \infty$.

Next we follow [3, 4].

We consider here the hyperbolic tangent function $\tanh x$, $x \in \mathbb{R}$:

$$\tanh x := \frac{e^x - e^{-x}}{e^x + e^{-x}}.$$

It has the properties $\tanh 0 = 0$, $-1 < \tanh x < 1$, $\forall\, x \in \mathbb{R}$, and $\tanh(-x) = -\tanh x$. Furthermore $\tanh x \to 1$ as $x \to \infty$, and $\tanh x \to -1$, as $x \to -\infty$, and it is strictly increasing on $\mathbb{R}$.

This function plays the role of an activation function in the hidden layer of neural networks.

We further consider

$$\Psi(x) := \frac{1}{4}(\tanh(x+1) - \tanh(x-1)) > 0, \quad \forall\, x \in \mathbb{R}.$$

We easily see that $\Psi(-x) = \Psi(x)$, that is Ψ is even on $\mathbb{R}$. Obviously Ψ is differentiable, thus continuous.

Proposition 6.1 ([3]) *$\Psi(x)$ for $x \geq 0$ is strictly decreasing.*

Obviously $\Psi(x)$ is strictly increasing for $x \leq 0$. Also it holds $\lim_{x\to-\infty}\Psi(x) = 0 = \lim_{x\to\infty}\Psi(x)$.

Infact Ψ has the bell shape with horizontal asymptote the x-axis. So the maximum of Ψ is zero, $\Psi(0) = 0.3809297$.

Theorem 6.2 ([3]) *We have that $\sum_{i=-\infty}^{\infty}\Psi(x-i) = 1$, $\forall\, x \in \mathbb{R}$.*

Thus

$$\sum_{i=-\infty}^{\infty}\Psi(nx-i) = 1, \quad \forall\, n \in \mathbb{N},\ \forall\, x \in \mathbb{R}.$$

Also it holds

$$\sum_{i=-\infty}^{\infty}\Psi(x+i) = 1, \quad \forall x \in \mathbb{R}.$$

Theorem 6.3 ([3]) *It holds $\int_{-\infty}^{\infty}\Psi(x)\,dx = 1$.*

So $\Psi(x)$ is a density function on $\mathbb{R}$.

Theorem 6.4 ([3]) *Let $0 < \alpha < 1$ and $n \in \mathbb{N}$. It holds*

$$\sum_{\begin{cases} k=-\infty \\ : |nx-k| \geq n^{1-\alpha} \end{cases}}^{\infty}\Psi(nx-k) \leq e^4 \cdot e^{-2n^{(1-\alpha)}}.$$

Theorem 6.5 ([3]) *Let $x \in [a,b] \subset \mathbb{R}$ and $n \in \mathbb{N}$ so that $\lceil na \rceil \leq \lfloor nb \rfloor$. It holds*

$$\frac{1}{\sum_{k=\lceil na \rceil}^{\lfloor nb \rfloor}\Psi(nx-k)} < \frac{1}{\Psi(1)} = 4.1488766.$$

Also by [3] we get that

$$\lim_{n\to\infty} \sum_{k=\lceil na \rceil}^{\lfloor nb \rfloor} \Psi(nx-k) \neq 1,$$

for at least some $x \in [a,b]$.

In this chapter we will use

$$\Theta(x_1,\ldots,x_N) := \Theta(x) := \prod_{i=1}^{N} \Psi(x_i), \quad x = (x_1,\ldots,x_N) \in \mathbb{R}^N, \ N \in \mathbb{N}. \tag{6.4}$$

It has the properties:

(i)* $\Theta(x) > 0, \ \forall\, x \in \mathbb{R}^N$,

(ii)*

$$\sum_{k=-\infty}^{\infty} \Theta(x-k) := \sum_{k_1=-\infty}^{\infty} \sum_{k_2=-\infty}^{\infty} \ldots \sum_{k_N=-\infty}^{\infty} \Theta(x_1-k_1,\ldots,x_N-k_N) = 1,$$

where $k := (k_1,\ldots,k_N)$, $\forall\, x \in \mathbb{R}^N$.

(iii)*

$$\sum_{k=-\infty}^{\infty} \Theta(nx-k) :=$$

$$\sum_{k_1=-\infty}^{\infty} \sum_{k_2=-\infty}^{\infty} \ldots \sum_{k_N=-\infty}^{\infty} \Theta(nx_1-k_1,\ldots,nx_N-k_N) = 1,$$

$\forall\, x \in \mathbb{R}^N$; $n \in \mathbb{N}$.

(iv)*

$$\int_{\mathbb{R}^N} \Theta(x)\,dx = 1,$$

that is Θ is a multivariate density function.

We obviously see that

$$\sum_{k=\lceil na \rceil}^{\lfloor nb \rfloor} \Theta(nx-k) = \sum_{k=\lceil na \rceil}^{\lfloor nb \rfloor} \prod_{i=1}^{N} \Psi(nx_i-k_i) =$$

$$\sum_{k_1=\lceil na_1 \rceil}^{\lfloor nb_1 \rfloor} \ldots \sum_{k_N=\lceil na_N \rceil}^{\lfloor nb_N \rfloor} \prod_{i=1}^{N} \Psi(nx_i-k_i) = \prod_{i=1}^{N} \left(\sum_{k_i=\lceil na_i \rceil}^{\lfloor nb_i \rfloor} \Psi(nx_i-k_i) \right).$$

For $0 < \beta < 1$ and $n \in \mathbb{N}$, fixed $x \in \mathbb{R}^N$, we have that

$$\sum_{k=\lceil na\rceil}^{\lfloor nb\rfloor} \Theta(nx-k) = \sum_{\begin{cases} k=\lceil na\rceil \\ \left\|\frac{k}{n}-x\right\|_\infty \le \frac{1}{n^\beta}\end{cases}}^{\lfloor nb\rfloor} \Theta(nx-k) + \sum_{\begin{cases} k=\lceil na\rceil \\ \left\|\frac{k}{n}-x\right\|_\infty > \frac{1}{n^\beta}\end{cases}}^{\lfloor nb\rfloor} \Theta(nx-k).$$

In the last two sums the counting is over disjoint vector of k's, because the condition $\left\|\frac{k}{n}-x\right\|_\infty > \frac{1}{n^\beta}$ implies that there exists at least one $\left|\frac{k_r}{n}-x_r\right| > \frac{1}{n^\beta}$, $r \in \{1,\ldots,N\}$.

Il holds

(v)*

$$\sum_{\begin{cases} k=\lceil na\rceil \\ \left\|\frac{k}{n}-x\right\|_\infty > \frac{1}{n^\beta}\end{cases}}^{\lfloor nb\rfloor} \Theta(nx-k) \le e^4 \cdot e^{-2n^{(1-\beta)}},$$

$0 < \beta < 1, n \in \mathbb{N}, x \in \left(\prod_{i=1}^N [a_i, b_i]\right)$.

Also it holds

(vi)*

$$0 < \frac{1}{\sum_{k=\lceil na\rceil}^{\lfloor nb\rfloor} \Theta(nx-k)} < \frac{1}{(\Psi(1))^N} = (4.1488766)^N,$$

$\forall\, x \in \left(\prod_{i=1}^N [a_i, b_i]\right),\ n \in \mathbb{N}$.

It is clear that

(vii)*

$$\sum_{\begin{cases} k=-\infty \\ \left\|\frac{k}{n}-x\right\|_\infty > \frac{1}{n^\beta}\end{cases}}^{\infty} \Theta(nx-k) \le e^4 \cdot e^{-2n^{(1-\beta)}},$$

$0 < \beta < 1, n \in \mathbb{N}, x \in \mathbb{R}^N$.

Also we get

$$\lim_{n\to\infty} \sum_{k=\lceil na\rceil}^{\lfloor nb\rfloor} \Theta(nx-k) \neq 1,$$

for at least some $x \in \left(\prod_{i=1}^N [a_i, b_i]\right)$.

Let $f \in C\left(\prod_{i=1}^{N}[a_i, b_i]\right)$ and $n \in \mathbb{N}$ such that $\lceil na_i \rceil \leq \lfloor nb_i \rfloor$, $i = 1, \ldots, N$.

We define the multivariate positive linear neural network operator ($x := (x_1, \ldots, x_N) \in \left(\prod_{i=1}^{N}[a_i, b_i]\right)$)

$$F_n(f, x_1, \ldots, x_N) := F_n(f, x) := \frac{\sum_{k=\lceil na \rceil}^{\lfloor nb \rfloor} f\left(\frac{k}{n}\right) \Theta(nx - k)}{\sum_{k=\lceil na \rceil}^{\lfloor nb \rfloor} \Theta(nx - k)} \tag{6.5}$$

$$:= \frac{\sum_{k_1=\lceil na_1 \rceil}^{\lfloor nb_1 \rfloor} \sum_{k_2=\lceil na_2 \rceil}^{\lfloor nb_2 \rfloor} \cdots \sum_{k_N=\lceil na_N \rceil}^{\lfloor nb_N \rfloor} f\left(\frac{k_1}{n}, \ldots, \frac{k_N}{n}\right)\left(\prod_{i=1}^{N} \Psi(nx_i - k_i)\right)}{\prod_{i=1}^{N}\left(\sum_{k_i=\lceil na_i \rceil}^{\lfloor nb_i \rfloor} \Psi(nx_i - k_i)\right)}.$$

Our considered neural networks here are of one hidden layer.

In this chapter we find Voronovskaya type asymptotic expansions for the above described neural networks quasi-interpolation normalized operators $G_n(f, x)$, $F_n(f, x)$, where $x \in \left(\prod_{i=1}^{N}[a_i, b_i]\right)$ is fixed but arbitrary. For other neural networks related work, see [2–6, 8]. For neural networks in general, see [9–11].

Next we follow [1], pp. 284–286.

About Taylor formula -Multivariate Case and Estimates

Let Q be a compact convex subset of $\mathbb{R}^N$; $N \geq 2$; $z := (z_1, \ldots, z_N)$, $x_0 := (x_{01}, \ldots, x_{0N}) \in Q$.

Let $f : Q \to \mathbb{R}$ be such that all partial derivatives of order $(m-1)$ are coordinatewise absolutely continuous functions, $m \in \mathbb{N}$. Also $f \in C^{m-1}(Q)$. That is $f \in AC^m(Q)$. Each mth order partial derivative is denoted by $f_\alpha := \frac{\partial^\alpha f}{\partial x^\alpha}$, where $\alpha := (\alpha_1, \ldots, \alpha_N)$, $\alpha_i \in \mathbb{Z}^+$, $i = 1, \ldots, N$ and $|\alpha| := \sum_{i=1}^{N} \alpha_i = m$. Consider $g_z(t) := f(x_0 + t(z - x_0))$, $t \geq 0$. Then

$$g_z^{(j)}(t) = \left[\left(\sum_{i=1}^{N}(z_i - x_{0i})\frac{\partial}{\partial x_i}\right)^j f\right](x_{01} + t(z_1 - x_{01}), \ldots, x_{0N} + t(z_N - x_{0N})), \tag{6.6}$$

for all $j = 0, 1, 2, \ldots, m$.

Example 6.6 Let $m = N = 2$. Then

$$g_z(t) = f(x_{01} + t(z_1 - x_{01}), x_{02} + t(z_2 - x_{02})), \ t \in \mathbb{R},$$

and

$$g_z'(t) = (z_1 - x_{01})\frac{\partial f}{\partial x_1}(x_0 + t(z - x_0)) + (z_2 - x_{02})\frac{\partial f}{\partial x_2}(x_0 + t(z - x_0)). \tag{6.7}$$

Setting

$$(*) = (x_{01} + t(z_1 - x_{01}), x_{02} + t(z_2 - x_{02})) = (x_0 + t(z - x_0)),$$

we get

$$g_z''(t) = (z_1 - x_{01})^2 \frac{\partial f^2}{\partial x_1^2}(*) + (z_1 - x_{01})(z_2 - x_{02}) \frac{\partial f^2}{\partial x_2 \partial x_1}(*) +$$

$$(z_1 - x_{01})(z_2 - x_{02}) \frac{\partial f^2}{\partial x_1 \partial x_2}(*) + (z_2 - x_{02})^2 \frac{\partial f^2}{\partial x_2^2}(*). \tag{6.8}$$

Similarly, we have the general case of $m, N \in \mathbb{N}$ for $g_z^{(m)}(t)$.

We mention the following multivariate Taylor theorem.

Theorem 6.7 *Under the above assumptions we have*

$$f(z_1, \dots, z_N) = g_z(1) = \sum_{j=0}^{m-1} \frac{g_z^{(j)}(0)}{j!} + R_m(z, 0), \tag{6.9}$$

where

$$R_m(z, 0) := \int_0^1 \left(\int_0^{t_1} \dots \left(\int_0^{t_{m-1}} g_z^{(m)}(t_m)\, dt_m \right) \dots \right) dt_1, \tag{6.10}$$

or

$$R_m(z, 0) = \frac{1}{(m-1)!} \int_0^1 (1 - \theta)^{m-1} g_z^{(m)}(\theta)\, d\theta. \tag{6.11}$$

Notice that $g_z(0) = f(x_0)$.

We make

Remark 6.8 Assume here that

$$\| f_\alpha \|_{\infty, Q, m}^{\max} := \max_{|\alpha| = m} \| f_\alpha \|_{\infty, Q} < \infty.$$

Then

$$\left\| g_z^{(m)} \right\|_{\infty,[0,1]} = \left\| \left[\left(\sum_{i=1}^N (z_i - x_{0i}) \frac{\partial}{\partial x_i} \right)^m f \right] (x_0 + t(z - x_0)) \right\|_{\infty,[0,1]} \leq \tag{6.12}$$

$$\left(\sum_{i=1}^N |z_i - x_{0i}| \right)^m \| f_\alpha \|_{\infty, Q, m}^{\max},$$

that is

$$\left\| g_z^{(m)} \right\|_{\infty,[0,1]} \leq \left(\|z - x_0\|_{l_1} \right)^m \|f_\alpha\|_{\infty,Q,m}^{\max} < \infty. \tag{6.13}$$

Hence we get by (6.11) that

$$|R_m(z,0)| \leq \frac{\left\| g_z^{(m)} \right\|_{\infty,[0,1]}}{m!} < \infty. \tag{6.14}$$

And it holds

$$|R_m(z,0)| \leq \frac{\left(\|z - x_0\|_{l_1} \right)^m}{m!} \|f_\alpha\|_{\infty,Q,m}^{\max}, \tag{6.15}$$

$\forall\, z, x_0 \in Q$.

Inequality (6.15) will be an important tool in proving our main results.

6.2 Main Results

We present our first main result

Theorem 6.9 *Let* $0 < \beta < 1$, $x \in \prod_{i=1}^N [a_i, b_i]$, $n \in \mathbb{N}$ *large enough,* $f \in AC^m\left(\prod_{i=1}^N [a_i, b_i]\right)$, $m, N \in \mathbb{N}$. *Assume further that* $\|f_\alpha\|_{\infty,m}^{\max} < \infty$. *Then*

$$G_n(f,x) - f(x) =$$

$$\sum_{j=1}^{m-1} \left(\sum_{|\alpha|=j} \left(\frac{f_\alpha(x)}{\prod_{i=1}^N \alpha_i!} \right) G_n \left(\prod_{i=1}^N (\cdot - x_i)^{\alpha_i}, x \right) \right) + o\left(\frac{1}{n^{\beta(m-\varepsilon)}} \right), \tag{6.16}$$

where $0 < \varepsilon \leq m$.

If $m = 1$, *the sum in (6.16) collapses.*

The last (6.16) implies that

$$n^{\beta(m-\varepsilon)} \left[G_n(f,x) - f(x) - \sum_{j=1}^{m-1} \left(\sum_{|\alpha|=j} \left(\frac{f_\alpha(x)}{\prod_{i=1}^N \alpha_i!} \right) G_n \left(\prod_{i=1}^N (\cdot - x_i)^{\alpha_i}, x \right) \right) \right] \tag{6.17}$$

$$\to 0, \text{ as } n \to \infty,\ 0 < \varepsilon \leq m.$$

When $m = 1$, *or* $f_\alpha(x) = 0$, *for* $|\alpha| = j$, $j = 1, \ldots, m-1$, *then we derive that*

$$n^{\beta(m-\varepsilon)}\left[G_n(f,x) - f(x)\right] \to 0,$$

as $n \to \infty$, $0 < \varepsilon \leq m$.

Proof Consider $g_z(t) := f(x_0 + t(z - x_0))$, $t \geq 0$; $x_0, z \in \prod_{i=1}^N [a_i, b_i]$. Then

$$g_z^{(j)}(t) = \left[\left(\sum_{i=1}^N (z_i - x_{0i}) \frac{\partial}{\partial x_i}\right)^j f\right](x_{01} + t(z_1 - x_{01}), \ldots, x_{0N} + t(z_N - x_{0N})), \tag{6.18}$$

for all $j = 0, 1, \ldots, m$.

By (6.9) we have the multivariate Taylor's formula

$$f(z_1, \ldots, z_N) = g_z(1) = \sum_{j=0}^{m-1} \frac{g_z^{(j)}(0)}{j!} + \frac{1}{(m-1)!}\int_0^1 (1-\theta)^{m-1} g_z^{(m)}(\theta)\, d\theta. \tag{6.19}$$

Notice $g_z(0) = f(x_0)$. Also for $j = 0, 1, \ldots, m-1$, we have

$$g_z^{(j)}(0) = \sum_{\substack{\alpha:=(\alpha_1,\ldots,\alpha_N),\ \alpha_i \in \mathbb{Z}^+, \\ i=1,\ldots,N,\ |\alpha|:=\sum_{i=1}^N \alpha_i = j}} \left(\frac{j!}{\prod_{i=1}^N \alpha_i!}\right)\left(\prod_{i=1}^N (z_i - x_{0i})^{\alpha_i}\right) f_\alpha(x_0). \tag{6.20}$$

Furthermore

$$g_z^{(m)}(\theta) = \sum_{\substack{\alpha:=(\alpha_1,\ldots,\alpha_N),\ \alpha_i \in \mathbb{Z}^+, \\ i=1,\ldots,N,\ |\alpha|:=\sum_{i=1}^N \alpha_i = m}} \left(\frac{m!}{\prod_{i=1}^N \alpha_i!}\right)\left(\prod_{i=1}^N (z_i - x_{0i})^{\alpha_i}\right) f_\alpha(x_0 + \theta(z - x_0)), \tag{6.21}$$

$0 \leq \theta \leq 1$.

So we treat $f \in AC^m\left(\prod_{i=1}^N [a_i, b_i]\right)$ with $\|f_\alpha\|_{\infty,m}^{\max} < \infty$.

Thus, by (6.19) we have for $\frac{k}{n}, x \in \left(\prod_{i=1}^N [a_i, b_i]\right)$ that

$$f\left(\frac{k_1}{n}, \ldots, \frac{k_N}{n}\right) - f(x) =$$

$$\sum_{j=1}^{m-1} \sum_{\substack{\alpha:=(\alpha_1,\ldots,\alpha_N),\ \alpha_i \in \mathbb{Z}^+, \\ i=1,\ldots,N,\ |\alpha|:=\sum_{i=1}^N \alpha_i = j}} \left(\frac{1}{\prod_{i=1}^N \alpha_i!}\right)\left(\prod_{i=1}^N \left(\frac{k_i}{n} - x_i\right)^{\alpha_i}\right) f_\alpha(x) + R, \tag{6.22}$$

where

$$R := m \int_0^1 (1-\theta)^{m-1} \sum_{\substack{\alpha:=(\alpha_1,\ldots,\alpha_N),\, \alpha_i \in \mathbb{Z}^+,\\ i=1,\ldots,N,\ |\alpha|:=\sum_{i=1}^N \alpha_i = m}} \left(\frac{1}{\prod_{i=1}^N \alpha_i!} \right) \cdot$$

$$\left(\prod_{i=1}^N \left(\frac{k_i}{n} - x_i \right)^{\alpha_i} \right) f_\alpha \left(x + \theta \left(\frac{k}{n} - x \right) \right) d\theta. \tag{6.23}$$

By (6.15) we obtain

$$|R| \leq \frac{\left(\left\| x - \frac{k}{n} \right\|_{l_1} \right)^m}{m!} \left\| f_\alpha \right\|_{\infty,m}^{\max}. \tag{6.24}$$

Notice here that

$$\left\| \frac{k}{n} - x \right\|_\infty \leq \frac{1}{n^\beta} \Leftrightarrow \left| \frac{k_i}{n} - x_i \right| \leq \frac{1}{n^\beta},\ i = 1, \ldots, N. \tag{6.25}$$

So, if $\left\| \frac{k}{n} - x \right\|_\infty \leq \frac{1}{n^\beta}$ we get that $\left\| x - \frac{k}{n} \right\|_{l_1} \leq \frac{N}{n^\beta}$, and

$$|R| \leq \frac{N^m}{n^{m\beta} m!} \left\| f_\alpha \right\|_{\infty,m}^{\max}. \tag{6.26}$$

Also we see that

$$\left\| x - \frac{k}{n} \right\|_{l_1} = \sum_{i=1}^N \left| x_i - \frac{k_i}{n} \right| \leq \sum_{i=1}^N (b_i - a_i) = \|b - a\|_{l_1},$$

therefore in general it holds

$$|R| \leq \frac{\left(\|b - a\|_{l_1} \right)^m}{m!} \left\| f_\alpha \right\|_{\infty,m}^{\max}. \tag{6.27}$$

Call

$$V(x) := \sum_{k=\lceil na \rceil}^{\lfloor nb \rfloor} \Phi(nx - k).$$

Hence we have

$$U_n(x) := \frac{\sum_{k=\lceil na \rceil}^{\lfloor nb \rfloor} \Phi(nx-k) R}{V(x)} = \tag{6.28}$$

$$\frac{\sum_{\begin{cases} k = \lceil na \rceil \\ : \left\| \frac{k}{n} - x \right\|_\infty \le \frac{1}{n^\beta} \end{cases}}^{\lfloor nb \rfloor} \Phi(nx-k) R}{V(x)} + \frac{\sum_{\begin{cases} k = \lceil na \rceil \\ : \left\| \frac{k}{n} - x \right\|_\infty > \frac{1}{n^\beta} \end{cases}}^{\lfloor nb \rfloor} \Phi(nx-k) R}{V(x)}.$$

Consequently we obtain

$$|U_n(x)| \le \left(\frac{\sum_{\begin{cases} k = \lceil na \rceil \\ : \left\| \frac{k}{n} - x \right\|_\infty \le \frac{1}{n^\beta} \end{cases}}^{\lfloor nb \rfloor} \Phi(nx-k)}{V(x)} \right) \left(\frac{N^m}{n^{m\beta} m!} \| f_\alpha \|_{\infty,m}^{\max} \right) +$$

$$\frac{1}{V(x)} \left(\sum_{\begin{cases} k = \lceil na \rceil \\ : \left\| \frac{k}{n} - x \right\|_\infty > \frac{1}{n^\beta} \end{cases}}^{\lfloor nb \rfloor} \Phi(nx-k) \right) \frac{\left(\|b-a\|_{l_1} \right)^m}{m!} \| f_\alpha \|_{\infty,m}^{\max} \overset{\text{(by (v)', (vi)')}}{\le}$$

$$\frac{N^m}{n^{m\beta} m!} \| f_\alpha \|_{\infty,m}^{\max} + (5.250312578)^N (3.1992) e^{-n^{(1-\beta)}} \frac{\left(\|b-a\|_{l_1} \right)^m}{m!} \| f_\alpha \|_{\infty,m}^{\max}. \tag{6.29}$$

Therefore we have found

$$|U_n(x)| \le \frac{\| f_\alpha \|_{\infty,m}^{\max}}{m!} \left\{ \frac{N^m}{n^{m\beta}} + (5.250312578)^N (3.1992) e^{-n^{(1-\beta)}} \left(\|b-a\|_{l_1} \right)^m \right\}. \tag{6.30}$$

For large enough $n \in \mathbb{N}$ we get

$$|U_n(x)| \le \left(\frac{2 \| f_\alpha \|_{\infty,m}^{\max} N^m}{m!} \right) \left(\frac{1}{n^{m\beta}} \right). \tag{6.31}$$

That is

$$|U_n(x)| = O\left(\frac{1}{n^{m\beta}} \right), \tag{6.32}$$

and

$$|U_n(x)| = o(1). \tag{6.33}$$

And, letting $0 < \varepsilon \leq m$, we derive

$$\frac{|U_n(x)|}{\left(\frac{1}{n^{\beta(m-\varepsilon)}}\right)} \leq \left(\frac{2 \|f_\alpha\|_{\infty,m}^{\max} N^m}{m!}\right) \frac{1}{n^{\beta\varepsilon}} \to 0, \tag{6.34}$$

as $n \to \infty$.

I.e.

$$|U_n(x)| = o\left(\frac{1}{n^{\beta(m-\varepsilon)}}\right). \tag{6.35}$$

By (6.22) we observe that

$$\frac{\sum_{k=\lceil na \rceil}^{\lfloor nb \rfloor} f\left(\frac{k}{n}\right) \Phi(nx-k)}{V(x)} - f(x) =$$

$$\sum_{j=1}^{m-1} \left(\sum_{|\alpha|=j} \left(\frac{f_\alpha(x)}{\prod_{i=1}^N \alpha_i!}\right)\right) \frac{\left(\sum_{k=\lceil na \rceil}^{\lfloor nb \rfloor} \Phi(nx-k) \left(\prod_{i=1}^N \left(\frac{k_i}{n} - x_i\right)^{\alpha_i}\right)\right)}{V(x)} +$$

$$\frac{\sum_{k=\lceil na \rceil}^{\lfloor nb \rfloor} \Phi(nx-k) R}{V(x)}. \tag{6.36}$$

The last says

$$G_n(f,x) - f(x) - \sum_{j=1}^{m-1} \left(\sum_{|\alpha|=j} \left(\frac{f_\alpha(x)}{\prod_{i=1}^N \alpha_i!}\right) G_n\left(\prod_{i=1}^N (\cdot - x_i)^{\alpha_i}, x\right)\right) = U_n(x). \tag{6.37}$$

The proof of the theorem is complete. ■

We present our second main result

Theorem 6.10 *Let* $0 < \beta < 1$, $x \in \prod_{i=1}^N [a_i, b_i]$, $n \in \mathbb{N}$ *large enough,* $f \in AC^m\left(\prod_{i=1}^N [a_i, b_i]\right)$, $m, N \in \mathbb{N}$. *Assume further that* $\|f_\alpha\|_{\infty,m}^{\max} < \infty$. *Then*

$$F_n(f,x) - f(x) =$$

$$\sum_{j=1}^{m-1}\left(\sum_{|\alpha|=j}\left(\frac{f_\alpha(x)}{\prod_{i=1}^N \alpha_i!}\right)F_n\left(\prod_{i=1}^N(\cdot - x_i)^{\alpha_i}, x\right)\right) + o\left(\frac{1}{n^{\beta(m-\varepsilon)}}\right), \qquad (6.38)$$

where $0 < \varepsilon \leq m$.

If $m = 1$, *the sum in (6.38) collapses.*

The last (6.38) implies that

$$n^{\beta(m-\varepsilon)}\left[F_n(f,x) - f(x) - \sum_{j=1}^{m-1}\left(\sum_{|\alpha|=j}\left(\frac{f_\alpha(x)}{\prod_{i=1}^N \alpha_i!}\right)F_n\left(\prod_{i=1}^N(\cdot - x_i)^{\alpha_i}, x\right)\right)\right] \qquad (6.39)$$

$$\to 0, \text{ as } n \to \infty, \ 0 < \varepsilon \leq m.$$

When $m = 1$, *or* $f_\alpha(x) = 0$, *for* $|\alpha| = j$, $j = 1, \ldots, m-1$, *then we derive that*

$$n^{\beta(m-\varepsilon)}\left[F_n(f,x) - f(x)\right] \to 0,$$

as $n \to \infty$, $0 < \varepsilon \leq m$.

Proof Similar to Theorem 6.9, using the properties of $\Theta(x)$, see (6.4), (i)*–(vii)* and (6.5). ■

References

1. G.A. Anastassiou, *Advanced Inequalities* (World Scientific Publishing Company, Singapore, 2011)
2. G.A. Anastassiou, *Intelligent Systems: Approximation by Artificial Neural Networks*, Intelligent Systems Reference Library, vol. 19 (Springer, Heidelberg, 2011)
3. G.A. Anastassiou, Univariate hyperbolic tangent neural network approximation. Math. Comput. Model. **53**, 1111–1132 (2011)
4. G.A. Anastassiou, Multivariate hyperbolic tangent neural network approximation. Comput. Math. **61**, 809–821 (2011)
5. G.A. Anastassiou, Multivariate sigmoidal neural network approximation. Neural Netw. **24**, 378–386 (2011)
6. G.A. Anastassiou, Univariate sigmoidal neural network approximation. J. Comput. Anal. Appl. **14**(4), 659–690 (2012)
7. G.A. Anastassiou, Voronovskaya type asymptotic expansions for multivariate quasi-interpolation neural network operators. Cubo **16**(2), 33–47 (2014)
8. Z. Chen, F. Cao, The approximation operators with sigmoidal functions. Comput. Math. Appl. **58**, 758–765 (2009)
9. S. Haykin, *Neural Networks: A Comprehensive Foundation*, 2nd edn. (Prentice Hall, New York, 1998)
10. W. McCulloch, W. Pitts, A logical calculus of the ideas immanent in nervous activity. Bull. Math. Biophys. **7**, 115–133 (1943)
11. T.M. Mitchell, *Machine Learning* (WCB-McGraw-Hill, New York, 1997)

Chapter 7
Fractional Approximation by Normalized Bell and Squashing Type Neural Networks

This chapter deals with the determination of the fractional rate of convergence to the unit of some neural network operators, namely, the normalized bell and "squashing" type operators. This is given through the moduli of continuity of the involved right and left Caputo fractional derivatives of the approximated function and they appear in the right-hand side of the associated Jackson type inequalities. It follows [7].

7.1 Introduction

The Cardaliaguet-Euvrard operators were studied extensively in [8], where the authors among many other things proved that these operators converge uniformly on compacta, to the unit over continuous and bounded functions. Our "normalized bell and squashing type operators" (see (7.22), (7.63)) were motivated and inspired by the "bell" and "squashing functions" of [8]. The work in [8] is qualitative where the used bell-shaped function is general. However, our work, though greatly motivated by [8], is quantitative and the used bell-shaped and "squashing" functions are of compact support. We produce a series of Jackson type inequalities giving close upper bounds to the errors in approximating the unit operator by the above neural network induced operators. All involved constants there are well determined. These are pointwise, uniform and L_p, $p \geq 1$, estimates involving the first moduli of continuity of the engaged right and left Caputo fractional derivatives of the function under approximation. We give all necessary background of fractional calculus.

Initial work of the subject was done in [1], where we involved only ordinary derivatives. Article [1] motivated the current chapter.

G.A. Anastassiou, *Intelligent Systems II: Complete Approximation by Neural Network Operators*, Studies in Computational Intelligence 608,
DOI 10.1007/978-3-319-20505-2_7

7.2 Background

We need

Definition 7.1 Let $f \in C(\mathbb{R})$ which is bounded or uniformly continuous, $h > 0$. We define the first modulus of continuity of f at h as follows

$$\omega_1(f, h) = \sup\{|f(x) - f(y)| ; x, y \in \mathbb{R}, \ |x - y| \leq h\} \tag{7.1}$$

Notice that $\omega_1(f, h)$ is finite for any $h > 0$, and

$$\lim_{h \to 0} \omega_1(f, h) = 0.$$

We also need

Definition 7.2 Let $f : \mathbb{R} \to \mathbb{R}$, $\nu \geq 0$, $n = \lceil \nu \rceil$ ($\lceil \cdot \rceil$ is the ceiling of the number), $f \in AC^n([a, b])$ (space of functions f with $f^{(n-1)} \in AC([a, b])$, absolutely continuous functions), $\forall [a, b] \subset \mathbb{R}$. We call left Caputo fractional derivative (see [9], pp. 49–52) the function

$$D_{*a}^{\nu} f(x) = \frac{1}{\Gamma(n - \nu)} \int_a^x (x - t)^{n-\nu-1} f^{(n)}(t) \, dt, \tag{7.2}$$

$\forall x \geq a$, where Γ is the gamma function $\Gamma(\nu) = \int_0^\infty e^{-t} t^{\nu-1} dt$, $\nu > 0$. Notice $D_{*a}^{\nu} f \in L_1([a, b])$ and $D_{*a}^{\nu} f$ exists a.e. on $[a, b]$, $\forall b > a$. We set $D_{*a}^{0} f(x) = f(x)$, $\forall x \in [a, \infty)$.

Lemma 7.3 ([5]) *Let $\nu > 0$, $\nu \notin \mathbb{N}$, $n = \lceil \nu \rceil$, $f \in C^{n-1}(\mathbb{R})$ and $f^{(n)} \in L_\infty(\mathbb{R})$. Then $D_{*a}^{\nu} f(a) = 0$, $\forall a \in \mathbb{R}$.*

Definition 7.4 (*see also* [2, 10, 11]) Let $f : \mathbb{R} \to \mathbb{R}$, such that $f \in AC^m([a, b])$, $\forall [a, b] \subset \mathbb{R}$, $m = \lceil \alpha \rceil$, $\alpha > 0$. The right Caputo fractional derivative of order $\alpha > 0$ is given by

$$D_{b-}^{\alpha} f(x) = \frac{(-1)^m}{\Gamma(m - \alpha)} \int_x^b (J - x)^{m-\alpha-1} f^{(m)}(J) \, dJ, \tag{7.3}$$

$\forall x \leq b$. We set $D_{b-}^{0} f(x) = f(x)$, $\forall x \in (-\infty, b]$. Notice that $D_{b-}^{\alpha} f \in L_1([a, b])$ and $D_{b-}^{\alpha} f$ exists a.e. on $[a, b]$, $\forall a < b$.

Lemma 7.5 ([5]) *Let $f \in C^{m-1}(\mathbb{R})$, $f^{(m)} \in L_\infty(\mathbb{R})$, $m = \lceil \alpha \rceil$, $\alpha > 0$. Then $D_{b-}^{\alpha} f(b) = 0$, $\forall b \in \mathbb{R}$.*

Convention 7.6 *We assume that*

$$D^{\alpha}_{*x_0} f(x) = 0, \text{ for } x < x_0, \tag{7.4}$$

and

$$D^{\alpha}_{x_0 -} f(x) = 0, \text{ for } x > x_0, \tag{7.5}$$

for all $x, x_0 \in \mathbb{R}$.

We mention

Proposition 7.7 (by [3]) *Let* $f \in C^n(\mathbb{R})$, *where* $n = \lceil \nu \rceil$, $\nu > 0$. *Then* $D^{\nu}_{*a} f(x)$ *is continuous in* $x \in [a, \infty)$.

Also we have

Proposition 7.8 (by [3]) *Let* $f \in C^m(\mathbb{R})$, $m = \lceil \alpha \rceil$, $\alpha > 0$. *Then* $D^{\alpha}_{b-} f(x)$ *is continuous in* $x \in (-\infty, b]$.

We further mention

Proposition 7.9 (by [3]) *Let* $f \in C^{m-1}(\mathbb{R})$, $f^{(m)} \in L_{\infty}(\mathbb{R})$, $m = \lceil \alpha \rceil$, $\alpha > 0$ *and*

$$D^{\alpha}_{*x_0} f(x) = \frac{1}{\Gamma(m-\alpha)} \int_{x_0}^{x} (x-t)^{m-\alpha-1} f^{(m)}(t) \, dt, \tag{7.6}$$

for all $x, x_0 \in \mathbb{R} : x \geq x_0$.

Then $D^{\alpha}_{*x_0} f(x)$ *is continuous in* x_0.

Proposition 7.10 (by [3]) *Let* $f \in C^{m-1}(\mathbb{R})$, $f^{(m)} \in L_{\infty}(\mathbb{R})$, $m = \lceil \alpha \rceil$, $\alpha > 0$ *and*

$$D^{\alpha}_{x_0 -} f(x) = \frac{(-1)^m}{\Gamma(m-\alpha)} \int_{x}^{x_0} (J-x)^{m-\alpha-1} f^{(m)}(J) \, dJ, \tag{7.7}$$

for all $x, x_0 \in \mathbb{R} : x_0 \geq x$.

Then $D^{\alpha}_{x_0 -} f(x)$ *is continuous in* x_0.

Proposition 7.11 ([5]) *Let* $g \in C_b(\mathbb{R})$ *(continuous and bounded),* $0 < c < 1$, $x, x_0 \in \mathbb{R}$. *Define*

$$L(x, x_0) = \int_{x_0}^{x} (x-t)^{c-1} g(t) \, dt, \text{ for } x \geq x_0, \tag{7.8}$$

and $L(x, x_0) = 0$, *for* $x < x_0$.

Then L *is jointly continuous in* $(x, x_0) \in \mathbb{R}^2$.

We mention

Proposition 7.12 ([5]) *Let $g \in C_b(\mathbb{R})$, $0 < c < 1$, $x, x_0 \in \mathbb{R}$. Define*

$$K(x, x_0) = \int_x^{x_0} (J - x)^{c-1} g(J)\, dJ, \text{ for } x \leq x_0, \tag{7.9}$$

and $K(x, x_0) = 0$, for $x > x_0$.

Then $K(x, x_0)$ is jointly continuous in $(x, x_0) \in \mathbb{R}^2$.

Based on Propositions 7.11, 7.12 we derive

Corollary 7.13 ([5]) *Let $f \in C^m(\mathbb{R})$, $f^{(m)} \in L_\infty(\mathbb{R})$, $m = \lceil \alpha \rceil$, $\alpha > 0$, $\alpha \notin \mathbb{N}$, $x, x_0 \in \mathbb{R}$. Then $D^\alpha_{*x_0} f(x)$, $D^\alpha_{x_0-} f(x)$ are jointly continuous functions in (x, x_0) from $\mathbb{R}^2$ into $\mathbb{R}$.*

We need

Proposition 7.14 ([5]) *Let $f : \mathbb{R}^2 \to \mathbb{R}$ be jointly continuous. Consider*

$$G(x) = \omega_1 (f(\cdot, x), \delta)_{[x,+\infty)}, \ \delta > 0, x \in \mathbb{R}. \tag{7.10}$$

(Here ω_1 is defined over $[x, +\infty)$ instead of $\mathbb{R}$.)

Then G is continuous on $\mathbb{R}$.

Proposition 7.15 ([5]) *Let $f : \mathbb{R}^2 \to \mathbb{R}$ be jointly continuous. Consider*

$$H(x) = \omega_1 (f(\cdot, x), \delta)_{(-\infty,x]}, \ \delta > 0, x \in \mathbb{R}. \tag{7.11}$$

(Here ω_1 is defined over $(-\infty, x]$ instead of $\mathbb{R}$.)

Then H is continuous on $\mathbb{R}$.

By Propositions 7.14, 7.15 and Corollary 7.13 we derive

Proposition 7.16 ([5]) *Let $f \in C^m(\mathbb{R})$, $\left\| f^{(m)} \right\|_\infty < \infty$, $m = \lceil \alpha \rceil$, $\alpha \notin \mathbb{N}$, $\alpha > 0$, $x \in \mathbb{R}$. Then $\omega_1 \left(D^\alpha_{*x} f, h \right)_{[x,+\infty)}$, $\omega_1 \left(D^\alpha_{x-} f, h \right)_{(-\infty,x]}$ are continuous functions of $x \in \mathbb{R}$, $h > 0$ fixed.*

We make

Remark 7.17 Let g be continuous and bounded from $\mathbb{R}$ to $\mathbb{R}$. Then

$$\omega_1 (g, t) \leq 2 \|g\|_\infty < \infty. \tag{7.12}$$

Assuming that $\left(D_{*x}^{\alpha} f\right)(t)$, $\left(D_{x-}^{\alpha} f\right)(t)$, are both continuous and bounded in $(x,t) \in \mathbb{R}^2$, i.e.

$$\left\| D_{*x}^{\alpha} f \right\|_{\infty} \le K_1, \forall\, x \in \mathbb{R}; \tag{7.13}$$

$$\left\| D_{x-}^{\alpha} f \right\|_{\infty} \le K_2, \forall\, x \in \mathbb{R}, \tag{7.14}$$

where $K_1, K_2 > 0$, we get

$$\begin{aligned} &\omega_1\left(D_{*x}^{\alpha} f, \xi\right)_{[x,+\infty)} \le 2K_1; \\ &\omega_1\left(D_{x-}^{\alpha} f, \xi\right)_{(-\infty,x]} \le 2K_2, \forall\, \xi \ge 0, \end{aligned} \tag{7.15}$$

for each $x \in \mathbb{R}$.

Therefore, for any $\xi \ge 0$,

$$\sup_{x\in\mathbb{R}} \left[\max\left(\omega_1\left(D_{*x}^{\alpha} f, \xi\right)_{[x,+\infty)}, \omega_1\left(D_{x-}^{\alpha} f, \xi\right)_{(-\infty,x]}\right)\right] \le 2\max\left(K_1, K_2\right) < \infty. \tag{7.16}$$

So in our setting for $f \in C^m(\mathbb{R})$, $\left\| f^{(m)} \right\|_{\infty} < \infty$, $m = \lceil \alpha \rceil$, $\alpha \notin \mathbb{N}$, $\alpha > 0$, by Corollary 7.13 both $\left(D_{*x}^{\alpha} f\right)(t)$, $\left(D_{x-}^{\alpha} f\right)(t)$ are jointly continuous in (t,x) on $\mathbb{R}^2$. Assuming further that they are both bounded on $\mathbb{R}^2$ we get (7.16) valid. In particular, each of $\omega_1\left(D_{*x}^{\alpha} f, \xi\right)_{[x,+\infty)}$, $\omega_1\left(D_{x-}^{\alpha} f, \xi\right)_{(-\infty,x]}$ is finite for any $\xi \ge 0$.

Let us now assume only that $f \in C^{m-1}(\mathbb{R})$, $f^{(m)} \in L_\infty(\mathbb{R})$, $m = \lceil \alpha \rceil$, $\alpha > 0$, $\alpha \notin \mathbb{N}$, $x \in \mathbb{R}$. Then, by Proposition 15.114, p. 388 of [4], we find that $D_{*x}^{\alpha} f \in C([x,+\infty))$, and by [6] we obtain that $D_{x-}^{\alpha} f \in C((-\infty,x])$.

We make

Remark 7.18 Again let $f \in C^m(\mathbb{R})$, $m = \lceil \alpha \rceil$, $\alpha \notin \mathbb{N}$, $\alpha > 0$; $f^{(m)}(x) = 1$, $\forall\, x \in \mathbb{R}$; $x_0 \in \mathbb{R}$. Notice $0 < m - \alpha < 1$. Then

$$D_{*x_0}^{\alpha} f(x) = \frac{(x - x_0)^{m-\alpha}}{\Gamma(m - \alpha + 1)}, \forall\, x \ge x_0. \tag{7.17}$$

Let us consider $x, y \ge x_0$, then

$$\begin{aligned} \left| D_{*x_0}^{\alpha} f(x) - D_{*x_0}^{\alpha} f(y) \right| &= \frac{1}{\Gamma(m-\alpha+1)} \left| (x - x_0)^{m-\alpha} - (y - x_0)^{m-\alpha} \right| \\ &\le \frac{|x - y|^{m-\alpha}}{\Gamma(m-\alpha+1)}. \end{aligned} \tag{7.18}$$

So it is not strange to assume that

$$\left| D_{*x_0}^{\alpha} f(x_1) - D_{*x_0}^{\alpha} f(x_2) \right| \le K\, |x_1 - x_2|^{\beta}, \tag{7.19}$$

$K > 0$, $0 < \beta \leq 1$, $\forall\, x_1, x_2 \in \mathbb{R}$, $x_1, x_2 \geq x_0 \in \mathbb{R}$, where more generally it is $\left\| f^{(m)} \right\|_{\infty} < \infty$. Thus, one may assume

$$\omega_1 \left(D_{x-}^{\alpha} f, \xi \right)_{(-\infty,x]} \leq M_1 \xi^{\beta_1}, \text{ and} \tag{7.20}$$

$$\omega_1 \left(D_{*x}^{\alpha} f, \xi \right)_{[x,+\infty)} \leq M_2 \xi^{\beta_2},$$

where $0 < \beta_1, \beta_2 \leq 1$, $\forall\, \xi > 0$, $M_1, M_2 > 0$; any $x \in \mathbb{R}$.

Setting $\beta = \min(\beta_1, \beta_2)$ and $M = \max(M_1, M_2)$, in that case we obtain

$$\sup_{x \in \mathbb{R}} \left\{ \max \left(\omega_1 \left(D_{x-}^{\alpha} f, \xi \right)_{(-\infty,x]}, \omega_1 \left(D_{*x}^{\alpha} f, \xi \right)_{[x,+\infty)} \right) \right\} \leq M \xi^{\beta} \to 0, \text{ as } \xi \to 0+. \tag{7.21}$$

7.3 Results

7.3.1 Fractional Convergence with Rates of the Normalized Bell Type Neural Network Operators

We need the following (see [8]).

Definition 7.19 A function $b : \mathbb{R} \to \mathbb{R}$ is said to be bell-shaped if b belongs to L^1 and its integral is nonzero, if it is nondecreasing on $(-\infty, a)$ and nonincreasing on $[a, +\infty)$, where a belongs to $\mathbb{R}$. In particular $b(x)$ is a nonnegative number and at a b takes a global maximum; it is the center of the bell-shaped function. A bell-shaped function is said to be centered if its center is zero. The function $b(x)$ may have jump discontinuities. In this work we consider only centered bell-shaped functions of compact support $[-T, T]$, $T > 0$.

We follow [1, 8].

Example 7.20 (1) $b(x)$ can be the characteristic function over $[-1, 1]$.

(2) $b(x)$ can be the hat function over $[-1, 1]$, i.e.,

$$b(x) = \begin{cases} 1 + x, & -1 \leq x \leq 0, \\ 1 - x, & 0 < x \leq 1 \\ 0, & \text{elsewhere.} \end{cases}$$

These are centered bell-shaped functions of compact support.

Here we consider functions $f : \mathbb{R} \to \mathbb{R}$ that are continuous.

In this chapter we study the fractional convergence with rates over the real line, to the unit operator, of the "normalized bell type neural network operators",

$$\left(H_n\left(f\right)\right)\left(x\right) := \frac{\sum_{k=-n^2}^{n^2} f\left(\frac{k}{n}\right) \cdot b\left(n^{1-\alpha} \cdot \left(x - \frac{k}{n}\right)\right)}{\sum_{k=-n^2}^{n^2} b\left(n^{1-\alpha} \cdot \left(x - \frac{k}{n}\right)\right)}, \tag{7.22}$$

where $0 < \alpha < 1$ and $x \in \mathbb{R}$, $n \in \mathbb{N}$. The terms in the ratio of sums (7.22) can be nonzero iff

$$\left| n^{1-\alpha}\left(x - \frac{k}{n}\right)\right| \leq T, \text{ i.e. } \left| x - \frac{k}{n}\right| \leq \frac{T}{n^{1-\alpha}}$$

iff

$$nx - Tn^{\alpha} \leq k \leq nx + Tn^{\alpha}. \tag{7.23}$$

In order to have the desired order of numbers

$$-n^2 \leq nx - Tn^{\alpha} \leq nx + Tn^{\alpha} \leq n^2, \tag{7.24}$$

it is sufficient enough to assume that

$$n \geq T + |x|. \tag{7.25}$$

When $x \in [-T, T]$ it is enough to assume $n \geq 2T$ which implies (7.24).

Proposition 7.21 *Let $a \leq b$, $a, b \in \mathbb{R}$. Let $card\left(k\right)$ (≥ 0) be the maximum number of integers contained in $[a, b]$. Then*

$$\max\left(0, (b-a) - 1\right) \leq card\left(k\right) \leq (b-a) + 1. \tag{7.26}$$

Remark 7.22 We would like to establish a lower bound on $card\left(k\right)$ over the interval $[nx - Tn^{\alpha}, nx + Tn^{\alpha}]$. From Proposition 7.21 we get that

$$card\left(k\right) \geq \max\left(2Tn^{\alpha} - 1, 0\right).$$

We obtain $card\left(k\right) \geq 1$, if

$$2Tn^{\alpha} - 1 \geq 1 \text{ iff } n \geq T^{-\frac{1}{\alpha}}.$$

So to have the desired order (7.24) and $card\left(k\right) \geq 1$ over $[nx - Tn^{\alpha}, nx + Tn^{\alpha}]$, we need to consider

$$n \geq \max\left(T + |x|, T^{-\frac{1}{\alpha}}\right). \tag{7.27}$$

Also notice that $card\left(k\right) \to +\infty$, as $n \to +\infty$.

Denote by $[\cdot]$ the integral part of a number.

We make

Remark 7.23 Clearly we have that

$$nx - Tn^{\alpha} \leq nx \leq nx + Tn^{\alpha}. \tag{7.28}$$

We prove in general that

$$nx - Tn^{\alpha} \leq [nx] \leq nx \leq \lceil nx \rceil \leq nx + Tn^{\alpha}. \tag{7.29}$$

Indeed we have that, if $[nx] < nx - Tn^{\alpha}$, then $[nx] + Tn^{\alpha} < nx$, and $[nx] + [Tn^{\alpha}] \leq [nx]$, resulting into $[Tn^{\alpha}] = 0$, which for large enough n is not true. Therefore $nx - Tn^{\alpha} \leq [nx]$. Similarly, if $\lceil nx \rceil > nx + Tn^{\alpha}$, then $nx + Tn^{\alpha} \geq nx + [Tn^{\alpha}]$, and $\lceil nx \rceil - [Tn^{\alpha}] > nx$, thus $\lceil nx \rceil - [Tn^{\alpha}] \geq \lceil nx \rceil$, resulting into $[Tn^{\alpha}] = 0$, which again for large enough n is not true.

Therefore without loss of generality we may assume that

$$nx - Tn^{\alpha} \leq [nx] \leq nx \leq \lceil nx \rceil \leq nx + Tn^{\alpha}. \tag{7.30}$$

Hence $\lceil nx - Tn^{\alpha} \rceil \leq [nx]$ and $\lceil nx \rceil \leq [nx + Tn^{\alpha}]$. Also if $[nx] \neq \lceil nx \rceil$, then $\lceil nx \rceil = [nx] + 1$. If $[nx] = \lceil nx \rceil$, then $nx \in \mathbb{Z}$; and by assuming $n \geq T^{-\frac{1}{\alpha}}$, we get $Tn^{\alpha} \geq 1$ and $nx + Tn^{\alpha} \geq nx + 1$, so that $[nx + Tn^{\alpha}] \geq nx + 1 = [nx] + 1$.

We present our first main result

Theorem 7.24 *We consider* $f : \mathbb{R} \to \mathbb{R}$. *Let* $\beta > 0$, $N = \lceil \beta \rceil$, $\beta \notin \mathbb{N}$, $f \in AC^N([a,b])$, $\forall\, [a,b] \subset \mathbb{R}$, *with* $f^{(N)} \in L_{\infty}(\mathbb{R})$. *Let also* $x \in \mathbb{R}$, $T > 0$, $n \in \mathbb{N} : n \geq \max\left(T + |x|, T^{-\frac{1}{\alpha}}\right)$. *We further assume that* $D_{*x}^{\beta} f$, $D_{x-}^{\beta} f$ *are uniformly continuous functions or continuous and bounded on* $[x, +\infty)$, $(-\infty, x]$, *respectively.*

Then

(1)

$$|H_n(f)(x) - f(x)| \leq \left(\sum_{j=1}^{N-1} \frac{\left| f^{(j)}(x) \right| T^j}{j! n^{(1-\alpha) j}} \right) + \frac{T^{\beta}}{\Gamma(\beta+1) n^{(1-\alpha)\beta}} \cdot \tag{7.31}$$

$$\left\{ \omega_1 \left(D_{*x}^{\beta} f, \frac{T}{n^{1-\alpha}} \right)_{[x,+\infty)} + \omega_1 \left(D_{x-}^{\beta} f, \frac{T}{n^{1-\alpha}} \right)_{(-\infty,x]} \right\},$$

above $\sum_{j=1}^{0} \cdot = 0$,

(2)

$$\left|(H_n(f))(x)-\sum_{j=0}^{N-1}\frac{f^{(j)}(x)}{j!}\left(H_n\left((\cdot-x)^j\right)\right)(x)\right|\leq\frac{T^\beta}{\Gamma(\beta+1)\,n^{(1-\alpha)\beta}}\cdot \tag{7.32}$$

$$\left\{\omega_1\left(D^\beta_{*x}f,\frac{T}{n^{1-\alpha}}\right)_{[x,+\infty)}+\omega_1\left(D^\beta_{x-}f,\frac{T}{n^{1-\alpha}}\right)_{(-\infty,x]}\right\}=:\lambda_n(x),$$

(3) assume further that $f^{(j)}(x)=0$, for $j=1,\ldots,N-1$, we get

$$|H_n(f)(x)-f(x)|\leq\lambda_n(x), \tag{7.33}$$

(4) in case of $N=1$, we obtain again

$$|H_n(f)(x)-f(x)|\leq\lambda_n(x). \tag{7.34}$$

Here we get fractionally with rates the pointwise convergence of $(H_n(f))(x)\to f(x)$, as $n\to\infty$, $x\in\mathbb{R}$.

Proof Let $x\in\mathbb{R}$. We have that

$$D^\beta_{x-}f(x)=D^\beta_{*x}f(x)=0. \tag{7.35}$$

From [9], p. 54, we get by the left Caputo fractional Taylor formula that

$$f\left(\frac{k}{n}\right)=\sum_{j=0}^{N-1}\frac{f^{(j)}(x)}{j!}\left(\frac{k}{n}-x\right)^j+ \tag{7.36}$$

$$\frac{1}{\Gamma(\beta)}\int_x^{\frac{k}{n}}\left(\frac{k}{n}-J\right)^{\beta-1}\left(D^\beta_{*x}f(J)-D^\beta_{*x}f(x)\right)dJ,$$

for all $x\leq\frac{k}{n}\leq x+Tn^{\alpha-1}$, iff $\lceil nx\rceil\leq k\leq[nx+Tn^\alpha]$, where $k\in\mathbb{Z}$.

Also from [2], using the right Caputo fractional Taylor formula we get

$$f\left(\frac{k}{n}\right)=\sum_{j=0}^{N-1}\frac{f^{(j)}(x)}{j!}\left(\frac{k}{n}-x\right)^j+ \tag{7.37}$$

$$\frac{1}{\Gamma(\beta)}\int_{\frac{k}{n}}^{x}\left(J-\frac{k}{n}\right)^{\beta-1}\left(D^\beta_{x-}f(J)-D^\beta_{x-}f(x)\right)dJ,$$

for all $x-Tn^{\alpha-1}\leq\frac{k}{n}\leq x$, iff $\lceil nx-Tn^\alpha\rceil\leq k\leq[nx]$, where $k\in\mathbb{Z}$.

Notice that $\lceil nx \rceil \leq [nx] + 1$.
Call

$$V(x) := \sum_{k=\lceil nx - Tn^{\alpha} \rceil}^{[nx+Tn^{\alpha}]} b\left(n^{1-\alpha}\left(x - \frac{k}{n}\right)\right).$$

Hence we have

$$\frac{f\left(\frac{k}{n}\right) b\left(n^{1-\alpha}\left(x-\frac{k}{n}\right)\right)}{V(x)} = \sum_{j=0}^{N-1} \frac{f^{(j)}(x)}{j!}\left(\frac{k}{n}-x\right)^{j} \frac{b\left(n^{1-\alpha}\left(x-\frac{k}{n}\right)\right)}{V(x)} + \tag{7.38}$$

$$\frac{b\left(n^{1-\alpha}\left(x-\frac{k}{n}\right)\right)}{V(x)\Gamma(\beta)} \int_{x}^{\frac{k}{n}} \left(\frac{k}{n}-J\right)^{\beta-1}\left(D_{*x}^{\beta} f(J) - D_{*x}^{\beta} f(x)\right) dJ,$$

and

$$\frac{f\left(\frac{k}{n}\right) b\left(n^{1-\alpha}\left(x-\frac{k}{n}\right)\right)}{V(x)} = \sum_{j=0}^{N-1} \frac{f^{(j)}(x)}{j!}\left(\frac{k}{n}-x\right)^{j} \frac{b\left(n^{1-\alpha}\left(x-\frac{k}{n}\right)\right)}{V(x)} + \tag{7.39}$$

$$\frac{b\left(n^{1-\alpha}\left(x-\frac{k}{n}\right)\right)}{V(x)\Gamma(\beta)} \int_{\frac{k}{n}}^{x} \left(J-\frac{k}{n}\right)^{\beta-1}\left(D_{x-}^{\beta} f(J) - D_{x-}^{\beta} f(x)\right) dJ.$$

Therefore we obtain

$$\frac{\sum_{k=[nx]+1}^{[nx+Tn^{\alpha}]} f\left(\frac{k}{n}\right) b\left(n^{1-\alpha}\left(x-\frac{k}{n}\right)\right)}{V(x)} = \tag{7.40}$$

$$\sum_{j=0}^{N-1} \frac{f^{(j)}(x)}{j!}\left(\frac{\sum_{k=[nx]+1}^{[nx+Tn^{\alpha}]}\left(\frac{k}{n}-x\right)^{j} b\left(n^{1-\alpha}\left(x-\frac{k}{n}\right)\right)}{V(x)}\right) +$$

$$\sum_{k=[nx]+1}^{[nx+Tn^{\alpha}]} \frac{b\left(n^{1-\alpha}\left(x-\frac{k}{n}\right)\right)}{V(x)\Gamma(\beta)} \int_{x}^{\frac{k}{n}} \left(\frac{k}{n}-J\right)^{\beta-1}\left(D_{*x}^{\beta} f(J) - D_{*x}^{\beta} f(x)\right) dJ,$$

and

$$\frac{\sum_{k=\lceil nx - Tn^{\alpha} \rceil}^{[nx]} f\left(\frac{k}{n}\right) b\left(n^{1-\alpha}\left(x-\frac{k}{n}\right)\right)}{V(x)} = \tag{7.41}$$

$$\sum_{j=0}^{N-1} \frac{f^{(j)}(x)}{j!} \frac{\sum_{k=\lceil nx - Tn^{\alpha} \rceil}^{[nx]}\left(\frac{k}{n}-x\right)^{j} b\left(n^{1-\alpha}\left(x-\frac{k}{n}\right)\right)}{V(x)} +$$

$$\frac{\sum_{k=\lceil nx-Tn^{\alpha}\rceil}^{[nx]} b\left(n^{1-\alpha}\left(x-\frac{k}{n}\right)\right)}{V(x)\Gamma(\beta)} \int_{\frac{k}{n}}^{x}\left(J-\frac{k}{n}\right)^{\beta-1}\left(D_{x-}^{\beta}f(J)-D_{x-}^{\beta}f(x)\right)dJ.$$

We notice here that

$$(H_n(f))(x) := \frac{\sum_{k=-n^2}^{n^2} f\left(\frac{k}{n}\right) b\left(n^{1-\alpha}\left(x-\frac{k}{n}\right)\right)}{\sum_{k=-n^2}^{n^2} b\left(n^{1-\alpha}\left(x-\frac{k}{n}\right)\right)} = \tag{7.42}$$

$$\frac{\sum_{k=\lceil nx-Tn^{\alpha}\rceil}^{[nx+Tn^{\alpha}]} f\left(\frac{k}{n}\right) b\left(n^{1-\alpha}\left(x-\frac{k}{n}\right)\right)}{V(x)}.$$

Adding the two equalities (7.40) and (7.41) we obtain

$$(H_n(f))(x) =$$

$$\sum_{j=0}^{N-1}\frac{f^{(j)}(x)}{j!}\left(\frac{\sum_{k=\lceil nx-Tn^{\alpha}\rceil}^{[nx+Tn^{\alpha}]}\left(\frac{k}{n}-x\right)^{j} b\left(n^{1-\alpha}\left(x-\frac{k}{n}\right)\right)}{V(x)}\right)+\theta_n(x), \tag{7.43}$$

where

$$\theta_n(x) := \frac{\sum_{k=\lceil nx-Tn^{\alpha}\rceil}^{[nx]} b\left(n^{1-\alpha}\left(x-\frac{k}{n}\right)\right)}{V(x)\Gamma(\beta)}.$$

$$\int_{\frac{k}{n}}^{x}\left(J-\frac{k}{n}\right)^{\beta-1}\left(D_{x-}^{\beta}f(J)-D_{x-}^{\beta}f(x)\right)dJ +$$

$$\sum_{k=[nx]+1}^{[nx+Tn^{\alpha}]}\frac{b\left(n^{1-\alpha}\left(x-\frac{k}{n}\right)\right)}{V(x)\Gamma(\beta)}\int_{x}^{\frac{k}{n}}\left(\frac{k}{n}-J\right)^{\beta-1}\left(D_{*x}^{\beta}f(J)-D_{*x}^{\beta}f(x)\right)dJ. \tag{7.44}$$

We call

$$\theta_{1n}(x) := \frac{\sum_{k=\lceil nx-Tn^{\alpha}\rceil}^{[nx]} b\left(n^{1-\alpha}\left(x-\frac{k}{n}\right)\right)}{V(x)\Gamma(\beta)}.$$

$$\int_{\frac{k}{n}}^{x}\left(J-\frac{k}{n}\right)^{\beta-1}\left(D_{x-}^{\beta}f(J)-D_{x-}^{\beta}f(x)\right)dJ, \tag{7.45}$$

and

$$\theta_{2n}(x) := \sum_{k=[nx]+1}^{[nx+Tn^{\alpha}]}\frac{b\left(n^{1-\alpha}\left(x-\frac{k}{n}\right)\right)}{V(x)\Gamma(\beta)}.$$

$$\int_x^{\frac{k}{n}} \left(\frac{k}{n} - J\right)^{\beta-1} \left(D_{*x}^{\beta} f(J) - D_{*x}^{\beta} f(x)\right) dJ. \tag{7.46}$$

I.e.

$$\theta_n(x) = \theta_{1n}(x) + \theta_{2n}(x). \tag{7.47}$$

We further have

$$(H_n(f))(x) - f(x) = \tag{7.48}$$

$$\sum_{j=1}^{N-1} \frac{f^{(j)}(x)}{j!} \left(\frac{\sum_{k=\lceil nx - Tn^{\alpha} \rceil}^{[nx+Tn^{\alpha}]} \left(\frac{k}{n} - x\right)^j b\left(n^{1-\alpha}\left(x - \frac{k}{n}\right)\right)}{V(x)} \right) + \theta_n(x),$$

and

$$|(H_n(f))(x) - f(x)| \leq$$

$$\sum_{j=1}^{N-1} \frac{\left|f^{(j)}(x)\right|}{j!} \left(\frac{\sum_{k=\lceil nx - Tn^{\alpha} \rceil}^{[nx+Tn^{\alpha}]} \left|x - \frac{k}{n}\right|^j b\left(n^{1-\alpha}\left(x - \frac{k}{n}\right)\right)}{V(x)} \right) + |\theta_n(x)| \leq \tag{7.49}$$

$$\sum_{j=1}^{N-1} \frac{\left|f^{(j)}(x)\right|}{j!} \frac{T^j}{n^{(1-\alpha)j}} + |\theta_n(x)| =: (*).$$

Next we see that

$$\gamma_{1n} := \frac{1}{\Gamma(\beta)} \left| \int_{\frac{k}{n}}^{x} \left(J - \frac{k}{n}\right)^{\beta-1} \left(D_{x-}^{\beta} f(J) - D_{x-}^{\beta} f(x)\right) dJ \right| \leq \tag{7.50}$$

$$\frac{1}{\Gamma(\beta)} \int_{\frac{k}{n}}^{x} \left(J - \frac{k}{n}\right)^{\beta-1} \left| D_{x-}^{\beta} f(J) - D_{x-}^{\beta} f(x)\, dJ \right| \leq$$

$$\frac{1}{\Gamma(\beta)} \int_{\frac{k}{n}}^{x} \left(J - \frac{k}{n}\right)^{\beta-1} \omega_1\left(D_{x-}^{\beta} f, |J - x|\right)_{(-\infty,x]} dJ \leq$$

$$\frac{1}{\Gamma(\beta)} \omega_1\left(D_{x-}^{\beta} f, \left|x - \frac{k}{n}\right|\right)_{(-\infty,x]} \int_{\frac{k}{n}}^{x} \left(J - \frac{k}{n}\right)^{\beta-1} dJ \leq$$

$$\frac{1}{\Gamma(\beta)} \omega_1\left(D_{x-}^{\beta} f, \frac{T}{n^{1-\alpha}}\right)_{(-\infty,x]} \frac{\left(x - \frac{k}{n}\right)^{\beta}}{\beta} \leq$$

$$\frac{1}{\Gamma(\beta+1)}\omega_1\left(D_{x-}^{\beta}f,\frac{T}{n^{1-\alpha}}\right)_{(-\infty,x]}\frac{T^{\beta}}{n^{(1-\alpha)\beta}}.$$

That is

$$\gamma_{1n}\leq\frac{T^{\beta}}{\Gamma(\beta+1)\,n^{(1-\alpha)\beta}}\omega_1\left(D_{x-}^{\beta}f,\frac{T}{n^{1-\alpha}}\right)_{(-\infty,x]}. \tag{7.51}$$

Furthermore

$$|\theta_{1n}(x)|\leq\sum_{k=\lceil nx-Tn^{\alpha}\rceil}^{[nx]}\frac{b\left(n^{1-\alpha}\left(x-\frac{k}{n}\right)\right)}{V(x)}\gamma_{1n}\leq \tag{7.52}$$

$$\left(\sum_{k=\lceil nx-Tn^{\alpha}\rceil}^{[nx]}\frac{b\left(n^{1-\alpha}\left(x-\frac{k}{n}\right)\right)}{V(x)}\right)\frac{T^{\beta}}{\Gamma(\beta+1)\,n^{(1-\alpha)\beta}}\omega_1\left(D_{x-}^{\beta}f,\frac{T}{n^{1-\alpha}}\right)_{(-\infty,x]}\leq$$

$$\left(\sum_{k=\lceil nx-Tn^{\alpha}\rceil}^{[nx+Tn^{\alpha}]}\frac{b\left(n^{1-\alpha}\left(x-\frac{k}{n}\right)\right)}{V(x)}\right)\frac{T^{\beta}}{\Gamma(\beta+1)\,n^{(1-\alpha)\beta}}\omega_1\left(D_{x-}^{\beta}f,\frac{T}{n^{1-\alpha}}\right)_{(-\infty,x]}=$$

$$\frac{T^{\beta}}{\Gamma(\beta+1)\,n^{(1-\alpha)\beta}}\omega_1\left(D_{x-}^{\beta}f,\frac{T}{n^{1-\alpha}}\right)_{(-\infty,x]}.$$

So that

$$|\theta_{1n}(x)|\leq\frac{T^{\beta}}{\Gamma(\beta+1)\,n^{(1-\alpha)\beta}}\omega_1\left(D_{x-}^{\beta}f,\frac{T}{n^{1-\alpha}}\right)_{(-\infty,x]}. \tag{7.53}$$

Similarly we derive

$$\gamma_{2n}:=\frac{1}{\Gamma(\beta)}\left|\int_x^{\frac{k}{n}}\left(\frac{k}{n}-J\right)^{\beta-1}\left(D_{*x}^{\beta}f(J)-D_{*x}^{\beta}f(x)\right)dJ\right|\leq \tag{7.54}$$

$$\frac{1}{\Gamma(\beta)}\int_x^{\frac{k}{n}}\left(\frac{k}{n}-J\right)^{\beta-1}\left|D_{*x}^{\beta}f(J)-D_{*x}^{\beta}f(x)\right|dJ\leq$$

$$\frac{\omega_1\left(D_{*x}^{\beta}f,\frac{T}{n^{1-\alpha}}\right)_{[x,+\infty)}}{\Gamma(\beta+1)}\left(\frac{k}{n}-x\right)^{\beta}\leq$$

$$\frac{\omega_1\left(D_{*x}^{\beta}f,\frac{T}{n^{1-\alpha}}\right)_{[x,+\infty)}}{\Gamma(\beta+1)}\frac{T^{\beta}}{n^{(1-\alpha)\beta}}.$$

That is

$$\gamma_{2n} \leq \frac{T^{\beta}}{\Gamma(\beta+1) n^{(1-\alpha)\beta}} \omega_1 \left(D_{*x}^{\beta} f, \frac{T}{n^{1-\alpha}} \right)_{[x,+\infty)} . \tag{7.55}$$

Consequently we find

$$|\theta_{2n}(x)| \leq \left(\sum_{k=[nx]+1}^{[nx+Tn^{\alpha}]} \frac{b\left(n^{1-\alpha}\left(x-\frac{k}{n}\right)\right)}{V(x)} \right) \cdot$$

$$\frac{T^{\beta}}{\Gamma(\beta+1) n^{(1-\alpha)\beta}} \omega_1 \left(D_{*x}^{\beta} f, \frac{T}{n^{1-\alpha}} \right)_{[x,+\infty)} \leq \tag{7.56}$$

$$\frac{T^{\beta}}{\Gamma(\beta+1) n^{(1-\alpha)\beta}} \omega_1 \left(D_{*x}^{\beta} f, \frac{T}{n^{1-\alpha}} \right)_{[x,+\infty)} .$$

So we have proved that

$$|\theta_n(x)| \leq \frac{T^{\beta}}{\Gamma(\beta+1) n^{(1-\alpha)\beta}} \cdot \tag{7.57}$$

$$\left\{ \omega_1 \left(D_{*x}^{\beta} f, \frac{T}{n^{1-\alpha}} \right)_{[x,+\infty)} + \omega_1 \left(D_{x-}^{\beta} f, \frac{T}{n^{1-\alpha}} \right)_{(-\infty,x]} \right\} .$$

Combining (7.49) and (7.57) we have (7.31). ■

As an application of Theorem 7.24 we give

Theorem 7.25 *Let $\beta > 0$, $N = \lceil \beta \rceil$, $\beta \notin \mathbb{N}$, $f \in C^N(\mathbb{R})$, with $f^{(N)} \in L_{\infty}(\mathbb{R})$. Let also $T > 0$, $n \in \mathbb{N} : n \geq \max\left(2T, T^{-\frac{1}{\alpha}}\right)$. We further assume that $D_{*x}^{\beta} f(t)$, $D_{x-}^{\beta} f(t)$ are both bounded in $(x,t) \in \mathbb{R}^2$. Then*

(1)

$$\| H_n(f) - f \|_{\infty,[-T,T]} \leq \tag{7.58}$$

$$\left(\sum_{j=1}^{N-1} \frac{\| f^{(j)} \|_{\infty,[-T,T]} T^j}{j! n^{(1-\alpha)j}} \right) + \frac{T^{\beta}}{\Gamma(\beta+1) n^{(1-\alpha)\beta}} \cdot$$

$$\left\{ \sup_{x\in[-T,T]} \omega_1 \left(D_{*x}^{\beta} f, \frac{T}{n^{1-\alpha}} \right)_{[x,+\infty)} + \sup_{x\in[-T,T]} \omega_1 \left(D_{x-}^{\beta} f, \frac{T}{n^{1-\alpha}} \right)_{(-\infty,x]} \right\},$$

(2) in case of $N = 1$, we obtain

$$\|H_n(f) - f\|_{\infty,[-T,T]} \leq \frac{T^{\beta}}{\Gamma(\beta+1)\, n^{(1-\alpha)\beta}} \cdot \tag{7.59}$$

$$\left\{ \sup_{x\in[-T,T]} \omega_1 \left(D_{*x}^{\beta} f, \frac{T}{n^{1-\alpha}} \right)_{[x,+\infty)} + \sup_{x\in[-T,T]} \omega_1 \left(D_{x-}^{\beta} f, \frac{T}{n^{1-\alpha}} \right)_{(-\infty,x]} \right\}.$$

An interesting case is when $\beta = \frac{1}{2}$.

Here we get fractionally with rates the uniform convergence of $H_n(f) \to f$, as $n \to \infty$.

Proof From (7.31), (7.34) of Theorem 7.24, and by Remark 7.17. ∎

One can also apply Remark 7.18 to the last Theorem 7.25, to get interesting and simplified results.

We make

Remark 7.26 Let $b(x)$ be a centered bell-shaped continuous function on $\mathbb{R}$ of compact support $[-T, T]$, $T > 0$. Let $x \in [-T^*, T^*]$, $T^* > 0$, and $n \in \mathbb{N} : n \geq \max\left(T + T^*, T^{-\frac{1}{\alpha}}\right)$, $0 < \alpha < 1$. Consider $p \geq 1$.

Let also $\beta > 0$, $N = \lceil \beta \rceil$, $\beta \notin \mathbb{N}$, $f \in C^N(\mathbb{R})$, $f^{(N)} \in L_{\infty}(\mathbb{R})$. Here both $D_{*x}^{\alpha} f(t)$, $D_{x-}^{\alpha} f(t)$ are bounded in $(x,t) \in \mathbb{R}^2$.

By Theorem 7.24 we have

$$|H_n(f)(x) - f(x)| \leq \tag{7.60}$$

$$\left(\sum_{j=1}^{N-1} \frac{\left|f^{(j)}(x)\right| T^j}{j! n^{(1-\alpha)j}} \right) + \frac{T^{\beta}}{\Gamma(\beta+1)\, n^{(1-\alpha)\beta}} \cdot$$

$$\left\{ \sup_{x\in[-T^*,T^*]} \omega_1 \left(D_{*x}^{\beta} f, \frac{T}{n^{1-\alpha}} \right)_{[x,+\infty)} + \sup_{x\in[-T^*,T^*]} \omega_1 \left(D_{x-}^{\beta} f, \frac{T}{n^{1-\alpha}} \right)_{(-\infty,x]} \right\}.$$

Applying to the last inequality (7.60) the monotonicity and subadditive property of $\|\cdot\|_p$, we derive the following L_p, $p \geq 1$, interesting result.

Theorem 7.27 *Let $b(x)$ be a centered bell-shaped continuous function on $\mathbb{R}$ of compact support $[-T, T]$, $T > 0$. Let $x \in [-T^*, T^*]$, $T^* > 0$, and $n \in \mathbb{N} : n \geq \max\left(T + T^*, T^{-\frac{1}{\alpha}}\right)$, $0 < \alpha < 1$, $p \geq 1$. Let $\beta > 0$, $N = \lceil \beta \rceil$, $\beta \notin \mathbb{N}$, $f \in C^N(\mathbb{R})$, with $f^{(N)} \in L_{\infty}(\mathbb{R})$. Here both $D_{*x}^{\beta} f(t)$, $D_{x-}^{\beta} f(t)$ are bounded in $(x,t) \in \mathbb{R}^2$. Then*

(1)

$$\left\| H_n f - f \right\|_{p,[-T^*,T^*]} \leq \tag{7.61}$$

$$\left(\sum_{j=1}^{N-1} \frac{\left\| f^{(j)} \right\|_{p,[-T^*,T^*]} T^j}{j! n^{(1-\alpha)j}} \right) + 2^{\frac{1}{p}} T^{*\frac{1}{p}} \frac{T^\beta}{\Gamma(\beta+1) n^{(1-\alpha)\beta}} \cdot$$

$$\left\{ \sup_{x\in[-T^*,T^*]} \omega_1 \left(D_{*x}^\beta f, \frac{T}{n^{1-\alpha}} \right)_{[x,+\infty)} + \sup_{x\in[-T^*,T^*]} \omega_1 \left(D_{x-}^\beta f, \frac{T}{n^{1-\alpha}} \right)_{(-\infty,x]} \right\},$$

(2) When $N = 1$, we derive

$$\left\| H_n f - f \right\|_{p,[-T^*,T^*]} \leq 2^{\frac{1}{p}} T^{*\frac{1}{p}} \frac{T^\beta}{\Gamma(\beta+1) n^{(1-\alpha)\beta}} \cdot \tag{7.62}$$

$$\left\{ \sup_{x\in[-T^*,T^*]} \omega_1 \left(D_{*x}^\beta f, \frac{T}{n^{1-\alpha}} \right)_{[x,+\infty)} + \sup_{x\in[-T^*,T^*]} \omega_1 \left(D_{x-}^\beta f, \frac{T}{n^{1-\alpha}} \right)_{(-\infty,x]} \right\}.$$

By (7.61), (7.62) we derive the fractional L_p, $p \geq 1$, convergence with rates of $H_n f$ to f.

7.3.2 The "Normalized Squashing Type Operators" and Their Fractional Convergence to the Unit with Rates

We need (see also [1, 8]).

Definition 7.28 Let the nonnegative function $S : \mathbb{R} \to \mathbb{R}$, S has compact support $[-T, T]$, $T > 0$, and is nondecreasing there and it can be continuous only on either $(-\infty, T]$ or $[-T, T]$. S can have jump discontinuities. We call S the "squashing function".

Let $f : \mathbb{R} \to \mathbb{R}$ be continuous.
For $x \in \mathbb{R}$ we define the "normalized squashing type operator"

$$(K_n(f))(x) := \frac{\sum_{k=-n^2}^{n^2} f\left(\frac{k}{n}\right) \cdot S\left(n^{1-\alpha} \cdot \left(x - \frac{k}{n}\right)\right)}{\sum_{k=-n^2}^{n^2} S\left(n^{1-\alpha} \cdot \left(x - \frac{k}{n}\right)\right)}, \tag{7.63}$$

$0 < \alpha < 1$ and $n \in \mathbb{N} : n \geq \max\left(T + |x|, T^{-\frac{1}{\alpha}}\right)$. It is clear that

$$(K_n(f))(x) = \frac{\sum_{k=\lceil nx-Tn^{\alpha}\rceil}^{[nx+Tn^{\alpha}]} f\left(\frac{k}{n}\right) \cdot S\left(n^{1-\alpha} \cdot \left(x - \frac{k}{n}\right)\right)}{\sum_{k=\lceil nx-Tn^{\alpha}\rceil}^{[nx+Tn^{\alpha}]} S\left(n^{1-\alpha} \cdot \left(x - \frac{k}{n}\right)\right)}. \tag{7.64}$$

Here we study the fractional convergence with rates of $(K_n f)(x) \to f(x)$, as $n \to +\infty$, $x \in \mathbb{R}$.

We present our second main result

Theorem 7.29 *We consider $f : \mathbb{R} \to \mathbb{R}$. Let $\beta > 0$, $N = \lceil \beta \rceil$, $\beta \notin \mathbb{N}$, $f \in AC^N([a,b])$, $\forall [a,b] \subset \mathbb{R}$, with $f^{(N)} \in L_\infty(\mathbb{R})$. Let also $x \in \mathbb{R}$, $T > 0$, $n \in \mathbb{N} : n \geq \max\left(T + |x|, T^{-\frac{1}{\alpha}}\right)$. We further assume that $D_{*x}^{\beta} f$, $D_{x-}^{\beta} f$ are uniformly continuous functions or continuous and bounded on $[x, +\infty)$, $(-\infty, x]$, respectively.*

Then

(1)

$$|K_n(f)(x) - f(x)| \leq \tag{7.65}$$

$$\left(\sum_{j=1}^{N-1} \frac{\left|f^{(j)}(x)\right| T^j}{j! n^{(1-\alpha)j}}\right) + \frac{T^\beta}{\Gamma(\beta+1) n^{(1-\alpha)\beta}} \cdot$$

$$\left\{\omega_1\left(D_{*x}^{\beta} f, \frac{T}{n^{1-\alpha}}\right)_{[x,+\infty)} + \omega_1\left(D_{x-}^{\beta} f, \frac{T}{n^{1-\alpha}}\right)_{(-\infty,x]}\right\},$$

above $\sum_{j=1}^{0} \cdot = 0$,

(2)

$$\left|(K_n(f))(x) - \sum_{j=0}^{N-1} \frac{f^{(j)}(x)}{j!} \left(K_n\left((\cdot - x)^j\right)\right)(x)\right| \leq \frac{T^\beta}{\Gamma(\beta+1) n^{(1-\alpha)\beta}} \cdot \tag{7.66}$$

$$\left\{\omega_1\left(D_{*x}^{\beta} f, \frac{T}{n^{1-\alpha}}\right)_{[x,+\infty)} + \omega_1\left(D_{x-}^{\beta} f, \frac{T}{n^{1-\alpha}}\right)_{(-\infty,x]}\right\} =: \lambda_n^*(x),$$

(3) assume further that $f^{(j)}(x) = 0$, for $j = 1, \ldots, N-1$, we get

$$|K_n(f)(x) - f(x)| \leq \lambda_n^*(x), \tag{7.67}$$

(4) in case of $N = 1$, we obtain also

$$|K_n(f)(x) - f(x)| \leq \lambda_n^*(x). \tag{7.68}$$

Here we get fractionally with rates the pointwise convergence of $(K_n(f))(x) \to f(x)$, as $n \to \infty$, $x \in \mathbb{R}$.

Proof Let $x \in \mathbb{R}$. We have that

$$D_{x-}^{\beta} f(x) = D_{*x}^{\beta} f(x) = 0.$$

From [9], p. 54, we get by the left Caputo fractional Taylor formula that

$$f\left(\frac{k}{n}\right) = \sum_{j=0}^{N-1} \frac{f^{(j)}(x)}{j!} \left(\frac{k}{n} - x\right)^j + \tag{7.69}$$

$$\frac{1}{\Gamma(\beta)} \int_x^{\frac{k}{n}} \left(\frac{k}{n} - J\right)^{\beta-1} \left(D_{*x}^{\beta} f(J) - D_{*x}^{\beta} f(x)\right) dJ,$$

for all $x \leq \frac{k}{n} \leq x + Tn^{\alpha-1}$, iff $\lceil nx \rceil \leq k \leq [nx + Tn^{\alpha}]$, where $k \in \mathbb{Z}$.

Also from [2], using the right Caputo fractional Taylor formula we get

$$f\left(\frac{k}{n}\right) = \sum_{j=0}^{N-1} \frac{f^{(j)}(x)}{j!} \left(\frac{k}{n} - x\right)^j + \tag{7.70}$$

$$\frac{1}{\Gamma(\beta)} \int_{\frac{k}{n}}^{x} \left(J - \frac{k}{n}\right)^{\beta-1} \left(D_{x-}^{\beta} f(J) - D_{x-}^{\beta} f(x)\right) dJ,$$

for all $x - Tn^{\alpha-1} \leq \frac{k}{n} \leq x$, iff $\lceil nx - Tn^{\alpha} \rceil \leq k \leq [nx]$, where $k \in \mathbb{Z}$.

Call

$$W(x) := \sum_{k=\lceil nx - Tn^{\alpha} \rceil}^{[nx+Tn^{\alpha}]} S\left(n^{1-\alpha}\left(x - \frac{k}{n}\right)\right).$$

Hence we have

$$\frac{f\left(\frac{k}{n}\right) S\left(n^{1-\alpha}\left(x - \frac{k}{n}\right)\right)}{W(x)} = \sum_{j=0}^{N-1} \frac{f^{(j)}(x)}{j!} \left(\frac{k}{n} - x\right)^j \frac{S\left(n^{1-\alpha}\left(x - \frac{k}{n}\right)\right)}{W(x)} + \tag{7.71}$$

$$\frac{S\left(n^{1-\alpha}\left(x - \frac{k}{n}\right)\right)}{W(x)\Gamma(\beta)} \int_x^{\frac{k}{n}} \left(\frac{k}{n} - J\right)^{\beta-1} \left(D_{*x}^{\beta} f(J) - D_{*x}^{\beta} f(x)\right) dJ,$$

and

$$\frac{f\left(\frac{k}{n}\right) S\left(n^{1-\alpha}\left(x - \frac{k}{n}\right)\right)}{W(x)} = \sum_{j=0}^{N-1} \frac{f^{(j)}(x)}{j!} \left(\frac{k}{n} - x\right)^j \frac{S\left(n^{1-\alpha}\left(x - \frac{k}{n}\right)\right)}{W(x)} + \tag{7.72}$$

$$\frac{S\left(n^{1-\alpha}\left(x-\frac{k}{n}\right)\right)}{W\left(x\right)\Gamma\left(\beta\right)}\int_{\frac{k}{n}}^{x}\left(J-\frac{k}{n}\right)^{\beta-1}\left(D_{x-}^{\beta}f\left(J\right)-D_{x-}^{\beta}f\left(x\right)\right)dJ.$$

Therefore we obtain

$$\frac{\sum_{k=[nx]+1}^{[nx+Tn^{\alpha}]}f\left(\frac{k}{n}\right)S\left(n^{1-\alpha}\left(x-\frac{k}{n}\right)\right)}{W\left(x\right)}= \tag{7.73}$$

$$\sum_{j=0}^{N-1}\frac{f^{(j)}\left(x\right)}{j!}\left(\frac{\sum_{k=[nx]+1}^{[nx+Tn^{\alpha}]}\left(\frac{k}{n}-x\right)^{j}S\left(n^{1-\alpha}\left(x-\frac{k}{n}\right)\right)}{W\left(x\right)}\right)+$$

$$\sum_{k=[nx]+1}^{[nx+Tn^{\alpha}]}\frac{S\left(n^{1-\alpha}\left(x-\frac{k}{n}\right)\right)}{W\left(x\right)\Gamma\left(\beta\right)}\int_{x}^{\frac{k}{n}}\left(\frac{k}{n}-J\right)^{\beta-1}\left(D_{*x}^{\beta}f\left(J\right)-D_{*x}^{\beta}f\left(x\right)\right)dJ,$$

and

$$\frac{\sum_{k=\lceil nx-Tn^{\alpha}\rceil}^{[nx]}f\left(\frac{k}{n}\right)S\left(n^{1-\alpha}\left(x-\frac{k}{n}\right)\right)}{W\left(x\right)}= \tag{7.74}$$

$$\sum_{j=0}^{N-1}\frac{f^{(j)}\left(x\right)}{j!}\frac{\sum_{k=\lceil nx-Tn^{\alpha}\rceil}^{[nx]}\left(\frac{k}{n}-x\right)^{j}S\left(n^{1-\alpha}\left(x-\frac{k}{n}\right)\right)}{W\left(x\right)}+$$

$$\frac{\sum_{k=\lceil nx-Tn^{\alpha}\rceil}^{[nx]}S\left(n^{1-\alpha}\left(x-\frac{k}{n}\right)\right)}{W\left(x\right)\Gamma\left(\beta\right)}\int_{\frac{k}{n}}^{x}\left(J-\frac{k}{n}\right)^{\beta-1}\left(D_{x-}^{\beta}f\left(J\right)-D_{x-}^{\beta}f\left(x\right)\right)dJ.$$

Adding the two equalities (7.73) and (7.74) we obtain

$$\left(K_{n}\left(f\right)\right)\left(x\right)=$$

$$\sum_{j=0}^{N-1}\frac{f^{(j)}\left(x\right)}{j!}\left(\frac{\sum_{k=\lceil nx-Tn^{\alpha}\rceil}^{[nx+Tn^{\alpha}]}\left(\frac{k}{n}-x\right)^{j}S\left(n^{1-\alpha}\left(x-\frac{k}{n}\right)\right)}{W\left(x\right)}\right)+M_{n}\left(x\right), \tag{7.75}$$

where

$$M_{n}\left(x\right):=\frac{\sum_{k=\lceil nx-Tn^{\alpha}\rceil}^{[nx]}S\left(n^{1-\alpha}\left(x-\frac{k}{n}\right)\right)}{W\left(x\right)\Gamma\left(\beta\right)}\cdot$$

$$\int_{\frac{k}{n}}^{x}\left(J-\frac{k}{n}\right)^{\beta-1}\left(D_{x-}^{\beta}f\left(J\right)-D_{x-}^{\beta}f\left(x\right)\right)dJ+$$

$$\sum_{k=[nx]+1}^{[nx+Tn^{\alpha}]} \frac{S\left(n^{1-\alpha}\left(x-\frac{k}{n}\right)\right)}{W(x)\Gamma(\beta)} \int_{x}^{\frac{k}{n}} \left(\frac{k}{n}-J\right)^{\beta-1} \left(D_{*x}^{\beta}f(J)-D_{*x}^{\beta}f(x)\right) dJ. \tag{7.76}$$

We call

$$M_{1n}(x) := \sum_{k=\lceil nx-Tn^{\alpha}\rceil}^{[nx]} \frac{S\left(n^{1-\alpha}\left(x-\frac{k}{n}\right)\right)}{W(x)\Gamma(\beta)} \cdot$$

$$\int_{\frac{k}{n}}^{x} \left(J-\frac{k}{n}\right)^{\beta-1} \left(D_{x-}^{\beta}f(J)-D_{x-}^{\beta}f(x)\right) dJ, \tag{7.77}$$

and

$$M_{2n}(x) := \sum_{k=[nx]+1}^{[nx+Tn^{\alpha}]} \frac{S\left(n^{1-\alpha}\left(x-\frac{k}{n}\right)\right)}{W(x)\Gamma(\beta)} \cdot$$

$$\int_{x}^{\frac{k}{n}} \left(\frac{k}{n}-J\right)^{\beta-1} \left(D_{*x}^{\beta}f(J)-D_{*x}^{\beta}f(x)\right) dJ. \tag{7.78}$$

I.e.

$$M_n(x) = M_{1n}(x) + M_{2n}(x). \tag{7.79}$$

We further have

$$(K_n(f))(x) - f(x) = \tag{7.80}$$

$$\sum_{j=1}^{N-1} \frac{f^{(j)}(x)}{j!} \left(\frac{\sum_{k=\lceil nx-Tn^{\alpha}\rceil}^{[nx+Tn^{\alpha}]} \left(\frac{k}{n}-x\right)^{j} S\left(n^{1-\alpha}\left(x-\frac{k}{n}\right)\right)}{W(x)} \right) + M_n(x),$$

and

$$|(K_n(f))(x) - f(x)| \leq$$

$$\sum_{j=1}^{N-1} \frac{\left|f^{(j)}(x)\right|}{j!} \left(\frac{\sum_{k=\lceil nx-Tn^{\alpha}\rceil}^{[nx+Tn^{\alpha}]} \left|x-\frac{k}{n}\right|^{j} S\left(n^{1-\alpha}\left(x-\frac{k}{n}\right)\right)}{W(x)} \right) + |M_n(x)| \leq$$

$$\sum_{j=1}^{N-1} \frac{\left|f^{(j)}(x)\right|}{j!} \frac{T^j}{n^{(1-\alpha)j}} \left(\frac{\sum_{k=\lceil nx-Tn^{\alpha}\rceil}^{[nx+Tn^{\alpha}]} S\left(n^{1-\alpha}\left(x-\frac{k}{n}\right)\right)}{W(x)} \right) + |M_n(x)| =: (*). \tag{7.81}$$

Therefore we obtain

$$|(K_n(f))(x) - f(x)| \leq \left(\sum_{j=1}^{N-1} \frac{\left|f^{(j)}(x)\right| T^j}{j! n^{(1-\alpha)j}} \right) + |M_n(x)|. \tag{7.82}$$

We call

$$\gamma_{1n} := \frac{1}{\Gamma(\beta)} \left| \int_{\frac{k}{n}}^{x} \left(J - \frac{k}{n} \right)^{\beta-1} \left(D_{x-}^{\beta} f(J) - D_{x-}^{\beta} f(x) \right) dJ \right|. \tag{7.83}$$

As in the proof of Theorem 7.24 we have

$$\gamma_{1n} \leq \frac{T^{\beta}}{\Gamma(\beta+1) n^{(1-\alpha)\beta}} \omega_1 \left(D_{x-}^{\beta} f, \frac{T}{n^{1-\alpha}} \right)_{(-\infty,x]}. \tag{7.84}$$

Furthermore

$$|M_{1n}(x)| \leq \sum_{k=\lceil nx-Tn^{\alpha} \rceil}^{[nx]} \frac{S\left(n^{1-\alpha}\left(x-\frac{k}{n}\right)\right)}{W(x)} \gamma_{1n} \leq \tag{7.85}$$

$$\left(\sum_{k=\lceil nx-Tn^{\alpha} \rceil}^{[nx]} \frac{S\left(n^{1-\alpha}\left(x-\frac{k}{n}\right)\right)}{W(x)} \right) \frac{T^{\beta}}{\Gamma(\beta+1) n^{(1-\alpha)\beta}} \omega_1 \left(D_{x-}^{\beta} f, \frac{T}{n^{1-\alpha}} \right)_{(-\infty,x]} \leq$$

$$\frac{T^{\beta}}{\Gamma(\beta+1) n^{(1-\alpha)\beta}} \omega_1 \left(D_{x-}^{\beta} f, \frac{T}{n^{1-\alpha}} \right)_{(-\infty,x]}.$$

So that

$$|M_{1n}(x)| \leq \frac{T^{\beta}}{\Gamma(\beta+1) n^{(1-\alpha)\beta}} \omega_1 \left(D_{x-}^{\beta} f, \frac{T}{n^{1-\alpha}} \right)_{(-\infty,x]}. \tag{7.86}$$

We also call

$$\gamma_{2n} := \frac{1}{\Gamma(\beta)} \left| \int_{x}^{\frac{k}{n}} \left(\frac{k}{n} - J \right)^{\beta-1} \left(D_{*x}^{\beta} f(J) - D_{*x}^{\beta} f(x) \right) dJ \right|. \tag{7.87}$$

As in the proof of Theorem 7.24 we get

$$\gamma_{2n} \leq \frac{T^{\beta}}{\Gamma(\beta+1) n^{(1-\alpha)\beta}} \omega_1 \left(D_{*x}^{\beta} f, \frac{T}{n^{1-\alpha}} \right)_{[x,+\infty)}. \tag{7.88}$$

Consequently we find

$$|M_{2n}(x)| \leq \left(\sum_{k=[nx]+1}^{[nx+Tn^{\alpha}]} \frac{S\left(n^{1-\alpha}\left(x - \frac{k}{n}\right)\right)}{W(x)} \right) \cdot$$

$$\frac{T^{\beta}}{\Gamma(\beta+1) n^{(1-\alpha)\beta}} \omega_1 \left(D_{*x}^{\beta} f, \frac{T}{n^{1-\alpha}} \right)_{[x,+\infty)} \leq \tag{7.89}$$

$$\frac{T^{\beta}}{\Gamma(\beta+1) n^{(1-\alpha)\beta}} \omega_1 \left(D_{*x}^{\beta} f, \frac{T}{n^{1-\alpha}} \right)_{[x,+\infty)}.$$

So we have proved that

$$|M_n(x)| \leq \frac{T^{\beta}}{\Gamma(\beta+1) n^{(1-\alpha)\beta}} \cdot \tag{7.90}$$

$$\left\{ \omega_1 \left(D_{*x}^{\beta} f, \frac{T}{n^{1-\alpha}} \right)_{[x,+\infty)} + \omega_1 \left(D_{x-}^{\beta} f, \frac{T}{n^{1-\alpha}} \right)_{(-\infty,x]} \right\}.$$

Combining (7.82) and (7.90) we have (7.65). ■

As an application of Theorem 7.29 we give

Theorem 7.30 *Let $\beta > 0$, $N = \lceil \beta \rceil$, $\beta \notin \mathbb{N}$, $f \in C^N(\mathbb{R})$, with $f^{(N)} \in L_{\infty}(\mathbb{R})$. Let also $T > 0$, $n \in \mathbb{N} : n \geq \max\left(2T, T^{-\frac{1}{\alpha}}\right)$. We further assume that $D_{*x}^{\beta} f(t)$, $D_{x-}^{\beta} f(t)$ are both bounded in $(x,t) \in \mathbb{R}^2$. Then*

(1)

$$\|K_n(f) - f\|_{\infty,[-T,T]} \leq \tag{7.91}$$

$$\left(\sum_{j=1}^{N-1} \frac{\left\| f^{(j)} \right\|_{\infty,[-T,T]} T^j}{j! n^{(1-\alpha)j}} \right) + \frac{T^{\beta}}{\Gamma(\beta+1) n^{(1-\alpha)\beta}} \cdot$$

$$\left\{ \sup_{x \in [-T,T]} \omega_1 \left(D_{*x}^{\beta} f, \frac{T}{n^{1-\alpha}} \right)_{[x,+\infty)} + \sup_{x \in [-T,T]} \omega_1 \left(D_{x-}^{\beta} f, \frac{T}{n^{1-\alpha}} \right)_{(-\infty,x]} \right\},$$

(2) in case of $N = 1$, we obtain

$$\|K_n(f) - f\|_{\infty,[-T,T]} \leq \frac{T^{\beta}}{\Gamma(\beta+1) n^{(1-\alpha)\beta}} \cdot \tag{7.92}$$

$$\left\{ \sup_{x\in[-T,T]} \omega_1 \left(D_{*x}^{\beta} f, \frac{T}{n^{1-\alpha}} \right)_{[x,+\infty)} + \sup_{x\in[-T,T]} \omega_1 \left(D_{x-}^{\beta} f, \frac{T}{n^{1-\alpha}} \right)_{(-\infty,x]} \right\}.$$

An interesting case is when $\beta = \frac{1}{2}$.

Here we get fractionally with rates the uniform convergence of $K_n(f) \to f$, *as* $n \to \infty$.

Proof From (7.65), (7.68) of Theorem 7.29, and by Remark 7.17. ■

One can also apply Remark 7.18 to the last Theorem 7.30, to get interesting and simplified results.

Note 7.31 *The maps* H_n, K_n, $n \in \mathbb{N}$, *are positive linear operators reproducing constants.*

References

1. G.A. Anastassiou, Rate of convergence of some neural network operators to the unit-univariate case. J. Math. Anal. Appl. **212**, 237–262 (1997)
2. G.A. Anastassiou, On right fractional calculus. Chaos, Solitons Fractals **42**, 365–376 (2009)
3. G.A. Anastassiou, Fractional korovkin theory. Chaos, Solitons Fractals **42**(4), 2080–2094 (2009)
4. G.A. Anastassiou, *Fractional Differentiation Inequalities* (Springer, New York, 2009)
5. G.A. Anastassiou, Quantitative approximation by fractional smooth picard singular operators. Math. Eng. Sci. Aerosp. **2**(1), 71–87 (2011)
6. G.A. Anastassiou, Fractional representation formulae and right fractional inequalities. Math. Comput. Model. **54**(11–12), 3098–3115 (2011)
7. G.A. Anastassiou, Fractional approximation by normalized bell and squashing type neural network operators. New Math. Nat. Comput. **9**(1), 43–63 (2013)
8. P. Cardaliaguet, G. Euvrard, Approximation of a function and its derivative with a neural network. Neural Netw. **5**, 207–220 (1992)
9. K. Diethelm, *The Analysis of Fractional Differential Equations*, Lecture Notes in Mathematics, vol. 2004 (Springer, Heidelberg, 2010)
10. A.M.A. El-Sayed, M. Gaber, On the finite Caputo and finite Riesz derivatives. Electron. J. Theor. Phys. **3**(12), 81–95 (2006)
11. G.S. Frederico, D.F.M. Torres, Fractional optimal control in the sense of Caputo and the fractional Noether's theorem. Int. Math. Forum **3**(10), 479–493 (2008)

Chapter 8
Fractional Voronovskaya Type Asymptotic Expansions for Bell and Squashing Type Neural Networks

Here we introduce the normalized bell and squashing type neural network operators of one hidden layer. Based on fractional calculus theory we derive fractional Voronovskaya type asymptotic expansions for the error of approximation of these operators to the unit operator. It follows [7].

8.1 Background

We need

Definition 8.1 Let $f : \mathbb{R} \to \mathbb{R}$, $\nu > 0$, $n = \lceil \nu \rceil$ ($\lceil \cdot \rceil$ is the ceiling of the number), such that $f \in AC^n([a, b])$ (space of functions f with $f^{(n-1)} \in AC([a, b])$, absolutely continuous functions), $\forall [a, b] \subset \mathbb{R}$. We call left Caputo fractional derivative (see [9], pp. 49–52) the function

$$D_{*a}^{\nu} f(x) = \frac{1}{\Gamma(n-\nu)} \int_a^x (x-t)^{n-\nu-1} f^{(n)}(t)\, dt, \tag{8.1}$$

$\forall x \geq a$, where Γ is the gamma function $\Gamma(\nu) = \int_0^{\infty} e^{-t} t^{\nu-1} dt$, $\nu > 0$. Notice $D_{*a}^{\nu} f \in L_1([a, b])$ and $D_{*a}^{\nu} f$ exists a.e. on $[a, b]$, $\forall b > a$.

We set $D_{*a}^{0} f(x) = f(x)$, $\forall x \in [a, +\infty)$.

We also need

Definition 8.2 (*see also* [2, 10, 11]). Let $f : \mathbb{R} \to \mathbb{R}$, such that $f \in AC^m([a, b])$, $\forall [a, b] \subset \mathbb{R}$, $m = \lceil \alpha \rceil$, $\alpha > 0$. The right Caputo fractional derivative of order $\alpha > 0$ is given by

$$D_{b-}^{\alpha} f(x) = \frac{(-1)^m}{\Gamma(m-\alpha)} \int_x^b (J-x)^{m-\alpha-1} f^{(m)}(J)\, dJ, \tag{8.2}$$

G.A. Anastassiou, *Intelligent Systems II: Complete Approximation by Neural Network Operators*, Studies in Computational Intelligence 608,
DOI 10.1007/978-3-319-20505-2_8

$\forall x \leq b$. We set $D_{b-}^{0} f(x) = f(x), \forall x \in (-\infty, b]$. Notice that $D_{b-}^{\alpha} f \in L_1([a, b])$ and $D_{b-}^{\alpha} f$ exists a.e. on $[a, b]$, $\forall a < b$.

We mention the left Caputo fractional Taylor formula with integral remainder.

Theorem 8.3 ([9], p. 54) *Let $f \in AC^m([a, b])$, $\forall [a, b] \subset \mathbb{R}$, $m = \lceil \alpha \rceil$, $\alpha > 0$. Then*

$$f(x) = \sum_{k=0}^{m-1} \frac{f^{(k)}(x_0)}{k!}(x - x_0)^k + \frac{1}{\Gamma(\alpha)} \int_{x_0}^{x} (x - J)^{\alpha-1} D_{*x_0}^{\alpha} f(J) \, dJ, \quad (8.3)$$

$\forall x \geq x_0$.

Also we mention the right Caputo fractional Taylor formula.

Theorem 8.4 ([2]) *Let $f \in AC^m([a, b])$, $\forall [a, b] \subset \mathbb{R}$, $m = \lceil \alpha \rceil$, $\alpha > 0$. Then*

$$f(x) = \sum_{k=0}^{m-1} \frac{f^{(k)}(x_0)}{k!}(x - x_0)^k + \frac{1}{\Gamma(\alpha)} \int_{x}^{x_0} (J - x)^{\alpha-1} D_{x_0-}^{\alpha} f(J) \, dJ, \quad (8.4)$$

$\forall x \leq x_0$.

Convention 8.5 *We assume that*

$$D_{*x_0}^{\alpha} f(x) = 0, \text{ for } x < x_0,$$

and

$$D_{x_0-}^{\alpha} f(x) = 0, \text{ for } x > x_0,$$

for all $x, x_0 \in \mathbb{R}$.

We mention

Proposition 8.6 (by [3]) *(i) Let $f \in C^n(\mathbb{R})$, where $n = \lceil \nu \rceil$, $\nu > 0$. Then $D_{*a}^{\nu} f(x)$ is continuous in $x \in [a, \infty)$.*

(ii) Let $f \in C^m(\mathbb{R})$, $m = \lceil \alpha \rceil$, $\alpha > 0$. Then $D_{b-}^{\alpha} f(x)$ is continuous in $x \in (-\infty, b]$.

We also mention

Theorem 8.7 ([5]) *Let $f \in C^m(\mathbb{R})$, $f^{(m)} \in L_\infty(\mathbb{R})$, $m = \lceil \alpha \rceil$, $\alpha > 0$, $\alpha \notin \mathbb{N}$, $x, x_0 \in \mathbb{R}$. Then $D_{*x_0}^{\alpha} f(x)$, $D_{x_0-}^{\alpha} f(x)$ are jointly continuous in (x, x_0) from $\mathbb{R}^2$ into $\mathbb{R}$.*

For more see [4, 6].

We need the following (see [8]).

Definition 8.8 A function $b : \mathbb{R} \to \mathbb{R}$ is said to be bell-shaped if b belongs to L^1 and its integral is nonzero, if it is nondecreasing on $(-\infty, a)$ and nonincreasing on $[a, +\infty)$, where a belongs to $\mathbb{R}$. In particular $b(x)$ is a nonnegative number and at a b takes a global maximum; it is the center of the bell-shaped function. A bell-shaped function is said to be centered if its center is zero. The function $b(x)$ may have jump discontinuities. In this work we consider only centered bell-shaped functions of compact support $[-T, T]$, $T > 0$.

Example 8.9 (1) $b(x)$ can be the characteristic function over $[-1, 1]$.

(2) $b(x)$ can be the hat function over $[-1, 1]$, i.e.,

$$b(x) = \begin{cases} 1 + x, \ -1 \le x \le 0, \\ 1 - x, \ 0 < x \le 1 \\ 0, \textit{elsewhere}. \end{cases}$$

Here we consider functions $f \in C(\mathbb{R})$.

We study the following "normalized bell type neural network operators" (see also related [1, 8])

$$(H_n(f))(x) := \frac{\sum_{k=-n^2}^{n^2} f\left(\frac{k}{n}\right) b\left(n^{1-\alpha}\left(x - \frac{k}{n}\right)\right)}{\sum_{k=-n^2}^{n^2} b\left(n^{1-\alpha}\left(x - \frac{k}{n}\right)\right)}, \tag{8.5}$$

where $0 < \alpha < 1$ and $x \in \mathbb{R}$, $n \in \mathbb{N}$.

We find a fractional Voronovskaya type asymptotic expansion for $H_n(f)(x)$.

The terms in $H_n(f)(x)$ are nonzero iff

$$\left|n^{1-\alpha}\left(x - \frac{k}{n}\right)\right| \le T \text{ , i.e. } \left|x - \frac{k}{n}\right| \le \frac{T}{n^{1-\alpha}}$$

iff

$$nx - Tn^{\alpha} \le k \le nx + Tn^{\alpha}. \tag{8.6}$$

In order to have the desired order of numbers

$$-n^2 \le nx - Tn^{\alpha} \le nx + Tn^{\alpha} \le n^2, \tag{8.7}$$

it is sufficient enough to assume that

$$n \ge T + |x|. \tag{8.8}$$

When $x \in [-T, T]$ it is enough to assume $n \ge 2T$ which implies (8.7).

Proposition 8.10 (see [1]) *Let $a \le b$, $a, b \in \mathbb{R}$. Let $card(k)$ (≥ 0) be the maximum number of integers contained in $[a, b]$. Then*

$$\max(0, (b-a)-1) \leq card\,(k) \leq (b-a)+1. \tag{8.9}$$

Remark 8.11 We would like to establish a lower bound on $card\,(k)$ over the interval $[nx - Tn^\alpha, nx + Tn^\alpha]$. From Proposition 8.10 we get that

$$card\,(k) \geq \max\left(2Tn^\alpha - 1, 0\right).$$

We obtain $card\,(k) \geq 1$, if

$$2Tn^\alpha - 1 \geq 1 \text{ iff } n \geq T^{-\frac{1}{\alpha}}.$$

So to have the desired order (8.7) and $card\,(k) \geq 1$ over $[nx - Tn^\alpha, nx + Tn^\alpha]$, we need to consider

$$n \geq \max\left(T + |x|, T^{-\frac{1}{\alpha}}\right). \tag{8.10}$$

Also notice that $card\,(k) \to +\infty$, as $n \to +\infty$.

Denote by $[\cdot]$ the integral part of a number.

Remark 8.12 Clearly we have that

$$nx - Tn^\alpha \leq nx \leq nx + Tn^\alpha. \tag{8.11}$$

We prove in general that

$$nx - Tn^\alpha \leq [nx] \leq nx \leq \lceil nx \rceil \leq nx + Tn^\alpha. \tag{8.12}$$

Indeed we have that, if $[nx] < nx - Tn^\alpha$, then $[nx] + Tn^\alpha < nx$, and $[nx] + [Tn^\alpha] \leq [nx]$, resulting into $[Tn^\alpha] = 0$, which for large enough n is not true. Therefore $nx - Tn^\alpha \leq [nx]$. Similarly, if $\lceil nx \rceil > nx + Tn^\alpha$, then $nx + Tn^\alpha \geq nx + [Tn^\alpha]$, and $\lceil nx \rceil - [Tn^\alpha] > nx$, thus $\lceil nx \rceil - [Tn^\alpha] \geq \lceil nx \rceil$, resulting into $[Tn^\alpha] = 0$, which again for large enough n is not true.

Therefore without loss of generality we may assume that

$$nx - Tn^\alpha \leq [nx] \leq nx \leq \lceil nx \rceil \leq nx + Tn^\alpha. \tag{8.13}$$

Hence $\lceil nx - Tn^\alpha \rceil \leq [nx]$ and $\lceil nx \rceil \leq [nx + Tn^\alpha]$. Also if $[nx] \neq \lceil nx \rceil$, then $\lceil nx \rceil = [nx] + 1$. If $[nx] = \lceil nx \rceil$, then $nx \in \mathbb{Z}$; and by assuming $n \geq T^{-\frac{1}{\alpha}}$, we get $Tn^\alpha \geq 1$ and $nx + Tn^\alpha \geq nx + 1$, so that $[nx + Tn^\alpha] \geq nx + 1 = [nx] + 1$.

We need also

Definition 8.13 Let the nonnegative function $S : \mathbb{R} \to \mathbb{R}$, S has compact support $[-T, T]$, $T > 0$, and is nondecreasing there and it can be continuous only on either $(-\infty, T]$ or $[-T, T]$, S can have jump discontinuites. We call S the "squashing function", see [1, 8].

Let $f \in C(\mathbb{R})$. For $x \in \mathbb{R}$ we define the following "normalized squashing type neural network operators" (see also related [1])

$$(K_n(f))(x) := \frac{\sum_{k=-n^2}^{n^2} f\left(\frac{k}{n}\right) S\left(n^{1-\alpha}\left(x-\frac{k}{n}\right)\right)}{\sum_{k=-n^2}^{n^2} S\left(n^{1-\alpha}\left(x-\frac{k}{n}\right)\right)}, \tag{8.14}$$

$0 < \alpha < 1$ and $n \in \mathbb{N} : n \geq \max\left(T + |x|, T^{-\frac{1}{\alpha}}\right)$.

It is clear that

$$(K_n(f))(x) := \frac{\sum_{k=\lceil nx-Tn^{\alpha}\rceil}^{[nx+Tn^{\alpha}]} f\left(\frac{k}{n}\right) S\left(n^{1-\alpha}\left(x-\frac{k}{n}\right)\right)}{\sum_{k=\lceil nx-Tn^{\alpha}\rceil}^{[nx+Tn^{\alpha}]} S\left(n^{1-\alpha}\left(x-\frac{k}{n}\right)\right)}. \tag{8.15}$$

We find a fractional Voronovskaya type asymptotic expansion for $(K_n(f))(x)$.

8.2 Main Results

We present our first main result.

Theorem 8.14 *Let $\beta > 0$, $N \in \mathbb{N}$, $N = \lceil \beta \rceil$, $f \in AC^N([a,b])$, $\forall [a,b] \subset \mathbb{R}$, with $\left\| D_{x_0-}^{\beta} f \right\|_{\infty}$, $\left\| D_{*x_0}^{\beta} f \right\|_{\infty} \leq M$, $M > 0$, $x_0 \in \mathbb{R}$. Let $T > 0$, $n \in \mathbb{N} : n \geq \max\left(T + |x_0|, T^{-\frac{1}{\alpha}}\right)$ Then*

$$(H_n(f))(x_0) - f(x_0) = \sum_{j=1}^{N-1} \frac{f^{(j)}(x_0)}{j!} H_n\left((\cdot - x_0)^j\right)(x_0) + o\left(\frac{1}{n^{(1-\alpha)(\beta-\varepsilon)}}\right), \tag{8.16}$$

where $0 < \varepsilon \leq \beta$.

If $N = 1$, the sum in (8.16) *disappears.*

The last (8.16) *implies that*

$$n^{(1-\alpha)(\beta-\varepsilon)} \left[(H_n(f))(x_0) - f(x_0) - \sum_{j=1}^{N-1} \frac{f^{(j)}(x_0)}{j!} H_n\left((\cdot - x_0)^j\right)(x_0) \right] \to 0, \tag{8.17}$$

as $n \to \infty$, $0 < \varepsilon \leq \beta$.

When $N = 1$, or $f^{(j)}(x_0) = 0$, $j = 1, \ldots, N-1$, then we derive

$$n^{(1-\alpha)(\beta-\varepsilon)} \left[(H_n(f))(x_0) - f(x_0) \right] \to 0$$

as $n \to \infty$, $0 < \varepsilon \leq \beta$. Of great interest is the case of $\beta = \frac{1}{2}$.

Proof From [9], p. 54; (3), we get by the left Caputo fractional Taylor formula that

$$f\left(\frac{k}{n}\right)=\sum_{j=0}^{N-1}\frac{f^{(j)}(x_0)}{j!}\left(\frac{k}{n}-x_0\right)^j+\frac{1}{\Gamma(\beta)}\int_{x_0}^{\frac{k}{n}}\left(\frac{k}{n}-J\right)^{\beta-1}D_{*x_0}^{\beta}f(J)\,dJ, \tag{8.18}$$

for all $x_0\leq\frac{k}{n}\leq x_0+Tn^{\alpha-1}$, iff $\lceil nx_0\rceil\leq k\leq[nx_0+Tn^{\alpha}]$, where $k\in\mathbb{Z}$.

Also from [2]; (4), using the right Caputo fractional Taylor formula we get

$$f\left(\frac{k}{n}\right)=\sum_{j=0}^{N-1}\frac{f^{(j)}(x_0)}{j!}\left(\frac{k}{n}-x_0\right)^j+\frac{1}{\Gamma(\beta)}\int_{\frac{k}{n}}^{x_0}\left(J-\frac{k}{n}\right)^{\beta-1}D_{x_0-}^{\beta}f(J)\,dJ, \tag{8.19}$$

for all $x_0-Tn^{\alpha-1}\leq\frac{k}{n}\leq x_0$, iff $\lceil nx_0-Tn^{\alpha}\rceil\leq k\leq[nx_0]$, where $k\in\mathbb{Z}$. Notice that $\lceil nx_0\rceil\leq[nx_0]+1$.

Call

$$V(x_0):=\sum_{k=\lceil nx_0-Tn^{\alpha}\rceil}^{[nx_0+Tn^{\alpha}]}b\left(n^{1-\alpha}\left(x_0-\frac{k}{n}\right)\right).$$

Hence we have

$$\frac{f\left(\frac{k}{n}\right)b\left(n^{1-\alpha}\left(x_0-\frac{k}{n}\right)\right)}{V(x_0)}=\sum_{j=0}^{N-1}\frac{f^{(j)}(x_0)}{j!}\left(\frac{k}{n}-x_0\right)^j\frac{b\left(n^{1-\alpha}\left(x_0-\frac{k}{n}\right)\right)}{V(x_0)}+$$
$$\frac{b\left(n^{1-\alpha}\left(x_0-\frac{k}{n}\right)\right)}{V(x_0)\Gamma(\beta)}\int_{x_0}^{\frac{k}{n}}\left(\frac{k}{n}-J\right)^{\beta-1}D_{*x_0}^{\beta}f(J)\,dJ, \tag{8.20}$$

and

$$\frac{f\left(\frac{k}{n}\right)b\left(n^{1-\alpha}\left(x_0-\frac{k}{n}\right)\right)}{V(x_0)}=\sum_{j=0}^{N-1}\frac{f^{(j)}(x_0)}{j!}\left(\frac{k}{n}-x_0\right)^j\frac{b\left(n^{1-\alpha}\left(x_0-\frac{k}{n}\right)\right)}{V(x_0)}+$$
$$\frac{b\left(n^{1-\alpha}\left(x_0-\frac{k}{n}\right)\right)}{V(x_0)\Gamma(\beta)}\int_{\frac{k}{n}}^{x_0}\left(J-\frac{k}{n}\right)^{\beta-1}D_{x_0-}^{\beta}f(J)\,dJ, \tag{8.21}$$

Therefore we obtain

$$\frac{\sum_{k=[nx_0]+1}^{[nx_0+Tn^{\alpha}]}f\left(\frac{k}{n}\right)b\left(n^{1-\alpha}\left(x_0-\frac{k}{n}\right)\right)}{V(x_0)}=$$
$$\sum_{j=0}^{N-1}\frac{f^{(j)}(x_0)}{j!}\left(\frac{\sum_{k=[nx_0]+1}^{[nx_0+Tn^{\alpha}]}\left(\frac{k}{n}-x_0\right)^j b\left(n^{1-\alpha}\left(x_0-\frac{k}{n}\right)\right)}{V(x_0)}\right)+ \tag{8.22}$$

$$\sum_{k=[nx_0]+1}^{[nx_0+Tn^\alpha]} \frac{b\left(n^{1-\alpha}\left(x_0-\frac{k}{n}\right)\right)}{V(x_0)\,\Gamma(\beta)} \int_{x_0}^{\frac{k}{n}} \left(\frac{k}{n}-J\right)^{\beta-1} D_{*x_0}^{\beta} f(J)\,dJ,$$

and

$$\frac{\sum_{k=\lceil nx_0-Tn^\alpha\rceil}^{[nx_0]} f\left(\frac{k}{n}\right) b\left(n^{1-\alpha}\left(x_0-\frac{k}{n}\right)\right)}{V(x_0)} =$$

$$\sum_{j=0}^{N-1} \frac{f^{(j)}(x_0)}{j!} \frac{\sum_{k=\lceil nx_0-Tn^\alpha\rceil}^{[nx_0]} \left(\frac{k}{n}-x_0\right)^j b\left(n^{1-\alpha}\left(x_0-\frac{k}{n}\right)\right)}{V(x_0)} + \tag{8.23}$$

$$\frac{\sum_{k=\lceil nx_0-Tn^\alpha\rceil}^{[nx_0]} b\left(n^{1-\alpha}\left(x_0-\frac{k}{n}\right)\right)}{V(x_0)\,\Gamma(\beta)} \int_{\frac{k}{n}}^{x_0} \left(J-\frac{k}{n}\right)^{\beta-1} D_{x_0-}^{\beta} f(J)\,dJ.$$

We notice here that

$$(H_n(f))(x) := \frac{\sum_{k=-n^2}^{n^2} f\left(\frac{k}{n}\right) b\left(n^{1-\alpha}\left(x-\frac{k}{n}\right)\right)}{\sum_{k=-n^2}^{n^2} b\left(n^{1-\alpha}\left(x-\frac{k}{n}\right)\right)} \tag{8.24}$$

$$= \frac{\sum_{k=\lceil nx-Tn^\alpha\rceil}^{[nx+Tn^\alpha]} f\left(\frac{k}{n}\right) b\left(n^{1-\alpha}\left(x-\frac{k}{n}\right)\right)}{\sum_{k=\lceil nx-Tn^\alpha\rceil}^{[nx+Tn^\alpha]} b\left(n^{1-\alpha}\left(x-\frac{k}{n}\right)\right)}, \ \forall\, x \in \mathbb{R}.$$

Adding the two equalities (8.22), (8.23) and rewriting it, we obtain

$$T(x_0) := (H_n(f))(x_0) - f(x_0) - \sum_{j=1}^{N-1} \frac{f^{(j)}(x_0)}{j!} H_n\left((\cdot - x_0)^j\right)(x_0) = \theta_n^*(x_0), \tag{8.25}$$

where

$$\theta_n^*(x_0) := \frac{\sum_{k=\lceil nx_0-Tn^\alpha\rceil}^{[nx_0]} b\left(n^{1-\alpha}\left(x_0-\frac{k}{n}\right)\right)}{V(x_0)\,\Gamma(\beta)} \int_{\frac{k}{n}}^{x_0} \left(J-\frac{k}{n}\right)^{\beta-1} D_{x_0-}^{\beta} f(J)\,dJ$$

$$+ \sum_{k=[nx_0]+1}^{[nx_0+Tn^\alpha]} \frac{b\left(n^{1-\alpha}\left(x_0-\frac{k}{n}\right)\right)}{V(x_0)\,\Gamma(\beta)} \int_{x_0}^{\frac{k}{n}} \left(\frac{k}{n}-J\right)^{\beta-1} D_{*x_0}^{\beta} f(J)\,dJ. \tag{8.26}$$

We observe that

$$\left|\theta_n^*(x_0)\right| \leq \frac{1}{V(x_0)\Gamma(\beta)}.$$

$$\left\{\sum_{k=\lceil nx_0 - Tn^{\alpha}\rceil}^{[nx_0]} b\left(n^{1-\alpha}\left(x_0 - \frac{k}{n}\right)\right)\int_{\frac{k}{n}}^{x_0}\left(J - \frac{k}{n}\right)^{\beta-1}\left|D_{x_0-}^{\beta} f(J)\right| dJ \right. \tag{8.27}$$

$$\left. + \sum_{k=[nx_0]+1}^{[nx_0+Tn^{\alpha}]} b\left(n^{1-\alpha}\left(x_0 - \frac{k}{n}\right)\right)\int_{x_0}^{\frac{k}{n}}\left(\frac{k}{n} - J\right)^{\beta-1}\left|D_{*x_0}^{\beta} f(J)\right| dJ\right\} \leq$$

$$\frac{M}{V(x_0)\Gamma(\beta)}\left\{\sum_{k=\lceil nx_0 - Tn^{\alpha}\rceil}^{[nx_0]} b\left(n^{1-\alpha}\left(x_0 - \frac{k}{n}\right)\right)\frac{\left(x_0 - \frac{k}{n}\right)^{\beta}}{\beta} + \right.$$

$$\left.\sum_{k=[nx_0]+1}^{[nx_0+Tn^{\alpha}]} b\left(n^{1-\alpha}\left(x_0 - \frac{k}{n}\right)\right)\frac{\left(\frac{k}{n} - x_0\right)^{\beta}}{\beta}\right\} \leq$$

$$\frac{M}{V(x_0)\Gamma(\beta+1)}\left\{\left(\sum_{k=\lceil nx_0 - Tn^{\alpha}\rceil}^{[nx_0]} b\left(n^{1-\alpha}\left(x_0 - \frac{k}{n}\right)\right)\right)\left(\frac{T}{n^{1-\alpha}}\right)^{\beta} + \right.$$

$$\left.\left(\sum_{k=[nx_0]+1}^{[nx_0+Tn^{\alpha}]} b\left(n^{1-\alpha}\left(x_0 - \frac{k}{n}\right)\right)\right)\left(\frac{T}{n^{1-\alpha}}\right)^{\beta}\right\} = \frac{M}{\Gamma(\beta+1)}\frac{T^{\beta}}{n^{(1-\alpha)\beta}}. \tag{8.28}$$

So we have proved that

$$|T(x_0)| = \left|\theta_n^*(x_0)\right| \leq \left(\frac{MT^{\beta}}{\Gamma(\beta+1)}\right)\left(\frac{1}{n^{(1-\alpha)\beta}}\right), \tag{8.29}$$

resulting to

$$|T(x_0)| = O\left(\frac{1}{n^{(1-\alpha)\beta}}\right), \tag{8.30}$$

and

$$|T(x_0)| = o(1). \tag{8.31}$$

And, letting $0 < \varepsilon \leq \beta$, we derive

$$\frac{|T(x_0)|}{\left(\frac{1}{n^{(1-\alpha)(\beta-\varepsilon)}}\right)} \leq \frac{MT^{\beta}}{\Gamma(\beta+1)} \left(\frac{1}{n^{(1-\alpha)\varepsilon}}\right) \to 0, \tag{8.32}$$

as $n \to \infty$.

I.e.

$$|T(x_0)| = o\left(\frac{1}{n^{(1-\alpha)(\beta-\varepsilon)}}\right), \tag{8.33}$$

proving the claim. ■

Our second main result follows

Theorem 8.15 *Same assumptions as in Theorem 8.14. Then*

$$(K_n(f))(x_0) - f(x_0) = \sum_{j=1}^{N-1} \frac{f^{(j)}(x_0)}{j!} K_n\left((\cdot - x_0)^j\right)(x_0) + o\left(\frac{1}{n^{(1-\alpha)(\beta-\varepsilon)}}\right), \tag{8.34}$$

where $0 < \varepsilon \leq \beta$.

If $N = 1$, *the sum in (8.34) disappears.*

The last (8.34) implies that

$$n^{(1-\alpha)(\beta-\varepsilon)} \left[(K_n(f))(x_0) - f(x_0) - \sum_{j=1}^{N-1} \frac{f^{(j)}(x_0)}{j!} K_n\left((\cdot - x_0)^j\right)(x_0)\right] \to 0, \tag{8.35}$$

as $n \to \infty$, $0 < \varepsilon \leq \beta$.

When $N = 1$, *or* $f^{(j)}(x_0) = 0$, $j = 1, \ldots, N-1$, *then we derive*

$$n^{(1-\alpha)(\beta-\varepsilon)} \left[(K_n(f))(x_0) - f(x_0)\right] \to 0 \tag{8.36}$$

as $n \to \infty$, $0 < \varepsilon \leq \beta$. *Of great interest is the case of* $\beta = \frac{1}{2}$.

Proof As in Theorem 8.14. ■

References

1. G.A. Anastassiou, Rate of convergence of some neural network operators to the unit-univariate case. J. Math. Anal. Appl. **212**, 237–262 (1997)
2. G.A. Anastassiou, On right fractional calculus. Chaos, Solitons Fractals **42**, 365–376 (2009)
3. G.A. Anastassiou, Fractional Korovkin theory. Chaos, Solitons Fractals **42**(4), 2080–2094 (2009)
4. G.A. Anastassiou, *Fractional Differentiation Inequalities* (Springer, New York, 2009)

5. G.A. Anastassiou, Quantitative approximation by fractional smooth picard singular operators. Math. Eng. Sci. Aerosp. **2**(1), 71–87 (2011)
6. G.A. Anastassiou, Fractional representation formulae and right fractional inequalities. Math. Comput. Model. **54**(11–12), 3098–3115 (2011)
7. G.A. Anastassiou, Fractional Voronovskaya type asymptotic expansions for bell and squashing type neural network operators. J. Comput. Anal. Appl. **15**(7), 1231–1239 (2013)
8. P. Cardaliaguet, G. Euvrard, Approximation of a function and its derivative with a neural network. Neural Netw. **5**, 207–220 (1992)
9. K. Diethelm, *The Analysis of Fractional Differential Equations*, Lecture Notes in Mathematics 2004 (Springer, Berlin, 2010)
10. A.M.A. El-Sayed, M. Gaber, On the finite Caputo and finite Riesz derivatives. Electron. J. Theor. Phys. **3**(12), 81–95 (2006)
11. G.S. Frederico, D.F.M. Torres, Fractional optimal control in the sense of Caputo and the fractional Noether's theorem. Int. Math. Forum **3**(10), 479–493 (2008)

Chapter 9
Multivariate Voronovskaya Type Asymptotic Expansions for Normalized Bell and Squashing Type Neural Networks

Here we introduce the multivariate normalized bell and squashing type neural network operators of one hidden layer. We derive multivariate Voronovskaya type asymptotic expansions for the error of approximation of these operators to the unit operator. It follows [6].

9.1 Background

In [7] the authors presented for the fist time approximation of functions by specific completely described neural network operators. However their approach was only qualitative. The author in [1, 2] continued the work of [7] by presenting for the first time quantitative approximation by determining the rate of convergence and involving the modulus of continuity of the function under approximation. In this chapter we engage very flexible neural network operators that derive by normalization of operators of [7], so we are able to produce asymptotic expansions of Voronovkaya type regarding the approximation of these operators to the unit operator.

We use the following (see [7]).

Definition 9.1 A function $b : \mathbb{R} \to \mathbb{R}$ is said to be bell-shaped if b belongs to L^1 and its integral is nonzero, if it is nondecreasing on $(-\infty, a)$ and nonincreasing on $[a, +\infty)$, where a belongs to $\mathbb{R}$. In particular $b(x)$ is a nonnegative number and at a, b takes a global maximum; it is the center of the bell-shaped function. A bell-shaped function is said to be centered if its center is zero.

Definition 9.2 (*see* [7]) A function $b : \mathbb{R}^d \to \mathbb{R}$ $(d \geq 1)$ is said to be a d-dimensional bell-shaped function if it is integrable and its integral is not zero, and for all $i = 1, \ldots, d$,

$$t \to b(x_1, \ldots, t, \ldots, x_d)$$

is a centered bell-shaped function, where $\overrightarrow{x} := (x_1, \ldots, x_d) \in \mathbb{R}^d$ arbitrary.

G.A. Anastassiou, *Intelligent Systems II: Complete Approximation by Neural Network Operators*, Studies in Computational Intelligence 608,
DOI 10.1007/978-3-319-20505-2_9

Example 9.3 (from [7]) Let b be a centered bell-shaped function over $\mathbb{R}$, then $(x_1, \ldots, x_d) \to b(x_1) \ldots b(x_d)$ is a d-dimensional bell-shaped function.

Assumption 9.4 Here $b\left(\overrightarrow{x}\right)$ is of compact support $\mathcal{B} := \prod_{i=1}^{d} [-T_i, T_i]$, $T_i > 0$ and it may have jump discontinuities there.

Let $f \in C\left(\mathbb{R}^d\right)$.

In this chapter we find a multivariate Voronovskaya type asymptotic expansion for the multivariate normalized bell type neural network operators,

$$M_n(f)\left(\overrightarrow{x}\right) :=$$

$$\frac{\sum_{k_1=-n^2}^{n^2} \cdots \sum_{k_d=-n^2}^{n^2} f\left(\frac{k_1}{n}, \ldots, \frac{k_d}{n}\right) b\left(n^{1-\beta}\left(x_1 - \frac{k_1}{n}\right), \ldots, n^{1-\beta}\left(x_d - \frac{k_d}{n}\right)\right)}{\sum_{k_1=-n^2}^{n^2} \cdots \sum_{k_d=-n^2}^{n^2} b\left(n^{1-\beta}\left(x_1 - \frac{k_1}{n}\right), \ldots, n^{1-\beta}\left(x_d - \frac{k_d}{n}\right)\right)}, \tag{9.1}$$

where $0 < \beta < 1$ and $\overrightarrow{x} := (x_1, \ldots, x_d) \in \mathbb{R}^d$, $n \in \mathbb{N}$. Clearly M_n is a positive linear operator.

The terms in the ratio of multiple sums (9.1) can be nonzero iff simultaneously

$$\left|n^{1-\beta}\left(x_i - \frac{k_i}{n}\right)\right| \leq T_i, \text{ all } i = 1, \ldots, d$$

i.e., $\left|x_i - \frac{k_i}{n}\right| \leq \frac{T_i}{n^{1-\beta}}$, all $i = 1, \ldots, d$, iff

$$nx_i - T_i n^{\beta} \leq k_i \leq nx_i + T_i n^{\beta}, \text{ all } i = 1, \ldots, d. \tag{9.2}$$

To have the order

$$-n^2 \leq nx_i - T_i n^{\beta} \leq k_i \leq nx_i + T_i n^{\beta} \leq n^2, \tag{9.3}$$

we need $n \geq T_i + |x_i|$, all $i = 1, \ldots, d$. So (9.3) is true when we consider

$$n \geq \max_{i \in \{1,\ldots,d\}} (T_i + |x_i|). \tag{9.4}$$

When $\overrightarrow{x} \in \mathcal{B}$ in order to have (9.3) it is enough to suppose that $n \geq 2T^*$, where $T^* := \max\{T_1, \ldots, T_d\} > 0$. Take

$$\widetilde{I}_i := \left[nx_i - T_i n^{\beta}, nx_i + T_i n^{\beta}\right], \ i = 1, \ldots, d, n \in \mathbb{N}.$$

The length of $\widetilde{I}_i$ is $2T_i n^{\beta}$. By Proposition 2.1, p. 61 of [3], we obtain that the cardinality of $k_i \in \mathbb{Z}$ that belong to $\widetilde{I}_i := card(k_i) \geq \max\left(2T_i n^{\beta} - 1, 0\right)$, any $i \in \{1, \ldots, d\}$. In order to have $card(k_i) \geq 1$ we need $2T_i n^{\beta} - 1 \geq 1$ iff $n \geq T_i^{-\frac{1}{\beta}}$, any $i \in \{1, \ldots, d\}$.

Therefore, a sufficient condition for causing the order (9.3) along with the interval $\widetilde{I}_i$ to contain at least one integer for all $i = 1, \ldots, d$ is that

$$n \geq \max_{i \in \{1,\ldots,d\}} \left\{ T_i + |x_i|, T_i^{-\frac{1}{\beta}} \right\}. \tag{9.5}$$

Clearly as $n \to +\infty$ we get that $card(k_i) \to +\infty$, all $i = 1, \ldots, d$. Also notice that $card(k_i)$ equals to the cardinality of integers in $\left[\lceil nx_i - T_i n^\beta \rceil, [nx_i + T_i n^\beta]\right]$ for all $i = 1, \ldots, d$.

Here we denote by $\lceil \cdot \rceil$ the ceiling of the number, and by $[\cdot]$ we denote the integral part.

From now on in this chapter we assume (9.5). Therefore

$$\left(M_n(f)\right)\left(\overrightarrow{x}\right) = \tag{9.6}$$

$$\frac{\sum_{k_1=\lceil nx_1-T_1n^\beta\rceil}^{[nx_1+T_1n^\beta]} \cdots \sum_{k_d=\lceil nx_d-T_dn^\beta\rceil}^{[nx_d+T_dn^\beta]} f\left(\frac{k_1}{n}, \ldots, \frac{k_d}{n}\right) b\left(n^{1-\beta}\left(x_1 - \frac{k_1}{n}\right), \ldots, n^{1-\beta}\left(x_d - \frac{k_d}{n}\right)\right)}{\sum_{k_1=\lceil nx_1-T_1n^\beta\rceil}^{[nx_1+T_1n^\beta]} \cdots \sum_{k_d=\lceil nx_d-T_dn^\beta\rceil}^{[nx_d+T_dn^\beta]} b\left(n^{1-\beta}\left(x_1 - \frac{k_1}{n}\right), \ldots, n^{1-\beta}\left(x_d - \frac{k_d}{n}\right)\right)}$$

all $\overrightarrow{x} := (x_1, \ldots, x_d) \in \mathbb{R}^d$.

In brief we write

$$\left(M_n(f)\right)\left(\overrightarrow{x}\right) = \frac{\sum_{\overrightarrow{k}=\lceil n\overrightarrow{x}-\overrightarrow{T}n^\beta\rceil}^{[n\overrightarrow{x}+\overrightarrow{T}n^\beta]} f\left(\frac{\overrightarrow{k}}{n}\right) b\left(n^{1-\beta}\left(\overrightarrow{x} - \frac{\overrightarrow{k}}{n}\right)\right)}{\sum_{\overrightarrow{k}=\lceil n\overrightarrow{x}-\overrightarrow{T}n^\beta\rceil}^{[n\overrightarrow{x}+\overrightarrow{T}n^\beta]} b\left(n^{1-\beta}\left(\overrightarrow{x} - \frac{\overrightarrow{k}}{n}\right)\right)}, \tag{9.7}$$

all $\overrightarrow{x} \in \mathbb{R}^d$.

Denote by $\|\cdot\|_\infty$ the maximum norm on $\mathbb{R}^d$, $d \geq 1$. So if $\left|n^{1-\beta}\left(x_i - \frac{k_i}{n}\right)\right| \leq T_i$, all $i = 1, \ldots, d$, we find that

$$\left\|\overrightarrow{x} - \frac{\overrightarrow{k}}{n}\right\|_\infty \leq \frac{T^*}{n^{1-\beta}}, \tag{9.8}$$

where $\overrightarrow{k} := (k_1, \ldots, k_d)$.

We also need

Definition 9.5 Let the nonnegative function $S : \mathbb{R}^d \to \mathbb{R}$, $d \geq 1$, S has compact support $\mathcal{B} := \prod_{i=1}^d [-T_i, T_i]$, $T_i > 0$ and is nondecreasing for each coordinate. S can be continuous only on either $\prod_{i=1}^d (-\infty, T_i]$ or $\mathcal{B}$ and can have jump discontinuities. We call S the multivariate "squashing function" (see also [7]).

Example 9.6 Let $\widehat{S}$ as above when $d = 1$. Then

$$S\left(\overrightarrow{x}\right) := \widehat{S}(x_1)\ldots\widehat{S}(x_d), \ \overrightarrow{x} := (x_1, \ldots, x_d) \in \mathbb{R}^d,$$

is a multivariate "squashing function".

Let $f \in C\left(\mathbb{R}^d\right)$.

For $\overrightarrow{x} \in \mathbb{R}^d$ we define also the "multivariate normalized squashing type neural network operators",

$$L_n(f)\left(\overrightarrow{x}\right) :=$$

$$\frac{\sum_{k_1=-n^2}^{n^2}\cdots\sum_{k_d=-n^2}^{n^2} f\left(\frac{k_1}{n},\ldots,\frac{k_d}{n}\right) S\left(n^{1-\beta}\left(x_1-\frac{k_1}{n}\right),\ldots,n^{1-\beta}\left(x_d-\frac{k_d}{n}\right)\right)}{\sum_{k_1=-n^2}^{n^2}\cdots\sum_{k_d=-n^2}^{n^2} S\left(n^{1-\beta}\left(x_1-\frac{k_1}{n}\right),\ldots,n^{1-\beta}\left(x_d-\frac{k_d}{n}\right)\right)}. \quad (9.9)$$

We also here find a multivariate Voronovskaya type asymptotic expansion for $(L_n(f))\left(\overrightarrow{x}\right)$.

Here again $0 < \beta < 1$ and $n \in \mathbb{N}$:

$$n \geq \max_{i \in \{1,\ldots,d\}} \left\{T_i + |x_i|, T_i^{-\frac{1}{\beta}}\right\},$$

and L_n is a positive linear operator. It is clear that

$$\left(L_n(f)\right)\left(\overrightarrow{x}\right) = \frac{\sum_{\overrightarrow{k}=\left\lceil n\overrightarrow{x}-\overrightarrow{T}n^{\beta}\right\rceil}^{\left[n\overrightarrow{x}+\overrightarrow{T}n^{\beta}\right]} f\left(\frac{\overrightarrow{k}}{n}\right) S\left(n^{1-\beta}\left(\overrightarrow{x}-\frac{\overrightarrow{k}}{n}\right)\right)}{\sum_{\overrightarrow{k}=\left\lceil n\overrightarrow{x}-\overrightarrow{T}n^{\beta}\right\rceil}^{\left[n\overrightarrow{x}+\overrightarrow{T}n^{\beta}\right]} S\left(n^{1-\beta}\left(\overrightarrow{x}-\frac{\overrightarrow{k}}{n}\right)\right)}. \quad (9.10)$$

For related articles on neural networks approximation, see [1–3, 5]. For neural networks in general, see [8–10].

Next we follow [4], pp. 284–286.

About Multivariate Taylor Formula and Estimates

Let $\mathbb{R}^d$; $d \geq 2$; $z := (z_1, \ldots, z_d)$, $x_0 := (x_{01}, \ldots, x_{0d}) \in \mathbb{R}^d$. We consider the space of functions $AC^N\left(\mathbb{R}^d\right)$ with $f : \mathbb{R}^d \to \mathbb{R}$ be such that all partial derivatives of order $(N-1)$ are coordinatewise absolutely continuous functions on compacta, $N \in \mathbb{N}$. Also $f \in C^{N-1}\left(\mathbb{R}^d\right)$. Each Nth order partial derivative is denoted by $f_\alpha := \frac{\partial^\alpha f}{\partial x^\alpha}$, where $\alpha := (\alpha_1, \ldots, \alpha_d)$, $\alpha_i \in \mathbb{Z}^+$, $i = 1, \ldots, d$ and $|\alpha| := \sum_{i=1}^d \alpha_i = N$. Consider $g_z(t) := f(x_0 + t(z - x_0))$, $t \geq 0$. Then

$$g_z^{(j)}(t) = \left[\left(\sum_{i=1}^{d}(z_i - x_{0i})\frac{\partial}{\partial x_i}\right)^j f\right](x_{01} + t(z_1 - x_{01}), \ldots, x_{0d} + t(z_N - x_{0d})), \tag{9.11}$$

for all $j = 0, 1, 2, \ldots, N$.

Example 9.7 Let $d = N = 2$. Then

$$g_z(t) = f(x_{01} + t(z_1 - x_{01}), x_{02} + t(z_2 - x_{02})), \ t \in \mathbb{R},$$

and

$$g_z'(t) = (z_1 - x_{01})\frac{\partial f}{\partial x_1}(x_0 + t(z - x_0)) + (z_2 - x_{02})\frac{\partial f}{\partial x_2}(x_0 + t(z - x_0)). \tag{9.12}$$

Setting

$$(*) = (x_{01} + t(z_1 - x_{01}), x_{02} + t(z_2 - x_{02})) = (x_0 + t(z - x_0)),$$

we get

$$g_z''(t) = (z_1 - x_{01})^2\frac{\partial f^2}{\partial x_1^2}(*) + (z_1 - x_{01})(z_2 - x_{02})\frac{\partial f^2}{\partial x_2 \partial x_1}(*) +$$

$$(z_1 - x_{01})(z_2 - x_{02})\frac{\partial f^2}{\partial x_1 \partial x_2}(*) + (z_2 - x_{02})^2\frac{\partial f^2}{\partial x_2^2}(*). \tag{9.13}$$

Similarly, we have the general case of $d, N \in \mathbb{N}$ for $g_z^{(N)}(t)$.

We mention the following multivariate Taylor theorem.

Theorem 9.8 *Under the above assumptions we have*

$$f(z_1, \ldots, z_d) = g_z(1) = \sum_{j=0}^{N-1}\frac{g_z^{(j)}(0)}{j!} + R_N(z, 0), \tag{9.14}$$

where

$$R_N(z, 0) := \int_0^1\left(\int_0^{t_1} \ldots \left(\int_0^{t_{N-1}} g_z^{(N)}(t_N)\,dt_N\right)\ldots\right)dt_1, \tag{9.15}$$

or

$$R_N(z, 0) = \frac{1}{(N-1)!}\int_0^1 (1 - \theta)^{N-1} g_z^{(N)}(\theta)\,d\theta. \tag{9.16}$$

Notice that $g_z(0) = f(x_0)$.

We make

Remark 9.9 Assume here that

$$\|f_\alpha\|_{\infty,\mathbb{R}^d,N}^{\max} := \max_{|\alpha|=N} \|f_\alpha\|_{\infty,\mathbb{R}^d} < \infty.$$

Then

$$\left\|g_z^{(N)}\right\|_{\infty,[0,1]} = \left\|\left[\left(\sum_{i=1}^d (z_i - x_{0i}) \frac{\partial}{\partial x_i}\right)^N f\right](x_0 + t(z - x_0))\right\|_{\infty,[0,1]} \leq \tag{9.17}$$

$$\left(\sum_{i=1}^d |z_i - x_{0i}|\right)^N \|f_\alpha\|_{\infty,\mathbb{R}^d,N}^{\max},$$

that is

$$\left\|g_z^{(N)}\right\|_{\infty,[0,1]} \leq \left(\|z - x_0\|_{l_1}\right)^N \|f_\alpha\|_{\infty,\mathbb{R}^d,N}^{\max} < \infty. \tag{9.18}$$

Hence we get by (9.16) that

$$|R_N(z,0)| \leq \frac{\left\|g_z^{(N)}\right\|_{\infty,[0,1]}}{N!} < \infty. \tag{9.19}$$

And it holds

$$|R_N(z,0)| \leq \frac{\left(\|z - x_0\|_{l_1}\right)^N}{N!} \|f_\alpha\|_{\infty,\mathbb{R}^d,N}^{\max}, \tag{9.20}$$

$\forall\, z, x_0 \in \mathbb{R}^d$.

Inequality (9.20) will be an important tool in proving our main results.

9.2 Main Results

We present our first main result.

Theorem 9.10 *Let* $f \in AC^N(\mathbb{R}^d)$, $d \in \mathbb{N} - \{1\}$, $N \in \mathbb{N}$, *with* $\|f_\alpha\|_{\infty,\mathbb{R}^d,N}^{\max} < \infty$. *Here* $n \geq \max_{i \in \{1,\dots,d\}} \left\{T_i + |x_i|, T_i^{-\frac{1}{\beta}}\right\}$, *where* $\overrightarrow{x} \in \mathbb{R}^d$, $0 < \beta < 1$, $n \in \mathbb{N}$, $T_i > 0$. *Then*

$$\left(M_n(f)\right)\left(\overrightarrow{x}\right) - f\left(\overrightarrow{x}\right) =$$

$$\sum_{j=1}^{N-1}\left(\sum_{|\alpha|=j}\left(\frac{f_\alpha\left(\overrightarrow{x}\right)}{\prod_{i=1}^{d}\alpha_i!}\right)M_n\left(\prod_{i=1}^{d}(\cdot - x_i)^{\alpha_i}, \overrightarrow{x}\right)\right) + o\left(\frac{1}{n^{(N-\varepsilon)(1-\beta)}}\right), \quad (9.21)$$

where $0 < \varepsilon \leq N$.

If $N = 1$, *the sum in (9.21) collapses.*

The last (9.21) implies that

$$n^{(N-\varepsilon)(1-\beta)}\left[\left(M_n(f)\right)\left(\overrightarrow{x}\right) - f\left(\overrightarrow{x}\right) - \right. \quad (9.22)$$

$$\left.\sum_{j=1}^{N-1}\left(\sum_{|\alpha|=j}\left(\frac{f_\alpha\left(\overrightarrow{x}\right)}{\prod_{i=1}^{d}\alpha_i!}\right)M_n\left(\prod_{i=1}^{d}(\cdot - x_i)^{\alpha_i}, \overrightarrow{x}\right)\right)\right] \to 0, \text{ as } n \to \infty,$$

$0 < \varepsilon \leq N$.

When $N = 1$, *or* $f_\alpha\left(\overrightarrow{x}\right) = 0$, *all* $\alpha : |\alpha| = j = 1, \ldots, N-1$, *then we derive*

$$n^{(N-\varepsilon)(1-\beta)}\left[\left(M_n(f)\right)\left(\overrightarrow{x}\right) - f\left(\overrightarrow{x}\right)\right] \to 0,$$

as $n \to \infty$, $0 < \varepsilon \leq N$.

Proof Put

$$g_{\frac{\overrightarrow{k}}{n}}(t) := f\left(\overrightarrow{x} + t\left(\frac{\overrightarrow{k}}{n} - \overrightarrow{x}\right)\right), \quad 0 \leq t \leq 1.$$

Then

$$g_{\frac{\overrightarrow{k}}{n}}^{(j)}(t) =$$

$$\left[\left(\sum_{i=1}^{d}\left(\frac{k_i}{n} - x_i\right)\frac{\partial}{\partial x_i}\right)^j f\right]\left(x_1 + t\left(\frac{k_1}{n} - x_1\right), \ldots, x_d + t\left(\frac{k_d}{n} - x_d\right)\right), \quad (9.23)$$

and $g_{\frac{\overrightarrow{k}}{n}}(0) = f\left(\overrightarrow{x}\right)$. By Taylor's formula (9.14), (9.16) we obtain

$$f\left(\frac{k_1}{n}, \ldots, \frac{k_d}{n}\right) = g_{\frac{\overrightarrow{k}}{n}}(1) = \sum_{j=0}^{N-1}\frac{g_{\frac{\overrightarrow{k}}{n}}^{(j)}(0)}{j!} + R_N\left(\frac{\overrightarrow{k}}{n}, 0\right), \quad (9.24)$$

where

$$R_N\left(\frac{\overrightarrow{k}}{n}, 0\right) = \frac{1}{(N-1)!}\int_0^1 (1-\theta)^{N-1} g_{\frac{\overrightarrow{k}}{n}}^{(N)}(\theta)\, d\theta. \quad (9.25)$$

More precisely we can rewrite

$$f\left(\frac{\overrightarrow{k}}{n}\right)-f\left(\overrightarrow{x}\right)=$$

$$\sum_{j=1}^{N-1}\sum_{\substack{\alpha:=(\alpha_1,\dots,\alpha_d),\alpha_i\in\mathbb{Z}^+,\\ i=1,\dots,d,|\alpha|:=\sum_{i=1}^{d}\alpha_i=j}}\left(\frac{1}{\prod_{i=1}^{d}\alpha_i!}\right)\left(\prod_{i=1}^{d}\left(\frac{k_i}{n}-x_i\right)^{\alpha_i}\right)f_\alpha\left(\overrightarrow{x}\right)+R_N\left(\frac{\overrightarrow{k}}{n},0\right), \tag{9.26}$$

where

$$R_N\left(\frac{\overrightarrow{k}}{n},0\right)=N\int_0^1(1-\theta)^{N-1}\sum_{\substack{\alpha:=(\alpha_1,\dots,\alpha_d),\alpha_i\in\mathbb{Z}^+,\\ i=1,\dots,d,|\alpha|:=\sum_{i=1}^{d}\alpha_i=N}}\left(\frac{1}{\prod_{i=1}^{d}\alpha_i!}\right)\cdot$$

$$\left(\prod_{i=1}^{d}\left(\frac{k_i}{n}-x_i\right)^{\alpha_i}\right)f_\alpha\left(\overrightarrow{x}+\theta\left(\frac{\overrightarrow{k}}{n}-\overrightarrow{x}\right)\right)d\theta. \tag{9.27}$$

By (9.20) we get

$$\left|R_N\left(\frac{\overrightarrow{k}}{n},0\right)\right|\leq\frac{\left(\left\|\frac{\overrightarrow{k}}{n}-\overrightarrow{x}\right\|_{l_1}\right)^N}{N!}\left\|f_\alpha\right\|_{\infty,\mathbb{R}^d,N}^{\max}. \tag{9.28}$$

So, since here it holds

$$\left\|\overrightarrow{x}-\frac{\overrightarrow{k}}{n}\right\|_\infty\leq\frac{T^*}{n^{1-\beta}},$$

then

$$\left\|\overrightarrow{x}-\frac{\overrightarrow{k}}{n}\right\|_{l_1}\leq\frac{dT^*}{n^{1-\beta}},$$

and

$$\left|R_N\left(\frac{\overrightarrow{k}}{n},0\right)\right|\leq\frac{d^NT^{*N}}{n^{N(1-\beta)}N!}\left\|f_\alpha\right\|_{\infty,\mathbb{R}^d,N}^{\max}, \tag{9.29}$$

for all $\overrightarrow{k}\in\left\{\left\lceil n\overrightarrow{x}-\overrightarrow{T}n^\beta\right\rceil,\dots,\left[n\overrightarrow{x}+\overrightarrow{T}n^\beta\right]\right\}$.

Call

$$V\left(\overrightarrow{x}\right) := \sum_{\overrightarrow{k}=\left\lceil n\overrightarrow{x}-\overrightarrow{T}n^{\beta}\right\rceil}^{\left[n\overrightarrow{x}+\overrightarrow{T}n^{\beta}\right]} b\left(n^{1-\beta}\left(\overrightarrow{x}-\frac{\overrightarrow{k}}{n}\right)\right). \tag{9.30}$$

We observe for

$$U_n\left(\overrightarrow{x}\right) := \frac{\sum_{\overrightarrow{k}=\left\lceil n\overrightarrow{x}-\overrightarrow{T}n^{\beta}\right\rceil}^{\left[n\overrightarrow{x}+\overrightarrow{T}n^{\beta}\right]} R_N\left(\frac{\overrightarrow{k}}{n},0\right) b\left(n^{1-\beta}\left(\overrightarrow{x}-\frac{\overrightarrow{k}}{n}\right)\right)}{V\left(\overrightarrow{x}\right)}, \tag{9.31}$$

that

$$\left|U_n\left(\overrightarrow{x}\right)\right| \overset{\text{(by (9.29))}}{\leq} \frac{d^N T^{*N}}{n^{N(1-\beta)}N!}\left\|f_\alpha\right\|_{\infty,\mathbb{R}^d,N}^{\max}. \tag{9.32}$$

That is

$$\left|U_n\left(\overrightarrow{x}\right)\right| = O\left(\frac{1}{n^{N(1-\beta)}}\right), \tag{9.33}$$

and

$$\left|U_n\left(\overrightarrow{x}\right)\right| = o\left(1\right). \tag{9.34}$$

And, letting $0 < \varepsilon \leq N$, we derive

$$\frac{\left|U_n\left(\overrightarrow{x}\right)\right|}{\left(\frac{1}{n^{(N-\varepsilon)(1-\beta)}}\right)} \leq \left(\frac{d^N T^{*N}\left\|f_\alpha\right\|_{\infty,\mathbb{R}^d,N}^{\max}}{N!}\right)\frac{1}{n^{\varepsilon(1-\beta)}} \to 0, \tag{9.35}$$

as $n \to \infty$.

I.e.

$$\left|U_n\left(\overrightarrow{x}\right)\right| = o\left(\frac{1}{n^{(N-\varepsilon)(1-\beta)}}\right). \tag{9.36}$$

By (9.26) we get

$$\frac{\sum_{\overrightarrow{k}=\left\lceil n\overrightarrow{x}-\overrightarrow{T}n^{\beta}\right\rceil}^{\left[n\overrightarrow{x}+\overrightarrow{T}n^{\beta}\right]} f\left(\frac{\overrightarrow{k}}{n}\right) b\left(n^{1-\beta}\left(\overrightarrow{x}-\frac{\overrightarrow{k}}{n}\right)\right)}{V\left(\overrightarrow{x}\right)} - f\left(\overrightarrow{x}\right) =$$

$$\sum_{j=1}^{N-1} \sum_{\substack{\alpha:=(\alpha_1,\ldots,\alpha_d),\alpha_i\in\mathbb{Z}^+,\\ i=1,\ldots,d,|\alpha|:=\sum_{i=1}^d \alpha_i=j}} \left(\frac{f_\alpha\left(\overrightarrow{x}\right)}{\prod_{i=1}^d \alpha_i!}\right). \tag{9.37}$$

$$\frac{\left(\sum_{\overrightarrow{k}=\lceil n\overrightarrow{x}-\overrightarrow{T}n^{\beta}\rceil}^{[n\overrightarrow{x}+\overrightarrow{T}n^{\beta}]}\left(\prod_{i=1}^{d}\left(\frac{k_i}{n}-x_i\right)^{\alpha_i}\right)\right)b\left(n^{1-\beta}\left(\overrightarrow{x}-\frac{\overrightarrow{k}}{n}\right)\right)}{V\left(\overrightarrow{x}\right)}+U_n\left(\overrightarrow{x}\right).$$

The last says

$$(M_n(f))\left(\overrightarrow{x}\right)-f\left(\overrightarrow{x}\right)-$$

$$\sum_{j=1}^{N-1}\left(\sum_{|\alpha|=j}\left(\frac{f_{\alpha}\left(\overrightarrow{x}\right)}{\prod_{i=1}^{d}\alpha_i!}\right)M_n\left(\prod_{i=1}^{d}(\cdot-x_i)^{\alpha_i},\overrightarrow{x}\right)\right)=U_n\left(\overrightarrow{x}\right). \tag{9.38}$$

The proof of the theorem is complete. ■

We present our second main result

Theorem 9.11 *Let* $f\in AC^N\left(\mathbb{R}^d\right)$, $d\in\mathbb{N}-\{1\}$, $N\in\mathbb{N}$, *with* $\|f_\alpha\|_{\infty,\mathbb{R}^d,N}^{\max}<\infty$. *Here* $n\geq\max_{i\in\{1,\dots,d\}}\left\{T_i+|x_i|,T_i^{-\frac{1}{\beta}}\right\}$, *where* $\overrightarrow{x}\in\mathbb{R}^d$, $0<\beta<1$, $n\in\mathbb{N}$, $T_i>0$. *Then*

$$(L_n(f))\left(\overrightarrow{x}\right)-f\left(\overrightarrow{x}\right)=$$

$$\sum_{j=1}^{N-1}\left(\sum_{|\alpha|=j}\left(\frac{f_{\alpha}\left(\overrightarrow{x}\right)}{\prod_{i=1}^{d}\alpha_i!}\right)L_n\left(\prod_{i=1}^{d}(\cdot-x_i)^{\alpha_i},\overrightarrow{x}\right)\right)+o\left(\frac{1}{n^{(N-\varepsilon)(1-\beta)}}\right), \tag{9.39}$$

where $0<\varepsilon\leq N$.

If $N=1$, *the sum in (9.39) collapses.*

The last (9.39) implies that

$$n^{(N-\varepsilon)(1-\beta)}\left[(L_n(f))\left(\overrightarrow{x}\right)-f\left(\overrightarrow{x}\right)-\right. \tag{9.40}$$

$$\left.\sum_{j=1}^{N-1}\left(\sum_{|\alpha|=j}\left(\frac{f_{\alpha}\left(\overrightarrow{x}\right)}{\prod_{i=1}^{d}\alpha_i!}\right)L_n\left(\prod_{i=1}^{d}(\cdot-x_i)^{\alpha_i},\overrightarrow{x}\right)\right)\right]\to 0,\ as\ n\to\infty,$$

$0<\varepsilon\leq N$.

When $N=1$, *or* $f_\alpha\left(\overrightarrow{x}\right)=0$, *all* $\alpha:|\alpha|=j=1,\dots,N-1$, *then we derive that*

$$n^{(N-\varepsilon)(1-\beta)}\left[(L_n(f))\left(\overrightarrow{x}\right)-f\left(\overrightarrow{x}\right)\right]\to 0,$$

as $n\to\infty$, $0<\varepsilon\leq N$.

Proof As similar to Theorem 9.10 is omitted. ■

References

1. G.A. Anastassiou, Rate of convergence of some neural network operators to the unit-univariate case. J. Math. Anal. Appl. **212**, 237–262 (1997)
2. G.A. Anastassiou, Rate of convergence of some multivariate neural network operators to the unit. Comput. Math. Appl. **40**(1), 1–19 (2000)
3. G.A. Anastassiou, *Quantitative Approximations* (Chapman & Hall/CRC, Boca Raton, 2001)
4. G.A. Anastassiou, *Advanced Inequalities* (World Scientific Publishing Co., Singapore, 2011)
5. G.A. Anastassiou, *Inteligent Systems: Approximation by Artificial Neural Networks*, Intelligent Systems Reference Library, vol. 19 (Springer, Heidelberg, 2011)
6. G.A. Anastassiou, Multivariate Voronovskaya type asymptotic expansions for normalized bell and squashing type neural network operators. Neural Parallel Sci. Comput. **20**, 1–10 (2012)
7. P. Cardaliaguet, G. Euvrard, Approximation of a function and its derivative with a neural network. Neural Netw. **5**, 207–220 (1992)
8. S. Haykin, *Neural Networks: A Comprehensive Foundation*, 2nd edn. (Prentice Hall, New York, 1998)
9. W. McCulloch, W. Pitts, A logical calculus of the ideas immanent in nervous activity. Bull. Math. Biophy. **7**, 115–133 (1943)
10. T.M. Mitchell, *Machine Learning* (WCB-McGraw-Hill, New York, 1997)

Chapter 10
Multivariate Fuzzy-Random Normalized Neural Network Approximation

In this chapter we study the rate of multivariate pointwise convergence in the q-mean to the Fuzzy-Random unit operator or its perturbation of very precise multivariate normalized Fuzzy-Random neural network operators of Cardaliaguet-Euvrard and "Squashing" types. These multivariate Fuzzy-Ranfom operators arise in a natural and common way among multivariate Fuzzy-Random neural network. These rates are given through multivariate Probabilistic-Jackson type inequalities involving the multivariate Fuzzy-Random modulus of continuity of the engaged multivariate Fuzzy-Random function or its Fuzzy partial derivatives. Also several interesting results in multivariate Fuzzy-Random Analysis are given of independent merit, which are used then in the proof of the main results of the chapter. It follows [7].

10.1 Introduction

Let $(X, \mathcal{B}, P)$ be a probability space. Consider the set of all fuzzy-random variables $\mathcal{L}_F(X, \mathcal{B}, P)$. Let $f : \mathbb{R}^d \to \mathcal{L}_F(X, \mathcal{B}, P)$, $d \in \mathbb{N}$, be a multivariate fuzzy-random function or fuzzy-stochastic process. Here for $\overrightarrow{t} \in \mathbb{R}^d$, $s \in X$ we denote $\left(f\left(\overrightarrow{t}\right)\right)(s) = f\left(\overrightarrow{t}, s\right)$ and actually we have $f : \mathbb{R}^d \times X \to \mathbb{R}_{\mathcal{F}}$, where $\mathbb{R}_{\mathcal{F}}$ is the set of fuzzy real numbers. Let $1 \leq q < +\infty$. Here we consider only multivariate fuzzy-random functions f which are (q-mean) uniformly continuous over $\mathbb{R}^d$. For each $n \in \mathbb{N}$, the multivariate fuzzy-random neural network we deal with has the following structure.

It is a three-layer feed forward network with one hidden layer. It has one input unit and one output unit. The hidden layer has $\left(2n^2 + 1\right)$ processing units. To each pair of connecting units (input to each processing unit) we assign the same weight $n^{1-\alpha}$, $0 < \alpha < 1$. The threshold values $\frac{\overrightarrow{k}}{n^\alpha}$ are one for each processing unit $\overrightarrow{k}$, $\overrightarrow{k} \in \mathbb{Z}^d$. The activation function b (or S) is the same for each processing unit. The Fuzzy-Random

G.A. Anastassiou, *Intelligent Systems II: Complete Approximation by Neural Network Operators*, Studies in Computational Intelligence 608,
DOI 10.1007/978-3-319-20505-2_10

weights associated with the output unit are $f\left(\overrightarrow{\frac{k}{n}}, s\right) \odot \frac{1}{\sum_{\vec{k}=-\vec{n}^2}^{\vec{n}^2} b\left(n^{1-\alpha}\left(\vec{x}-\frac{\vec{k}}{n}\right)\right)}$, one for each processing unit $\vec{k}$, $\odot$ denotes the scalar Fuzzy multiplication.

The above precisely described multivariate Fuzzy-Random neural networks induce some completely described multivariate Fuzzy-Random neural network operators of normalized Cardaliaguet-Euvrard and "Squashing" types.

We study here thoroughly the multivariate Fuzzy-Random pointwise convergence (in q-mean) of these operators to the unit operator and its perturbation. See Theorems 10.35, 10.38 and Comment 10.41. This is done with rates through multivariate Probabilistic-Jackson type inequalities involving Fuzzy-Random moduli of continuity of the engaged Fuzzy-Random function and its Fuzzy partial derivatives.

On the way to establish these main results we produce some new independent and interesting results for multivariate Fuzzy-Random Analysis. The real ordinary theory of the above mentioned operators was presented earlier in [1, 2, 10]. And the fuzzy case was treated in [3]. The fuzzy random case was studied first in [4]. Of course this chapter is strongly motivated from there and is a continuation.

The monumental revolutionizing work of Zadeh [14] is the foundation of this work, as well as another strong motivation. Fuzzyness in Computer Science and Engineering seems one of the main trends today. Also Fuzzyness has penetrated many areas of Mathematics and Statistics. These are other strong reasons for this work.

Our approach is quantitative and recent on the topic, started in [1, 2] and continued in [3, 6, 8]. It determines precisely the rates of convergence through natural very tight inequalities using the measurement of smoothness of the engaged multivariate Fuzzy-Random functions.

10.2 Background

We begin with

Definition 10.1 (*see* [13]) Let $\mu : \mathbb{R} \to [0, 1]$ with the following properties:

(i) is normal, i.e., $\exists\, x_0 \in \mathbb{R} : \mu(x_0) = 1$.
(ii) $\mu(\lambda x + (1-\lambda) y) \geq \min\{\mu(x), \mu(y)\}$, $\forall\, x, y \in \mathbb{R}$, $\forall\, \lambda \in [0, 1]$ (μ is called a convex fuzzy subset).
(iii) μ is upper semicontinuous on $\mathbb{R}$, i.e., $\forall\, x_0 \in \mathbb{R}$ and $\forall\, \varepsilon > 0$, $\exists$ neighborhood $V(x_0) : \mu(x) \leq \mu(x_0) + \varepsilon$, $\forall\, x \in V(x_0)$.
(iv) the set $\overline{\operatorname{supp}(\mu)}$ is compact in $\mathbb{R}$ (where $\operatorname{supp}(\mu) := \{x \in \mathbb{R}; \mu(x) > 0\}$).

We call μ a fuzzy real number. Denote the set of all μ with $\mathbb{R}_{\mathcal{F}}$.

E.g., $\chi_{\{x_0\}} \in \mathbb{R}_{\mathcal{F}}$, for any $x_0 \in \mathbb{R}$, where $\chi_{\{x_0\}}$ is the characteristic function at x_0.

For $0 < r \leq 1$ and $\mu \in \mathbb{R}_{\mathcal{F}}$ define $[\mu]^r := \{x \in \mathbb{R} : \mu(x) \geq r\}$ and $[\mu]^0 := \overline{\{x \in \mathbb{R} : \mu(x) > 0\}}$.

Then it is well known that for each $r \in [0,1]$, $[\mu]^r$ is a closed and bounded interval of $\mathbb{R}$. For $u, v \in \mathbb{R}_{\mathcal{F}}$ and $\lambda \in \mathbb{R}$, we define uniquely the sum $u \oplus v$ and the product $\lambda \odot u$ by

$$[u \oplus v]^r = [u]^r + [v]^r, \quad [\lambda \odot u]^r = \lambda [u]^r, \quad \forall\, r \in [0,1],$$

where $[u]^r + [v]^r$ means the usual addition of two intervals (as subsets of $\mathbb{R}$) and $\lambda [u]^r$ means the usual product between a scalar and a subset of $\mathbb{R}$ (see, e.g., [13]). Notice $1 \odot u = u$ and it holds $u \oplus v = v \oplus u$, $\lambda \odot u = u \odot \lambda$. If $0 \le r_1 \le r_2 \le 1$ then $[u]^{r_2} \subseteq [u]^{r_1}$. Actually $[u]^r = \left[u_-^{(r)}, u_+^{(r)}\right]$, where $u_-^{(r)} < u_+^{(r)}$, $u_-^{(r)}, u_+^{(r)} \in \mathbb{R}$, $\forall\, r \in [0,1]$.

Define

$$D : \mathbb{R}_{\mathcal{F}} \times \mathbb{R}_{\mathcal{F}} \to \mathbb{R}_+ \cup \{0\}$$

by

$$D(u,v) := \sup_{r \in [0,1]} \max \left\{ \left| u_-^{(r)} - v_-^{(r)} \right|, \left| u_+^{(r)} - v_+^{(r)} \right| \right\},$$

where $[v]^r = \left[v_-^{(r)}, v_+^{(r)}\right]$; $u, v \in \mathbb{R}_{\mathcal{F}}$. We have that D is a metric on $\mathbb{R}_{\mathcal{F}}$. Then $(\mathbb{R}_{\mathcal{F}}, D)$ is a complete metric space, see [13], with the properties

$$D(u \oplus w, v \oplus w) = D(u,v), \quad \forall\, u, v, w \in \mathbb{R}_{\mathcal{F}}, \tag{10.1}$$
$$D(k \odot u, k \odot v) = |k|\, D(u,v), \quad \forall\, u, v \in \mathbb{R}_{\mathcal{F}}, \forall\, k \in \mathbb{R},$$
$$D(u \oplus v, w \oplus e) \le D(u,w) + D(v,e), \quad \forall\, u, v, w, e \in \mathbb{R}_{\mathcal{F}}.$$

Let $f, g : \mathbb{R} \to \mathbb{R}_{\mathcal{F}}$ be fuzzy real number valued functions. The distance between f, g is defined by

$$D^*(f,g) := \sup_{x \in \mathbb{R}} D(f(x), g(x)).$$

On $\mathbb{R}_{\mathcal{F}}$ we define a partial order by "$\le$" : $u, v \in \mathbb{R}_{\mathcal{F}}$, $u \le v$ iff $u_-^{(r)} \le v_-^{(r)}$ and $u_+^{(r)} \le v_+^{(r)}$, $\forall\, r \in [0,1]$.

We need

Lemma 10.2 ([9]) *For any $a, b \in \mathbb{R} : a \cdot b \ge 0$ and any $u \in \mathbb{R}_{\mathcal{F}}$ we have*

$$D(a \odot u, b \odot u) \le |a - b| \cdot D(u, \widetilde{o}), \tag{10.2}$$

where $\widetilde{o} \in \mathbb{R}_{\mathcal{F}}$ is defined by $\widetilde{o} := \chi_{\{0\}}$.

Lemma 10.3 ([9])

(i) If we denote $\widetilde{o} := \chi_{\{0\}}$, then $\widetilde{o} \in \mathbb{R}_{\mathcal{F}}$ is the neutral element with respect to $\oplus$, i.e., $u \oplus \widetilde{o} = \widetilde{o} \oplus u = u$, $\forall\, u \in \mathbb{R}_{\mathcal{F}}$.

(ii) With respect to $\widetilde{o}$, none of $u \in \mathbb{R}_{\mathcal{F}}$, $u \ne \widetilde{o}$ has opposite in $\mathbb{R}_{\mathcal{F}}$.

(iii) *Let* $a, b \in \mathbb{R} : a \cdot b \geq 0$, *and any* $u \in \mathbb{R}_{\mathcal{F}}$, *we have* $(a + b) \odot u = a \odot u \oplus b \odot u$. *For general* $a, b \in \mathbb{R}$, *the above property is false.*
(iv) *For any* $\lambda \in \mathbb{R}$ *and any* $u, v \in \mathbb{R}_{\mathcal{F}}$, *we have* $\lambda \odot (u \oplus v) = \lambda \odot u \oplus \lambda \odot v$.
(v) *For any* $\lambda, \mu \in \mathbb{R}$ *and* $u \in \mathbb{R}_{\mathcal{F}}$, *we have* $\lambda \odot (\mu \odot u) = (\lambda \cdot \mu) \odot u$.
(vi) *If we denote* $\|u\|_{\mathcal{F}} := D(u, \tilde{o})$, $\forall\, u \in \mathbb{R}_{\mathcal{F}}$, *then* $\|\cdot\|_{\mathcal{F}}$ *has the properties of a usual norm on* $\mathbb{R}_{\mathcal{F}}$, *i.e.,*

$$\|u\|_{\mathcal{F}} = 0 \text{ iff } u = \tilde{o}, \quad \|\lambda \odot u\|_{\mathcal{F}} = |\lambda| \cdot \|u\|_{\mathcal{F}},$$
$$\|u \oplus v\|_{\mathcal{F}} \leq \|u\|_{\mathcal{F}} + \|v\|_{\mathcal{F}}, \quad \|u\|_{\mathcal{F}} - \|v\|_{\mathcal{F}} \leq D(u, v). \tag{10.3}$$

Notice that $(\mathbb{R}_{\mathcal{F}}, \oplus, \odot)$ is not a linear space over $\mathbb{R}$; and consequently $(\mathbb{R}_{\mathcal{F}}, \|\cdot\|_{\mathcal{F}})$ is not a normed space.

As in Remark 4.4 [9] one can show easily that a sequence of operators of the form

$$L_n(f)(x) := \sum_{k=0}^{n*} f\left(x_{k_n}\right) \odot w_{n,k}(x), \quad n \in \mathbb{N}, \tag{10.4}$$

($\sum^{*}$ denotes the fuzzy summation) where $f : \mathbb{R}^d \to \mathbb{R}_{\mathcal{F}}$, $x_{k_n} \in \mathbb{R}^d$, $d \in \mathbb{N}$, $w_{n,k}(x)$ real valued weights, are linear over $\mathbb{R}^d$, i.e.,

$$L_n(\lambda \odot f \oplus \mu \odot g)(x) = \lambda \odot L_n(f)(x) \oplus \mu \odot L_n(g)(x), \tag{10.5}$$

$\forall \lambda, \mu \in \mathbb{R}$, any $x \in \mathbb{R}^d$; $f, g : \mathbb{R}^d \to \mathbb{R}_{\mathcal{F}}$. (Proof based on Lemma 10.3 (iv).)

We need

Definition 10.4 (*see* [13]) Let $x, y \in \mathbb{R}_{\mathcal{F}}$. If there exists a $z \in \mathbb{R}_{\mathcal{F}}$ such that $x = y + z$, then we call z the H-difference of x and y, denoted by $z := x - y$.

Definition 10.5 (*see* [13]) Let $T := [x_0, x_0 + \beta] \subset \mathbb{R}$, with $\beta > 0$. A function $f : T \to \mathbb{R}_{\mathcal{F}}$ is differentiable at $x \in T$ if there exists a $f'(x) \in \mathbb{R}_{\mathcal{F}}$ such that the limits

$$\lim_{h \to 0+} \frac{f(x+h) - f(x)}{h}, \quad \lim_{h \to 0+} \frac{f(x) - f(x-h)}{h}$$

exist and are equal to $f'(x)$. We call f' the derivative of f or H-derivative of f at x. If f is differentiable at any $x \in T$, we call f differentiable or H-differentiable and it has derivative over T the function f'.

We need also a particular case of the Fuzzy Henstock integral ($\delta(x) = \frac{\delta}{2}$) introduced in [13], Definition 2.1 there.

That is,

Definition 10.6 (*see* [12], *p. 644*) Let $f : [a, b] \to \mathbb{R}_{\mathcal{F}}$. We say that f is Fuzzy-Riemann integrable to $I \in \mathbb{R}_{\mathcal{F}}$ if for any $\varepsilon > 0$, there exists $\delta > 0$ such that for any division $P = \{[u, v]; \xi\}$ of $[a, b]$ with the norms $\Delta(P) < \delta$, we have

$$D\left(\sum_{P}^{*}(v-u)\odot f(\xi), I\right)<\varepsilon. \tag{10.6}$$

We choose to write

$$I := (FR)\int_a^b f(x)\,dx. \tag{10.7}$$

We also call an f as above (FR)-integrable.

We mention the following fundamental theorem of Fuzzy Calculus:

Corollary 10.7 ([3]) *If $f : [a, b] \to \mathbb{R}_{\mathcal{F}}$ has a fuzzy continuous derivative f' on $[a, b]$, then $f'(x)$ is (FR)-integrable over $[a, b]$ and*

$$f(s) = f(t) \oplus (FR)\int_t^s f'(x)\,dx, \quad \text{for any } s \geq t,\ s, t \in [a, b]. \tag{10.8}$$

Note In Corollary 10.7 when $s < t$ the formula is invalid! Since fuzzy real numbers correspond to closed intervals etc.

We need also

Lemma 10.8 ([3]) *If $f, g : [a, b] \subseteq \mathbb{R} \to \mathbb{R}_{\mathcal{F}}$ are fuzzy continuous (with respect to metric D), then the function $F : [a, b] \to \mathbb{R}_+ \cup \{0\}$ defined by $F(x) := D(f(x), g(x))$ is continuous on $[a, b]$, and*

$$D\left((FR)\int_a^b f(u)\,du, (FR)\int_a^b g(u)\,du\right) \leq \int_a^b D(f(x), g(x))\,dx. \tag{10.9}$$

Lemma 10.9 ([3]) *Let $f : [a, b] \to \mathbb{R}_{\mathcal{F}}$ fuzzy continuous (with respect to metric D), then $D(f(x), \widetilde{o}) \leq M$, $\forall\, x \in [a, b]$, $M > 0$, that is f is fuzzy bounded.*

We mention

Lemma 10.10 ([4]) *Let $f : [a, b] \to \mathbb{R}_{\mathcal{F}}$ have an existing fuzzy derivative f' at $c \in [a, b]$. Then f is fuzzy continuous at c.*

Note Higher order fuzzy derivatives and all fuzzy partial derivatives are defined the obvious and analogous way to the real derivatives, all based on Definitions 10.4, 10.5, here.

We need the fuzzy multivariate Taylor formula.

Theorem 10.11 ([6], p. 54) *Let U be an open convex subset on $\mathbb{R}^n$, $n \in \mathbb{N}$ and $f : U \to \mathbb{R}_{\mathcal{F}}$ be a fuzzy continuous function. Assume that all H-fuzzy partial derivatives of f up to order $m \in \mathbb{N}$ exist and are fuzzy continuous. Let $z := (z_1, \ldots, z_n)$, $x_0 := (x_{01}, \ldots, x_{0n}) \in U$ such that $z_i \geq x_{0i}$, $i = 1, \ldots, n$. Let $0 \leq t \leq 1$, we define $x_i := x_{0i} + t(z_i - z_{0i})$, $i = 1, 2, \ldots, n$ and $g_z(t) := f(x_0 + t(z - x_0))$. (Clearly $x_0 + t(z - x_0) \in U$). Then for $N = 1, \ldots, m$ we obtain*

$$g_z^{(N)}(t) = \left[\left(\sum_{i=1}^{n*}(z_i - x_{0i}) \odot \frac{\partial}{\partial x_i}\right)^N f\right](x_1, x_2, \ldots, x_n). \tag{10.10}$$

Furthermore it holds the following fuzzy multivariate Taylor formula

$$f(z) = f(x_0) \oplus \sum_{N=1}^{m-1*} \frac{g_z^{(N)}(0)}{N!} \oplus \mathcal{R}_m(0, 1), \tag{10.11}$$

where

$$\mathcal{R}_m(0, 1) := \frac{1}{(m-1)!} \odot (FR) \int_0^1 (1-s)^{m-1} \odot g_z^{(m)}(s)\, ds. \tag{10.12}$$

Comment 10.12 *(explaining formula (10.10)) When $N = n = 2$ we have ($z_i \geq x_{0i}$, $i = 1, 2$)*

$$g_z(t) = f(x_{01} + t(z_1 - x_{01}), x_{02} + t(z_2 - x_{02})), \quad 0 \leq t \leq 1.$$

We apply Theorems 2.18, 2.19, 2.21 of [6] repeatedly, etc. Thus we find

$$g_z'(t) = (z_1 - x_{01}) \odot \frac{\partial f}{\partial x_1}(x_1, x_2) \oplus (z_2 - x_{02}) \odot \frac{\partial f}{\partial x_2}(x_1, x_2).$$

Furthermore it holds

$$g_z''(t) = (z_1 - x_{01})^2 \odot \frac{\partial^2 f}{\partial x_1^2}(x_1, x_2) \oplus 2(z_1 - x_{01})(z_2 - x_{02}) \odot \frac{\partial^2 f(x_1, x_2)}{\partial x_1 \partial x_2}$$
$$\oplus (z_2 - x_{02})^2 \odot \frac{\partial^2 f}{\partial x_2^2}(x_1, x_2). \tag{10.13}$$

When $n = 2$ and $N = 3$ we obtain

$$g_z'''(t) = (z_1 - x_{01})^3 \odot \frac{\partial^3 f}{\partial x_1^3}(x_1, x_2) \oplus 3(z_1 - x_{01})^2 (z_2 - x_{02}) \odot \frac{\partial^3 f(x_1, x_2)}{\partial x_1^2 \partial x_2}$$

$$\oplus 3(z_1 - x_{01})(z_2 - x_{02})^2 \odot \frac{\partial^3 f(x_1, x_2)}{\partial x_1 \partial x_2^2} \oplus (z_2 - x_{02})^3 \odot \frac{\partial^3 f}{\partial x_2^3}(x_1, x_2). \tag{10.14}$$

When $n = 3$ and $N = 2$ we get ($z_i \geq x_{0i}$, $i = 1, 2, 3$)

$$g_z''(t) = (z_1 - x_{01})^2 \odot \frac{\partial^2 f}{\partial x_1^2}(x_1, x_2, x_3) \oplus (z_2 - x_{02})^2 \odot \frac{\partial^2 f}{\partial x_2^2}(x_1, x_2, x_3)$$

$$\oplus \left(z_3 - x_{03}\right)^2 \odot \frac{\partial^2 f}{\partial x_3^2}\left(x_1, x_2, x_3\right) \oplus 2\left(z_1 - x_{01}\right)\left(z_2 - x_{02}\right) \tag{10.15}$$

$$\odot \frac{\partial^2 f\left(x_1, x_2, x_3\right)}{\partial x_1 \partial x_2} \oplus 2\left(z_2 - x_{02}\right)\left(z_3 - x_{03}\right) \odot \frac{\partial^2 f\left(x_1, x_2, x_3\right)}{\partial x_2 \partial x_3}$$

$$\oplus 2\left(z_3 - x_{03}\right)\left(z_1 - x_{01}\right) \odot \frac{\partial^2 f\left(x_1, x_2, x_3\right)}{\partial x_3 \partial x_1},$$

etc.

10.3 Basic Properties

We need

Definition 10.13 (*see also* [12], *Definition 13.16, p. 654*) Let $(X, \mathcal{B}, P)$ be a probability space. A fuzzy-random variable is a $\mathcal{B}$-measurable mapping $g : X \to \mathbb{R}_{\mathcal{F}}$ (i.e., for any open set $U \subseteq \mathbb{R}_{\mathcal{F}}$, in the topology of $\mathbb{R}_{\mathcal{F}}$ generated by the metric D, we have

$$g^{-1}(U) = \{s \in X; g(s) \in U\} \in \mathcal{B}). \tag{10.16}$$

The set of all fuzzy-random variables is denoted by $\mathcal{L}_{\mathcal{F}}(X, \mathcal{B}, P)$. Let $g_n, g \in \mathcal{L}_{\mathcal{F}}(X, \mathcal{B}, P)$, $n \in \mathbb{N}$ and $0 < q < +\infty$. We say $g_n(s) \overset{\text{"}q\text{-mean"}}{\underset{n\to+\infty}{\rightarrow}} g(s)$ if

$$\lim_{n\to+\infty} \int_X D\left(g_n(s), g(s)\right)^q P(ds) = 0. \tag{10.17}$$

Remark 10.14 (see [12], p. 654) If $f, g \in \mathcal{L}_{\mathcal{F}}(X, \mathcal{B}, P)$, let us denote $F : X \to \mathbb{R}_+ \cup \{0\}$ by $F(s) = D(f(s), g(s))$, $s \in X$. Here, F is $\mathcal{B}$-measurable, because $F = G \circ H$, where $G(u, v) = D(u, v)$ is continuous on $\mathbb{R}_{\mathcal{F}} \times \mathbb{R}_{\mathcal{F}}$, and $H : X \to \mathbb{R}_{\mathcal{F}} \times \mathbb{R}_{\mathcal{F}}$, $H(s) = (f(s), g(s))$, $s \in X$, is $\mathcal{B}$-measurable. This shows that the above convergence in q-mean makes sense.

Definition 10.15 (*see* [12], *p. 654, Definition 13.17*) Let $(T, \mathcal{T})$ be a topological space. A mapping $f : T \to \mathcal{L}_{\mathcal{F}}(X, \mathcal{B}, P)$ will be called fuzzy-random function (or fuzzy-stochastic process) on T. We denote $f(t)(s) = f(t, s)$, $t \in T$, $s \in X$.

Remark 10.16 (see [12], p. 655) Any usual fuzzy real function $f : T \to \mathbb{R}_{\mathcal{F}}$ can be identified with the degenerate fuzzy-random function $f(t, s) = f(t)$, $\forall\, t \in T$, $s \in X$.

Remark 10.17 (see [12], p. 655) Fuzzy-random functions that coincide with probability one for each $t \in T$ will be consider equivalent.

Remark 10.18 (see [12], p. 655) Let $f, g : T \to \mathcal{L}_{\mathcal{F}}(X, \mathcal{B}, P)$. Then $f \oplus g$ and $k \odot f$ are defined pointwise, i.e.,

$$(f \oplus g)(t, s) = f(t, s) \oplus g(t, s),$$
$$(k \odot f)(t, s) = k \odot f(t, s), \quad t \in T, s \in X.$$

Definition 10.19 (*see also Definition 13.18, pp. 655–656,* [12]) For a fuzzy-random function $f : \mathbb{R}^d \to \mathcal{L}_{\mathcal{F}}(X, \mathcal{B}, P)$, $d \in \mathbb{N}$, we define the (first) fuzzy-random modulus of continuity

$$\Omega_1^{(\mathcal{F})}(f, \delta)_{L^q} =$$

$$\sup\left\{\left(\int_X D^q(f(x, s), f(y, s)) P(ds)\right)^{\frac{1}{q}} : x, y \in \mathbb{R}^d, \ \|x - y\|_{l_1} \le \delta\right\},$$

$0 < \delta$, $1 \le q < \infty$.

Definition 10.20 Here $1 \le q < +\infty$. Let $f : \mathbb{R}^d \to \mathcal{L}_{\mathcal{F}}(X, \mathcal{B}, P)$, $d \in \mathbb{N}$, be a fuzzy random function. We call f a (q-mean) uniformly continuous fuzzy random function over $\mathbb{R}^d$, iff $\forall\, \varepsilon > 0\, \exists\, \delta > 0$:whenever $\|x - y\|_{l_1} \le \delta$, $x, y \in \mathbb{R}^d$, implies that

$$\int_X (D(f(x, s), f(y, s)))^q P(ds) \le \varepsilon.$$

We denote it as $f \in C_{FR}^{U_q}(\mathbb{R}^d)$.

Proposition 10.21 *Let* $f \in C_{FR}^{U_q}(\mathbb{R}^d)$. *Then* $\Omega_1^{(\mathcal{F})}(f, \delta)_{L^q} < \infty$, *any* $\delta > 0$.

Proof Let $\varepsilon_0 > 0$ be arbitrary but fixed. Then there exists $\delta_0 > 0 : \|x - y\|_{l_1} \le \delta_0$ implies

$$\int_X (D(f(x, s), f(y, s)))^q P(ds) \le \varepsilon_0 < \infty.$$

That is $\Omega_1^{(\mathcal{F})}(f, \delta_0)_{L^q} \le \varepsilon_0^{\frac{1}{q}} < \infty$. Let now $\delta > 0$ arbitrary, $x, y \in \mathbb{R}^d$ such that $\|x - y\|_{l_1} \le \delta$. Choose $n \in \mathbb{N} : n\delta_0 \ge \delta$ and set $x_i := x + \frac{i}{n}(y - x)$, $0 \le i \le n$. Then

$$D(f(x, s), f(y, s)) \le D(f(x, s), f(x_1, s)) + D(f(x_1, s), f(x_2, s)) + \cdots + D(f(x_{n-1}, s), f(y, s)).$$

Consequently

$$\left(\int_X (D(f(x,s), f(y,s)))^q P(ds)\right)^{\frac{1}{q}} \leq \left(\int_X (D(f(x,s), f(x_1,s)))^q P(ds)\right)^{\frac{1}{q}}$$

$$+\cdots+\left(\int_X (D(f(x_{n-1},s), f(y,s)))^q P(ds)\right)^{\frac{1}{q}} \leq n\Omega_1^{(\mathcal{F})}(f,\delta_0)_{L^q} \leq n\varepsilon_0^{\frac{1}{q}} < \infty,$$

since $\|x_i - x_{i+1}\|_{l_1} = \frac{1}{n}\|x-y\|_{l_1} \leq \frac{1}{n}\delta \leq \delta_0$, $0 \leq i \leq n$.

Therefore $\Omega_1^{(\mathcal{F})}(f,\delta)_{L^q} \leq n\varepsilon_0^{\frac{1}{q}} < \infty$. ∎

Proposition 10.22 *Let $f, g : \mathbb{R}^d \to \mathcal{L}_{\mathcal{F}}(X, \mathcal{B}, P)$, $d \in \mathbb{N}$, be fuzzy random functions. It holds*

(i) $\Omega_1^{(\mathcal{F})}(f,\delta)_{L^q}$ *is nonnegative and nondecreasing in* $\delta > 0$.

(ii) $\lim_{\delta\downarrow 0}\Omega_1^{(\mathcal{F})}(f,\delta)_{L^q} = \Omega_1^{(\mathcal{F})}(f,0)_{L^q} = 0$, *iff* $f \in C_{FR}^{U_q}(\mathbb{R}^d)$.

(iii) $\Omega_1^{(\mathcal{F})}(f,\delta_1+\delta_2)_{L^q} \leq \Omega_1^{(\mathcal{F})}(f,\delta_1)_{L^q} + \Omega_1^{(\mathcal{F})}(f,\delta_2)_{L^q}$, $\delta_1, \delta_2 > 0$.

(iv) $\Omega_1^{(\mathcal{F})}(f,n\delta)_{L^q} \leq n\Omega_1^{(\mathcal{F})}(f,\delta)_{L^q}$, $\delta > 0$, $n \in \mathbb{N}$.

(v) $\Omega_1^{(\mathcal{F})}(f,\lambda\delta)_{L^q} \leq \lceil\lambda\rceil\,\Omega_1^{(\mathcal{F})}(f,\delta)_{L^q} \leq (\lambda+1)\,\Omega_1^{(\mathcal{F})}(f,\delta)_{L^q}$, $\lambda > 0$, $\delta > 0$, *where* $\lceil\cdot\rceil$ *is the ceilling of the number.*

(vi) $\Omega_1^{(\mathcal{F})}(f \oplus g,\delta)_{L^q} \leq \Omega_1^{(\mathcal{F})}(f,\delta)_{L^q} + \Omega_1^{(\mathcal{F})}(g,\delta)_{L^q}$, $\delta > 0$. *Here* $f \oplus g$ *is a fuzzy random function.*

(vii) $\Omega_1^{(\mathcal{F})}(f,\cdot)_{L^q}$ *is continuous on* $\mathbb{R}_+$, *for* $f \in C_{FR}^{U_q}(\mathbb{R}^d)$.

Proof (i) is obvious.

(ii) $\Omega_1(f,0)_{L^q} = 0$.

(⇒) Let $\lim_{\delta\downarrow 0}\Omega_1(f,\delta)_{L^q} = 0$. Then $\forall\,\varepsilon > 0$, $\varepsilon^{\frac{1}{q}} > 0$ and $\exists\,\delta > 0$, $\Omega_1(f,\delta)_{L^q} \leq \varepsilon^{\frac{1}{q}}$. I.e. for any $x, y \in \mathbb{R}^d : \|x-y\|_{l_1} \leq \delta$ we get

$$\int_X D^q(f(x,s), f(y,s))\,P(ds) \leq \varepsilon.$$

That is $f \in C_{FR}^{U_q}(\mathbb{R}^d)$.

(⇐) Let $f \in C_{FR}^{U_q}(\mathbb{R}^d)$. Then $\forall\,\varepsilon > 0\;\exists\,\delta > 0$: whenever $\|x-y\|_{l_1} \leq \delta$, $x, y \in \mathbb{R}^d$, it implies

$$\int_X D^q(f(x,s), f(y,s))\,P(ds) \leq \varepsilon.$$

I.e. $\forall\,\varepsilon > 0\;\exists\,\delta > 0 : \Omega_1(f,\delta)_{L^q} \leq \varepsilon^{\frac{1}{q}}$. That is $\Omega_1(f,\delta)_{L^q} \to 0$ as $\delta \downarrow 0$.

(iii) Let $x_1, x_2 \in \mathbb{R}^d : \|x_1 - x_2\|_{l_1} \le \delta_1 + \delta_2$. Set $x = \frac{\delta_2}{\delta_1+\delta_2}x_1 + \frac{\delta_1}{\delta_1+\delta_2}x_2$, so that $x \in \overline{x_1x_2}$. Hence $\|x - x_1\|_{l_1} \le \delta_1$ and $\|x_2 - x\|_{l_1} \le \delta_2$. We have

$$\left(\int_X D^q\left(f\left(x_1,s\right), f\left(x_2,s\right)\right) P\left(ds\right)\right)^{\frac{1}{q}}$$

$$\le \left(\int_X D^q\left(f\left(x_1,s\right), f\left(x,s\right)\right) P\left(ds\right)\right)^{\frac{1}{q}} + \left(\int_X D^q\left(f\left(x,s\right), f\left(x_2,s\right)\right) P\left(ds\right)\right)^{\frac{1}{q}} \le$$

$$\Omega_1\left(f, \|x_1 - x\|_{l_1}\right)_{L^q} + \Omega_1\left(f, \|x_2 - x\|_{l_1}\right)_{L^q} \le$$

$$\Omega_1\left(f, \delta_1\right)_{L^q} + \Omega_2\left(f, \delta_2\right)_{L^q} .$$

Therefore (iii) is true.

(iv) and (v) are obvious.

(vi) Notice that

$$\left(\int_X D^q\left(\left(f \oplus g\right)\left(x,s\right), \left(f \oplus g\right)\left(y,s\right)\right) P\left(ds\right)\right)^{\frac{1}{q}} \le$$

$$\left(\int_X D^q\left(f\left(x,s\right), f\left(y,s\right)\right) P\left(ds\right)\right)^{\frac{1}{q}} + \left(\int_X D^q\left(g\left(x,s\right), g\left(y,s\right)\right) P\left(ds\right)\right)^{\frac{1}{q}} .$$

That is (vi) is now clear.

(vii) By (iii) we get

$$\left|\Omega_1^{(\mathcal{F})}\left(f, \delta_1 + \delta_2\right)_{L^q} - \Omega_1^{(\mathcal{F})}\left(f, \delta_1\right)_{L^q}\right| \le \Omega_1^{(\mathcal{F})}\left(f, \delta_2\right)_{L^q} .$$

Let now $f \in C_{FR}^{U_q}\left(\mathbb{R}^d\right)$, then by (ii) $\lim_{\delta_2 \downarrow 0} \Omega_1^{(\mathcal{F})}\left(f, \delta_2\right)_{L^q} = 0$. That is proving the continuity of $\Omega_1^{(\mathcal{F})}\left(f, \cdot\right)_{L^q}$ on $\mathbb{R}_+$. ■

We give

Definition 10.23 ([5]) Let $f(t, s)$ be a stochastic process from $\mathbb{R}^d \times (X, \mathcal{B}, P)$ into $\mathbb{R}$, $d \in \mathbb{N}$, where $(X, \mathcal{B}, P)$ is a probability space. We define the q-mean multivariate first moduli of continuity of f by

$$\Omega_1\left(f, \delta\right)_{L^q} :=$$

$$\sup\left\{\left(\int_X \left|f\left(x,s\right) - f\left(y,s\right)\right|^q P\left(ds\right)\right)^{\frac{1}{q}} : x, y \in \mathbb{R}^d,\ \|x - y\|_{l_1} \le \delta\right\}, \quad (10.18)$$

$\delta > 0$, $1 \le q < \infty$.

For more see [5].

We also give

Proposition 10.24 *Assume that $\Omega_1^{(\mathcal{F})}(f,\delta)_{L^q}$ is finite, $\delta>0$, $1\leq q<\infty$. Then*

$$\Omega_1^{(\mathcal{F})}(f,\delta)_{L^q}\geq\sup_{r\in[0,1]}\max\left\{\Omega_1\left(f_-^{(r)},\delta\right)_{L^q},\Omega_1\left(f_+^{(r)},\delta\right)_{L^q}\right\}. \tag{10.19}$$

The reverse direction "$\leq$" is not possible.

Proof We observe that

$$D(f(x,s),f(y,s))=$$

$$\sup_{r\in[0,1]}\max\left\{\left|f_-^{(r)}(x,s)-f_-^{(r)}(y,s)\right|,\left|f_+^{(r)}(x,s)-f_+^{(r)}(y,s)\right|\right\}$$

$$\geq\left|f_\pm^{(r)}(x,s)-f_\pm^{(r)}(y,s)\right|,$$

respectively in $+,-$.

Hence

$$\left(\int_X D^q(f(x,s),f(y,s))P(ds)\right)^{\frac{1}{q}}\geq\left(\int_X\left|f_\pm^{(r)}(x,s)-f_\pm^{(r)}(y,s)\right|^qP(ds)\right)^{\frac{1}{q}},$$

respectively in $+,-$.

Therefore it holds

$$\sup_{\substack{x,y\in\mathbb{R}^d\\ \|x-y\|_{l_1}\leq\delta}}\left(\int_X D^q(f(x,s),f(y,s))P(ds)\right)^{\frac{1}{q}}\geq$$

$$\sup_{r\in[0,1]}\max_{\{+,-\}}\left\{\sup_{\substack{x,y\in\mathbb{R}^d\\ \|x-y\|_{l_1}\leq\delta}}\left(\int_X\left|f_\pm^{(r)}(x,s)-f_\pm^{(r)}(y,s)\right|^qP(ds)\right)^{\frac{1}{q}}\right\},$$

proving the claim. ■

Remark 10.25 For each $s\in X$ we define the usual first modulus of continuity of $f(\cdot,s)$ by

$$\omega_1^{(\mathcal{F})}(f(\cdot,s),\delta):=\sup_{\substack{x,y\in\mathbb{R}^d\\ \|x-y\|_{l_1}\leq\delta}}D(f(x,s),f(y,s)),\quad\delta>0. \tag{10.20}$$

Therefore

$$D^q(f(x,s), f(y,s)) \leq \left(\omega_1^{(\mathcal{F})}(f(\cdot,s),\delta)\right)^q,$$

$\forall s \in X$ and $x, y \in \mathbb{R}^d : \|x - y\|_{l_1} \leq \delta$, $\delta > 0$.

Hence it holds

$$\left(\int_X D^q(f(x,s), f(y,s)) P(ds)\right)^{\frac{1}{q}} \leq \left(\int_X \left(\omega_1^{(\mathcal{F})}(f(\cdot,s),\delta)\right)^q P(ds)\right)^{\frac{1}{q}},$$

$\forall x, y \in \mathbb{R}^d : \|x - y\|_{l_1} \leq \delta$.

We have that

$$\Omega_1^{(\mathcal{F})}(f,\delta)_{L^q} \leq \left(\int_X \left(\omega_1^{(\mathcal{F})}(f(\cdot,s),\delta)\right)^q P(ds)\right)^{\frac{1}{q}}, \tag{10.21}$$

under the assumption that the right hand side of (10.21) is finite.

The reverse "$\geq$" of the last (10.21) is not true.

Also we have

Proposition 10.26 ([4])

(i) Let $Y(t,\omega)$ be a real valued stochastic process such that Y is continuous in $t \in [a,b]$. Then Y is jointly measurable in (t,ω).

(ii) Further assume that the expectation $(E|Y|)(t) \in C([a,b])$, or more generally $\int_a^b (E|Y|)(t)\,dt$ makes sense and is finite. Then

$$E\left(\int_a^b Y(t,\omega)\,dt\right) = \int_a^b (EY)(t)\,dt. \tag{10.22}$$

According to [11], p. 94 we have the following

Definition 10.27 Let $(Y, \mathcal{T})$ be a topological space, with its σ-algebra of Borel sets $\mathcal{B} := \mathcal{B}(Y,\mathcal{T})$ generated by $\mathcal{T}$. If $(X, \mathcal{S})$ is a measurable space, a function $f : X \to Y$ is called measurable iff $f^{-1}(B) \in \mathcal{S}$ for all $B \in \mathcal{B}$.

By Theorem 4.1.6 of [11], p. 89 f as above is measurable iff

$$f^{-1}(C) \in \mathcal{S} \text{ for all } C \in \mathcal{T}.$$

We would need

Theorem 10.28 (see [11], p. 95) *Let $(X,\mathcal{S})$ be a measurable space and (Y,d) be a metric space. Let f_n be measurable functions from X into Y such that for all $x \in X$, $f_n(x) \to f(x)$ in Y. Then f is measurable. I.e., $\lim_{n\to\infty} f_n = f$ is measurable.*

We need also

Proposition 10.29 *Let f, g be fuzzy random variables from $\mathcal{S}$ into $\mathbb{R}_{\mathcal{F}}$. Then*

(i) Let $c \in \mathbb{R}$, then $c \odot f$ is a fuzzy random variable.
(ii) $f \oplus g$ is a fuzzy random variable.

Finally we need

Proposition 10.30 ([4]) *Let $f : [a, b] \to L_{\mathcal{F}}(X, \mathcal{B}, P)$, $a, b \in \mathbb{R}$, be a fuzzy-random function. We assume that $f(t, s)$ is fuzzy continuous in $t \in [a, b]$, $s \in X$. Then $(FR) \int_a^b f(t, s)\,dt$ exists and is a fuzzy-random variable.*

10.4 Main Results

We need the following (see [10]) definitions.

Definition 10.31 A function $b : \mathbb{R} \to \mathbb{R}$ is said to be bell-shaped if b belongs to L_1 and its integral is nonzero, if it is nondecreasing on $(-\infty, a)$ and nonincreasing on $[a, +\infty)$, where a belongs to $\mathbb{R}$. In particular $b(x)$ is a nonnegative number and at a, b takes a global maximum; it is the center of the bell-shaped function. A bell-shaped function is said to be centered if its center is zero.

Definition 10.32 (*see* [10]) A function $b : \mathbb{R}^d \to \mathbb{R}$ $(d \geq 1)$ is said to be a d-dimensional bell-shaped function if it is integrable and its integral is not zero, and for all $i = 1, \ldots, d$,

$$t \to b(x_1, \ldots, t, \ldots, x_d)$$

is a centered bell-shaped function, where $\overrightarrow{x} := (x_1, \ldots, x_d) \in \mathbb{R}^d$ arbitrary.

Example 10.33 (from [10]) Let b be a centered bell-shaped function over $\mathbb{R}$, then $(x_1, \ldots, x_d) \to b(x_1) \ldots b(x_d)$ is a d-dimensional bell-shaped function, e.g. b could be the characteristic function or the hat function on $[-1, 1]$.

Assumption 10.34 Here $b\left(\overrightarrow{x}\right)$ is of compact support $\mathcal{B}^* := \prod_{i=1}^{d} [-T_i, T_i]$, $T_i > 0$ and it may have jump discontinuities there. Here we consider functions $f \in C_{FR}^{U_q}\left(\mathbb{R}^d\right)$.

In this chapter we study among others in q-mean $(1 \leq q < \infty)$ the pointwise convergence with rates over $\mathbb{R}^d$, to the fuzzy-random unit operator or a perturbation of it, of the following fuzzy-random multivariate neural network operators, $(0 < \alpha < 1$, $\overrightarrow{x} := (x_1, \ldots, x_d) \in \mathbb{R}^d$, $s \in X$, $(X, \mathcal{B}, P)$ a probability space, $n \in \mathbb{N})$

$$(M_n(f))(\overrightarrow{x}, s) =$$

$$\frac{\sum_{k_1=-n^2}^{n^2*} \cdots \sum_{k_d=-n^2}^{n^2*} f\left(\frac{k_1}{n}, \ldots, \frac{k_d}{n}, s\right) \odot b\left(n^{1-\alpha}\left(x_1 - \frac{k_1}{n}\right), \ldots, n^{1-\alpha}\left(x_d - \frac{k_d}{n}\right)\right)}{\sum_{k_1=-n^2}^{n^2} \cdots \sum_{k_d=-n^2}^{n^2} b\left(n^{1-\alpha}\left(x_1 - \frac{k_1}{n}\right), \ldots, n^{1-\alpha}\left(x_d - \frac{k_d}{n}\right)\right)}. \tag{10.23}$$

In short, we can write

$$(M_n(f))(\overrightarrow{x}, s) = \frac{\sum_{\overrightarrow{k}=-\overrightarrow{n^2}}^{\overrightarrow{n^2}*} f\left(\frac{\overrightarrow{k}}{n}, s\right) \odot b\left(n^{1-\alpha}\left(\overrightarrow{x} - \frac{\overrightarrow{k}}{n}\right)\right)}{\sum_{\overrightarrow{k}=-\overrightarrow{n^2}}^{\overrightarrow{n^2}} b\left(n^{1-\alpha}\left(\overrightarrow{x} - \frac{\overrightarrow{k}}{n}\right)\right)}. \tag{10.24}$$

In this chapter we assume that

$$n \geq \max_{i \in \{1, \ldots, d\}} \left\{ T_i + |x_i|, T_i^{-\frac{1}{\alpha}} \right\}, \tag{10.25}$$

see also [2], p. 91.

So, by (10.25) we can rewrite ($[\cdot]$ is the integral part of a number, while $\lceil \cdot \rceil$ is the ceiling of a number)

$$(M_n(f))(\overrightarrow{x}, s) =$$

$$\frac{\sum_{k_1=\lceil nx_1 - T_1 n^\alpha \rceil}^{[nx_1 + T_1 n^\alpha]*} \cdots \sum_{k_d=\lceil nx_d - T_d n^\alpha \rceil}^{[nx_d + T_d n^\alpha]*} f\left(\frac{k_1}{n}, \ldots, \frac{k_d}{n}, s\right) \odot b\left(n^{1-\alpha}\left(x_1 - \frac{k_1}{n}\right), \ldots, n^{1-\alpha}\left(x_d - \frac{k_d}{n}\right)\right)}{\sum_{k_1=\lceil nx_1 - T_1 n^\alpha \rceil}^{[nx_1 + T_1 n^\alpha]} \cdots \sum_{k_d=\lceil nx_d - T_d n^\alpha \rceil}^{[nx_d + T_d n^\alpha]} b\left(n^{1-\alpha}\left(x_1 - \frac{k_1}{n}\right), \ldots, n^{1-\alpha}\left(x_d - \frac{k_d}{n}\right)\right)}. \tag{10.26}$$

In short we can write

$$(M_n(f))(\overrightarrow{x}, s) = \frac{\sum_{\overrightarrow{k}=\lceil n\overrightarrow{x} - \overrightarrow{T} n^\alpha \rceil}^{[n\overrightarrow{x} + \overrightarrow{T} n^\alpha]*} f\left(\frac{\overrightarrow{k}}{n}, s\right) \odot b\left(n^{1-\alpha}\left(\overrightarrow{x} - \frac{\overrightarrow{k}}{n}\right)\right)}{\sum_{\overrightarrow{k}=\lceil n\overrightarrow{x} - \overrightarrow{T} n^\alpha \rceil}^{[n\overrightarrow{x} + \overrightarrow{T} n^\alpha]} b\left(n^{1-\alpha}\left(\overrightarrow{x} - \frac{\overrightarrow{k}}{n}\right)\right)}. \tag{10.27}$$

Denoting

$$V\left(\overrightarrow{x}\right) := \sum_{\overrightarrow{k}=\left\lceil n\overrightarrow{x}-\overrightarrow{T}n^{\alpha}\right\rceil}^{\left[n\overrightarrow{x}+\overrightarrow{T}n^{\alpha}\right]} b\left(n^{1-\alpha}\left(\overrightarrow{x}-\frac{\overrightarrow{k}}{n}\right)\right), \tag{10.28}$$

we will write and use from now on that

$$\left(M_n(f)\right)\left(\overrightarrow{x},s\right) = \sum_{\overrightarrow{k}=\left\lceil n\overrightarrow{x}-\overrightarrow{T}n^{\alpha}\right\rceil}^{\left[n\overrightarrow{x}+\overrightarrow{T}n^{\alpha}\right]*} f\left(\frac{\overrightarrow{k}}{n},s\right) \odot \frac{b\left(n^{1-\alpha}\left(\overrightarrow{x}-\frac{\overrightarrow{k}}{n}\right)\right)}{V\left(\overrightarrow{x}\right)}. \tag{10.29}$$

The above M_n are linear operators over $\mathbb{R}^d \times X$.

Related works were done in [6, 8].

We present

Theorem 10.35 *Here all as above. Then*

$$\left(\int_X D^q\left(\left(M_n(f)\right)\left(\overrightarrow{x},s\right), f\left(\overrightarrow{x},s\right)\right) P(ds)\right)^{\frac{1}{q}} \leq \Omega_1^{(\mathcal{F})}\left(f, \frac{\sum_{i=1}^{d} T_i}{n^{1-\alpha}}\right)_{L^q}. \tag{10.30}$$

As $n \to \infty$, we get that

$$\left(M_n f\right)\left(\overrightarrow{x},s\right) \overset{\text{“}q\text{-mean”}}{\rightarrow} f\left(\overrightarrow{x},s\right)$$

with rates.

Proof We observe that

$$D\left(\left(M_n(f)\right)\left(\overrightarrow{x},s\right), f\left(\overrightarrow{x},s\right)\right) =$$

$$D\left(\sum_{\overrightarrow{k}=\left\lceil n\overrightarrow{x}-\overrightarrow{T}n^{\alpha}\right\rceil}^{\left[n\overrightarrow{x}+\overrightarrow{T}n^{\alpha}\right]*} f\left(\frac{\overrightarrow{k}}{n},s\right) \odot \frac{b\left(n^{1-\alpha}\left(\overrightarrow{x}-\frac{\overrightarrow{k}}{n}\right)\right)}{V\left(\overrightarrow{x}\right)}, f\left(\overrightarrow{x},s\right) \odot 1\right) =$$

$$D\left(\sum_{\overrightarrow{k}=\left\lceil n\overrightarrow{x}-\overrightarrow{T}n^{\alpha}\right\rceil}^{\left[n\overrightarrow{x}+\overrightarrow{T}n^{\alpha}\right]*} f\left(\frac{\overrightarrow{k}}{n},s\right) \odot \frac{b\left(n^{1-\alpha}\left(\overrightarrow{x}-\frac{\overrightarrow{k}}{n}\right)\right)}{V\left(\overrightarrow{x}\right)}, f\left(\overrightarrow{x},s\right) \odot \frac{V\left(\overrightarrow{x}\right)}{V\left(\overrightarrow{x}\right)}\right) =$$

$$D\left(\sum_{\vec{k}=\lceil n\vec{x}-\vec{T}n^{\alpha}\rceil}^{[n\vec{x}+\vec{T}n^{\alpha}]*} f\left(\frac{\vec{k}}{n},s\right)\odot\frac{b\left(n^{1-\alpha}\left(\vec{x}-\frac{\vec{k}}{n}\right)\right)}{V(\vec{x})},\right.$$

$$\left.\sum_{\vec{k}=\lceil n\vec{x}-\vec{T}n^{\alpha}\rceil}^{[n\vec{x}+\vec{T}n^{\alpha}]*} f(\vec{x},s)\odot\frac{b\left(n^{1-\alpha}\left(\vec{x}-\frac{\vec{k}}{n}\right)\right)}{V(\vec{x})}\right)\leq$$

$$\sum_{\vec{k}=\lceil n\vec{x}-\vec{T}n^{\alpha}\rceil}^{[n\vec{x}+\vec{T}n^{\alpha}]}\frac{b\left(n^{1-\alpha}\left(\vec{x}-\frac{\vec{k}}{n}\right)\right)}{V(\vec{x})}D\left(f\left(\frac{\vec{k}}{n},s\right),f(\vec{x},s)\right).$$

That is it holds

$$D\left((M_n(f))(\vec{x},s),f(\vec{x},s)\right)\leq \tag{10.31}$$

$$\sum_{\vec{k}=\lceil n\vec{x}-\vec{T}n^{\alpha}\rceil}^{[n\vec{x}+\vec{T}n^{\alpha}]}\frac{b\left(n^{1-\alpha}\left(\vec{x}-\frac{\vec{k}}{n}\right)\right)}{V(\vec{x})}D\left(f\left(\frac{\vec{k}}{n},s\right),f(\vec{x},s)\right).$$

Hence

$$\left(\int_X D^q\left((M_n(f))(\vec{x},s),f(\vec{x},s)\right)P(ds)\right)^{\frac{1}{q}}\leq$$

$$\sum_{\vec{k}=\lceil n\vec{x}-\vec{T}n^{\alpha}\rceil}^{[n\vec{x}+\vec{T}n^{\alpha}]}\frac{b\left(n^{1-\alpha}\left(\vec{x}-\frac{\vec{k}}{n}\right)\right)}{V(\vec{x})}\left(\int_X D^q\left(f\left(\frac{\vec{k}}{n},s\right),f(\vec{x},s)\right)P(ds)\right)^{\frac{1}{q}}\leq$$

$$\sum_{\vec{k}=\lceil n\vec{x}-\vec{T}n^{\alpha}\rceil}^{[n\vec{x}+\vec{T}n^{\alpha}]}\frac{b\left(n^{1-\alpha}\left(\vec{x}-\frac{\vec{k}}{n}\right)\right)}{V(\vec{x})}\Omega_1^{(\mathcal{F})}\left(f,\left\|\vec{x}-\frac{\vec{k}}{n}\right\|_{l_1}\right)_{L^q}\leq \tag{10.32}$$

$$\sum_{\vec{k}=\lceil n\vec{x}-\vec{T}n^{\alpha}\rceil}^{[n\vec{x}+\vec{T}n^{\alpha}]}\frac{b\left(n^{1-\alpha}\left(\vec{x}-\frac{\vec{k}}{n}\right)\right)}{V(\vec{x})}\Omega_1^{(\mathcal{F})}\left(f,\frac{\sum_{i=1}^{d}T_i}{n^{1-\alpha}}\right)_{L^q}=$$

$$\Omega_1^{(\mathcal{F})}\left(f, \frac{\sum_{i=1}^{d} T_i}{n^{1-\alpha}}\right)_{L^q}. \tag{10.33}$$

Condition (10.25) implies that

$$\left|x_i - \frac{k_i}{n}\right| \leq \frac{T_i}{n^{1-\alpha}}, \quad \text{all } i = 1, \ldots, d. \tag{10.34}$$

The proof of (10.30) is now finished. ■

Remark 10.36 Consider the fuzzy-random perturbed unit operator

$$(T_n(f))\left(\overrightarrow{x}, s\right) := f\left(\overrightarrow{x} - \frac{\overrightarrow{T}}{n^{1-\alpha}}, s\right),$$

$\forall \left(\overrightarrow{x}, s\right) \in \mathbb{R}^d \times X$; $\overrightarrow{T} := (T_1, \ldots, T_d)$, $n \in \mathbb{N}$, $1 \leq q < \infty$.

We observe that

$$\int_X D^q\left(\left(T_n(f)\left(\overrightarrow{x}, s\right), f\left(\overrightarrow{x}, s\right)\right) P(ds)\right)^{\frac{1}{q}} =$$

$$\left(\int_X D^q\left(f\left(\overrightarrow{x} - \frac{\overrightarrow{T}}{n^{1-\alpha}}, s\right), f\left(\overrightarrow{x}, s\right)\right) P(ds)\right)^{\frac{1}{q}} \leq \Omega_1^{(\mathcal{F})}\left(f, \frac{\sum_{i=1}^{d} T_i}{n^{1-\alpha}}\right)_{L^q}. \tag{10.35}$$

Given that $f \in C_{FR}^{U_q}\left(\mathbb{R}^d\right)$, we get that $(T_n(f))\left(\overrightarrow{x}, s\right) \underset{n\to\infty}{\overset{\text{"}q\text{-mean"}}{\rightarrow}} f\left(\overrightarrow{x}, s\right)$, $\forall \left(\overrightarrow{x}, s\right) \in \mathbb{R}^d \times X$.

Next we estimate in high order with rates the 1-mean difference

$$\int_X D\left((M_n(f))\left(\overrightarrow{x}, s\right), (T_n(f))\left(\overrightarrow{x}, s\right)\right) P(ds), \quad n \in \mathbb{N}.$$

We make

Assumption 10.37 Let $\overrightarrow{x} \in \mathbb{R}^d$, $d \in \mathbb{N}$, $s \in X$; where $(X, \mathcal{B}, P)$ is a probability space, n as in (10.25), b of compact support $\mathcal{B}^*$, $0 < \alpha < 1$, M_n as in (10.29).

Let $f : \mathbb{R}^d \to L_{\mathcal{F}}(X, \mathcal{B}, P)$ be a fuzzy continuous in $\overrightarrow{x} \in \mathbb{R}^d$ random function.

We assume that all H-fuzzy partial derivatives of f up to order $N \in \mathbb{N}$ exist and are fuzzy continuous in $\overrightarrow{x} \in \mathbb{R}^d$ and all belong to $C_{FR}^{U_1}\left(\mathbb{R}^d\right)$.

Furthermore we assume that

$$\int_X D\left(f_\alpha\left(\overrightarrow{x}, s\right), \widetilde{o}\right) P(ds) < \infty, \tag{10.36}$$

for all $\alpha : |\alpha| = j$, $j = 1, \ldots, N$.

Call

$$\Omega_{1,N}^{(\mathcal{F})}\left(f_\alpha^{\max}, \delta\right)_{L^1} := \max_{|\alpha|=N} \Omega_1^{(\mathcal{F})}\left(f_\alpha, \delta\right)_{L^1}, \quad \delta > 0. \tag{10.37}$$

We give

Theorem 10.38 *All here as in Assumption 10.37. Then*

$$\int_X D\left(\left(M_n(f)\right)\left(\overrightarrow{x}, s\right), \left(T_n(f)\right)\left(\overrightarrow{x}, s\right)\right) P(ds) \leq$$

$$\sum_{j=1}^{N}\left[\sum_{|\alpha|=j}\left(\frac{1}{\prod_{i=1}^{d}\alpha_i!}\right)\left(\prod_{i=1}^{d}\left(\frac{2T_i}{n^{1-\alpha}}\right)^{\alpha_i}\right)\right.$$

$$\left.\left\{\int_X D\left(f_\alpha\left(\overrightarrow{x}, s\right), \widetilde{o}\right) P(ds) + \Omega_1^{(\mathcal{F})}\left(f_\alpha, \frac{\sum_{i=1}^{d} T_i}{n^{1-\alpha}}\right)_{L^1}\right\}\right]$$

$$+\left[\sum_{|\alpha|=N}\left(\frac{1}{\prod_{i=1}^{d}\alpha_i!}\right)\left(\prod_{i=1}^{d}\left(\frac{2T_i}{n^{1-\alpha}}\right)^{\alpha_i}\right)\right]\Omega_{1,N}^{(\mathcal{F})}\left(f_\alpha^{\max}, \frac{2\left(\sum_{i=1}^{d} T_i\right)}{n^{1-\alpha}}\right)_{L^1}. \tag{10.38}$$

By (10.38), as $n \to \infty$, we get

$$\int_X D\left(\left(M_n(f)\right)\left(\overrightarrow{x}, s\right), \left(T_n(f)\right)\left(\overrightarrow{x}, s\right)\right) P(ds) \to 0,$$

with rates in high order.

Proof By (10.25) we get that

$$-\frac{T_i}{n^{1-\alpha}} \leq x_i - \frac{k_i}{n} \leq \frac{T_i}{n^{1-\alpha}}, \quad \text{all } i = 1, \ldots, d, \; x_i \in \mathbb{R}. \tag{10.39}$$

We consider the case

$$x_i - \frac{T_i}{n^{1-\alpha}} \leq \frac{k_i}{n}, \quad i = 1, \ldots, d. \tag{10.40}$$

Set

$$g_{\frac{\vec{k}}{n}}(t, s) := f\left(\vec{x} - \frac{\vec{T}}{n^{1-\alpha}} + t\left(\frac{\vec{k}}{n} - \vec{x} + \frac{\vec{T}}{n^{1-\alpha}}\right), s\right), \tag{10.41}$$

$0 \leq t \leq 1, \forall\, s \in X$.

We apply Theorem 10.11, and we have by the fuzzy multivariate Taylor formula that

$$f\left(\frac{\vec{k}}{n}, s\right) = f\left(\vec{x} - \frac{\vec{T}}{n^{1-\alpha}}, s\right) \oplus \sum_{j=1}^{N-1*} \frac{g^{(j)}_{\frac{\vec{k}}{n}}(0, s)}{j!} \oplus \mathcal{R}_N(0, 1, s), \tag{10.42}$$

where

$$\mathcal{R}_N(0, 1, s) := \frac{1}{(N-1)!} \odot (FR) \int_0^1 (1-\theta)^{N-1} \odot g^{(N)}_{\frac{\vec{k}}{n}}(\theta, s)\, d\theta, \tag{10.43}$$

$\forall\, s \in X$.

Here for $j = 1, \ldots, N$, we obtain

$$g^{(j)}_{\frac{\vec{k}}{n}}(\theta, s) = \left[\left(\sum_{i=1}^{d*}\left(\frac{k_i}{n} - x_i + \frac{T_i}{n^{1-\alpha}}\right) \odot \frac{\partial}{\partial x_i}\right)^j f\right](x_1, x_2, \ldots, x_d, s), \tag{10.44}$$

$0 \leq \theta \leq 1, \forall\, s \in X$.

More precisely we have for $j = 1, \ldots, N$, that

$$g^{(j)}_{\frac{\vec{k}}{n}}(0, s) = \sum_{\substack{\alpha := (\alpha_1, \ldots, \alpha_d),\ \alpha_i \in \mathbb{Z}^+ \\ i = 1, \ldots, d,\ |\alpha| := \sum_{i=1}^{d} \alpha_i = j}}^{*} \left(\frac{j!}{\prod_{i=1}^{d} \alpha_i!}\right)\left(\prod_{i=1}^{d}\left(\frac{k_i}{n} - x_i + \frac{T_i}{n^{1-\alpha}}\right)^{\alpha_i}\right)$$

$$\odot f_\alpha\left(\vec{x} - \frac{\vec{T}}{n^{1-\alpha}}, s\right), \tag{10.45}$$

$\forall\, s \in X$, and

$$g^{(N)}_{\frac{\vec{k}}{n}}(\theta, s) = \sum^{*}_{\substack{\alpha:=(\alpha_1,\ldots,\alpha_d),\ \alpha_i \in \mathbb{Z}^+ \\ i=1,\ldots,d,\ |\alpha|:=\sum_{i=1}^{d}\alpha_i = N}} \left(\frac{N!}{\prod_{i=1}^{d}\alpha_i!} \right) \left(\prod_{i=1}^{d} \left(\frac{k_i}{n} - x_i + \frac{T_i}{n^{1-\alpha}} \right)^{\alpha_i} \right) \tag{10.46}$$

$$\odot f_{\alpha}\left(\vec{x} - \frac{\vec{T}}{n^{1-\alpha}} + \theta\left(\frac{\vec{k}}{n} - \vec{x} + \frac{\vec{T}}{n^{1-\alpha}} \right), s \right),$$

$0 \leq \theta \leq 1$, $\forall\, s \in X$.

Multiplying (10.42) by $\frac{b\left(n^{1-\alpha}\left(\vec{x} - \frac{\vec{k}}{n}\right)\right)}{V(\vec{x})}$ and applying $\sum^{\lfloor n\vec{x} + \vec{T} n^{\alpha} \rfloor *}_{\vec{k} = \lceil n\vec{x} - \vec{T} n^{\alpha} \rceil}$ to both sides, we obtain

$$(M_n(f))(\vec{x}, s) = f\left(\vec{x} - \frac{\vec{T}}{n^{1-\alpha}}, s \right) \oplus \sum^{\lfloor n\vec{x} + \vec{T} n^{\alpha} \rfloor *}_{\vec{k} = \lceil n\vec{x} - \vec{T} n^{\alpha} \rceil} \sum_{j=1}^{N-1 *} \frac{g^{(j)}_{\frac{\vec{k}}{n}}(0, s)}{j!} \tag{10.47}$$

$$\odot \frac{b\left(n^{1-\alpha}\left(\vec{x} - \frac{\vec{k}}{n}\right)\right)}{V(\vec{x})} \oplus \sum^{\lfloor n\vec{x} + \vec{T} n^{\alpha} \rfloor *}_{\vec{k} = \lceil n\vec{x} - \vec{T} n^{\alpha} \rceil} \mathcal{R}_N(0, 1, s) \odot \frac{b\left(n^{1-\alpha}\left(\vec{x} - \frac{\vec{k}}{n}\right)\right)}{V(\vec{x})},$$

$\forall\, s \in X$.

Next we observe

$$D\left((M_n(f))(\vec{x}, s), f\left(\vec{x} - \frac{\vec{T}}{n^{1-\alpha}}, s \right) \right) \overset{((10.47),\,(10.1))}{=}$$

$$D\left(\sum^{\lfloor n\vec{x} + \vec{T} n^{\alpha} \rfloor *}_{\vec{k} = \lceil n\vec{x} - \vec{T} n^{\alpha} \rceil} \sum_{j=1}^{N-1 *} \frac{g^{(j)}_{\frac{\vec{k}}{n}}(0, s)}{j!} \odot \frac{b\left(n^{1-\alpha}\left(\vec{x} - \frac{\vec{k}}{n}\right)\right)}{V(\vec{x})} \oplus \right.$$

$$\sum_{\overrightarrow{k}=\lceil n\overrightarrow{x}-\overrightarrow{T}n^{\alpha}\rceil}^{[n\overrightarrow{x}+\overrightarrow{T}n^{\alpha}]*} \mathcal{R}_N(0,1,s)\odot\frac{b\left(n^{1-\alpha}\left(\overrightarrow{x}-\frac{\overrightarrow{k}}{n}\right)\right)}{V\left(\overrightarrow{x}\right)},\widetilde{o}\right)\overset{(10.1)}{=} \tag{10.48}$$

$$D\left(\sum_{\overrightarrow{k}=\lceil n\overrightarrow{x}-\overrightarrow{T}n^{\alpha}\rceil}^{[n\overrightarrow{x}+\overrightarrow{T}n^{\alpha}]*}\sum_{j=1}^{N-1*}\frac{g_{\frac{\overrightarrow{k}}{n}}^{(j)}(0,s)}{j!}\odot\frac{b\left(n^{1-\alpha}\left(\overrightarrow{x}-\frac{\overrightarrow{k}}{n}\right)\right)}{V\left(\overrightarrow{x}\right)}\right.$$

$$\oplus\sum_{\overrightarrow{k}=\lceil n\overrightarrow{x}-\overrightarrow{T}n^{\alpha}\rceil}^{[n\overrightarrow{x}+\overrightarrow{T}n^{\alpha}]*}\mathcal{R}_N(0,1,s)\odot\frac{b\left(n^{1-\alpha}\left(\overrightarrow{x}-\frac{\overrightarrow{k}}{n}\right)\right)}{V\left(\overrightarrow{x}\right)},$$

$$\left.\sum_{\overrightarrow{k}=\lceil n\overrightarrow{x}-\overrightarrow{T}n^{\alpha}\rceil}^{[n\overrightarrow{x}+\overrightarrow{T}n^{\alpha}]*}\frac{g_{\frac{\overrightarrow{k}}{n}}^{(N)}(0,s)}{N!}\odot\frac{b\left(n^{1-\alpha}\left(\overrightarrow{x}-\frac{\overrightarrow{k}}{n}\right)\right)}{V\left(\overrightarrow{x}\right)}\right)\overset{(10.1)}{\leq}$$

$$\sum_{j=1}^{N}\sum_{\overrightarrow{k}=\lceil n\overrightarrow{x}-\overrightarrow{T}n^{\alpha}\rceil}^{[n\overrightarrow{x}+\overrightarrow{T}n^{\alpha}]}\frac{b\left(n^{1-\alpha}\left(\overrightarrow{x}-\frac{\overrightarrow{k}}{n}\right)\right)}{V\left(\overrightarrow{x}\right)j!}D\left(g_{\frac{\overrightarrow{k}}{n}}^{(j)}(0,s),\widetilde{o}\right)+ \tag{10.49}$$

$$\sum_{\overrightarrow{k}=\lceil n\overrightarrow{x}-\overrightarrow{T}n^{\alpha}\rceil}^{[n\overrightarrow{x}+\overrightarrow{T}n^{\alpha}]}\frac{b\left(n^{1-\alpha}\left(\overrightarrow{x}-\frac{\overrightarrow{k}}{n}\right)\right)}{V\left(\overrightarrow{x}\right)}D\left(\mathcal{R}_N(0,1,s),\frac{g_{\frac{\overrightarrow{k}}{n}}^{(N)}(0,s)}{N!}\right).$$

So we obtain

$$D\left((M_n(f))\left(\overrightarrow{x},s\right),(T_n(f))\left(\overrightarrow{x},s\right)\right)\leq$$

$$\sum_{j=1}^{N}\sum_{\overrightarrow{k}=\lceil n\overrightarrow{x}-\overrightarrow{T}n^{\alpha}\rceil}^{[n\overrightarrow{x}+\overrightarrow{T}n^{\alpha}]}\frac{b\left(n^{1-\alpha}\left(\overrightarrow{x}-\frac{\overrightarrow{k}}{n}\right)\right)}{V\left(\overrightarrow{x}\right)j!}D\left(g_{\frac{\overrightarrow{k}}{n}}^{(j)}(0,s),\widetilde{o}\right)+ \tag{10.50}$$

$$\sum_{\overrightarrow{k}=\lceil n\overrightarrow{x}-\overrightarrow{T}n^{\alpha}\rceil}^{[n\overrightarrow{x}+\overrightarrow{T}n^{\alpha}]}\frac{b\left(n^{1-\alpha}\left(\overrightarrow{x}-\frac{\overrightarrow{k}}{n}\right)\right)}{V\left(\overrightarrow{x}\right)}D\left(\mathcal{R}_N(0,1,s),\frac{g_{\frac{\overrightarrow{k}}{n}}^{(N)}(0,s)}{N!}\right),$$

$\forall\, s\in X$.

Notice that

$$\frac{g_{\frac{\vec{k}}{n}}^{(N)}(0,s)}{N!} = \frac{1}{(N-1)!} \odot (FR) \int_0^1 (1-\theta)^{N-1} \odot g_{\frac{\vec{k}}{n}}^{(N)}(0,s)\, d\theta. \tag{10.51}$$

We estimate

$$D\left(\mathcal{R}_N(0,1,s), \frac{g_{\frac{\vec{k}}{n}}^{(N)}(0,s)}{N!}\right) \overset{(10.51)}{=}$$

$$D\left(\frac{1}{(N-1)!} \odot (FR) \int_0^1 (1-\theta)^{N-1} \odot g_{\frac{\vec{k}}{n}}^{(N)}(\theta,s)\, d\theta,\right.$$

$$\left.\frac{1}{(N-1)!} \odot (FR) \int_0^1 (1-\theta)^{N-1} \odot g_{\frac{\vec{k}}{n}}^{(N)}(0,s)\, d\theta\right) \overset{(10.9)}{\leq} \tag{10.52}$$

$$\frac{1}{(N-1)!} \int_0^1 (1-\theta)^{N-1} D\left(g_{\frac{\vec{k}}{n}}^{(N)}(\theta,s), g_{\frac{\vec{k}}{n}}^{(N)}(0,s)\right) d\theta \overset{(10.46)}{\leq}$$

$$\frac{1}{(N-1)!} \int_0^1 (1-\theta)^{N-1} \left[\sum_{|\alpha|=N} \left(\frac{N!}{\prod_{i=1}^{d} \alpha_i!}\right) \left(\prod_{i=1}^{d} \left(\frac{k_i}{n} - x_i + \frac{T_i}{n^{1-\alpha}}\right)^{\alpha_i}\right)\right.$$

$$\left. D\left(f_\alpha\left(\vec{x} - \frac{\vec{T}}{n^{1-\alpha}} + \theta\left(\frac{\vec{k}}{n} - \vec{x} + \frac{\vec{T}}{n^{1-\alpha}}\right), s\right), f_\alpha\left(\vec{x} - \frac{\vec{T}}{n^{1-\alpha}}, s\right)\right)\right] d\theta \leq$$

$$\frac{1}{(N-1)!} \left[\sum_{|\alpha|=N} \left(\frac{N!}{\prod_{i=1}^{d} \alpha_i!}\right) \left(\prod_{i=1}^{d} \left(\frac{2T_i}{n^{1-\alpha}}\right)^{\alpha_i}\right) \int_0^1 (1-\theta)^{N-1}\right.$$

$$\left. D\left(f_\alpha\left(\vec{x} - \frac{\vec{T}}{n^{1-\alpha}} + \theta\left(\frac{\vec{k}}{n} - \vec{x} + \frac{\vec{T}}{n^{1-\alpha}}\right), s\right), f_\alpha\left(\vec{x} - \frac{\vec{T}}{n^{1-\alpha}}, s\right)\right) d\theta\right]. \tag{10.53}$$

Therefore it holds

$$\int_X D\left(\mathcal{R}_N(0,1,s), \frac{g^{(N)}_{\frac{\vec{k}}{n}}(0,s)}{N!}\right) P(ds) \overset{\text{(by (10.53) and Tonelli's (10.11) theorem)}}{\leq}$$

$$\frac{1}{(N-1)!}\left[\sum_{|\alpha|=N}\left(\frac{N!}{\prod_{i=1}^{d}\alpha_i!}\right)\left(\prod_{i=1}^{d}\left(\frac{2T_i}{n^{1-\alpha}}\right)^{\alpha_i}\right)\int_0^1 (1-\theta)^{N-1}\right. \tag{10.54}$$

$$\left(\int_X D\left(f_\alpha\left(\vec{x}-\frac{\vec{T}}{n^{1-\alpha}}+\theta\left(\frac{\vec{k}}{n}-\vec{x}+\frac{\vec{T}}{n^{1-\alpha}}\right),s\right),\right.\right.$$

$$\left.\left.f_\alpha\left(\vec{x}-\frac{\vec{T}}{n^{1-\alpha}},s\right)\right)P(ds)\right)d\theta\Bigg] \leq$$

$$\frac{1}{(N-1)!}\left[\sum_{|\alpha|=N}\left(\frac{N!}{\prod_{i=1}^{d}\alpha_i!}\right)\left(\prod_{i=1}^{d}\left(\frac{2T_i}{n^{1-\alpha}}\right)^{\alpha_i}\right)\right.$$

$$\left.\int_0^1 (1-\theta)^{N-1}\,\Omega_1^{(\mathcal{F})}\left(f_\alpha, \theta\left(\sum_{i=1}^{d}\left(\frac{k_i}{n}-x_i+\frac{T_i}{n^{1-\alpha}}\right)\right)\right)_{L^1} d\theta\right] \leq$$

$$\sum_{|\alpha|=N}\left(\frac{1}{\prod_{i=1}^{d}\alpha_i!}\right)\left(\prod_{i=1}^{d}\left(\frac{2T_i}{n^{1-\alpha}}\right)^{\alpha_i}\right)\Omega_1^{(\mathcal{F})}\left(f_\alpha, \frac{2\left(\sum_{i=1}^{d}T_i\right)}{n^{1-\alpha}}\right)_{L^1} \leq \tag{10.55}$$

$$\left[\sum_{|\alpha|=N}\left(\frac{1}{\prod_{i=1}^{d}\alpha_i!}\right)\left(\prod_{i=1}^{d}\left(\frac{2T_i}{n^{1-\alpha}}\right)^{\alpha_i}\right)\right]\Omega_{1,N}^{(\mathcal{F})}\left(f_{\alpha}^{\max},\frac{2\left(\sum_{i=1}^{d}T_i\right)}{n^{1-\alpha}}\right)_{L^1}.$$

We have proved that

$$\sum_{\vec{k}=\lceil n\vec{x}-\vec{T}n^{\alpha}\rceil}^{[n\vec{x}+\vec{T}n^{\alpha}]}\frac{b\left(n^{1-\alpha}\left(\vec{x}-\frac{\vec{k}}{n}\right)\right)}{V\left(\vec{x}\right)}\left(\int_X D\left(\mathcal{R}_N(0,1,s),\frac{g_{\frac{\vec{k}}{n}}^{(N)}(0,s)}{N!}\right)P(ds)\right)$$

$$\leq\left[\sum_{|\alpha|=N}\left(\frac{1}{\prod_{i=1}^{d}\alpha_i!}\right)\left(\prod_{i=1}^{d}\left(\frac{2T_i}{n^{1-\alpha}}\right)^{\alpha_i}\right)\right]\Omega_{1,N}^{(\mathcal{F})}\left(f_{\alpha}^{\max},\frac{2\left(\sum_{i=1}^{d}T_i\right)}{n^{1-\alpha}}\right)_{L^1}. \tag{10.56}$$

Next we extimate

$$\frac{1}{j!}D\left(g_{\frac{\vec{k}}{n}}^{(j)}(0,s),\widetilde{o}\right)=$$

$$\frac{1}{j!}D\left(\sum_{|\alpha|=j}^{*}\left(\frac{j!}{\prod_{i=1}^{d}\alpha_i!}\right)\left(\prod_{i=1}^{d}\left(\frac{k_i}{n}-x_i+\frac{T_i}{n^{1-\alpha}}\right)^{\alpha_i}\right)\odot f_{\alpha}\left(\vec{x}-\frac{\vec{T}}{n^{1-\alpha}},s\right),\widetilde{o}\right)\leq \tag{10.57}$$

$$\sum_{|\alpha|=j}\left(\frac{1}{\prod_{i=1}^{d}\alpha_i!}\right)\left(\prod_{i=1}^{d}\left(\frac{k_i}{n}-x_i+\frac{T_i}{n^{1-\alpha}}\right)^{\alpha_i}\right)D\left(f_{\alpha}\left(\vec{x}-\frac{\vec{T}}{n^{1-\alpha}},s\right),\widetilde{o}\right)\leq$$

$$\sum_{|\alpha|=j}\left(\frac{1}{\prod_{i=1}^{d}\alpha_i!}\right)\left(\prod_{i=1}^{d}\left(\frac{2T_i}{n^{1-\alpha}}\right)^{\alpha_i}\right)$$

$$\left[D\left(f_{\alpha}\left(\overrightarrow{x}, s\right), \widetilde{o}\right) + D\left(f_{\alpha}\left(\overrightarrow{x} - \frac{\overrightarrow{T}}{n^{1-\alpha}}, s\right), f_{\alpha}\left(\overrightarrow{x}, s\right)\right)\right]. \tag{10.58}$$

Consequently we have

$$\frac{1}{j!}\int_{X} D\left(g_{\frac{k}{n}}^{(j)}(0, s), \widetilde{o}\right) P(ds) \leq \sum_{|\alpha|=j} \left(\frac{1}{\prod_{i=1}^{d} \alpha_i!}\right)\left(\prod_{i=1}^{d}\left(\frac{2T_i}{n^{1-\alpha}}\right)^{\alpha_i}\right)$$

$$\left\{\int_{X} D\left(f_{\alpha}\left(\overrightarrow{x}, s\right), \widetilde{o}\right) P(ds) + \Omega_1^{(\mathcal{F})}\left(f_{\alpha}, \frac{\sum_{i=1}^{d} T_i}{n^{1-\alpha}}\right)_{L^1}\right\}. \tag{10.59}$$

Furthermore it holds

$$\sum_{j=1}^{N} \sum_{\overrightarrow{k}=\lceil n\overrightarrow{x}-\overrightarrow{T}n^{\alpha}\rceil}^{[n\overrightarrow{x}+\overrightarrow{T}n^{\alpha}]} \frac{b\left(n^{1-\alpha}\left(\overrightarrow{x} - \frac{\overrightarrow{k}}{n}\right)\right)}{V\left(\overrightarrow{x}\right) j!} \int_{X} D\left(g_{\frac{\overrightarrow{k}}{n}}^{(j)}(0, s), \widetilde{o}\right) P(ds) \leq$$

$$\sum_{j=1}^{N}\left[\sum_{|\alpha|=j} \left(\frac{1}{\prod_{i=1}^{d} \alpha_i!}\right)\left(\prod_{i=1}^{d}\left(\frac{2T_i}{n^{1-\alpha}}\right)^{\alpha_i}\right)\right.$$

$$\left.\left\{\int_{X} D\left(f_{\alpha}\left(\overrightarrow{x}, s\right), \widetilde{o}\right) P(ds) + \Omega_1^{(\mathcal{F})}\left(f_{\alpha}, \frac{\sum_{i=1}^{d} T_i}{n^{1-\alpha}}\right)_{L^1}\right\}\right]. \tag{10.60}$$

By (10.50) we get

$$\int_{X} D\left(\left(M_n(f)\right)\left(\overrightarrow{x}, s\right), \left(T_n(f)\right)\left(\overrightarrow{x}, s\right)\right) P(ds) \leq$$

$$\sum_{j=1}^{N} \sum_{\vec{k}=\lceil n\vec{x}-\vec{T}n^{\alpha}\rceil}^{[n\vec{x}+\vec{T}n^{\alpha}]} \frac{b\left(n^{1-\alpha}\left(\vec{x}-\frac{\vec{k}}{n}\right)\right)}{V\left(\vec{x}\right) j!} \int_X D\left(g_{\frac{\vec{k}}{n}}^{(j)}(0,s), \tilde{o}\right) P(ds) + \quad (10.61)$$

$$\sum_{\vec{k}=\lceil n\vec{x}-\vec{T}n^{\alpha}\rceil}^{[n\vec{x}+\vec{T}n^{\alpha}]} \frac{b\left(n^{1-\alpha}\left(\vec{x}-\frac{\vec{k}}{n}\right)\right)}{V\left(\vec{x}\right)} \left(\int_X D\left(\mathcal{R}_N(0,1,s), \frac{g_{\frac{\vec{k}}{n}}^{(N)}(0,s)}{N!}\right) P(ds)\right)$$

$$\overset{((10.56),(10.60))}{\leq} \sum_{j=1}^{N} \left[\sum_{|\alpha|=j} \left(\frac{1}{\prod_{i=1}^{d} \alpha_i!} \right) \left(\prod_{i=1}^{d} \left(\frac{2T_i}{n^{1-\alpha}} \right)^{\alpha_i} \right) \right.$$

$$\left. \left\{ \int_X D\left(f_\alpha\left(\vec{x}, s\right), \tilde{o}\right) P(ds) + \Omega_1^{(\mathcal{F})} \left(f_\alpha, \frac{\sum_{i=1}^{d} T_i}{n^{1-\alpha}} \right)_{L^1} \right\} \right] + \quad (10.62)$$

$$\left[\sum_{|\alpha|=j} \left(\frac{1}{\prod_{i=1}^{d} \alpha_i!} \right) \left(\prod_{i=1}^{d} \left(\frac{2T_i}{n^{1-\alpha}} \right)^{\alpha_i} \right) \right] \Omega_{1,N}^{(\mathcal{F})} \left(f_\alpha^{\max}, \frac{2\left(\sum_{i=1}^{d} T_i\right)}{n^{1-\alpha}} \right)_{L^1},$$

proving (10.38). ■

Remark 10.39 Inequality (10.38) reveals that the operators M_n behave in good approximation like the simple operators T_n. So T_n is a good simplification of M_n.

We give the following definition

Definition 10.40 Let the nonnegative function $S : \mathbb{R}^d \to \mathbb{R}$, $d \geq 1$, S has compact support $\mathcal{B}^* := \prod_{i=1}^{d} [-T_i, T_i]$, $T_i > 0$ and is nondecreasing there for each coordinate. S can be continuous only on either $\prod_{i=1}^{d} (-\infty, T_i]$ or $\mathcal{B}^*$ and can have jump discontinuities. We call S the multivariate "squashing function" (see also [10]).

Comment 10.41 *If the operators M_n, see (10.23), replace b by S, we derive the normalized "squashing" operators K_n. Then Theorems 10.35, 10.38 remain valid for K_n, just replace M_n by K_n there.*

References

1. G.A. Anastassiou, Rate of convergence of some neural network operators to the unit univariate case. J. Math. Anal. Appl. **212**, 237–262 (1997)
2. G.A. Anastassiou, *Quantitative Approximation* (Chapman & Hall/CRC, Boca Raton, 2001)
3. G.A. Anastassiou, Rate of convergence of fuzzy neural network operators, univariate case. J. Fuzzy Math. **10**(3), 755–780 (2002)
4. G.A. Anastassiou, *Univariate Fuzzy-random Neural Network Approximation operators*, Computers and Mathematics with Applications, Special issue/ Proceedings, ed. by G. Anastassiou. Computational Methods in Analysis, vol. 48, AMS meeting in Orlando. Florida, 2004, pp. 1263–1283, Nov 2002
5. G.A. Anastassiou, Multivariate stochastic korovkin theory given quantitatively. Math. Comput. Model. **48**, 558–580 (2008)
6. G.A. Anastassiou, *Fuzzy Mathematics: Approximation Theory* (Springer, New York, 2010)
7. G.A. Anastassiou, Multivariate fuzzy-random normalized neural network approximation operators. Ann. Fuzzy Math. Inform. **6**(1), 191–212 (2013)
8. G.A. Anastassiou, Higher order multivariate fuzzy approximation by basic neural network operators. Cubo **16**(03), 21–35 (2014)
9. G.A. Anastassiou, S. Gal, On a fuzzy trigonometric approximation theorem of Weierstrass-type. J. Fuzzy Math. **9**(3), 701–708 (2001)
10. P. Cardaliaguet, G. Euvrard, Approximation of a function and its derivative with a neural network. Neural Netw. **5**, 207–220 (1992)
11. R.M. Dudley, *Real Analysis and Probability* (Wadsworth & Brooks/Cole Mathematics Series, Pacific Grove, 1989)
12. S. Gal, *Approximation Theory in Fuzzy Setting*, ed. by G. Anastassiou. Chapter 13 in Handbook of Analytic-Computational Methods in Applied Mathematics (Chapman & Hall/CRC, Boca Raton, 2000), pp. 617–666
13. C. Wu, Z. Gong, On Henstock integral of fuzzy number valued functions (I). Fuzzy Sets Syst. **115**(3), 377–391 (2000)
14. L.A. Zadeh, Fuzzy sets. Inform. Control **8**, 338–353 (1965)

Chapter 11
Fuzzy Fractional Approximations by Fuzzy Normalized Bell and Squashing Type Neural Networks

This chapter deals with the determination of the fuzzy fractional rate of convergence to the unit to some fuzzy neural network operators, namely, the fuzzy normalized bell and "squashing" type operators. This is given through the fuzzy moduli of continuity of the involved right and left fuzzy Caputo fractional derivatives of the approximated function and they appear in the right-hand side of the associated fuzzy Jackson type inequalities. It follows [14].

11.1 Introduction

The Cardaliaguet-Euvrard operators were studied extensively in [15], where the authors among many other things proved that these operators converge uniformly on compacta, to the unit over continuous and bounded functions. Our fuzzy "normalized bell and squashing type operators" are motivated and inspired by the "bell" and "squashing functions" of [15]. The work in [15] is qualitative where the used bell-shaped function is general. However, our work, though greatly motivated by [15], is quantitative and the used bell-shaped and "squashing" functions are of compact support. We produce a series of fuzzy Jackson type inequalities giving close upper bounds to the errors in approximating the unit operator by the above fuzzy neural network induced operators. All involved constants there are well determined. These are poitwise estimates involving the first fuzzy moduli of continuity of the engaged right and left fuzzy Caputo fractional derivatives of the function under approximation. We give all necessary background of fuzzy and fractional calculus.

Initial work of the subject was done in [1, 8, 13]. These works motivated the current chapter.

G.A. Anastassiou, *Intelligent Systems II: Complete Approximation by Neural Network Operators*, Studies in Computational Intelligence 608,
DOI 10.1007/978-3-319-20505-2_11

11.2 Fuzzy Mathematical Analysis Background

We need the following basic background

Definition 11.1 (*see* [24]) Let $\mu : \mathbb{R} \to [0, 1]$ with the following properties:

(i) is normal, i.e., $\exists\, x_0 \in \mathbb{R} : \mu(x_0) = 1$.
(ii) $\mu(\lambda x + (1-\lambda) y) \geq \min\{\mu(x), \mu(y)\}$, $\forall\, x, y \in \mathbb{R}$, $\forall\, \lambda \in [0, 1]$ (μ is called a convex fuzzy subset).
(iii) μ is upper semicontinuous on $\mathbb{R}$, i.e., $\forall\, x_0 \in \mathbb{R}$ and $\forall\, \varepsilon > 0$, $\exists$ neighborhood $V(x_0) : \mu(x) \leq \mu(x_0) + \varepsilon$, $\forall\, x \in V(x_0)$.
(iv) the set $\overline{\text{supp}(\mu)}$ is compact in $\mathbb{R}$ (where $\text{supp}(\mu) := \{x \in \mathbb{R} : \mu(x) > 0\}$).

We call μ a fuzzy real number. Denote the set of all μ with $\mathbb{R}_{\mathcal{F}}$.
E.g., $\chi_{\{x_0\}} \in \mathbb{R}_{\mathcal{F}}$, for any $x_0 \in \mathbb{R}$, where $\chi_{\{x_0\}}$ is the characteristic function at x_0.
For $0 < r \leq 1$ and $\mu \in \mathbb{R}_{\mathcal{F}}$ define $[\mu]^r := \{x \in \mathbb{R} : \mu(x) \geq r\}$ and $[\mu]^0 := \overline{\{x \in \mathbb{R} : \mu(x) \geq 0\}}$.

Then it is well known that for each $r \in [0, 1]$, $[\mu]^r$ is a closed and bounded interval of $\mathbb{R}$ [20].

For $u, v \in \mathbb{R}_{\mathcal{F}}$ and $\lambda \in \mathbb{R}$, we define uniquely the sum $u \oplus v$ and the product $\lambda \odot u$ by

$$[u \oplus v]^r = [u]^r + [v]^r, \quad [\lambda \odot u]^r = \lambda [u]^r, \quad \forall\, r \in [0, 1],$$

where

$[u]^r + [v]^r$ means the usual addition of two intervals (as subsets of $\mathbb{R}$) and $\lambda [u]^r$ means the usual product between a scalar and a subset of $\mathbb{R}$ (see, e.g., [24]).
Notice $1 \odot u = u$ and it holds

$$u \oplus v = v \oplus u, \quad \lambda \odot u = u \odot \lambda.$$

If $0 \leq r_1 \leq r_2 \leq 1$ then $[u]^{r_2} \subseteq [u]^{r_1}$. Actually $[u]^r = \left[u_-^{(r)}, u_+^{(r)}\right]$, where $u_-^{(r)} \leq u_+^{(r)}$, $u_-^{(r)}, u_+^{(r)} \in \mathbb{R}$, $\forall\, r \in [0, 1]$.

For $\lambda > 0$ one has $\lambda u_\pm^{(r)} = (\lambda \odot u)_\pm^{(r)}$, respectively.

Define

$$D : \mathbb{R}_{\mathcal{F}} \times \mathbb{R}_{\mathcal{F}} \to \mathbb{R}_+$$

by

$$D(u, v) := \sup_{r \in [0,1]} \max\left\{\left|u_-^{(r)} - v_-^{(r)}\right|, \left|u_+^{(r)} - v_+^{(r)}\right|\right\},$$

where

$$[v]^r = \left[v_-^{(r)}, v_+^{(r)}\right]; \quad u, v \in \mathbb{R}_{\mathcal{F}}.$$

We have that D is a metric on $\mathbb{R}_{\mathcal{F}}$.

Then $(\mathbb{R}_{\mathcal{F}}, D)$ is a complete metric space, see [24, 25].

Here $\overset{*}{\sum}$ stands for fuzzy summation and $\widetilde{0} : \chi_{\{0\}} \in \mathbb{R}_{\mathcal{F}}$ is the neural element with respect to $\oplus$, i.e.,

$$u \oplus \widetilde{0} = \widetilde{0} \oplus u = u, \forall\, u \in \mathbb{R}_{\mathcal{F}}.$$

Denote

$$D^* (f, g) := \sup_{x \in X \subseteq \mathbb{R}} D (f, g),$$

where $f, g : X \to \mathbb{R}_{\mathcal{F}}$.

We mention

Definition 11.2 Let $f : X \subseteq \mathbb{R} \to \mathbb{R}_{\mathcal{F}}$, X interval, we define the (first) fuzzy modulus of continuity of f by

$$\omega_1^{(\mathcal{F})} (f, \delta)_X = \sup_{x, y \in X,\ |x - y| \le \delta} D (f (x), f (y)), \delta > 0.$$

We define by $C_{\mathcal{F}}^U (\mathbb{R})$ the space of fuzzy uniformly continuous functions from $\mathbb{R} \to \mathbb{R}_{\mathcal{F}}$, also $C_{\mathcal{F}} (\mathbb{R})$ is the space of fuzzy continuous functions on $\mathbb{R}$, and $C_b (\mathbb{R}, \mathbb{R}_{\mathcal{F}})$ is the fuzzy continuous and bounded functions.

We mention

Proposition 11.3 ([4]) *Let $f \in C_{\mathcal{F}}^U (X)$. Then $\omega_1^{(\mathcal{F})} (f, \delta)_X < \infty$, for any $\delta > 0$.*

Proposition 11.4 ([4]) *It holds*

$$\lim_{\delta \to 0} \omega_1^{(\mathcal{F})} (f, \delta)_X = \omega_1^{(\mathcal{F})} (f, 0)_X = 0,$$

iff $f \in C_{\mathcal{F}}^U (X)$.

Proposition 11.5 ([4]) *Here $[f]^r = \left[f_-^{(r)}, f_+^{(r)}\right], r \in [0, 1]$. Let $f \in C_{\mathcal{F}} (\mathbb{R})$. Then $f_{\pm}^{(r)}$ are equicontinuous with respect to $r \in [0, 1]$ over $\mathbb{R}$, respectively in $\pm$.*

Note 11.6 *It is clear by Propositions 11.4 and 11.5, that if $f \in C_{\mathcal{F}}^U (\mathbb{R})$, then $f_{\pm}^{(r)} \in C_U (\mathbb{R})$ (uniformly continuous on $\mathbb{R}$).*

Proposition 11.7 *Let $f : \mathbb{R} \to \mathbb{R}_{\mathcal{F}}$. Assume that $\omega_1^{\mathcal{F}} (f, \delta)_X$, $\omega_1 \left(f_-^{(r)}, \delta\right)_X$, $\omega_1 \left(f_+^{(r)}, \delta\right)_X$ are finite for any $\delta > 0$, $r \in [0, 1]$, where X any interval of $\mathbb{R}$.*

Then

$$\omega_1^{(\mathcal{F})}(f,\delta)_X = \sup_{r\in[0,1]} \max\left\{\omega_1\left(f_-^{(r)},\delta\right)_X, \omega_1\left(f_+^{(r)},\delta\right)_X\right\}.$$

Proof Similar to Proposition 14.15, p. 246 of [8]. ■

We need

Remark 11.8 ([2]). Here $r \in [0,1]$, $x_i^{(r)}, y_i^{(r)} \in \mathbb{R}$, $i = 1,\ldots,m \in \mathbb{N}$. Suppose that

$$\sup_{r\in[0,1]} \max\left(x_i^{(r)}, y_i^{(r)}\right) \in \mathbb{R}, \text{ for } i = 1,\ldots,m.$$

Then one sees easily that

$$\sup_{r\in[0,1]} \max\left(\sum_{i=1}^m x_i^{(r)}, \sum_{i=1}^m y_i^{(r)}\right) \le \sum_{i=1}^m \sup_{r\in[0,1]} \max\left(x_i^{(r)}, y_i^{(r)}\right). \tag{11.1}$$

We need

Definition 11.9 Let $x, y \in \mathbb{R}_{\mathcal{F}}$. If there exists $z \in \mathbb{R}_{\mathcal{F}} : x = y \oplus z$, then we call z the H-difference on x and y, denoted $x - y$.

Definition 11.10 ([23]) Let $T := [x_0, x_0 + \beta] \subset \mathbb{R}$, with $\beta > 0$. A function $f : T \to \mathbb{R}_{\mathcal{F}}$ is H-difference at $x \in T$ if there exists an $f'(x) \in \mathbb{R}_{\mathcal{F}}$ such that the limits (with respect to D)

$$\lim_{h\to 0+} \frac{f(x+h) - f(x)}{h}, \lim_{h\to 0+} \frac{f(x) - f(x-h)}{h} \tag{11.2}$$

exist and are equal to $f'(x)$.

We call f' the H-derivative or fuzzy derivative of f at x.

Above is assumed that the H-differences $f(x+h) - f(x)$, $f(x) - f(x-h)$ exist in $\mathbb{R}_{\mathcal{F}}$ in a neighborhood of x.

Higher order H-fuzzy derivatives are defined the obvious way, like in the real case.

We denote by $C_{\mathcal{F}}^N(\mathbb{R})$, $N \ge 1$, the space of all N-times continuously H-fuzzy differentiable functions from $\mathbb{R}$ into $\mathbb{R}_{\mathcal{F}}$.

We mention

Theorem 11.11 ([21]) *Let $f : \mathbb{R} \to \mathbb{R}_{\mathcal{F}}$ be H-fuzzy differentiable. Let $t \in \mathbb{R}$, $0 \le r \le 1$. Clearly*

$$[f(t)]^r = \left[f(t)_-^{(r)}, f(t)_+^{(r)}\right] \subseteq \mathbb{R}.$$

Then $(f(t))_{\pm}^{(r)}$ *are differentiable and*

$$\left[f'(t)\right]^r = \left[\left(f(t)_-^{(r)}\right)', \left(f(t)_+^{(r)}\right)'\right].$$

I.e.

$$(f')_{\pm}^{(r)} = \left(f_{\pm}^{(r)}\right)', \forall r \in [0,1].$$

Remark 11.12 ([3]) Let $f \in C_{\mathcal{F}}^N(\mathbb{R})$, $N \geq 1$. Then by Theorem 11.11 we obtain

$$\left[f^{(i)}(t)\right]^r = \left[\left(f(t)_-^{(r)}\right)^{(i)}, \left(f(t)_+^{(r)}\right)^{(i)}\right],$$

for $i = 0, 1, 2, \ldots, N$, and in particular we have that

$$\left(f^{(i)}\right)_{\pm}^{(r)} = \left(f_{\pm}^{(r)}\right)^{(i)},$$

for any $r \in [0,1]$, all $i = 0, 1, 2, \ldots, N$.

Note 11.13 ([3]) *Let* $f \in C_{\mathcal{F}}^N(\mathbb{R})$, $N \geq 1$. *Then by Theorem 11.11 we have* $f_{\pm}^{(r)} \in C^N(\mathbb{R})$, *for any* $r \in [0,1]$.

We need also a particular case of the Fuzzy Henstock integral ($\delta(x) = \frac{\delta}{2}$), see [24].

Definition 11.14 ([19], p. 644) Let $f : [a,b] \to \mathbb{R}_{\mathcal{F}}$. We say that f is Fuzzy-Riemann integrable to $I \in \mathbb{R}_{\mathcal{F}}$ if for any $\varepsilon > 0$, there exists $\delta > 0$ such that for any division $P = \{[u,v];\xi\}$ of $[a,b]$ with the norms $\Delta(P) < \delta$, we have

$$D\left(\sum_{P}^{*}(v-u) \odot f(\xi), I\right) < \varepsilon.$$

We write

$$I := (FR)\int_a^b f(x)\,dx. \tag{11.3}$$

We mention

Theorem 11.15 ([20]) *Let* $f : [a,b] \to \mathbb{R}_{\mathcal{F}}$ *be fuzzy continuous. Then*

$$(FR)\int_a^b f(x)\,dx$$

exists and belongs to $\mathbb{R}_{\mathcal{F}}$, *furthermore it holds*

$$\left[(FR)\int_a^b f(x)\,dx\right]^r = \left[\int_a^b (f)_-^{(r)}(x)\,dx, \int_a^b (f)_+^{(r)}(x)\,dx\right],$$

$\forall\, r \in [0, 1]$.

For the definition of general fuzzy integral we follow [22] next.

Definition 11.16 Let (Ω, Σ, μ) be a complete σ-finite measure space. We call $F : \Omega \to R_{\mathcal{F}}$ measurable iff $\forall$ closed $B \subseteq \mathbb{R}$ the function $F^{-1}(B) : \Omega \to [0, 1]$ defined by

$$F^{-1}(B)(w) := \sup_{x \in B} F(w)(x), \text{ all } w \in \Omega$$

is measurable, see [22].

Theorem 11.17 ([22]) *For* $F : \Omega \to \mathbb{R}_{\mathcal{F}}$,

$$F(w) = \left\{\left(F_-^{(r)}(w), F_+^{(r)}(w)\right) | 0 \le r \le 1\right\},$$

the following are equivalent

(1) F *is measurable,*
(2) $\forall\, r \in [0, 1]$, $F_-^{(r)}$, $F_+^{(r)}$ *are measurable.*

Following [22], given that for each $r \in [0, 1]$, $F_-^{(r)}$, $F_+^{(r)}$ are integrable we have that the parametrized representation

$$\left\{\left(\int_A F_-^{(r)} d\mu, \int_A F_+^{(r)} d\mu\right) | 0 \le r \le 1\right\} \tag{11.4}$$

is a fuzzy real number for each $A \in \Sigma$.

The last fact leads to

Definition 11.18 ([22]) A measurable function $F : \Omega \to \mathbb{R}_{\mathcal{F}}$,

$$F(w) = \left\{\left(F_-^{(r)}(w), F_+^{(r)}(w)\right) | 0 \le r \le 1\right\}$$

is integrable if for each $r \in [0, 1]$, $F_\pm^{(r)}$ are integrable, or equivalently, if $F_\pm^{(0)}$ are integrable.

In this case, the fuzzy integral of F over $A \in \Sigma$ is defined by

$$\int_A F d\mu := \left\{\left(\int_A F_-^{(r)} d\mu, \int_A F_+^{(r)} d\mu\right) | 0 \le r \le 1\right\}.$$

By [22], F is integrable iff $w \to \|F(w)\|_{\mathcal{F}}$ is real-valued integrable.

Here denote

$$\|u\|_{\mathcal{F}} := D\left(u, \widetilde{0}\right), \forall\, u \in \mathbb{R}_{\mathcal{F}}.$$

We need also

Theorem 11.19 ([22]) *Let $F, G : \Omega \to \mathbb{R}_{\mathcal{F}}$ be integrable. Then*

(1). Let $a, b \in \mathbb{R}$, then $aF + bG$ is integrable and for each $A \in \Sigma$,

$$\int_A (aF + bG)\, d\mu = a \int_A F d\mu + b \int_A G d\mu;$$

(2) $D(F, G)$ is a real-valued integrable function and for each $A \in \Sigma$,

$$D\left(\int_A F d\mu, \int_A G d\mu\right) \le \int_A D(F, G)\, d\mu.$$

In particular,

$$\left\|\int_A F d\mu\right\|_{\mathcal{F}} \le \int_A \|F\|_{\mathcal{F}}\, d\mu.$$

Above μ could be the Lebesgue measure, with all the basic properties valid here too.

Basically here we have

$$\left[\int_A F d\mu\right]^r = \left[\int_A F_-^{(r)} d\mu, \int_A F_+^{(r)} d\mu\right], \tag{11.5}$$

i.e.

$$\left(\int_A F d\mu\right)_{\pm}^{(r)} = \int_A F_{\pm}^{(r)} d\mu, \; \forall\, r \in [0, 1].$$

We need

Definition 11.20 Let $f : \mathbb{R} \to \mathbb{R}$, $\nu \ge 0$, $n = \lceil \nu \rceil$ ($\lceil \cdot \rceil$ is the ceiling of the number), $f \in AC^n([a, b])$ (space of functions f with $f^{(n-1)} \in AC([a, b])$, absolutely continuous functions), $\forall\, [a, b] \subset \mathbb{R}$. We call left Caputo fractional derivative (see [16], pp. 49–52) the function

$$D_{*a}^{\nu} f(x) = \frac{1}{\Gamma(n - \nu)} \int_a^x (x - t)^{n-\nu-1} f^{(n)}(t)\, dt, \tag{11.6}$$

$\forall\, x \ge a$, where Γ is the gamma function $\Gamma(\nu) = \int_0^\infty e^{-t} t^{\nu-1} dt$, $\nu > 0$. Notice $D_{*a}^{\nu} f \in L_1([a, b])$ and $D_{*a}^{\nu} f$ exists a.e. on $[a, b]$, $\forall\, b > a$. We set $D_{*a}^{0} f(x) = f(x)$, $\forall\, x \in [a, \infty)$.

Lemma 11.21 ([10]) *Let* $\nu > 0$, $\nu \notin \mathbb{N}$, $n = \lceil \nu \rceil$, $f \in C^{n-1}(\mathbb{R})$ *and* $f^{(n)} \in L_{\infty}(\mathbb{R})$. *Then* $D_{*a}^{\nu} f(a) = 0$, $\forall\, a \in \mathbb{R}$.

Definition 11.22 (*see also* [7, 17, 18]) *Let* $f : \mathbb{R} \to \mathbb{R}$, *such that* $f \in AC^m([a,b])$, $\forall\,[a,b] \subset \mathbb{R}$, $m = \lceil \alpha \rceil$, $\alpha > 0$. The right Caputo fractional derivative of order $\alpha > 0$ is given by

$$D_{b-}^{\alpha} f(x) = \frac{(-1)^m}{\Gamma(m-\alpha)} \int_x^b (J - x)^{m-\alpha-1} f^{(m)}(J)\, dJ, \tag{11.7}$$

$\forall\, x \leq b$. We set $D_{b-}^{0} f(x) = f(x)$, $\forall\, x \in (-\infty, b]$. Notice that $D_{b-}^{\alpha} f \in L_1([a,b])$ and $D_{b-}^{\alpha} f$ exists a.e. on $[a,b]$, $\forall\, a < b$.

Lemma 11.23 ([10]) *Let* $f \in C^{m-1}(R)$, $f^{(m)} \in L_{\infty}(\mathbb{R})$, $m = \lceil \alpha \rceil$, $\alpha > 0$. *Then* $D_{b-}^{\alpha} f(b) = 0$, $\forall\, b \in \mathbb{R}$.

Convention 11.24 *We assume that*

$$D_{*x_0}^{\alpha} f(x) = 0, \text{ for } x < x_0, \tag{11.8}$$

and

$$D_{x_0-}^{\alpha} f(x) = 0, \text{ for } x > x_0, \tag{11.9}$$

for all $x, x_0 \in \mathbb{R}$.

We mention

Proposition 11.25 (by [5]) *Let* $f \in C^n(\mathbb{R})$, *where* $n = \lceil \nu \rceil$, $\nu > 0$. *Then* $D_{*a}^{\nu} f(x)$ *is continuous in* $x \in [a, \infty)$.

Also we have

Proposition 11.26 (by [5]) *Let* $f \in C^m(\mathbb{R})$, $m = \lceil \alpha \rceil$, $\alpha > 0$. *Then* $D_{b-}^{\alpha} f(x)$ *is continuous in* $x \in (-\infty, b]$.

We further mention

Proposition 11.27 (by [5]) *Let* $f \in C^{m-1}(\mathbb{R})$, $f^{(m)} \in L_{\infty}(\mathbb{R})$, $m = \lceil \alpha \rceil$, $\alpha > 0$ *and*

$$D_{*x_0}^{\alpha} f(x) = \frac{1}{\Gamma(m-\alpha)} \int_{x_0}^{x} (x-t)^{m-\alpha-1} f^{(m)}(t)\, dt, \tag{11.10}$$

for all $x, x_0 \in \mathbb{R} : x \geq x_0$.

Then $D_{*x_0}^{\alpha} f(x)$ *is continuous in* x_0.

Proposition 11.28 (by [5]) *Let* $f \in C^{m-1}(\mathbb{R})$, $f^{(m)} \in L_{\infty}(\mathbb{R})$, $m = \lceil \alpha \rceil$, $\alpha > 0$ *and*

$$D_{x_0-}^{\alpha} f(x) = \frac{(-1)^m}{\Gamma(m-\alpha)} \int_x^{x_0} (J-x)^{m-\alpha-1} f^{(m)}(J)\, dJ, \tag{11.11}$$

for all $x, x_0 \in \mathbb{R} : x_0 \geq x$.

Then $D_{x_0-}^{\alpha} f(x)$ *is continuous in* x_0.

Corollary 11.29 ([10]) *Let* $f \in C^m(\mathbb{R})$, $f^{(m)} \in L_\infty(\mathbb{R})$, $m = \lceil \alpha \rceil$, $\alpha > 0$, $\alpha \notin \mathbb{N}$, $x, x_0 \in \mathbb{R}$. *Then* $D_{*x_0}^{\alpha} f(x)$, $D_{x_0-}^{\alpha} f(x)$ *are jointly continuous functions in* (x, x_0) *from* $\mathbb{R}^2$ *into* $\mathbb{R}$.

We need

Proposition 11.30 ([10]) *Let* $f : \mathbb{R}^2 \to \mathbb{R}$ *be jointly continuous. Consider*

$$G(x) = \omega_1 (f(\cdot, x), \delta)_{[x,+\infty)}, \quad \delta > 0, \, x \in \mathbb{R}. \tag{11.12}$$

(Here ω_1 *is defined over* $[x, +\infty)$ *instead of* $\mathbb{R}$*).*

Then G *is continuous on* $\mathbb{R}$.

Proposition 11.31 ([10]) *Let* $f : \mathbb{R}^2 \to \mathbb{R}$ *be jointly continous. Consider*

$$H(x) = \omega_1 (f(\cdot, x), \delta)_{(-\infty,x]}, \quad \delta > 0, \, x \in \mathbb{R}. \tag{11.13}$$

(Here ω_1 *is defined over* $(-\infty, x]$ *instead of* $\mathbb{R}$*).*

Then H *is continuous on* $\mathbb{R}$.

By Propositions 11.30, 11.31 and Corollary 11.29 we derive

Proposition 11.32 ([10]) *Let* $f \in C^m(\mathbb{R})$, $\left\| f^{(m)} \right\|_\infty < \infty$, $m = \lceil \alpha \rceil$, $\alpha \notin \mathbb{N}$, $\alpha > 0$, $x \in \mathbb{R}$. *Then* $\omega_1 \left(D_{*x}^{\alpha} f, h \right)_{[x,+\infty)}$, $\omega_1 \left(D_{x-}^{\alpha} f, h \right)_{(-\infty,x]}$ *are continuous functions of* $x \in \mathbb{R}$, $h > 0$ *fixed.*

We make

Remark 11.33 Let g be continuous and bounded from $\mathbb{R}$ to $\mathbb{R}$. Then

$$\omega_1 (g, t) \leq 2 \|g\|_\infty < \infty. \tag{11.14}$$

Assuming that $\left(D_{*x}^{\alpha} f \right)(t)$, $\left(D_{x-}^{\alpha} f \right)(t)$, are both continuous and bounded in $(x, t) \in \mathbb{R}^2$, i.e.

$$\left\| D_{*x}^{\alpha} f \right\|_\infty \leq K_1, \forall \, x \in \mathbb{R}; \tag{11.15}$$

$$\left\| D_{x-}^{\alpha} f \right\|_\infty \leq K_2, \forall \, x \in \mathbb{R}; \tag{11.16}$$

where $K_1, K_2 > 0$, we get

$$\omega_1\left(D^{\alpha}_{*x} f, \xi\right)_{[x,+\infty)} \leq 2K_1; \qquad (11.17)$$
$$\omega_1\left(D^{\alpha}_{x-} f, \xi\right)_{(-\infty,x]} \leq 2K_2, \forall\, \xi \geq 0,$$

for each $x \in \mathbb{R}$.

Therefore, for any $\xi \geq 0$,

$$\sup_{x\in\mathbb{R}}\left[\max\left(\omega_1\left(D^{\alpha}_{*x} f, \xi\right)_{[x,+\infty)}, \omega_1\left(D^{\alpha}_{x-} f, \xi\right)_{(-\infty,x]}\right)\right] \leq 2\max\left(K_1, K_2\right) < \infty. \qquad (11.18)$$

So in our setting for $f \in C^m(\mathbb{R})$, $\left\|f^{(m)}\right\|_{\infty} < \infty$, $m = \lceil\alpha\rceil$, $\alpha \notin \mathbb{N}$, $\alpha > 0$, by Corollary 11.29 both $\left(D^{\alpha}_{*x} f\right)(t)$, $\left(D^{\alpha}_{x-} f\right)(t)$ are jointly continuous in (t, x) on $\mathbb{R}^2$. Assuming further that they are both bounded on $\mathbb{R}^2$ we get (11.18) valid. In particular, each of $\omega_1\left(D^{\alpha}_{*x} f, \xi\right)_{[x,+\infty)}, \omega_1\left(D^{\alpha}_{x-} f, \xi\right)_{(-\infty,x]}$ is finite for any $\xi \geq 0$.

Let us now assume only that $f \in C^{m-1}(\mathbb{R})$, $f^{(m)} \in L_{\infty}(\mathbb{R})$, $m = \lceil\alpha\rceil$, $\alpha > 0$, $\alpha \notin \mathbb{N}$, $x \in \mathbb{R}$. Then, by Proposition 15.114, p. 388 of [6], we find that $D^{\alpha}_{*x} f \in C([x, +\infty))$, and by [9] we obtain that $D^{\alpha}_{x-} f \in C((-\infty, x])$.

We make

Remark 11.34 Again let $f \in C^m(\mathbb{R})$, $m = \lceil\alpha\rceil$, $\alpha \notin \mathbb{N}$, $\alpha > 0$; $f^{(m)}(x) = 1$, $\forall$, $x \in \mathbb{R}$; $x_0 \in \mathbb{R}$. Notice $0 < m - \alpha < 1$. Then

$$D^{\alpha}_{*x_0} f(x) = \frac{(x - x_0)^{m-\alpha}}{\Gamma(m - \alpha + 1)}, \quad \forall\, x \geq x_0. \qquad (11.19)$$

Let us consider $x, y \geq x_0$, then

$$\left|D^{\alpha}_{*x_0} f(x) - D^{\alpha}_{*x_0} f(y)\right| = \frac{1}{\Gamma(m - \alpha + 1)}\left|(x - x_0)^{m-\alpha} - (y - x_0)^{m-\alpha}\right|$$
$$\leq \frac{|x - y|^{m-\alpha}}{\Gamma(m - \alpha + 1)}. \qquad (11.20)$$

So it is not strange to assume that

$$\left|D^{\alpha}_{*x_0} f(x_1) - D^{\alpha}_{*x_0} f(x_2)\right| \leq K\left|x_1 - x_2\right|^{\beta}, \qquad (11.21)$$

$K > 0$, $0 < \beta \leq 1$, $\forall\, x_1, x_2 \in \mathbb{R}$, $x_1, x_2 \geq x_0 \in \mathbb{R}$, where more generally it is $\left\|f^{(m)}\right\|_{\infty} < \infty$. Thus, one may assume

$$\omega_1\left(D^{\alpha}_{x-} f, \xi\right)_{(-\infty,x]} \leq M_1 \xi^{\beta_1}, \text{ and} \qquad (11.22)$$

$$\omega_1\left(D^{\alpha}_{*x} f, \xi\right)_{[x,+\infty)} \leq M_2 \xi^{\beta_2},$$

where $0 < \beta_1, \beta_2 \leq 1, \forall\, \xi > 0, M_1, M_2 > 0$; any $x \in \mathbb{R}$.

Setting $\beta = \min(\beta_1, \beta_2)$ and $M = \max(M_1, M_2)$, in that case we obtain

$$\sup_{x \in \mathbb{R}} \left\{ \max \left(\omega_1 \left(D_{x-}^{\alpha} f, \xi \right)_{(-\infty, x]}, \omega_1 \left(D_{*x}^{\alpha} f, \xi \right)_{[x, +\infty)} \right) \right\} \leq M \xi^{\beta} \to 0, \text{ as } \xi \to 0+. \tag{11.23}$$

We need

Definition 11.35 ([11]) Let $f \in C_{\mathcal{F}}([a,b])$ (fuzzy continuous on $[a,b] \subset \mathbb{R}$), $\nu > 0$.

We define the Fuzzy Fractional left Riemann-Liouville operator as

$$J_a^{\nu} f(x) := \frac{1}{\Gamma(\nu)} \odot \int_a^x (x-t)^{\nu-1} \odot f(t)\, dt, \ x \in [a,b], \tag{11.24}$$

$$J_a^0 f := f.$$

Also, we define the Fuzzy Fractional right Riemann-Liouville operator as

$$I_{b-}^{\nu} f(x) := \frac{1}{\Gamma(\nu)} \odot \int_x^b (t-x)^{\nu-1} \odot f(t)\, dt, \ x \in [a,b], \tag{11.25}$$

$$I_{b-}^0 f := f.$$

We mention

Definition 11.36 ([11]) Let $f : [a,b] \to \mathbb{R}_{\mathcal{F}}$ is called fuzzy absolutely continuous iff $\forall\, \epsilon > 0, \exists\, \delta > 0$ for every finite, pairwise disjoint, family

$$(c_k, d_k)_{k=1}^n \subseteq (a,b) \text{ with } \sum_{k=1}^n (d_k - c_k) < \delta$$

we get

$$\sum_{k=1}^n D(f(d_k), f(c_k)) < \epsilon. \tag{11.26}$$

We denote the related space of functions by $AC_{\mathcal{F}}([a,b])$.

If $f \in AC_{\mathcal{F}}([a,b])$, then $f \in C_{\mathcal{F}}([a,b])$.

It holds

Proposition 11.37 ([11]) $f \in AC_{\mathcal{F}}([a,b]) \Leftrightarrow f_{\pm}^{(r)} \in AEC([a,b]), \forall\, r \in [0,1]$ *(absolutely equicontinuous).*

We need

Definition 11.38 ([11]) We define the Fuzzy Fractional left Caputo derivative, $x \in [a,b]$.

Let $f \in C_{\mathcal{F}}^{n}([a,b])$, $n = \lceil \nu \rceil$, $\nu > 0$ ($\lceil \cdot \rceil$ denotes the ceiling). We define

$$D_{*a}^{\nu \mathcal{F}} f(x) := \frac{1}{\Gamma(n-\nu)} \odot \int_a^x (x-t)^{n-\nu-1} \odot f^{(n)}(t)\, dt$$

$$= \left\{ \left(\frac{1}{\Gamma(n-\nu)} \int_a^x (x-t)^{n-\nu-1} \left(f^{(n)}\right)_-^{(r)}(t)\, dt, \right.\right.$$

$$\left.\left. \frac{1}{\Gamma(n-\nu)} \int_a^x (x-t)^{n-\nu-1} \left(f^{(n)}\right)_+^{(r)}(t)\, dt \right) | 0 \le r \le 1 \right\}$$

$$= \left\{ \left(\frac{1}{\Gamma(n-\nu)} \int_a^x (x-t)^{n-\nu-1} \left(f_-^{(r)}\right)^{(n)}(t)\, dt, \right.\right.$$

$$\left.\left. \frac{1}{\Gamma(n-\nu)} \int_a^x (x-t)^{n-\nu-1} \left(f_+^{(r)}\right)^{(n)}(t)\, dt \right) | 0 \le r \le 1 \right\}. \tag{11.27}$$

So, we get

$$\left[D_{*a}^{\nu \mathcal{F}} f(x) \right]^r = \left[\left(\frac{1}{\Gamma(n-\nu)} \int_a^x (x-t)^{n-\nu-1} \left(f_-^{(r)}\right)^{(n)}(t)\, dt, \right.\right.$$

$$\left.\left. \frac{1}{\Gamma(n-\nu)} \int_a^x (x-t)^{n-\nu-1} \left(f_+^{(r)}\right)^{(n)}(t)\, dt \right) \right], \quad 0 \le r \le 1. \tag{11.28}$$

That is

$$\left(D_{*a}^{\nu \mathcal{F}} f(x) \right)_{\pm}^{(r)} = \frac{1}{\Gamma(n-\nu)} \int_a^x (x-t)^{n-\nu-1} \left(f_{\pm}^{(r)}\right)^{(n)}(t)\, dt = \left(D_{*a}^{\nu} \left(f_{\pm}^{(r)} \right) \right)(x),$$

see [6, 16].

I.e. we get that

$$\left(D_{*a}^{\nu \mathcal{F}} f(x) \right)_{\pm}^{(r)} = \left(D_{*a}^{\nu} \left(f_{\pm}^{(r)} \right) \right)(x), \tag{11.29}$$

$\forall\, x \in [a,b]$, in short

$$\left(D_{*a}^{\nu \mathcal{F}} f \right)_{\pm}^{(r)} = D_{*a}^{\nu} \left(f_{\pm}^{(r)} \right), \ \forall\, r \in [0,1]. \tag{11.30}$$

We need

Lemma 11.39 ([11]) *$D_{*a}^{\nu \mathcal{F}} f(x)$ is fuzzy continuous in $x \in [a,b]$.*

We need

Definition 11.40 ([11]) We define the Fuzzy Fractional right Caputo derivative, $x \in [a, b]$.

Let $f \in C^n_{\mathcal{F}}([a, b])$, $n = \lceil \nu \rceil$, $\nu > 0$. We define

$$D_{b-}^{\nu\mathcal{F}} f(x) := \frac{(-1)^n}{\Gamma(n-\nu)} \odot \int_x^b (t-x)^{n-\nu-1} \odot f^{(n)}(t)\, dt$$

$$= \left\{ \left(\frac{(-1)^n}{\Gamma(n-\nu)} \int_x^b (t-x)^{n-\nu-1} \left(f^{(n)} \right)_-^{(r)} (t)\, dt, \right. \right.$$

$$\left. \left. \frac{(-1)^n}{\Gamma(n-\nu)} \int_x^b (t-x)^{n-\nu-1} \left(f^{(n)} \right)_+^{(r)} (t)\, dt \right) | 0 \le r \le 1 \right\} \tag{11.31}$$

$$= \left\{ \left(\frac{(-1)^n}{\Gamma(n-\nu)} \int_x^b (t-x)^{n-\nu-1} \left(f_-^{(r)} \right)^{(n)} (t)\, dt, \right. \right.$$

$$\left. \left. \frac{(-1)^n}{\Gamma(n-\nu)} \int_x^b (t-x)^{n-\nu-1} \left(f_+^{(r)} \right)^{(n)} (t)\, dt \right) | 0 \le r \le 1 \right\}.$$

We get

$$\left[D_{b-}^{\nu\mathcal{F}} f(x) \right]^r = \left[\left(\frac{(-1)^n}{\Gamma(n-\nu)} \int_x^b (t-x)^{n-\nu-1} \left(f_-^{(r)} \right)^{(n)} (t)\, dt, \right. \right.$$

$$\left. \left. \frac{(-1)^n}{\Gamma(n-\nu)} \int_x^b (t-x)^{n-\nu-1} \left(f_+^{(r)} \right)^{(n)} (t)\, dt \right) \right], \ 0 \le r \le 1.$$

That is

$$\left(D_{b-}^{\nu\mathcal{F}} f(x) \right)_\pm^{(r)} = \frac{(-1)^n}{\Gamma(n-\nu)} \int_x^b (t-x)^{n-\nu-1} \left(f_\pm^{(r)} \right)^{(n)} (t)\, dt = \left(D_{b-}^{\nu} \left(f_\pm^{(r)} \right) \right)(x),$$

see [7].

I.e. we get that

$$\left(D_{b-}^{\nu\mathcal{F}} f(x) \right)_\pm^{(r)} = \left(D_{b-}^{\nu} \left(f_\pm^{(r)} \right) \right)(x), \tag{11.32}$$

$\forall\, x \in [a, b]$, in short

$$\left(D_{b-}^{\nu\mathcal{F}} f \right)_\pm^{(r)} = D_{b-}^{\nu} \left(f_\pm^{(r)} \right), \ \forall\, r \in [0, 1]. \tag{11.33}$$

Clearly,

$$D_{b-}^{\nu}\left(f_{-}^{(r)}\right) \leq D_{b-}^{\nu}\left(f_{+}^{(r)}\right), \ \forall r \in [0, 1].$$

We need

Lemma 11.41 ([11]) *$D_{b-}^{\nu\mathcal{F}} f(x)$ is fuzzy continuous in $x \in [a, b]$.*

11.3 Fractional Convergence with Rates of Real Normalized Bell Type Neural Network Operators

We need the following (see [15]).

Definition 11.42 A function $b : \mathbb{R} \to \mathbb{R}$ is said to be bell-shaped if b belongs to L^1 and its integral is nonzero, if it is nondecreasing on $(-\infty, a)$ and nonincreasing on $[a, +\infty)$, where a belongs to $\mathbb{R}$. In particular $b(x)$ is a nonnegative number and at a, b takes a global maximum; it is the center of the bell-shaped function. A bell-shaped function is said to be centered if its center is zero. The function $b(x)$ may have jump discontinuities. In this work we consider only centered bell-shaped functions of compact support $[-T, T]$, $T > 0$.

We follow [1, 8, 13, 15].

Example 11.43 (1) $b(x)$ can be the characteristic function over $[-1, 1]$.
(2) $b(x)$ can be the hat function over $[-1, 1]$, i.e.,

$$b(x) = \begin{cases} 1 + x, & -1 \leq x \leq 0, \\ 1 - x, & 0 < x \leq 1 \\ 0, & elsewhere. \end{cases}$$

These are centered bell-shaped functions of compact support.
Here we consider functions $f : \mathbb{R} \to \mathbb{R}$ that are continuous.

In [13] we studied the fractional convergence with rates over the real line, to the unit operator, of the real "normalized bell type neural network operators",

$$(H_n(f))(x) := \frac{\sum_{k=-n^2}^{n^2} f\left(\frac{k}{n}\right) b\left(n^{1-\alpha}\left(x - \frac{k}{n}\right)\right)}{\sum_{k=-n^2}^{n^2} b\left(n^{1-\alpha}\left(x - \frac{k}{n}\right)\right)}, \tag{11.34}$$

where $0 < \alpha < 1$ and $x \in \mathbb{R}$, $n \in \mathbb{N}$. The terms in the ratio of sums (11.34) can be nonzero iff

$$\left|n^{1-\alpha}\left(x - \frac{k}{n}\right)\right| \leq T, \text{ i.e. } \left|x - \frac{k}{n}\right| \leq \frac{T}{n^{1-\alpha}},$$

iff

$$nx - Tn^{\alpha} \leq k \leq nx + Tn^{\alpha}. \tag{11.35}$$

In order to have the desired order of numbers

$$-n^2 \leq nx - Tn^{\alpha} \leq nx + Tn^{\alpha} \leq n^2, \tag{11.36}$$

and $card(k)$:=cardinality$(k) \geq 1$ over $[nx - Tn^{\alpha}, nx + Tn^{\alpha}]$, we need to consider

$$n \geq \max\left(T + |x|, T^{-\frac{1}{\alpha}}\right). \tag{11.37}$$

Also notice that $card(k) \to +\infty$, as $n \to +\infty$.

Denote by $[\cdot]$ the integral part of a number, and by $\lceil \cdot \rceil$ the ceiling of the number.

Without loss of generality [13] we may assume that

$$nx - Tn^{\alpha} \leq [nx] \leq nx \leq \lceil nx \rceil \leq nx + Tn^{\alpha}. \tag{11.38}$$

So here we assume (11.37) and we have

$$(H_n(f))(x) := \frac{\sum_{k=\lceil nx - Tn^{\alpha}\rceil}^{[nx+Tn^{\alpha}]} f\left(\frac{k}{n}\right) b\left(n^{1-\alpha}\left(x - \frac{k}{n}\right)\right)}{\sum_{k=\lceil nx - Tn^{\alpha}\rceil}^{[nx+Tn^{\alpha}]} b\left(n^{1-\alpha}\left(x - \frac{k}{n}\right)\right)}. \tag{11.39}$$

We need

Theorem 11.44 ([13]) *We consider $f : \mathbb{R} \to \mathbb{R}$. Let $\beta > 0$, $N = \lceil \beta \rceil$, $\beta \notin \mathbb{N}$, $f \in AC^N([a,b])$, $\forall [a,b] \subset \mathbb{R}$, with $f^{(N)} \in L_{\infty}(\mathbb{R})$. Let also $x \in \mathbb{R}$, $T > 0$, $n \in \mathbb{N} : n \geq \max\left(T + |x|, T^{-\frac{1}{\alpha}}\right)$. We further assume that $D_{*x}^{\beta} f$, $D_{x-}^{\beta} f$ are uniformly continuous functions or continuous and bounded on $[x, +\infty)$, $(-\infty, x]$, respectively.*

Then

(1)

$$|H_n(f)(x) - f(x)| \leq \left(\sum_{j=1}^{N-1} \frac{\left|f^{(j)}(x)\right| T^j}{j! n^{(1-\alpha)j}}\right) + \frac{T^{\beta}}{\Gamma(\beta+1) n^{(1-\alpha)\beta}} \cdot \tag{11.40}$$

$$\left\{\omega_1\left(D_{*x}^{\beta} f, \frac{T}{n^{1-\alpha}}\right)_{[x,+\infty)} + \omega_1\left(D_{x-}^{\beta} f, \frac{T}{n^{1-\alpha}}\right)_{(-\infty,x]}\right\},$$

above $\sum_{j=1}^{0} \cdot = 0,$

(2)

$$\left| H_n(f)(x) - \sum_{j=1}^{N-1} \frac{f^{(j)}(x)}{j!} \left(H_n \left((\cdot - x)^j \right) \right)(x) \right| \leq \frac{T^\beta}{\Gamma(\beta+1) n^{(1-\alpha)\beta}} \cdot \tag{11.41}$$

$$\left\{ \omega_1 \left(D_{*x}^\beta f, \frac{T}{n^{1-\alpha}} \right)_{[x,+\infty)} + \omega_1 \left(D_{x-}^\beta f, \frac{T}{n^{1-\alpha}} \right)_{(-\infty,x]} \right\} =: \lambda_n(x),$$

(3) assume further that $f^{(j)}(x) = 0$, for $j = 1, \ldots, N-1$, we get

$$|H_n(f)(x) - f(x)| \leq \lambda_n(x), \tag{11.42}$$

(4) in case of $N = 1$, we obtain again

$$|H_n(f)(x) - f(x)| \leq \lambda_n(x). \tag{11.43}$$

Here we get fractionally with rates the poitwise convergence of $(H_n(f))(x) \to f(x)$, as $n \to \infty$, $x \in \mathbb{R}$.

We will use (11.40).

11.4 Fractional Convergence with Rates of Fuzzy Normalized Neural Network Operators

Here b as Definition 11.42, $n \in \mathbb{N}$ as in (11.37), and $f \in C_{\mathcal{F}}(\mathbb{R})$.

Based on (11.34), (11.39) we define the corresponding fuzzy operators

$$\left(H_n^{\mathcal{F}}(f) \right)(x) := \frac{\sum_{k=\lceil nx-Tn^\alpha \rceil}^{[nx+Tn^\alpha]*} f\left(\frac{k}{n}\right) \odot b\left(n^{1-\alpha}\left(x - \frac{k}{n}\right)\right)}{\sum_{k=\lceil nx-Tn^\alpha \rceil}^{[nx+Tn^\alpha]} b\left(n^{1-\alpha}\left(x - \frac{k}{n}\right)\right)}, \tag{11.44}$$

where $0 < \alpha < 1$, and $x \in \mathbb{R}$.

We notice that ($r \in [0, 1]$)

$$\left[\left(H_n^{\mathcal{F}}(f) \right)(x) \right]^r = \sum_{k=\lceil nx-Tn^\alpha \rceil}^{[nx+Tn^\alpha]} \left[f\left(\frac{k}{n}\right) \right]^r \frac{b\left(n^{1-\alpha}\left(x - \frac{k}{n}\right)\right)}{\sum_{k=\lceil nx-Tn^\alpha \rceil}^{[nx+Tn^\alpha]} b\left(n^{1-\alpha}\left(x - \frac{k}{n}\right)\right)} =$$

$$\sum_{k=\lceil nx-Tn^\alpha \rceil}^{[nx+Tn^\alpha]} \left[f_-^{(r)}\left(\frac{k}{n}\right), f_+^{(r)}\left(\frac{k}{n}\right) \right]^r \frac{b\left(n^{1-\alpha}\left(x - \frac{k}{n}\right)\right)}{\sum_{k=\lceil nx-Tn^\alpha \rceil}^{[nx+Tn^\alpha]} b\left(n^{1-\alpha}\left(x - \frac{k}{n}\right)\right)} = \tag{11.45}$$

$$\left[\sum_{k=\lceil nx-Tn^{\alpha}\rceil}^{[nx+Tn^{\alpha}]} f_{-}^{(r)}\left(\frac{k}{n}\right)\frac{b\left(n^{1-\alpha}\left(x-\frac{k}{n}\right)\right)}{\sum_{k=\lceil nx-Tn^{\alpha}\rceil}^{[nx+Tn^{\alpha}]} b\left(n^{1-\alpha}\left(x-\frac{k}{n}\right)\right)},\right.$$

$$\left.\sum_{k=\lceil nx-Tn^{\alpha}\rceil}^{[nx+Tn^{\alpha}]} f_{+}^{(r)}\left(\frac{k}{n}\right)\frac{b\left(n^{1-\alpha}\left(x-\frac{k}{n}\right)\right)}{\sum_{k=\lceil nx-Tn^{\alpha}\rceil}^{[nx+Tn^{\alpha}]} b\left(n^{1-\alpha}\left(x-\frac{k}{n}\right)\right)}\right] = \tag{11.46}$$

$$\left[\left(H_n\left(f_{-}^{(r)}\right)\right)(x), \left(H_n\left(f_{+}^{(r)}\right)\right)(x)\right].$$

We have proved that

$$\left(H_n^{\mathcal{F}}(f)\right)_{\pm}^{(r)} = H_n\left(f_{\pm}^{(r)}\right), \ \forall\, r \in [0,1], \tag{11.47}$$

respectively.

We present

Theorem 11.45 *Let $x \in \mathbb{R}$; then*

$$D\left(\left(H_n^{\mathcal{F}}(f)\right)(x), f(x)\right) \leq \omega_1^{(\mathcal{F})}\left(f, \frac{T}{n^{1-\alpha}}\right)_{\mathbb{R}}. \tag{11.48}$$

Notice that (11.48) gives $H_n^{\mathcal{F}}(f) \stackrel{D}{\rightarrow} f$ pointwise and uniformly, as $n \to \infty$, when $f \in C_{\mathcal{F}}^{U}(\mathbb{R})$.

Proof We observe that

$$D\left(\left(H_n^{\mathcal{F}}(f)\right)(x), f(x)\right) = \sup_{r\in[0,1]} \max\left\{\left|\left(H_n^{\mathcal{F}}(f)\right)_{-}^{(r)}(x) - f_{-}^{(r)}(x)\right|,\right.$$

$$\left.\left|\left(H_n^{\mathcal{F}}(f)\right)_{+}^{(r)}(x) - f_{+}^{(r)}(x)\right|\right\} \stackrel{(11.47)}{=}$$

$$\sup_{r\in[0,1]} \max\left\{\left|\left(H_n\left(f_{-}^{(r)}\right)\right)(x) - f_{-}^{(r)}(x)\right|, \left|\left(H_n\left(f_{+}^{(r)}\right)\right)(x) - f_{+}^{(r)}(x)\right|\right\} \stackrel{\text{(by [12])}}{\leq} \tag{11.49}$$

$$\sup_{r\in[0,1]} \max\left\{\omega_1\left(f_{-}^{(r)}, \frac{T}{n^{1-\alpha}}\right)_{\mathbb{R}}, \omega_1\left(f_{+}^{(r)}, \frac{T}{n^{1-\alpha}}\right)_{\mathbb{R}}\right\} \tag{11.50}$$

(by Proposition 11.7)

$$= \omega_1^{(\mathcal{F})}\left(f, \frac{T}{n^{1-\alpha}}\right)_{\mathbb{R}}.$$

■

We also give

Theorem 11.46 *We consider* $f : \mathbb{R} \to \mathbb{R}_{\mathcal{F}}$. *Let* $\beta > 0$, $N = \lceil \beta \rceil$, $\beta \notin \mathbb{N}$, $f \in C_{\mathcal{F}}^{N}(\mathbb{R})$, *with* $D\left(f^{(N)}(\cdot), \widetilde{o}\right) \in L_{\infty}(\mathbb{R})$. *Let also* $x \in \mathbb{R}$, $T > 0$, $n \in \mathbb{N}$: $n \geq \max\left(T + |x|, T^{-\frac{1}{\alpha}}\right)$. *We further assume that* $D_{*x}^{\beta\mathcal{F}} f$, $D_{x-}^{\beta\mathcal{F}} f$ *are fuzzy uniformly continuous functions or fuzzy continuous and bounded on* $[x, +\infty)$, $(-\infty, x]$, *respectively.*

Then

(1)

$$D\left(\left(H_{n}^{\mathcal{F}}(f)\right)(x), f(x)\right) \leq \sum_{j=1}^{N-1} \frac{T^{j}}{j! n^{(1-\alpha)j}} D\left(f^{(j)}(x), \widetilde{o}\right) + \frac{T^{\beta}}{\Gamma(\beta+1) n^{(1-\alpha)\beta}} \cdot \tag{11.51}$$

$$\left\{\omega_{1}^{(\mathcal{F})}\left(D_{*x}^{\beta\mathcal{F}} f, \frac{T}{n^{1-\alpha}}\right)_{[x,+\infty)} + \omega_{1}^{(\mathcal{F})}\left(D_{x-}^{\beta\mathcal{F}} f, \frac{T}{n^{1-\alpha}}\right)_{(-\infty,x]}\right\} =: \lambda_{n}^{\mathcal{F}}(x),$$

above $\sum_{j=1}^{0} \cdot = 0$,

(2) in case of $N = 1$ *or* $D\left(f^{(j)}(x), \widetilde{o}\right) = 0$, *all* $j = 1, \ldots, N-1$, $N \geq 2$, *we get that*

$$D\left(\left(H_{n}^{\mathcal{F}}(f)\right)(x), f(x)\right) \leq \frac{T^{\beta}}{\Gamma(\beta+1) n^{(1-\alpha)\beta}} \cdot \tag{11.52}$$

$$\left\{\omega_{1}^{(\mathcal{F})}\left(D_{*x}^{\beta\mathcal{F}} f, \frac{T}{n^{1-\alpha}}\right)_{[x,+\infty)} + \omega_{1}^{(\mathcal{F})}\left(D_{x-}^{\beta\mathcal{F}} f, \frac{T}{n^{1-\alpha}}\right)_{(-\infty,x]}\right\}.$$

Here we get fractionally with high rates the fuzzy pointwise convergence of $\left(H_{n}^{\mathcal{F}}(f)\right)(x) \xrightarrow{D} f(x)$, *as* $n \to \infty$, $x \in \mathbb{R}$.

Proof We have

$$D\left(f^{(N)}(x), \widetilde{o}\right) = \sup_{r\in[0,1]} \max\left\{\left|(f_{-}^{(r)})^{(N)}(x)\right|, \left|(f_{+}^{(r)})^{(N)}(x)\right|\right\},$$

so that

$$\left|(f_{\pm}^{(r)})^{(N)}(x)\right| \leq D\left(f^{(N)}(x), \widetilde{o}\right), \forall\, r \in [0,1], \text{ any } x \in \mathbb{R}. \tag{11.53}$$

Thus

$$(f_{\pm}^{(r)})^{(N)} \in L_{\infty}(\mathbb{R}), \ \forall\, r \in [0,1].$$

Also we have $f_{\pm}^{(r)} \in C^{N}(\mathbb{R})$, hence $f_{\pm}^{(r)} \in AC^{N}([a,b])$, $\forall\,[a,b] \subset \mathbb{R}$; $\forall\, r \in [0,1]$.

By assumptions we get that $\left(D_{*x}^{\beta\mathcal{F}} f\right)_{\pm}^{(r)} \in C_U([x,+\infty))$ or in $C_b([x,+\infty))$ (bounded and continuous on $[x,+\infty)$ functions), also it holds $\left(D_{x-}^{\beta\mathcal{F}} f\right)_{\pm}^{(r)} \in C_U((-\infty,x])$ or in $C_b((-\infty,x])$, $\forall\, r \in [0,1]$.

By (11.30) we have

$$\left(D_{*x}^{\beta\mathcal{F}} f\right)_{\pm}^{(r)} = D_{*x}^{\beta}\left(f_{\pm}^{(r)}\right), \ \forall r \in [0,1]. \tag{11.54}$$

And by (11.33) we have that

$$\left(D_{x-}^{\beta\mathcal{F}} f\right)_{\pm}^{(r)} = D_{x-}^{\beta}\left(f_{\pm}^{(r)}\right), \ \forall r \in [0,1]. \tag{11.55}$$

Therefore all asumptions of Theorem 11.44 are fulfilled by each of $f_{\pm}^{(r)}$, $\forall r \in [0,1]$; thus (11.40) is valid for these functions.

We observe that

$$D\left(\left(H_n^{\mathcal{F}}(f)\right)(x), f(x)\right) = \sup_{r\in[0,1]} \max\left\{\left|\left(H_n^{\mathcal{F}}(f)\right)_{-}^{(r)}(x) - f_{-}^{(r)}(x)\right|,\right.$$

$$\left.\left|\left(H_n^{\mathcal{F}}(f)\right)_{+}^{(r)}(x) - f_{+}^{(r)}(x)\right|\right\} \overset{(11.47)}{=}$$

$$\sup_{r\in[0,1]} \max\left\{\left|\left(H_n\left(f_{-}^{(r)}\right)\right)(x) - f_{-}^{(r)}(x)\right|, \left|\left(H_n\left(f_{+}^{(r)}\right)\right)(x) - f_{+}^{(r)}(x)\right|\right\} \overset{(11.40)}{\leq} \tag{11.56}$$

$$\sup_{r\in[0,1]} \max\left\{\sum_{j=1}^{N-1} \frac{\left|\left(f_{-}^{(r)}\right)^{(j)}(x)\right| T^j}{j!\,n^{(1-\alpha)j}} + \frac{T^{\beta}}{\Gamma(\beta+1)\, n^{(1-\alpha)\beta}}\cdot\right.$$

$$\left\{\omega_1\left(D_{*x}^{\beta}\left(f_{-}^{(r)}\right), \frac{T}{n^{1-\alpha}}\right)_{[x,+\infty)} + \omega_1\left(D_{x-}^{\beta}\left(f_{-}^{(r)}\right), \frac{T}{n^{1-\alpha}}\right)_{(-\infty,x]}\right\},$$

$$\sum_{j=1}^{N-1} \frac{\left|\left(f_{+}^{(r)}\right)^{(j)}(x)\right| T^j}{j!\,n^{(1-\alpha)j}} + \frac{T^{\beta}}{\Gamma(\beta+1)\, n^{(1-\alpha)\beta}}\cdot \tag{11.57}$$

$$\left.\left\{\omega_1\left(D_{*x}^{\beta}\left(f_{+}^{(r)}\right), \frac{T}{n^{1-\alpha}}\right)_{[x,+\infty)} + \omega_1\left(D_{x-}^{\beta}\left(f_{+}^{(r)}\right), \frac{T}{n^{1-\alpha}}\right)_{(-\infty,x]}\right\}\right\} =$$

(by Remark 11.12, (11.54), (11.55))

$$\sup_{r\in[0,1]} \max\left\{\sum_{j=1}^{N-1} \frac{\left|\left(f^{(j)}\right)_{-}^{(r)}(x)\right| T^{j}}{j! n^{(1-\alpha) j}}+\frac{T^{\beta}}{\Gamma(\beta+1) n^{(1-\alpha) \beta}} \cdot\right.$$

$$\left\{\omega_{1}\left(\left(D_{*x}^{\beta\mathcal{F}} f\right)_{-}^{(r)}, \frac{T}{n^{1-\alpha}}\right)_{[x,+\infty)}+\omega_{1}\left(\left(D_{x-}^{\beta\mathcal{F}} f\right)_{-}^{(r)}, \frac{T}{n^{1-\alpha}}\right)_{(-\infty, x]}\right\},$$

$$\sum_{j=1}^{N-1} \frac{\left|\left(f^{(j)}\right)_{+}^{(r)}(x)\right| T^{j}}{j! n^{(1-\alpha) j}}+\frac{T^{\beta}}{\Gamma(\beta+1) n^{(1-\alpha) \beta}} \cdot \tag{11.58}$$

$$\left.\left\{\omega_{1}\left(\left(D_{*x}^{\beta\mathcal{F}} f\right)_{+}^{(r)}, \frac{T}{n^{1-\alpha}}\right)_{[x,+\infty)}+\omega_{1}\left(\left(D_{x-}^{\beta\mathcal{F}} f\right)_{+}^{(r)}, \frac{T}{n^{1-\alpha}}\right)_{(-\infty, x]}\right\}\right\} \stackrel{(11.1)}{\leq}$$

$$\sum_{j=1}^{N-1} \frac{T^{j}}{j! n^{(1-\alpha) j}} \sup_{r\in[0,1]} \max\left\{\left|\left(f^{(j)}\right)_{-}^{(r)}(x)\right|,\left|\left(f^{(j)}\right)_{+}^{(r)}(x)\right|\right\}+ \tag{11.59}$$

$$\frac{T^{\beta}}{\Gamma(\beta+1) n^{(1-\alpha) \beta}} \sup_{r\in[0,1]} \max\left\{\omega_{1}\left(\left(D_{*x}^{\beta\mathcal{F}} f\right)_{-}^{(r)}, \frac{T}{n^{1-\alpha}}\right)_{[x,+\infty)}+\right.$$

$$\omega_{1}\left(\left(D_{x-}^{\beta\mathcal{F}} f\right)_{-}^{(r)}, \frac{T}{n^{1-\alpha}}\right)_{(-\infty, x]}, \omega_{1}\left(\left(D_{*x}^{\beta\mathcal{F}} f\right)_{+}^{(r)}, \frac{T}{n^{1-\alpha}}\right)_{[x,+\infty)}+$$

$$\left.\omega_{1}\left(\left(D_{x-}^{\beta\mathcal{F}} f\right)_{+}^{(r)}, \frac{T}{n^{1-\alpha}}\right)_{(-\infty, x]}\right\} \stackrel{\text{(by Definition 11.1, (11.1))}}{\leq}$$

$$\sum_{j=1}^{N-1} \frac{T^{j}}{j! n^{(1-\alpha) j}} D\left(f^{(j)}(x), \widetilde{o}\right)+\frac{T^{\beta}}{\Gamma(\beta+1) n^{(1-\alpha) \beta}} \cdot$$

$$\left\{\sup_{r\in[0,1]} \max\left\{\omega_{1}\left(\left(D_{*x}^{\beta\mathcal{F}} f\right)_{-}^{(r)}, \frac{T}{n^{1-\alpha}}\right)_{[x,+\infty)}, \omega_{1}\left(\left(D_{*x}^{\beta\mathcal{F}} f\right)_{+}^{(r)}, \frac{T}{n^{1-\alpha}}\right)_{[x,+\infty)}\right\}+\right.$$

$$\left.\sup_{r\in[0,1]} \max\left\{\omega_{1}\left(\left(D_{x-}^{\beta\mathcal{F}} f\right)_{-}^{(r)}, \frac{T}{n^{1-\alpha}}\right)_{(-\infty, x]}, \omega_{1}\left(\left(D_{x-}^{\beta\mathcal{F}} f\right)_{+}^{(r)}, \frac{T}{n^{1-\alpha}}\right)_{(-\infty, x]}\right\}\right\}$$

$$\overset{\text{(by Proposition 11.7)}}{=} \sum_{j=1}^{N-1} \frac{T^j}{j!n^{(1-\alpha)j}} D\left(f^{(j)}(x), \widetilde{o}\right) + \frac{T^\beta}{\Gamma(\beta+1)\, n^{(1-\alpha)\beta}} \cdot$$

$$\left\{ \omega_1^{(\mathcal{F})}\left(\left(D_{*x}^{\beta\mathcal{F}} f\right), \frac{T}{n^{1-\alpha}}\right)_{[x,+\infty)} + \omega_1^{(\mathcal{F})}\left(\left(D_{x-}^{\beta\mathcal{F}} f\right), \frac{T}{n^{1-\alpha}}\right)_{(-\infty,x]} \right\}, \tag{11.60}$$

proving the claim. ■

We need

Definition 11.47 ([1, 15]) Let the nonnegative function $S : \mathbb{R} \to \mathbb{R}$, S has compact support $[-T, T]$, $T > 0$, and is nondecreasing there and it can be continuous only on either $(-\infty, T]$ or $[-T, T]$. S can have jump discontinuities. We call S the "squashing function".

Remark 11.48 If in (11.44) we replace b by S we derive the so called normalized fuzzy squashing operator $K_n^{\mathcal{F}}$. Then under the same terms and assumptions as in Theorems 11.45 and 11.46, we get

$$D\left(\left(K_n^{\mathcal{F}}(f)\right)(x), f(x)\right) \leq \omega_1^{(\mathcal{F})}\left(f, \frac{T}{n^{1-\alpha}}\right)_{\mathbb{R}}, \tag{11.61}$$

(see (11.48)), and

$$D\left(\left(K_n^{\mathcal{F}}(f)\right)(x), f(x)\right) \leq \lambda_n^{\mathcal{F}}(x), \tag{11.62}$$

(see (11.51)), respectively.

References

1. G.A. Anastassiou, Rate of convergence of some neural network operators to the unit-univariate case. J. Math. Anal. Appl. **212**, 237–262 (1997)
2. G.A. Anastassiou, Fuzzy approximation by fuzzy convolution type operators. Comput. Math. **48**, 1369–1386 (2004)
3. G.A. Anastassiou, Higher order fuzzy korovkin theory via inequalities. Comm. Appl. Anal. **10**(2), 359–392 (2006)
4. G.A. Anastassiou, Fuzzy korovkin theorems and Inequalities. J. Fuzzy Math. **15**(1), 169–205 (2007)
5. G.A. Anastassiou, Fractional korovkin theory. Chaos Solitons Fractals **42**(4), 2080–2094 (2009)
6. G.A. Anastassiou, *Fractional Differentiation Inequalities* (Springer, New York, 2009)
7. G.A. Anastassiou, On right fractional calculus. Chaos Solitons Fractals **42**, 365–376 (2009)
8. G.A. Anastassiou, *Fuzzy Mathematics: Approximation Theory* (Springer, Heidelberg, 2010)
9. G.A. Anastassiou, Fractional representation formulae and right fractional inequalities. Math. Comput. Model. **54**(11–12), 3098–3115 (2011)

10. G.A. Anastassiou, Quantitative approximation by fractional smooth picard singular operators. Math. Eng. Sci. Aerosp. **2**(1), 71–87 (2011)
11. G.A. Anastassiou, Fuzzy fractional Calculus and Ostrowski inequality. J. Fuzzy Math. **19**(3), 577–590 (2011)
12. G.A. Anastassiou, Rate of Convergence of some neural network operators to the Unit-univariate case, revisited. Matematicki Vesnik **65**(4), 511–518 (2013)
13. G.A. Anastassiou, Fractional approximation by normalized Bell and Squashing type neural network operators. New Math. Nat. Comput. **9**(1), 43–63 (2013)
14. G.A. Anastassiou, Fuzzy fractional approximations by fuzzy normalized bell and squashing type neural network operators. J. Fuzzy Math. **22**(1), 139–156 (2014)
15. P. Cardaliaguet, G. Euvrard, Approximation of a function and its derivative with a neural network. Neural Netw. **5**, 207–220 (1992)
16. K. Diethelm, *The Analysis of Fractional Differential Equations*, vol. 2004, Lecture Notes in Mathematics (Springer, Berlin, 2010)
17. A.M.A. El-Sayed, M. Gaber, On the finite Caputo and finite Riesz derivatives. Electron. J. Theoret. Phys. **3**(12), 81–95 (2006)
18. G.S. Frederico, D.F.M. Torres, Fractional optimal control in the sense of Caputo and the fractional Noether's theorem. Int. Math. Forum **3**(10), 479–493 (2008)
19. S. Gal, in *Approximation Theory in Fuzzy Setting*. ed. by G. Anastassiou. Handbook of Analytic-Computational Methods in Applied Mathematics, Chap. 13 (Chapman & Hall/CRC, New York, 2000), pp. 617–666
20. R. Goetschel Jr, W. Voxman, Elementary fuzzy calculus. Fuzzy Sets Syst. **18**, 31–43 (1986)
21. O. Kaleva, Fuzzy differential equations. Fuzzy Sets Syst. **24**, 301–317 (1987)
22. Y.K. Kim, B.M. Ghil, Integrals of fuzzy-number-valued functions. Fuzzy Sets Syst. **86**, 213–222 (1997)
23. C. Wu, Z. Gong, On Henstock integrals of interval-valued functions and fuzzy-valued functions. Fuzzy Sets Syst. **115**(3), 377–391 (2000)
24. C. Wu, Z. Gong, On Henstock integral of fuzzy number valued functions (I). Fuzzy Sets Syst. **120**(3), 523–532 (2001)
25. C. Wu, M. Ma, On embedding problem of fuzzy number spaces: part 1. Fuzzy Sets Syst. **44**, 33–38 (1991)

Chapter 12
Fuzzy Fractional Neural Network Approximation Using Fuzzy Quasi-interpolation Operators

Here we consider the univariate fuzzy fractional quantitative approximation of fuzzy real valued functions on a compact interval by quasi-interpolation sigmoidal and hyperbolic tangent fuzzy neural network operators. These approximations are derived by establishing fuzzy Jackson type inequalities involving the fuzzy moduli of continuity of the right and left Caputo fuzzy fractional derivatives of the engaged function. The approximations are fuzzy pointwise and fuzzy uniform. The related feed-forward fuzzy neural networks are with one hidden layer. Our fuzzy fractional approximation results into higher order converges better than the fuzzy ordinary ones. It follows [19].

12.1 Introduction

The author in [1, 2], see Chaps. 2–5, was the first to establish neural network approximations to continuous functions with rates by very specifically defined neural network operators of Cardaliaguet-Euvrard and "Squashing" types, by employing the modulus of continuity of the engaged function or its high order derivative, and producing very tight Jackson type inequalities. He treats there both the univariate and multivariate cases. The defining these operators "bell-shaped" and "squashing" function are assumed to be of compact support.

The author motivated by [23], continued his studies on neural networks approximation by introducing and using the proper quasi-interpolation operators of sigmoidal and hyperbolic tangent type which resulted into [12, 14–18], by treating both the univariate and multivariate cases.

Continuation of the author's last work [18] is this chapter where fuzzy neural network approximation is taken at the fractional level resulting into higher rates of approximation. We involve the right and left Caputo fuzzy fractional derivatives of the fuzzy function under approximation and we establish tight fuzzy Jackson type inequalities. An extensive background is given on fuzzyness, fractional calculus and neural networks, all needed to expose our work.

G.A. Anastassiou, *Intelligent Systems II: Complete Approximation by Neural Network Operators*, Studies in Computational Intelligence 608,
DOI 10.1007/978-3-319-20505-2_12

Our fuzzy feed-forward neural networks (FFNNs) are with one hidden layer. About neural networks in general, see [29, 32, 33].

12.2 Fuzzy Mathematical Analysis Background

We need the following basic background

Definition 12.1 (*see* [36]) Let $\mu : \mathbb{R} \to [0, 1]$ with the following properties:

(i) is normal, i.e., $\exists\, x_0 \in \mathbb{R};\ \mu(x_0) = 1$.
(ii) $\mu(\lambda x + (1-\lambda) y) \geq \min\{\mu(x), \mu(y)\}$, $\forall\, x, y \in \mathbb{R}$, $\forall\, \lambda \in [0, 1]$ (μ is called a convex fuzzy subset).
(iii) μ is upper semicontinuous on $\mathbb{R}$, i.e. $\forall\, x_0 \in \mathbb{R}$ and $\forall\, \varepsilon > 0$, $\exists$ neighborhood $V(x_0) : \mu(x) \leq \mu(x_0) + \varepsilon$, $\forall\, x \in V(x_0)$.
(iv) The set $\overline{\text{supp}(\mu)}$ is compact in $\mathbb{R}$ (where $\text{supp}(\mu) := \{x \in \mathbb{R} : \mu(x) > 0\}$).

We call μ a fuzzy real number. Denote the set of all μ with $\mathbb{R}_{\mathcal{F}}$.

E.g. $\chi_{\{x_0\}} \in \mathbb{R}_{\mathcal{F}}$, for any $x_0 \in \mathbb{R}$, where $\chi_{\{x_0\}}$ is the characteristic function at x_0.

For $0 < r \leq 1$ and $\mu \in \mathbb{R}_{\mathcal{F}}$ define

$$[\mu]^r := \{x \in \mathbb{R} : \mu(x) \geq r\}$$

and

$$[\mu]^0 := \overline{\{x \in \mathbb{R} : \mu(x) \geq 0\}}.$$

Then it is well known that for each $r \in [0, 1]$, $[\mu]^r$ is a closed and bounded interval on $\mathbb{R}$ [28].

For $u, v \in \mathbb{R}_{\mathcal{F}}$ and $\lambda \in \mathbb{R}$, we define uniquely the sum $u \oplus v$ and the product $\lambda \odot u$ by

$$[u \oplus v]^r = [u]^r + [v]^r\ ,\quad [\lambda \odot u]^r = \lambda [u]^r\ ,\ \forall\, r \in [0, 1],$$

where $[u]^r + [v]^r$ means the usual addition of two intervals (as substes of $\mathbb{R}$) and $\lambda [u]^r$ means the usual product between a scalar and a subset of $\mathbb{R}$ (see, e.g. [36]).

Notice $1 \odot u = u$ and it holds

$$u \oplus v = v \oplus u,\ \lambda \odot u = u \odot \lambda.$$

If $0 \leq r_1 \leq r_2 \leq 1$ then

$$[u]^{r_2} \subseteq [u]^{r_1}.$$

Actually $[u]^r = \left[u_-^{(r)}, u_+^{(r)}\right]$, where $u_-^{(r)} \leq u_+^{(r)}$, $u_-^{(r)}, u_+^{(r)} \in \mathbb{R}$, $\forall\, r \in [0, 1]$.

For $\lambda > 0$ one has $\lambda u_{\pm}^{(r)} = (\lambda \odot u)_{\pm}^{(r)}$, respectively.

Define $D : \mathbb{R}_{\mathcal{F}} \times \mathbb{R}_{\mathcal{F}} \to \mathbb{R}_{\mathcal{F}}$ by

$$D(u, v) := \sup_{r \in [0,1]} \max \left\{ \left| u_{-}^{(r)} - v_{-}^{(r)} \right|, \left| u_{+}^{(r)} - v_{+}^{(r)} \right| \right\},$$

where

$$[v]^r = \left[v_{-}^{(r)}, v_{+}^{(r)} \right]; \; u, v \in \mathbb{R}_{\mathcal{F}}.$$

We have that D is a metric on $\mathbb{R}_{\mathcal{F}}$.

Then $(\mathbb{R}_{\mathcal{F}}, D)$ is a complete metric space, see [36, 37].

Here $\sum^{*}$ stands for fuzzy summation and $\widetilde{o} := \chi_{\{0\}} \in \mathbb{R}_{\mathcal{F}}$ is the neural element with respect to $\oplus$, i.e.,

$$u \oplus \widetilde{0} = \widetilde{0} \oplus u = u, \; \forall \, u \in \mathbb{R}_{\mathcal{F}}.$$

Denote

$$D^{*}(f, g) = \sup_{x \in X \subseteq \mathbb{R}} D(f, g),$$

where $f, g : X \to \mathbb{R}_{\mathcal{F}}$.

We mention

Definition 12.2 Let $f : X \subseteq \mathbb{R} \to \mathbb{R}_{\mathcal{F}}$, X interval, we define the (first) fuzzy modulus of continuity of f by

$$\omega_1^{(\mathcal{F})}(f, \delta)_X = \sup_{x, y \in X, \, |x-y| \le \delta} D(f(x), f(y)), \; \delta > 0.$$

When $g : X \subseteq \mathbb{R} \to \mathbb{R}$, we define

$$\omega_1(g, \delta)_X = \sup_{x, y \in X, \, |x-y| \le \delta} |g(x) - g(y)|.$$

We define by $C_{\mathcal{F}}^{U}(\mathbb{R})$ the space of fuzzy uniformly continuous functions from $\mathbb{R} \to \mathbb{R}_{\mathcal{F}}$, also $C_{\mathcal{F}}(\mathbb{R})$ is the space of fuzzy continuous functions on $\mathbb{R}$, and $C_b(\mathbb{R}, \mathbb{R}_{\mathcal{F}})$ is the fuzzy continuous and bounded functions.

We mention

Proposition 12.3 ([5]) *Let $f \in C_{\mathcal{F}}^{U}(X)$. Then $\omega_1^{\mathcal{F}}(f, \delta)_X < \infty$, for any $\delta > 0$.*

Proposition 12.4 ([5]) *It holds*

$$\lim_{\delta \to 0} \omega_1^{(\mathcal{F})}(f, \delta)_X = \omega_1^{(\mathcal{F})}(f, 0)_X = 0,$$

iff $f \in C_{\mathcal{F}}^{U}(X)$.

Proposition 12.5 ([5]) *Here* $[f]^r = \left[f_-^{(r)}, f_+^{(r)}\right]$, $r \in [0,1]$. *Let* $f \in C_{\mathcal{F}}(\mathbb{R})$. *Then* $f_\pm^{(r)}$ *are equicontinuous with respect to* $r \in [0,1]$ *over* $\mathbb{R}$, *respectively in* $\pm$.

Note 12.6 It is clear by Propositions 12.4 and 12.5, that if $f \in C_{\mathcal{F}}^U(\mathbb{R})$, then $f_\pm^{(r)} \in C_U(\mathbb{R})$ (uniformly continuous on $\mathbb{R}$).

Proposition 12.7 *Let* $f : \mathbb{R} \to \mathbb{R}_{\mathcal{F}}$. *Assume that* $\omega_1^{\mathcal{F}}(f,\delta)_X$, $\omega_1\left(f_-^{(r)},\delta\right)_X$, $\omega_1\left(f_+^{(r)},\delta\right)_X$ *are finite for any* $\delta > 0$, $r \in [0,1]$, *where* X *any interval of* $\mathbb{R}$.
Then

$$\omega_1^{(\mathcal{F})}(f,\delta)_X = \sup_{r\in[0,1]} \max\left\{\omega_1\left(f_-^{(r)},\delta\right)_X, \omega_1\left(f_+^{(r)},\delta\right)_X\right\}.$$

Proof Similar to Proposition 14.15, p. 246 of [9]. ■

We need

Remark 12.8 ([3]) Here $r \in [0,1]$, $x_i^{(r)}, y_i^{(r)} \in \mathbb{R}$, $i = 1,\ldots,m \in \mathbb{N}$. Suppose that

$$\sup_{r\in[0,1]} \max\left(x_i^{(r)}, y_i^{(r)}\right) \in \mathbb{R}, \text{ for } i = 1,\ldots,m.$$

Then one sees easily that

$$\sup_{r\in[0,1]} \max\left(\sum_{i=1}^m x_i^{(r)}, \sum_{i=1}^m y_i^{(r)}\right) \le \sum_{i=1}^m \sup_{r\in[0,1]} \max\left(x_i^{(r)}, y_i^{(r)}\right). \tag{12.1}$$

We need

Definition 12.9 Let $x, y \in \mathbb{R}_{\mathcal{F}}$. If there exists $z \in \mathbb{R}_{\mathcal{F}} : x = y \oplus z$, then we call z the H-difference on x and y, denoted $x - y$.

Definition 12.10 ([34]) Let $T := [x_0, x_0 + \beta] \subset \mathbb{R}$, with $\beta > 0$. A function $f : T \to \mathbb{R}_{\mathcal{F}}$ is H-difference at $x \in T$ if there exists an $f'(x) \in \mathbb{R}_{\mathcal{F}}$ such that the limits (with respect to D)

$$\lim_{h\to 0+} \frac{f(x+h) - f(x)}{h}, \quad \lim_{h\to 0+} \frac{f(x) - f(x-h)}{h} \tag{12.2}$$

exist and are equal to $f'(x)$.
We call f' the H-derivative or fuzzy derivative of f at x.

Above is assumed that the H-differences $f(x+h) - f(x)$, $f(x) - f(x-h)$ exists in $\mathbb{R}_{\mathcal{F}}$ in a neighborhood of x.

Higher order H-fuzzy derivatives are defined the obvious way, like in the real case.

We denote by $C_{\mathcal{F}}^{N}(\mathbb{R})$, $N \geq 1$, the space of all N-times continuously H-fuzzy differentiable functions from $\mathbb{R}$ into $\mathbb{R}_{\mathcal{F}}$.

We mention

Theorem 12.11 ([30]) *Let $f : \mathbb{R} \to \mathbb{R}_{\mathcal{F}}$ be H-fuzzy differentiable. Let $t \in \mathbb{R}$, $0 \leq r \leq 1$. Clearly*

$$[f(t)]^r = \left[f(t)_-^{(r)}, f(t)_+^{(r)}\right] \subseteq \mathbb{R}.$$

Then $(f(t))_{\pm}^{(r)}$ are differentiable and

$$[f'(t)]^r = \left[\left(f(t)_-^{(r)}\right)', \left(f(t)_+^{(r)}\right)'\right].$$

I.e.

$$(f')_{\pm}^{(r)} = \left(f_{\pm}^{(r)}\right)', \quad \forall r \in [0,1].$$

Remark 12.12 ([4]) Let $f \in C_{\mathcal{F}}^{N}(\mathbb{R})$, $N \geq 1$. Then by Theorem 12.11 we obtain

$$\left[f^{(i)}(t)\right]^r = \left[\left(f(t)_-^{(r)}\right)^{(i)}, \left(f(t)_+^{(r)}\right)^{(i)}\right],$$

for $i = 0, 1, 2, \ldots, N$, and in particular we have that

$$\left(f^{(i)}\right)_{\pm}^{(r)} = \left(f_{\pm}^{(r)}\right)^{(i)},$$

for any $r \in [0,1]$, all $i = 0, 1, 2, \ldots, N$.

Note 12.13 ([4]) Let $f \in C_{\mathcal{F}}^{N}(\mathbb{R})$, $N \geq 1$. Then by Theorem 12.11 we have $f_{\pm}^{(r)} \in C^N(\mathbb{R})$, for any $r \in [0,1]$.

We need also a particular case of the Fuzzy Henstock integral ($\delta(x) = \frac{\delta}{2}$), see [36].

Definition 12.14 ([27], *p. 644*) Let $f : [a,b] \to \mathbb{R}_{\mathcal{F}}$. We say that f is Fuzzy-Riemann integrable to $I \in \mathbb{R}_{\mathcal{F}}$ if for any $\varepsilon > 0$, there exists $\delta > 0$ such that for any division $P = \{[u,v]; \xi\}$ of $[a,b]$ with the norms $\Delta(P) < \delta$, we have

$$D\left(\sum_{P}^{*} (v-u) \odot f(\xi), I\right) < \varepsilon.$$

We write

$$I := (FR)\int_a^b f(x)\,dx. \tag{12.3}$$

We mention

Theorem 12.15 ([28]) *Let $f : [a, b] \to \mathbb{R}_{\mathcal{F}}$ be fuzzy continuous. Then*

$$(FR) \int_a^b f(x)\, dx$$

exists and belongs to $\mathbb{R}_{\mathcal{F}}$, furthermore it holds

$$\left[(FR) \int_a^b f(x)\, dx \right]^r = \left[\int_a^b (f)_-^{(r)}(x)\, dx, \int_a^b (f)_+^{(r)}(x)\, dx \right],$$

$\forall\, r \in [0, 1]$.

For the definition of general fuzzy integral we follow [31] next.

Definition 12.16 Let (Ω, Σ, μ) be a complete σ-finite measure space. We call $F : \Omega \to R_{\mathcal{F}}$ measurable iff $\forall$ closed $B \subseteq \mathbb{R}$ the function $F^{-1}(B) : \Omega \to [0, 1]$ defined by

$$F^{-1}(B)(w) := \sup_{x \in B} F(w)(x), \text{ all } w \in \Omega$$

is measurable, see [31].

Theorem 12.17 ([31]) *For $F : \Omega \to \mathbb{R}_{\mathcal{F}}$,*

$$F(w) = \left\{ \left(F_-^{(r)}(w), F_+^{(r)}(w) \right) | 0 \le r \le 1 \right\},$$

the following are equivalent

(1) F is measurable,
(2) $\forall\, r \in [0, 1]$, $F_-^{(r)}$, $F_+^{(r)}$ are measurable.

Following [31], given that for each $r \in [0, 1]$, $F_-^{(r)}$, $F_+^{(r)}$ are integrable we have that the parametrized representation

$$\left\{ \left(\int_A F_-^{(r)} d\mu, \int_A F_+^{(r)} d\mu \right) | 0 \le r \le 1 \right\} \tag{12.4}$$

is a fuzzy real number for each $A \in \Sigma$.

The last fact leads to

Definition 12.18 ([31]) A measurable function $F : \Omega \to \mathbb{R}_{\mathcal{F}}$,

$$F(w) = \left\{ \left(F_-^{(r)}(w), F_+^{(r)}(w) \right) | 0 \le r \le 1 \right\}$$

is integrable if for each $r \in [0, 1]$, $F_\pm^{(r)}$ are integrable, or equivalently, if $F_\pm^{(0)}$ are integrable.

In this case, the fuzzy integral of F over $A \in \Sigma$ is defined by

$$\int_A F d\mu := \left\{ \left(\int_A F_-^{(r)} d\mu, \int_A F_+^{(r)} d\mu \right) | 0 \le r \le 1 \right\}.$$

By [31], F is integrable iff $w \to \|F(w)\|_{\mathcal{F}}$ is real-valued integrable.

Here denote

$$\|u\|_{\mathcal{F}} := D\left(u, \widetilde{0}\right), \forall\, u \in \mathbb{R}_{\mathcal{F}}.$$

We need also

Theorem 12.19 ([31]) *Let $F, G : \Omega \to \mathbb{R}_{\mathcal{F}}$ be integrable. Then*

(1) Let $a, b \in \mathbb{R}$, then $aF + bG$ is integrable and for each $A \in \Sigma$,

$$\int_A (aF + bG)\, d\mu = a \int_A F d\mu + b \int_A G d\mu;$$

(2) $D(F, G)$ is a real- valued integrable function and for each $A \in \Sigma$,

$$D\left(\int_A F d\mu, \int_A G d\mu \right) \le \int_A D(F, G)\, d\mu.$$

In particular,

$$\left\| \int_A F d\mu \right\|_{\mathcal{F}} \le \int_A \|F\|_{\mathcal{F}}\, d\mu.$$

Above μ could be the Lebesgue measure, with all the basic properties valid here too.

Basically here we have

$$\left[\int_A F d\mu \right]^r = \left[\int_A F_-^{(r)} d\mu, \int_A F_+^{(r)} d\mu \right], \tag{12.5}$$

i.e.

$$\left(\int_A F d\mu \right)_{\pm}^{(r)} = \int_A F_{\pm}^{(r)} d\mu, \quad \forall\, r \in [0, 1].$$

We need

Definition 12.20 Let $\nu \ge 0$, $n = \lceil \nu \rceil$ ($\lceil \cdot \rceil$ is the ceiling of the number), $f \in AC^n([a, b])$ (space of functions f with $f^{(n-1)} \in AC([a, b])$, absolutely continuous functions). We call left Caputo fractional derivative (see [24], pp. 49–52, [26, 34]) the function

$$D_{*a}^{\nu} f(x) = \frac{1}{\Gamma(n - \nu)} \int_a^x (x - t)^{n-\nu-1} f^{(n)}(t)\, dt, \tag{12.6}$$

$\forall\, x \in [a, b]$, where Γ is the gamma function $\Gamma(\nu) = \int_0^\infty e^{-t} t^{\nu-1} dt$, $\nu > 0$.

Notice $D_{*a}^{\nu} f \in L_1([a, b])$ and $D_{*a}^{\nu} f$ exists a.e. on $[a, b]$.

We set $D_{*a}^{0} f(x) = f(x)$, $\forall\, x \in [a, b]$.

Lemma 12.21 ([8]) *Let* $\nu > 0$, $\nu \notin \mathbb{N}$, $n = \lceil \nu \rceil$, $f \in C^{n-1}([a, b])$ *and* $f^{(n)} \in L_\infty([a, b])$. *Then* $D_{*a}^{\nu} f(a) = 0$.

Definition 12.22 (*see also* [6, 25, 26]) Let $f \in AC^m([a, b])$, $m = \lceil \alpha \rceil$, $\alpha > 0$. The right Caputo fractional derivative of order $\alpha > 0$ is given by

$$D_{b-}^{\alpha} f(x) = \frac{(-1)^m}{\Gamma(m-\alpha)} \int_x^b (\zeta - x)^{m-\alpha-1} f^{(m)}(\zeta)\, d\zeta, \tag{12.7}$$

$\forall\, x \in [a, b]$. We set $D_{b-}^{0} f(x) = f(x)$. Notice that $D_{b-}^{\alpha} f \in L_1([a, b])$ and $D_{b-}^{\alpha} f$ exists a.e. on $[a, b]$.

Lemma 12.23 ([8]) *Let* $f \in C^{m-1}([a, b])$, $f^{(m)} \in L_\infty([a, b])$, $m = \lceil \alpha \rceil$, $\alpha > 0$, $\alpha \notin \mathbb{N}$. *Then* $D_{b-}^{\alpha} f(b) = 0$.

Convention 12.24 *We assume that*

$$D_{*x_0}^{\alpha} f(x) = 0, \textit{ for } x < x_0, \tag{12.8}$$

and

$$D_{x_0-}^{\alpha} f(x) = 0, \textit{ for } x > x_0, \tag{12.9}$$

for all $x, x_0 \in (a, b]$.

We mention

Proposition 12.25 ([8]) *Let* $f \in C^n([a, b])$, $n = \lceil \nu \rceil$, $\nu > 0$. *Then* $D_{*a}^{\nu} f(x)$ *is continuous in* $x \in [a, b]$.

Also we have

Proposition 12.26 ([8]) *Let* $f \in C^m([a, b])$, $m = \lceil \alpha \rceil$, $\alpha > 0$. *Then* $D_{b-}^{\alpha} f(x)$ *is continuous in* $x \in [a, b]$.

We further mention

Proposition 12.27 ([8]) *Let* $f \in C^{m-1}([a, b])$, $f^{(m)} \in L_\infty([a, b])$, $m = \lceil \alpha \rceil$, $\alpha > 0$ *and*

$$D_{*x_0}^{\alpha} f(x) = \frac{1}{\Gamma(m-\alpha)} \int_{x_0}^{x} (x - t)^{m-\alpha-1} f^{(m)}(t)\, dt, \tag{12.10}$$

for all $x, x_0 \in [a, b] : x \geq x_0$.

Then $D_{*x_0}^{\alpha} f(x)$ *is continuous in* x_0.

Proposition 12.28 ([8]) *Let* $f \in C^{m-1}([a,b])$, $f^{(m)} \in L_{\infty}([a,b])$, $m = \lceil \alpha \rceil$, $\alpha > 0$ *and*

$$D_{x_0-}^{\alpha} f(x) = \frac{(-1)^m}{\Gamma(m-\alpha)} \int_x^{x_0} (\zeta - x)^{m-\alpha-1} f^{(m)}(\zeta)\, d\zeta, \tag{12.11}$$

for all $x, x_0 \in [a,b] : x \leq x_0$.

Then $D_{x_0-}^{\alpha} f(x)$ *is continuous in* x_0.

We need

Proposition 12.29 ([8]) *Let* $g \in C([a,b])$, $0 < c < 1$, $x, x_0 \in [a,b]$. *Define*

$$L(x, x_0) = \int_{x_0}^{x} (x - t)^{c-1} g(t)\, dt, \text{ for } x \geq x_0, \tag{12.12}$$

and $L(x, x_0) = 0$, *for* $x < x_0$.

Then L *is jointly continuous in* (x, x_0) *on* $[a,b]^2$.

We mention

Proposition 12.30 ([8]) *Let* $g \in C([a,b])$, $0 < c < 1$, $x, x_0 \in [a,b]$. *Define*

$$K(x, x_0) = \int_{x_0}^{x} (\zeta - x)^{c-1} g(\zeta)\, d\zeta, \text{ for } x \leq x_0, \tag{12.13}$$

and $K(x, x_0) = 0$, *for* $x > x_0$.

Then $K(x, x_0)$ *is jointly continuous from* $[a,b]^2$ *into* $\mathbb{R}$.

Based on Propositions 12.29 and 12.30 we derive

Corollary 12.31 ([8]) *Let* $f \in C^m([a,b])$, $m = \lceil \alpha \rceil$, $\alpha > 0$, $\alpha \notin \mathbb{N}$, $x, x_0 \in [a,b]$. *Then* $D_{*x_0}^{\alpha} f(x)$, $D_{x_0-}^{\alpha} f(x)$ *are jointly continuous functions in* (x, x_0) *from* $[a,b]^2$ *into* $\mathbb{R}$.

We need

Theorem 12.32 ([8]) *Let* $f : [a,b]^2 \to \mathbb{R}$ *be jointly continuous. Consider*

$$G(x) = \omega_1(f(\cdot, x), \delta)_{[x,b]}, \tag{12.14}$$

$\delta > 0$, $x \in [a,b]$.

Then G *is continuous in* $x \in [a,b]$.

Also it holds

Theorem 12.33 ([8]) *Let $f : [a, b]^2 \to \mathbb{R}$ be jointly continuous. Then*

$$H(x) = \omega_1 (f(\cdot, x), \delta)_{[a,x]}, \tag{12.15}$$

$x \in [a, b]$, is continuous in $x \in [a, b]$, $\delta > 0$.

So that for $f \in C^m([a, b])$, $m = \lceil \alpha \rceil$, $\alpha > 0$, $\alpha \notin \mathbb{N}$, $x, x_0 \in [a, b]$, we have that $\omega_1 \left(D_{*x}^{\alpha} f, h\right)_{[x,b]}$, $\omega_1 \left(D_{x-}^{\alpha} f, h\right)_{[a,x]}$ are continuous functions in $x \in [a, b]$, $h > 0$ is fixed.

We make

Remark 12.34 ([8]) Let $f \in C^{n-1}([a, b])$, $f^{(n)} \in L_\infty([a, b])$, $n = \lceil \nu \rceil$, $\nu > 0$, $\nu \notin \mathbb{N}$. Then we have

$$\left|D_{*a}^{\nu} f(x)\right| \leq \frac{\left\| f^{(n)} \right\|_\infty}{\Gamma(n - \nu + 1)} (x - a)^{n-\nu}, \ \forall\, x \in [a, b]. \tag{12.16}$$

Thus we observe

$$\omega_1 \left(D_{*a}^{\nu} f, \delta\right) = \sup_{\substack{x,y\in[a,b] \\ |x-y|\leq\delta}} \left|D_{*a}^{\nu} f(x) - D_{*a}^{\nu} f(y)\right| \tag{12.17}$$

$$\leq \sup_{\substack{x,y\in[a,b] \\ |x-y|\leq\delta}} \left(\frac{\left\| f^{(n)} \right\|_\infty}{\Gamma(n - \nu + 1)} (x - a)^{n-\nu} + \frac{\left\| f^{(n)} \right\|_\infty}{\Gamma(n - \nu + 1)} (y - a)^{n-\nu} \right)$$

$$\leq \frac{2 \left\| f^{(n)} \right\|_\infty}{\Gamma(n - \nu + 1)} (b - a)^{n-\nu}. \tag{12.18}$$

Consequently

$$\omega_1 \left(D_{*a}^{\nu} f, \delta\right) \leq \frac{2 \left\| f^{(n)} \right\|_\infty}{\Gamma(n - \nu + 1)} (b - a)^{n-\nu}. \tag{12.19}$$

Similarly, let $f \in C^{m-1}([a, b])$, $f^{(m)} \in L_\infty([a, b])$, $m = \lceil \alpha \rceil$, $\alpha > 0$, $\alpha \notin \mathbb{N}$, then

$$\omega_1 \left(D_{b-}^{\alpha} f, \delta\right) \leq \frac{2 \left\| f^{(m)} \right\|_\infty}{\Gamma(m - \alpha + 1)} (b - a)^{m-\alpha}. \tag{12.20}$$

So for $f \in C^{m-1}([a, b])$, $f^{(m)} \in L_\infty([a, b])$, $m = \lceil \alpha \rceil$, $\alpha > 0$, $\alpha \notin \mathbb{N}$, we find

$$s_1(\delta) := \sup_{x_0\in[a,b]} \omega_1 \left(D_{*x_0}^{\alpha} f, \delta\right)_{[x_0,b]} \leq \frac{2 \left\| f^{(m)} \right\|_\infty}{\Gamma(m - \alpha + 1)} (b - a)^{m-\alpha}, \tag{12.21}$$

and

$$s_2(\delta) := \sup_{x_0 \in [a,b]} \omega_1 \left(D^{\alpha}_{x_0-} f, \delta \right)_{[a,x_0]} \leq \frac{2 \left\| f^{(m)} \right\|_{\infty}}{\Gamma(m - \alpha + 1)} (b - a)^{m-\alpha}. \quad (12.22)$$

By Proposition 15.114, p. 388 of [7], we get here that $D^{\alpha}_{*x_0} f \in C([x_0, b])$, and by [13] we obtain that $D^{\alpha}_{x_0-} f \in C([a, x_0])$.

We need

Definition 12.35 ([11]) Let $f \in C_{\mathcal{F}}([a, b])$ (fuzzy continuous on $[a, b] \subset \mathbb{R}$), $\nu > 0$.

We define the Fuzzy Fractional left Riemann-Liouville operator as

$$J^{\nu}_{a} f(x) := \frac{1}{\Gamma(\nu)} \odot \int_{a}^{x} (x - t)^{\nu-1} \odot f(t) \, dt, \quad x \in [a, b], \quad (12.23)$$

$$J^{0}_{a} f := f. \quad (12.24)$$

Also, we define the Fuzzy Fractional right Riemann-Liouville operator as

$$I^{\nu}_{b-} f(x) := \frac{1}{\Gamma(\nu)} \odot \int_{x}^{b} (t - x)^{\nu-1} \odot f(t) \, dt, \quad x \in [a, b], \quad (12.25)$$

$$I^{0}_{b-} f := f.$$

We mention

Definition 12.36 ([11]) Let $f : [a, b] \to \mathbb{R}_{\mathcal{F}}$ is called fuzzy absolutely continuous iff $\forall \epsilon > 0$, $\exists \delta > 0$ for every finite, pairwise disjoint, family

$$(c_k, d_k)_{k=1}^{n} \subseteq (a, b) \text{ with } \sum_{k=1}^{n} (d_k - c_k) < \delta$$

we get

$$\sum_{k=1}^{n} D(f(d_k), f(c_k)) < \epsilon. \quad (12.26)$$

We denote the related space of functions by $AC_{\mathcal{F}}([a, b])$.

If $f \in AC_{\mathcal{F}}([a, b])$, then $f \in C_{\mathcal{F}}([a, b])$.

It holds

Proposition 12.37 ([11]) $f \in AC_{\mathcal{F}}([a, b]) \Leftrightarrow f^{(r)}_{\pm} \in AEC([a, b])$, $\forall r \in [0, 1]$ *(absolutely equicontinuous).*

We need

Definition 12.38 ([11]) We define the Fuzzy Fractional left Caputo derivative, $x \in [a, b]$.

Let $f \in C^n_{\mathcal{F}}([a, b])$, $n = \lceil \nu \rceil$, $\nu > 0$ ($\lceil \cdot \rceil$ denotes the ceiling). We define

$$D^{\nu \mathcal{F}}_{*a} f(x) := \frac{1}{\Gamma(n-\nu)} \odot \int_a^x (x-t)^{n-\nu-1} \odot f^{(n)}(t)\, dt$$

$$= \left\{ \left(\frac{1}{\Gamma(n-\nu)} \int_a^x (x-t)^{n-\nu-1} \left(f^{(n)}\right)_-^{(r)}(t)\, dt, \right. \right.$$

$$\left. \left. \frac{1}{\Gamma(n-\nu)} \int_a^x (x-t)^{n-\nu-1} \left(f^{(n)}\right)_+^{(r)}(t)\, dt \right) | 0 \le r \le 1 \right\}$$

$$= \left\{ \left(\frac{1}{\Gamma(n-\nu)} \int_a^x (x-t)^{n-\nu-1} \left(f_-^{(r)}\right)^{(n)}(t)\, dt, \right. \right.$$

$$\left. \left. \frac{1}{\Gamma(n-\nu)} \int_a^x (x-t)^{n-\nu-1} \left(f_+^{(r)}\right)^{(n)}(t)\, dt \right) | 0 \le r \le 1 \right\}. \tag{12.27}$$

So, we get

$$\left[D^{\nu \mathcal{F}}_{*a} f(x) \right]^r = \left[\left(\frac{1}{\Gamma(n-\nu)} \int_a^x (x-t)^{n-\nu-1} \left(f_-^{(r)}\right)^{(n)}(t)\, dt, \right. \right.$$

$$\left. \left. \frac{1}{\Gamma(n-\nu)} \int_a^x (x-t)^{n-\nu-1} \left(f_+^{(r)}\right)^{(n)}(t)\, dt \right) \right], \quad 0 \le r \le 1. \tag{12.28}$$

That is

$$\left(D^{\nu \mathcal{F}}_{*a} f(x) \right)_\pm^{(r)} = \frac{1}{\Gamma(n-\nu)} \int_a^x (x-t)^{n-\nu-1} \left(f_\pm^{(r)}\right)^{(n)}(t)\, dt = \left(D^\nu_{*a} \left(f_\pm^{(r)} \right) \right)(x),$$

see [7, 24].

I.e. we get that

$$\left(D^{\nu \mathcal{F}}_{*a} f(x) \right)_\pm^{(r)} = \left(D^\nu_{*a} \left(f_\pm^{(r)} \right) \right)(x), \tag{12.29}$$

$\forall\, x \in [a, b]$, in short

$$\left(D^{\nu \mathcal{F}}_{*a} f \right)_\pm^{(r)} = D^\nu_{*a} \left(f_\pm^{(r)} \right), \quad \forall\, r \in [0, 1]. \tag{12.30}$$

We need

Lemma 12.39 ([11]) *$D^{\nu \mathcal{F}}_{*a} f(x)$ is fuzzy continuous in $x \in [a, b]$.*

We need

Definition 12.40 ([11]) We define the Fuzzy Fractional right Caputo derivative, $x \in [a, b]$.

Let $f \in C_{\mathcal{F}}^{n}([a, b])$, $n = \lceil \nu \rceil$, $\nu > 0$. We define

$$D_{b-}^{\nu \mathcal{F}} f(x) := \frac{(-1)^n}{\Gamma(n-\nu)} \odot \int_x^b (t-x)^{n-\nu-1} \odot f^{(n)}(t)\, dt$$

$$= \left\{ \left(\frac{(-1)^n}{\Gamma(n-\nu)} \int_x^b (t-x)^{n-\nu-1} \left(f^{(n)}\right)_-^{(r)}(t)\, dt, \right. \right.$$

$$\left. \left. \frac{(-1)^n}{\Gamma(n-\nu)} \int_x^b (t-x)^{n-\nu-1} \left(f^{(n)}\right)_+^{(r)}(t)\, dt \right) | 0 \le r \le 1 \right\} \tag{12.31}$$

$$= \left\{ \left(\frac{(-1)^n}{\Gamma(n-\nu)} \int_x^b (t-x)^{n-\nu-1} \left(f_-^{(r)}\right)^{(n)}(t)\, dt, \right. \right.$$

$$\left. \left. \frac{(-1)^n}{\Gamma(n-\nu)} \int_x^b (t-x)^{n-\nu-1} \left(f_+^{(r)}\right)^{(n)}(t)\, dt \right) | 0 \le r \le 1 \right\}.$$

We get

$$\left[D_{b-}^{\nu \mathcal{F}} f(x) \right]^r = \left[\left(\frac{(-1)^n}{\Gamma(n-\nu)} \int_x^b (t-x)^{n-\nu-1} \left(f_-^{(r)}\right)^{(n)}(t)\, dt, \right. \right.$$

$$\left. \left. \frac{(-1)^n}{\Gamma(n-\nu)} \int_x^b (t-x)^{n-\nu-1} \left(f_+^{(r)}\right)^{(n)}(t)\, dt \right) \right], \quad 0 \le r \le 1.$$

That is

$$\left(D_{b-}^{\nu \mathcal{F}} f(x) \right)_{\pm}^{(r)} = \frac{(-1)^n}{\Gamma(n-\nu)} \int_x^b (t-x)^{n-\nu-1} \left(f_{\pm}^{(r)}\right)^{(n)}(t)\, dt = \left(D_{b-}^{\nu} \left(f_{\pm}^{(r)} \right) \right)(x),$$

see [6].

I.e. we get that

$$\left(D_{b-}^{\nu \mathcal{F}} f(x) \right)_{\pm}^{(r)} = \left(D_{b-}^{\nu} \left(f_{\pm}^{(r)} \right) \right)(x), \tag{12.32}$$

$\forall\, x \in [a, b]$, in short

$$\left(D_{b-}^{\nu \mathcal{F}} f \right)_{\pm}^{(r)} = D_{b-}^{\nu} \left(f_{\pm}^{(r)} \right), \quad \forall\, r \in [0, 1]. \tag{12.33}$$

Clearly,

$$D_{b-}^{\nu}\left(f_{-}^{(r)}\right)\leq D_{b-}^{\nu}\left(f_{+}^{(r)}\right),\quad\forall\, r\in[0,1].$$

We need

Lemma 12.41 ([11]) *$D_{b-}^{\nu\mathcal{F}}f(x)$ is fuzzy continuous in $x\in[a,b]$.*

12.3 Fractional Neural Network Approximation by Quasi-interpolation Operators

We consider here the sigmoidal function of logarithmic type

$$s(x)=\frac{1}{1+e^{-x}},\quad x\in\mathbb{R}.$$

It has the properties $\lim_{x\to+\infty}s(x)=1$ and $\lim_{x\to-\infty}s(x)=0$.

This function plays the role of an activation function in the hidden layer of neural networks.

As in [22], we consider

$$\Phi(x):=\frac{1}{2}\left(s(x+1)-s(x-1)\right),\quad x\in\mathbb{R}. \tag{12.34}$$

We notice the following properties:

(i) $\Phi(x)>0$, $\forall\, x\in\mathbb{R}$,
(ii) $\sum_{k=-\infty}^{\infty}\Phi(x-k)=1$, $\forall\, x\in\mathbb{R}$,
(iii) $\sum_{k=-\infty}^{\infty}\Phi(nx-k)=1$, $\forall\, x\in\mathbb{R};\ n\in\mathbb{N}$,
(iv) $\int_{-\infty}^{\infty}\Phi(x)\,dx=1$,
(v) Φ is a density function,
(vi) Φ is even: $\Phi(-x)=\Phi(x)$, $x\geq0$.

We see that ([23])

$$\Phi(x)=\left(\frac{e^2-1}{2e^2}\right)\frac{1}{\left(1+e^{x-1}\right)\left(1+e^{-x-1}\right)}. \tag{12.35}$$

(vii) By [23] Φ is decreasing on $\mathbb{R}_+$, and increasing on $\mathbb{R}_-$.
(viii) By [17] for $n\in\mathbb{N}$, $0<\beta<1$, we get

$$\sum_{\substack{k=-\infty \\ :\,|nx-k|>n^{1-\beta}}}^{\infty}\Phi(nx-k)<\left(\frac{e^2-1}{2}\right)e^{-n^{(1-\beta)}}=3.1992e^{-n^{(1-\beta)}}. \tag{12.36}$$

Denote by $\lfloor \cdot \rfloor$ the integral part of a number. Consider $x \in [a, b] \subset \mathbb{R}$ and $n \in \mathbb{N}$ such that $\lceil na \rceil \leq \lfloor nb \rfloor$.

(ix) By [17] it holds

$$\frac{1}{\sum_{k=\lceil na \rceil}^{\lfloor nb \rfloor} \Phi(nx-k)} < \frac{1}{\Phi(1)} = 5.250312578, \ \forall\, x \in [a, b]. \tag{12.37}$$

(x) By [17] it holds $\lim_{n\to\infty} \sum_{k=\lceil na \rceil}^{\lfloor nb \rfloor} \Phi(nx-k) \neq 1$, for at least some $x \in [a, b]$.

Let $f \in C([a, b])$ and $n \in \mathbb{N}$ such that $\lceil na \rceil \leq \lfloor nb \rfloor$.
We consider further (see also [17]) the positive linear neural network operator

$$G_n(f, x) := \frac{\sum_{k=\lceil na \rceil}^{\lfloor nb \rfloor} f\left(\frac{k}{n}\right) \Phi(nx-k)}{\sum_{k=\lceil na \rceil}^{\lfloor nb \rfloor} \Phi(nx-k)}, \quad x \in [a, b]. \tag{12.38}$$

For large enough n we always obtain $\lceil na \rceil \leq \lfloor nb \rfloor$. Also $a \leq \frac{k}{n} \leq b$, iff $\lceil na \rceil \leq k \leq \lfloor nb \rfloor$.

We present here at fractional level the pointwise and uniform convergence of $G_n(f, x)$ to $f(x)$ with rates.

We also consider here the hyperbolic tangent function $\tanh x$, $x \in \mathbb{R}$:

$$\tanh x := \frac{e^x - e^{-x}}{e^x + e^{-x}} = \frac{e^{2x} - 1}{e^{2x} + 1}.$$

It has the properties $\tanh 0 = 0$, $-1 < \tanh x < 1$, $\forall\, x \in \mathbb{R}$, and $\tanh(-x) = -\tanh x$. Furthermore $\tanh x \to 1$ as $x \to \infty$, and $\tanh x \to -1$, as $x \to -\infty$, and it is strictly increasing on $\mathbb{R}$. Furthermore it holds $\frac{d}{dx} \tanh x = \frac{1}{\cosh^2 x} > 0$.

This function plays also the role of an activation function in the hidden layer of neural networks.

We further consider

$$\Psi(x) := \frac{1}{4}(\tanh(x+1) - \tanh(x-1)) > 0, \quad \forall\ x \in \mathbb{R}. \tag{12.39}$$

We easily see that $\Psi(-x) = \Psi(x)$, that is Ψ is even on $\mathbb{R}$. Obviously Ψ is differentiable, thus continuous.

Here we follow [14].

Proposition 12.42 $\Psi(x)$ *for* $x \geq 0$ *is strictly decreasing.*

Obviously $\Psi(x)$ is strictly increasing for $x \leq 0$. Also it holds $\lim_{x\to-\infty} \Psi(x) = 0 = \lim_{x\to\infty} \Psi(x)$.

Infact Ψ has the bell shape with horizontal asymptote the x-axis. So the maximum of Ψ is at zero, $\Psi(0) = 0.3809297$.

Theorem 12.43 *We have that* $\sum_{i=-\infty}^{\infty} \Psi(x-i) = 1, \ \forall\, x \in \mathbb{R}$.

Thus

$$\sum_{i=-\infty}^{\infty} \Psi(nx-i) = 1, \quad \forall\, n \in \mathbb{N},\ \forall\, x \in \mathbb{R}.$$

Furthermore we get:
Since Ψ is even it holds $\sum_{i=-\infty}^{\infty} \Psi(i-x) = 1, \ \forall x \in \mathbb{R}$.
Hence $\sum_{i=-\infty}^{\infty} \Psi(i+x) = 1, \ \forall\, x \in \mathbb{R}$, and $\sum_{i=-\infty}^{\infty} \Psi(x+i) = 1, \ \forall\, x \in \mathbb{R}$.

Theorem 12.44 *It holds* $\int_{-\infty}^{\infty} \Psi(x)\, dx = 1$.

So $\Psi(x)$ is a density function on $\mathbb{R}$.

Theorem 12.45 *Let* $0 < \beta < 1$ *and* $n \in \mathbb{N}$. *It holds*

$$\sum_{\begin{cases} k = -\infty \\ : |nx-k| \geq n^{1-\beta} \end{cases}}^{\infty} \Psi(nx-k) \leq e^4 \cdot e^{-2n^{(1-\beta)}}. \tag{12.40}$$

Theorem 12.46 *Let* $x \in [a,b] \subset \mathbb{R}$ *and* $n \in \mathbb{N}$ *so that* $\lceil na \rceil \leq \lfloor nb \rfloor$. *It holds*

$$\frac{1}{\sum_{k=\lceil na \rceil}^{\lfloor nb \rfloor} \Psi(nx-k)} < 4.1488766 = \frac{1}{\Psi(1)}. \tag{12.41}$$

Also by [14], we obtain

$$\lim_{n \to \infty} \sum_{k=\lceil na \rceil}^{\lfloor nb \rfloor} \Psi(nx-k) \neq 1,$$

for at least some $x \in [a,b]$.

Definition 12.47 Let $f \in C([a,b])$ and $n \in \mathbb{N}$ such that $\lceil na \rceil \leq \lfloor nb \rfloor$.
We further treat, as in [14], the positive linear neural network operator

$$F_n(f,x) := \frac{\sum_{k=\lceil na \rceil}^{\lfloor nb \rfloor} f\left(\frac{k}{n}\right) \Psi(nx-k)}{\sum_{k=\lceil na \rceil}^{\lfloor nb \rfloor} \Psi(nx-k)}, \quad x \in [a,b]. \tag{12.42}$$

We consider F_n similarly to G_n.
Here $\|\cdot\|_\infty$ stands for the supremum norm.
From [18] we need

Theorem 12.48 *Let* $\alpha > 0$, $N = \lceil \alpha \rceil$, $\alpha \notin \mathbb{N}$, $f \in AC^N([a,b])$, *with* $f^{(N)} \in L_\infty([a,b])$, $0 < \beta < 1$, $x \in [a,b]$, $n \in \mathbb{N}$. *Then*

(i)

$$\left| G_n(f,x) - \sum_{j=1}^{N-1} \frac{f^{(j)}(x)}{j!} G_n\left((\cdot - x)^j \right)(x) - f(x) \right| \leq \tag{12.43}$$

$$\frac{(5.250312578)}{\Gamma(\alpha+1)} \cdot \left\{ \frac{\left(\omega_1 \left(D_{x-}^{\alpha} f, \frac{1}{n^{\beta}} \right)_{[a,x]} + \omega_1 \left(D_{*x}^{\alpha} f, \frac{1}{n_{\beta}} \right)_{[x,b]} \right)}{n^{\alpha\beta}} + \right.$$

$$\left. 3.1992 e^{-n^{(1-\beta)}} \left(\left\| D_{x-}^{\alpha} f \right\|_{\infty,[a,x]} (x-a)^{\alpha} + \left\| D_{*x}^{\alpha} f \right\|_{\infty,[x,b]} (b-x)^{\alpha} \right) \right\},$$

(ii) if $f^{(j)}(x) = 0$, for $j = 1, \ldots, N-1$, we have

$$|G_n(f,x) - f(x)| \leq \frac{(5.250312578)}{\Gamma(\alpha+1)} \cdot \tag{12.44}$$

$$\left\{ \frac{\left(\omega_1 \left(D_{x-}^{\alpha} f, \frac{1}{n^{\beta}} \right)_{[a,x]} + \omega_1 \left(D_{*x}^{\alpha} f, \frac{1}{n_{\beta}} \right)_{[x,b]} \right)}{n^{\alpha\beta}} + \right.$$

$$\left. 3.1992 e^{-n^{(1-\beta)}} \left(\left\| D_{x-}^{\alpha} f \right\|_{\infty,[a,x]} (x-a)^{\alpha} + \left\| D_{*x}^{\alpha} f \right\|_{\infty,[x,b]} (b-x)^{\alpha} \right) \right\},$$

when $\alpha > 1$ notice here the extremely high rate of convergence at $n^{-(\alpha+1)\beta}$,

(iii)

$$|G_n(f,x) - f(x)| \leq (5.250312578) \cdot \tag{12.45}$$

$$\left\{ \sum_{j=1}^{N-1} \frac{\left| f^{(j)}(x) \right|}{j!} \left\{ \frac{1}{n^{\beta j}} + (b-a)^j (3.1992) e^{-n^{(1-\beta)}} \right\} + \right.$$

$$\frac{1}{\Gamma(\alpha+1)}\left\{\frac{\left(\omega_1\left(D_{x-}^{\alpha}f,\frac{1}{n^{\beta}}\right)_{[a,x]}+\omega_1\left(D_{*x}^{\alpha}f,\frac{1}{n_{\beta}}\right)_{[x,b]}\right)}{n^{\alpha\beta}}+\right.$$

$$\left.\left.3.1992e^{-n^{(1-\beta)}}\left(\left\|D_{x-}^{\alpha}f\right\|_{\infty,[a,x]}(x-a)^{\alpha}+\left\|D_{*x}^{\alpha}f\right\|_{\infty,[x,b]}(b-x)^{\alpha}\right)\right\}\right\},$$

$\forall\, x \in [a,b]$,
and

(iv)

$$\|G_n f - f\|_{\infty} \leq (5.250312578)\cdot \tag{12.46}$$

$$\left\{\sum_{j=1}^{N-1}\frac{\|f^{(j)}\|_{\infty}}{j!}\left\{\frac{1}{n^{\beta j}}+(b-a)^{j}(3.1992)\,e^{-n^{(1-\beta)}}\right\}+\right.$$

$$\frac{1}{\Gamma(\alpha+1)}\left\{\frac{\left(\sup_{x\in[a,b]}\omega_1\left(D_{x-}^{\alpha}f,\frac{1}{n^{\beta}}\right)_{[a,x]}+\sup_{x\in[a,b]}\omega_1\left(D_{*x}^{\alpha}f,\frac{1}{n_{\beta}}\right)_{[x,b]}\right)}{n^{\alpha\beta}}+\right.$$

$$\left.\left.3.1992e^{-n^{(1-\beta)}}(b-a)^{\alpha}\left(\sup_{x\in[a,b]}\left\|D_{x-}^{\alpha}f\right\|_{\infty,[a,x]}+\sup_{x\in[a,b]}\left\|D_{*x}^{\alpha}f\right\|_{\infty,[x,b]}\right)\right\}\right\}.$$

Above, when $N=1$ the sum $\sum_{j=1}^{N-1}\cdot=0$.

As we see here we obtain fractionally type pointwise and uniform convergence with rates of $G_n \to I$ the unit operator, as $n\to\infty$.

Also from [18] we need

Theorem 12.49 *Let $\alpha>0$, $N=\lceil\alpha\rceil$, $\alpha\notin\mathbb{N}$, $f\in AC^{N}([a,b])$, with $f^{(N)}\in L_{\infty}([a,b])$, $0<\beta<1$, $x\in[a,b]$, $n\in\mathbb{N}$. Then*

(i)

$$\left| F_n(f,x) - \sum_{j=1}^{N-1} \frac{f^{(j)}(x)}{j!} F_n\left((\cdot - x)^j\right)(x) - f(x) \right| \leq$$

$$\frac{(4.1488766)}{\Gamma(\alpha+1)} \cdot \left\{ \frac{\left(\omega_1\left(D_{x-}^{\alpha} f, \frac{1}{n^{\beta}}\right)_{[a,x]} + \omega_1\left(D_{*x}^{\alpha} f, \frac{1}{n_{\beta}}\right)_{[x,b]}\right)}{n^{\alpha\beta}} + \right.$$

$$\left. e^4 e^{-2n^{(1-\beta)}} \left(\left\| D_{x-}^{\alpha} f \right\|_{\infty,[a,x]} (x-a)^{\alpha} + \left\| D_{*x}^{\alpha} f \right\|_{\infty,[x,b]} (b-x)^{\alpha} \right) \right\}, \tag{12.47}$$

(ii) if $f^{(j)}(x) = 0$, for $j = 1, \ldots, N-1$, we have

$$|F_n(f,x) - f(x)| \leq \frac{(4.1488766)}{\Gamma(\alpha+1)} \cdot$$

$$\left\{ \frac{\left(\omega_1\left(D_{x-}^{\alpha} f, \frac{1}{n^{\beta}}\right)_{[a,x]} + \omega_1\left(D_{*x}^{\alpha} f, \frac{1}{n_{\beta}}\right)_{[x,b]}\right)}{n^{\alpha\beta}} + \right.$$

$$\left. e^4 e^{-2n^{(1-\beta)}} \left(\left\| D_{x-}^{\alpha} f \right\|_{\infty,[a,x]} (x-a)^{\alpha} + \left\| D_{*x}^{\alpha} f \right\|_{\infty,[x,b]} (b-x)^{\alpha} \right) \right\}, \tag{12.48}$$

when $\alpha > 1$ notice here the extremely high rate of convergence of $n^{-(\alpha+1)\beta}$,

(iii)

$$|F_n(f,x) - f(x)| \leq (4.1488766) \cdot$$

$$\left\{ \sum_{j=1}^{N-1} \frac{\left|f^{(j)}(x)\right|}{j!} \left\{ \frac{1}{n^{\beta j}} + (b-a)^j e^4 e^{-2n^{(1-\beta)}} \right\} + \right.$$

$$\frac{1}{\Gamma(\alpha+1)} \left\{ \frac{\left(\omega_1\left(D_{x-}^{\alpha} f, \frac{1}{n^{\beta}}\right)_{[a,x]} + \omega_1\left(D_{*x}^{\alpha} f, \frac{1}{n_{\beta}}\right)_{[x,b]}\right)}{n^{\alpha\beta}} + \right.$$

$$e^4 e^{-2n^{(1-\beta)}} \left(\left\| D^{\alpha}_{x-} f \right\|_{\infty,[a,x]} (x-a)^{\alpha} + \left\| D^{\alpha}_{*x} f \right\|_{\infty,[x,b]} (b-x)^{\alpha} \right) \Bigg\} \Bigg] \Bigg], \tag{12.49}$$

$\forall\, x \in [a,b]$,
and

(iv)

$$\|F_n f - f\|_{\infty} \leq (4.1488766) \cdot$$

$$\Bigg\{ \sum_{j=1}^{N-1} \frac{\|f^{(j)}\|_{\infty}}{j!} \left\{ \frac{1}{n^{\beta j}} + (b-a)^j e^4 e^{-2n^{(1-\beta)}} \right\} +$$

$$\frac{1}{\Gamma(\alpha+1)} \Bigg\{ \frac{\left(\sup_{x\in[a,b]} \omega_1 \left(D^{\alpha}_{x-} f, \frac{1}{n^{\beta}} \right)_{[a,x]} + \sup_{x\in[a,b]} \omega_1 \left(D^{\alpha}_{*x} f, \frac{1}{n_{\beta}} \right)_{[x,b]} \right)}{n^{\alpha\beta}} +$$

$$e^4 e^{-2n^{(1-\beta)}} (b-a)^{\alpha} \left(\sup_{x\in[a,b]} \left\| D^{\alpha}_{x-} f \right\|_{\infty,[a,x]} + \sup_{x\in[a,b]} \left\| D^{\alpha}_{*x} f \right\|_{\infty,[x,b]} \right) \Bigg\} \Bigg\}. \tag{12.50}$$

Above, when $N = 1$ the sum $\sum_{j=1}^{N-1} \cdot = 0$.

As we see here we obtain fractionally type pointwise and uniform convergence with rates of $F_n \to I$ the unit operator, as $n \to \infty$.

Also from [18] we mention for $\alpha = \frac{1}{2}$ the next

Corollary 12.50 *Let $0 < \beta < 1$, $f \in AC([a,b])$, $f' \in L_{\infty}([a,b])$, $n \in \mathbb{N}$. Then*

(i)

$$\|G_n f - f\|_{\infty} \leq \frac{(10.50062516)}{\sqrt{\pi}} \cdot$$

$$\Bigg\{ \frac{\left(\sup_{x\in[a,b]} \omega_1 \left(D^{\frac{1}{2}}_{x-} f, \frac{1}{n^{\beta}} \right)_{[a,x]} + \sup_{x\in[a,b]} \omega_1 \left(D^{\frac{1}{2}}_{*x} f, \frac{1}{n^{\beta}} \right)_{[x,b]} \right)}{n^{\frac{\beta}{2}}} +$$

$$3.1992 e^{-n^{(1-\beta)}} \sqrt{b-a} \left(\sup_{x \in [a,b]} \left\| D_{x-}^{\frac{1}{2}} f \right\|_{\infty,[a,x]} + \sup_{x \in [a,b]} \left\| D_{*x}^{\frac{1}{2}} f \right\|_{\infty,[x,b]} \right) \Bigg\}, \tag{12.51}$$

and

(ii)

$$\| F_n f - f \|_\infty \leq \frac{(8.2977532)}{\sqrt{\pi}} \cdot$$

$$\left\{ \frac{\left(\sup_{x \in [a,b]} \omega_1 \left(D_{x-}^{\frac{1}{2}} f, \frac{1}{n^\beta} \right)_{[a,x]} + \sup_{x \in [a,b]} \omega_1 \left(D_{*x}^{\frac{1}{2}} f, \frac{1}{n^\beta} \right)_{[x,b]} \right)}{n^{\frac{\beta}{2}}} + \right.$$

$$e^4 e^{-2n^{(1-\beta)}} \sqrt{b-a} \left(\sup_{x \in [a,b]} \left\| D_{x-}^{\frac{1}{2}} f \right\|_{\infty,[a,x]} + \sup_{x \in [a,b]} \left\| D_{*x}^{\frac{1}{2}} f \right\|_{\infty,[x,b]} \right) \Bigg\}. \tag{12.52}$$

Denote $\omega_1^{(\mathcal{F})} (f, \delta)_{[a,b]} = \omega_1^{(\mathcal{F})} (f, \delta)$ and $\omega_1 (f, \delta)_{[a,b]} = \omega_1 (f, \delta)$.
From [12] we need

Theorem 12.51 *Let $f \in C([a, b])$, $0 < \beta < 1$, $n \in \mathbb{N}$, $x \in [a, b]$. Then*

(i)

$$|G_n (f, x) - f (x)| \leq (5.250312578) \cdot$$

$$\left\{ \omega_1 \left(f, \frac{1}{n^\beta} \right) + (6.3984) \| f \|_\infty e^{-n^{(1-\beta)}} \right\} =: \lambda, \tag{12.53}$$

(ii)

$$\| G_n (f) - f \|_\infty \leq \lambda, \tag{12.54}$$

where $\|\cdot\|_\infty$ is the supremum norm.

Finally from [12] we need

Theorem 12.52 *Let $f \in C([a, b])$, $0 < \beta < 1$, $n \in \mathbb{N}$, $x \in [a, b]$. Then*

(i)

$$|F_n (f, x) - f (x)| \leq (4.1488766) \cdot$$

$$\left\{\omega_1\left(f, \frac{1}{n^{\beta}}\right) + 2e^4 \|f\|_{\infty} e^{-2n^{(1-\beta)}}\right\} =: \lambda^*, \tag{12.55}$$

(ii)

$$\|F_n(f) - f\|_{\infty} \leq \lambda^*. \tag{12.56}$$

12.4 Fractional Approximation by Fuzzy Quasi-interpolation Neural Network Operators

Let $f \in C_{\mathcal{F}}([a, b])$, $n \in \mathbb{N}$. We define the Fuzzy Quasi-Interpolation Neural Network operators

$$G_n^{\mathcal{F}}(f, x) := \sum_{k=\lceil na \rceil}^{\lfloor nb \rfloor *} f\left(\frac{k}{n}\right) \odot \frac{\Phi(nx-k)}{\sum_{k=\lceil na \rceil}^{\lfloor nb \rfloor} \Phi(nx-k)}, \tag{12.57}$$

$\forall\, x \in [a, b]$, see also (12.38).

Similarly, we define

$$F_n^{\mathcal{F}}(f, x) := \sum_{k=\lceil na \rceil}^{\lfloor nb \rfloor *} f\left(\frac{k}{n}\right) \odot \frac{\Psi(nx-k)}{\sum_{k=\lceil na \rceil}^{\lfloor nb \rfloor} \Psi(nx-k)}, \tag{12.58}$$

$\forall\, x \in [a, b]$, see also (12.42).

The fuzzy sums in (12.57) and (12.58) are finite.

Let $r \in [0, 1]$, we observe that

$$\left[G_n^{\mathcal{F}}(f, x)\right]^r = \sum_{k=\lceil na \rceil}^{\lfloor nb \rfloor} \left[f\left(\frac{k}{n}\right)\right]^r \left(\frac{\Phi(nx-k)}{\sum_{k=\lceil na \rceil}^{\lfloor nb \rfloor} \Phi(nx-k)}\right) =$$

$$\sum_{k=\lceil na \rceil}^{\lfloor nb \rfloor} \left[f_-^{(r)}\left(\frac{k}{n}\right), f_+^{(r)}\left(\frac{k}{n}\right)\right] \left(\frac{\Phi(nx-k)}{\sum_{k=\lceil na \rceil}^{\lfloor nb \rfloor} \Phi(nx-k)}\right) \tag{12.59}$$

$$= \left[\sum_{k=\lceil na \rceil}^{\lfloor nb \rfloor} f_-^{(r)} \left(\frac{k}{n} \right) \left(\frac{\Phi (nx-k)}{\sum_{k=\lceil na \rceil}^{\lfloor nb \rfloor} \Phi (nx-k)} \right), \sum_{k=\lceil na \rceil}^{\lfloor nb \rfloor} f_+^{(r)} \left(\frac{k}{n} \right) \left(\frac{\Phi (nx-k)}{\sum_{k=\lceil na \rceil}^{\lfloor nb \rfloor} \Phi (nx-k)} \right) \right]$$

$$= \left[G_n \left(f_-^{(r)}, x \right), G_n \left(f_+^{(r)}, x \right) \right]. \tag{12.60}$$

We have proved that

$$\left(G_n^{\mathcal{F}} (f, x) \right)_{\pm}^{(r)} = G_n \left(f_{\pm}^{(r)}, x \right), \tag{12.61}$$

respectively, $\forall\, r \in [0, 1]$, $\forall\, x \in [a, b]$.

Similarly, it holds

$$\left(F_n^{\mathcal{F}} (f, x) \right)_{\pm}^{(r)} = F_n \left(f_{\pm}^{(r)}, x \right), \tag{12.62}$$

respectively, $\forall\, r \in [0, 1]$, $\forall\, x \in [a, b]$.

Therefore we get

$$D \left(G_n^{\mathcal{F}} (f, x), f (x) \right) =$$

$$\sup_{r \in [0,1]} \max \left\{ \left| \left(G_n \left(f_-^{(r)}, x \right) \right) - f_-^{(r)} (x) \right|, \left| \left(G_n \left(f_+^{(r)}, x \right) \right) - f_+^{(r)} (x) \right| \right\}, \tag{12.63}$$

and

$$D \left(F_n^{\mathcal{F}} (f, x), f (x) \right) =$$

$$\sup_{r \in [0,1]} \max \left\{ \left| \left(F_n \left(f_-^{(r)}, x \right) \right) - f_-^{(r)} (x) \right|, \left| \left(F_n \left(f_+^{(r)}, x \right) \right) - f_+^{(r)} (x) \right| \right\}, \tag{12.64}$$

$\forall\, x \in [a, b]$.

We present

Theorem 12.53 *Let $f \in C_{\mathcal{F}} ([a, b])$, $0 < \beta < 1$, $n \in \mathbb{N}$, $x \in [a, b]$. Then*

(1)

$$D \left(G_n^{\mathcal{F}} (f, x), f (x) \right) \leq (5.250312578) \cdot \tag{12.65}$$

$$\left[\omega_1^{(\mathcal{F})} \left(f, \frac{1}{n^{\beta}} \right) + (6.3984)\, D^* (f, \widetilde{o})\, e^{-n^{(1-\beta)}} \right] =: \lambda^{(\mathcal{F})},$$

(2)

$$D^* \left(G_n^{\mathcal{F}} (f), f \right) \leq \lambda^{(\mathcal{F})}. \tag{12.66}$$

Proof Since $f \in C_{\mathcal{F}}([a,b])$ we have that $f_{\pm}^{(r)} \in C([a,b])$, $\forall\, r \in [0,1]$. Hence by (12.53) we obtain

$$\left| G_n\left(f_{\pm}^{(r)}, x\right) - f_{\pm}^{(r)}(x)\right| \leq (5.250312578) \cdot$$

$$\left[\omega_1\left(f_{\pm}^{(r)}, \frac{1}{n^{\beta}}\right) + (6.3984)\left\| f_{\pm}^{(r)}\right\|_{\infty} e^{-n^{(1-\beta)}}\right]$$

(by Proposition 12.7 and $\left\| f_{\pm}^{(r)}\right\|_{\infty} \leq D^*(f, \widetilde{o})$)

$$\leq (5.250312578)\left[\omega_1^{(\mathcal{F})}\left(f, \frac{1}{n^{\beta}}\right) + (6.3984)\, D^*(f, \widetilde{o})\, e^{-n^{(1-\beta)}}\right].$$

Taking into account (12.63) the proof of the claim is completed. ■

We also give

Theorem 12.54 *Let $f \in C_{\mathcal{F}}([a,b])$, $0 < \beta < 1$, $n \in \mathbb{N}$, $x \in [a,b]$. Then*

(1)

$$D\left(F_n^{\mathcal{F}}(f,x), f(x)\right) \leq (4.1488766) \cdot \tag{12.67}$$

$$\left[\omega_1^{(\mathcal{F})}\left(f, \frac{1}{n^{\beta}}\right) + 2e^4 D^*(f, \widetilde{o})\, e^{-2n^{(1-\beta)}}\right] =: \lambda_*^{(\mathcal{F})},$$

(2)

$$D^*\left(F_n^{\mathcal{F}}(f), f\right) \leq \lambda_*^{(\mathcal{F})}. \tag{12.68}$$

Proof Similar to Theorem 12.53 based on Theorem 12.52. ■

Next we present

Theorem 12.55 *Let $\alpha > 0$, $N = \lceil \alpha \rceil$, $\alpha \notin \mathbb{N}$, $f \in C_{\mathcal{F}}^N([a,b])$, $0 < \beta < 1$, $x \in [a,b]$, $n \in \mathbb{N}$. Then*

(i)

$$D\left(G_n^{\mathcal{F}}(f,x), f(x)\right) \leq (5.250312578) \cdot \tag{12.69}$$

$$\left\{ \sum_{j=1}^{N-1} \frac{D\left(f^{(j)}(x), \widetilde{o}\right)}{j!} \left\{ \frac{1}{n^{\beta j}} + (b-a)^j \cdot (3.1992)\, e^{-n^{(1-\beta)}} \right\} \right.$$

$$+\frac{1}{\Gamma(\alpha+1)}\left\{\frac{\left(\omega_1^{(\mathcal{F})}\left(\left(D_{x-}^{\alpha\mathcal{F}}f\right),\frac{1}{n^\beta}\right)_{[a,x]}+\omega_1^{(\mathcal{F})}\left(\left(D_{*x}^{\alpha\mathcal{F}}f\right),\frac{1}{n^\beta}\right)_{[x,b]}\right)}{n^{\alpha\beta}}\right.$$

$$+(3.1992)\,e^{-n^{(1-\beta)}}\cdot$$

$$\left.\left.\left(D^*\left(\left(D_{x-}^{\alpha\mathcal{F}}f\right),\widetilde{o}\right)_{[a,x]}(x-a)^\alpha+D^*\left(\left(D_{*x}^{\alpha\mathcal{F}}f\right),\widetilde{o}\right)_{[x,b]}(b-x)^\alpha\right)\right]\right\},$$

(ii) if $f^{(j)}(x_0)=0$, $j=1,\ldots,N-1$, *we have*

$$D\left(G_n^{\mathcal{F}}(f,x_0),f(x_0)\right)\leq\frac{(5.250312578)}{\Gamma(\alpha+1)}\cdot \tag{12.70}$$

$$\left\{\frac{\left(\omega_1^{(\mathcal{F})}\left(\left(D_{x_0-}^{\alpha\mathcal{F}}f\right),\frac{1}{n^\beta}\right)_{[a,x_0]}+\omega_1^{(\mathcal{F})}\left(\left(D_{*x_0}^{\alpha\mathcal{F}}f\right),\frac{1}{n^\beta}\right)_{[x_0,b]}\right)}{n^{\alpha\beta}}\right.$$

$$+(3.1992)\,e^{-n^{(1-\beta)}}\cdot$$

$$\left.\left(D^*\left(\left(D_{x_0-}^{\alpha\mathcal{F}}f\right),\widetilde{o}\right)_{[a,x_0]}(x_0-a)^\alpha+D^*\left(\left(D_{*x_0}^{\alpha\mathcal{F}}f\right),\widetilde{o}\right)_{[x_0,b]}(b-x_0)^\alpha\right)\right\},$$

when $\alpha>1$ *notice here the extremely high rate of convergence at* $n^{-(\alpha+1)\beta}$, *and*

(iii)

$$D^*\left(G_n^{\mathcal{F}}(f),f\right)\leq(5.250312578)\cdot \tag{12.71}$$

$$\left\{\sum_{j=1}^{N-1}\frac{D^*\left(f^{(j)}(x),\widetilde{o}\right)}{j!}\left\{\frac{1}{n^{\beta j}}+(b-a)^j\cdot(3.1992)\,e^{-n^{(1-\beta)}}\right\}\right.$$

$$+\frac{1}{\Gamma(\alpha+1)}\left\{\frac{\left(\sup_{x\in[a,b]}\omega_1^{(\mathcal{F})}\left(\left(D_{x-}^{\alpha\mathcal{F}}f\right),\frac{1}{n^{\beta}}\right)_{[a,x]}+\sup_{x\in[a,b]}\omega_1^{(\mathcal{F})}\left(\left(D_{*x}^{\alpha\mathcal{F}}f\right),\frac{1}{n^{\beta}}\right)_{[x,b]}\right)}{n^{\alpha\beta}}\right.$$

$$+(3.1992)\,e^{-n^{(1-\beta)}}(b-a)^{\alpha}\cdot$$

$$\left.\left.\left(\sup_{x\in[a,b]}D^{*}\left(\left(D_{x-}^{\alpha\mathcal{F}}f\right),\widetilde{o}\right)_{[a,x]}+\sup_{x\in[a,b]}D^{*}\left(\left(D_{*x}^{\alpha\mathcal{F}}f\right),\widetilde{o}\right)_{[x,b]}\right)\right\}\right\},$$

above, when $N=1$ *the sum* $\sum_{j=1}^{N-1}\cdot=0$.

As we see here we obtain fractionally the fuzzy pointwise and uniform convergence with rates of $G_n^{\mathcal{F}}\to I$ *the unit operator, as* $n\to\infty$.

Proof Here $f_{\pm}^{(r)}\in C^{N}([a,b])$, $\forall\, r\in[0,1]$, and $D_{x-}^{\alpha\mathcal{F}}f$, $D_{*x}^{\alpha\mathcal{F}}f$ are fuzzy continuous over $[a,b]$, $\forall\, x\in[a,b]$, so that $\left(D_{x-}^{\alpha\mathcal{F}}f\right)_{\pm}^{(r)}$, $\left(D_{*x}^{\alpha\mathcal{F}}f\right)_{\pm}^{(r)}\in C([a,b])$, $\forall\, r\in[0,1]$, $\forall\, x\in[a,b]$.

We observe by (12.45) that (respectively in $\pm$)

$$\left|G_n\left(f_{\pm}^{(r)},x\right)-f_{\pm}^{(r)}(x)\right|\le(5.250312578)\cdot$$

$$\left\{\sum_{j=1}^{N-1}\frac{\left|\left(f_{\pm}^{(r)}\right)^{(j)}(x)\right|}{j!}\left\{\frac{1}{n^{\beta j}}+(b-a)^{j}\cdot(3.1992)\,e^{-n^{(1-\beta)}}\right\}\right.\tag{12.72}$$

$$+\frac{1}{\Gamma(\alpha+1)}\left\{\frac{\left(\omega_1\left(D_{x-}^{\alpha}\left(f_{\pm}^{(r)}\right),\frac{1}{n^{\beta}}\right)_{[a,x]}+\omega_1\left(D_{*x}^{\alpha}\left(f_{\pm}^{(r)}\right),\frac{1}{n^{\beta}}\right)_{[x,b]}\right)}{n^{\alpha\beta}}\right.$$

$$+(3.1992)\,e^{-n^{(1-\beta)}}\cdot$$

$$\left.\left.\left(\left\|D_{x-}^{\alpha}\left(f_{\pm}^{(r)}\right)\right\|_{\infty,[a,x]}(x-a)^{\alpha}+\left\|D_{*x}^{\alpha}\left(f_{\pm}^{(r)}\right)\right\|_{\infty,[x,b]}(b-x)^{\alpha}\right)\right\}\right\}=$$

$$(5.250312578)\left\{\sum_{j=1}^{N-1}\frac{\left|\left(f^{(j)}(x)\right)_{\pm}^{(r)}\right|}{j!}\left\{\frac{1}{n^{\beta j}}+(b-a)^{j}\cdot(3.1992)\,e^{-n^{(1-\beta)}}\right\}\right.$$

$$+\frac{1}{\Gamma(\alpha+1)}\left\{\frac{\left(\omega_1\left(\left(D_{x-}^{\alpha\mathcal{F}}f\right)_{\pm}^{(r)},\frac{1}{n^{\beta}}\right)_{[a,x]}+\omega_1\left(\left(D_{*x}^{\alpha\mathcal{F}}f\right)_{\pm}^{(r)},\frac{1}{n^{\beta}}\right)_{[x,b]}\right)}{n^{\alpha\beta}}\right.$$

$$+(3.1992)\,e^{-n^{(1-\beta)}}\cdot \tag{12.73}$$

$$\left.\left.\left.\left(\left\|\left(D_{x-}^{\alpha\mathcal{F}}f\right)_{\pm}^{(r)}\right\|_{\infty,[a,x]}(x-a)^{\alpha}+\left\|\left(D_{*x}^{\alpha\mathcal{F}}f\right)_{\pm}^{(r)}\right\|_{\infty,[x,b]}(b-x)^{\alpha}\right)\right\}\right]\right\}\leq$$

$$(5.250312578)\left\{\sum_{j=1}^{N-1}\frac{D\left(f^{(j)}(x),\widetilde{o}\right)}{j!}\left\{\frac{1}{n^{\beta j}}+(b-a)^{j}\cdot(3.1992)\,e^{-n^{(1-\beta)}}\right\}\right.$$

$$+\frac{1}{\Gamma(\alpha+1)}\left\{\frac{\left(\omega_1^{(\mathcal{F})}\left(\left(D_{x-}^{\alpha\mathcal{F}}f\right),\frac{1}{n^{\beta}}\right)_{[a,x]}+\omega_1^{(\mathcal{F})}\left(\left(D_{*x}^{\alpha\mathcal{F}}f\right),\frac{1}{n^{\beta}}\right)_{[x,b]}\right)}{n^{\alpha\beta}}\right.$$

$$+(3.1992)\,e^{-n^{(1-\beta)}}\cdot \tag{12.74}$$

$$\left.\left.\left.\left(D^{*}\left(\left(D_{x-}^{\alpha\mathcal{F}}f\right),\widetilde{o}\right)_{[a,x]}(x-a)^{\alpha}+D^{*}\left(\left(D_{*x}^{\alpha\mathcal{F}}f\right),\widetilde{o}\right)_{[x,b]}(b-x)^{\alpha}\right)\right\}\right]\right\},$$

along with (12.63) proving all inequalities of theorem.

Here we notice that

$$\left(D_{x-}^{\alpha\mathcal{F}}f\right)_{\pm}^{(r)}(t)=\left(D_{x-}^{\alpha}\left(f_{\pm}^{(r)}\right)\right)(t)$$

$$=\frac{(-1)^{N}}{\Gamma(N-\alpha)}\int_{t}^{x}(s-t)^{N-\alpha-1}\left(f_{\pm}^{(r)}\right)^{(N)}(s)\,ds, \tag{12.75}$$

where $a\leq t\leq x$.

Hence

$$\left|\left(D_{x-}^{\alpha\mathcal{F}} f\right)_{\pm}^{(r)}(t)\right| \leq \frac{1}{\Gamma(N-\alpha)} \int_{t}^{x} (s-t)^{N-\alpha-1} \left|\left(f_{\pm}^{(r)}\right)^{(N)}(s)\right| ds \leq \tag{12.76}$$

$$\frac{\left\|\left(f^{(N)}\right)_{\pm}^{(r)}\right\|_{\infty}}{\Gamma(N-\alpha+1)} (x-t)^{N-\alpha} \leq \frac{\left\|\left(f^{(N)}\right)_{\pm}^{(r)}\right\|_{\infty}}{\Gamma(N-\alpha+1)} (b-a)^{N-\alpha}$$

$$\leq \frac{D^{*}\left(f^{(N)}, \widetilde{o}\right)}{\Gamma(N-\alpha+1)} (b-a)^{N-\alpha}.$$

So we have

$$\left|\left(D_{x-}^{\alpha\mathcal{F}} f\right)_{\pm}^{(r)}(t)\right| \leq \frac{D^{*}\left(f^{(N)}, \widetilde{o}\right)}{\Gamma(N-\alpha+1)} (b-a)^{N-\alpha}, \tag{12.77}$$

all $a \leq t \leq x$.

And it holds

$$\left\|\left(D_{x-}^{\alpha\mathcal{F}} f\right)_{\pm}^{(r)}\right\|_{\infty,[a,x]} \leq \frac{D^{*}\left(f^{(N)}, \widetilde{o}\right)}{\Gamma(N-\alpha+1)} (b-a)^{N-\alpha}, \tag{12.78}$$

that is

$$D^{*}\left(\left(D_{x-}^{\alpha\mathcal{F}} f\right), \widetilde{o}\right)_{[a,x]} \leq \frac{D^{*}\left(f^{(N)}, \widetilde{o}\right)}{\Gamma(N-\alpha+1)} (b-a)^{N-\alpha},$$

and

$$\sup_{x\in[a,b]} D^{*}\left(\left(D_{x-}^{\alpha\mathcal{F}} f\right), \widetilde{o}\right)_{[a,x]} \leq \frac{D^{*}\left(f^{(N)}, \widetilde{o}\right)}{\Gamma(N-\alpha+1)} (b-a)^{N-\alpha} < \infty. \tag{12.79}$$

Similarly we have

$$\left(D_{*x}^{\alpha\mathcal{F}} f\right)_{\pm}^{(r)}(t) = \left(D_{*x}^{\alpha}\left(f_{\pm}^{(r)}\right)\right)(t)$$

$$= \frac{1}{\Gamma(N-\alpha)} \int_{x}^{t} (t-s)^{N-\alpha-1} \left(f_{\pm}^{(r)}\right)^{(N)}(s)\, ds, \tag{12.80}$$

where $x \leq t \leq b$.

Hence

$$\left|\left(D_{*x}^{\alpha\mathcal{F}} f\right)_{\pm}^{(r)}(t)\right| \leq \frac{1}{\Gamma(N-\alpha)} \int_{x}^{t} (t-s)^{N-\alpha-1} \left|\left(f^{(N)}\right)_{\pm}^{(r)}(s)\right| ds \leq \tag{12.81}$$

$$\frac{\left\| \left(f^{(N)}\right)_{\pm}^{(r)} \right\|_{\infty}}{\Gamma(N-\alpha+1)} (b-a)^{N-\alpha} \leq \frac{D^*\left(f^{(N)}, \widetilde{o}\right)}{\Gamma(N-\alpha+1)} (b-a)^{N-\alpha},$$

$x \leq t \leq b$.

So we have

$$\left\| \left(D_{*x}^{\alpha\mathcal{F}} f\right)_{\pm}^{(r)} \right\|_{\infty,[x,b]} \leq \frac{D^*\left(f^{(N)}, \widetilde{o}\right)}{\Gamma(N-\alpha+1)} (b-a)^{N-\alpha}, \tag{12.82}$$

that is

$$D^*\left(\left(D_{*x}^{\alpha\mathcal{F}} f\right), \widetilde{o}\right)_{[x,b]} \leq \frac{D^*\left(f^{(N)}, \widetilde{o}\right)}{\Gamma(N-\alpha+1)} (b-a)^{N-\alpha}, \tag{12.83}$$

and

$$\sup_{x\in[a,b]} D^*\left(\left(D_{*x}^{\alpha\mathcal{F}} f\right), \widetilde{o}\right)_{[x,b]} \leq \frac{D^*\left(f^{(N)}, \widetilde{o}\right)}{\Gamma(N-\alpha+1)} (b-a)^{N-\alpha} < +\infty. \tag{12.84}$$

Furthermore we notice

$$\omega_1^{(\mathcal{F})}\left(\left(D_{x-}^{\alpha\mathcal{F}} f\right), \frac{1}{n^\beta}\right)_{[a,x]} = \sup_{\substack{s,t\in[a,x] \\ |s-t|\leq\frac{1}{n^\beta}}} D\left(\left(D_{x-}^{\alpha\mathcal{F}} f\right)(s), \left(D_{x-}^{\alpha\mathcal{F}} f\right)(t)\right) \leq$$

$$\sup_{\substack{s,t\in[a,x] \\ |s-t|\leq\frac{1}{n^\beta}}} \left\{D\left(\left(D_{x-}^{\alpha\mathcal{F}} f\right)(s), \widetilde{o}\right), D\left(\left(D_{x-}^{\alpha\mathcal{F}} f\right)(t), \widetilde{o}\right)\right\} \leq 2D^*\left(\left(D_{x-}^{\alpha\mathcal{F}} f\right), \widetilde{o}\right)_{[a,x]} \tag{12.85}$$

$$\leq \frac{2D^*\left(f^{(N)}, \widetilde{o}\right)}{\Gamma(N-\alpha+1)} (b-a)^{N-\alpha}.$$

Therefore it holds

$$\sup_{x\in[a,b]} \omega_1^{(\mathcal{F})}\left(\left(D_{x-}^{\alpha\mathcal{F}} f\right), \frac{1}{n^\beta}\right)_{[a,x]} \leq \frac{2D^*\left(f^{(N)}, \widetilde{o}\right)}{\Gamma(N-\alpha+1)} (b-a)^{N-\alpha} < +\infty. \tag{12.86}$$

Similarly we observe

$$\omega_1^{(\mathcal{F})}\left(\left(D_{*x}^{\alpha\mathcal{F}} f\right), \frac{1}{n^\beta}\right)_{[x,b]} = \sup_{\substack{s,t\in[x,b] \\ |s-t|\leq\frac{1}{n^\beta}}} D\left(\left(D_{*x}^{\alpha\mathcal{F}} f\right)(s), \left(D_{*x}^{\alpha\mathcal{F}} f\right)(t)\right) \leq$$

$$2D^*\left(\left(D_{*x}^{\alpha\mathcal{F}}f\right),\widetilde{o}\right)_{[x,b]}\leq\frac{2D^*\left(f^{(N)},\widetilde{o}\right)}{\Gamma\left(N-\alpha+1\right)}\left(b-a\right)^{N-\alpha}. \tag{12.87}$$

Consequently it holds

$$\sup_{x\in[a,b]}\omega_1^{(\mathcal{F})}\left(\left(D_{*x}^{\alpha\mathcal{F}}f\right),\frac{1}{n^{\beta}}\right)_{[x,b]}\leq\frac{2D^*\left(f^{(N)},\widetilde{o}\right)}{\Gamma\left(N-\alpha+1\right)}\left(b-a\right)^{N-\alpha}<+\infty. \tag{12.88}$$

So everything in the statements of the theorem makes sense.

The proof of the Theorem is now completed. ■

We also give

Theorem 12.56 *Let* $\alpha>0$, $N=\lceil\alpha\rceil$, $\alpha\notin\mathbb{N}$, $f\in C_{\mathcal{F}}^{N}([a,b])$, $0<\beta<1$, $x\in[a,b]$, $n\in\mathbb{N}$. *Then*

(i)

$$D\left(F_n^{\mathcal{F}}\left(f,x\right),f\left(x\right)\right)\leq\left(4.1488766\right)\cdot \tag{12.89}$$

$$\left\{\sum_{j=1}^{N-1}\frac{D\left(f^{(j)}\left(x\right),\widetilde{o}\right)}{j!}\left\{\frac{1}{n^{\beta j}}+\left(b-a\right)^{j}\cdot e^{4}e^{-2n^{(1-\beta)}}\right\}\right.$$

$$+\frac{1}{\Gamma\left(\alpha+1\right)}\left\{\frac{\left(\omega_1^{(\mathcal{F})}\left(\left(D_{x-}^{\alpha\mathcal{F}}f\right),\frac{1}{n^{\beta}}\right)_{[a,x]}+\omega_1^{(\mathcal{F})}\left(\left(D_{*x}^{\alpha\mathcal{F}}f\right),\frac{1}{n^{\beta}}\right)_{[x,b]}\right)}{n^{\alpha\beta}}\right.$$

$$\left.\left.+e^{4}e^{-2n^{(1-\beta)}}\left(D^*\left(\left(D_{x-}^{\alpha\mathcal{F}}f\right),\widetilde{o}\right)_{[a,x]}\left(x-a\right)^{\alpha}+D^*\left(\left(D_{*x}^{\alpha\mathcal{F}}f\right),\widetilde{o}\right)_{[x,b]}\left(b-x\right)^{\alpha}\right)\right\}\right\},$$

(ii) if $f^{(j)}\left(x_0\right)=0$, $j=1,\ldots,N-1$, *we have*

$$D\left(F_n^{\mathcal{F}}\left(f,x_0\right),f\left(x_0\right)\right)\leq\frac{\left(4.1488766\right)}{\Gamma\left(\alpha+1\right)}\cdot \tag{12.90}$$

$$\left\{\frac{\left(\omega_1^{(\mathcal{F})}\left(\left(D_{x_0-}^{\alpha\mathcal{F}}f\right),\frac{1}{n^{\beta}}\right)_{[a,x_0]}+\omega_1^{(\mathcal{F})}\left(\left(D_{*x_0}^{\alpha\mathcal{F}}f\right),\frac{1}{n^{\beta}}\right)_{[x_0,b]}\right)}{n^{\alpha\beta}}+e^{4}e^{-2n^{(1-\beta)}}\cdot\right.$$

$$\left(D^*\left(\left(D_{x_0-}^{\alpha\mathcal{F}} f\right), \widetilde{o}\right)_{[a,x_0]} (x_0-a)^\alpha + D^*\left(\left(D_{*x_0}^{\alpha\mathcal{F}} f\right), \widetilde{o}\right)_{[x_0,b]} (b-x_0)^\alpha\right)\Bigg\},$$

when $\alpha > 1$ notice here the extremely high rate of convergence at $n^{-(\alpha+1)\beta}$,

(iii)

$$D^*\left(F_n^{\mathcal{F}}(f), f\right) \leq (4.1488766) \cdot \tag{12.91}$$

$$\Bigg\{\sum_{j=1}^{N-1} \frac{D^*\left(f^{(j)}, \widetilde{o}\right)}{j!} \left\{\frac{1}{n^{\beta j}} + (b-a)^j \cdot e^4 e^{-2n^{(1-\beta)}}\right\} +$$

$$\frac{1}{\Gamma(\alpha+1)} \Bigg\{ \frac{\left(\sup_{x\in[a,b]} \omega_1^{(\mathcal{F})}\left(\left(D_{x-}^{\alpha\mathcal{F}} f\right), \frac{1}{n^\beta}\right)_{[a,x]} + \sup_{x\in[a,b]} \omega_1^{(\mathcal{F})}\left(\left(D_{*x}^{\alpha\mathcal{F}} f\right), \frac{1}{n^\beta}\right)_{[x,b]}\right)}{n^{\alpha\beta}}$$

$$+ e^4 e^{-2n^{(1-\beta)}} (b-a)^\alpha \cdot$$

$$\left(\sup_{x\in[a,b]} D^*\left(\left(D_{x-}^{\alpha\mathcal{F}} f\right), \widetilde{o}\right)_{[a,x]} + \sup_{x\in[a,b]} D^*\left(\left(D_{*x}^{\alpha\mathcal{F}} f\right), \widetilde{o}\right)_{[x,b]}\right)\Bigg\}\Bigg\}.$$

Above, when $N = 1$ the sum $\sum_{j=1}^{N-1} \cdot = 0$.

As we see here we obtain fractionally the fuzzy pointwise and uniform convergence with rates of $F_n^{\mathcal{F}} \to I$, as $n \to \infty$.

Proof Similar to Theorem 12.55, using Theorem 12.49. ■

We make

Remark 12.57 Looking at (12.63) and (12.64), and applying the principle of iterated suprema we obtain

$$D^*\left(G_n^{\mathcal{F}}(f), f\right) = \sup_{r\in[0,1]} \max\left\{\left\|G_n\left(f_-^{(r)}\right) - f_-^{(r)}\right\|_\infty, \left\|G_n\left(f_+^{(r)}\right) - f_+^{(r)}\right\|_\infty\right\}, \tag{12.92}$$

$$D^*\left(F_n^{\mathcal{F}}(f), f\right) = \sup_{r\in[0,1]} \max\left\{\left\|F_n\left(f_-^{(r)}\right) - f_-^{(r)}\right\|_\infty, \left\|F_n\left(f_+^{(r)}\right) - f_+^{(r)}\right\|_\infty\right\}, \tag{12.93}$$

where $f \in C_{\mathcal{F}}([a, b])$, see also proof of Theorem 12.53.

We finish with

Corollary 12.58 *Let* $0 < \beta < 1$, $f \in C_{\mathcal{F}}^1([a, b])$, $n \in \mathbb{N}$. *Then*

(i)

$$D^*\left(G_n^{\mathcal{F}}(f), f\right) \leq \left(\frac{10.50062516}{\sqrt{\pi}}\right) \cdot \tag{12.94}$$

$$\left\{\frac{\left(\sup_{x\in[a,b]} \omega_1^{(\mathcal{F})}\left(\left(D_{x-}^{\frac{1}{2}\mathcal{F}} f\right), \frac{1}{n^\beta}\right)_{[a,x]} + \sup_{x\in[a,b]} \omega_1^{(\mathcal{F})}\left(\left(D_{*x}^{\frac{1}{2}\mathcal{F}} f\right), \frac{1}{n^\beta}\right)_{[x,b]}\right)}{n^{\frac{\beta}{2}}}\right.$$

$$+ (3.1992)\, e^{-n^{(1-\beta)}} \sqrt{b-a} \cdot$$

$$\left.\left(\sup_{x\in[a,b]} D^*\left(\left(D_{x-}^{\frac{1}{2}\mathcal{F}} f\right), \widetilde{o}\right)_{[a,x]} + \sup_{x\in[a,b]} D^*\left(\left(D_{*x}^{\frac{1}{2}\mathcal{F}} f\right), \widetilde{o}\right)_{[x,b]}\right)\right\},$$

and

(ii)

$$D^*\left(F_n^{\mathcal{F}}(f), f\right) \leq \left(\frac{8.2977532}{\sqrt{\pi}}\right) \cdot \tag{12.95}$$

$$\left\{\frac{\left(\sup_{x\in[a,b]} \omega_1^{(\mathcal{F})}\left(\left(D_{x-}^{\frac{1}{2}\mathcal{F}} f\right), \frac{1}{n^\beta}\right)_{[a,x]} + \sup_{x\in[a,b]} \omega_1^{(\mathcal{F})}\left(\left(D_{*x}^{\frac{1}{2}\mathcal{F}} f\right), \frac{1}{n^\beta}\right)_{[x,b]}\right)}{n^{\frac{\beta}{2}}}\right.$$

$$+ e^4 e^{-2n^{(1-\beta)}} \sqrt{b-a} \cdot$$

$$\left.\left(\sup_{x\in[a,b]} D^*\left(\left(D_{x-}^{\frac{1}{2}\mathcal{F}} f\right), \widetilde{o}\right)_{[a,x]} + \sup_{x\in[a,b]} D^*\left(\left(D_{*x}^{\frac{1}{2}\mathcal{F}} f\right), \widetilde{o}\right)_{[x,b]}\right)\right\}.$$

Proof By (12.51) we get

$$\left\| G_n^{\mathcal{F}}\left(f_{\pm}^{(r)}\right) - \left(f_{\pm}^{(r)}\right)\right\|_{\infty} \leq \left(\frac{10.50062516}{\sqrt{\pi}}\right) \cdot \tag{12.96}$$

$$\left\{ \frac{\left(\sup_{x\in[a,b]} \omega_1 \left(D_{x-}^{\frac{1}{2}} \left(f_{\pm}^{(r)}\right), \frac{1}{n^{\beta}} \right)_{[a,x]} + \sup_{x\in[a,b]} \omega_1 \left(D_{*x}^{\frac{1}{2}} \left(f_{\pm}^{(r)}\right), \frac{1}{n^{\beta}} \right)_{[x,b]} \right)}{n^{\frac{\beta}{2}}} \right.$$

$$+ (3.1992)\, e^{-n^{(1-\beta)}} \sqrt{b-a} \cdot$$

$$\left. \left(\sup_{x\in[a,b]} \left\| D_{x-}^{\frac{1}{2}} \left(f_{\pm}^{(r)}\right) \right\|_{\infty,[a,x]} + \sup_{x\in[a,b]} \left\| D_{*x}^{\frac{1}{2}} \left(f_{\pm}^{(r)}\right) \right\|_{\infty,[x,b]} \right) \right\} =$$

$$\left(\frac{10.50062516}{\sqrt{\pi}}\right) \cdot$$

$$\left\{ \frac{\left(\sup_{x\in[a,b]} \omega_1 \left(\left(D_{x-}^{\frac{1}{2}} f\right)_{\pm}^{(r)}, \frac{1}{n^{\beta}} \right)_{[a,x]} + \sup_{x\in[a,b]} \omega_1 \left(\left(D_{*x}^{\frac{1}{2}} f\right)_{\pm}^{(r)}, \frac{1}{n^{\beta}} \right)_{[x,b]} \right)}{n^{\frac{\beta}{2}}} \right.$$

$$+ (3.1992)\, e^{-n^{(1-\beta)}} \sqrt{b-a} \cdot \tag{12.97}$$

$$\left. \left(\sup_{x\in[a,b]} \left\| \left(D_{x-}^{\frac{1}{2}} f\right)_{\pm}^{(r)} \right\|_{\infty,[a,x]} + \sup_{x\in[a,b]} \left\| \left(D_{*x}^{\frac{1}{2}} f\right)_{\pm}^{(r)} \right\|_{\infty,[x,b]} \right) \right\} \leq$$

$$\left(\frac{10.50062516}{\sqrt{\pi}}\right) \cdot$$

$$\left\{\frac{\left(\sup_{x\in[a,b]}\omega_1^{(\mathcal{F})}\left(\left(D_{x-}^{\frac{1}{2}\mathcal{F}}f\right),\frac{1}{n^{\beta}}\right)_{[a,x]}+\sup_{x\in[a,b]}\omega_1^{(\mathcal{F})}\left(\left(D_{*x}^{\frac{1}{2}\mathcal{F}}f\right),\frac{1}{n^{\beta}}\right)_{[x,b]}\right)}{n^{\frac{\beta}{2}}}\right.$$

$$+(3.1992)\,e^{-n^{(1-\beta)}}\sqrt{b-a}\cdot \tag{12.98}$$

$$\left.\left(\sup_{x\in[a,b]}D^*\left(\left(D_{x-}^{\frac{1}{2}\mathcal{F}}f\right),\widetilde{o}\right)_{[a,x]}+\sup_{x\in[a,b]}D^*\left(\left(D_{*x}^{\frac{1}{2}\mathcal{F}}f\right),\widetilde{o}\right)_{[x,b]}\right)\right\},$$

proving the claim with the use of Remark 12.57.

Part (ii) follows similarly. ■

References

1. G.A. Anastassiou, Rate of convergence of some neural network operators to the unit-univariate case. J. Math. Anal. Appl. **212**, 237–262 (1997)
2. G.A. Anastassiou, *Quantitative Approximation* (Chapmann & Hall/CRC, Boca Raton, 2001)
3. G.A. Anastassiou, Fuzzy approximation by fuzzy convolution type operators. Comput. Math. **48**, 1369–1386 (2004)
4. G.A. Anastassiou, Higher order fuzzy Korovkin theory via inequalities. Commun. Appl. Anal. **10**(2), 359–392 (2006)
5. G.A. Anastassiou, Fuzzy Korovkin theorems and inequalities. J. Fuzzy Math. **15**(1), 169–205 (2007)
6. G.A. Anastassiou, On right fractional calculus. Chaos, Solitons Fractals **42**, 365–376 (2009)
7. G.A. Anastassiou, *Fractional Differentiation Inequalities* (Springer, New York, 2009)
8. G.A. Anastassiou, Fractional korovkin theory. Chaos, Solitons Fractals **42**(4), 2080–2094 (2009)
9. G.A. Anastassiou, *Fuzzy Mathematics: Approximation Theory* (Springer, New York, 2010)
10. G.A. Anastassiou, Quantitative approximation by fractional smooth picard singular operators. Math. Eng. Sci. Aerosp. **2**(1), 71–87 (2011)
11. G.A. Anastassiou, Fuzzy fractional calculus and Ostrowski inequality. J. Fuzzy Math. **19**(3), 577–590 (2011)
12. G.A. Anastassiou, *Inteligent Systems: Approximation by Artificial Neural Networks*, Intelligent Systems reference Library, vol. 19 (Springer, Heidelberg, 2011)
13. G.A. Anastassiou, Fractional representation formulae and right fractional inequalities. Math. Comput. Model. **54**(11–12), 3098–3115 (2011)
14. G.A. Anastassiou, Univariate hyperbolic tangent neural network approximation. Math. Comput. Model. **53**, 1111–1132 (2011)
15. G.A. Anastassiou, Multivariate hyperbolic tangent neural network approximation. Comput. Math. **61**, 809–821 (2011)
16. G.A. Anastassiou, Multivariate sigmoidal neural network approximation. Neural Netw. **24**, 378–386 (2011)

17. G.A. Anastassiou, Univariate sigmoidal neural network approximation. J. Comput. Anal. Appl. **14**(4), 659–690 (2012)
18. G.A. Anastassiou, Fractional neural network approximation. Comput. Math. Appl. **64**(6), 1655–1676 (2012)
19. G.A. Anastassiou, Fuzzy fractional neural network approximation by fuzzy quasi-interpolation operators. J. Appl. Non Linear Dyn. **2**(3), 235–259 (2013)
20. G.A. Anastassiou, Rate of convergence of some neural network operators to the unit-univariate case, revisited. Matematicki Vesnik **65**(4), 511–518 (2013)
21. G.A. Anastassiou, Fractional approximation by normalized bell and squashing type neural network operators. New Math. Nat. Comput. **9**(1), 43–63 (2013)
22. P. Cardaliaguet, G. Euvrard, Approximation of a function and its derivative with a neural network. Neural Netw. **5**, 207–220 (1992)
23. Z. Chen, F. Cao, The approximation operators with sigmoidal functions. Comput. Math. Appl. **58**, 758–765 (2009)
24. K. Diethelm, *The Analysis of Fractional Differential Equations*, Lecture Notes in Mathematics, vol. 2004 (Springer, Heidelberg, 2010)
25. A.M.A. El-Sayed, M. Gaber, On the finite Caputo and finite Riesz derivatives. Electron. J. Theor. Phys. **3**(12), 81–95 (2006)
26. G.S. Frederico, D.F.M. Torres, Fractional optimal control in the sense of Caputo and the fractional Noether's theorem. Int. Math. Forum **3**(10), 479–493 (2008)
27. S. Gal, in *Approximation Theory in Fuzzy Setting*, ed. by G. Anastassiou. Handbook of Analytic-Computational Methods in Applied Mathematics, Chap. 13 (Chapman & Hall/CRC, Boca Raton, 2000), pp. 617–666
28. R. Goetschel Jr, W. Voxman, Elementary fuzzy calculus. Fuzzy Sets Syst. **18**, 31–43 (1986)
29. S. Haykin, *Neural Networks: A Comprehensive Foundation*, 2nd edn. (Prentice Hall, New York, 1998)
30. O. Kaleva, Fuzzy differential equations. Fuzzy Sets Syst. **24**, 301–317 (1987)
31. Y.K. Kim, B.M. Ghil, Integrals of fuzzy-number-valued functions. Fuzzy Sets Syst. **86**, 213–222 (1997)
32. W. McCulloch, W. Pitts, A logical calculus of the ideas immanent in nervous activity. Bull. Math. Biophys. **7**, 115–133 (1943)
33. T.M. Mitchell, *Machine Learning* (WCB-McGraw-Hill, New York, 1997)
34. S.G. Samko, A.A. Kilbas, O.I. Marichev, *Fractional Integrals and Derivatives, Theory and Applications* (Gordon and Breach, Amsterdam, 1993) [English translation from the Russian, Integrals and Derivatives of Fractional Order and Some of Their Applications (Nauka i Tekhnika, Minsk, 1987)]
35. C. Wu, Z. Gong, On Henstock integrals of interval-valued functions and fuzzy valued functions. Fuzzy Sets Syst. **115**(3), 377–391 (2000)
36. C. Wu, Z. Gong, On Henstock integral of fuzzy-number-valued functions (I). Fuzzy Sets Syst. **120**(3), 523–532 (2001)
37. C. Wu, M. Ma, On embedding problem of fuzzy number space: part 1. Fuzzy Sets Syst. **44**, 33–38 (1991)

Chapter 13
Higher Order Multivariate Fuzzy Approximation Using Basic Neural Network Operators

Here are studied in terms of multivariate fuzzy high approximation to the multivariate unit basic sequences of multivariate fuzzy neural network operators. These operators are multivariate fuzzy analogs of earlier studied multivariate real ones. The produced results generalize earlier real ones into the fuzzy setting. Here the high order multivariate fuzzy pointwise convergence with rates to the multivariate fuzzy unit operator is established through multivariate fuzzy inequalities involving the multivariate fuzzy moduli of continuity of the Nth order ($N \geq 1$) H-fuzzy partial derivatives, of the engaged multivariate fuzzy number valued function. It follows [10].

13.1 Fuzzy Real Analysis Background

We need the following background

Definition 13.1 (*see* [15]) Let $\mu : \mathbb{R} \to [0, 1]$ with the following properties

(i) is normal, i.e., $\exists\, x_0 \in \mathbb{R}$; $\mu(x_0) = 1$.
(ii) $\mu(\lambda x + (1-\lambda) y) \geq \min\{\mu(x), \mu(y)\}$, $\forall\, x, y \in \mathbb{R}$, $\forall\, \lambda \in [0, 1]$ (μ is called a convex fuzzy subset).
(iii) μ is upper semicontinuous on $\mathbb{R}$, i.e. $\forall\, x_0 \in \mathbb{R}$ and $\forall\, \varepsilon > 0$, $\exists$ neighborhood $V(x_0) : \mu(x) \leq \mu(x_0) + \varepsilon$, $\forall\, x \in V(x_0)$.
(iv) The set $\overline{\sup p(\mu)}$ is compact in $\mathbb{R}$, (where $\sup p(\mu) := \{x \in \mathbb{R} : \mu(x) > 0\}$).

We call μ a fuzzy real number. Denote the set of all μ with $\mathbb{R}_{\mathcal{F}}$.

E.g. $\chi_{\{x_0\}} \in \mathbb{R}_{\mathcal{F}}$, for any $x_0 \in \mathbb{R}$, where $\chi_{\{x_0\}}$ is the characteristic function at x_0.

For $0 < r \leq 1$ and $\mu \in \mathbb{R}_{\mathcal{F}}$ define

$$[\mu]^r := \{x \in \mathbb{R} : \mu(x) \geq r\} \tag{13.1}$$

G.A. Anastassiou, *Intelligent Systems II: Complete Approximation by Neural Network Operators*, Studies in Computational Intelligence 608,
DOI 10.1007/978-3-319-20505-2_13

and

$$[\mu]^0 := \overline{\{x \in \mathbb{R} : \mu(x) \geq 0\}}.$$

Then it is well known that for each $r \in [0,1]$, $[\mu]^r$ is a closed and bounded interval on $\mathbb{R}$ [12].

For $u, v \in \mathbb{R}_{\mathcal{F}}$ and $\lambda \in \mathbb{R}$, we define uniquely the sum $u \oplus v$ and the product $\lambda \odot u$ by

$$[u \oplus v]^r = [u]^r + [v]^r, \quad [\lambda \odot u]^r = \lambda [u]^r, \ \forall\, r \in [0,1],$$

where $[u]^r + [v]^r$ means the usual addition of two intervals (as subsets of $\mathbb{R}$) and $\lambda [u]^r$ means the usual product between a scalar and a subset of $\mathbb{R}$ (see, e.g. [15]).

Notice $1 \odot u = u$ and it holds

$$u \oplus v = v \oplus u, \lambda \odot u = u \odot \lambda.$$

If $0 \leq r_1 \leq r_2 \leq 1$ then

$$[u]^{r_2} \subseteq [u]^{r_1}.$$

Actually $[u]^r = \left[u_-^{(r)}, u_+^{(r)}\right]$, where $u_-^{(r)} \leq u_+^{(r)}$, $u_-^{(r)}, u_+^{(r)} \in \mathbb{R}$, $\forall\, r \in [0,1]$.

For $\lambda > 0$ one has $\lambda u_\pm^{(r)} = (\lambda \odot u)_\pm^{(r)}$, respectively.

Define $D : \mathbb{R}_{\mathcal{F}} \times \mathbb{R}_{\mathcal{F}} \to \mathbb{R}_{\mathcal{F}}$ by

$$D(u,v) := \sup_{r \in [0,1]} \max \left\{ \left| u_-^{(r)} - v_-^{(r)} \right|, \left| u_+^{(r)} - v_+^{(r)} \right| \right\}, \tag{13.2}$$

where

$$[v]^r = \left[v_-^{(r)}, v_+^{(r)}\right]; \ u, v \in \mathbb{R}_{\mathcal{F}}.$$

We have that D is a metric on $\mathbb{R}_{\mathcal{F}}$.

Then $(\mathbb{R}_{\mathcal{F}}, D)$ is a complete metric space, see [15, 16].

Let $f, g : \mathbb{R}^m \to \mathbb{R}_{\mathcal{F}}$. We define the distance

$$D^*(f,g) = \sup_{x \in \mathbb{R}^m} D(f(x), g(x)).$$

Here Σ^* stands for fuzzy summation and $\widetilde{0} := \chi_{\{0\}} \in \mathbb{R}_{\mathcal{F}}$ is the neutral element with respect to $\oplus$, i.e.,

$$u \oplus \widetilde{0} = \widetilde{0} \oplus u = u, \ \forall\, u \in \mathbb{R}_{\mathcal{F}}.$$

We need

Remark 13.2 ([5]). Here $r \in [0,1]$, $x_i^{(r)}, y_i^{(r)} \in \mathbb{R}$, $i = 1, \ldots, m \in \mathbb{N}$. Suppose that

$$\sup_{r\in[0,1]} \max\left(x_i^{(r)}, y_i^{(r)}\right) \in \mathbb{R}, \text{ for } i = 1, \ldots, m.$$

Then one sees easily that

$$\sup_{r\in[0,1]} \max\left(\sum_{i=1}^{m} x_i^{(r)}, \sum_{i=1}^{m} y_i^{(r)}\right) \le \sum_{i=1}^{m} \sup_{r\in[0,1]} \max\left(x_i^{(r)}, y_i^{(r)}\right). \tag{13.3}$$

Definition 13.3 Let $f \in C(\mathbb{R}^m)$, $m \in \mathbb{N}$, which is bounded or uniformly continuous, we define ($h > 0$)

$$\omega_1(f, h) := \sup_{\text{all } x_i, x_i' \in \mathbb{R},\ |x_i - x_i'| \le h, \text{ for } i=1,\ldots,m} \left| f(x_1, \ldots, x_m) - f\left(x_1', \ldots, x_m'\right)\right|. \tag{13.4}$$

Definition 13.4 Let $f : \mathbb{R}^m \to \mathbb{R}_{\mathcal{F}}$, we define the fuzzy modulus of continuity of f by

$$\omega_1^{(\mathcal{F})}(f, \delta) = \sup_{x,y\in\mathbb{R},\ |x_i - y_i| \le \delta, \text{ for } i=1,\ldots,m} D(f(x), f(y)),\ \delta > 0, \tag{13.5}$$

where $x = (x_1, \ldots, x_m)$, $y = (y_1, \ldots, y_m)$.

For $f : \mathbb{R}^m \to \mathbb{R}_{\mathcal{F}}$, we use

$$[f]^r = \left[f_-^{(r)}, f_+^{(r)}\right], \tag{13.6}$$

where $f_\pm^{(r)} : \mathbb{R}^m \to \mathbb{R}$, $\forall\, r \in [0, 1]$.

We need

Proposition 13.5 *Let $f : \mathbb{R}^m \to \mathbb{R}_{\mathcal{F}}$. Assume that $\omega_1^{\mathcal{F}}(f, \delta)$, $\omega_1\left(f_-^{(r)}, \delta\right)$, $\omega_1\left(f_+^{(r)}, \delta\right)$ are finite for any $\delta > 0$, $r \in [0, 1]$.*

Then

$$\omega_1^{(\mathcal{F})}(f, \delta) = \sup_{r\in[0,1]} \max\left\{\omega_1\left(f_-^{(r)}, \delta\right), \omega_1\left(f_+^{(r)}, \delta\right)\right\}. \tag{13.7}$$

Proof By Proposition 1 of [8]. ■

We define by $C_{\mathcal{F}}^U(\mathbb{R}^m)$ the space of fuzzy uniformly continuous functions from $\mathbb{R}^m \to \mathbb{R}_{\mathcal{F}}$, also $C_{\mathcal{F}}(\mathbb{R}^m)$ is the space of fuzzy continuous functions on $\mathbb{R}^m$, and $C_b(\mathbb{R}^m, \mathbb{R}_{\mathcal{F}})$ is the fuzzy continuous and bounded functions.

We mention

Proposition 13.6 ([7]) *Let $f \in C_{\mathcal{F}}^U(\mathbb{R}^m)$. Then $\omega_1^{(\mathcal{F})}(f, \delta) < \infty$, for any $\delta > 0$.*

Proposition 13.7 ([7]) *It holds*

$$\lim_{\delta \to 0} \omega_1^{(\mathcal{F})}(f, \delta) = \omega_1^{(\mathcal{F})}(f, 0) = 0, \tag{13.8}$$

iff $f \in C_{\mathcal{F}}^{U}(\mathbb{R}^m)$.

Proposition 13.8 ([7]) *Let* $f \in C_{\mathcal{F}}(\mathbb{R}^m)$. *Then* $f_{\pm}^{(r)}$ *are equicontinuous with respect to* $r \in [0, 1]$ *over* $\mathbb{R}^m$, *respectively in* $\pm$.

Note 13.9 It is clear by Propositions 13.5, 13.7, that if $f \in C_{\mathcal{F}}^{U}(\mathbb{R}^m)$, then $f_{\pm}^{(r)} \in C_U(\mathbb{R}^m)$ (uniformly continuous on $\mathbb{R}^m$).

We need

Definition 13.10 Let $x, y \in \mathbb{R}_{\mathcal{F}}$. If there exists $z \in \mathbb{R}_{\mathcal{F}} : x = y \oplus z$, then we call z the H-difference on x and y, denoted $x - y$.

Definition 13.11 ([15]) Let $T := [x_0, x_0 + \beta] \subset \mathbb{R}$, with $\beta > 0$. A function $f : T \to \mathbb{R}_{\mathcal{F}}$ is H-difference at $x \in T$ if there exists an $f'(x) \in \mathbb{R}_{\mathcal{F}}$ such that the limits (with respect to D)

$$\lim_{h \to 0+} \frac{f(x+h) - f(x)}{h}, \quad \lim_{h \to 0+} \frac{f(x) - f(x-h)}{h} \tag{13.9}$$

exist and are equal to $f'(x)$.

We call f' the H-derivative or fuzzy derivative of f at x.

Above is assumed that the H-differences $f(x+h) - f(x)$, $f(x) - f(x-h)$ exists in $\mathbb{R}_{\mathcal{F}}$ in a neighborhood of x.

Definition 13.12 We denote by $C_{\mathcal{F}}^{N}(\mathbb{R}^m)$, $N \in \mathbb{N}$, the space of all N-times fuzzy continuously differentiable functions from $\mathbb{R}^m$ into $\mathbb{R}_{\mathcal{F}}$.

Here fuzzy partial derivatives are defined via Definition 13.11 in the obvious way as in the ordinary real case.

We mention

Theorem 13.13 ([13]) *Let* $f : [a, b] \subseteq \mathbb{R} \to \mathbb{R}_{\mathcal{F}}$ *be* H-*fuzzy differentiable. Let* $t \in [a, b]$, $0 \leq r \leq 1$. *Clearly*

$$[f(t)]^r = \left[f(t)_{-}^{(r)}, f(t)_{+}^{(r)} \right] \subseteq \mathbb{R}.$$

Then $(f(t))_{\pm}^{(r)}$ *are differentiable and*

$$[f'(t)]^r = \left[\left(f(t)_{-}^{(r)} \right)', \left(f(t)_{+}^{(r)} \right)' \right].$$

I.e.

$$\left(f'\right)_{\pm}^{(r)} = \left(f_{\pm}^{(r)}\right)', \quad \forall\, r \in [0, 1]. \tag{13.10}$$

Remark 13.14 (se also [6]) Let $f \in C^N(\mathbb{R}, \mathbb{R}_{\mathcal{F}})$, $N \geq 1$. Then by Theorem 13.13 we obtain $f_{\pm}^{(r)} \in C^N(\mathbb{R})$ and

$$\left[f^{(i)}(t)\right]^r = \left[\left(f(t)_{-}^{(r)}\right)^{(i)}, \left(f(t)_{+}^{(r)}\right)^{(i)}\right],$$

for $i = 0, 1, 2, \ldots, N$, and in particular we have

$$\left(f^{(i)}\right)_{\pm}^{(r)} = \left(f_{\pm}^{(r)}\right)^{(i)}, \tag{13.11}$$

for any $r \in [0, 1]$.

Let $f \in C_{\mathcal{F}}^N(\mathbb{R}^m)$, denote $f_{\widetilde{\alpha}} := \frac{\partial^{\widetilde{\alpha}} f}{\partial x^{\widetilde{\alpha}}}$, where $\widetilde{\alpha} := (\widetilde{\alpha_1}, \ldots, \widetilde{\alpha_m})$, $\widetilde{\alpha_i} \in \mathbb{Z}^+$, $i = 1, \ldots, m$ and

$$0 < |\widetilde{\alpha}| := \sum_{i=1}^{m} \widetilde{\alpha_i} \leq N, \quad N > 1.$$

Then by Theorem 13.13 we get that

$$\left(f_{\pm}^{(r)}\right)_{\widetilde{\alpha}} = \left(f_{\widetilde{\alpha}}\right)_{\pm}^{(r)}, \quad \forall\, r \in [0, 1], \tag{13.12}$$

and any $\widetilde{\alpha} : |\widetilde{\alpha}| \leq N$. Here $f_{\pm}^{(r)} \in C^N(\mathbb{R}^m)$.

For the definition of general fuzzy integral we follow [14] next.

Definition 13.15 Let (Ω, Σ, μ) be a complete σ-finite measure space. We call $F : \Omega \to \mathbb{R}_{\mathcal{F}}$ measurable iff $\forall$ closed $B \subseteq \mathbb{R}$ the function $F^{-1}(B) : \Omega \to [0, 1]$ defined by

$$F^{-1}(B)(w) := \sup_{x \in B} F(w)(x), \text{ all } w \in \Omega$$

is measurable, see [14].

Theorem 13.16 ([14]) *For* $F : \Omega \to \mathbb{R}_{\mathcal{F}}$,

$$F(w) = \{(F_{-}^{(r)}(w), F_{+}^{(r)}(w)) | 0 \leq r \leq 1\},$$

the following are equivalent

(1) F *is measurable,*
(2) $\forall\, r \in [0, 1]$, $F_{-}^{(r)}$, $F_{+}^{(r)}$ *are measurable.*

Following [14], given that for each $r \in [0,1]$, $F_{-}^{(r)}, F_{+}^{(r)}$ are integrable we have that the parametrized representation

$$\left\{ \left(\int_A F_{-}^{(r)} d\mu, \int_A F_{+}^{(r)} d\mu \right) | 0 \leq r \leq 1 \right\}$$

is a fuzzy real number for each $A \in \Sigma$.

The last fact leads to

Definition 13.17 ([14]) A measurable function $F : \Omega \to \mathbb{R}_{\mathcal{F}}$,

$$F(w) = \{ (F_{-}^{(r)}(w), F_{+}^{(r)}(w)) | 0 \leq r \leq 1 \}$$

is integrable if for each $r \in [0,1]$, $F_{\pm}^{(r)}$ are integrable, or equivalently, if $F_{\pm}^{(0)}$ are integrable.

In this case, the fuzzy integral of F over $A \in \Sigma$ is defined by

$$\int_A F d\mu := \left\{ \left(\int_A F_{-}^{(r)} d\mu, \int_A F_{+}^{(r)} d\mu \right) | 0 \leq r \leq 1 \right\}. \tag{13.13}$$

By [14] F is integrable iff $w \to \|F(w)\|_{\mathcal{F}}$ is real-valeud integrable. Here

$$\|u\|_{\mathcal{F}} := D\left(u, \widetilde{0}\right), \ \forall\, u \in \mathbb{R}_{\mathcal{F}}.$$

We need also

Theorem 13.18 ([14]) *Let $F, G : \Omega \to \mathbb{R}_{\mathcal{F}}$ be integrable. Then*

(1) Let $a, b \in \mathbb{R}$, then $aF + bG$ is integrable and for each $A \in \Sigma$,

$$\int_A (aF + bG) d\mu = a \int_A F d\mu + b \int_A G d\mu;$$

(2) $D(F, G)$ is a real-valued integrable function and for each $A \in \Sigma$,

$$D\left(\int_A F d\mu, \int_A G d\mu \right) \leq \int_A D(F, G) d\mu. \tag{13.14}$$

In particular,

$$\left\| \int_A F d\mu \right\|_{\mathcal{F}} \leq \int_A \|F\|_{\mathcal{F}} d\mu.$$

Above μ could be the Lebesgue measure, with all the basic properties valid here too.

Basically here we have

$$\left[\int_A F d\mu\right]^r := \left[\int_A F_-^{(r)} d\mu, \int_A F_+^{(r)} d\mu\right], \tag{13.15}$$

i.e.

$$\left(\int_A F d\mu\right)_\pm^{(r)} = \int_A F_\pm^{(r)} d\mu, \tag{13.16}$$

$\forall\, r \in [0, 1]$, respectively.

We use

Notation 13.19 *We denote*

$$\left(\sum_{i=1}^{2} D\left(\frac{\partial}{\partial x_i}, \widetilde{o}\right)\right)^2 f\left(\overrightarrow{x}\right) := \tag{13.17}$$

$$D\left(\frac{\partial^2 f(x_1, x_2)}{\partial x_1^2}, \widetilde{o}\right) + D\left(\frac{\partial^2 f(x_1, x_2)}{\partial x_2^2}, \widetilde{o}\right) + 2D\left(\frac{\partial^2 f(x_1, x_2)}{\partial x_1 \partial x_2}, \widetilde{o}\right).$$

In general we denote $(j = 1, \ldots, N)$

$$\left(\sum_{i=1}^{m} D\left(\frac{\partial}{\partial x_i}, \widetilde{o}\right)\right)^j f\left(\overrightarrow{x}\right) := \tag{13.18}$$

$$\sum_{(j_1,\ldots,j_m)\in\mathbb{Z}_+^m : \sum_{i=1}^m j_i = j} \frac{j!}{j_1! j_2! \ldots j_m!} D\left(\frac{\partial^j f(x_1, \ldots, x_m)}{\partial x_1^{j_1} \partial x_2^{j_2} \ldots \partial x_m^{j_m}}, \widetilde{o}\right).$$

13.2 Convergence with Rates of Real Multivariate Neural Network Operators

Here we follow [9].

We need the following (see [11]) definitions.

Definition 13.20 A function $b : \mathbb{R} \to \mathbb{R}$ is said to be bell-shaped if b belongs to L^1 and its integral is nonzero, if it is nondecreasing on $(-\infty, a)$ and nonincreasing on $[a, +\infty)$, where a belongs to $\mathbb{R}$. In particular $b(x)$ is a nonnegative number and at a, b takes a global maximum; it is the center of the bell-shaped function. A bell-shaped function is said to be centered if its center is zero.

Definition 13.21 (*see* [11]) A function $b : \mathbb{R}^d \to \mathbb{R}$ $(d \geq 1)$ is said to be a d-dimensional bell-shaped function if it is integrable and its integral is not zero, and for all $i = 1, \ldots, d$,

$$t \to b(x_1, \ldots, t, \ldots, x_d)$$

is a centered bell-shaped function, where $\overrightarrow{x} := (x_1, \ldots, x_d) \in \mathbb{R}^d$ arbitrary.

Example 13.22 (from [11]) Let b be a centered bell-shaped function over $\mathbb{R}$, then $(x_1, \ldots, x_d) \to b(x_1) \ldots b(x_d)$ is a d-dimensional bell-shaped function.

Assumption 13.23 Here $b\left(\overrightarrow{x}\right)$ is of compact support $\mathcal{B} := \prod_{i=1}^{d} [-T_i, T_i]$, $T_i > 0$ and it may have jump discontinuities there. Let $f : \mathbb{R}^d \to \mathbb{R}$ be a continuous and bounded function or a uniformly continuous function.

Here we mention the study [9] of poitwise convergence with rates over $\mathbb{R}^d$, to the unit operator I, of the "normalized bell" real multivariate neural network operators

$$M_n(f)\left(\overrightarrow{x}\right) := \tag{13.19}$$

$$\frac{\sum_{k_1=-n^2}^{n^2} \cdots \sum_{k_d=-n^2}^{n^2} f\left(\frac{k_1}{n}, \ldots \frac{k_d}{n}\right) b\left(n^{1-\alpha}\left(x_1 - \frac{k_1}{n}\right), \ldots, n^{1-\alpha}\left(x_d - \frac{k_d}{n}\right)\right)}{\sum_{k_1=-n^2}^{n^2} \cdots \sum_{k_d=-n^2}^{n^2} b\left(n^{1-\alpha}\left(x_1 - \frac{k_1}{n}\right), \ldots, n^{1-\alpha}\left(x_d - \frac{k_d}{n}\right)\right)},$$

where $0 < \alpha < 1$ and $\overrightarrow{x} := (x_1, \ldots, x_d) \in \mathbb{R}^d$, $n \in \mathbb{N}$. Clearly, M_n is a positive linear operator.

The terms in the ratio of multiple sums (13.19) can be nonzero iff simultaneously

$$\left|n^{1-\alpha}\left(x_i - \frac{k_i}{n}\right)\right| \leq T_i, \quad \text{all } i = 1, \ldots, d,$$

i.e., $\left|x_i - \frac{k_i}{n}\right| \leq \frac{T_i}{n^{1-\alpha}}$, all $i = 1, \ldots, d$, iff

$$nx_i - T_i n^{\alpha} \leq k_i \leq nx_i + T_i n^{\alpha}, \text{ all } i = 1, \ldots, d. \tag{13.20}$$

To have the order

$$-n^2 \leq nx_i - T_i n^{\alpha} \leq k_i \leq nx_i + T_i n^{\alpha} \leq n^2, \tag{13.21}$$

we need $n \geq T_i + |x_i|$, all $i = 1, \ldots, d$. So (13.21) is true when we take

$$n \geq \max_{i \in \{1, \ldots, d\}} (T_i + |x_i|). \tag{13.22}$$

When $\overrightarrow{x} \in \mathcal{B}$ in order to have (13.21) it is enough to assume that $n \geq 2T^*$, where $T^* := \max\{T_1, \ldots, T_d\} > 0$. Consider

$$\widetilde{I}_i := \left[nx_i - T_i n^{\alpha}, nx_i + T_i n^{\alpha}\right], \ i = 1, \ldots, d, \ n \in \mathbb{N}.$$

The length of $\widetilde{I}_i$ is $2T_i n^{\alpha}$. By Proposition 1 of [1], we get that the cardinality of $k_i \in \mathbb{Z}$ that belong to $\widetilde{I}_i := card\,(k_i) \geq \max\,(2T_i n^{\alpha} - 1, 0)$, any $i \in \{1, \ldots, d\}$. In order to have $card\,(k_i) \geq 1$, we need $2T_i n^{\alpha} - 1 \geq 1$ iff $n \geq T_i^{-\frac{1}{\alpha}}$, any $i \in \{1, \ldots, d\}$.

Therefore, a sufficient condition in order to obtain the order (13.21) along with the interval $\widetilde{I}_i$ to contain at least one integer for all $i = 1, \ldots, d$ is that

$$n \geq \max_{i \in \{1, \ldots, d\}} \left\{ T_i + |x_i|, T_i^{-\frac{1}{\alpha}} \right\}. \tag{13.23}$$

Clearly as $n \to +\infty$ we get that $card\,(k_i) \to +\infty$, all $i = 1, \ldots, d$. Also notice that $card\,(k_i)$ equals to the cardinality of integers in $[\lceil nx_i - T_i n^{\alpha} \rceil, [nx_i + T_i n^{\alpha}]]$ for all $i = 1, \ldots, d$. Here, $[\cdot]$ denotes the integral part of the number, while $\lceil \cdot \rceil$ denotes its ceiling.

From now on, in this chapter we will assume (13.23). Furthermore it holds

$$(M_n(f))\left(\overrightarrow{x}\right) := \frac{\sum_{k_1=\lceil nx_1 - T_1 n^{\alpha}\rceil}^{[nx_1 + T_1 n^{\alpha}]} \cdots \sum_{k_d=\lceil nx_d - T_d n^{\alpha}\rceil}^{[nx_d + T_d n^{\alpha}]} f\left(\frac{k_1}{n}, \ldots \frac{k_d}{n}\right)}{V\left(\overrightarrow{x}\right)} \tag{13.24}$$

$$\cdot b\left(n^{1-\alpha}\left(x_1 - \frac{k_1}{n}\right), \ldots, n^{1-\alpha}\left(x_d - \frac{k_d}{n}\right)\right)$$

all $\overrightarrow{x} := (x_1, \ldots, x_d) \in \mathbb{R}^d$, where

$$V\left(\overrightarrow{x}\right) :=$$

$$\sum_{k_1=\lceil nx_1 - T_1 n^{\alpha}\rceil}^{[nx_1 + T_1 n^{\alpha}]} \cdots \sum_{k_d=\lceil nx_d - T_d n^{\alpha}\rceil}^{[nx_d + T_d n^{\alpha}]} b\left(n^{1-\alpha}\left(x_1 - \frac{k_1}{n}\right), \ldots, n^{1-\alpha}\left(x_d - \frac{k_d}{n}\right)\right). \tag{13.25}$$

From [9], we need and mention

Theorem 13.24 *Let $\overrightarrow{x} \in \mathbb{R}^d$; then*

$$\left|(M_n(f))\left(\overrightarrow{x}\right) - f\left(\overrightarrow{x}\right)\right| \leq \omega_1\left(f, \frac{T^*}{n^{1-\alpha}}\right). \tag{13.26}$$

Inequality (13.26) *is attained by constant functions.*

Inequalities (13.26) *gives* $M_n(f)(\overrightarrow{x}) \to f(\overrightarrow{x})$, *pointwise with rates, as* $n \to +\infty$, *where* $\overrightarrow{x} \in \mathbb{R}^d$, $d \geq 1$, *provided that* f *is uniformly continuous on* $\mathbb{R}^d$. *In the last case it is clear that* $M_n \to I$, *uniformly.*

From [10], we also need and mention

Theorem 13.25 *Let* $\overrightarrow{x} \in \mathbb{R}^d$, $f \in C^N(\mathbb{R}^d)$, $N \in \mathbb{N}$, *such that all of its partial derivatives* $f_{\widetilde{\alpha}}$ *of order* N, $\widetilde{\alpha} : |\widetilde{\alpha}| = N$, *are uniformly continuous or continuous are bounded. Then*

$$\left|(M_n(f))(\overrightarrow{x}) - f(\overrightarrow{x})\right| \leq \tag{13.27}$$

$$\left\{\sum_{j=1}^{N} \frac{(T^*)^j}{j! n^{j(1-\alpha)}} \left(\left(\sum_{i=1}^{d} \left|\frac{\partial}{\partial x_i}\right|\right)^j f(\overrightarrow{x})\right)\right\} + \frac{(T^*)^N d^N}{N! n^{N(1-\alpha)}} \cdot \max_{\widetilde{\alpha}:|\widetilde{\alpha}|=N} \omega_1\left(f_{\widetilde{\alpha}}, \frac{T^*}{n^{1-\alpha}}\right).$$

Inequality (13.27) *is attained by constant functions. Also,* (13.27) *gives us with rates the pointwise convergences of* $M_n(f) \to f$ *over* $\mathbb{R}^d$, *as* $n \to +\infty$.

13.3 Main Results—Convergence with Rates of Fuzzy Multivariate Neural Networks

Here b is as in Definition 13.21.

Assumption 13.26 We suppose that $b(\overrightarrow{x})$ is of compact support $B := \prod_{i=1}^{d}[-T_i, T_i]$, $T_i > 0$, and it may have jump discontinuities there. We consider $f : \mathbb{R}^d \to \mathbb{R}_{\mathcal{F}}$ to be fuzzy continuous and fuzzy bounded function or fuzzy uniformly continuous function.

In this section we study the D-metric pointwise convergence with rates over $\mathbb{R}^d$, to the fuzzy unit operator $I_{\mathcal{F}}$, of the fuzzy multivariate neural network operators ($0 < \alpha < 1$, $\overrightarrow{x} := (x_1, \ldots, x_d) \in \mathbb{R}^d$, $n \in \mathbb{N}$)

$$M_n^{\mathcal{F}}(f)(\overrightarrow{x}) := \tag{13.28}$$

$$\frac{\sum_{k_1=-n^2}^{n^2 *} \cdots \sum_{k_d=-n^2}^{n^2 *} f\left(\frac{k_1}{n}, \ldots \frac{k_d}{n}\right) \odot b\left(n^{1-\alpha}\left(x_1 - \frac{k_1}{n}\right), \ldots, n^{1-\alpha}\left(x_d - \frac{k_d}{n}\right)\right)}{\sum_{k_1=-n^2}^{n^2} \cdots \sum_{k_d=-n^2}^{n^2} b\left(n^{1-\alpha}\left(x_1 - \frac{k_1}{n}\right), \ldots, n^{1-\alpha}\left(x_d - \frac{k_d}{n}\right)\right)}$$

$$= \sum_{k_1=\lceil nx_1 - T_1 n^{\alpha}\rceil}^{[nx_1 + T_1 n^{\alpha}]*} \cdots \sum_{k_d=\lceil nx_d - T_d n^{\alpha}\rceil}^{[nx_d + T_d n^{\alpha}]*} f\left(\frac{k_1}{n}, \ldots \frac{k_d}{n}\right) \tag{13.29}$$

$$\odot \frac{b\left(n^{1-\alpha}\left(x_1-\frac{k_1}{n}\right),\ldots,n^{1-\alpha}\left(x_d-\frac{k_d}{n}\right)\right)}{V\left(\overrightarrow{x}\right)},$$

where $V\left(\overrightarrow{x}\right)$ as in (13.25) and under the assumption (13.23).

We notice for $r\in[0,1]$ that

$$\left[M_n^{\mathcal{F}}(f)\left(\overrightarrow{x}\right)\right]^r=\sum_{k_1=\lceil nx_1-T_1n^{\alpha}\rceil}^{[nx_1+T_1n^{\alpha}]}\cdots\sum_{k_d=\lceil nx_d-T_dn^{\alpha}\rceil}^{[nx_d+T_dn^{\alpha}]}\left[f\left(\frac{k_1}{n},\ldots\frac{k_d}{n}\right)\right]^r$$

$$\cdot\frac{b\left(n^{1-\alpha}\left(x_1-\frac{k_1}{n}\right),\ldots,n^{1-\alpha}\left(x_d-\frac{k_d}{n}\right)\right)}{V\left(\overrightarrow{x}\right)} \tag{13.30}$$

$$=\sum_{k_1=\lceil nx_1-T_1n^{\alpha}\rceil}^{[nx_1+T_1n^{\alpha}]}\cdots\sum_{k_d=\lceil nx_d-T_dn^{\alpha}\rceil}^{[nx_d+T_dn^{\alpha}]}\left[f_-^{(r)}\left(\frac{k_1}{n},\ldots\frac{k_d}{n}\right),f_+^{(r)}\left(\frac{k_1}{n},\ldots\frac{k_d}{n}\right)\right]$$

$$\cdot\frac{b\left(n^{1-\alpha}\left(x_1-\frac{k_1}{n}\right),\ldots,n^{1-\alpha}\left(x_d-\frac{k_d}{n}\right)\right)}{V\left(\overrightarrow{x}\right)}$$

$$=\left[\sum_{k_1=\lceil nx_1-T_1n^{\alpha}\rceil}^{[nx_1+T_1n^{\alpha}]}\cdots\sum_{k_d=\lceil nx_d-T_dn^{\alpha}\rceil}^{[nx_d+T_dn^{\alpha}]}f_-^{(r)}\left(\frac{k_1}{n},\ldots\frac{k_d}{n}\right)\right.$$

$$\cdot\frac{b\left(n^{1-\alpha}\left(x_1-\frac{k_1}{n}\right),\ldots,n^{1-\alpha}\left(x_d-\frac{k_d}{n}\right)\right)}{V\left(\overrightarrow{x}\right)},$$

$$\sum_{k_1=\lceil nx_1-T_1n^{\alpha}\rceil}^{[nx_1+T_1n^{\alpha}]}\cdots\sum_{k_d=\lceil nx_d-T_dn^{\alpha}\rceil}^{[nx_d+T_dn^{\alpha}]}f_+^{(r)}\left(\frac{k_1}{n},\ldots\frac{k_d}{n}\right)$$

$$\left.\cdot\frac{b\left(n^{1-\alpha}\left(x_1-\frac{k_1}{n}\right),\ldots,n^{1-\alpha}\left(x_d-\frac{k_d}{n}\right)\right)}{V\left(\overrightarrow{x}\right)}\right] \tag{13.31}$$

$$=\left[\left(M_n\left(f_-^{(r)}\right)\right)\left(\overrightarrow{x}\right),\left(M_n\left(f_+^{(r)}\right)\right)\left(\overrightarrow{x}\right)\right].$$

We have proved that

$$\left(M_n^{\mathcal{F}}(f)\right)_{\pm}^{(r)} = M_n\left(f_{\pm}^{(r)}\right),\ \forall\, r \in [0,1], \tag{13.32}$$

respectively.

We present

Theorem 13.27 *Let $\overrightarrow{x} \in \mathbb{R}^d$; then*

$$D\left(\left(M_n^{\mathcal{F}}(f)\right)(\overrightarrow{x}), f(\overrightarrow{x})\right) \leq \omega_1^{(\mathcal{F})}\left(f, \frac{T^*}{n^{1-\alpha}}\right). \tag{13.33}$$

Notice that (13.33) *gives $M_n^{\mathcal{F}} \overset{D}{\rightarrow} I_{\mathcal{F}}$ pointwise and uniformly, as $n \rightarrow \infty$, when $f \in C_{\mathcal{F}}^U(\mathbb{R}^d)$.*

Proof We observe that

$$D\left(\left(M_n^{\mathcal{F}}(f)\right)(\overrightarrow{x}), f(\overrightarrow{x})\right) =$$

$$\sup_{r\in[0,1]} \max\left\{\left|\left(M_n^{\mathcal{F}}(f)\right)_{-}^{(r)}(\overrightarrow{x}) - f_{-}^{(r)}(\overrightarrow{x})\right|, \left|\left(M_n^{\mathcal{F}}(f)\right)_{+}^{(r)}(\overrightarrow{x}) - f_{+}^{(r)}(\overrightarrow{x})\right|\right\} \overset{(13.32)}{=}$$

$$\sup_{r\in[0,1]} \max\left\{\left|\left(M_n\left(f_{-}^{(r)}\right)\right)(\overrightarrow{x}) - f_{-}^{(r)}(\overrightarrow{x})\right|, \left|\left(M_n\left(f_{+}^{(r)}\right)\right)(\overrightarrow{x}) - f_{+}^{(r)}(\overrightarrow{x})\right|\right\} \overset{(13.26)}{\leq}$$

$$\sup_{r\in[0,1]} \max\left\{\omega_1\left(f_{-}^{(r)}, \frac{T^*}{n^{1-\alpha}}\right), \omega_1\left(f_{+}^{(r)}, \frac{T^*}{n^{1-\alpha}}\right)\right\} \overset{(13.7)}{=} \omega_1^{(\mathcal{F})}\left(f, \frac{T^*}{n^{1-\alpha}}\right),$$

proving the claim. ■

We continue with

Theorem 13.28 *Let $\overrightarrow{x} \in \mathbb{R}^d$, $f \in C_{\mathcal{F}}^N(\mathbb{R}^d)$, $N \in \mathbb{N}$, such that all of its fuzzy partial derivatives $f_{\widetilde{\alpha}}$ of order N, $\widetilde{\alpha} : |\widetilde{\alpha}| = N$, are fuzzy uniformly continuous or fuzzy continuous and fuzzy bounded. Then*

$$D\left(\left(M_n^{\mathcal{F}}(f)\right)(\overrightarrow{x}), f(\overrightarrow{x})\right) \leq \tag{13.34}$$

$$\left\{\sum_{j=1}^{N} \frac{(T^*)^j}{j!n^{j(1-\alpha)}}\left[\left(\sum_{i=1}^{d} D\left(\frac{\partial}{\partial x_i}, \widetilde{0}\right)\right)^j f(\overrightarrow{x})\right]\right\}$$

$$+\frac{(T^*)^N d^N}{N!n^{N(1-\alpha)}} \max_{\widetilde{\alpha}:|\widetilde{\alpha}|=N} \omega_1^{(\mathcal{F})}\left(f_{\widetilde{\alpha}}, \frac{T^*}{n^{1-\alpha}}\right).$$

As $n \to \infty$, we get $D\left(\left(M_n^{\mathcal{F}}(f)\right)(\overrightarrow{x}), f(\overrightarrow{x})\right) \to 0$ pointwise with rates.

Proof As before we have

$$D\left(\left(M_n^{\mathcal{F}}(f)\right)(\overrightarrow{x}), f(\overrightarrow{x})\right) \overset{(13.32)}{=}$$

$$\sup_{r\in[0,1]} \max\left\{\left|\left(M_n\left(f_-^{(r)}\right)\right)(\overrightarrow{x}) - f_-^{(r)}(\overrightarrow{x})\right|, \left|\left(M_n\left(f_+^{(r)}\right)\right)(\overrightarrow{x}) - f_+^{(r)}(\overrightarrow{x})\right|\right\} \overset{(13.27)}{\leq}$$

$$\sup_{r\in[0,1]} \max\left\{\left\{\sum_{j=1}^{N} \frac{(T^*)^j}{j!n^{j(1-\alpha)}}\left[\left(\sum_{i=1}^{d}\left|\frac{\partial}{\partial x_i}\right|\right)^j f_-^{(r)}(\overrightarrow{x})\right]\right\}\right.$$

$$+\frac{(T^*)^N d^N}{N!n^{N(1-\alpha)}} \max_{\widetilde{\alpha}:|\widetilde{\alpha}|=N} \omega_1\left(\left(f_-^{(r)}\right)_{\widetilde{\alpha}}, \frac{T^*}{n^{1-\alpha}}\right),$$

$$\left\{\sum_{j=1}^{N} \frac{(T^*)^j}{j!n^{j(1-\alpha)}}\left[\left(\sum_{i=1}^{d}\left|\frac{\partial}{\partial x_i}\right|\right)^j f_+^{(r)}(\overrightarrow{x})\right]\right\}$$

$$\left.+\frac{(T^*)^N d^N}{N!n^{N(1-\alpha)}} \max_{\widetilde{\alpha}:|\widetilde{\alpha}|=N} \omega_1\left(\left(f_+^{(r)}\right)_{\widetilde{\alpha}}, \frac{T^*}{n^{1-\alpha}}\right)\right\} \overset{(13.3)}{\leq} \sum_{j=1}^{N} \frac{(T^*)^j}{j!n^{j(1-\alpha)}} \cdot \tag{13.36}$$

$$\sup_{r\in[0,1]} \max\left\{\left(\left(\sum_{i=1}^{d}\left|\frac{\partial}{\partial x_i}\right|\right)^j f_-^{(r)}(\overrightarrow{x})\right), \left(\left(\sum_{i=1}^{d}\left|\frac{\partial}{\partial x_i}\right|\right)^j f_+^{(r)}(\overrightarrow{x})\right)\right\} +$$

$$\frac{(T^*)^N d^N}{N!n^{N(1-\alpha)}} \max_{\widetilde{\alpha}:|\widetilde{\alpha}|=N} \sup_{r\in[0,1]} \max\left\{\omega_1\left(\left(f_-^{(r)}\right)_{\widetilde{\alpha}}, \frac{T^*}{n^{1-\alpha}}\right), \omega_1\left(\left(f_+^{(r)}\right)_{\widetilde{\alpha}}, \frac{T^*}{n^{1-\alpha}}\right)\right\}$$

$$\overset{\text{(by (13.3), (13.7), (13.12), (13.18))}}{\leq} \left\{\sum_{j=1}^{N} \frac{(T^*)^j}{j!n^{j(1-\alpha)}}\left[\left(\sum_{i=1}^{d} D\left(\frac{\partial}{\partial x_i}, \widetilde{o}\right)\right)^j f(\overrightarrow{x})\right]\right\} + \tag{13.37}$$

$$\frac{(T^*)^N d^N}{N!n^{N(1-\alpha)}} \max_{\widetilde{\alpha}:|\widetilde{\alpha}|=N} \omega_1^{(\mathcal{F})}\left(f_{\widetilde{\alpha}}, \frac{T^*}{n^{1-\alpha}}\right),$$

proving the claim. ■

13.4 Main Results—The Fuzzy Multivariate "Normalized Squashing Type Operators" and Their Fuzzy Convergence to the Fuzzy Unit with Rates

We give the following definition

Definition 13.29 Let the nonnegative function $S : \mathbb{R}^d \to \mathbb{R}$, $d \geq 1$, S has compact support $\mathcal{B} := \prod_{i=1}^{d} [-T_i, T_i]$, $T_i > 0$ and is nondecreasing there for each coordinate. S can be continuous only on either $\prod_{i=1}^{d} (-\infty, T_i]$ or $\mathcal{B}$ and can have jump discontinuities. We call S the multivariate "squashing function" (see also [11]).

Let $f : \mathbb{R}^d \to \mathbb{R}_{\mathcal{F}}$ be either fuzzy uniformly continuous or fuzzy continuous and fuzzy bounded function.

For $\overrightarrow{x} \in \mathbb{R}^d$, we define the fuzzy multivariate "normalized squashing type operator",

$$L_n^{\mathcal{F}} (f) \left(\overrightarrow{x}\right) := \tag{13.38}$$

$$\frac{\sum_{k_1=-n^2}^{n^2*} \cdots \sum_{k_d=-n^2}^{n^2*} f\left(\frac{k_1}{n}, \ldots \frac{k_d}{n}\right) \odot S\left(n^{1-\alpha}\left(x_1 - \frac{k_1}{n}\right), \ldots, n^{1-\alpha}\left(x_d - \frac{k_d}{n}\right)\right)}{W\left(\overrightarrow{x}\right)},$$

where $0 < \alpha < 1$ and $n \in \mathbb{N}$:

$$n \geq \max_{i \in \{1,\ldots,d\}} \left\{T_i + |x_i|, T_i^{-\frac{1}{\alpha}}\right\}, \tag{13.39}$$

and

$$W\left(\overrightarrow{x}\right) := \sum_{k_1=-n^2}^{n^2*} \cdots \sum_{k_d=-n^2}^{n^2*} S\left(n^{1-\alpha}\left(x_1 - \frac{k_1}{n}\right), \ldots, n^{1-\alpha}\left(x_d - \frac{k_d}{n}\right)\right). \tag{13.40}$$

It is clear that

$$\left(L_n^{\mathcal{F}}(f)\right)\left(\overrightarrow{x}\right) := \sum_{\overrightarrow{k}=\left\lceil n\overrightarrow{x} - \overrightarrow{T} n^{\alpha}\right\rceil}^{\left[n\overrightarrow{x} + \overrightarrow{T} n^{\alpha}\right]*} \frac{f\left(\frac{\overrightarrow{k}}{n}\right) \odot S\left(n^{1-\alpha}\left(\overrightarrow{x} - \frac{\overrightarrow{k}}{n}\right)\right)}{\Phi\left(\overrightarrow{x}\right)}, \tag{13.41}$$

where

$$\Phi\left(\overrightarrow{x}\right) := \sum_{\overrightarrow{k}=\lceil n\overrightarrow{x}-\overrightarrow{T}n^{\alpha}\rceil}^{\lfloor n\overrightarrow{x}+\overrightarrow{T}n^{\alpha}\rfloor} S\left(n^{1-\alpha}\left(\overrightarrow{x}-\frac{\overrightarrow{k}}{n}\right)\right). \quad (13.42)$$

Here, we study the D-metric pointwise convergence with rates of $\left(L_n^{\mathcal{F}}(f)\right)\left(\overrightarrow{x}\right) \to f\left(\overrightarrow{x}\right)$, as $n \to +\infty$, $\overrightarrow{x} \in \mathbb{R}^d$.

This is given first by the next result.

Theorem 13.30 *Under the above terms and asumptions, we find that*

$$D\left(\left(L_n^{\mathcal{F}}(f)\right)\left(\overrightarrow{x}\right), f\left(\overrightarrow{x}\right)\right) \leq \omega_1^{(\mathcal{F})}\left(f, \frac{T^*}{n^{1-\alpha}}\right). \quad (13.43)$$

Notice that (13.43) *gives* $L_n^{\mathcal{F}} \xrightarrow{D} I_{\mathcal{F}}$ *pointwise and uniformly, as* $n \to \infty$, *when* $f \in C_{\mathcal{F}}^U\left(\mathbb{R}^d\right)$.

Proof Similar to (13.33). ∎

We also give

Theorem 13.31 *Let* $\overrightarrow{x} \in \mathbb{R}^d$, $f \in C_{\mathcal{F}}^N\left(\mathbb{R}^d\right)$, $N \in \mathbb{N}$, *such that all of its fuzzy partial derivatives* $f_{\widetilde{\alpha}}$ *of order* N, $\widetilde{\alpha} : |\widetilde{\alpha}| = N$, *are fuzzy uniformly continuous or fuzzy continuous and fuzzy bounded. Then*

$$D\left(\left(L_n^{\mathcal{F}}(f)\right)\left(\overrightarrow{x}\right), f\left(\overrightarrow{x}\right)\right) \leq \quad (13.44)$$

$$\left\{\sum_{j=1}^{N} \frac{(T^*)^j}{j! n^{j(1-\alpha)}} \left[\left(\sum_{i=1}^{d} D\left(\frac{\partial}{\partial x_i}, \widetilde{0}\right)\right)^j f\left(\overrightarrow{x}\right)\right]\right\}$$

$$+\frac{(T^*)^N d^N}{N! n^{N(1-\alpha)}} \max_{\widetilde{\alpha}:|\widetilde{\alpha}|=N} \omega_1^{(\mathcal{F})}\left(f_{\widetilde{\alpha}}, \frac{T^*}{n^{1-\alpha}}\right).$$

Inequality (13.44) *gives us with rates the poitwise convergence of* $D\left(\left(L_n^{\mathcal{F}}(f)\right)\left(\overrightarrow{x}\right), f\left(\overrightarrow{x}\right)\right) \to 0$ *over* $\mathbb{R}^d$, *as* $n \to \infty$.

Proof Similar to (13.34). ∎

References

1. G.A. Anastassiou, Rate of convergence of some neural network operators to the unit-univariate case. J. Math. Anal. Appl. **212**, 237–262 (1997)
2. G.A. Anastassiou, Rate of convergence of some multivariate neural network operators to the unit. Comput. Math. **40**, 1–19 (2000)
3. G.A. Anastassiou, *Quantitative Approximation* (Chapmann and Hall/CRC, Boca Raton, 2001)
4. G.A. Anastassiou, Higher order fuzzy approximation by fuzzy wavelet type and neural network operators. Comput. Math. **48**, 1387–1401 (2004)
5. G.A. Anastassiou, Fuzzy approximation by fuzzy convolution type operators. Comput. Math. **48**, 1369–1386 (2004)
6. G.A. Anastassiou, Higher order fuzzy Korovkin theory via inequalities. Commun. Appl. Anal. **10**(2), 359–392 (2006)
7. G.A. Anastassiou, Fuzzy Korovkin theorems and inequalities. J. Fuzzy Math. **15**(1), 169–205 (2007)
8. G.A. Anastassiou, Higher order multivariate fuzzy approximation by multivariate fuzzy wavelet type and neural network operators. J. Fuzzy Math. **19**(3), 601–618 (2011)
9. G.A. Anastassiou, Rate of convergence of some multivariate neural network operators to the unit, revisited. J. Comput. Anal. Appl. **15**(7), 1300–1309 (2013)
10. G.A. Anastassiou, Higher order multivariate fuzzy approximation by basic neural network operators. Cubo **16**(0.3), 21–35 (2014)
11. P. Cardaliaguet, G. Euvrard, Approximation of a function and its derivative with a neural network. Neural Networks **5**, 207–220 (1992)
12. R. Goetschel Jr, W. Voxman, Elementary fuzzy calculus. Fuzzy Sets Syst. **18**, 31–43 (1986)
13. O. Kaleva, Fuzzy differential equations. Fuzzy Sets Syst. **24**, 301–317 (1987)
14. Y.K. Kim, B.M. Ghil, Integrals of fuzzy-number-valued functions. Fuzzy Sets Syst. **86**, 213–222 (1997)
15. C. Wu, Z. Gong, On Henstock integral of fuzzy-number-valued functions (I). Fuzzy Sets Syst. **120**(3), 523–532 (2001)
16. C. Wu, M. Ma, On embedding problem of fuzzy numer spaces: part 1. Fuzzy Sets Syst. **44**, 33–38 (1991)

Chapter 14
High Order Multivariate Fuzzy Approximation Using Quasi-interpolation Neural Networks

Here are considered in terms of multivariate fuzzy high approximation to the multivariate unit sequences of multivariate fuzzy quasi-interpolation neural network operators. These operators are multivariate fuzzy analogs of earlier considered multivariate real ones. The derived results generalize earlier real ones into the fuzzy setting. Here the high degree multivariate fuzzy pointwise and uniform convergences with rates to the multivariate fuzzy unit operator are given through multivariate fuzzy inequalities involving the multivariate fuzzy moduli of continuity of the Nth order ($N \geq 1$) H-fuzzy partial derivatives, of the involved multivariate fuzzy number valued function. It follows [16].

14.1 Fuzzy Real Analysis Background

We need the following background

Definition 14.1 (see [24]) Let $\mu : \mathbb{R} \rightarrow [0, 1]$ with the following properties

(i) is normal, i.e., $\exists\, x_0 \in \mathbb{R}$; $\mu(x_0) = 1$.

(ii) $\mu(\lambda x + (1-\lambda) y) \geq \min\{\mu(x), \mu(y)\}$, $\forall\, x, y \in \mathbb{R}$, $\forall\, \lambda \in [0, 1]$ (μ is called a convex fuzzy subset).

(iii) μ is upper semicontinuous on $\mathbb{R}$, i.e. $\forall x_0 \in \mathbb{R}$ and $\forall \varepsilon > 0$, $\exists$ neighborhood $V(x_0) : \mu(x) \leq \mu(x_0) + \varepsilon$, $\forall x \in V(x_0)$.

(iv) The set $\overline{\sup p(\mu)}$ is compact in $\mathbb{R}$, (where $\sup p(\mu) := \{x \in \mathbb{R} : \mu(x) > 0\}$).

We call μ a fuzzy real number. Denote the set of all μ with $\mathbb{R}_{\mathcal{F}}$.
E.g. $\chi_{\{x_0\}} \in \mathbb{R}_{\mathcal{F}}$, for any $x_0 \in \mathbb{R}$, where $\chi_{\{x_0\}}$ is the characteristic function at x_0.

For $0 < r \leq 1$ and $\mu \in \mathbb{R}_{\mathcal{F}}$ define

$$[\mu]^r := \{x \in \mathbb{R} : \mu(x) \geq r\} \tag{14.1}$$

G.A. Anastassiou, *Intelligent Systems II: Complete Approximation by Neural Network Operators*, Studies in Computational Intelligence 608,
DOI 10.1007/978-3-319-20505-2_14

and

$$[\mu]^0 := \overline{\{x \in \mathbb{R} : \mu(x) \geq 0\}}.$$

Then it is well known that for each $r \in [0, 1]$, $[\mu]^r$ is a closed and bounded interval on $\mathbb{R}$ [18].

For $u, v \in \mathbb{R}_{\mathcal{F}}$ and $\lambda \in \mathbb{R}$, we define uniquely the sum $u \oplus v$ and the product $\lambda \odot u$ by

$$[u \oplus v]^r = [u]^r + [v]^r, [\lambda \odot u]^r = \lambda [u]^r, \ \forall \, r \in [0, 1],$$

where $[u]^r + [v]^r$ means the usual addition of two intervals (as substes of $\mathbb{R}$) and $\lambda [u]^r$ means the usual product between a scalar and a subset of $\mathbb{R}$ (see, e.g. [24]).

Notice $1 \odot u = u$ and it holds

$$u \oplus v = v \oplus u, \lambda \odot u = u \odot \lambda.$$

If $0 \leq r_1 \leq r_2 \leq 1$ then

$$[u]^{r_2} \subseteq [u]^{r_1}.$$

Actually $[u]^r = \left[u_-^{(r)}, u_+^{(r)}\right]$, where $u_-^{(r)} \leq u_+^{(r)}, u_-^{(r)}, u_+^{(r)} \in \mathbb{R}, \forall \, r \in [0, 1]$.

For $\lambda > 0$ one has $\lambda u_\pm^{(r)} = (\lambda \odot u)_\pm^{(r)}$, respectively.

Define $D : \mathbb{R}_{\mathcal{F}} \times \mathbb{R}_{\mathcal{F}} \to \mathbb{R}_{\mathcal{F}}$ by

$$D(u, v) := \sup_{r \in [0,1]} \max \left\{ \left| u_-^{(r)} - v_-^{(r)} \right|, \left| u_+^{(r)} - v_+^{(r)} \right| \right\}, \tag{14.2}$$

where

$$[v]^r = \left[v_-^{(r)}, v_+^{(r)}\right]; u, v \in \mathbb{R}_{\mathcal{F}}.$$

We have that D is a metric on $\mathbb{R}_{\mathcal{F}}$.

Then $(\mathbb{R}_{\mathcal{F}}, D)$ is a complete metric space, see [24, 25].

Let $f, g : W \subseteq \mathbb{R}^m \to \mathbb{R}_{\mathcal{F}}$. We define the distance

$$D^*(f, g) = \sup_{x \in W} D(f(x), g(x)).$$

Remark 14.2 We try to determine better and use

$$D^*(f, \widetilde{o}) = \sup_{x \in W} D(f(x), \widetilde{o}) =$$

$$\sup_{x \in W} \sup_{r \in [0,1]} \max \left\{ \left| f_-^{(r)}(x) \right|, \left| f_+^{(r)}(x) \right| \right\}.$$

By the principle of iterated suprema we find that

$$D^{*}\left(f,\widetilde{o}\right)=\sup_{r\in[0,1]}\max\left\{\left\|f_{-}^{(r)}\right\|_{\infty},\left\|f_{+}^{(r)}\right\|_{\infty}\right\}, \tag{14.3}$$

under the assumption $D^{*}\left(f,\widetilde{o}\right)<\infty$, that is f is a fuzzy bounded function.

Above $\|\cdot\|_{\infty}$ is the supremum norm of the function over $W\subseteq\mathbb{R}^{m}$.

Another direct proof of (14.3) follows:

We easily see that

$$D^{*}\left(f,\widetilde{o}\right)\leq\sup_{r\in[0,1]}\max\left\{\left\|f_{-}^{(r)}\right\|_{\infty},\left\|f_{+}^{(r)}\right\|_{\infty}\right\}.$$

On the other hand we observe that $\forall\, x\in W$: each

$$\left|f_{\pm}^{(r)}(x)\right|\leq\max\left\{\left|f_{\pm}^{(r)}(x)\right|\right\}\leq\sup_{r\in[0,1]}\max\left\{\left|f_{\pm}^{(r)}(x)\right|\right\}\leq$$

$$\sup_{x\in W}\sup_{r\in[0,1]}\max\left\{\left|f_{\pm}^{(r)}(x)\right|\right\}=D^{*}\left(f,\widetilde{o}\right).$$

That is, each

$$\left|f_{\pm}^{(r)}(x)\right|\leq D^{*}\left(f,\widetilde{o}\right),\ \forall\, x\in W,$$

hence each

$$\left\|f_{\pm}^{(r)}\right\|_{\infty}\leq D^{*}\left(f,\widetilde{o}\right),$$

and

$$\max\left\{\left\|f_{-}^{(r)}\right\|_{\infty},\left\|f_{+}^{(r)}\right\|_{\infty}\right\}\leq D^{*}\left(f,\widetilde{o}\right),$$

and

$$\sup_{r\in[0,1]}\max\left\{\left\|f_{-}^{(r)}\right\|_{\infty},\left\|f_{+}^{(r)}\right\|_{\infty}\right\}\leq D^{*}\left(f,\widetilde{o}\right),$$

proving (14.3).

The assumption $D^{*}\left(f,\widetilde{o}\right)<\infty$ implies $\left\|f_{\pm}^{(r)}\right\|_{\infty}<\infty$, $\forall\, r\in[0,1]$.

Here Σ^{*} stands for fuzzy summation and $\widetilde{0}:=\chi_{\{0\}}\in\mathbb{R}_{\mathcal{F}}$ is the neutral element with respect to $\oplus$, i.e.,

$$u\oplus\widetilde{0}=\widetilde{0}\oplus u=u,\ \forall\, u\in\mathbb{R}_{\mathcal{F}}.$$

We need

Remark 14.3 ([5]). Here $r \in [0,1]$, $x_i^{(r)}, y_i^{(r)} \in \mathbb{R}$, $i = 1, \ldots, m \in \mathbb{N}$. Suppose that

$$\sup_{r\in[0,1]} \max\left(x_i^{(r)}, y_i^{(r)}\right) \in \mathbb{R}, \quad for \quad i = 1, \ldots, m.$$

Then one sees easily that

$$\sup_{r\in[0,1]} \max\left(\sum_{i=1}^{m} x_i^{(r)}, \sum_{i=1}^{m} y_i^{(r)}\right) \le \sum_{i=1}^{m} \sup_{r\in[0,1]} \max\left(x_i^{(r)}, y_i^{(r)}\right). \tag{14.4}$$

Definition 14.4 Let $f \in C(W)$, $W \subseteq \mathbb{R}^m$, $m \in \mathbb{N}$, which is bounded or uniformly continuous, we define $(h > 0)$

$$\omega_1(f, h) := \sup_{x,y\in W,\ \|x-y\|_\infty \le h} |f(x) - f(y)|, \tag{14.5}$$

where $x = (x_1, \ldots, x_m)$, $y = (y_1, \ldots, y_m)$.

Definition 14.5 Let $f : W \to \mathbb{R}_{\mathcal{F}}$, $W \subseteq \mathbb{R}^m$, we define the fuzzy modulus of continuity of f by

$$\omega_1^{(\mathcal{F})}(f, h) = \sup_{x,y\in W,\ \|x-y\|_\infty \le h} D(f(x), f(y)),\ h > 0. \tag{14.6}$$

where $x = (x_1, \ldots, x_m)$, $y = (y_1, \ldots, y_m)$.

For $f : W \to \mathbb{R}_{\mathcal{F}}$, $W \subseteq \mathbb{R}^m$, we use

$$[f]^r = \left[f_-^{(r)}, f_+^{(r)}\right], \tag{14.7}$$

where $f_\pm^{(r)} : W \to \mathbb{R}$, $\forall\, r \in [0,1]$.
We need

Proposition 14.6 *Let $f : W \to \mathbb{R}_{\mathcal{F}}$. Assume that $\omega_1^{\mathcal{F}}(f, \delta)$, $\omega_1\left(f_-^{(r)}, \delta\right)$, $\omega_1\left(f_+^{(r)}, \delta\right)$ are finite for any $\delta > 0$, $r \in [0,1]$.*
Then

$$\omega_1^{(\mathcal{F})}(f, \delta) = \sup_{r\in[0,1]} \max\left\{\omega_1\left(f_-^{(r)}, \delta\right), \omega_1\left(f_+^{(r)}, \delta\right)\right\}. \tag{14.8}$$

Proof As in [5]. ■

We define by $C_{\mathcal{F}}^U(W)$ the space of fuzzy uniformly continuous functions from $W \to \mathbb{R}_{\mathcal{F}}$, also $C_{\mathcal{F}}(W)$ is the space of fuzzy continuous functions on $W \subseteq \mathbb{R}^m$, and $C_b(W, \mathbb{R}_{\mathcal{F}})$ is the fuzzy continuous and bounded functions.

We mention

Proposition 14.7 ([7]) *Let $f \in C^U_{\mathcal{F}}(W)$, where $W \subseteq \mathbb{R}^m$ is convex. Then $\omega_1^{(\mathcal{F})}(f, \delta) < \infty$, for any $\delta > 0$.*

Proposition 14.8 ([7]) *It holds*

$$\lim_{\delta \to 0} \omega_1^{(\mathcal{F})}(f, \delta) = \omega_1^{(\mathcal{F})}(f, 0) = 0, \tag{14.9}$$

iff $f \in C^U_{\mathcal{F}}(W)$, $W \subseteq \mathbb{R}^m$.

Proposition 14.9 ([7]) *Let $f \in C_{\mathcal{F}}(W)$, $W \subseteq \mathbb{R}^m$ open or compact. Then $f_{\pm}^{(r)}$ are equicontinuous with respect to $r \in [0, 1]$ over W, respectively in $\pm$.*

Notation 14.10 *It is clear by Propositions 14.6, 14.8, that if $f \in C^U_{\mathcal{F}}(W)$, then $f_{\pm}^{(r)} \in C_U(W)$ (uniformly continuous on W).*

We need

Definition 14.11 Let $x, y \in \mathbb{R}_{\mathcal{F}}$. If there exists $z \in \mathbb{R}_{\mathcal{F}} : x = y \oplus z$, then we call z the H-difference on x and y, denoted $x - y$.

Definition 14.12 ([24]) Let $T := [x_0, x_0 + \beta] \subset \mathbb{R}$, with $\beta > 0$. A function $f : T \to \mathbb{R}_{\mathcal{F}}$ is H-difference at $x \in T$ if there exists an $f'(x) \in \mathbb{R}_{\mathcal{F}}$ such that the limits (with respect to D)

$$\lim_{h \to 0+} \frac{f(x+h) - f(x)}{h}, \quad \lim_{h \to 0+} \frac{f(x) - f(x-h)}{h} \tag{14.10}$$

exist and are equal to $f'(x)$.

We call f' the H-derivative or fuzzy derivative of f at x.

Above is assumed that the H-differences $f(x+h) - f(x)$, $f(x) - f(x-h)$ exists in $\mathbb{R}_{\mathcal{F}}$ in a neighborhood of x.

Definition 14.13 We denote by $C^N_{\mathcal{F}}(W)$, $N \in \mathbb{N}$, the space of all N-times fuzzy continuously differentiable functions from W into $\mathbb{R}_{\mathcal{F}}$, $W \subseteq \mathbb{R}^m$ open or compact which is convex.

Here fuzzy partial derivatives are defined via Definition 14.12 in the obvious way as in the ordinary real case.

We mention

Theorem 14.14 ([20]) *Let $f : [a, b] \subseteq \mathbb{R} \to \mathbb{R}_{\mathcal{F}}$ be H-fuzzy differentiable. Let $t \in [a, b]$, $0 \le r \le 1$. Clearly*

$$[f(t)]^r = \left[f(t)_-^{(r)}, f(t)_+^{(r)} \right] \subseteq \mathbb{R}.$$

Then $(f(t))_{\pm}^{(r)}$ *are differentiable and*

$$[f'(t)]^r = \left[\left(f(t)_-^{(r)}\right)', \left(f(t)_+^{(r)}\right)'\right].$$

I.e.

$$(f')_{\pm}^{(r)} = \left(f_{\pm}^{(r)}\right)', \ \forall r \in [0,1]. \tag{14.11}$$

Remark 14.15 (se also [6]) Let $f \in C^N([a,b], \mathbb{R}_{\mathcal{F}})$, $N \geq 1$. Then by Theorem 14.14 we obtain $f_{\pm}^{(r)} \in C^N([a,b])$ and

$$\left[f^{(i)}(t)\right]^r = \left[\left(f(t)_-^{(r)}\right)^{(i)}, \left(f(t)_+^{(r)}\right)^{(i)}\right],$$

for $i = 0, 1, 2, \ldots, N$, and in particular we have

$$\left(f^{(i)}\right)_{\pm}^{(r)} = \left(f_{\pm}^{(r)}\right)^{(i)}, \tag{14.12}$$

for any $r \in [0,1]$.

Let $f \in C_{\mathcal{F}}^N(W)$, $W \subseteq \mathbb{R}^m$, open or compact, which is convex, denote $f_{\widetilde{\alpha}} := \frac{\partial^{\widetilde{\alpha}} f}{\partial x^{\widetilde{\alpha}}}$, where $\widetilde{\alpha} := (\widetilde{\alpha_1}, \ldots, \widetilde{\alpha_m})$, $\widetilde{\alpha_i} \in \mathbb{Z}^+$, $i = 1, \ldots, m$ and

$$0 < |\widetilde{\alpha}| := \sum_{i=1}^m \widetilde{\alpha_i} \leq N, \ N > 1.$$

Then by Theorem 14.14 we get that

$$\left(f_{\pm}^{(r)}\right)_{\widetilde{\alpha}} = (f_{\widetilde{\alpha}})_{\pm}^{(r)}, \ \forall r \in [0,1], \tag{14.13}$$

and any $\widetilde{\alpha} : |\widetilde{\alpha}| \leq N$. Here $f_{\pm}^{(r)} \in C^N(W)$.

Notation 14.16 *We denote*

$$\left(\sum_{i=1}^2 D\left(\frac{\partial}{\partial x_i}, \widetilde{o}\right)\right)^2 f(\overrightarrow{x}) := \tag{14.14}$$

$$D\left(\frac{\partial^2 f(x_1,x_2)}{\partial x_1^2}, \widetilde{o}\right) + D\left(\frac{\partial^2 f(x_1,x_2)}{\partial x_2^2}, \widetilde{o}\right) + 2D\left(\frac{\partial^2 f(x_1,x_2)}{\partial x_1 \partial x_2}, \widetilde{o}\right).$$

In general we denote ($j = 1, \ldots, N$)

$$\left(\sum_{i=1}^{m} D\left(\frac{\partial}{\partial x_i}, \widetilde{o}\right)\right)^{j} f\left(\overrightarrow{x}\right) := \tag{14.15}$$

$$\sum_{(j_1,\ldots,j_m)\in\mathbb{Z}_+^m:\sum_{i=1}^m j_i=j} \frac{j!}{j_1!j_2!\ldots j_m!} D\left(\frac{\partial^j f(x_1,\ldots,x_m)}{\partial x_1^{j_1}\partial x_2^{j_2}\ldots\partial x_m^{j_m}}, \widetilde{o}\right).$$

14.2 Basic on Real Quasi-interpolation Neural Network Operators Approximation

(I) Here all come from [8, 17].

We consider the sigmoidal function of logarithmic type

$$s_i(x_i) = \frac{1}{1+e^{-x_i}}, \quad x_i \in \mathbb{R}, i = 1, \ldots, N; \ x := (x_1, \ldots, x_N) \in \mathbb{R}^N,$$

each has the properties $\lim_{x_i \to +\infty} s_i(x_i) = 1$ and $\lim_{x_i \to -\infty} s_i(x_i) = 0$, $i = 1, \ldots, N$.

These functions play the role of activation functions in the hidden layer of neural networks.

As in [17], we consider

$$\Phi_i(x_i) := \frac{1}{2}\left(s_i(x_i+1) - s_i(x_i-1)\right), \quad x_i \in \mathbb{R}, i = 1, \ldots, N.$$

We notice the following properties:

(i) $\Phi_i(x_i) > 0, \ \forall \, x_i \in \mathbb{R}$,

(ii) $\sum_{k_i=-\infty}^{\infty} \Phi_i(x_i - k_i) = 1, \ \forall \, x_i \in \mathbb{R}$,

(iii) $\sum_{k_i=-\infty}^{\infty} \Phi_i(nx_i - k_i) = 1, \ \forall \, x_i \in \mathbb{R}; n \in \mathbb{N}$,

(iv) $\int_{-\infty}^{\infty} \Phi_i(x_i)\, dx_i = 1$,

(v) Φ_i is a density function,

(vi) Φ_i is even: $\Phi_i(-x_i) = \Phi_i(x_i), \ x_i \geq 0$, for $i = 1, \ldots, N$.

We see that [12]

$$\Phi_i(x_i) = \left(\frac{e^2-1}{2e^2}\right) \frac{1}{\left(1+e^{x_i-1}\right)\left(1+e^{-x_i-1}\right)}, \quad i = 1, \ldots, N.$$

(vii) Φ_i is decreasing on $\mathbb{R}_+$, and increasing on $\mathbb{R}_-$, $i = 1, \ldots, N$.
Notice that $\Phi_i(x_i) = \Phi_i(0) = 0.231$.

Let $0 < \beta < 1$, $n \in \mathbb{N}$. Then as in [12] we get

(viii)

$$\sum_{\begin{cases} k_i = -\infty \\ : |nx_i - k_i| > n^{1-\beta} \end{cases}}^{\infty} \Phi_i(nx_i - k_i) \leq 3.1992 e^{-n^{(1-\beta)}}, \quad i = 1, \ldots, N.$$

Denote by $\lceil \cdot \rceil$ the ceiling of a number, and by $\lfloor \cdot \rfloor$ the integral part of a number. Consider here $x \in \left(\prod_{i=1}^{N} [a_i, b_i] \right) \subset \mathbb{R}^N$, $N \in \mathbb{N}$ such that $\lceil na_i \rceil \leq \lfloor nb_i \rfloor$, $i = 1, \ldots, N$; $a := (a_1, \ldots, a_N)$, $b := (b_1, \ldots, b_N)$.

As in [12] we obtain

(ix)

$$0 < \frac{1}{\sum_{k_i = \lceil na_i \rceil}^{\lfloor nb_i \rfloor} \Phi_i(nx_i - k_i)} < \frac{1}{\Phi_i(1)} = 5.250312578,$$

$\forall\, x_i \in [a_i, b_i]$, $i = 1, \ldots, N$.

(x) As in [12], we see that

$$\lim_{n \to \infty} \sum_{k_i = \lceil na_i \rceil}^{\lfloor nb_i \rfloor} \Phi_i(nx_i - k_i) \neq 1,$$

for at least some $x_i \in [a_i, b_i]$, $i = 1, \ldots, N$.

We use here

$$\Phi(x_1, \ldots, x_N) := \Phi(x) := \prod_{i=1}^{N} \Phi_i(x_i), \quad x \in \mathbb{R}^N. \tag{14.16}$$

It has the properties:

(i)' $\Phi(x) > 0$, $\forall\, x \in \mathbb{R}^N$,

(ii)'

$$\sum_{k=-\infty}^{\infty} \Phi(x - k) := \sum_{k_1=-\infty}^{\infty} \sum_{k_2=-\infty}^{\infty} \cdots \sum_{k_N=-\infty}^{\infty} \Phi(x_1 - k_1, \ldots, x_N - k_N) = 1, \tag{14.17}$$

$k := (k_1, \ldots, k_N)$, $\forall\, x \in \mathbb{R}^N$.

(iii)'

$$\sum_{k=-\infty}^{\infty} \Phi(nx-k) :=$$

$$\sum_{k_1=-\infty}^{\infty} \sum_{k_2=-\infty}^{\infty} \dots \sum_{k_N=-\infty}^{\infty} \Phi(nx_1-k_1, \dots, nx_N-k_N) = 1, \tag{14.18}$$

$\forall\, x \in \mathbb{R}^N; n \in \mathbb{N}.$

(iv)'

$$\int_{\mathbb{R}^N} \Phi(x)\, dx = 1,$$

that is Φ is a multivariate density function.

Here $\|x\|_\infty := \max\{|x_1|, \dots, |x_N|\}$, $x \in \mathbb{R}^N$, also set $\infty := (\infty, \dots, \infty)$, $-\infty := (-\infty, \dots, -\infty)$ upon the multivariate context, and

$$\lceil na \rceil := (\lceil na_1 \rceil, \dots, \lceil na_N \rceil),$$
$$\lfloor nb \rfloor := (\lfloor nb_1 \rfloor, \dots, \lfloor nb_N \rfloor).$$

We also have

(v)'

$$\sum_{\begin{cases} k = \lceil na \rceil \\ \left\| \frac{k}{n} - x \right\|_\infty > \frac{1}{n^\beta} \end{cases}}^{\lfloor nb \rfloor} \Phi(nx-k) \leq 3.1992 e^{-n^{(1-\beta)}},$$

$0 < \beta < 1, n \in \mathbb{N}, x \in \left(\prod_{i=1}^N [a_i, b_i]\right).$

(vi)'

$$0 < \frac{1}{\sum_{k=\lceil na \rceil}^{\lfloor nb \rfloor} \Phi(nx-k)} < (5.250312578)^N,$$

$\forall\, x \in \left(\prod_{i=1}^N [a_i, b_i]\right),\ n \in \mathbb{N}.$

(vii)'

$$\sum_{\begin{cases} k = -\infty \\ \left\| \frac{k}{n} - x \right\|_\infty > \frac{1}{n^\beta} \end{cases}}^{\infty} \Phi(nx-k) \leq 3.1992 e^{-n^{(1-\beta)}},$$

$0 < \beta < 1, n \in \mathbb{N}, x \in \mathbb{R}^N.$

(viii)'

$$\lim_{n\to\infty}\sum_{k=\lceil na\rceil}^{\lfloor nb\rfloor}\Phi(nx-k)\neq 1 \tag{14.19}$$

for at least some $x\in\left(\prod_{i=1}^{N}[a_i,b_i]\right)$.

Let $f\in C\left(\prod_{i=1}^{N}[a_i,b_i]\right)$ and $n\in\mathbb{N}$ such that $\lceil na_i\rceil\leq\lfloor nb_i\rfloor$, $i=1,\ldots,N$.

We introduce and define [11] the multivariate positive linear neural network operator ($x:=(x_1,\ldots,x_N)\in\left(\prod_{i=1}^{N}[a_i,b_i]\right)$)

$$G_n(f,x_1,\ldots,x_N):=G_n(f,x):=\frac{\sum_{k=\lceil na\rceil}^{\lfloor nb\rfloor}f\left(\frac{k}{n}\right)\Phi(nx-k)}{\sum_{k=\lceil na\rceil}^{\lfloor nb\rfloor}\Phi(nx-k)} \tag{14.20}$$

$$:=\frac{\sum_{k_1=\lceil na_1\rceil}^{\lfloor nb_1\rfloor}\sum_{k_2=\lceil na_2\rceil}^{\lfloor nb_2\rfloor}\cdots\sum_{k_N=\lceil na_N\rceil}^{\lfloor nb_N\rfloor}f\left(\frac{k_1}{n},\ldots,\frac{k_N}{n}\right)\left(\prod_{i=1}^{N}\Phi_i(nx_i-k_i)\right)}{\prod_{i=1}^{N}\left(\sum_{k_i=\lceil na_i\rceil}^{\lfloor nb_i\rfloor}\Phi_i(nx_i-k_i)\right)}.$$

For large enough n we always obtain $\lceil na_i\rceil\leq\lfloor nb_i\rfloor$, $i=1,\ldots,N$. Also $a_i\leq\frac{k_i}{n}\leq b_i$, iff $\lceil na_i\rceil\leq k_i\leq\lfloor nb_i\rfloor$, $i=1,\ldots,N$.

When $f\in C_B\left(\mathbb{R}^N\right)$ (continuous and bounded functions on $\mathbb{R}^N$) we define [11]

$$\overline{G}_n(f,x):=\overline{G}_n(f,x_1,\ldots,x_N):=\sum_{k=-\infty}^{\infty}f\left(\frac{k}{n}\right)\Phi(nx-k) \tag{14.21}$$

$$:=\sum_{k_1=-\infty}^{\infty}\sum_{k_2=-\infty}^{\infty}\cdots\sum_{k_N=-\infty}^{\infty}f\left(\frac{k_1}{n},\frac{k_2}{n},\ldots,\frac{k_N}{n}\right)\left(\prod_{i=1}^{N}\Phi_i(nx_i-k_i)\right),$$

$n\in\mathbb{N}$, $\forall\ x\in\mathbb{R}^N$, $N\geq 1$, the multivariate quasi-interpolation neural network operator.

We mention from [11]:

Theorem 14.17 *Let* $f\in C\left(\prod_{i=1}^{N}[a_i,b_i]\right)$, $0<\beta<1$, $x\in\left(\prod_{i=1}^{N}[a_i,b_i]\right)$, $n,N\in\mathbb{N}$. *Then*

(i)

$$|G_n(f,x)-f(x)|\leq(5.250312578)^N\cdot$$

$$\left\{\omega_1\left(f,\frac{1}{n^{\beta}}\right)+(6.3984)\|f\|_{\infty}e^{-n^{(1-\beta)}}\right\}=:\lambda_1, \tag{14.22}$$

(ii)

$$\|G_n(f) - f\|_\infty \leq \lambda_1. \tag{14.23}$$

Theorem 14.18 *Let $f \in C_B(\mathbb{R}^N)$, $0 < \beta < 1$, $x \in \mathbb{R}^N$, $n, N \in \mathbb{N}$. Then*

(i)

$$\left|\overline{G}_n(f, x) - f(x)\right| \leq \omega_1\left(f, \frac{1}{n^\beta}\right) + (6.3984)\|f\|_\infty e^{-n^{(1-\beta)}} =: \lambda_2, \tag{14.24}$$

(ii)

$$\left\|\overline{G}_n(f) - f\right\|_\infty \leq \lambda_2. \tag{14.25}$$

(II) Here we follow [8, 10].

We also consider here the hyperbolic tangent function $\tanh x$, $x \in \mathbb{R}$:

$$\tanh x := \frac{e^x - e^{-x}}{e^x + e^{-x}}.$$

It has the properties $\tanh 0 = 0$, $-1 < \tanh x < 1$, $\forall\, x \in \mathbb{R}$, and $\tanh(-x) = -\tanh x$. Furthermore $\tanh x \to 1$ as $x \to \infty$, and $\tanh x \to -1$, as $x \to -\infty$, and it is strictly increasing on $\mathbb{R}$.

This function plays the role of an activation function in the hidden layer of neural networks.

We further consider

$$\Psi(x) := \frac{1}{4}(\tanh(x+1) - \tanh(x-1)) > 0, \quad \forall\, x \in \mathbb{R}. \tag{14.26}$$

We easily see that $\Psi(-x) = \Psi(x)$, that is Ψ is even on $\mathbb{R}$. Obviously Ψ is differentiable, thus continuous.

Proposition 14.19 ([9]) *$\Psi(x)$ for $x \geq 0$ is strictly decreasing.*

Obviously $\Psi(x)$ is strictly increasing for $x \leq 0$. Also it holds $\lim_{x\to-\infty} \Psi(x) = 0 = \lim_{x\to\infty} \Psi(x)$.

Infact Ψ has the bell shape with horizontal asymptote the x-axis. So the maximum of Ψ is zero, $\Psi(0) = 0.3809297$.

Theorem 14.20 ([9]) *We have that $\sum_{i=-\infty}^{\infty} \Psi(x-i) = 1$, $\forall\, x \in \mathbb{R}$.*

Thus

$$\sum_{i=-\infty}^{\infty} \Psi(nx - i) = 1, \quad \forall\, n \in \mathbb{N},\ \forall\, x \in \mathbb{R}.$$

Also it holds

$$\sum_{i=-\infty}^{\infty} \Psi(x+i) = 1, \quad \forall x \in \mathbb{R}.$$

Theorem 14.21 ([9]) *It holds* $\int_{-\infty}^{\infty} \Psi(x)\, dx = 1$.

So $\Psi(x)$ is a density function on $\mathbb{R}$.

Theorem 14.22 ([9]) *Let* $0 < \alpha < 1$ *and* $n \in \mathbb{N}$. *It holds*

$$\sum_{\substack{k=-\infty \\ : |nx-k| \geq n^{1-\alpha}}}^{\infty} \Psi(nx-k) \leq e^4 \cdot e^{-2n^{(1-\alpha)}}.$$

Theorem 14.23 ([9]) *Let* $x \in [a,b] \subset \mathbb{R}$ *and* $n \in \mathbb{N}$ *so that* $\lceil na \rceil \leq \lfloor nb \rfloor$. *It holds*

$$\frac{1}{\sum_{k=\lceil na \rceil}^{\lfloor nb \rfloor} \Psi(nx-k)} < \frac{1}{\Psi(1)} = 4.1488766.$$

Also by [9] we get that

$$\lim_{n \to \infty} \sum_{k=\lceil na \rceil}^{\lfloor nb \rfloor} \Psi(nx-k) \neq 1,$$

for at least some $x \in [a,b]$.

We use (see [10])

$$\Theta(x_1, \ldots, x_N) := \Theta(x) := \prod_{i=1}^{N} \Psi(x_i), \quad x = (x_1, \ldots, x_N) \in \mathbb{R}^N, \quad N \in \mathbb{N}. \tag{14.27}$$

It has the properties:

(i)* $\Theta(x) > 0$, $\forall\, x \in \mathbb{R}^N$,

(ii)*

$$\sum_{k=-\infty}^{\infty} \Theta(x-k) := \sum_{k_1=-\infty}^{\infty} \sum_{k_2=-\infty}^{\infty} \cdots \sum_{k_N=-\infty}^{\infty} \Theta(x_1-k_1, \ldots, x_N-k_N) = 1, \tag{14.28}$$

where $k := (k_1, \ldots, k_N)$, $\forall\, x \in \mathbb{R}^N$.

(iii)*

$$\sum_{k=-\infty}^{\infty} \Theta(nx-k) :=$$

$$\sum_{k_1=-\infty}^{\infty} \sum_{k_2=-\infty}^{\infty} \cdots \sum_{k_N=-\infty}^{\infty} \Theta\left(nx_1 - k_1, \ldots, nx_N - k_N\right) = 1, \tag{14.29}$$

$\forall\, x \in \mathbb{R}^N;\ n \in \mathbb{N}.$

(iv)*

$$\int_{\mathbb{R}^N} \Theta(x)\, dx = 1,$$

that is Θ is a multivariate density function.

(v)*

$$\sum_{\substack{k = \lceil na \rceil \\ \left\| \frac{k}{n} - x \right\|_{\infty} > \frac{1}{n^{\beta}}}}^{\lfloor nb \rfloor} \Theta(nx - k) \leq e^4 \cdot e^{-2n^{(1-\beta)}}, \tag{14.30}$$

$0 < \beta < 1,\ n \in \mathbb{N},\ x \in \left(\prod_{i=1}^{N} [a_i, b_i]\right).$

(vi)*

$$0 < \frac{1}{\sum_{k=\lceil na \rceil}^{\lfloor nb \rfloor} \Theta(nx - k)} < \frac{1}{(\Psi(1))^N} = (4.1488766)^N, \tag{14.31}$$

$\forall\, x \in \left(\prod_{i=1}^{N} [a_i, b_i]\right),\ n \in \mathbb{N}.$

(vii)*

$$\sum_{\substack{k = -\infty \\ \left\| \frac{k}{n} - x \right\|_{\infty} > \frac{1}{n^{\beta}}}}^{\infty} \Theta(nx - k) \leq e^4 \cdot e^{-2n^{(1-\beta)}}, \tag{14.32}$$

$0 < \beta < 1,\ n \in \mathbb{N},\ x \in \mathbb{R}^N.$

Also we get that

$$\lim_{n \to \infty} \sum_{k=\lceil na \rceil}^{\lfloor nb \rfloor} \Theta(nx - k) \neq 1, \tag{14.33}$$

for at least some $x \in \left(\prod_{i=1}^{N} [a_i, b_i]\right)$.

Let $f \in C\left(\prod_{i=1}^{N} [a_i, b_i]\right)$ and $n \in \mathbb{N}$ such that $\lceil na_i \rceil \leq \lfloor nb_i \rfloor,\ i = 1, \ldots, N$.

We introduce and define the multivariate positive linear neural network operator [10] ($x := (x_1, \ldots, x_N) \in \left(\prod_{i=1}^{N} [a_i, b_i]\right)$)

$$F_n(f, x_1, \ldots, x_N) := F_n(f, x) := \frac{\sum_{k=\lceil na \rceil}^{\lfloor nb \rfloor} f\left(\frac{k}{n}\right) \Theta(nx - k)}{\sum_{k=\lceil na \rceil}^{\lfloor nb \rfloor} \Theta(nx - k)} \tag{14.34}$$

$$:= \frac{\sum_{k_1=\lceil na_1 \rceil}^{\lfloor nb_1 \rfloor} \sum_{k_2=\lceil na_2 \rceil}^{\lfloor nb_2 \rfloor} \cdots \sum_{k_N=\lceil na_N \rceil}^{\lfloor nb_N \rfloor} f\left(\frac{k_1}{n}, \ldots, \frac{k_N}{n}\right) \left(\prod_{i=1}^{N} \Psi(nx_i - k_i)\right)}{\prod_{i=1}^{N} \left(\sum_{k_i=\lceil na_i \rceil}^{\lfloor nb_i \rfloor} \Psi(nx_i - k_i)\right)}.$$

When $f \in C_B(\mathbb{R}^N)$ we define [10],

$$\overline{F}_n(f, x) := \overline{F}_n(f, x_1, \ldots, x_N) := \sum_{k=-\infty}^{\infty} f\left(\frac{k}{n}\right) \Theta(nx - k) := \tag{14.35}$$

$$\sum_{k_1=-\infty}^{\infty} \sum_{k_2=-\infty}^{\infty} \cdots \sum_{k_N=-\infty}^{\infty} f\left(\frac{k_1}{n}, \frac{k_2}{n}, \ldots, \frac{k_N}{n}\right) \left(\prod_{i=1}^{N} \Psi(nx_i - k_i)\right),$$

$n \in \mathbb{N}$, $\forall\, x \in \mathbb{R}^N$, $N \geq 1$, the multivariate quasi-interpolation neural network operator.

We mention from [10]:

Theorem 14.24 *Let* $f \in C\left(\prod_{i=1}^{N} [a_i, b_i]\right)$, $0 < \beta < 1$, $x \in \left(\prod_{i=1}^{N} [a_i, b_i]\right)$, $n, N \in \mathbb{N}$. *Then*

(i)

$$|F_n(f, x) - f(x)| \leq (4.1488766)^N \cdot$$

$$\left\{\omega_1\left(f, \frac{1}{n^\beta}\right) + 2e^4 \|f\|_\infty e^{-2n^{(1-\beta)}}\right\} =: \lambda_1, \tag{14.36}$$

(ii)

$$\|F_n(f) - f\|_\infty \leq \lambda_1. \tag{14.37}$$

Theorem 14.25 *Let* $f \in C_B(\mathbb{R}^N)$, $0 < \beta < 1$, $x \in \mathbb{R}^N$, $n, N \in \mathbb{N}$. *Then*

(i)

$$\left|\overline{F}_n(f, x) - f(x)\right| \leq \omega_1\left(f, \frac{1}{n^\beta}\right) + 2e^4 \|f\|_\infty e^{-2n^{(1-\beta)}} =: \lambda_2, \tag{14.38}$$

(ii)

$$\left\|\overline{F}_n(f) - f\right\|_\infty \leq \lambda_2. \tag{14.39}$$

Notation 14.26 *Let $f \in C^m \left(\prod_{i=1}^N [a_i, b_i] \right)$, $m, N \in \mathbb{N}$. Here f_α denotes a partial derivative of f, $\alpha := (\alpha_1, \ldots, \alpha_N)$, $\alpha_i \in \mathbb{Z}^+$, $i = 1, \ldots, N$, and $|\alpha| := \sum_{i=1}^N \alpha_i = l$, where $l = 0, 1, \ldots, m$. We write also $f_\alpha := \frac{\partial^\alpha f}{\partial x^\alpha}$ and we say it is of order l.*

We denote

$$\omega_{1,m}^{\max} (f_\alpha, h) := \max_{\alpha : |\alpha| = m} \omega_1 (f_\alpha, h) . \tag{14.40}$$

Call also

$$\| f_\alpha \|_{\infty,m}^{\max} := \max_{|\alpha| = m} \left\{ \| f_\alpha \|_\infty \right\}, \tag{14.41}$$

In the next we mention the high order of approximation by using the smoothness of f.

We give

Theorem 14.27 ([8]) *Let $f \in C^m \left(\prod_{i=1}^N [a_i, b_i] \right)$, $0 < \beta < 1$, $n, m, N \in \mathbb{N}$, $x \in \left(\prod_{i=1}^N [a_i, b_i] \right)$. Then*

(i)

$$\left| G_n (f, x) - f (x) - \sum_{j=1}^m \left(\sum_{|\alpha| = j} \left(\frac{f_\alpha (x)}{\prod_{i=1}^N \alpha_i !} \right) G_n \left(\prod_{i=1}^N (\cdot - x_i)^{\alpha_i} , x \right) \right) \right| \leq \tag{14.42}$$
$$(5.250312578)^N \cdot \left\{ \frac{N^m}{m! n^{m\beta}} \omega_{1,m}^{\max} \left(f_\alpha, \frac{1}{n^\beta} \right) + \right.$$
$$\left. \left(\frac{(6.3984) \| b - a \|_\infty^m \| f_\alpha \|_{\infty,m}^{\max} N^m}{m!} \right) e^{-n^{(1-\beta)}} \right\},$$

(ii)

$$|G_n (f, x) - f (x)| \leq (5.250312578)^N \cdot \tag{14.43}$$
$$\left\{ \sum_{j=1}^m \left(\sum_{|\alpha| = j} \left(\frac{|f_\alpha (x)|}{\prod_{i=1}^N \alpha_i !} \right) \left[\frac{1}{n^{\beta j}} + \left(\prod_{i=1}^N (b_i - a_i)^{\alpha_i} \right) \cdot (3.1992) e^{-n^{(1-\beta)}} \right] \right) + \right.$$
$$\left. \frac{N^m}{m! n^{m\beta}} \omega_{1,m}^{\max} \left(f_\alpha, \frac{1}{n^\beta} \right) + \left(\frac{(6.3984) \| b - a \|_\infty^m \| f_\alpha \|_{\infty,m}^{\max} N^m}{m!} \right) e^{-n^{(1-\beta)}} \right\},$$

(iii)

$$\| G_n (f) - f \|_\infty \leq (5.250312578)^N \cdot \tag{14.44}$$

$$\left\{\sum_{j=1}^{N}\left(\sum_{|\alpha|=j}\left(\frac{\|f_\alpha\|_\infty}{\prod_{i=1}^{N}\alpha_i!}\right)\left[\frac{1}{n^{\beta j}}+\left(\prod_{i=1}^{N}(b_i-a_i)^{\alpha_i}\right)(3.1992)\,e^{-n^{(1-\beta)}}\right]\right)+\right.$$

$$\left.\frac{N^m}{m!n^{m\beta}}\omega_{1,m}^{\max}\left(f_\alpha,\frac{1}{n^\beta}\right)+\left(\frac{(6.3984)\,\|b-a\|_\infty^m\,\|f_\alpha\|_{\infty,m}^{\max}\,N^m}{m!}\right)e^{-n^{(1-\beta)}}\right\},$$

(iv) Assume $f_\alpha(x_0)=0$*, for all* $\alpha:|\alpha|=1,\ldots,m$; $x_0\in\left(\prod_{i=1}^{N}[a_i,b_i]\right)$*. Then*

$$|G_n(f,x_0)-f(x_0)|\le(5.250312578)^N\cdot \tag{14.45}$$

$$\left\{\frac{N^m}{m!n^{m\beta}}\omega_1^{\max}\left(f_\alpha,\frac{1}{n^\beta}\right)+\left(\frac{(6.3984)\,\|b-a\|_\infty^m\,\|f_\alpha\|_{\infty,m}^{\max}\,N^m}{m!}\right)e^{-n^{(1-\beta)}}\right\},$$

notice in the last the extremely high rate of convergence at $n^{-\beta(m+1)}$.

We also mention

Theorem 14.28 ([8]) *Let* $f\in C^m\left(\prod_{i=1}^{N}[a_i,b_i]\right)$, $0<\beta<1$, $n,m,N\in\mathbb{N}$, $x\in\left(\prod_{i=1}^{N}[a_i,b_i]\right)$*. Then*

(i)

$$\left|F_n(f,x)-f(x)-\sum_{j=1}^{m}\left(\sum_{|\alpha|=j}\left(\frac{f_\alpha(x)}{\prod_{i=1}^{N}\alpha_i!}\right)F_n\left(\prod_{i=1}^{N}(\cdot-x_i)^{\alpha_i},x\right)\right)\right|\le \tag{14.46}$$

$$(4.1488766)^N\cdot\left\{\frac{N^m}{m!n^{m\beta}}\omega_{1,m}^{\max}\left(f_\alpha,\frac{1}{n^\beta}\right)+\right.$$

$$\left.\left(\frac{2e^4\,\|b-a\|_\infty^m\,\|f_\alpha\|_{\infty,m}^{\max}\,N^m}{m!}\right)e^{-2n^{(1-\beta)}}\right\},$$

(ii)

$$|F_n(f,x)-f(x)|\le(4.1488766)^N\cdot \tag{14.47}$$

$$\left\{\sum_{j=1}^{m}\left(\sum_{|\alpha|=j}\left(\frac{|f_\alpha(x)|}{\prod_{i=1}^{N}\alpha_i!}\right)\left[\frac{1}{n^{\beta j}}+\left(\prod_{i=1}^{N}(b_i-a_i)^{\alpha_i}\right)\cdot e^4e^{-2n^{(1-\beta)}}\right]\right)+\right.$$

$$\left.\frac{N^m}{m!n^{m\beta}}\omega_{1,m}^{\max}\left(f_\alpha,\frac{1}{n^\beta}\right)+\left(\frac{2e^4\,\|b-a\|_\infty^m\,\|f_\alpha\|_{\infty,m}^{\max}\,N^m}{m!}\right)e^{-2n^{(1-\beta)}}\right\},$$

(iii)

$$\|F_n(f) - f\|_\infty \leq (4.1488766)^N \cdot \tag{14.48}$$

$$\left\{ \sum_{j=1}^{m} \left(\sum_{|\alpha|=j} \left(\frac{\|f_\alpha\|_\infty}{\prod_{i=1}^{N} \alpha_i!} \right) \left[\frac{1}{n^{\beta j}} + \left(\prod_{i=1}^{N} (b_i - a_i)^{\alpha_i} \right) e^4 e^{-2n^{(1-\beta)}} \right] \right) + \right.$$

$$\left. \frac{N^m}{m! n^{m\beta}} \omega_{1,m}^{\max} \left(f_\alpha, \frac{1}{n^\beta} \right) + \left(\frac{2e^4 \|b-a\|_\infty^m \|f_\alpha\|_{\infty,m}^{\max} N^m}{m!} \right) e^{-2n^{(1-\beta)}} \right\},$$

(iv) Assume $f_\alpha(x_0) = 0$, *for all* $\alpha : |\alpha| = 1, \ldots, m$; $x_0 \in \left(\prod_{i=1}^{N} [a_i, b_i] \right)$. *Then*

$$|F_n(f, x_0) - f(x_0)| \leq (4.1488766)^N \cdot \tag{14.49}$$

$$\left\{ \frac{N^m}{m! n^{m\beta}} \omega_1^{\max} \left(f_\alpha, \frac{1}{n^\beta} \right) + \left(\frac{2e^4 \|b-a\|_\infty^m \|f_\alpha\|_{\infty,m}^{\max} N^m}{m!} \right) e^{-2n^{(1-\beta)}} \right\},$$

notice in the last the extremely high rate of convergence at $n^{-\beta(m+1)}$.

We need

Notation 14.29 ([15]) *Call* $L_n = G_n, \overline{G}_n, F_n, \overline{F}_n$.
Denote by

$$c_N = \begin{cases} (5.250312578)^N, \text{ if } L_n = G_n, \\ (4.1488766)^N, \text{ if } L_n = F_n, \\ 1, \text{ if } L_n = \overline{G}_n, \overline{F}_n, \end{cases} \tag{14.50}$$

$$\mu = \begin{cases} 6.3984, \text{ if } L_n = G_n, \overline{G}_n, \\ 2e^4, \text{ if } L_n = F_n, \overline{F}_n, \end{cases} \tag{14.51}$$

and

$$\gamma = \begin{cases} 1, \text{ when } L_n = G_n, \overline{G}_n, \\ 2 \text{ when } L_n = F_n, \overline{F}_n. \end{cases} \tag{14.52}$$

Based on the above notations Theorems 14.17, 14.18, 14.24 and 14.25 can put in a unified way as follows.

Theorem 14.30 ([15]) *Let* $f \in C\left(\prod_{i=1}^{N} [a_i, b_i] \right)$ *or* $f \in C_B\left(\mathbb{R}^N\right)$; $n, N \in \mathbb{N}$, $0 < \beta < 1$, $x \in \left(\prod_{i=1}^{N} [a_i, b_i] \right)$ *or* $x \in \mathbb{R}^N$. *Then*

(i)

$$|L_n(f, x) - f(x)| \leq c_N \left\{ \omega_1 \left(f, \frac{1}{n^\beta} \right) + \mu \|f\|_\infty e^{-\gamma n^{(1-\beta)}} \right\} =: \rho_n, \tag{14.53}$$

(ii)

$$\|L_n(f) - f\|_\infty \le \rho_n. \tag{14.54}$$

For basic neural networks knowledge we refer to [22, 23].

14.3 Approximation by Fuzzy Quasi-interpolation Neural Network Operators

Let $f \in C_{\mathcal{F}}\left(\prod_{i=1}^{N}[a_i, b_i]\right)$, $n \in \mathbb{N}$. We define the Fuzzy Quasi-Interpolation Neural Network operators

$$G_n^{\mathcal{F}}(f, x_1, \dots, x_N) := G_n^{\mathcal{F}}(f, x) := \sum_{k=\lceil na \rceil}^{\lfloor nb \rfloor *} f\left(\frac{k}{n}\right) \odot \frac{\Phi(nx - k)}{\sum_{k=\lceil na \rceil}^{\lfloor nb \rfloor} \Phi(nx - k)}, \tag{14.55}$$

$\forall\, x := (x_1, \dots, x_N) \in \left(\prod_{i=1}^{N}[a_i, b_i]\right)$, see also (14.20).

Similarly we define

$$F_n^{\mathcal{F}}(f, x_1, \dots, x_N) := F_n^{\mathcal{F}}(f, x) := \sum_{k=\lceil na \rceil}^{\lfloor nb \rfloor *} f\left(\frac{k}{n}\right) \odot \frac{\Theta(nx - k)}{\sum_{k=\lceil na \rceil}^{\lfloor nb \rfloor} \Theta(nx - k)}, \tag{14.56}$$

$\forall\, x \in \left(\prod_{i=1}^{N}[a_i, b_i]\right)$, see also (14.34).

Let $f \in C_b\left(\mathbb{R}^N, \mathbb{R}_{\mathcal{F}}\right)$. We define

$$\overline{G}_n^{\mathcal{F}}(f, x) := \overline{G}_n^{\mathcal{F}}(f, x_1, \dots, x_N) := \sum_{k=-\infty}^{\infty *} f\left(\frac{k}{n}\right) \odot \Phi(nx - k) \tag{14.57}$$

$$:= \sum_{k_1=-\infty}^{\infty} \sum_{k_2=-\infty}^{\infty} \cdots \sum_{k_N=-\infty}^{\infty *} f\left(\frac{k_1}{n}, \frac{k_2}{n}, \dots, \frac{k_N}{n}\right) \odot \left(\prod_{i=1}^{N} \Phi_i(nx_i - k_i)\right),$$

and

$$\overline{F}_n^{\mathcal{F}}(f, x) := \sum_{k=-\infty}^{\infty *} f\left(\frac{k}{n}\right) \odot \Theta(nx - k), \tag{14.58}$$

$\forall\, x \in \mathbb{R}^N$, $N \geq 1$.

The sum in (14.55), (14.56) are finite.

Let $r \in [0, 1]$, we observe that

$$\left[G_n^{\mathcal{F}}(f, x)\right]^r = \sum_{k=\lceil na \rceil}^{\lfloor nb \rfloor} \left[f\left(\frac{k}{n}\right)\right]^r \left(\frac{\Phi(nx - k)}{\sum_{k=\lceil na \rceil}^{\lfloor nb \rfloor} \Phi(nx - k)}\right) =$$

$$\sum_{k=\lceil na \rceil}^{\lfloor nb \rfloor} \left[f_-^{(r)}\left(\frac{k}{n}\right), f_+^{(r)}\left(\frac{k}{n}\right)\right] \left(\frac{\Phi(nx - k)}{\sum_{k=\lceil na \rceil}^{\lfloor nb \rfloor} \Phi(nx - k)}\right) = \tag{14.59}$$

$$\left[\sum_{k=\lceil na \rceil}^{\lfloor nb \rfloor} f_-^{(r)}\left(\frac{k}{n}\right) \left(\frac{\Phi(nx - k)}{\sum_{k=\lceil na \rceil}^{\lfloor nb \rfloor} \Phi(nx - k)}\right), \sum_{k=\lceil na \rceil}^{\lfloor nb \rfloor} f_+^{(r)}\left(\frac{k}{n}\right) \left(\frac{\Phi(nx - k)}{\sum_{k=\lceil na \rceil}^{\lfloor nb \rfloor} \Phi(nx - k)}\right)\right]$$

$$= \left[G_n\left(f_-^{(r)}, x\right), G_n\left(f_+^{(r)}, x\right)\right]. \tag{14.60}$$

We have proved that

$$\left(G_n^{\mathcal{F}}(f, x)\right)_{\pm}^{(r)} = G_n\left(f_{\pm}^{(r)}, x\right), \tag{14.61}$$

respectively, $\forall\, r \in [0, 1]$, $\forall\, x \in \left(\prod_{i=1}^{N} [a_i, b_i]\right)$.

Similarly, it holds

$$\left(F_n^{\mathcal{F}}(f, x)\right)_{\pm}^{(r)} = F_n\left(f_{\pm}^{(r)}, x\right), \tag{14.62}$$

respectively, $\forall\, r \in [0, 1]$, $\forall\, x \in \left(\prod_{i=1}^{N} [a_i, b_i]\right)$.

We will prove also that

$$\left(\overline{G}_n^{\mathcal{F}}(f,x)\right)_{\pm}^{(r)} = \overline{G}_n\left(f_{\pm}^{(r)},x\right), \tag{14.63}$$

$$\left(\overline{F}_n^{\mathcal{F}}(f,x)\right)_{\pm}^{(r)} = \overline{F}_n\left(f_{\pm}^{(r)},x\right), \tag{14.64}$$

respectively, $\forall\, r \in [0,1]$, $\forall\, x \in \mathbb{R}^N$.

The sums in (14.63), (14.64) are doubly infinite and their proof is more complicated and follows.

We need

Remark 14.31 (see also [21]) (1) Here $k = (k_1,k_2) \in \mathbb{Z}^2$, $m = (m_1,m_2) \in \mathbb{Z}_-^2$, $n = (n_1,n_2) \in \mathbb{Z}_+^2$, $\infty = (\infty,\infty)$, $-\infty = (-\infty,-\infty)$, $\sum$ is a double sum.

Let $(u_k)_{k\in\mathbb{Z}^2} \in \mathbb{R}_{\mathcal{F}}$. We denote the fuzzy double infinite series by $\sum\limits_{k=-\infty}^{\infty *} u_k$ and we say that it converges to $u \in \mathbb{R}_{\mathcal{F}}$ iff $\lim\limits_{m\to\infty}\lim\limits_{n\to\infty} D\left(\sum\limits_{k=m}^{n*} u_k, u\right) = 0$. We denote the last by $\sum\limits_{k=-\infty}^{\infty *} u_k = u$.

Let $(u_k)_{k\in\mathbb{Z}^2}, (v_k)_{k\in\mathbb{Z}^2}, u, v, \mathbb{R}_{\mathcal{F}}$ such that $\sum\limits_{k=-\infty}^{\infty *} u_k = u$, $\sum\limits_{k=-\infty}^{\infty *} v_k = v$. Then

$$\sum_{k=-\infty}^{\infty *} (u_k \oplus v_k) = u \oplus v = \sum_{k=-\infty}^{\infty *} u_k \oplus \sum_{k=-\infty}^{\infty *} v_k. \tag{14.65}$$

The last is true since

$$\lim_{m\to\infty}\lim_{n\to\infty} D\left(\sum_{k=m}^{n*} (u_k \oplus v_k), u \oplus v\right) =$$

$$\lim_{m\to\infty}\lim_{n\to\infty} D\left(\left(\sum_{k=m}^{n*} u_k\right) \oplus \left(\sum_{k=m}^{n*} v_k\right), u \oplus v\right)$$

$$\leq \lim_{m\to\infty}\lim_{n\to\infty}\left[D\left(\sum_{k=m}^{n*} u_k, u\right) + D\left(\sum_{k=m}^{n*} v_k, v\right)\right] = 0.$$

Let $\sum\limits_{k=-\infty}^{\infty *} u_k = u \in \mathbb{R}_{\mathcal{F}}$, then one has that

$$\sum_{k=-\infty}^{\infty} (u_k)_-^{(r)} = u_-^{(r)} = \left(\sum_{k=-\infty}^{\infty *} u_k \right)_-^{(r)}, \tag{14.66}$$

and

$$\sum_{k=-\infty}^{\infty} (u_k)_+^{(r)} = u_+^{(r)} = \left(\sum_{k=-\infty}^{\infty *} u_k \right)_+^{(r)}, \tag{14.67}$$

$\forall\, r \in [0, 1]$.

We prove the last claim:

We have that

$$0 = \lim_{m\to\infty} \lim_{n\to\infty} D\left(\sum_{k=m}^{n} u_k, u \right) =$$

$$\lim_{m\to\infty} \lim_{n\to\infty} \sup_{r\in[0,1]} \max \left\{ \left| \sum_{k=m}^{n} (u_k)_-^{(r)} - u_-^{(r)} \right|, \left| \sum_{k=m}^{n} (u_k)_+^{(r)} - u_+^{(r)} \right| \right\} \geq$$

$$\lim_{m\to\infty} \lim_{n\to\infty} \left\{ \left| \sum_{k=m}^{n} (u_k)_-^{(r)} - u_-^{(r)} \right|, \left| \sum_{k=m}^{n} (u_k)_+^{(r)} - u_+^{(r)} \right| \right\},$$

$\forall\, r \in [0, 1]$, proving the claim.

Also we need: let $(u_k)_{k\in\mathbb{Z}^2} \in \mathbb{R}_{\mathcal{F}}$ with $\sum_{k=-\infty}^{\infty *} u_k = u \in \mathbb{R}_{\mathcal{F}}$, then clearly one has for any $\lambda \in \mathbb{R}$ that $\sum_{k=-\infty}^{\infty *} \lambda u_k = \lambda u$.

Clearly also here $\sum_{k=-\infty}^{\infty *} u_k = u \in \mathbb{R}_{\mathcal{F}}$, here $(u_k)_{k\in\mathbb{Z}^2} \in \mathbb{R}_{\mathcal{F}}$, iff $\sum_{k=-\infty}^{\infty} (u_k)_-^{(r)} = (u)_-^{(r)}$ and $\sum_{k=-\infty}^{\infty} (u_k)_+^{(r)} = (u)_+^{(r)}$, uniformly in $r \in [0, 1]$, see also [21].

(2) By [21] we see: Let $(k \in \mathbb{Z}^2)$ $u_k := \left\{ \left((u_k)_-^{(r)}, (u_k)_+^{(r)} \right) | 0 \leq r \leq 1 \right\} \in \mathbb{R}_{\mathcal{F}}$ such that $\sum_{k=-\infty}^{\infty} (u_k)_-^{(r)} = (u)_-^{(r)}$ and $\sum_{k=-\infty}^{\infty} (u_k)_+^{(r)} = (u)_+^{(r)}$ converge uniformly in $r \in [0, 1]$, then $u := \left\{ \left((u)_-^{(r)}, (u)_+^{(r)} \right) | 0 \leq r \leq 1 \right\} \in \mathbb{R}_{\mathcal{F}}$ and $u = \sum_{k=-\infty}^{\infty *} u_k$.

I.e. we have

$$\sum_{k=-\infty}^{\infty *} \left\{ \left((u_k)_-^{(r)}, (u_k)_+^{(r)} \right) | 0 \leq r \leq 1 \right\} =$$

$$\left\{\left(\sum_{k=-\infty}^{\infty *}(u_k)_-^{(r)},\ \sum_{k=-\infty}^{\infty *}(u_k)_+^{(r)}\right)|0\leq r\leq 1\right\}. \tag{14.68}$$

All the content of Remark 14.31 goes through and is valid for $k \in \mathbb{Z}^N$, $N \geq 2$.

Proof of (14.63) and (14.64).

The proof of (14.64) is totally similar to the proof of (14.63). So we prove only (14.63).

Let $f \in C_b\left(\mathbb{R}^N, \mathbb{R}_{\mathcal{F}}\right)$, then $f_{\pm}^{(r)} \in C_B\left(\mathbb{R}^N\right)$, $\forall\, r \in [0,1]$. We have

$$(f(x))_-^{(0)} \leq (f(x))_-^{(r)} \leq (f(x))_-^{(1)} \leq (f(x))_+^{(1)}$$

$$\leq (f(x))_+^{(r)} \leq (f(x))_+^{(0)},\ \forall\, x \in \mathbb{R}^N. \tag{14.69}$$

We get that

$$\left|(f(x))_-^{(r)}\right| \leq \max\left\{\left|(f(x))_-^{(0)}\right|, \left|(f(x))_-^{(1)}\right|\right\} =$$

$$\frac{1}{2}\left\{\left|(f(x))_-^{(0)}\right| + \left|(f(x))_-^{(1)}\right| - \left|\left|(f(x))_-^{(0)}\right| - \left|(f(x))_-^{(1)}\right|\right|\right\}$$

$$=: A_-^{(0,1)}(x) \in C_B\left(\mathbb{R}^N\right). \tag{14.70}$$

Also it holds

$$\left|(f(x))_+^{(r)}\right| \leq \max\left\{\left|(f(x))_+^{(0)}\right|, \left|(f(x))_+^{(1)}\right|\right\} =$$

$$\frac{1}{2}\left\{\left|(f(x))_+^{(0)}\right| + \left|(f(x))_+^{(1)}\right| - \left|\left|(f(x))_+^{(0)}\right| - \left|(f(x))_+^{(1)}\right|\right|\right\}$$

$$=: A_+^{(0,1)}(x) \in C_B\left(\mathbb{R}^N\right). \tag{14.71}$$

I.e. we have obtained that

$$0 \leq \left|(f(x))_-^{(r)}\right| \leq A_-^{(0,1)}(x), \tag{14.72}$$

$$0 \leq \left|(f(x))_+^{(r)}\right| \leq A_+^{(0,1)}(x),\ \forall\, r \in [0,1], \forall x \in \mathbb{R}^N.$$

Hence by positivity of the operators $\overline{G}_n$ we get

$$0 \leq \left|\overline{G}_n\left(f_{\pm}^{(r)}, x\right)\right| \leq \overline{G}_n\left(\left|f_{\pm}^{(r)}\right|, x\right) \leq \overline{G}_n\left(A_{\pm}^{(0,1)}, x\right), \tag{14.73}$$

respectively in $\pm$, $\forall\, r \in [0,1]$, $\forall\, x \in \mathbb{R}^N$.

In detail one has

$$\left| \left(f\left(\frac{k}{n}\right)\right)_{\pm}^{(r)} \right| \Phi(nx-k) \leq A_{\pm}^{(0,1)}\left(\frac{k}{n}\right)\Phi(nx-k), \tag{14.74}$$

$\forall\, r \in [0,1]$, $k \in \mathbb{Z}^N$, $n \in \mathbb{N}$ fixed, $\forall\, x \in \mathbb{R}^N$ fixed, respectively in $\pm$.

We notice that

$$\sum_{k=-\infty}^{\infty} A_{\pm}^{(0,1)}\left(\frac{k}{n}\right)\Phi(nx-k) \leq \left\| A_{\pm}^{(0,1)} \right\|_{\infty} < \infty, \tag{14.75}$$

i.e. $\sum\limits_{k=-\infty}^{\infty} A_{\pm}^{(0,1)}\left(\frac{k}{n}\right)\Phi(nx-k)$ converges, respectively in $\pm$.

Thus by Weierstrass M-test we obtain that $\overline{G}_n\left(\left|f_{\pm}^{(r)}\right|, x\right)$ as series converges uniformly in $r \in [0,1]$, respectively in $\pm$, $\forall\, n \in \mathbb{N}$, $\forall\, x \in \mathbb{R}^N$.

And by Cauchy criterion for series of uniform convergence we get that $\overline{G}_n\left(f_{\pm}^{(r)}, x\right)$ as series converges uniformly in $r \in [0,1]$, $\forall\, n \in \mathbb{N}$, $\forall\, x \in \mathbb{R}^N$, respectively in $\pm$.

Here for $k \in \mathbb{Z}^N$, $f\left(\frac{k}{n}\right) = \left\{\left(f\left(\frac{k}{n}\right)\right)_{-}^{(r)}, \left(f\left(\frac{k}{n}\right)\right)_{+}^{(r)} | 0 \leq r \leq 1\right\} \in \mathbb{R}_{\mathcal{F}}$, and also

$$u_k := \left\{ \left(f\left(\frac{k}{n}\right)\right)_{-}^{(r)} \Phi(nx-k), \left(f\left(\frac{k}{n}\right)\right)_{+}^{(r)} \Phi(nx-k) \,|\, 0 \leq r \leq 1 \right\} \tag{14.76}$$

$$= f\left(\frac{k}{n}\right) \odot \Phi(nx-k) \in \mathbb{R}_{\mathcal{F}}.$$

That is $(u_k)_{\pm}^{(r)} = \left(f\left(\frac{k}{n}\right)\right)_{\pm}^{(r)} \Phi(nx-k)$, respectively in $\pm$.

But we proved that

$$\sum_{k=-\infty}^{\infty} (u_k)_{\pm}^{(r)} = \overline{G}_n\left(f_{\pm}^{(r)}, x\right) =: (u)_{\pm}^{(r)},$$

converges uniformly in $r \in [0,1]$, respectively in $\pm$.

Then by Remark 14.31 (2) we get

$$u := \left\{ \left((u)_{-}^{(r)}, (u)_{+}^{(r)} \right) | 0 \leq r \leq 1 \right\}$$

$$= \left\{ \left(\overline{G}_n\left(f_{-}^{(r)}, x\right), \overline{G}_n\left(f_{+}^{(r)}, x\right)\right) | 0 \leq r \leq 1 \right\} \in \mathbb{R}_{\mathcal{F}} \tag{14.77}$$

and

$$u=\sum_{k=-\infty}^{\infty *}u_k=$$

$$\sum_{k=-\infty}^{\infty *}\left\{\left(f\left(\frac{k}{n}\right)\right)_{-}^{(r)}\Phi(nx-k),\left(f\left(\frac{k}{n}\right)\right)_{+}^{(r)}\Phi(nx-k)\,|0\leq r\leq 1\right\} \quad (14.78)$$

$$=\sum_{k=-\infty}^{\infty *}\left\{\left(\left(f\left(\frac{k}{n}\right)\right)_{-}^{(r)},\left(f\left(\frac{k}{n}\right)\right)_{+}^{(r)}\right)|0\leq r\leq 1\right\}\odot\Phi(nx-k) \quad (14.79)$$

$$=\sum_{k=-\infty}^{\infty *}f\left(\frac{k}{n}\right)\odot\Phi(nx-k)=\overline{G}_n^{\mathcal{F}}(f,x).$$

So we have proved (14.63). ∎

Conclusion 14.32 *Call $L_n^{\mathcal{F}}$ any of the operators $G_n^{\mathcal{F}}$, $F_n^{\mathcal{F}}$, $\overline{G}_n^{\mathcal{F}}$, $\overline{F}_n^{\mathcal{F}}$.*

Let also L_n be any of real operators G_n, F_n, $\overline{G}_n$, $\overline{F}_n$.
We have proved that

$$\left(L_n^{\mathcal{F}}(f,x)\right)_{\pm}^{(r)}=L_n\left(f_{\pm}^{(r)},x\right), \quad (14.80)$$

respectively in $\pm$ and operator matching couples, $\forall\, r\in[0,1]$, $\forall\, x\in\left(\prod_{i=1}^{N}([a_i,b_i])\right)$ or $\forall\, x\in\mathbb{R}^N$, $N\in\mathbb{N}$.

We present our first main result

Theorem 14.33 *Let $f\in C_{\mathcal{F}}\left(\prod_{i=1}^{N}[a_i,b_i]\right)$ or $f\in C_b\left(\mathbb{R}^N,\mathbb{R}_{\mathcal{F}}\right)$; $n,N\in\mathbb{N}$, $0<\beta<1$, $x\in\prod_{i=1}^{N}[a_i,b_i]$ or $x\in\mathbb{R}^N$. Then*

(i)

$$D(L_n(f,x),f(x))\leq c_N\left\{\omega_1^{(\mathcal{F})}\left(f,\frac{1}{n^\beta}\right)+\mu D^*(f,\widetilde{o})\,e^{-\gamma n(1-\beta)}\right\}=:\rho_n^{\mathcal{F}}. \quad (14.81)$$

(ii)

$$D^*(L_n(f),f)\leq\rho_n^{\mathcal{F}}, \quad (14.82)$$

when c_N, μ, γ as in Notation 14.29.

Proof We see that

$$D\left(L_n\left(f,x\right),f\left(x\right)\right)=$$

$$\sup_{r\in[0,1]}\max\left\{\left|\left(L_n\left(f,x\right)\right)_-^{(r)}-f_-^{(r)}\left(x\right)\right|,\left|\left(L_n\left(f,x\right)\right)_+^{(r)}-f_+^{(r)}\left(x\right)\right|\right\}$$

$$\overset{(14.80)}{=}\sup_{r\in[0,1]}\max\left\{\left|\left(L_n\left(f_-^{(r)},x\right)\right)-f_-^{(r)}\left(x\right)\right|,\left|\left(L_n\left(f_+^{(r)},x\right)\right)-f_+^{(r)}\left(x\right)\right|\right\},\tag{14.83}$$

$\forall\, x\in\prod_{i=1}^{N}\left[a_i,b_i\right]$ or $\forall\, x\in\mathbb{R}^N$.

The assumption implies $f_\pm^{(r)}\in C\left(\prod_{i=1}^{N}\left[a_i,b_i\right]\right)$ or $f_\pm^{(r)}\in C_B\left(\mathbb{R}^N\right)$, respectively in $\pm$, $\forall\, r\in[0,1]$.

Hence by (14.53) we get

$$\left|\left(L_n\left(f_\pm^{(r)},x\right)\right)-f_\pm^{(r)}\left(x\right)\right|\leq c_N\left\{\omega_1\left(f_\pm^{(r)},\frac{1}{n^\beta}\right)+\mu\left\|f_\pm^{(r)}\right\|_\infty e^{-\gamma n^{(1-\beta)}}\right\}$$

(by (14.8) and (14.3))

$$\leq c_N\left\{\omega_1^{(\mathcal{F})}\left(f,\frac{1}{n^\beta}\right)+\mu D^*\left(f,\widetilde{o}\right)e^{-\gamma n^{(1-\beta)}}\right\},\tag{14.84}$$

respectively in $\pm$.

Now combining (14.83) and (14.84) we prove (14.81). ■

Remark 14.34 Let $f\in C_{\mathcal{F}}\left(\prod_{i=1}^{N}\left[a_i,b_i\right]\right)$ or $f\in\left(C_b\left(\mathbb{R}^N,\mathbb{R}_{\mathcal{F}}\right)\cap C_{\mathcal{F}}^U\left(\mathbb{R}^N\right)\right)$. Then $\lim_{n\to\infty}D^*\left(L_n\left(f\right),f\right)=0$, quantitatively with rates from (14.82).

Notation 14.35 *Let* $f\in C_{\mathcal{F}}^m\left(\prod_{i=1}^{N}\left[a_i,b_i\right]\right)$, $m,N\in\mathbb{N}$. *Here* f_α *denotes a fuzzy partial derivative as in Notation 14.26. We denote*

$$\omega_{1,m}^{(\mathcal{F})\max}\left(f_\alpha,h\right):=\max_{\alpha:|\alpha|=m}\omega_1^{(\mathcal{F})}\left(f_\alpha,h\right).\tag{14.85}$$

Call also

$$D_m^{*\max}\left(f_\alpha,\widetilde{o}\right):=\max_{\alpha:|\alpha|=m}\left\{D^*\left(f_\alpha,\widetilde{o}\right)\right\}.\tag{14.86}$$

We present

Theorem 14.36 *Let* $f \in C_{\mathcal{F}}^{m}\left(\prod_{i=1}^{N}[a_i, b_i]\right)$, $0 < \beta < 1$, $n, m, N \in \mathbb{N}$, $x \in \left(\prod_{i=1}^{N}[a_i, b_i]\right)$. *Then*

(1)

$$D\left(G_n^{\mathcal{F}}(f, x), f(x)\right) \leq (5.250312578)^N \cdot \tag{14.87}$$

$$\left\{\sum_{j=1}^{m}\left(\sum_{|\alpha|=j}\left(\frac{D\left(f_\alpha(x), \widetilde{o}\right)}{\prod_{i=1}^{N}\alpha_i!}\right)\left[\frac{1}{n^{\beta j}} + \left(\prod_{i=1}^{N}(b_i - a_i)^{\alpha_i}\right)\cdot(3.1992)\, e^{-n^{(1-\beta)}}\right]\right)\right.$$

$$+\frac{N^m}{m!n^{m\beta}}\omega_{1,m}^{(\mathcal{F})\max}\left(f_\alpha, \frac{1}{n^\beta}\right) +$$

$$\left.\left(\frac{(6.3984)\,\|b-a\|_\infty^m\, D_m^{*\max}\left(f_\alpha, \widetilde{o}\right)N^m}{m!}\right)e^{-n^{(1-\beta)}}\right\},$$

(2)

$$D^*\left(G_n^{\mathcal{F}}(f), f\right) \leq (5.250312578)^N \cdot \tag{14.88}$$

$$\left\{\sum_{j=1}^{m}\left(\sum_{|\alpha|=j}\left(\frac{D^*\left(f_\alpha, \widetilde{o}\right)}{\prod_{i=1}^{N}\alpha_i!}\right)\left[\frac{1}{n^{\beta j}} + \left(\prod_{i=1}^{N}(b_i - a_i)^{\alpha_i}\right)\cdot(3.1992)\, e^{-n^{(1-\beta)}}\right]\right)\right.$$

$$+\frac{N^m}{m!n^{m\beta}}\omega_{1,m}^{(\mathcal{F})\max}\left(f_\alpha, \frac{1}{n^\beta}\right) +$$

$$\left.\left(\frac{(6.3984)\,\|b-a\|_\infty^m\, D_m^{*\max}\left(f_\alpha, \widetilde{o}\right)N^m}{m!}\right)e^{-n^{(1-\beta)}}\right\},$$

(3) assume that $f_\alpha(x_0) = \widetilde{o}$, *for all* $\alpha : |\alpha| = 1, \ldots, m$; $x_0 \in \left(\prod_{i=1}^N [a_i, b_i]\right)$, *then*

$$D\left(G_n^{\mathcal{F}}(f, x_0), f(x_0)\right) \leq (5.250312578)^N \cdot \tag{14.89}$$

$$\left\{\frac{N^m}{m!n^{m\beta}}\omega_{1,m}^{(\mathcal{F})\max}\left(f_\alpha, \frac{1}{n^\beta}\right) + \left(\frac{(6.3984)\|b-a\|_\infty^m D_m^{*\max}(f_\alpha, \widetilde{o}) N^m}{m!}\right)e^{-n^{(1-\beta)}}\right\},$$

notice in the last the extremely high rate of convergence at $n^{-\beta(m+1)}$.

Proof We observe that

$$\left|G_n\left(f_\pm^{(r)}, x\right) - f_\pm^{(r)}(x)\right| \overset{(14.43)}{\leq} (5.250312578)^N \cdot$$

$$\left\{\sum_{j=1}^m\left(\sum_{|\alpha|=j}\left(\frac{\left|\left(f_\pm^{(r)}\right)_\alpha(x)\right|}{\prod_{i=1}^N \alpha_i!}\right)\left[\frac{1}{n^{\beta j}} + \left(\prod_{i=1}^N (b_i - a_i)^{\alpha_i}\right)\cdot (3.1992)\, e^{-n^{(1-\beta)}}\right]\right)\right.$$

$$+\frac{N^m}{m!n^{m\beta}}\omega_{1,m}^{\max}\left(\left(f_\pm^{(r)}\right)_\alpha, \frac{1}{n^\beta}\right) + \tag{14.90}$$

$$\left.\left(\frac{(6.3984)\|b-a\|_\infty^m \left\|\left(f_\pm^{(r)}\right)_\alpha\right\|_{\infty,m}^{\max} N^m}{m!}\right)e^{-n^{(1-\beta)}}\right\} \overset{(14.13)}{=} (5.250312578)^N \cdot$$

$$\left\{\sum_{j=1}^m\left(\sum_{|\alpha|=j}\left(\frac{\left|(f_\alpha)_\pm^{(r)}(x)\right|}{\prod_{i=1}^N \alpha_i!}\right)\left[\frac{1}{n^{\beta j}} + \left(\prod_{i=1}^N (b_i - a_i)^{\alpha_i}\right)\cdot (3.1992)\, e^{-n^{(1-\beta)}}\right]\right)\right.$$

$$+\frac{N^m}{m!n^{m\beta}}\omega_{1,m}^{\max}\left((f_\alpha)_\pm^{(r)}, \frac{1}{n^\beta}\right) + \tag{14.91}$$

$$\left.\left(\frac{(6.3984)\,\|b-a\|_{\infty}^{m}\left\|(f_{\alpha})_{\pm}^{(r)}\right\|_{\infty,m}^{\max}N^{m}}{m!}\right)e^{-n^{(1-\beta)}}\right\}\overset{((14.8),(14.3))}{\leq}(5.250312578)^{N}\cdot$$

$$\left\{\sum_{j=1}^{m}\left(\sum_{|\alpha|=j}\left(\frac{D\left(f_{\alpha}\left(x\right),\widetilde{o}\right)}{\prod_{i=1}^{N}\alpha_{i}!}\right)\left[\frac{1}{n^{\beta j}}+\left(\prod_{i=1}^{N}(b_{i}-a_{i})^{\alpha_{i}}\right)\cdot(3.1992)\,e^{-n^{(1-\beta)}}\right]\right)\right.$$

$$+\frac{N^{m}}{m!n^{m\beta}}\omega_{1,m}^{(\mathcal{F})\max}\left(f_{\alpha},\frac{1}{n^{\beta}}\right)+ \tag{14.92}$$

$$\left.\left(\frac{(6.3984)\,\|b-a\|_{\infty}^{m}\,D_{m}^{*\max}\left(f_{\alpha},\widetilde{o}\right)N^{m}}{m!}\right)e^{-n^{(1-\beta)}}\right\},$$

We have proved that

$$\left|G_{n}\left(f_{\pm}^{(r)},x\right)-f_{\pm}^{(r)}\left(x\right)\right|\leq(5.250312578)^{N}\cdot \tag{14.93}$$

$$\left\{\sum_{j=1}^{m}\left(\sum_{|\alpha|=j}\left(\frac{D\left(f_{\alpha}\left(x\right),\widetilde{o}\right)}{\prod_{i=1}^{N}\alpha_{i}!}\right)\left[\frac{1}{n^{\beta j}}+\left(\prod_{i=1}^{N}(b_{i}-a_{i})^{\alpha_{i}}\right)\cdot(3.1992)\,e^{-n^{(1-\beta)}}\right]\right)\right.$$

$$+\frac{N^{m}}{m!n^{m\beta}}\omega_{1,m}^{(\mathcal{F})\max}\left(f_{\alpha},\frac{1}{n^{\beta}}\right)+$$

$$\left.\left(\frac{(6.3984)\,\|b-a\|_{\infty}^{m}\,D_{m}^{*\max}\left(f_{\alpha},\widetilde{o}\right)N^{m}}{m!}\right)e^{-n^{(1-\beta)}}\right\},$$

respectively in $\pm$.

By (14.61) we get

$$D\left(G_n^{\mathcal{F}}(f,x), f(x)\right) =$$

$$\sup_{r\in[0,1]} \max\left\{\left|G_n\left(f_-^{(r)},x\right) - f_-^{(r)}(x)\right|, \left|G_n\left(f_+^{(r)},x\right) - f_+^{(r)}(x)\right|\right\}. \tag{14.94}$$

Combining (14.94) and (14.93) we have established claim. ∎

Using (14.47), (14.62), (14.13), (14.8), (14.3) and acting similarly as in the proof of Theorem 14.36 we derive the last result of this chapter, coming next.

Theorem 14.37 *All assumptions and terminology as in Theorem 14.36 and Notation 14.35. Then*

(1)

$$D\left(F_n^{\mathcal{F}}(f,x), f(x)\right) \leq (4.1488766)^N \cdot \tag{14.95}$$

$$\left\{\sum_{j=1}^{m}\left(\sum_{|\alpha|=j}\left(\frac{D(f_\alpha(x),\widetilde{o})}{\prod_{i=1}^N \alpha_i!}\right)\left[\frac{1}{n^{\beta j}} + \left(\prod_{i=1}^N (b_i-a_i)^{\alpha_i}\right)\cdot e^4 e^{-2n^{(1-\beta)}}\right]\right) +\right.$$

$$\left.\frac{N^m}{m!n^{m\beta}}\omega_{1,m}^{(\mathcal{F})\max}\left(f_\alpha,\frac{1}{n^\beta}\right) + \left(\frac{2e^4\|b-a\|_\infty^m D_m^{*\max}(f_\alpha,\widetilde{o})N^m}{m!}\right)e^{-2n^{(1-\beta)}}\right\},$$

(2)

$$D^*\left(F_n^{\mathcal{F}}(f), f\right) \leq (4.1488766)^N \cdot \tag{14.96}$$

$$\left\{\sum_{j=1}^{m}\left(\sum_{|\alpha|=j}\left(\frac{D^*(f_\alpha,\widetilde{o})}{\prod_{i=1}^N \alpha_i!}\right)\left[\frac{1}{n^{\beta j}} + \left(\prod_{i=1}^N (b_i-a_i)^{\alpha_i}\right)\cdot e^4 e^{-2n^{(1-\beta)}}\right]\right) +\right.$$

$$\left.\frac{N^m}{m!n^{m\beta}}\omega_{1,m}^{(\mathcal{F})\max}\left(f_\alpha,\frac{1}{n^\beta}\right) + \left(\frac{2e^4\|b-a\|_\infty^m D_m^{*\max}(f_\alpha,\widetilde{o})N^m}{m!}\right)e^{-2n^{(1-\beta)}}\right\},$$

(3) assume that $f_\alpha(x_0) = \widetilde{o}$, for all $\alpha : |\alpha| = 1, \ldots, m$; $x_0 \in \left(\prod_{i=1}^N [a_i,b_i]\right)$. Then

$$D\left(F_n^{\mathcal{F}}(f,x_0), f(x_0)\right) \leq (4.1488766)^N \cdot \tag{14.97}$$

$$\left\{\frac{N^m}{m!n^{m\beta}}\omega_{1,m}^{(\mathcal{F})\max}\left(f_\alpha,\frac{1}{n^\beta}\right) + \left(\frac{2e^4\|b-a\|_\infty^m D_m^{*\max}(f_\alpha,\widetilde{o})N^m}{m!}\right)e^{-2n^{(1-\beta)}}\right\},$$

notice in the last the extremely high rate of convergence at $n^{-\beta(m+1)}$.

Conclusion 14.38 *In Theorems 14.36, 14.37 we studied quantitatively with rates the high speed approximation of*

$$D\left(G_n^{\mathcal{F}}(f,x), f(x)\right) \to 0, \tag{14.98}$$

$$D\left(F_n^{\mathcal{F}}(f,x), f(x)\right) \to 0,$$

as $n \to \infty$.

Also we proved with rates that

$$D^*\left(G_n^{\mathcal{F}}(f), f\right) \to 0, \tag{14.99}$$

$$D^*\left(F_n^{\mathcal{F}}(f), f\right) \to 0,$$

as $n \to \infty$, involving smoothness of f.

References

1. G.A. Anastassiou, Rate of convergence of some neural network operators to the unit-univariate case. J. Math. Anal. Appl. **212**, 237–262 (1997)
2. G.A. Anastassiou, Rate of convergence of some multivariate neural network operators to the unit. Comput. Math. **40**, 1–19 (2000)
3. G.A. Anastassiou, *Quantitative Approximation* (Chapmann and Hall/CRC, Boca Raton, New York, 2001)
4. G.A. Anastassiou, Higher order fuzzy approximation by fuzzy wavelet type and neural network operators. Comput. Math. **48**, 1387–1401 (2004)
5. G.A. Anastassiou, Fuzzy approximation by fuzzy convolution type operators. Comput. Math. **48**, 1369–1386 (2004)
6. G.A. Anastassiou, Higher order fuzzy Korovkin theory via inequalities. Commun. Appl. Anal. **10**(2), 359–392 (2006)
7. G.A. Anastassiou, Fuzzy Korovkin theorems and inequalities. J. Fuzzy Math. **15**(1), 169–205 (2007)
8. G.A. Anastassiou, *Inteligent Systems: Approximation by Artificial Neural Networks* (Springer, Heidelberg, 2011)
9. G.A. Anastassiou, Univariate hyperbolic tangent neural network approximation. Math. Comput. Model. **53**, 1111–1132 (2011)
10. G.A. Anastassiou, Multivariate hyperbolic tangent neural network approximation. Comput. Math. **61**, 809–821 (2011)
11. G.A. Anastassiou, Multivariate sigmoidal neural network approximation. Neural Netw. **24**, 378–386 (2011)
12. G.A. Anastassiou, Univariate sigmoidal neural network approximation. J. Comput. Anal. Appl. **14**(4), 659–690 (2012)
13. G.A. Anastassiou, Higher order multivariate fuzzy approximation by multivariate fuzzy wavelet type and neural network operators. J. Fuzzy Math. **19**(3), 601–618 (2011)
14. G.A. Anastassiou, Rate of convergence of some multivariate neural network operators to the unit, revisited. J. Comput. Anal. Appl. **15**(7), 1300–1309 (2013)

15. G.A. Anastassiou, Approximation by Neural Network Iterates, in *Advances in Applied Mathematics and Approximation Theory: Contributions from AMAT 2012*, ed. by G. Anastassiou, O. Duman (Springer, New York, 2013), pp. 1–20
16. G.A. Anastassiou, High degree multivariate fuzzy approximation by quasi-interpolation neural network operators. Discontinuity Nonlinearity Complex. **2**(2), 125–146 (2013)
17. Z. Chen, F. Cao, The approximation operators with sigmoidal functions. Comput. Math. Appl. **58**, 758–765 (2009)
18. R. Goetschel Jr, W. Voxman, Elementary fuzzy calculus. Fuzzy Sets Syst. **18**, 31–43 (1986)
19. S. Haykin, *Neural Networks: a Comprehensive Foundation*, 2nd edn. (Prentice Hall, New York, 1998)
20. O. Kaleva, Fuzzy differential equations. Fuzzy Sets Syst. **24**, 301–317 (1987)
21. Y.K. Kim, B.M. Ghil, Integrals of fuzzy-number-valued functions. Fuzzy Sets Syst. **86**, 213–222 (1997)
22. T.M. Mitchell, *Machine Learning* (WCB-McGraw-Hill, New York, 1997)
23. W. McCulloch, W. Pitts, A logical calculus of the ideas immanent in nervous activity. Bull. Math. Biophys. **7**, 115–133 (1943)
24. C. Wu, Z. Gong, On Henstock integral of fuzzy-number-valued functions (I). Fuzzy Sets Syst. **120**(3), 523–532 (2001)
25. C. Wu, M. Ma, On embedding problem of fuzzy number space: part 1. Fuzzy Sets Syst. **44**, 33–38 (1991)

Chapter 15
Multivariate Fuzzy-Random Quasi-interpolation Neural Networks Approximation

In this chapter we study the rate of multivariate pointwise and uniform convergence in the q-mean to the Fuzzy-Random unit operator of multivariate Fuzzy-Random Quasi-Interpolation neural network operators. These multivariate Fuzzy-Random operators arise in a natural and common way among multivariate Fuzzy-Random neural networks. These rates are given through multivariate Probabilistic-Jackson type inequalities involving the multivariate Fuzzy-Random modulus of continuity of the engaged multivariate Fuzzy-Random function. The plain stochastic extreme analog of this theory is also presented. It follows [15].

15.1 Fuzzy-Random Functions Theory Background

We begin with

Definition 15.1 (*see* [23]) Let $\mu : \mathbb{R} \to [0, 1]$ with the following properties:

(i) is normal, i.e., $\exists\, x_0 \in \mathbb{R} : \mu(x_0) = 1$.
(ii) $\mu(\lambda x + (1-\lambda) y) \geq \min\{\mu(x), \mu(y)\}$, $\forall\, x, y \in \mathbb{R}$, $\forall\, \lambda \in [0, 1]$ (μ is called a convex fuzzy subset).
(iii) μ is upper semicontinuous on $\mathbb{R}$, i.e., $\forall\, x_0 \in \mathbb{R}$ and $\forall\, \varepsilon > 0$, $\exists$ neighborhood $V(x_0) : \mu(x) \leq \mu(x_0) + \varepsilon$, $\forall\, x \in V(x_0)$.
(iv) the set $\overline{\operatorname{supp}(\mu)}$ is compact in $\mathbb{R}$ (where $\operatorname{supp}(\mu) := \{x \in \mathbb{R};\, \mu(x) > 0\}$).

We call μ a fuzzy real number. Denote the set of all μ with $\mathbb{R}_{\mathcal{F}}$.

E.g., $\chi_{\{x_0\}} \in \mathbb{R}_{\mathcal{F}}$, for any $x_0 \in \mathbb{R}$, where $\chi_{\{x_0\}}$ is the characteristic function at x_0.

For $0 < r \leq 1$ and $\mu \in \mathbb{R}_{\mathcal{F}}$ define $[\mu]^r := \{x \in \mathbb{R} : \mu(x) \geq r\}$ and $[\mu]^0 := \overline{\{x \in \mathbb{R} : \mu(x) > 0\}}$.

Then it is well known that for each $r \in [0, 1]$, $[\mu]^r$ is a closed and bounded interval of $\mathbb{R}$. For $u, v \in \mathbb{R}_{\mathcal{F}}$ and $\lambda \in \mathbb{R}$, we define uniquely the sum $u \oplus v$ and the product $\lambda \odot u$ by

G.A. Anastassiou, *Intelligent Systems II: Complete Approximation by Neural Network Operators*, Studies in Computational Intelligence 608,
DOI 10.1007/978-3-319-20505-2_15

$$[u \oplus v]^r = [u]^r + [v]^r, \quad [\lambda \odot u]^r = \lambda [u]^r, \quad \forall\, r \in [0, 1],$$

where $[u]^r + [v]^r$ means the usual addition of two intervals (as subsets of $\mathbb{R}$) and $\lambda [u]^r$ means the usual product between a scalar and a subset of $\mathbb{R}$ (see, e.g., [23]). Notice $1 \odot u = u$ and it holds $u \oplus v = v \oplus u$, $\lambda \odot u = u \odot \lambda$. If $0 \le r_1 \le r_2 \le 1$ then $[u]^{r_2} \subseteq [u]^{r_1}$. Actually $[u]^r = \left[u_-^{(r)}, u_+^{(r)}\right]$, where $u_-^{(r)} < u_+^{(r)}$, $u_-^{(r)}, u_+^{(r)} \in \mathbb{R}$, $\forall\, r \in [0, 1]$.

Define

$$D : \mathbb{R}_{\mathcal{F}} \times \mathbb{R}_{\mathcal{F}} \to \mathbb{R}_+ \cup \{0\}$$

by

$$D(u, v) := \sup_{r \in [0,1]} \max \left\{ \left| u_-^{(r)} - v_-^{(r)} \right|, \left| u_+^{(r)} - v_+^{(r)} \right| \right\},$$

where $[v]^r = \left[v_-^{(r)}, v_+^{(r)}\right]$; $u, v \in \mathbb{R}_{\mathcal{F}}$. We have that D is a metric on $\mathbb{R}_{\mathcal{F}}$. Then $(\mathbb{R}_{\mathcal{F}}, D)$ is a complete metric space, see [23], with the properties

$$\begin{aligned} D(u \oplus w, v \oplus w) &= D(u, v), \quad \forall\, u, v, w \in \mathbb{R}_{\mathcal{F}}, \\ D(k \odot u, k \odot v) &= |k|\, D(u, v), \quad \forall\, u, v \in \mathbb{R}_{\mathcal{F}}, \forall\, k \in \mathbb{R}, \\ D(u \oplus v, w \oplus e) &\le D(u, w) + D(v, e), \quad \forall\, u, v, w, e \in \mathbb{R}_{\mathcal{F}}. \end{aligned} \tag{15.1}$$

Let (M, d) metric space and $f, g : M \to \mathbb{R}_{\mathcal{F}}$ be fuzzy real number valued functions. The distance between f, g is defined by

$$D^*(f, g) := \sup_{x \in M} D(f(x), g(x)).$$

On $\mathbb{R}_{\mathcal{F}}$ we define a partial order by "$\le$": $u, v \in \mathbb{R}_{\mathcal{F}}$, $u \le v$ iff $u_-^{(r)} \le v_-^{(r)}$ and $u_+^{(r)} \le v_+^{(r)}$, $\forall\, r \in [0, 1]$.

$\sum^{*}$ denotes the fuzzy summation, $\tilde{o} := \chi_{\{0\}} \in \mathbb{R}_{\mathcal{F}}$ the neutral element with respect to $\oplus$. For more see also [24, 25].

We need

Definition 15.2 (*see also* [19], *Definition* 13.16, *p. 654*) Let $(X, \mathcal{B}, P)$ be a probability space. A fuzzy-random variable is a $\mathcal{B}$-measurable mapping $g : X \to \mathbb{R}_{\mathcal{F}}$ (i.e., for any open set $U \subseteq \mathbb{R}_{\mathcal{F}}$, in the topology of $\mathbb{R}_{\mathcal{F}}$ generated by the metric D, we have

$$g^{-1}(U) = \{s \in X; g(s) \in U\} \in \mathcal{B}). \tag{15.2}$$

The set of all fuzzy-random variables is denoted by $\mathcal{L}_{\mathcal{F}}(X, \mathcal{B}, P)$. Let $g_n, g \in \mathcal{L}_{\mathcal{F}}(X, \mathcal{B}, P)$, $n \in \mathbb{N}$ and $0 < q < +\infty$. We say $g_n(s) \underset{n \to +\infty}{\overset{\text{“}q\text{-mean”}}{\to}} g(s)$ if

$$\lim_{n\to+\infty} \int_X D\left(g_n(s), g(s)\right)^q P(ds) = 0. \tag{15.3}$$

Remark 15.3 (see [19], p. 654) If $f, g \in \mathcal{L}_{\mathcal{F}}(X, \mathcal{B}, P)$, let us denote $F : X \to \mathbb{R}_+ \cup \{0\}$ by $F(s) = D(f(s), g(s))$, $s \in X$. Here, F is $\mathcal{B}$-measurable, because $F = G \circ H$, where $G(u, v) = D(u, v)$ is continuous on $\mathbb{R}_{\mathcal{F}} \times \mathbb{R}_{\mathcal{F}}$, and $H : X \to \mathbb{R}_{\mathcal{F}} \times \mathbb{R}_{\mathcal{F}}$, $H(s) = (f(s), g(s))$, $s \in X$, is $\mathcal{B}$-measurable. This shows that the above convergence in q -mean makes sense.

Definition 15.4 (*see* [19], *p. 654, Definition* 13.17) Let $(T, \mathcal{T})$ be a topological space. A mapping $f : T \to \mathcal{L}_{\mathcal{F}}(X, \mathcal{B}, P)$ will be called fuzzy-random function (or fuzzy-stochastic process) on T. We denote $f(t)(s) = f(t, s)$, $t \in T$, $s \in X$.

Remark 15.5 (see [19], p. 655) Any usual fuzzy real function $f : T \to \mathbb{R}_{\mathcal{F}}$ can be identified with the degenerate fuzzy-random function $f(t, s) = f(t)$, $\forall\, t \in T$, $s \in X$.

Remark 15.6 (see [19], p. 655) Fuzzy-random functions that coincide with probability one for each $t \in T$ will be consider equivalent.

Remark 15.7 (see [19], p. 655) Let $f, g : T \to \mathcal{L}_{\mathcal{F}}(X, \mathcal{B}, P)$. Then $f \oplus g$ and $k \odot f$ are defined pointwise, i.e.,

$$(f \oplus g)(t, s) = f(t, s) \oplus g(t, s),$$
$$(k \odot f)(t, s) = k \odot f(t, s)\ ,\ t \in T, s \in X.$$

Definition 15.8 (*see also Definition* 13.18, *pp. 655–656,* [19]) For a fuzzy-random function $f : W \subseteq \mathbb{R}^N \to \mathcal{L}_{\mathcal{F}}(X, \mathcal{B}, P)$, $N \in \mathbb{N}$, we define the (first) fuzzy-random modulus of continuity

$$\Omega_1^{(\mathcal{F})}(f, \delta)_{L^q} =$$

$$\sup\left\{\left(\int_X D^q\left(f(x, s), f(y, s)\right) P(ds)\right)^{\frac{1}{q}} : x, y \in W,\ \ \|x - y\|_\infty \le \delta\right\},$$

$0 < \delta$, $1 \le q < \infty$.

Definition 15.9 Here $1 \le q < +\infty$. Let $f : W \subseteq \mathbb{R}^N \to \mathcal{L}_{\mathcal{F}}(X, \mathcal{B}, P)$, $N \in \mathbb{N}$, be a fuzzy random function. We call f a (q-mean) uniformly continuous fuzzy random function over W, iff $\forall\, \varepsilon > 0\ \exists\, \delta > 0$:whenever $\|x - y\|_\infty \le \delta$, $x, y \in W$, implies that

$$\int_X \left(D\left(f(x, s), f(y, s)\right)\right)^q P(ds) \le \varepsilon.$$

We denote it as $f \in C_{FR}^{U_q}(W)$.

Proposition 15.10 *Let $f \in C_{FR}^{U_q}(W)$, where $W \subseteq \mathbb{R}^N$ is convex.*
Then $\Omega_1^{(\mathcal{F})}(f, \delta)_{L^q} < \infty$, any $\delta > 0$.

Proof Let $\varepsilon_0 > 0$ be arbitrary but fixed. Then there exists $\delta_0 > 0 : \|x - y\|_\infty \leq \delta_0$ implies

$$\int_X (D(f(x, s), f(y, s)))^q P(ds) \leq \varepsilon_0 < \infty.$$

That is $\Omega_1^{(\mathcal{F})}(f, \delta_0)_{L^q} \leq \varepsilon_0^{\frac{1}{q}} < \infty$. Let now $\delta > 0$ arbitrary, $x, y \in W$ such that $\|x - y\|_\infty \leq \delta$. Choose $n \in \mathbb{N} : n\delta_0 \geq \delta$ and set $x_i := x + \frac{i}{n}(y - x)$, $0 \leq i \leq n$. Then

$$D(f(x, s), f(y, s)) \leq D(f(x, s), f(x_1, s)) + D(f(x_1, s), f(x_2, s))$$

$$+ \cdots + D(f(x_{n-1}, s), f(y, s)).$$

Consequently

$$\left(\int_X (D(f(x, s), f(y, s)))^q P(ds)\right)^{\frac{1}{q}} \leq \left(\int_X (D(f(x, s), f(x_1, s)))^q P(ds)\right)^{\frac{1}{q}}$$

$$+ \cdots + \left(\int_X (D(f(x_{n-1}, s), f(y, s)))^q P(ds)\right)^{\frac{1}{q}} \leq n\Omega_1^{(\mathcal{F})}(f, \delta_0)_{L^q} \leq n\varepsilon_0^{\frac{1}{q}} < \infty,$$

since $\|x_i - x_{i+1}\|_\infty = \frac{1}{n}\|x - y\|_\infty \leq \frac{1}{n}\delta \leq \delta_0$, $0 \leq i \leq n$.

Therefore $\Omega_1^{(\mathcal{F})}(f, \delta)_{L^q} \leq n\varepsilon_0^{\frac{1}{q}} < \infty$. ∎

Proposition 15.11 *Let $f, g : W \subseteq \mathbb{R}^N \to \mathcal{L}_{\mathcal{F}}(X, \mathcal{B}, P)$, $N \in \mathbb{N}$, be fuzzy random functions. It holds*

(i) $\Omega_1^{(\mathcal{F})}(f, \delta)_{L^q}$ is nonnegative and nondecreasing in $\delta > 0$.
(ii) $\lim_{\delta \downarrow 0} \Omega_1^{(\mathcal{F})}(f, \delta)_{L^q} = \Omega_1^{(\mathcal{F})}(f, 0)_{L^q} = 0$, iff $f \in C_{FR}^{U_q}(W)$.

Proof (i) is obvious.
(ii) $\Omega_1(f, 0)_{L^q} = 0$.

($\Rightarrow$) Let $\lim_{\delta \downarrow 0} \Omega_1(f, \delta)_{L^q} = 0$. Then $\forall\, \varepsilon > 0$, $\varepsilon^{\frac{1}{q}} > 0$ and $\exists\, \delta > 0$, $\Omega_1(f, \delta)_{L^q} \leq \varepsilon^{\frac{1}{q}}$. I.e. for any $x, y \in W : \|x - y\|_\infty \leq \delta$ we get

$$\int_X D^q(f(x, s), f(y, s)) P(ds) \leq \varepsilon.$$

That is $f \in C_{FR}^{U_q}(W)$.

($\Leftarrow$) Let $f \in C_{FR}^{U_q}(W)$. Then $\forall\, \varepsilon > 0\ \exists\, \delta > 0$: whenever $\|x - y\|_\infty \leq \delta$, $x, y \in W$, it implies

$$\int_X D^q\left(f(x,s), f(y,s)\right) P(ds) \leq \varepsilon.$$

I.e. $\forall\, \varepsilon > 0\ \exists\, \delta > 0 : \Omega_1(f,\delta)_{L^q} \leq \varepsilon^{\frac{1}{q}}$. That is $\Omega_1(f,\delta)_{L^q} \to 0$ as $\delta \downarrow 0$. ■

We give

Definition 15.12 (*see also* [6]) Let $f(t,s)$ be a random function (stochastic process) from $W \times (X, \mathcal{B}, P)$ *into* $\mathbb{R}$, $W \subseteq \mathbb{R}^N$, where $(X, \mathcal{B}, P)$ is a probability space. We define the q-mean multivariate first modulus of continuity of f by

$$\Omega_1(f,\delta)_{L^q} :=$$

$$\sup\left\{\left(\int_X |f(x,s) - f(y,s)|^q P(ds)\right)^{\frac{1}{q}} : x, y \in W,\ \|x - y\|_\infty \leq \delta\right\}, \quad (15.4)$$

$\delta > 0$, $1 \leq q < \infty$.

The concept of f being (q-mean) uniformly continuous random real function is defined the same way as in Definition 15.9 just replace D by $|\cdot|$, etc. We denote it as $f \in C_{\mathbb{R}}^{U_q}(W)$.

Similar properties as in Propositions 15.10, 15.11 are valid for $\Omega_1(f,\delta)_{L^q}$.

Also we have

Proposition 15.13 ([3]) *Let $Y(t,\omega)$ be a real valued stochastic process such that Y is continuous in $t \in [a,b]$. Then Y is jointly measurable in (t,ω).*

According to [18], p. 94 we have the following

Definition 15.14 Let $(Y, \mathcal{T})$ be a topological space, with its σ-algebra of Borel sets $\mathcal{B} := \mathcal{B}(Y, \mathcal{T})$ generated by $\mathcal{T}$. If $(X, \mathcal{S})$ is a measurable space, a function $f : X \to Y$ is called measurable iff $f^{-1}(B) \in \mathcal{S}$ for all $B \in \mathcal{B}$.

By Theorem 4.1.6 of [18], p. 89 f as above is measurable iff

$$f^{-1}(C) \in \mathcal{S} \text{ for all } C \in \mathcal{T}.$$

We mention

Theorem 15.15 (see [18], p. 95) *Let $(X, \mathcal{S})$ be a measurable space and (Y, d) be a metric space. Let f_n be measurable functions from X into Y such that for all $x \in X$, $f_n(x) \to f(x)$ in Y. Then f is measurable. I.e., $\lim_{n\to\infty} f_n = f$ is measurable.*

We need also

Proposition 15.16 *Let f, g be fuzzy random variables from $\mathcal{S}$ into $\mathbb{R}_{\mathcal{F}}$. Then*

(i) Let $c \in \mathbb{R}$, then $c \odot f$ is a fuzzy random variable.
(ii) $f \oplus g$ is a fuzzy random variable.

15.2 Basics on Neural Network Operators

(I) Here all come from [8, 17]. We consider the sigmoidal function of logarithmic type

$$s_i(x_i) = \frac{1}{1+e^{-x_i}}, \quad x_i \in \mathbb{R}, i = 1, \ldots, N; \; x := (x_1, \ldots, x_N) \in \mathbb{R}^N,$$

each has the properties $\lim_{x_i \to +\infty} s_i(x_i) = 1$ and $\lim_{x_i \to -\infty} s_i(x_i) = 0$, $i = 1, \ldots, N$.

These functions play the role of activation functions in the hidden layer of neural networks.

As in [17], we consider

$$\Phi_i(x_i) := \frac{1}{2}\left(s_i(x_i+1) - s_i(x_i-1)\right), \quad x_i \in \mathbb{R}, i = 1, \ldots, N.$$

We notice the following properties:

(i) $\Phi_i(x_i) > 0, \ \forall \, x_i \in \mathbb{R}$,
(ii) $\sum_{k_i=-\infty}^{\infty} \Phi_i(x_i - k_i) = 1, \ \forall \, x_i \in \mathbb{R}$,
(iii) $\sum_{k_i=-\infty}^{\infty} \Phi_i(nx_i - k_i) = 1, \ \forall \, x_i \in \mathbb{R}; n \in \mathbb{N}$,
(iv) $\int_{-\infty}^{\infty} \Phi_i(x_i)\, dx_i = 1$,
(v) Φ_i is a density function,
(vi) Φ_i is even: $\Phi_i(-x_i) = \Phi_i(x_i)$, $x_i \geq 0$, for $i = 1, \ldots, N$.
We see that [13]

$$\Phi_i(x_i) = \left(\frac{e^2-1}{2e^2}\right) \frac{1}{\left(1+e^{x_i-1}\right)\left(1+e^{-x_i-1}\right)}, \quad i = 1, \ldots, N.$$

(vii) Φ_i is decreasing on $\mathbb{R}_+$, and increasing on $\mathbb{R}_-$, $i = 1, \ldots, N$.
Notice that $\Phi_i(x_i) = \Phi_i(0) = 0,231$.
Let $0 < \beta < 1$, $n \in \mathbb{N}$. Then as in [13] we get

(viii)

$$\sum_{\substack{k_i = -\infty \\ : |nx_i - k_i| > n^{1-\beta}}}^{\infty} \Phi_i(nx_i - k_i) \leq 3.1992 e^{-n^{(1-\beta)}}, \quad i = 1, \ldots, N.$$

Denote by $\lceil \cdot \rceil$ the ceiling of a number, and by $\lfloor \cdot \rfloor$ the integral part of a number. Consider here $x \in \left(\prod_{i=1}^{N} [a_i, b_i] \right) \subset \mathbb{R}^N$, $N \in \mathbb{N}$ such that $\lceil na_i \rceil \leq \lfloor nb_i \rfloor$, $i = 1, \ldots, N$; $a := (a_1, \ldots, a_N)$, $b := (b_1, \ldots, b_N)$.

As in [13] we obtain

(ix)

$$0 < \frac{1}{\sum_{k_i = \lceil na_i \rceil}^{\lfloor nb_i \rfloor} \Phi_i (nx_i - k_i)} < \frac{1}{\Phi_i (1)} = 5.250312578,$$

$\forall\, x_i \in [a_i, b_i]$, $i = 1, \ldots, N$.

(x) As in [13], we see that

$$\lim_{n \to \infty} \sum_{k_i = \lceil na_i \rceil}^{\lfloor nb_i \rfloor} \Phi_i (nx_i - k_i) \neq 1,$$

for at least some $x_i \in [a_i, b_i]$, $i = 1, \ldots, N$.

We will use here

$$\Phi (x_1, \ldots, x_N) := \Phi (x) := \prod_{i=1}^{N} \Phi_i (x_i), \quad x \in \mathbb{R}^N. \tag{15.5}$$

It has the properties:

(i)' $\Phi (x) > 0$, $\forall\, x \in \mathbb{R}^N$,

(ii)'

$$\sum_{k=-\infty}^{\infty} \Phi (x - k) := \sum_{k_1=-\infty}^{\infty} \sum_{k_2=-\infty}^{\infty} \cdots \sum_{k_N=-\infty}^{\infty} \Phi (x_1 - k_1, \ldots, x_N - k_N) = 1, \tag{15.6}$$

$k := (k_1, \ldots, k_N)$, $\forall\, x \in \mathbb{R}^N$.

(iii)'

$$\sum_{k=-\infty}^{\infty} \Phi (nx - k) :=$$

$$\sum_{k_1=-\infty}^{\infty} \sum_{k_2=-\infty}^{\infty} \cdots \sum_{k_N=-\infty}^{\infty} \Phi (nx_1 - k_1, \ldots, nx_N - k_N) = 1, \tag{15.7}$$

$\forall\, x \in \mathbb{R}^N$; $n \in \mathbb{N}$.

(iv)'

$$\int_{\mathbb{R}^N} \Phi (x)\, dx = 1,$$

that is Φ is a multivariate density function.
Here $\|x\|_\infty := \max\{|x_1|, \ldots, |x_N|\}$, $x \in \mathbb{R}^N$, also set $\infty := (\infty, \ldots, \infty)$, $-\infty := (-\infty, \ldots, -\infty)$ upon the multivariate context, and

$$\lceil na \rceil := (\lceil na_1 \rceil, \ldots, \lceil na_N \rceil),$$
$$\lfloor nb \rfloor := (\lfloor nb_1 \rfloor, \ldots, \lfloor nb_N \rfloor).$$

We also have

(v)'

$$\sum_{\substack{k = \lceil na \rceil \\ \left\| \frac{k}{n} - x \right\|_\infty > \frac{1}{n^\beta}}}^{\lfloor nb \rfloor} \Phi(nx - k) \leq 3.1992 e^{-n^{(1-\beta)}}, \tag{15.8}$$

$0 < \beta < 1$, $n \in \mathbb{N}$, $x \in \left(\prod_{i=1}^N [a_i, b_i]\right)$.

(vi)'

$$0 < \frac{1}{\sum_{k=\lceil na \rceil}^{\lfloor nb \rfloor} \Phi(nx - k)} < (5.250312578)^N, \tag{15.9}$$

$\forall\, x \in \left(\prod_{i=1}^N [a_i, b_i]\right)$, $n \in \mathbb{N}$.

(vii)'

$$\sum_{\substack{k = -\infty \\ \left\| \frac{k}{n} - x \right\|_\infty > \frac{1}{n^\beta}}}^{\infty} \Phi(nx - k) \leq 3.1992 e^{-n^{(1-\beta)}},$$

$0 < \beta < 1$, $n \in \mathbb{N}$, $x \in \mathbb{R}^N$.

(viii)'

$$\lim_{n \to \infty} \sum_{k=\lceil na \rceil}^{\lfloor nb \rfloor} \Phi(nx - k) \neq 1$$

for at least some $x \in \left(\prod_{i=1}^N [a_i, b_i]\right)$.

In general $\|.\|_\infty$ $s\tan ds\ for\ the\ \sup remum\ norm.$

Let $f \in C\left(\prod_{i=1}^N [a_i, b_i]\right)$ and $n \in \mathbb{N}$ such that $\lceil na_i \rceil \leq \lfloor nb_i \rfloor$, $i = 1, \ldots, N$.

We introduce and define [11] the multivariate positive linear neural network operator ($x := (x_1, \ldots, x_N) \in \left(\prod_{i=1}^N [a_i, b_i]\right)$)

$$G_n(f, x_1, \ldots, x_N) := G_n(f, x) := \frac{\sum_{k=\lceil na \rceil}^{\lfloor nb \rfloor} f\left(\frac{k}{n}\right) \Phi(nx - k)}{\sum_{k=\lceil na \rceil}^{\lfloor nb \rfloor} \Phi(nx - k)} \quad (15.10)$$

$$:= \frac{\sum_{k_1=\lceil na_1 \rceil}^{\lfloor nb_1 \rfloor} \sum_{k_2=\lceil na_2 \rceil}^{\lfloor nb_2 \rfloor} \cdots \sum_{k_N=\lceil na_N \rceil}^{\lfloor nb_N \rfloor} f\left(\frac{k_1}{n}, \ldots, \frac{k_N}{n}\right) \left(\prod_{i=1}^N \Phi_i(nx_i - k_i)\right)}{\prod_{i=1}^N \left(\sum_{k_i=\lceil na_i \rceil}^{\lfloor nb_i \rfloor} \Phi_i(nx_i - k_i)\right)}.$$

For large enough n we always obtain $\lceil na_i \rceil \leq \lfloor nb_i \rfloor$, $i = 1, \ldots, N$. Also $a_i \leq \frac{k_i}{n} \leq b_i$, iff $\lceil na_i \rceil \leq k_i \leq \lfloor nb_i \rfloor$, $i = 1, \ldots, N$.

We need, for $f \in C\left(\prod_{i=1}^N [a_i, b_i]\right)$ the first multivariate modulus of continuity

$$\omega_1(f, h) := \sup_{\substack{x, y \in \left(\prod_{i=1}^N [a_i, b_i]\right) \\ \|x - y\|_\infty \leq h}} |f(x) - f(y)|, \quad h > 0. \quad (15.11)$$

Similarly it is defined for $f \in C_B\left(\mathbb{R}^N\right)$ (continuous and bounded functions on $\mathbb{R}^N$). We have that $\lim_{h \to 0} \omega_1(f, h) = 0$ when f is uniformly continuous.

When $f \in C_B\left(\mathbb{R}^N\right)$ we define [11]

$$\overline{G}_n(f, x) := \overline{G}_n(f, x_1, \ldots, x_N) := \sum_{k=-\infty}^{\infty} f\left(\frac{k}{n}\right) \Phi(nx - k) \quad (15.12)$$

$$:= \sum_{k_1=-\infty}^{\infty} \sum_{k_2=-\infty}^{\infty} \cdots \sum_{k_N=-\infty}^{\infty} f\left(\frac{k_1}{n}, \frac{k_2}{n}, \ldots, \frac{k_N}{n}\right) \left(\prod_{i=1}^N \Phi_i(nx_i - k_i)\right),$$

$n \in \mathbb{N}$, $\forall\, x \in \mathbb{R}^N$, $N \geq 1$, the multivariate quasi-interpolation neural network operator.

We mention from [11]:

Theorem 15.17 *Let* $f \in C\left(\prod_{i=1}^N [a_i, b_i]\right)$, $0 < \beta < 1$, $x \in \left(\prod_{i=1}^N [a_i, b_i]\right)$, $n, N \in \mathbb{N}$. *Then*

(i)

$$|G_n(f, x) - f(x)| \leq (5.250312578)^N \cdot$$

$$\left\{\omega_1\left(f, \frac{1}{n^\beta}\right) + (6.3984)\|f\|_\infty e^{-n^{(1-\beta)}}\right\} =: \lambda_1, \quad (15.13)$$

(ii)

$$\|G_n(f) - f\|_\infty \leq \lambda_1. \quad (15.14)$$

Theorem 15.18 *Let $f \in C_B(\mathbb{R}^N)$, $0 < \beta < 1$, $x \in \mathbb{R}^N$, $n, N \in \mathbb{N}$. Then*

(i)

$$\left|\overline{G}_n(f, x) - f(x)\right| \leq \omega_1\left(f, \frac{1}{n^\beta}\right) + (6.3984)\|f\|_\infty e^{-n^{(1-\beta)}} =: \lambda_2, \quad (15.15)$$

(ii)

$$\left\|\overline{G}_n(f) - f\right\|_\infty \leq \lambda_2. \quad (15.16)$$

(II) Here we follow [8], [10].

We also consider here the hyperbolic tangent function $\tanh x$, $x \in \mathbb{R}$:

$$\tanh x := \frac{e^x - e^{-x}}{e^x + e^{-x}}.$$

It has the properties $\tanh 0 = 0$, $-1 < \tanh x < 1$, $\forall\, x \in \mathbb{R}$, and $\tanh(-x) = -\tanh x$. Furthermore $\tanh x \to 1$ as $x \to \infty$, and $\tanh x \to -1$, as $x \to -\infty$, and it is strictly increasing on $\mathbb{R}$.

This function plays the role of an activation function in the hidden layer of neural networks.

We further consider

$$\Psi(x) := \frac{1}{4}(\tanh(x+1) - \tanh(x-1)) > 0, \quad \forall\, x \in \mathbb{R}. \quad (15.26)$$

We easily see that $\Psi(-x) = \Psi(x)$, that is Ψ is even on $\mathbb{R}$. Obviously Ψ is differentiable, thus continuous.

Proposition 15.19 ([9]) *$\Psi(x)$ for $x \geq 0$ is strictly decreasing.*

Obviously $\Psi(x)$ is strictly increasing for $x \leq 0$. Also it holds $\lim_{x \to -\infty} \Psi(x) = 0 = \lim_{x \to \infty} \Psi(x)$.

Infact Ψ has the bell shape with horizontal asymptote the x-axis. So the maximum of Ψ is zero, $\Psi(0) = 0.3809297$.

Theorem 15.20 ([9]) *We have that $\sum_{i=-\infty}^{\infty} \Psi(x - i) = 1$, $\forall\, x \in \mathbb{R}$.*

Thus

$$\sum_{i=-\infty}^{\infty} \Psi(nx - i) = 1, \quad \forall\, n \in \mathbb{N},\ \forall\, x \in \mathbb{R}.$$

Also it holds

$$\sum_{i=-\infty}^{\infty} \Psi(x + i) = 1, \quad \forall x \in \mathbb{R}.$$

Theorem 15.21 ([9]) *It holds* $\int_{-\infty}^{\infty} \Psi(x)\,dx = 1$.

So $\Psi(x)$ is a density function on $\mathbb{R}$.

Theorem 15.22 ([9]) *Let* $0 < \alpha < 1$ *and* $n \in \mathbb{N}$. *It holds*

$$\sum_{\substack{k=-\infty \\ : |nx-k| \geq n^{1-\alpha}}}^{\infty} \Psi(nx-k) \leq e^4 \cdot e^{-2n^{(1-\alpha)}}.$$

Theorem 15.23 ([9]) *Let* $x \in [a,b] \subset \mathbb{R}$ *and* $n \in \mathbb{N}$ *so that* $\lceil na \rceil \leq \lfloor nb \rfloor$. *It holds*

$$\frac{1}{\sum_{k=\lceil na \rceil}^{\lfloor nb \rfloor} \Psi(nx-k)} < \frac{1}{\Psi(1)} = 4.1488766.$$

Also by [9] we get that

$$\lim_{n\to\infty} \sum_{k=\lceil na \rceil}^{\lfloor nb \rfloor} \Psi(nx-k) \neq 1,$$

for at least some $x \in [a,b]$.

In this chapter we use (see [10])

$$\Theta(x_1,\ldots,x_N) := \Theta(x) := \prod_{i=1}^{N} \Psi(x_i), \quad x = (x_1,\ldots,x_N) \in \mathbb{R}^N,\ N \in \mathbb{N}. \tag{15.17}$$

It has the properties:

(i) $\Theta(x) > 0$, $\forall\, x \in \mathbb{R}^N$,

(ii)

$$\sum_{k=-\infty}^{\infty} \Theta(x-k) := \sum_{k_1=-\infty}^{\infty} \sum_{k_2=-\infty}^{\infty} \ldots \sum_{k_N=-\infty}^{\infty} \Theta(x_1-k_1,\ldots,x_N-k_N) = 1, \tag{15.18}$$

where $k := (k_1,\ldots,k_N)$, $\forall\, x \in \mathbb{R}^N$.

(iii)

$$\sum_{k=-\infty}^{\infty} \Theta(nx-k) :=$$

$$\sum_{k_1=-\infty}^{\infty} \sum_{k_2=-\infty}^{\infty} \ldots \sum_{k_N=-\infty}^{\infty} \Theta(nx_1-k_1,\ldots,nx_N-k_N) = 1, \tag{15.19}$$

$\forall\, x \in \mathbb{R}^N;\ n \in \mathbb{N}.$

(iv)

$$\int_{\mathbb{R}^N} \Theta(x)\, dx = 1,$$

that is Θ is a multivariate density function.

(v)

$$\sum_{\begin{cases} k = \lceil na \rceil \\ \left\| \frac{k}{n} - x \right\|_\infty > \frac{1}{n^\beta} \end{cases}}^{\lfloor nb \rfloor} \Theta(nx - k) \leq e^4 \cdot e^{-2n^{(1-\beta)}}, \tag{15.20}$$

$0 < \beta < 1, n \in \mathbb{N}, x \in \left(\prod_{i=1}^N [a_i, b_i]\right).$

(vi)

$$0 < \frac{1}{\sum_{k=\lceil na \rceil}^{\lfloor nb \rfloor} \Theta(nx - k)} < \frac{1}{(\Psi(1))^N} = (4.1488766)^N, \tag{15.21}$$

$\forall\, x \in \left(\prod_{i=1}^N [a_i, b_i]\right),\ n \in \mathbb{N}.$

(vii)

$$\sum_{\begin{cases} k = -\infty \\ \left\| \frac{k}{n} - x \right\|_\infty > \frac{1}{n^\beta} \end{cases}}^{\infty} \Theta(nx - k) \leq e^4 \cdot e^{-2n^{(1-\beta)}}, \tag{15.22}$$

$0 < \beta < 1, n \in \mathbb{N}, x \in \mathbb{R}^N.$

Also we get that

$$\lim_{n \to \infty} \sum_{k=\lceil na \rceil}^{\lfloor nb \rfloor} \Theta(nx - k) \neq 1,$$

for at least some $x \in \left(\prod_{i=1}^N [a_i, b_i]\right)$.

Let $f \in C\left(\prod_{i=1}^N [a_i, b_i]\right)$ and $n \in \mathbb{N}$ such that $\lceil na_i \rceil \leq \lfloor nb_i \rfloor, i = 1, \ldots, N$.

We introduce and define the multivariate positive linear neural network operator ([10]) ($x := (x_1, \ldots, x_N) \in \left(\prod_{i=1}^N [a_i, b_i]\right)$)

$$F_n(f, x_1, \ldots, x_N) := F_n(f, x) := \frac{\sum_{k=\lceil na \rceil}^{\lfloor nb \rfloor} f\left(\frac{k}{n}\right) \Theta(nx - k)}{\sum_{k=\lceil na \rceil}^{\lfloor nb \rfloor} \Theta(nx - k)} \tag{15.23}$$

$$:= \frac{\sum_{k_1=\lceil na_1 \rceil}^{\lfloor nb_1 \rfloor} \sum_{k_2=\lceil na_2 \rceil}^{\lfloor nb_2 \rfloor} \cdots \sum_{k_N=\lceil na_N \rceil}^{\lfloor nb_N \rfloor} f\left(\frac{k_1}{n}, \ldots, \frac{k_N}{n}\right)\left(\prod_{i=1}^{N} \Psi\left(nx_i - k_i\right)\right)}{\prod_{i=1}^{N}\left(\sum_{k_i=\lceil na_i \rceil}^{\lfloor nb_i \rfloor} \Psi\left(nx_i - k_i\right)\right)}.$$

When $f \in C_B\left(\mathbb{R}^N\right)$ we define [10],

$$\overline{F}_n(f, x) := \overline{F}_n(f, x_1, \ldots, x_N) := \sum_{k=-\infty}^{\infty} f\left(\frac{k}{n}\right)\Theta(nx - k) := \tag{15.24}$$

$$\sum_{k_1=-\infty}^{\infty} \sum_{k_2=-\infty}^{\infty} \cdots \sum_{k_N=-\infty}^{\infty} f\left(\frac{k_1}{n}, \frac{k_2}{n}, \ldots, \frac{k_N}{n}\right)\left(\prod_{i=1}^{N} \Psi\left(nx_i - k_i\right)\right),$$

$n \in \mathbb{N}$, $\forall\, x \in \mathbb{R}^N$, $N \geq 1$, the multivariate quasi-interpolation neural network operator.

We mention from [10]:

Theorem 15.24 *Let* $f \in C\left(\prod_{i=1}^{N}[a_i, b_i]\right)$, $0 < \beta < 1$, $x \in \left(\prod_{i=1}^{N}[a_i, b_i]\right)$, $n, N \in \mathbb{N}$. *Then*

(i)

$$|F_n(f, x) - f(x)| \leq (4.1488766)^N \cdot$$

$$\left\{\omega_1\left(f, \frac{1}{n^\beta}\right) + 2e^4 \|f\|_\infty e^{-2n^{(1-\beta)}}\right\} =: \lambda_1, \tag{15.25}$$

(ii)

$$\|F_n(f) - f\|_\infty \leq \lambda_1. \tag{15.26}$$

Theorem 15.25 *Let* $f \in C_B\left(\mathbb{R}^N\right)$, $0 < \beta < 1$, $x \in \mathbb{R}^N$, $n, N \in \mathbb{N}$. *Then*

(i)

$$\left|\overline{F}_n(f, x) - f(x)\right| \leq \omega_1\left(f, \frac{1}{n^\beta}\right) + 2e^4 \|f\|_\infty e^{-2n^{(1-\beta)}} =: \lambda_2, \tag{15.27}$$

(ii)

$$\left\|\overline{F}_n(f) - f\right\|_\infty \leq \lambda_2. \tag{15.28}$$

We are also motivated by [1, 2, 4–7, 12, 14, 16]. For general knowledge on neural networks we recommend [20–22].

15.3 Main Results

(I) q-mean Approximation by Fuzzy-Random Quasi-interpolation Neural Network Operators

All terms and assumptions here as in Sects. 15.1, 15.2.

Let $f \in C_{\mathcal{FR}}^{U_q}\left(\prod_{i=1}^{N}[a_i,b_i]\right)$, $1 \leq q < +\infty$, $n, N \in \mathbb{N}$, $0 < \beta < 1$, $\overrightarrow{x} \in \left(\prod_{i=1}^{N}[a_i,b_i]\right)$, $(X, \mathcal{B}, P)$ probability space, $s \in X$.

We define the following multivariate fuzzy random quasi-interpolation linear neural network operators

$$\left(G_n^{\mathcal{FR}}(f)\right)(\overrightarrow{x}, s) := \sum_{\overrightarrow{k}=\lceil na \rceil}^{\lfloor nb \rfloor *} f\left(\frac{\overrightarrow{k}}{n}, s\right) \odot \frac{\Phi(nx-k)}{\sum_{k=\lceil na \rceil}^{\lfloor nb \rfloor} \Phi(nx-k)}, \tag{15.29}$$

(see also (15.10)), and

$$\left(F_n^{\mathcal{FR}}(f)\right)(\overrightarrow{x}, s) := \sum_{\overrightarrow{k}=\lceil na \rceil}^{\lfloor nb \rfloor *} f\left(\frac{\overrightarrow{k}}{n}, s\right) \odot \frac{\theta(nx-k)}{\sum_{k=\lceil na \rceil}^{\lfloor nb \rfloor} \theta(nx-k)}, \tag{15.30}$$

(see also (15.23)).

We present

Theorem 15.26 *Let $f \in C_{\mathcal{FR}}^{U_q}\left(\prod_{i=1}^{N}[a_i,b_i]\right)$, $0 < \beta < 1$, $\overrightarrow{x} \in \left(\prod_{i=1}^{N}[a_i,b_i]\right)$, $n, N \in \mathbb{N}$, $1 \leq q < +\infty$. Assume that $\int_X \left(D^*\left(f(\cdot, s), \widetilde{o}\right)\right)^q P(ds) < \infty$. Then*

(1)

$$\left(\int_X D^q\left(\left(G_n^{\mathcal{FR}}(f)\right)(\overrightarrow{x}, s), f(\overrightarrow{x}, s)\right) P(ds)\right)^{\frac{1}{q}} \leq (5.250312578)^N \cdot \tag{15.31}$$

$$\left\{\Omega_1^{(\mathcal{F})}\left(f, \frac{1}{n^{\beta}}\right)_{L^q} + (6.3984)\left(\int_X \left(D^*\left(f(\cdot, s), \widetilde{o}\right)\right)^q P(ds)\right)^{\frac{1}{q}} e^{-n^{(1-\beta)}}\right\} =: \lambda_1^{(\mathcal{FR})},$$

(2)

$$\left\| \left(\int_X D^q \left(\left(G_n^{\mathcal{FR}} (f) \right) (\overrightarrow{x}, s), f (\overrightarrow{x}, s) \right) P(ds) \right)^{\frac{1}{q}} \right\|_{\infty, \left(\prod_{i=1}^{N} [a_i, b_i] \right)} \leq \lambda_1^{(\mathcal{FR})}. \tag{15.32}$$

(see also Theorem 15.17).

Proof We notice that

$$D\left(f\left(\frac{\overrightarrow{k}}{n}, s \right), f (\overrightarrow{x}, s) \right) \leq D\left(f\left(\frac{\overrightarrow{k}}{n}, s \right), \widetilde{o} \right) + D\left(f (\overrightarrow{x}, s), \widetilde{o} \right) \tag{15.33}$$

$$\leq 2D^* \left(f (\cdot, s), \widetilde{o} \right).$$

Hence

$$D^q \left(f\left(\frac{\overrightarrow{k}}{n}, s \right), f (\overrightarrow{x}, s) \right) \leq 2^q D^{*q} \left(f (\cdot, s), \widetilde{o} \right), \tag{15.34}$$

and

$$\left(\int_X D^q \left(f\left(\frac{\overrightarrow{k}}{n}, s \right), f (\overrightarrow{x}, s) \right) P(ds) \right)^{\frac{1}{q}} \leq 2 \left(\int_X \left(D^* \left(f (\cdot, s), \widetilde{o} \right) \right)^q P(ds) \right)^{\frac{1}{q}}. \tag{15.35}$$

We observe that

$$D\left(\left(G_n^{\mathcal{FR}} (f) \right) (\overrightarrow{x}, s), f (\overrightarrow{x}, s) \right) = \tag{15.36}$$

$$D\left(\sum_{\overrightarrow{k} = \lceil na \rceil}^{\lfloor nb \rfloor *} f\left(\frac{\overrightarrow{k}}{n}, s \right) \odot \frac{\Phi (nx - k)}{\sum_{\overrightarrow{k} = \lceil na \rceil}^{\lfloor nb \rfloor} \Phi (nx - k)}, f (\overrightarrow{x}, s) \odot 1 \right) =$$

$$D\left(\sum_{\overrightarrow{k} = \lceil na \rceil}^{\lfloor nb \rfloor *} f\left(\frac{\overrightarrow{k}}{n}, s \right) \odot \frac{\Phi (nx - k)}{\sum_{\overrightarrow{k} = \lceil na \rceil}^{\lfloor nb \rfloor} \Phi (nx - k)}, f (\overrightarrow{x}, s) \odot \frac{\sum_{\overrightarrow{k} = \lceil na \rceil}^{\lfloor nb \rfloor} \Phi (nx - k)}{\sum_{\overrightarrow{k} = \lceil na \rceil}^{\lfloor nb \rfloor} \Phi (nx - k)} \right) = \tag{15.37}$$

$$D\left(\sum_{\overrightarrow{k}=\lceil na\rceil}^{\lfloor nb\rfloor *} f\left(\frac{\overrightarrow{k}}{n}, s\right) \odot \frac{\Phi(nx-k)}{\sum_{\overrightarrow{k}=\lceil na\rceil}^{\lfloor nb\rfloor} \Phi(nx-k)}, \sum_{\overrightarrow{k}=\lceil na\rceil}^{\lfloor nb\rfloor *} f(\overrightarrow{x}, s) \odot \frac{\Phi(nx-k)}{\sum_{\overrightarrow{k}=\lceil na\rceil}^{\lfloor nb\rfloor} \Phi(nx-k)}\right)$$

$$\leq \sum_{\overrightarrow{k}=\lceil na\rceil}^{\lfloor nb\rfloor} \left(\frac{\Phi(nx-k)}{\sum_{\overrightarrow{k}=\lceil na\rceil}^{\lfloor nb\rfloor} \Phi(nx-k)}\right) D\left(f\left(\frac{\overrightarrow{k}}{n}, s\right), f(\overrightarrow{x}, s)\right). \tag{15.38}$$

So that

$$D\left(\left(G_n^{\mathcal{FR}}(f)\right)(\overrightarrow{x}, s), f(\overrightarrow{x}, s)\right) \leq$$

$$\sum_{\overrightarrow{k}=\lceil na\rceil}^{\lfloor nb\rfloor} \left(\frac{\Phi(nx-k)}{\sum_{\overrightarrow{k}=\lceil na\rceil}^{\lfloor nb\rfloor} \Phi(nx-k)}\right) D\left(f\left(\frac{\overrightarrow{k}}{n}, s\right), f(\overrightarrow{x}, s)\right) = \tag{15.39}$$

$$\sum_{\substack{\overrightarrow{k}=\lceil na\rceil \\ \left\|\frac{\overrightarrow{k}}{n}-\overrightarrow{x}\right\|_\infty \leq \frac{1}{n^\beta}}}^{\lfloor nb\rfloor} \left(\frac{\Phi(nx-k)}{\sum_{\overrightarrow{k}=\lceil na\rceil}^{\lfloor nb\rfloor} \Phi(nx-k)}\right) D\left(f\left(\frac{\overrightarrow{k}}{n}, s\right), f(\overrightarrow{x}, s)\right) +$$

$$\sum_{\substack{\overrightarrow{k}=\lceil na\rceil \\ \left\|\frac{\overrightarrow{k}}{n}-\overrightarrow{x}\right\|_\infty > \frac{1}{n^\beta}}}^{\lfloor nb\rfloor} \left(\frac{\Phi(nx-k)}{\sum_{\overrightarrow{k}=\lceil na\rceil}^{\lfloor nb\rfloor} \Phi(nx-k)}\right) D\left(f\left(\frac{\overrightarrow{k}}{n}, s\right), f(\overrightarrow{x}, s)\right).$$

Hence it holds

$$\left(\int_X D^q\left(\left(G_n^{\mathcal{FR}}(f)\right)(\overrightarrow{x},s),f(\overrightarrow{x},s)\right)P(ds)\right)^{\frac{1}{q}}\leq \tag{15.40}$$

$$\sum_{\substack{\overrightarrow{k}=\lceil na\rceil\\ \left\|\frac{\overrightarrow{k}}{n}-\overrightarrow{x}\right\|_\infty\leq\frac{1}{n^\beta}}}^{\lfloor nb\rfloor}\left(\frac{\Phi(nx-k)}{\sum_{\overrightarrow{k}=\lceil na\rceil}^{\lfloor nb\rfloor}\Phi(nx-k)}\right)\left(\int_X D^q\left(f\left(\frac{\overrightarrow{k}}{n},s\right),f(\overrightarrow{x},s)\right)P(ds)\right)^{\frac{1}{q}}+$$

$$\sum_{\substack{\overrightarrow{k}=\lceil na\rceil\\ \left\|\frac{\overrightarrow{k}}{n}-\overrightarrow{x}\right\|_\infty>\frac{1}{n^\beta}}}^{\lfloor nb\rfloor}\left(\frac{\Phi(nx-k)}{\sum_{\overrightarrow{k}=\lceil na\rceil}^{\lfloor nb\rfloor}\Phi(nx-k)}\right)\left(\int_X D^q\left(f\left(\frac{\overrightarrow{k}}{n},s\right),f(\overrightarrow{x},s)\right)P(ds)\right)^{\frac{1}{q}}\leq$$

$$\left(\frac{1}{\sum_{\overrightarrow{k}=\lceil na\rceil}^{\lfloor nb\rfloor}\Phi(nx-k)}\right)\cdot\left\{\Omega_1^{(\mathcal{F})}\left(f,\frac{1}{n^\beta}\right)_{L^q}+\right. \tag{15.41}$$

$$\left.2\left(\int_X\left(D^*\left(f(\cdot,s),\widetilde{o}\right)\right)^q P(ds)\right)^{\frac{1}{q}}\left(\sum_{\substack{\overrightarrow{k}=\lceil na\rceil\\ \left\|\frac{\overrightarrow{k}}{n}-\overrightarrow{x}\right\|_\infty>\frac{1}{n^\beta}}}^{\lfloor nb\rfloor}\Phi(nx-k)\right)\right\}\leq$$

$$(5.250312578)^N\cdot$$

$$\left\{\Omega_1^{(\mathcal{F})}\left(f,\frac{1}{n^\beta}\right)_{L^q}+2\left(\int_X\left(D^*\left(f(\cdot,s),\widetilde{o}\right)\right)^q P(ds)\right)^{\frac{1}{q}}(3.1992)\,e^{-n^{(1-\beta)}}\right\}. \tag{15.42}$$

We have proved claim. ■

Similarly we give

Theorem 15.27 *All assumptions as in Theorem 15.26. Then*

(1)

$$\left(\int_X D^q\left(\left(F_n^{\mathcal{FR}}(f)\right)(\overrightarrow{x},s),f(\overrightarrow{x},s)\right)P(ds)\right)^{\frac{1}{q}} \leq (4.1488766)^N \cdot \tag{15.43}$$

$$\left\{\Omega_1^{(\mathcal{F})}\left(f,\frac{1}{n^\beta}\right)_{L^q}+2e^4\left(\int_X\left(D^*\left(f(\cdot,s),\widetilde{o}\right)\right)^q P(ds)\right)^{\frac{1}{q}}e^{-2n^{(1-\beta)}}\right\}=:\lambda_2^{(\mathcal{FR})},$$

(2)

$$\left\|\left(\int_X D^q\left(\left(F_n^{\mathcal{FR}}(f)\right)(\overrightarrow{x},s),f(\overrightarrow{x},s)\right)P(ds)\right)^{\frac{1}{q}}\right\|_{\infty,\left(\prod_{i=1}^N[a_i,b_i]\right)} \leq \lambda_2^{(\mathcal{FR})}. \tag{15.44}$$

(see also Theorem 15.24).

Proof Similar to the proof of Theorem 15.26. ■

Conclusion 15.28 *By Theorems 15.26, 15.27, we obtain the pointwise and uniform convergences with rates in the q-mean and D-metric of the operators $G_n^{\mathcal{FR}}$, $F_n^{\mathcal{FR}}$ to the unit operator for $f \in C_{\mathcal{FR}}^{U_q}\left(\prod_{i=1}^N [a_i,b_i]\right)$.*

(II) q-mean Approximation by Random Quasi-interpolation Neural Network Operators

Let $g \in C_{\mathcal{R}}^{U_1}(\mathbb{R}^N)$, $0<\beta<1$, $\overrightarrow{x} \in \mathbb{R}^N$, $n,N \in \mathbb{N}$, with $\|g\|_{\infty,\mathbb{R}^N,X}<\infty$, $(X,\mathcal{B},P)$ probability space, $s \in X$.

We define

$$\overline{G}_n^{(\mathcal{R})}(f,\overrightarrow{x}) := \sum_{\overrightarrow{k}=-\infty}^{\infty} f\left(\frac{\overrightarrow{k}}{n},s\right)\Phi(nx-k), \tag{15.45}$$

(see also (15.12)), also we define

$$\overline{F}_n^{(\mathcal{R})}(f,\overrightarrow{x}) := \sum_{\overrightarrow{k}=-\infty}^{\infty} f\left(\frac{\overrightarrow{k}}{n},s\right)\Theta(nx-k), \tag{15.46}$$

(see also (15.24)).

We give

Theorem 15.29 *Let $g \in C_{\mathcal{R}}^{U_1}(\mathbb{R}^N)$, $0<\beta<1$, $\overrightarrow{x} \in \mathbb{R}^N$, $n,N \in \mathbb{N}$, $\|g\|_{\infty,\mathbb{R}^N,X}<\infty$. Then*

(1)

$$\int_X \left| \left(\overline{G}_n^{(\mathcal{R})} (g) \right) \left(\overrightarrow{x}, s \right) - g \left(\overrightarrow{x}, s \right) \right| P (ds) \leq \tag{15.47}$$

$$\left\{ \Omega_1 \left(g, \frac{1}{n^{\beta}} \right)_{L^1} + (6.3984) \|g\|_{\infty, \mathbb{R}^N, X} \, e^{-n^{(1-\beta)}} \right\} =: \mu_1^{(\mathcal{R})},$$

(2)

$$\left\| \int_X \left| \left(\overline{G}_n^{(\mathcal{R})} (g) \right) \left(\overrightarrow{x}, s \right) - g \left(\overrightarrow{x}, s \right) \right| P (ds) \right\|_{\infty, \mathbb{R}^N} \leq \mu_1^{(\mathcal{R})}. \tag{15.48}$$

(see also Theorem 15.18).

Proof Since $\|g\|_{\infty, \mathbb{R}^N, X} < \infty$, then

$$\left| g \left(\frac{\overrightarrow{k}}{n}, s \right) - g \left(\overrightarrow{x}, s \right) \right| \leq 2 \|g\|_{\infty, \mathbb{R}^N, X} < \infty. \tag{15.49}$$

Hence

$$\int_X \left| g \left(\frac{\overrightarrow{k}}{n}, s \right) - g \left(\overrightarrow{x}, s \right) \right| P (ds) \leq 2 \|g\|_{\infty, \mathbb{R}^N, X} < \infty. \tag{15.50}$$

We observe that

$$\left(\overline{G}_n^{(\mathcal{R})} (g) \right) \left(\overrightarrow{x}, s \right) - g \left(\overrightarrow{x}, s \right) =$$

$$\sum_{\overrightarrow{k}=-\infty}^{\infty} g \left(\frac{\overrightarrow{k}}{n}, s \right) \Phi (nx - k) - g \left(\overrightarrow{x}, s \right) \sum_{\overrightarrow{k}=-\infty}^{\infty} \Phi (nx - k) = \tag{15.51}$$

$$\left(\sum_{\overrightarrow{k}=-\infty}^{\infty} g \left(\frac{\overrightarrow{k}}{n}, s \right) - g \left(\overrightarrow{x}, s \right) \right) \Phi (nx - k).$$

So that

$$\sum_{\overrightarrow{k}=-\infty}^{\infty} \left| g \left(\frac{\overrightarrow{k}}{n}, s \right) - g \left(\overrightarrow{x}, s \right) \right| \Phi (nx - k) \leq 2 \|g\|_{\infty, \mathbb{R}^N, X} < \infty. \tag{15.52}$$

Hence

$$\left| \left(\overline{G}_n^{(\mathcal{R})} (g) \right) \left(\overrightarrow{x}, s \right) - g \left(\overrightarrow{x}, s \right) \right| \leq \tag{15.53}$$

$$\sum_{\vec{k}=-\infty}^{\infty} \left| g\left(\frac{\vec{k}}{n}, s\right) - g\left(\vec{x}, s\right) \right| \Phi\left(nx - k\right) =$$

$$\sum_{\substack{\vec{k}=-\infty \\ \left\| \frac{\vec{k}}{n} - \vec{x} \right\|_{\infty} \leq \frac{1}{n^{\beta}}}}^{\infty} \left| g\left(\frac{\vec{k}}{n}, s\right) - g\left(\vec{x}, s\right) \right| \Phi\left(nx - k\right) +$$

$$\sum_{\substack{\vec{k}=-\infty \\ \left\| \frac{\vec{k}}{n} - \vec{x} \right\|_{\infty} > \frac{1}{n^{\beta}}}}^{\infty} \left| g\left(\frac{\vec{k}}{n}, s\right) - g\left(\vec{x}, s\right) \right| \Phi\left(nx - k\right).$$

Furthermore it holds

$$\left(\int_X \left| \left(\overline{G}_n^{(\mathcal{R})}(g) \right) \left(\vec{x}, s\right) - g\left(\vec{x}, s\right) \right| P\left(ds\right) \right) \leq$$

$$\sum_{\substack{\vec{k}=-\infty \\ \left\| \frac{\vec{k}}{n} - \vec{x} \right\|_{\infty} \leq \frac{1}{n^{\beta}}}}^{\infty} \left(\int_X \left| g\left(\frac{\vec{k}}{n}, s\right) - g\left(\vec{x}, s\right) \right| P\left(ds\right) \right) \Phi\left(nx - k\right) + \tag{15.55}$$

$$\sum_{\substack{\vec{k}=-\infty \\ \left\| \frac{\vec{k}}{n} - \vec{x} \right\|_{\infty} > \frac{1}{n^{\beta}}}}^{\infty} \left(\int_X \left| g\left(\frac{\vec{k}}{n}, s\right) - g\left(\vec{x}, s\right) \right| P\left(ds\right) \right) \Phi\left(nx - k\right) \leq$$

$$\Omega_1 \left(g, \frac{1}{n^{\beta}} \right)_{L^1} + 2 \left\| g \right\|_{\infty, \mathbb{R}^N, X} \sum_{\substack{\vec{k}=-\infty \\ \left\| \frac{\vec{k}}{n} - \vec{x} \right\|_{\infty} > \frac{1}{n^{\beta}}}}^{\infty} \Phi\left(nx - k\right) \leq$$

$$\Omega_1 \left(g, \frac{1}{n^{\beta}} \right)_{L^1} + \left\| g \right\|_{\infty, \mathbb{R}^N, X} (6.3984)\, e^{-n^{(1-\beta)}},$$

proving the claim. ■

We finish with

Theorem 15.30 *All as in Theorem 15.29. Then*

(1)

$$\int_X \left| \left(\overline{F}_n^{(\mathcal{R})}(g) \right) \left(\overrightarrow{x}, s \right) - g\left(\overrightarrow{x}, s \right) \right| P(ds) \leq \tag{15.57}$$

$$\left\{ \Omega_1 \left(g, \frac{1}{n^\beta} \right)_{L^1} + 2e^4 \|g\|_{\infty, \mathbb{R}^N, X} e^{-2n^{(1-\beta)}} \right\} =: \mu_2^{(\mathcal{R})},$$

(2)

$$\left\| \int_X \left| \left(\overline{F}_n^{(\mathcal{R})}(g) \right) \left(\overrightarrow{x}, s \right) - g\left(\overrightarrow{x}, s \right) \right| P(ds) \right\|_{\infty, \mathbb{R}^N} \leq \mu_2^{(\mathcal{R})}. \tag{15.58}$$

(see also Theorem 15.25).

Proof As similar to Theorem 15.29 is omitted. ■

Conclusion 15.31 *By Theorems 15.29, 15.30, we obtain pointwise and uniform convergences with rates in the q-mean of random operators $\overline{G}_n^{(\mathcal{R})}$, $\overline{F}_n^{(\mathcal{R})}$ to the unit operator for $g \in C_{\mathcal{R}}^{U_1}\left(\mathbb{R}^N\right)$.*

References

1. G.A. Anastassiou, Rate of convergence of Fuzzy neural network operators, univariate case. J. Fuzzy Math. **10**(3), 755–780 (2002)
2. G.A. Anastassiou, Higher order fuzzy approximation by fuzzy wavelet type and neural network operators. Comput. Math. **48**, 1387–1401 (2004)
3. G.A. Anastassiou, Univariate fuzzy-random neural network approximation operators, Computers and Mathematics with Applications, Special issue/ Proceedings edited by G. Anastassiou of special session "Computational Methods in Analysis", AMS meeting in Orlando. Florida, November 2002, Vol. 48 (2004), 1263–1283
4. G.A. Anastassiou, Higher order fuzzy Korovkin theory via inequalities. Commun. Appl. Anal. **10**(2), 359–392 (2006)
5. G.A. Anastassiou, Fuzzy Korovkin theorems and inequalities. J. Fuzzy Math. **15**(1), 169–205 (2007)
6. G.A. Anastassiou, Multivariate Stochastic Korovkin theory given quantitatively. Math. Comput. Model. **48**, 558–580 (2008)
7. G.A. Anastassiou, *Fuzzy Mathematics: Approximation Theory* (Springer, Heidelberg, 2010)
8. G.A. Anastassiou, *Inteligent Systems: Approximation by Artificial Neural Networks* (Springer, Heidelberg, 2011)
9. G.A. Anastassiou, Univariate hyperbolic tangent neural network approximation. Math. Comput. Model. **53**, 1111–1132 (2011)
10. G.A. Anastassiou, Multivariate hyperbolic tangent neural network approximation. Comput. Math. **61**, 809–821 (2011)
11. G.A. Anastassiou, Multivariate sigmoidal neural network approximation. Neural Networks **24**, 378–386 (2011)
12. G.A. Anastassiou, Higher order multivariate fuzzy approximation by multivariate fuzzy wavelet type and neural network operators. J. Fuzzy Math. **19**(3), 601–618 (2011)

13. G.A. Anastassiou, Univariate sigmoidal neural network approximation. J. Comput. Anal. Appl. **14**(4), 659–690 (2012)
14. G.A. Anastassiou, Rate of convergence of some multivariate neural network operators to the unit, revisited. J. Comput. Anal. Appl. **15**(7), 1300–1309 (2013)
15. G.A. Anastassiou, Multivariate fuzzy-random quasi-interpolation neural network approximation operators. J. Fuzzy Math. **22**(1), 167–184 (2014)
16. G.A. Anastassiou, Higher order multivariate fuzzy approximation by basic neural network operators. Cubo **16**(03), 21–35 (2014)
17. Z. Chen, F. Cao, The approximation operators with sigmoidal functions. Comput. Math. Appl. **58**, 758–765 (2009)
18. R.M. Dudley, *Real Analysis and Probability* (Wadsworth and Brooks/Cole Mathematics Series, Pacific Grove, 1989)
19. S. Gal, in *Approximation Theory in Fuzzy Setting, Chapter 13 in Handbook of Analytic-Computational Methods in Applied Mathematics*, pp. 617–666, ed. by G. Anastassiou, (Chapman and Hall/CRC, Boca Raton, 2000)
20. S. Haykin, *Neural Networks: A Comprehensive Foundation*, 2nd edn. (Prentice Hall, New York, 1998)
21. T.M. Mitchell, *Machine Learning* (WCB-McGraw-Hill, New York, 1997)
22. W. McCulloch, W. Pitts, A logical calculus of the ideas immanent in nervous activity. Bull. Math. Biophys. **7**, 115–133 (1943)
23. Wu Congxin, Gong Zengtai, On Henstock integrals of interval-valued functions and fuzzy valued functions. Fuzzy Sets Syst. **115**(3), 377–391 (2000)
24. C. Wu, Z. Gong, On Henstock integral of fuzzy-number-valued functions (I). Fuzzy Sets Syst. **120**(3), 523–532 (2001)
25. C. Wu, M. Ma, On embedding problem of fuzzy number space: part 1. Fuzzy Sets Syst. **44**, 33–38 (1991)

Chapter 16
Approximation by Kantorovich and Quadrature Type Quasi-interpolation Neural Networks

Here we present multivariate basic approximation by Kantorovich and Quadrature type quasi-interpolation neural network operators with respect to supremum norm. This is done with rates using the first multivariate modulus of continuity. We approximate continuous and bounded functions on $\mathbb{R}^N$. When they are also uniformly continuous we have pointwise and uniform convergences. It follows [6].

16.1 Background

We consider here the sigmoidal function of logarithmic type

$$s_i(x_i) = \frac{1}{1+e^{-x_i}}, \quad x_i \in \mathbb{R}, i = 1, \ldots, N; \; x := (x_1, \ldots, x_N) \in \mathbb{R}^N,$$

each has the properties $\lim_{x_i \to +\infty} s_i(x_i) = 1$ and $\lim_{x_i \to -\infty} s_i(x_i) = 0$, $i = 1, \ldots, N$.

These functions play the role of activation functions in the hidden layer of neural networks, also have applications in biology, demography, etc.

As in [8], we consider

$$\Phi_i(x_i) := \frac{1}{2}(s_i(x_i+1) - s_i(x_i-1)), \quad x_i \in \mathbb{R}, i = 1, \ldots, N.$$

We notice the following properties:

(i) $\Phi_i(x_i) > 0, \; \forall \, x_i \in \mathbb{R}$,
(ii) $\sum_{k_i=-\infty}^{\infty} \Phi_i(x_i - k_i) = 1, \; \forall \, x_i \in \mathbb{R}$,
(iii) $\sum_{k_i=-\infty}^{\infty} \Phi_i(nx_i - k_i) = 1, \; \forall \, x_i \in \mathbb{R}; n \in \mathbb{N}$,
(iv) $\int_{-\infty}^{\infty} \Phi_i(x_i)\, dx_i = 1$,
(v) Φ_i is a density function,

G.A. Anastassiou, *Intelligent Systems II: Complete Approximation by Neural Network Operators*, Studies in Computational Intelligence 608,
DOI 10.1007/978-3-319-20505-2_16

(vi) Φ_i is even: $\Phi_i(-x_i) = \Phi_i(x_i)$, $x_i \geq 0$, for $i = 1, \dots, N$.

We see that ([5])

$$\Phi_i(x_i) = \left(\frac{e^2-1}{2e^2}\right) \frac{1}{\left(1+e^{x_i-1}\right)\left(1+e^{-x_i-1}\right)}, \quad i = 1, \dots, N.$$

(vii) Φ_i is decreasing on $\mathbb{R}_+$, and increasing on $\mathbb{R}_-$, $i = 1, \dots, N$.

Let $0 < \beta < 1$, $n \in \mathbb{N}$. Then as in [5] we get

(viii)

$$\sum_{\substack{k_i = -\infty \\ : |nx_i - k_i| > n^{1-\beta}}}^{\infty} \Phi_i(nx_i - k_i) = \sum_{\substack{k_i = -\infty \\ : |nx_i - k_i| > n^{1-\beta}}}^{\infty} \Phi_i(|nx_i - k_i|)$$

$$\leq 3.1992 e^{-n^{(1-\beta)}}, \quad i = 1, \dots, N.$$

We use here the complete multivariate activation function ([4])

$$\Phi(x_1, \dots, x_N) := \Phi(x) := \prod_{i=1}^{N} \Phi_i(x_i), \quad x \in \mathbb{R}^N. \tag{16.1}$$

It has the properties ([4]):

(i)' $\Phi(x) > 0$, $\forall\, x \in \mathbb{R}^N$,

We see that

$$\sum_{k_1=-\infty}^{\infty} \sum_{k_2=-\infty}^{\infty} \dots \sum_{k_N=-\infty}^{\infty} \Phi(x_1 - k_1, x_2 - k_2, \dots, x_N - k_N) =$$

$$\sum_{k_1=-\infty}^{\infty} \sum_{k_2=-\infty}^{\infty} \dots \sum_{k_N=-\infty}^{\infty} \prod_{i=1}^{N} \Phi_i(x_i - k_i) = \prod_{i=1}^{N} \left(\sum_{k_i=-\infty}^{\infty} \Phi_i(x_i - k_i) \right) = 1. \tag{16.2}$$

That is

(ii)'

$$\sum_{k=-\infty}^{\infty} \Phi(x - k) := \sum_{k_1=-\infty}^{\infty} \sum_{k_2=-\infty}^{\infty} \dots \sum_{k_N=-\infty}^{\infty} \Phi(x_1 - k_1, \dots, x_N - k_N) = 1, \tag{16.3}$$

$k := (k_1, \dots, k_n)$, $\forall\, x \in \mathbb{R}^N$.

(iii)'

$$\sum_{k=-\infty}^{\infty} \Phi(nx-k) :=$$

$$\sum_{k_1=-\infty}^{\infty} \sum_{k_2=-\infty}^{\infty} \dots \sum_{k_N=-\infty}^{\infty} \Phi(nx_1-k_1,\dots,nx_N-k_N)=1, \tag{16.4}$$

$\forall\, x \in \mathbb{R}^N;\ n \in \mathbb{N}.$

(iv)'

$$\int_{\mathbb{R}^N} \Phi(x)\,dx = 1, \tag{16.5}$$

that is Φ is a multivariate density function.

Here $\|x\|_\infty := \max\{|x_1|,\dots,|x_N|\}$, $x \in \mathbb{R}^N$, also set $\infty := (\infty,\dots,\infty)$, $-\infty := (-\infty,\dots,-\infty)$ upon the multivariate context.

For $0<\beta<1$ and $n \in \mathbb{N}$, fixed $x \in \mathbb{R}^N$, we have proved ([4])

(v)'

$$\sum_{\substack{k=\lceil na\rceil \\ \left\|\frac{k}{n}-x\right\|_\infty > \frac{1}{n^\beta}}}^{\lfloor nb\rfloor} \Phi(nx-k) \leq 3.1992e^{-n^{(1-\beta)}}. \tag{16.6}$$

Let $f \in C_B\left(\mathbb{R}^N\right)$ (bounded and continuous functions on $\mathbb{R}^N$, $N \in \mathbb{N}$). We define the multivariate Kantorovich type neural network operators ($n \in \mathbb{N}$, $\forall\, x \in \mathbb{R}^N$)

$$K_n(f,x) := K_n(f,x_1,\dots,x_N) := \sum_{k=-\infty}^{\infty}\left(n^N \int_{\frac{k}{n}}^{\frac{k+1}{n}} f(t)\,dt\right)\Phi(nx-k) := \tag{16.7}$$

$$\sum_{k_1=-\infty}^{\infty}\dots\sum_{k_N=-\infty}^{\infty}\left(n^N \int_{\frac{k_1}{n}}^{\frac{k_1+1}{n}}\dots\int_{\frac{k_N}{n}}^{\frac{k_N+1}{n}} f(t_1,\dots,t_N)\,dt_1\dots dt_N\right)$$

$$\left(\prod_{i=1}^{N}\Phi_i(nx_i-k_i)\right).$$

We observe that

$$\int_{\frac{k}{n}}^{\frac{k+1}{n}} f(t)\,dt = \int_{\frac{k_1}{n}}^{\frac{k_1+1}{n}}\dots\int_{\frac{k_N}{n}}^{\frac{k_N+1}{n}} f(t_1,\dots,t_N)\,dt_1\dots dt_N =$$

$$\int_0^{\frac{1}{n}} \cdots \int_0^{\frac{1}{n}} f\left(t_1 + \frac{k_1}{n}, \ldots, t_N + \frac{k_N}{n}\right) dt_1 \ldots dt_N = \int_0^{\frac{1}{n}} f\left(t + \frac{k}{n}\right) dt. \quad (16.8)$$

Thus it holds

$$K_n(f, x) = \sum_{k=-\infty}^{\infty} \left(n^N \int_0^{\frac{1}{n}} f\left(t + \frac{k}{n}\right) dt\right) \Phi(nx - k). \quad (16.9)$$

Again for $f \in C_B(\mathbb{R}^N)$, $N \in \mathbb{N}$, we define the multivariate neural network operators of quadrature type $Q_n(f, x)$, $n \in \mathbb{N}$, as follows. Let $\theta = (\theta_1, \ldots, \theta_N) \in \mathbb{N}^N$, $r = (r_1, \ldots, r_N) \in \mathbb{Z}_+^N$, $w_r = w_{r_1,\ldots,r_N} \geq 0$, such that

$$\sum_{r=0}^{\theta} w_r = \sum_{r_1=0}^{\theta_1} \cdots \sum_{r_N=0}^{\theta_N} w_{r_1,\ldots,r_N} = 1; \ k \in \mathbb{Z}^N$$

and

$$\delta_{nk}(f) := \delta_{n,k_1,\ldots,k_N}(f) := \sum_{r=0}^{\theta} w_r f\left(\frac{k}{n} + \frac{r}{n\theta}\right) \quad (16.10)$$

$$:= \sum_{r_1=0}^{\theta_1} \cdots \sum_{r_N=0}^{\theta_N} w_{r_1,\ldots,r_N} f\left(\frac{k_1}{n} + \frac{r_1}{n\theta_1}, \ldots, \frac{k_N}{n} + \frac{r_N}{n\theta_N}\right), \quad (16.11)$$

where $\frac{r}{\theta} = \left(\frac{r_1}{\theta_1}, \ldots, \frac{r_N}{\theta_N}\right)$.

We define

$$Q_n(f, x) := Q_n(f, x_1, \ldots, x_N) := \sum_{k=-\infty}^{\infty} \delta_{nk}(f) \Phi(nx - k) \quad (16.12)$$

$$:= \sum_{k_1=-\infty}^{\infty} \cdots \sum_{k_N=-\infty}^{\infty} \delta_{n,k_1,\ldots,k_N}(f) \left(\prod_{i=1}^{N} \Phi_i(nx_i - k_i)\right), \quad \forall x \in \mathbb{R}^N.$$

We consider also here the hyperbolic tangent function $\tanh x$, $x \in \mathbb{R}$ (see also [2])

$$\tanh x := \frac{e^x - e^{-x}}{e^x + e^{-x}}. \quad (16.13)$$

It has the properties $\tanh 0 = 0$, $-1 < \tanh x < 1$, $\forall x \in \mathbb{R}$, and $\tanh(-x) = -\tanh x$. Furthermore $\tanh x \to 1$ as $x \to \infty$, and $\tanh x \to -1$, as $x \to -\infty$, and it is strictly increasing on $\mathbb{R}$.

This function plays the role of an activation function in the hidden layer of neural networks.

We further consider ([2])

$$\Psi(x) := \frac{1}{4}\left(\tanh(x+1) - \tanh(x-1)\right) > 0, \ \forall\, x \in \mathbb{R}. \tag{16.14}$$

We easily see that $\Psi(-x) = \Psi(x)$, that is Ψ is even on $\mathbb{R}$. Obviously Ψ is differentiable, thus continuous.

Proposition 16.1 ([2]) *$\Psi(x)$ for $x \geq 0$ is strictly decreasing.*

Obviously $\Psi(x)$ is strictly increasing for $x \leq 0$. Also it holds $\lim_{x\to-\infty} \Psi(x) = 0 = \lim_{x\to\infty} \Psi(x)$.

Infact Ψ has the bell shape with horizontal asymptote the x-axis. So the maximum of Ψ is zero, $\Psi(0) = 0.3809297$.

Theorem 16.2 ([2]) *We have that $\sum_{i=-\infty}^{\infty} \Psi(x-i) = 1, \ \forall\, x \in \mathbb{R}$.*

Thus

$$\sum_{i=-\infty}^{\infty} \Psi(nx - i) = 1, \quad \forall\, n \in \mathbb{N}, \ \forall\, x \in \mathbb{R}.$$

Also it holds

$$\sum_{i=-\infty}^{\infty} \Psi(x + i) = 1, \quad \forall x \in \mathbb{R}.$$

Theorem 16.3 ([2]) *It holds $\int_{-\infty}^{\infty} \Psi(x)\, dx = 1$.*

So $\Psi(x)$ is a density function on $\mathbb{R}$.

Theorem 16.4 ([2]) *Let $0 < \alpha < 1$ and $n \in \mathbb{N}$. It holds*

$$\sum_{\begin{cases} k = -\infty \\ : |nx - k| \geq n^{1-\alpha} \end{cases}}^{\infty} \Psi(nx - k) \leq e^4 \cdot e^{-2n^{(1-\alpha)}}.$$

In this chapter we also use the complete multivariate activation function

$$\Theta(x_1, \ldots, x_N) := \Theta(x) := \prod_{i=1}^{N} \Psi(x_i), \ \ x = (x_1, \ldots, x_N) \in \mathbb{R}^N, \ \ N \in \mathbb{N}. \tag{16.15}$$

It has the properties (see [3])

(i) $\Theta(x) > 0, \ \forall\, x \in \mathbb{R}^N$,

(ii)

$$\sum_{k=-\infty}^{\infty} \Theta(x-k) := \sum_{k_1=-\infty}^{\infty} \sum_{k_2=-\infty}^{\infty} \cdots \sum_{k_N=-\infty}^{\infty} \Theta(x_1-k_1, \ldots, x_N-k_N) = 1, \tag{16.16}$$

where $k := (k_1, \ldots, k_N)$, $\forall\, x \in \mathbb{R}^N$.

(iii)

$$\sum_{k=-\infty}^{\infty} \Theta(nx-k) :=$$

$$\sum_{k_1=-\infty}^{\infty} \sum_{k_2=-\infty}^{\infty} \cdots \sum_{k_N=-\infty}^{\infty} \Theta(nx_1-k_1, \ldots, nx_N-k_N) = 1, \tag{16.17}$$

$\forall\, x \in \mathbb{R}^N$; $n \in \mathbb{N}$.

(iv)

$$\int_{\mathbb{R}^N} \Theta(x)\, dx = 1, \tag{16.18}$$

that is Θ is a multivariate density function.

By [3] we get

(v)

$$\sum_{\begin{cases} k=-\infty \\ \left\| \frac{k}{n} - x \right\|_\infty > \frac{1}{n^\beta} \end{cases}}^{\infty} \Theta(nx-k) \le e^4 \cdot e^{-2n^{(1-\beta)}}, \tag{16.19}$$

$0 < \beta < 1$, $n \in \mathbb{N}$, $x \in \mathbb{R}^N$.

We also define the following Kantorovich type neural network operators, $f \in C_B(\mathbb{R}^N)$, $N \in \mathbb{N}$, $n \in \mathbb{N}$, $\forall\, x \in \mathbb{R}^N$, similarly to (16.7):

$$L_n(f,x) := L_n(f, x_1, \ldots, x_N) := \sum_{k=-\infty}^{\infty} \left(n^N \int_{\frac{k}{n}}^{\frac{k+1}{n}} f(t)\, dt \right) \Theta(nx-k) := \tag{16.20}$$

$$\sum_{k_1=-\infty}^{\infty} \cdots \sum_{k_N=-\infty}^{\infty} \left(n^N \int_{\frac{k_1}{n}}^{\frac{k_1+1}{n}} \cdots \int_{\frac{k_N}{n}}^{\frac{k_N+1}{n}} f(t_1, \ldots, t_N)\, dt_1 \ldots dt_N \right)$$

$$\left(\prod_{i=1}^{N} \Psi(nx_i - k_i) \right).$$

Similarly to (16.9) it holds

$$L_n(f,x) = \sum_{k=-\infty}^{\infty} \left(n^N \int_0^{\frac{1}{n}} f\left(t + \frac{k}{n}\right) dt \right) \Theta(nx - k). \tag{16.21}$$

Finally we define, similarly to (16.12), (for any $x \in \mathbb{R}^N$) the following quadrature type neural network operators

$$T_n(f,x) := T_n(f, x_1, \ldots, x_N) := \sum_{k=-\infty}^{\infty} \delta_{nk}(f)\,\Theta(nx - k) \tag{16.22}$$

$$:= \sum_{k_1=-\infty}^{\infty} \cdots \sum_{k_N=-\infty}^{\infty} \delta_{n,k_1,\ldots,k_N}(f) \left(\prod_{i=1}^{N} \Psi(nx_i - k_i) \right),$$

where $\delta_{nk}(f)$ is as in (16.10) and (16.11).

For $f \in C_B(\mathbb{R}^N)$ we define the first multivariate modulus of continuity

$$\omega_1(f,h) := \sup_{\substack{x,y \in \mathbb{R}^N \\ \|x-y\|_\infty \le h}} |f(x) - f(y)|, \; h > 0. \tag{16.23}$$

Given that $f \in C_U(\mathbb{R}^N)$ (uniformly continuous functions on $\mathbb{R}^N$) we get that $\lim_{h \to 0} \omega_1(f,h) = 0$, the same definition for ω_1.

In this chapter we study the pointwise and uniform convergence of operators K_n, Q_n, L_n and T_n to the unit operator I with rates. We are inspired by [1–5, 7, 8].

16.2 Main Results

We present

Theorem 16.5 *Let $f \in C_B(\mathbb{R}^N)$, $0 < \beta < 1$, $x \in \mathbb{R}^N$, $n, N \in \mathbb{N}$. Then*

(1)

$$|K_n(f,x) - f(x)| \le \omega_1\left(f, \frac{1}{n} + \frac{1}{n^\beta}\right) + (6.3984)\,\|f\|_\infty e^{-n^{(1-\beta)}} =: \rho_1 \tag{16.24}$$

(2)

$$\|K_n(f) - f\|_\infty \le \rho_1. \tag{16.25}$$

Proof We have that

$$K_n(f,x) - f(x) =$$

$$\sum_{k=-\infty}^{\infty} \left(n^N \int_0^{\frac{1}{n}} f\left(t + \frac{k}{n}\right) dt \right) \Phi(nx-k) - f(x) \sum_{k=-\infty}^{\infty} \Phi(nx-k) =$$

$$\sum_{k=-\infty}^{\infty} \left[\left(n^N \int_0^{\frac{1}{n}} f\left(t + \frac{k}{n}\right) dt \right) - f(x) \right] \Phi(nx-k) = \quad (16.26)$$

$$\sum_{k=-\infty}^{\infty} \left[n^N \int_0^{\frac{1}{n}} \left(f\left(t + \frac{k}{n}\right) - f(x) \right) dt \right] \Phi(nx-k).$$

Hence

$$|K_n(f,x) - f(x)| \leq \sum_{k=-\infty}^{\infty} \left[n^N \int_0^{\frac{1}{n}} \left| f\left(t + \frac{k}{n}\right) - f(x) \right| dt \right] \Phi(nx-k) \quad (16.27)$$

$$= \sum_{\substack{k=-\infty \\ \left\| \frac{k}{n} - x \right\|_\infty \leq \frac{1}{n^\beta}}}^{\infty} \left[n^N \int_0^{\frac{1}{n}} \left| f\left(t + \frac{k}{n}\right) - f(x) \right| dt \right] \Phi(nx-k) +$$

$$\sum_{\substack{k=-\infty \\ \left\| \frac{k}{n} - x \right\|_\infty > \frac{1}{n^\beta}}}^{\infty} \left[n^N \int_0^{\frac{1}{n}} \left| f\left(t + \frac{k}{n}\right) - f(x) \right| dt \right] \Phi(nx-k) \leq$$

$$\omega_1\left(f, \frac{1}{n} + \frac{1}{n^\beta}\right) + 2\|f\|_\infty \sum_{\substack{k=-\infty \\ \left\| \frac{k}{n} - x \right\|_\infty > \frac{1}{n^\beta}}}^{\infty} \Phi(nx-k) \overset{(16.6)}{\leq} \quad (16.28)$$

$$\omega_1\left(f, \frac{1}{n} + \frac{1}{n^\beta}\right) + (6.3984)\|f\|_\infty e^{-n^{(1-\beta)}},$$

proving the claim. ■

We continue with

Theorem 16.6 *Let* $f \in C_B(\mathbb{R}^N)$, $0 < \beta < 1$, $x \in \mathbb{R}^N$, $n, N \in \mathbb{N}$. *Then*
(1)

$$|Q_n(f,x) - f(x)| \leq \omega_1\left(f, \frac{1}{n} + \frac{1}{n^\beta}\right) + (6.3984)\,\|f\|_\infty\, e^{-n^{(1-\beta)}} = \rho_1, \quad (16.29)$$

(2)

$$\|Q_n(f) - f\|_\infty \leq \rho_1. \quad (16.30)$$

Proof We notice that

$$Q_n(f,x) - f(x) = \sum_{k=-\infty}^{\infty} \delta_{nk}(f)\,\Phi(nx-k) - f(x)\sum_{k=-\infty}^{\infty} \Phi(nx-k)$$

$$= \sum_{k=-\infty}^{\infty} (\delta_{nk}(f) - f(x))\,\Phi(nx-k)$$

$$= \sum_{k=-\infty}^{\infty}\left(\sum_{r=0}^{\theta} w_r\left(f\left(\frac{k}{n} + \frac{r}{n\theta}\right) - f(x)\right)\right)\Phi(nx-k). \quad (16.31)$$

Hence it holds

$$|Q_n(f,x) - f(x)| \leq \sum_{k=-\infty}^{\infty}\left(\sum_{r=0}^{\theta} w_r\left|f\left(\frac{k}{n} + \frac{r}{n\theta}\right) - f(x)\right|\right)\Phi(nx-k)$$

$$= \sum_{\substack{k=-\infty \\ \left\|\frac{k}{n} - x\right\|_\infty \leq \frac{1}{n^\beta}}}^{\infty}\left(\sum_{r=0}^{\theta} w_r\left|f\left(\frac{k}{n} + \frac{r}{n\theta}\right) - f(x)\right|\right)\Phi(nx-k) + \quad (16.32)$$

$$\sum_{\substack{k=-\infty \\ \left\|\frac{k}{n} - x\right\|_\infty > \frac{1}{n^\beta}}}^{\infty}\left(\sum_{r=0}^{\theta} w_r\left|f\left(\frac{k}{n} + \frac{r}{n\theta}\right) - f(x)\right|\right)\Phi(nx-k) \leq$$

$$\omega_1\left(f, \frac{1}{n} + \frac{1}{n^\beta}\right) + 2\,\|f\|_\infty \sum_{\substack{k=-\infty \\ \left\|\frac{k}{n} - x\right\|_\infty > \frac{1}{n^\beta}}}^{\infty} \Phi(nx-k) \overset{(16.6)}{\leq} \quad (16.33)$$

$$\omega_1\left(f, \frac{1}{n} + \frac{1}{n^\beta}\right) + (6.3984)\,\|f\|_\infty\, e^{-n^{(1-\beta)}},$$

proving the claim. ■

We further state

Theorem 16.7 *Same assumptions as in Theorem 16.5. Then*

(1)

$$|L_n(f,x) - f(x)| \leq \omega_1\left(f, \frac{1}{n} + \frac{1}{n^{\beta}}\right) + \|f\|_{\infty} 2e^4 e^{-2n^{(1-\beta)}} =: \rho_2, \tag{16.34}$$

and

(2)

$$\|L_n(f) - f\|_{\infty} \leq \rho_2. \tag{16.35}$$

Proof As in Theorem 16.5, using (16.19). ■

Theorem 16.8 *Same assumptions as in Theorem 16.5. Then*

(1)

$$|T_n(f,x) - f(x)| \leq \rho_2, \tag{16.36}$$

and

(2)

$$\|T_n(f) - f\|_{\infty} \leq \rho_2. \tag{16.37}$$

Proof As in Theorem 16.6, using (16.19). ■

Conclusion 16.9 *When* $f \in \left(C_B\left(\mathbb{R}^N\right) \cap C_U\left(\mathbb{R}^N\right)\right)$, *then* $K_n(f,x) \to f(x)$, $Q_n(f,x) \to f(x)$, $L_n(f,x) \to f(x)$, $T_n(f,x) \to f(x)$, *pointwise, as* $n \to \infty$, *and* $K_n(f) \to f$, $Q_n(f) \to f$, $L_n(f) \to f$, $T_n(f) \to f$, *uniformly, as* $n \to \infty$, *all at the speed of* $\frac{1}{n^{\beta}}$, $0 < \beta < 1$.

References

1. G.A. Anastassiou, *Inteligent Systems: Approximation by Artificial Neural Networks* (Springer, Heidelberg, 2011)
2. G.A. Anastassiou, Univariate hyperbolic tangent neural network approximation. Math. Comput. Model. **53**, 1111–1132 (2011)
3. G.A. Anastassiou, Multivariate hyperbolic tangent neural network approximation. Comput. Math. Appl. **61**, 809–821 (2011)
4. G.A. Anastassiou, Multivariate sigmoidal neural network approximation. Neural Netw. **24**, 378–386 (2011)
5. G.A. Anastassiou, Univariate sigmoidal neural network approximation. J. Comput. Anal. Appl. **14**(4), 659–690 (2012)
6. G.A. Anastassiou, Approximation by Kantorovich and Quadrature type quasi-interpolation neural network operators. J. Concrete Appl. Math. **13**(3–4), 242–251 (2015)
7. G.A. Anastassiou, Multivariate error function based neural network approximations, Rev. Anal. Numer. Theor. Approx. Accepted (2014)
8. Z. Chen, F. Cao, The approximation operators with sigmoidal functions. Comput. Math. Appl. **58**, 758–765 (2009)

Chapter 17
Univariate Error Function Based Neural Network Approximations

Here we study the univariate quantitative approximation of real and complex valued continuous functions on a compact interval or all the real line by quasi-interpolation, Baskakov type and quadrature type neural network operators. We perform also the related fractional approximation. These approximations are derived by establishing Jackson type inequalities involving the modulus of continuity of the engaged function or its high order derivative or fractional derivatives. Our operators are defined by using a density function induced by the error function. The approximations are pointwise and with respect to the uniform norm. The related feed-forward neural networks are with one hidden layer. It follows [14].

17.1 Introduction

The author in [2, 3], see Chaps. 2–5, was the first to establish neural network approximations to continuous functions with rates by very specifically defined neural network operators of Cardaliagnet-Euvrard and "Squashing" types, by employing the modulus of continuity of the engaged function or its high order derivative, and producing very tight Jackson type inequalities. He treats there both the univariate and multivariate cases. The defining these operators "bell-shaped" and "squashing" functions are assumed to be of compact support. Also in [3] he gives the Nth order asymptotic expansion for the error of weak approximation of these two operators to a special natural class of smooth functions, see Chaps. 4 and 5 there.

The author inspired by [16], continued his studies on neural networks approximation by introducing and using the proper quasi-interpolation operators of sigmoidal and hyperbolic tangent type which resulted into [7, 9–12], by treating both the univariate and multivariate cases. He did also the corresponding fractional case [13].

The author here performs univariate error function based neural network approximations to continuous functions over compact intervals of the real line or over the

G.A. Anastassiou, *Intelligent Systems II: Complete Approximation by Neural Network Operators*, Studies in Computational Intelligence 608,
DOI 10.1007/978-3-319-20505-2_17

whole $\mathbb{R}$, the he extends his results to complex valued functions. Finally he treats completely the related fractional approximation. All convergences here are with rates expressed via the modulus of continuity of the involved function or its high order derivative, or fractional derivatives and given by very tight Jackson type inequalities.

The author comes up with the "right" precisely defined quasi-interpolation, Baskakov type and quadrature neural networks operators, associated with the error function and related to a compact interval or real line. Our compact intervals are not necessarily symmetric to the origin. Some of our upper bounds to error quantity are very flexible and general. In preparation to prove our results we establish important properties of the basic density function defining our operators.

Feed-forward neural networks (FNNs) with one hidden layer, the only type of networks we deal with in this chapter, are mathematically expressed as

$$N_n(x) = \sum_{j=0}^{n} c_j \sigma\left(\langle a_j \cdot x \rangle + b_j\right), \quad x \in \mathbb{R}^s, \ s \in \mathbb{N},$$

where for $0 \leq j \leq n$, $b_j \in \mathbb{R}$ are the thresholds, $a_j \in \mathbb{R}^s$ are the connection weights, $c_j \in \mathbb{R}$ are the coefficients, $\langle a_j \cdot x \rangle$ is the inner product of a_j and x, and σ is the activation function of the network. In many fundamental neural network models, the activation function is the error. About neural networks in general read [20–22].

17.2 Basics

We consider here the (Gauss) error special function [1, 15]

$$\operatorname{erf}(x) = \frac{2}{\sqrt{\pi}} \int_0^x e^{-t^2} dt, \quad x \in \mathbb{R}, \tag{17.1}$$

which is a sigmoidal type function and a strictly increasing function.

It has the basic properties

$$\operatorname{erf}(0) = 0, \ \operatorname{erf}(-x) = -\operatorname{erf}(x), \quad \operatorname{erf}(+\infty) = 1, \ \operatorname{erf}(-\infty) = -1, \tag{17.2}$$

and

$$(\operatorname{erf}(x))' = \frac{2}{\sqrt{\pi}} e^{-x^2}, \quad x \in \mathbb{R}, \tag{17.3}$$

$$\int \operatorname{erf}(x)\, dx = x \operatorname{erf}(x) + \frac{e^{-x^2}}{\sqrt{\pi}} + C, \tag{17.4}$$

where C is a constant.

The error function is related to the cumulative probability distribution function of the standard normal distribution

$$\Phi(x) = \frac{1}{2} + \frac{1}{2}\operatorname{erf}\left(\frac{x}{\sqrt{2}}\right).$$

We consider the activation function

$$\chi(x) = \frac{1}{4}(\operatorname{erf}(x+1) - \operatorname{erf}(x-1)), \ x \in \mathbb{R}, \tag{17.5}$$

and we notice that

$$\chi(-x) = \frac{1}{4}(\operatorname{erf}(-x+1) - \operatorname{erf}(-x-1)) =$$

$$\frac{1}{4}(\operatorname{erf}(-(x-1)) - \operatorname{erf}(-(x+1))) = \frac{1}{4}(-\operatorname{erf}(x-1) + \operatorname{erf}(x+1)) = \chi(x), \tag{17.6}$$

thus χ is an even function.

Since $x+1 > x-1$, then $\operatorname{erf}(x+1) > \operatorname{erf}(x-1)$, and $\chi(x) > 0$, all $x \in \mathbb{R}$.

We see that

$$\chi(0) = \frac{\operatorname{erf}(1)}{2} \simeq \frac{0.843}{2} = 0.4215. \tag{17.7}$$

Let $x > 0$, we have

$$\chi'(x) = \frac{1}{4}\left(\frac{2}{\sqrt{\pi}}e^{-(x+1)^2} - \frac{2}{\sqrt{\pi}}e^{-(x-1)^2}\right) = \tag{17.8}$$

$$\frac{1}{2\sqrt{\pi}}\left(\frac{1}{e^{(x+1)^2}} - \frac{1}{e^{(x-1)^2}}\right) = \frac{1}{2\sqrt{\pi}}\left(\frac{e^{(x-1)^2} - e^{(x+1)^2}}{e^{(x+1)^2}e^{(x-1)^2}}\right) < 0,$$

proving $\chi'(x) < 0$, for $x > 0$.

That is χ is strictly decreasing on $[0, \infty)$ and is strictly increasing on $(-\infty, 0]$, and $\chi'(0) = 0$.

Clearly the x-axis is the horizontal asymptote on χ.

Conclusion, χ is a bell symmetric function with maximum $\chi(0) \simeq 0.4215$.

We further present

Theorem 17.1 *We have that*

$$\sum_{i=-\infty}^{\infty} \chi(x-i) = 1, \quad all\, x \in \mathbb{R}. \tag{17.9}$$

Proof We notice

$$\sum_{i=-\infty}^{\infty} \text{erf}(x-i) - \text{erf}(x-1-i) =$$

$$\sum_{i=0}^{\infty} (\text{erf}(x-i) - \text{erf}(x-1-i)) + \sum_{i=-\infty}^{-1} (\text{erf}(x-i) - \text{erf}(x-1-i)). \tag{17.10}$$

Furthermore ($\lambda \in \mathbb{Z}^+$) (telescoping sum)

$$\sum_{i=0}^{\infty} (\text{erf}(x-i) - \text{erf}(x-1-i)) =$$

$$\lim_{\lambda\to\infty} \sum_{i=0}^{\lambda} (\text{erf}(x-i) - \text{erf}(x-1-i)) =$$

$$\text{erf}(x) - \lim_{\lambda\to\infty} \text{erf}(x-1-\lambda) = 1 + \text{erf}(x). \tag{17.11}$$

Similarly we get

$$\sum_{i=-\infty}^{-1} (\text{erf}(x-i) - \text{erf}(x-1-i)) =$$

$$\lim_{\lambda\to\infty} \sum_{i=-\lambda}^{-1} (\text{erf}(x-i) - \text{erf}(x-1-i)) =$$

$$\lim_{\lambda\to\infty} (\text{erf}(x+\lambda) - \text{erf}(x)) = 1 - \text{erf}(x). \tag{17.12}$$

Adding (17.11) and (17.12), we get

$$\sum_{i=-\infty}^{\infty} (\text{erf}(x-i) - \text{erf}(x-1-i)) = 2, \quad \text{for any } x \in \mathbb{R}. \tag{17.13}$$

Hence (17.13) is true for $(x+1)$, giving us

$$\sum_{i=-\infty}^{\infty} (\text{erf}(x+1-i) - \text{erf}(x-i)) = 2, \quad \text{for any } x \in \mathbb{R}. \tag{17.14}$$

Adding (17.13) and (17.14) we obtain

$$\sum_{i=-\infty}^{\infty} (\text{erf}(x+1-i) - \text{erf}(x-1-i)) = 4, \quad \text{for any } x \in \mathbb{R}, \tag{17.15}$$

proving (17.9). ∎

Thus

$$\sum_{i=-\infty}^{\infty} \chi(nx-i) = 1, \quad \forall n \in \mathbb{N},\ \forall x \in \mathbb{R}. \tag{17.16}$$

Furthermore we get:
Since χ is even it holds $\sum_{i=-\infty}^{\infty} \chi(i-x) = 1$, for any $x \in \mathbb{R}$.
Hence $\sum_{i=-\infty}^{\infty} \chi(i+x) = 1$, $\forall x \in \mathbb{R}$, and $\sum_{i=-\infty}^{\infty} \chi(x+i) = 1$, $\forall x \in \mathbb{R}$.

Theorem 17.2 *It holds*

$$\int_{-\infty}^{\infty} \chi(x)\, dx = 1. \tag{17.17}$$

Proof We notice that

$$\int_{-\infty}^{\infty} \chi(x)\, dx = \sum_{j=-\infty}^{\infty} \int_{j}^{j+1} \chi(x)\, dx = \sum_{j=-\infty}^{\infty} \int_{0}^{1} \chi(x+j)\, dx =$$

$$\int_{0}^{1} \left(\sum_{j=-\infty}^{\infty} \chi(x+j) \right) dx = \int_{0}^{1} 1 dx = 1.$$

■

So $\chi(x)$ is a density function on $\mathbb{R}$.

Theorem 17.3 *Let* $0 < \alpha < 1$, *and* $n \in \mathbb{N}$ *with* $n^{1-\alpha} \geq 3$. *It holds*

$$\sum_{\begin{cases} k=-\infty \\ : |nx-k| \geq n^{1-\alpha} \end{cases}}^{\infty} \chi(nx-k) < \frac{1}{2\sqrt{\pi}\left(n^{1-\alpha}-2\right) e^{\left(n^{1-\alpha}-2\right)^2}}. \tag{17.18}$$

Proof Let $x \geq 1$. That is $0 \leq x-1 < x+1$. Applying the mean value theorem we get

$$\chi(x) = \frac{1}{4}\left(\operatorname{erf}(x+1) - \operatorname{erf}(x-1)\right) = \frac{1}{\sqrt{\pi}} e^{-\xi^2}, \tag{17.19}$$

where $x-1 < \xi < x+1$.

Hence

$$\chi(x) < \frac{e^{-(x-1)^2}}{\sqrt{\pi}}, \quad x \geq 1. \tag{17.20}$$

Thus we have

$$\sum_{\substack{k=-\infty \\ : |nx-k| \geq n^{1-\alpha}}}^{\infty} \chi(nx-k) = \sum_{\substack{k=-\infty \\ : |nx-k| \geq n^{1-\alpha}}}^{\infty} \chi(|nx-k|) <$$

$$\frac{1}{\sqrt{\pi}} \sum_{\substack{k=-\infty \\ : |nx-k| \geq n^{1-\alpha}}}^{\infty} e^{-(|nx-k|-1)^2} \leq \frac{1}{\sqrt{\pi}} \int_{(n^{1-\alpha}-1)}^{\infty} e^{-(x-1)^2} dx \tag{17.21}$$

$$= \frac{1}{\sqrt{\pi}} \int_{n^{1-\alpha}-2}^{\infty} e^{-z^2} dz$$

(see Sect. 3.7.3 of [23])

$$= \frac{1}{2\sqrt{\pi}} \left(\min \left(\sqrt{\pi}, \frac{1}{(n^{1-\alpha}-2)} \right) \right) e^{-(n^{1-\alpha}-2)^2}$$

(by $n^{1-\alpha} - 2 \geq 1$, hence $\frac{1}{n^{1-\alpha}-2} \leq 1 < \sqrt{\pi}$)

$$< \frac{1}{2\sqrt{\pi}\left(n^{1-\alpha}-2\right) e^{(n^{1-\alpha}-2)^2}}, \tag{17.22}$$

proving the claim. ■

Denote by $\lfloor \cdot \rfloor$ the integral part of the number and by $\lceil \cdot \rceil$ the ceiling of the number.

Theorem 17.4 *Let $x \in [a, b] \subset \mathbb{R}$ and $n \in \mathbb{N}$ so that $\lceil na \rceil \leq \lfloor nb \rfloor$. It holds*

$$\frac{1}{\sum_{k=\lceil na \rceil}^{\lfloor nb \rfloor} \chi(nx-k)} < \frac{1}{\chi(1)} \simeq 4.019, \ \forall\, x \in [a, b]. \tag{17.23}$$

Proof Let $x \in [a, b]$. We see that

$$1 = \sum_{k=-\infty}^{\infty} \chi(nx-k) > \sum_{k=\lceil na \rceil}^{\lfloor nb \rfloor} \chi(nx-k) = \tag{17.24}$$

$$\sum_{k=\lceil na \rceil}^{\lfloor nb \rfloor} \chi(|nx-k|) > \chi(|nx-k_0|),$$

$\forall\, k_0 \in [\lceil na \rceil, \lfloor nb \rfloor] \cap \mathbb{Z}$.

We can choose $k_0 \in [\lceil na \rceil, \lfloor nb \rfloor] \cap \mathbb{Z}$ such that $|nx - k_0| < 1$.

Therefore

$$\chi(|nx - k_0|) > \chi(1) = \frac{1}{4}(\text{erf}(2) - \text{erf}(0)) =$$

$$\frac{\text{erf}(2)}{4} = \frac{0.99533}{4} = 0.2488325. \quad (17.25)$$

Consequently we get

$$\sum_{k=\lceil na \rceil}^{\lfloor nb \rfloor} \chi(|nx - k|) > \chi(1) \simeq 0.2488325, \quad (17.26)$$

and

$$\frac{1}{\sum_{k=\lceil na \rceil}^{\lfloor nb \rfloor} \chi(|nx - k|)} < \frac{1}{\chi(1)} \simeq 4.019, \quad (17.27)$$

proving the claim. ■

Remark 17.5 We also notice that

$$1 - \sum_{k=\lceil na \rceil}^{\lfloor nb \rfloor} \chi(nb - k) = \sum_{k=-\infty}^{\lceil na \rceil - 1} \chi(nb - k) + \sum_{k=\lfloor nb \rfloor + 1}^{\infty} \chi(nb - k)$$

$$> \chi(nb - \lfloor nb \rfloor - 1)$$

(call $\varepsilon := nb - \lfloor nb \rfloor$, $0 \leq \varepsilon < 1$)

$$= \chi(\varepsilon - 1) = \chi(1 - \varepsilon) \geq \chi(1) > 0. \quad (17.28)$$

Therefore

$$\lim_{n \to \infty} \left(1 - \sum_{k=\lceil na \rceil}^{\lfloor nb \rfloor} \chi(nb - k) \right) > 0.$$

Similarly,

$$1 - \sum_{k=\lceil na \rceil}^{\lfloor nb \rfloor} \chi(na - k) = \sum_{k=-\infty}^{\lceil na \rceil - 1} \chi(na - k) + \sum_{k=\lfloor nb \rfloor + 1}^{\infty} \chi(na - k)$$

$$> \chi(na - \lceil na \rceil + 1)$$

(call $\eta := \lceil na \rceil - na,\ 0 \leq \eta < 1$)

$$= \chi(1-\eta) \geq \chi(1) > 0.$$

Therefore again

$$\lim_{n\to\infty} \left(1 - \sum_{k=\lceil na \rceil}^{\lfloor nb \rfloor} \chi(na-k)\right) > 0. \tag{17.29}$$

Hence we derive that

$$\lim_{n\to\infty} \sum_{k=\lceil na \rceil}^{\lfloor nb \rfloor} \chi(nx-k) \neq 1, \tag{17.30}$$

for at least some $x \in [a,b]$.

Note 17.6 *For large enough n we always obtain* $\lceil na \rceil \leq \lfloor nb \rfloor$. *Also* $a \leq \frac{k}{n} \leq b$, *iff* $\lceil na \rceil \leq k \leq \lfloor nb \rfloor$. *In general it holds (by (17.16)) that*

$$\sum_{k=\lceil na \rceil}^{\lfloor nb \rfloor} \chi(nx-k) \leq 1. \tag{17.31}$$

We give

Definition 17.7 Let $f \in C([a,b])$ $n \in \mathbb{N}$. We set

$$A_n(f,x) = \frac{\sum_{k=\lceil na \rceil}^{\lfloor nb \rfloor} f\left(\frac{k}{n}\right) \chi(nx-k)}{\sum_{k=\lceil na \rceil}^{\lfloor nb \rfloor} \chi(nx-k)}, \forall\ x \in [a.b], \tag{17.32}$$

A_n is a neural network operator.

Definition 17.8 Let $f \in C_B(\mathbb{R})$, (continuous and bounded functions on $\mathbb{R}$), $n \in \mathbb{N}$. We introduce the quasi-interpolation operator

$$B_n(f,x) := \sum_{k=-\infty}^{\infty} f\left(\frac{k}{n}\right) \chi(nx-k), \quad \forall\, x \in \mathbb{R}, \tag{17.33}$$

and the Kantorovich type operator

$$C_n(f,x) = \sum_{k=-\infty}^{\infty} \left(n \int_{\frac{k}{n}}^{\frac{k+1}{n}} f(t)\,dt\right) \chi(nx-k), \quad \forall\, x \in \mathbb{R}. \tag{17.34}$$

B_n, C_n are neural network operators.

Also we give

Definition 17.9 Let $f \in C_B(\mathbb{R})$, $n \in \mathbb{N}$. Let $\theta \in \mathbb{N}$, $w_r \geq 0$, $\sum_{r=0}^{\theta} w_r = 1$, $k \in \mathbb{Z}$, and

$$\delta_{nk}(f) = \sum_{r=0}^{\theta} w_r f\left(\frac{k}{n} + \frac{r}{n\theta}\right). \tag{17.35}$$

We put

$$D_n(f, x) = \sum_{k=-\infty}^{\infty} \delta_{nk}(f) \chi(nx - k), \quad \forall x \in \mathbb{R}. \tag{17.36}$$

D_n is a neural network operator of quadrature type.

We need

Definition 17.10 For $f \in C([a, b])$, the first modulus of continuity is given by

$$\omega_1(f, \delta) := \sup_{\substack{x, y \in [a, b] \\ |x - y| \leq \delta}} |f(x) - f(y)|, \quad \delta > 0. \tag{17.37}$$

We have that $\lim_{\delta \to 0} \omega_1(f, \delta) = 0$.

Similarly $\omega_1(f, \delta)$ is defined for $f \in C_B(\mathbb{R})$.

We know that, f is uniformly continuous on $\mathbb{R}$ iff $\lim_{\delta \to 0} \omega_1(f, \delta) = 0$.

We make

Remark 17.11 We notice the following, that

$$A_n(f, x) - f(x) \overset{(17.32)}{=} \frac{\sum_{k=\lceil na \rceil}^{\lfloor nb \rfloor} f\left(\frac{k}{n}\right) \chi(nx - k) - f(x) \sum_{k=\lceil na \rceil}^{\lfloor nb \rfloor} \chi(nx - k)}{\sum_{k=\lceil na \rceil}^{\lfloor nb \rfloor} \chi(nx - k)}, \tag{17.38}$$

using (17.23) we get,

$$|A_n(f, x) - f(x)| \leq (4.019) \left| \sum_{k=\lceil na \rceil}^{\lfloor nb \rfloor} f\left(\frac{k}{n}\right) \chi(nx - k) - f(x) \sum_{k=\lceil na \rceil}^{\lfloor nb \rfloor} \chi(nx - k) \right|. \tag{17.39}$$

Again here $0 < \alpha < 1$ and $n \in \mathbb{N}$ with $n^{1-\alpha} \geq 3$. Let the fixed $K, L > 0$; for the linear combination $\frac{K}{n^\alpha} + \frac{L}{(n^{1-\alpha}-2)e^{(n^{1-\alpha}-2)^2}}$, the dominant rate of convergence to zero, as $n \to \infty$, is $n^{-\alpha}$. The closer α is to 1, we get faster and better rate of convergence to zero.

In this chapter we study basic approximation properties of A_n, B_n, C_n, D_n neural network operators. That is, the quantitative pointwise and uniform convergence of these operators to the unit operator I.

17.3 Real Neural Network Approximations

Here we present a series of neural network approximations to a function given with rates.

We give

Theorem 17.12 *Let* $f \in C([a,b])$, $0 < \alpha < 1$, $x \in [a,b]$, $n \in \mathbb{N}$ *with* $n^{1-\alpha} \geq 3$, $\|\cdot\|_\infty$ *is the supremum norm. Then*

(i)

$$|A_n(f,x) - f(x)| \leq (4.019)\left[\omega_1\left(f, \frac{1}{n^\alpha}\right) + \frac{\|f\|_\infty}{\sqrt{\pi}\left(n^{1-\alpha}-2\right)e^{\left(n^{1-\alpha}-2\right)^2}}\right] =: \mu_{1n}, \tag{17.40}$$

(ii)

$$\|A_n(f) - f\|_\infty \leq \mu_{1n}. \tag{17.41}$$

We notice that $\lim_{n\to\infty} A_n(f) = f$, *pointwise and uniformly.*

Proof Using (17.39) we get

$$|A_n(f,x) - f(x)| \leq (4.019)\left[\sum_{k=\lceil na\rceil}^{\lfloor nb\rfloor}\left|f\left(\frac{k}{n}\right) - f(x)\right|\chi(nx-k)\right] \leq$$

$$(4.019)\left[\sum_{\substack{k=\lceil na\rceil \\ \left|\frac{k}{n}-x\right| \leq \frac{1}{n^\alpha}}}^{\lfloor nb\rfloor}\left|f\left(\frac{k}{n}\right) - f(x)\right|\chi(nx-k) + \right.$$

$$\left.\sum_{\substack{k=\lceil na\rceil \\ \left|\frac{k}{n}-x\right| > \frac{1}{n^\alpha}}}^{\lfloor nb\rfloor}\left|f\left(\frac{k}{n}\right) - f(x)\right|\chi(nx-k)\right] \leq \tag{17.42}$$

$$(4.019)\left[\omega_1\left(f, \frac{1}{n^\alpha}\right)\left(\sum_{k=\lceil na\rceil}^{\lfloor nb\rfloor}\chi(nx-k)\right) + \right.$$

$$2\left\|f\right\|_{\infty}\left(\sum_{\left\{\begin{array}{l}k=\lceil na\rceil\\ |nx-k|\geq n^{1-\alpha}\end{array}\right.}^{\lfloor nb\rfloor}\chi\left(nx-k\right)\right)\Bigg]\overset{(\text{by}(17.18),(17.31))}{\leq} \tag{17.43}$$

$$(4.019)\left[\omega_1\left(f,\frac{1}{n^{\alpha}}\right)+\frac{\|f\|_{\infty}}{\sqrt{\pi}\left(n^{1-\alpha}-2\right)e^{\left(n^{1-\alpha}-2\right)^2}}\right],$$

proving the claim. ■

We continue with

Theorem 17.13 *Let $f\in C_B(\mathbb{R})$, $0<\alpha<1$, $x\in\mathbb{R}$, $n\in\mathbb{N}$ with $n^{1-\alpha}\geq 3$. Then*

(i)

$$|B_n(f,x)-f(x)|\leq\omega_1\left(f,\frac{1}{n^{\alpha}}\right)+\frac{\|f\|_{\infty}}{\sqrt{\pi}\left(n^{1-\alpha}-2\right)e^{\left(n^{1-\alpha}-2\right)^2}}=:\mu_{2n}, \tag{17.44}$$

(ii)

$$\|B_n(f)-f\|_{\infty}\leq\mu_{2n}. \tag{17.45}$$

For $f\in(C_B(\mathbb{R})\cap C_u(\mathbb{R}))$ ($C_u(\mathbb{R})$ uniformly continuous functions on $\mathbb{R}$) we get $\lim_{n\to\infty}B_n(f)=f$, pointwise and uniformly.

Proof We see that

$$|B_n(f,x)-f(x)|\overset{(\text{by }(17.16),(17.33))}{=}\left|\sum_{k=-\infty}^{\infty}\left(f\left(\frac{k}{n}\right)-f(x)\right)\chi(nx-k)\right|\leq$$

$$\sum_{k=-\infty}^{\infty}\left|f\left(\frac{k}{n}\right)-f(x)\right|\chi(nx-k)\leq$$

$$\sum_{\left\{\begin{array}{l}k=-\infty\\ \left|\frac{k}{n}-x\right|\leq\frac{1}{n^{\alpha}}\end{array}\right.}^{\infty}\left|f\left(\frac{k}{n}\right)-f(x)\right|\chi(nx-k)+$$

$$\sum_{\left\{\begin{array}{l}k=-\infty\\ \left|\frac{k}{n}-x\right|\geq\frac{1}{n^{\alpha}}\end{array}\right.}^{\infty}\left|f\left(\frac{k}{n}\right)-f(x)\right|\chi(nx-k)\leq \tag{17.46}$$

$$\omega_1\left(f, \frac{1}{n^\alpha}\right)\left(\sum_{\substack{k=-\infty \\ \left\{ \left|\frac{k}{n}-x\right| \leq \frac{1}{n^\alpha} \right.}}^{\infty} \chi(nx-k)\right)+$$

$$2\|f\|_\infty\left(\sum_{\substack{k=-\infty \\ \left\{ |nx-k| \geq n^{1-\alpha} \right.}}^{\infty} \chi(nx-k)\right) \overset{\text{(by (17.16),(17.18))}}{\leq} \tag{17.47}$$

$$\omega_1\left(f, \frac{1}{n^\alpha}\right)+\frac{\|f\|_\infty}{\sqrt{\pi}\left(n^{1-\alpha}-2\right)e^{\left(n^{1-\alpha}-2\right)^2}}.$$

■

We continue with

Theorem 17.14 *Let $f \in C_B(\mathbb{R})$, $0<\alpha<1$, $x \in \mathbb{R}$, $n \in \mathbb{N}$ with $n^{1-\alpha} \geq 3$. Then*

(i)

$$|C_n(f,x)-f(x)| \leq \omega_1\left(f, \frac{1}{n}+\frac{1}{n^\alpha}\right)+\frac{\|f\|_\infty}{\sqrt{\pi}\left(n^{1-\alpha}-2\right)e^{\left(n^{1-\alpha}-2\right)^2}} =: \mu_{3n}, \tag{17.48}$$

(ii)

$$\|C_n(f)-f\|_\infty \leq \mu_{3n}. \tag{17.49}$$

For $f \in (C_B(\mathbb{R}) \cap C_u(\mathbb{R}))$ we get $\lim_{n\to\infty} C_n(f)=f$, pointwise and uniformly.

Proof We notice that

$$\int_{\frac{k}{n}}^{\frac{k+1}{n}} f(t)\,dt=\int_0^{\frac{1}{n}} f\left(t+\frac{k}{n}\right)dt. \tag{17.50}$$

Hence we can write

$$C_n(f,x)=\sum_{k=-\infty}^{\infty}\left(n\int_0^{\frac{1}{n}} f\left(t+\frac{k}{n}\right)dt\right)\chi(nx-k). \tag{17.51}$$

We observe that

$$|C_n(f,x) - f(x)| = \left| \sum_{k=-\infty}^{\infty} \left(\left(n \int_0^{\frac{1}{n}} f\left(t + \frac{k}{n}\right) dt \right) - f(x) \right) \chi(nx-k) \right| = \tag{17.52}$$

$$\left| \sum_{k=-\infty}^{\infty} \left(n \int_0^{\frac{1}{n}} \left(f\left(t + \frac{k}{n}\right) - f(x) \right) dt \right) \chi(nx-k) \right| \le$$

$$\sum_{k=-\infty}^{\infty} \left(n \int_0^{\frac{1}{n}} \left| f\left(t + \frac{k}{n}\right) - f(x) \right| dt \right) \chi(nx-k) \le$$

$$\sum_{\begin{cases} k=-\infty \\ \left|x - \frac{k}{n}\right| \le \frac{1}{n^\alpha} \end{cases}}^{\infty} \left(n \int_0^{\frac{1}{n}} \left| f\left(t + \frac{k}{n}\right) - f(x) \right| dt \right) \chi(nx-k) + \tag{17.53}$$

$$\sum_{\begin{cases} k=-\infty \\ \left|x - \frac{k}{n}\right| \ge \frac{1}{n^\alpha} \end{cases}}^{\infty} \left(n \int_0^{\frac{1}{n}} \left| f\left(t + \frac{k}{n}\right) - f(x) \right| dt \right) \chi(nx-k) \le$$

$$\sum_{\begin{cases} k=-\infty \\ \left|x - \frac{k}{n}\right| \le \frac{1}{n^\alpha} \end{cases}}^{\infty} \left(n \int_0^{\frac{1}{n}} \omega_1\left(f, \left|t + \frac{k}{n} - x\right|\right) dt \right) \chi(nx-k) +$$

$$2\|f\|_\infty \left(\sum_{\begin{cases} k=-\infty \\ |nx-k| \ge n^{1-\alpha} \end{cases}}^{\infty} \chi(|nx-k|) \right) \le \tag{17.54}$$

$$\sum_{\begin{cases} k=-\infty \\ |nx-k| \le n^{1-\alpha} \end{cases}}^{\infty} \left(n \int_0^{\frac{1}{n}} \omega_1\left(f, |t| + \frac{1}{n^\alpha}\right) dt \right) \chi(nx-k)$$

$$+ \frac{\|f\|_\infty}{\sqrt{\pi}\left(n^{1-\alpha} - 2\right) e^{\left(n^{1-\alpha}-2\right)^2}} \le$$

$$\omega_1\left(f, \frac{1}{n} + \frac{1}{n^{\alpha}}\right)\left(\sum_{\substack{k=-\infty \\ |nx-k| \leq n^{1-\alpha}}}^{\infty} \chi(nx-k)\right) \tag{17.55}$$

$$+\frac{\|f\|_{\infty}}{\sqrt{\pi}\left(n^{1-\alpha}-2\right) e^{\left(n^{1-\alpha}-2\right)^2}} \leq$$

$$\omega_1\left(f, \frac{1}{n} + \frac{1}{n^{\alpha}}\right) + \frac{\|f\|_{\infty}}{\sqrt{\pi}\left(n^{1-\alpha}-2\right) e^{\left(n^{1-\alpha}-2\right)^2}},$$

proving the claim. ■

We give next

Theorem 17.15 *Let $f \in C_B(\mathbb{R})$, $0 < \alpha < 1$, $x \in \mathbb{R}$, $n \in \mathbb{N}$ with $n^{1-\alpha} \geq 3$. Then*

(i)

$$|D_n(f, x) - f(x)| \leq \mu_{3n}, \tag{17.56}$$

and

(ii)

$$\|D_n(f) - f\|_{\infty} \leq \mu_{3n}, \tag{17.57}$$

where μ_{3n} as in (17.48).

For $f \in (C_B(\mathbb{R}) \cap C_u(\mathbb{R}))$ we get $\lim_{n\to\infty} D_n(f) = f$, pointwise and uniformly.

Proof We see that

$$|D_n(f, x) - f(x)| \overset{\text{(by (17.35),(17.36))}}{=}$$

$$\left|\sum_{k=-\infty}^{\infty}\left(\left(\sum_{r=0}^{\theta} w_r f\left(\frac{k}{n} + \frac{r}{n\theta}\right)\right) - f(x)\right)\chi(nx-k)\right| =$$

$$\left|\sum_{k=-\infty}^{\infty}\left(\sum_{r=0}^{\theta} w_r\left(f\left(\frac{k}{n} + \frac{r}{n\theta}\right) - f(x)\right)\right)\chi(nx-k)\right| \leq \tag{17.58}$$

$$\sum_{k=-\infty}^{\infty}\left(\sum_{r=0}^{\theta} w_r\left|f\left(\frac{k}{n} + \frac{r}{n\theta}\right) - f(x)\right|\right)\chi(nx-k) \leq$$

$$\sum_{\begin{cases} k=-\infty \\ \left|\frac{k}{n}-x\right| \le \frac{1}{n^\alpha} \end{cases}}^{\infty} \left(\sum_{r=0}^{\theta} w_r \left| f\left(\frac{k}{n}+\frac{r}{n\theta}\right) - f(x)\right|\right) \chi(nx-k) + \tag{17.59}$$

$$2\|f\|_\infty \sum_{\begin{cases} k=-\infty \\ |nx-k| \ge n^{1-\alpha} \end{cases}}^{\infty} \chi(|nx-k|) \le$$

(see that $\frac{r}{n\theta} \le \frac{1}{n}$)

$$\sum_{\begin{cases} k=-\infty \\ \left|\frac{k}{n}-x\right| \le \frac{1}{n^\alpha} \end{cases}}^{\infty} \left(\sum_{r=0}^{\theta} w_r \omega_1\left(f, \frac{1}{n^\alpha}+\frac{1}{n}\right)\right) \chi(nx-k) +$$

$$\frac{\|f\|_\infty}{\sqrt{\pi}\left(n^{1-\alpha}-2\right) e^{\left(n^{1-\alpha}-2\right)^2}} \le \tag{17.60}$$

$$\omega_1\left(f, \frac{1}{n^\alpha}+\frac{1}{n}\right) + \frac{\|f\|_\infty}{\sqrt{\pi}\left(n^{1-\alpha}-2\right) e^{\left(n^{1-\alpha}-2\right)^2}} = \mu_{3n},$$

proving the claim. ■

In the next we discuss high order of approximation by using the smoothness of f.

Theorem 17.16 *Let $f \in C^N([a,b])$, $n, N \in \mathbb{N}$, $n^{1-\alpha} \ge 3$, $0 < \alpha < 1$, $x \in [a,b]$. Then*

(i)

$$|A_n(f,x) - f(x)| \le (4.019) \cdot \tag{17.61}$$

$$\left\{\sum_{j=1}^{N} \frac{\left|f^{(j)}(x)\right|}{j!}\left[\frac{1}{n^{\alpha j}} + \frac{(b-a)^j}{2\sqrt{\pi}\left(n^{1-\alpha}-2\right) e^{\left(n^{1-\alpha}-2\right)^2}}\right] + \right.$$

$$\left.\left[\omega_1\left(f^{(N)}, \frac{1}{n^\alpha}\right)\frac{1}{n^{\alpha N} N!} + \frac{\left\|f^{(N)}\right\|_\infty (b-a)^N}{N!\sqrt{\pi}\left(n^{1-\alpha}-2\right) e^{\left(n^{1-\alpha}-2\right)^2}}\right]\right\},$$

(ii) assume further $f^{(j)}(x_0) = 0$, $j = 1, \ldots, N$, for some $x_0 \in [a,b]$, it holds

$$|A_n(f,x_0) - f(x_0)| \le (4.019) \cdot \tag{17.62}$$

$$\left[\omega_1\left(f^{(N)}, \frac{1}{n^\alpha}\right)\frac{1}{n^{\alpha N}N!} + \frac{\left\|f^{(N)}\right\|_\infty (b-a)^N}{N!\sqrt{\pi}\left(n^{1-\alpha}-2\right)e^{\left(n^{1-\alpha}-2\right)^2}}\right],$$

notice here the extremely high rate of convergence at $n^{-(N+1)\alpha}$,

(iii)

$$\|A_n(f) - f\|_\infty \leq (4.019)\cdot \tag{17.63}$$

$$\left\{\sum_{j=1}^{N}\frac{\left\|f^{(j)}\right\|_\infty}{j!}\left[\frac{1}{n^{\alpha j}} + \frac{(b-a)^j}{2\sqrt{\pi}\left(n^{1-\alpha}-2\right)e^{\left(n^{1-\alpha}-2\right)^2}}\right] + \right.$$

$$\left.\left[\omega_1\left(f^{(N)}, \frac{1}{n^\alpha}\right)\frac{1}{n^{\alpha N}N!} + \frac{\left\|f^{(N)}\right\|_\infty (b-a)^N}{N!\sqrt{\pi}\left(n^{1-\alpha}-2\right)e^{\left(n^{1-\alpha}-2\right)^2}}\right]\right\}.$$

Proof We use (17.39).

Call

$$A_n^*(f,x) := \sum_{k=\lceil na\rceil}^{\lfloor nb\rfloor} f\left(\frac{k}{n}\right)\chi(nx-k),$$

that is

$$A_n(f,x) = \frac{A_n^*(f,x)}{\sum_{k=\lceil na\rceil}^{\lfloor nb\rfloor}\chi(nx-k)}.$$

Next we apply Taylor's formula with integral remainder.

We have (here $\frac{k}{n}, x \in [a,b]$)

$$f\left(\frac{k}{n}\right) = \sum_{j=0}^{N}\frac{f^{(j)}(x)}{j!}\left(\frac{k}{n}-x\right)^j + \int_x^{\frac{k}{n}}\left(f^{(N)}(t) - f^{(N)}(x)\right)\frac{\left(\frac{k}{n}-t\right)^{N-1}}{(N-1)!}dt.$$

Then

$$f\left(\frac{k}{n}\right)\chi(nx-k) = \sum_{j=0}^{N}\frac{f^{(j)}(x)}{j!}\chi(nx-k)\left(\frac{k}{n}-x\right)^j +$$

$$\chi(nx-k)\int_x^{\frac{k}{n}}\left(f^{(N)}(t) - f^{(N)}(x)\right)\frac{\left(\frac{k}{n}-t\right)^{N-1}}{(N-1)!}dt.$$

Hence

$$\sum_{k=\lceil na\rceil}^{\lfloor nb\rfloor} f\left(\frac{k}{n}\right)\chi(nx-k) - f(x)\sum_{k=\lceil na\rceil}^{\lfloor nb\rfloor}\chi(nx-k)$$

$$= \sum_{j=1}^{N} \frac{f^{(j)}(x)}{j!} \sum_{k=\lceil na \rceil}^{\lfloor nb \rfloor} \chi(nx-k) \left(\frac{k}{n} - x \right)^{j} +$$

$$\sum_{k=\lceil na \rceil}^{\lfloor nb \rfloor} \chi(nx-k) \int_{x}^{\frac{k}{n}} \left(f^{(N)}(t) - f^{(N)}(x) \right) \frac{\left(\frac{k}{n} - t \right)^{N-1}}{(N-1)!} dt.$$

Thus

$$A_n^*(f,x) - f(x) \left(\sum_{k=\lceil na \rceil}^{\lfloor nb \rfloor} \chi(nx-k) \right) = \sum_{j=1}^{N} \frac{f^{(j)}(x)}{j!} A_n^* \left((\cdot - x)^j \right) + \Lambda_n(x), \tag{17.64}$$

where

$$\Lambda_n(x) := \sum_{k=\lceil na \rceil}^{\lfloor nb \rfloor} \chi(nx-k) \int_{x}^{\frac{k}{n}} \left(f^{(N)}(t) - f^{(N)}(x) \right) \frac{\left(\frac{k}{n} - t \right)^{N-1}}{(N-1)!} dt. \tag{17.65}$$

We assume that $b - a > \frac{1}{n^{\alpha}}$, which is always the case for large enough $n \in \mathbb{N}$, that is when $n > \left\lceil (b-a)^{-\frac{1}{\alpha}} \right\rceil$.

Thus $\left| \frac{k}{n} - x \right| \leq \frac{1}{n^{\alpha}}$ or $\left| \frac{k}{n} - x \right| > \frac{1}{n^{\alpha}}$.

As in [3], pp. 72–73 for

$$\gamma := \int_{x}^{\frac{k}{n}} \left(f^{(N)}(t) - f^{(N)}(x) \right) \frac{\left(\frac{k}{n} - t \right)^{N-1}}{(N-1)!} dt, \tag{17.66}$$

in case of $\left| \frac{k}{n} - x \right| \leq \frac{1}{n^{\alpha}}$, we find that

$$|\gamma| \leq \omega_1 \left(f^{(N)}, \frac{1}{n^{\alpha}} \right) \frac{1}{n^{\alpha N} N!}$$

(for $x \leq \frac{k}{n}$ or $x \geq \frac{k}{n}$).

Notice also for $x \leq \frac{k}{n}$ that

$$\left| \int_{x}^{\frac{k}{n}} \left(f^{(N)}(t) - f^{(N)}(x) \right) \frac{\left(\frac{k}{n} - t \right)^{N-1}}{(N-1)!} dt \right| \leq$$

$$\int_{x}^{\frac{k}{n}} \left| f^{(N)}(t) - f^{(N)}(x) \right| \frac{\left(\frac{k}{n} - t \right)^{N-1}}{(N-1)!} dt \leq$$

$$2\left\|f^{(N)}\right\|_{\infty}\int_{x}^{\frac{k}{n}}\frac{\left(\frac{k}{n}-t\right)^{N-1}}{(N-1)!}dt=2\left\|f^{(N)}\right\|_{\infty}\frac{\left(\frac{k}{n}-x\right)^{N}}{N!}\leq 2\left\|f^{(N)}\right\|_{\infty}\frac{(b-a)^{N}}{N!}.$$

Next assume $\frac{k}{n} \leq x$, then

$$\left|\int_{x}^{\frac{k}{n}}\left(f^{(N)}(t)-f^{(N)}(x)\right)\frac{\left(\frac{k}{n}-t\right)^{N-1}}{(N-1)!}dt\right| =$$

$$\left|\int_{\frac{k}{n}}^{x}\left(f^{(N)}(t)-f^{(N)}(x)\right)\frac{\left(\frac{k}{n}-t\right)^{N-1}}{(N-1)!}dt\right| \leq$$

$$\int_{\frac{k}{n}}^{x}\left|f^{(N)}(t)-f^{(N)}(x)\right|\frac{\left(t-\frac{k}{n}\right)^{N-1}}{(N-1)!}dt$$

$$\leq 2\left\|f^{(N)}\right\|_{\infty}\int_{\frac{k}{n}}^{x}\frac{\left(t-\frac{k}{n}\right)^{N-1}}{(N-1)!}dt=2\left\|f^{(N)}\right\|_{\infty}\frac{\left(x-\frac{k}{n}\right)^{N}}{N!}\leq 2\left\|f^{(N)}\right\|_{\infty}\frac{(b-a)^{N}}{N!}.$$

Thus

$$|\gamma| \leq 2\left\|f^{(N)}\right\|_{\infty}\frac{(b-a)^{N}}{N!}, \tag{17.67}$$

in all two cases.

Therefore

$$\Lambda_n(x) = \sum_{\substack{k=\lceil na\rceil \\ \left|\frac{k}{n}-x\right|\leq\frac{1}{n^{\alpha}}}}^{\lfloor nb\rfloor}\chi(nx-k)\gamma + \sum_{\substack{k=\lceil na\rceil \\ \left|\frac{k}{n}-x\right|>\frac{1}{n^{\alpha}}}}^{\lfloor nb\rfloor}\chi(nx-k)\gamma.$$

Hence

$$|\Lambda_n(x)| \leq \sum_{\substack{k=\lceil na\rceil \\ \left|\frac{k}{n}-x\right|\leq\frac{1}{n^{\alpha}}}}^{\lfloor nb\rfloor}\chi(nx-k)\left(\omega_1\left(f^{(N)},\frac{1}{n^{\alpha}}\right)\frac{1}{N!n^{N\alpha}}\right)+$$

$$\left(\sum_{\substack{k=\lceil na\rceil \\ \left|\frac{k}{n}-x\right|>\frac{1}{n^{\alpha}}}}^{\lfloor nb\rfloor}\chi(nx-k)\right)2\left\|f^{(N)}\right\|_{\infty}\frac{(b-a)^{N}}{N!}\leq$$

$$\omega_1\left(f^{(N)},\frac{1}{n^{\alpha}}\right)\frac{1}{N!n^{N\alpha}}+\left\|f^{(N)}\right\|_{\infty}\frac{(b-a)^N}{N!}\frac{1}{\sqrt{\pi}\left(n^{1-\alpha}-2\right)e^{\left(n^{1-\alpha}-2\right)^2}}.$$

Consequently we have

$$|\Lambda_n(x)|\leq\omega_1\left(f^{(N)},\frac{1}{n^{\alpha}}\right)\frac{1}{n^{\alpha N}N!}+\frac{\left\|f^{(N)}\right\|_{\infty}(b-a)^N}{N!\sqrt{\pi}\left(n^{1-\alpha}-2\right)e^{\left(n^{1-\alpha}-2\right)^2}}. \tag{17.68}$$

We further see that

$$A_n^*\left((\cdot-x)^j\right)=\sum_{k=\lceil na\rceil}^{\lfloor nb\rfloor}\chi(nx-k)\left(\frac{k}{n}-x\right)^j.$$

Therefore

$$\left|A_n^*\left((\cdot-x)^j\right)\right|\leq\sum_{k=\lceil na\rceil}^{\lfloor nb\rfloor}\chi(nx-k)\left|\frac{k}{n}-x\right|^j=$$

$$\sum_{\begin{cases}k=\lceil na\rceil\\ \left|\frac{k}{n}-x\right|\leq\frac{1}{n^{\alpha}}\end{cases}}^{\lfloor nb\rfloor}\chi(nx-k)\left|\frac{k}{n}-x\right|^j+\sum_{\begin{cases}k=\lceil na\rceil\\ \left|\frac{k}{n}-x\right|>\frac{1}{n^{\alpha}}\end{cases}}^{\lfloor nb\rfloor}\chi(nx-k)\left|\frac{k}{n}-x\right|^j\leq$$

$$\frac{1}{n^{\alpha j}}\sum_{\begin{cases}k=\lceil na\rceil\\ \left|\frac{k}{n}-x\right|\leq\frac{1}{n^{\alpha}}\end{cases}}^{\lfloor nb\rfloor}\chi(nx-k)+(b-a)^j\cdot\sum_{\begin{cases}k=\lceil na\rceil\\ |k-nx|>n^{1-\alpha}\end{cases}}^{\lfloor nb\rfloor}\chi(nx-k)$$

$$\leq\frac{1}{n^{\alpha j}}+(b-a)^j\cdot\frac{1}{2\sqrt{\pi}\left(n^{1-\alpha}-2\right)e^{\left(n^{1-\alpha}-2\right)^2}}.$$

Hence

$$\left|A_n^*\left((\cdot-x)^j\right)\right|\leq\frac{1}{n^{\alpha j}}+\frac{(b-a)^j}{2\sqrt{\pi}\left(n^{1-\alpha}-2\right)e^{\left(n^{1-\alpha}-2\right)^2}}, \tag{17.69}$$

for $j=1,\ldots,N$.

Putting things together we have proved

$$\left|A_n^*(f,x)-f(x)\sum_{k=\lceil na\rceil}^{\lfloor nb\rfloor}\chi(nx-k)\right|\leq$$

$$\sum_{j=1}^{N} \frac{\left|f^{(j)}(x)\right|}{j!} \left[\frac{1}{n^{\alpha j}} + \frac{(b-a)^j}{2\sqrt{\pi}\left(n^{1-\alpha}-2\right) e^{\left(n^{1-\alpha}-2\right)^2}} \right] + \tag{17.70}$$

$$\left[\omega_1 \left(f^{(N)}, \frac{1}{n^{\alpha}} \right) \frac{1}{n^{\alpha N} N!} + \frac{\left\| f^{(N)} \right\|_{\infty} (b-a)^N}{N! \sqrt{\pi} \left(n^{1-\alpha}-2\right) e^{\left(n^{1-\alpha}-2\right)^2}} \right],$$

that is establishing theorem. ■

17.4 Fractional Neural Network Approximation

We need

Definition 17.17 Let $\nu \geq 0$, $n = \lceil \nu \rceil$ ($\lceil \cdot \rceil$ is the ceiling of the number), $f \in AC^n([a,b])$ (space of functions f with $f^{(n-1)} \in AC([a,b])$, absolutely continuous functions). We call left Caputo fractional derivative (see [17], pp. 49–52, [19, 24]) the function

$$D_{*a}^{\nu} f(x) = \frac{1}{\Gamma(n-\nu)} \int_a^x (x-t)^{n-\nu-1} f^{(n)}(t)\, dt, \tag{17.71}$$

$\forall\, x \in [a,b]$, where Γ is the gamma function $\Gamma(\nu) = \int_0^{\infty} e^{-t} t^{\nu-1} dt$, $\nu > 0$.

Notice $D_{*a}^{\nu} f \in L_1([a,b])$ and $D_{*a}^{\nu} f$ exists a.e.on $[a,b]$.

We set $D_{*a}^{0} f(x) = f(x)$, $\forall\, x \in [a,b]$.

Lemma 17.18 ([6]) *Let $\nu > 0$, $\nu \notin \mathbb{N}$, $n = \lceil \nu \rceil$, $f \in C^{n-1}([a,b])$ and $f^{(n)} \in L_{\infty}([a,b])$. Then $D_{*a}^{\nu} f(a) = 0$.*

Definition 17.19 (*see also* [4, 18, 19]) Let $f \in AC^m([a,b])$, $m = \lceil \alpha \rceil$, $\alpha > 0$. The right Caputo fractional derivative of order $\alpha > 0$ is given by

$$D_{b-}^{\alpha} f(x) = \frac{(-1)^m}{\Gamma(m-\alpha)} \int_x^b (\zeta - x)^{m-\alpha-1} f^{(m)}(\zeta)\, d\zeta, \tag{17.72}$$

$\forall\, x \in [a,b]$. We set $D_{b-}^{0} f(x) = f(x)$. Notice $D_{b-}^{\alpha} f \in L_1([a,b])$ and $D_{b-}^{\alpha} f$ exists a.e.on $[a,b]$.

Lemma 17.20 ([6]) *Let $f \in C^{m-1}([a,b])$, $f^{(m)} \in L_{\infty}([a,b])$, $m = \lceil \alpha \rceil$, $\alpha > 0$. Then $D_{b-}^{\alpha} f(b) = 0$.*

Convention 17.21 *We assume that*

$$D_{*x_0}^{\alpha} f(x) = 0, \textit{ for } x < x_0, \tag{17.73}$$

and

$$D_{x_0-}^{\alpha} f(x) = 0, \text{ for } x > x_0, \tag{17.74}$$

for all $x, x_0 \in (a, b]$.

We mention

Proposition 17.22 ([6]) *Let* $f \in C^n([a, b])$, $n = \lceil \nu \rceil$, $\nu > 0$. *Then* $D_{*a}^{\nu} f(x)$ *is continuous in* $x \in [a, b]$.

Also we have

Proposition 17.23 ([6]) *Let* $f \in C^m([a, b])$, $m = \lceil \alpha \rceil$, $\alpha > 0$. *Then* $D_{b-}^{\alpha} f(x)$ *is continuous in* $x \in [a, b]$.

We further mention

Proposition 17.24 ([6]) *Let* $f \in C^{m-1}([a, b])$, $f^{(m)} \in L_\infty([a, b])$, $m = \lceil \alpha \rceil$, $\alpha > 0$ *and*

$$D_{*x_0}^{\alpha} f(x) = \frac{1}{\Gamma(m-\alpha)} \int_{x_0}^{x} (x-t)^{m-\alpha-1} f^{(m)}(t)\,dt, \tag{17.75}$$

for all $x, x_0 \in [a, b] : x \geq x_0$.

Then $D_{*x_0}^{\alpha} f(x)$ *is continous in* x_0.

Proposition 17.25 ([6]) *Let* $f \in C^{m-1}([a, b])$, $f^{(m)} \in L_\infty([a, b])$, $m = \lceil \alpha \rceil$, $\alpha > 0$ *and*

$$D_{x_0-}^{\alpha} f(x) = \frac{(-1)^m}{\Gamma(m-\alpha)} \int_{x}^{x_0} (\zeta-x)^{m-\alpha-1} f^{(m)}(\zeta)\,d\zeta, \tag{17.76}$$

for all $x, x_0 \in [a, b] : x \leq x_0$.

Then $D_{x_0-}^{\alpha} f(x)$ *is continuous in* x_0.

Proposition 17.26 ([6]) *Let* $f \in C^m([a, b])$, $m = \lceil \alpha \rceil$, $\alpha > 0$, $x, x_0 \in [a, b]$. *Then* $D_{*x_0}^{\alpha} f(x)$, $D_{x_0-}^{\alpha} f(x)$ *are jointly continuous functions in* (x, x_0) *from* $[a, b]^2$ *into* $\mathbb{R}$.

We recall

Theorem 17.27 ([6]) *Let* $f : [a, b]^2 \to \mathbb{R}$ *be jointly continuous. Consider*

$$G(x) = \omega_1(f(\cdot, x), \delta, [x, b]), \tag{17.77}$$

$\delta > 0$, $x \in [a, b]$.

Then G *is continuous in* $x \in [a, b]$.

Also it holds

Theorem 17.28 ([6]) *Let $f : [a, b]^2 \to \mathbb{R}$ be jointly continuous. Then*

$$H(x) = \omega_1 (f(\cdot, x), \delta, [a, x]), \tag{17.78}$$

$x \in [a, b]$, is continuous in $x \in [a, b]$, $\delta > 0$.

We need

Remark 17.29 ([6]) Let $f \in C^{n-1}([a, b])$, $f^{(n)} \in L_\infty([a, b])$, $n = \lceil \nu \rceil$, $\nu > 0$, $\nu \notin \mathbb{N}$. Then we have

$$\left| D_{*a}^{\nu} f(x) \right| \leq \frac{\| f^{(n)} \|_\infty}{\Gamma(n - \nu + 1)} (x - a)^{n-\nu}, \ \forall x \in [a, b]. \tag{17.79}$$

Thus we observe

$$\omega_1 \left(D_{*a}^{\nu} f, \delta \right) = \sup_{\substack{x,y \in [a,b] \\ |x-y| \leq \delta}} \left| D_{*a}^{\nu} f(x) - D_{*a}^{\nu} f(y) \right|$$

$$\leq \sup_{\substack{x,y \in [a,b] \\ |x-y| \leq \delta}} \left(\frac{\| f^{(n)} \|_\infty}{\Gamma(n - \nu + 1)} (x - a)^{n-\nu} + \frac{\| f^{(n)} \|_\infty}{\Gamma(n-\nu+1)} (y - a)^{n-\nu} \right)$$

$$\leq \frac{2 \| f^{(n)} \|_\infty}{\Gamma(n - \nu + 1)} (b - a)^{n-\nu}.$$

Consequently

$$\omega_1 \left(D_{*a}^{\nu} f, \delta \right) \leq \frac{2 \| f^{(n)} \|_\infty}{\Gamma(n - \nu + 1)} (b - a)^{n-\nu}. \tag{17.80}$$

Similarly, let $f \in C^{m-1}([a, b])$, $f^{(m)} \in L_\infty([a, b])$, $m = \lceil \alpha \rceil$, $\alpha > 0$, $\alpha \notin \mathbb{N}$, then

$$\omega_1 \left(D_{b-}^{\alpha} f, \delta \right) \leq \frac{2 \| f^{(m)} \|_\infty}{\Gamma(m - \alpha + 1)} (b - a)^{m-\alpha}. \tag{17.81}$$

So for $f \in C^{m-1}([a, b])$, $f^{(m)} \in L_\infty([a, b])$, $m = \lceil \alpha \rceil$, $\alpha > 0$, $\alpha \notin \mathbb{N}$, we find

$$\sup_{x_0 \in [a,b]} \omega_1 \left(D_{*x_0}^{\alpha} f, \delta \right)_{[x_0, b]} \leq \frac{2 \| f^{(m)} \|_\infty}{\Gamma(m - \alpha + 1)} (b - a)^{m-\alpha}, \tag{17.82}$$

and

$$\sup_{x_0 \in [a,b]} \omega_1 \left(D_{x_0-}^{\alpha} f, \delta \right)_{[a, x_0]} \leq \frac{2 \| f^{(m)} \|_\infty}{\Gamma(m - \alpha + 1)} (b - a)^{m-\alpha}. \tag{17.83}$$

By Proposition 15.114, p. 388 of [5], we get here that $D^{\alpha}_{*x_0} f \in C([x_0, b])$, and by [8] we obtain that $D^{\alpha}_{x_0-} f \in C([a, x_0])$.

Here comes our main fractional result

Theorem 17.30 *Let* $\alpha > 0$, $N = \lceil \alpha \rceil$, $\alpha \notin \mathbb{N}$, $f \in AC^N([a, b])$, *with* $f^{(N)} \in L_{\infty}([a, b])$, $0 < \beta < 1$, $x \in [a, b]$, $n \in \mathbb{N}$, $n^{1-\beta} \geq 3$. *Then*

(i)

$$\left| A_n(f, x) - \sum_{j=1}^{N-1} \frac{f^{(j)}(x)}{j!} A_n\left((\cdot - x)^j, x\right) - f(x) \right| \leq \tag{17.84}$$

$$\frac{(4.019)}{\Gamma(\alpha+1)} \cdot \left\{ \frac{\left(\omega_1\left(D^{\alpha}_{x-} f, \frac{1}{n^{\beta}}\right)_{[a,x]} + \omega_1\left(D^{\alpha}_{*x} f, \frac{1}{n_{\beta}}\right)_{[x,b]} \right)}{n^{\alpha\beta}} + \right.$$

$$\frac{1}{2\sqrt{\pi}\left(n^{1-\beta} - 2\right) e^{\left(n^{1-\beta}-2\right)^2}} \cdot$$

$$\left. \left(\left\| D^{\alpha}_{x-} f \right\|_{\infty,[a,x]} (x-a)^{\alpha} + \left\| D^{\alpha}_{*x} f \right\|_{\infty,[x,b]} (b-x)^{\alpha} \right) \right\},$$

(ii) if $f^{(j)}(x) = 0$, *for* $j = 1, \ldots, N-1$, *we have*

$$|A_n(f, x) - f(x)| \leq \frac{(4.019)}{\Gamma(\alpha+1)} \cdot \tag{17.85}$$

$$\left\{ \frac{\left(\omega_1\left(D^{\alpha}_{x-} f, \frac{1}{n^{\beta}}\right)_{[a,x]} + \omega_1\left(D^{\alpha}_{*x} f, \frac{1}{n_{\beta}}\right)_{[x,b]} \right)}{n^{\alpha\beta}} + \right.$$

$$\frac{1}{2\sqrt{\pi}\left(n^{1-\beta} - 2\right) e^{\left(n^{1-\beta}-2\right)^2}} \cdot$$

$$\left. \left(\left\| D^{\alpha}_{x-} f \right\|_{\infty,[a,x]} (x-a)^{\alpha} + \left\| D^{\alpha}_{*x} f \right\|_{\infty,[x,b]} (b-x)^{\alpha} \right) \right\},$$

when $\alpha > 1$ *notice here the extremely high rate of convergence at* $n^{-(\alpha+1)\beta}$,

(iii)

$$\left|A_n(f,x) - f(x)\right| \leq (4.019) \cdot \tag{17.86}$$

$$\left\{ \sum_{j=1}^{N-1} \frac{\left|f^{(j)}(x)\right|}{j!} \left\{ \frac{1}{n^{\beta j}} + (b-a)^j \frac{1}{2\sqrt{\pi}\left(n^{1-\beta}-2\right)e^{\left(n^{1-\beta}-2\right)^2}} \right\} + \right.$$

$$\frac{1}{\Gamma(\alpha+1)} \left\{ \frac{\left(\omega_1\left(D_{x-}^{\alpha} f, \frac{1}{n^{\beta}}\right)_{[a,x]} + \omega_1\left(D_{*x}^{\alpha} f, \frac{1}{n_{\beta}}\right)_{[x,b]} \right)}{n^{\alpha\beta}} + \right.$$

$$\frac{1}{2\sqrt{\pi}\left(n^{1-\beta}-2\right)e^{\left(n^{1-\beta}-2\right)^2}} \cdot$$

$$\left. \left. \left(\left\| D_{x-}^{\alpha} f \right\|_{\infty,[a,x]} (x-a)^{\alpha} + \left\| D_{*x}^{\alpha} f \right\|_{\infty,[x,b]} (b-x)^{\alpha} \right) \right\} \right\},$$

$\forall\, x \in [a,b]$, *and*

(iv)

$$\left\| A_n f - f \right\|_{\infty} \leq (4.019) \cdot$$

$$\left\{ \sum_{j=1}^{N-1} \frac{\left\|f^{(j)}\right\|_{\infty}}{j!} \left\{ \frac{1}{n^{\beta j}} + (b-a)^j \frac{1}{2\sqrt{\pi}\left(n^{1-\beta}-2\right)e^{\left(n^{1-\beta}-2\right)^2}} \right\} + \right.$$

$$\frac{1}{\Gamma(\alpha+1)} \left\{ \frac{\left(\sup_{x\in[a,b]} \omega_1\left(D_{x-}^{\alpha} f, \frac{1}{n^{\beta}}\right)_{[a,x]} + \sup_{x\in[a,b]} \omega_1\left(D_{*x}^{\alpha} f, \frac{1}{n_{\beta}}\right)_{[x,b]} \right)}{n^{\alpha\beta}} + \right.$$

$$\frac{1}{2\sqrt{\pi}\left(n^{1-\beta}-2\right)e^{\left(n^{1-\beta}-2\right)^2}} \cdot \tag{17.87}$$

$$\left. \left. (b-a)^{\alpha} \left(\sup_{x\in[a,b]} \left\| D_{x-}^{\alpha} f \right\|_{\infty,[a,x]} + \sup_{x\in[a,b]} \left\| D_{*x}^{\alpha} f \right\|_{\infty,[x,b]} \right) \right\} \right\}.$$

Above, when $N = 1$ the sum $\sum_{j=1}^{N-1} \cdot = 0$.
As we see here we obtain fractionally type pointwise and uniform convergence with rates of $A_n \to I$ the unit operator, as $n \to \infty$.

Proof Let $x \in [a, b]$. We have that $D_{x-}^{\alpha} f(x) = D_{*x}^{\alpha} f(x) = 0$.
From [17], p. 54, we get by the left Caputo fractional Taylor formula that

$$f\left(\frac{k}{n}\right) = \sum_{j=0}^{N-1} \frac{f^{(j)}(x)}{j!} \left(\frac{k}{n} - x\right)^j + \tag{17.88}$$

$$\frac{1}{\Gamma(\alpha)} \int_x^{\frac{k}{n}} \left(\frac{k}{n} - J\right)^{\alpha-1} \left(D_{*x}^{\alpha} f(J) - D_{*x}^{\alpha} f(x)\right) dJ,$$

for all $x \leq \frac{k}{n} \leq b$.
Also from [4], using the right Caputo fractional Taylor formula we get

$$f\left(\frac{k}{n}\right) = \sum_{j=0}^{N-1} \frac{f^{(j)}(x)}{j!} \left(\frac{k}{n} - x\right)^j + \tag{17.89}$$

$$\frac{1}{\Gamma(\alpha)} \int_{\frac{k}{n}}^{x} \left(J - \frac{k}{n}\right)^{\alpha-1} \left(D_{x-}^{\alpha} f(J) - D_{x-}^{\alpha} f(x)\right) dJ,$$

for all $a \leq \frac{k}{n} \leq x$.
Hence we have

$$f\left(\frac{k}{n}\right) \chi(nx - k) = \sum_{j=0}^{N-1} \frac{f^{(j)}(x)}{j!} \chi(nx - k) \left(\frac{k}{n} - x\right)^j + \tag{17.90}$$

$$\frac{\chi(nx - k)}{\Gamma(\alpha)} \int_x^{\frac{k}{n}} \left(\frac{k}{n} - J\right)^{\alpha-1} \left(D_{*x}^{\alpha} f(J) - D_{*x}^{\alpha} f(x)\right) dJ,$$

for all $x \leq \frac{k}{n} \leq b$, iff $\lceil nx \rceil \leq k \leq \lfloor nb \rfloor$, and

$$f\left(\frac{k}{n}\right) \chi(nx - k) = \sum_{j=0}^{N-1} \frac{f^{(j)}(x)}{j!} \chi(nx - k) \left(\frac{k}{n} - x\right)^j + \tag{17.91}$$

$$\frac{\chi(nx - k)}{\Gamma(\alpha)} \int_{\frac{k}{n}}^{x} \left(J - \frac{k}{n}\right)^{\alpha-1} \left(D_{x-}^{\alpha} f(J) - D_{x-}^{\alpha} f(x)\right) dJ,$$

for all $a \leq \frac{k}{n} \leq x$, iff $\lceil na \rceil \leq k \leq \lfloor nx \rfloor$.

We have that $\lceil nx \rceil \leq \lfloor nx \rfloor + 1$.
Therefore it holds

$$\sum_{k=\lfloor nx \rfloor+1}^{\lfloor nb \rfloor} f\left(\frac{k}{n}\right) \chi(nx-k) = \sum_{j=0}^{N-1} \frac{f^{(j)}(x)}{j!} \sum_{k=\lfloor nx \rfloor+1}^{\lfloor nb \rfloor} \chi(nx-k) \left(\frac{k}{n}-x\right)^j + \tag{17.92}$$

$$\frac{1}{\Gamma(\alpha)} \sum_{k=\lfloor nx \rfloor+1}^{\lfloor nb \rfloor} \chi(nx-k) \int_x^{\frac{k}{n}} \left(\frac{k}{n}-J\right)^{\alpha-1} \left(D_{*x}^{\alpha} f(J) - D_{*x}^{\alpha} f(x)\right) dJ,$$

and

$$\sum_{k=\lceil na \rceil}^{\lfloor nx \rfloor} f\left(\frac{k}{n}\right) \chi(nx-k) = \sum_{j=0}^{N-1} \frac{f^{(j)}(x)}{j!} \sum_{k=\lceil na \rceil}^{\lfloor nx \rfloor} \chi(nx-k) \left(\frac{k}{n}-x\right)^j + \tag{17.93}$$

$$\frac{1}{\Gamma(\alpha)} \sum_{k=\lceil na \rceil}^{\lfloor nx \rfloor} \chi(nx-k) \int_{\frac{k}{n}}^{x} \left(J-\frac{k}{n}\right)^{\alpha-1} \left(D_{x-}^{\alpha} f(J) - D_{x-}^{\alpha} f(x)\right) dJ.$$

Adding the last two equalities (17.92) and (17.93) we obtain

$$A_n^*(f,x) = \sum_{k=\lceil na \rceil}^{\lfloor nb \rfloor} f\left(\frac{k}{n}\right) \chi(nx-k) = \tag{17.94}$$

$$\sum_{j=0}^{N-1} \frac{f^{(j)}(x)}{j!} \sum_{k=\lceil na \rceil}^{\lfloor nb \rfloor} \chi(nx-k) \left(\frac{k}{n}-x\right)^j +$$

$$\frac{1}{\Gamma(\alpha)} \left\{ \sum_{k=\lceil na \rceil}^{\lfloor nx \rfloor} \chi(nx-k) \int_{\frac{k}{n}}^{x} \left(J-\frac{k}{n}\right)^{\alpha-1} \left(D_{x-}^{\alpha} f(J) - D_{x-}^{\alpha} f(x)\right) dJ + \right.$$

$$\left. \sum_{k=\lfloor nx \rfloor+1}^{\lfloor nb \rfloor} \chi(nx-k) \int_x^{\frac{k}{n}} \left(\frac{k}{n}-J\right)^{\alpha-1} \left(D_{*x}^{\alpha} f(J) - D_{*x}^{\alpha} f(x)\right) dJ \right\}.$$

So we have derived

$$A_n^*(f,x) - f(x) \left(\sum_{k=\lceil na \rceil}^{\lfloor nb \rfloor} \chi(nx-k) \right) = \tag{17.95}$$

$$\sum_{j=1}^{N-1} \frac{f^{(j)}(x)}{j!} A_n^*\left((\cdot - x)^j\right)(x) + \theta_n(x),$$

where

$$\theta_n(x) := \frac{1}{\Gamma(\alpha)} \left\{ \sum_{k=\lceil na \rceil}^{\lfloor nx \rfloor} \chi(nx-k) \int_{\frac{k}{n}}^{x} \left(J - \frac{k}{n}\right)^{\alpha-1} \left(D_{x-}^{\alpha} f(J) - D_{x-}^{\alpha} f(x)\right) dJ \right.$$

$$\left. + \sum_{k=\lfloor nx \rfloor+1}^{\lfloor nb \rfloor} \chi(nx-k) \int_{x}^{\frac{k}{n}} \left(\frac{k}{n} - J\right)^{\alpha-1} \left(D_{*x}^{\alpha} f(J) - D_{*x}^{\alpha} f(x)\right) dJ \right\}. \quad (17.96)$$

We set

$$\theta_{1n}(x) := \frac{1}{\Gamma(\alpha)} \sum_{k=\lceil na \rceil}^{\lfloor nx \rfloor} \chi(nx-k) \int_{\frac{k}{n}}^{x} \left(J - \frac{k}{n}\right)^{\alpha-1} \left(D_{x-}^{\alpha} f(J) - D_{x-}^{\alpha} f(x)\right) dJ, \quad (17.97)$$

and

$$\theta_{2n} := \frac{1}{\Gamma(\alpha)} \sum_{k=\lfloor nx \rfloor+1}^{\lfloor nb \rfloor} \chi(nx-k) \int_{x}^{\frac{k}{n}} \left(\frac{k}{n} - J\right)^{\alpha-1} \left(D_{*x}^{\alpha} f(J) - D_{*x}^{\alpha} f(x)\right) dJ, \quad (17.98)$$

i.e.

$$\theta_n(x) = \theta_{1n}(x) + \theta_{2n}(x). \quad (17.99)$$

We assume $b - a > \frac{1}{n^{\beta}}$, $0 < \beta < 1$, which is always the case for large enough $n \in \mathbb{N}$, that is when $n > \left\lceil (b-a)^{-\frac{1}{\beta}} \right\rceil$. It is always true that either $\left|\frac{k}{n} - x\right| \leq \frac{1}{n^{\beta}}$ or $\left|\frac{k}{n} - x\right| > \frac{1}{n^{\beta}}$.

For $k = \lceil na \rceil, \ldots, \lfloor nx \rfloor$, we consider

$$\gamma_{1k} := \left| \int_{\frac{k}{n}}^{x} \left(J - \frac{k}{n}\right)^{\alpha-1} \left(D_{x-}^{\alpha} f(J) - D_{x-}^{\alpha} f(x)\right) dJ \right| \quad (17.100)$$

$$= \left| \int_{\frac{k}{n}}^{x} \left(J - \frac{k}{n}\right)^{\alpha-1} D_{x-}^{\alpha} f(J)\, dJ \right| \leq \int_{\frac{k}{n}}^{x} \left(J - \frac{k}{n}\right)^{\alpha-1} \left|D_{x-}^{\alpha} f(J)\right| dJ$$

$$\leq \left\| D_{x-}^{\alpha} f \right\|_{\infty,[a,x]} \frac{\left(x - \frac{\kappa}{n}\right)^{\alpha}}{\alpha} \leq \left\| D_{x-}^{\alpha} f \right\|_{\infty,[a,x]} \frac{(x-a)^{\alpha}}{\alpha}. \quad (17.101)$$

That is

$$\gamma_{1k} \leq \left\| D_{x-}^{\alpha} f \right\|_{\infty,[a,x]} \frac{(x-a)^{\alpha}}{\alpha}, \quad (17.102)$$

for $k = \lceil na \rceil, \ldots, \lfloor nx \rfloor$.

Also we have in case of $\left|\frac{k}{n}-x\right| \leq \frac{1}{n^{\beta}}$ that

$$\gamma_{1k} \leq \int_{\frac{k}{n}}^{x} \left(J-\frac{k}{n}\right)^{\alpha-1} \left|D_{x-}^{\alpha} f(J) - D_{x-}^{\alpha} f(x)\right| dJ \tag{17.103}$$

$$\leq \int_{\frac{k}{n}}^{x} \left(J-\frac{k}{n}\right)^{\alpha-1} \omega_1 \left(D_{x-}^{\alpha} f, |J-x|\right)_{[a,x]} dJ$$

$$\leq \omega_1 \left(D_{x-}^{\alpha} f, \left|x-\frac{k}{n}\right|\right)_{[a,x]} \int_{\frac{k}{n}}^{x} \left(J-\frac{k}{n}\right)^{\alpha-1} dJ$$

$$\leq \omega_1 \left(D_{x-}^{\alpha} f, \frac{1}{n^{\beta}}\right)_{[a,x]} \frac{\left(x-\frac{k}{n}\right)^{\alpha}}{\alpha} \leq \omega_1 \left(D_{x-}^{\alpha} f, \frac{1}{n^{\beta}}\right)_{[a,x]} \frac{1}{\alpha n^{a\beta}}. \tag{17.104}$$

That is when $\left|\frac{k}{n}-x\right| \leq \frac{1}{n^{\beta}}$, then

$$\gamma_{1k} \leq \frac{\omega_1 \left(D_{x-}^{\alpha} f, \frac{1}{n^{\beta}}\right)_{[a,x]}}{\alpha n^{a\beta}}. \tag{17.105}$$

Consequently we obtain

$$|\theta_{1n}(x)| \leq \frac{1}{\Gamma(\alpha)} \sum_{k=\lceil na \rceil}^{\lfloor nx \rfloor} \chi(nx-k) \gamma_{1k} =$$

$$\frac{1}{\Gamma(\alpha)} \left\{ \sum_{\substack{k=\lceil na \rceil \\ : \left|\frac{k}{n}-x\right| \leq \frac{1}{n^{\beta}}}}^{\lfloor nx \rfloor} \chi(nx-k)\gamma_{1k} + \sum_{\substack{k=\lceil na \rceil \\ : \left|\frac{k}{n}-x\right| > \frac{1}{n^{\beta}}}}^{\lfloor nx \rfloor} \chi(nx-k)\gamma_{1k} \right\} \leq$$

$$\frac{1}{\Gamma(\alpha)} \left\{ \left(\sum_{\substack{k=\lceil na \rceil \\ : \left|\frac{k}{n}-x\right| \leq \frac{1}{n^{\beta}}}}^{\lfloor nx \rfloor} \chi(nx-k) \right) \frac{\omega_1 \left(D_{x-}^{\alpha} f, \frac{1}{n^{\beta}}\right)_{[a,x]}}{\alpha n^{\alpha\beta}} + \right.$$

$$\left(\sum_{\left\{\begin{array}{c} k=\lceil na\rceil \\ : \left|\frac{k}{n}-x\right| > \frac{1}{n^{\beta}} \end{array}\right.}^{\lfloor nx\rfloor} \chi(nx-k)\right) \left\| D_{x-}^{\alpha} f \right\|_{\infty,[a,x]} \frac{(x-a)^{\alpha}}{\alpha} \Bigg] \leq \tag{17.106}$$

$$\frac{1}{\Gamma(\alpha+1)} \left\{ \frac{\omega_1\left(D_{x-}^{\alpha} f, \frac{1}{n^{\beta}}\right)_{[a,x]}}{n^{\alpha\beta}} + \right.$$

$$\left(\sum_{\left\{\begin{array}{c} k=-\infty \\ : |nx-k| > n^{1-\beta} \end{array}\right.}^{\infty} \chi(nx-k)\right) \left\| D_{x-}^{\alpha} f \right\|_{\infty,[a,x]} (x-a)^{\alpha} \Bigg\} \overset{(17.18)}{\leq} \tag{17.107}$$

$$\frac{1}{\Gamma(\alpha+1)} \left\{ \frac{\omega_1\left(D_{x-}^{\alpha} f, \frac{1}{n^{\beta}}\right)_{[a,x]}}{n^{\alpha\beta}} + \right.$$

$$\left. \frac{1}{2\sqrt{\pi}\left(n^{1-\beta}-2\right) e^{\left(n^{1-\beta}-2\right)^2}} \left\| D_{x-}^{\alpha} f \right\|_{\infty,[a,x]} (x-a)^{\alpha} \right\}.$$

So we have proved that

$$|\theta_{1n}(x)| \leq \frac{1}{\Gamma(\alpha+1)} \left\{ \frac{\omega_1\left(D_{x-}^{\alpha} f, \frac{1}{n^{\beta}}\right)_{[a,x]}}{n^{\alpha\beta}} + \right. \tag{17.108}$$

$$\left. \frac{1}{2\sqrt{\pi}\left(n^{1-\beta}-2\right) e^{\left(n^{1-\beta}-2\right)^2}} \left\| D_{x-}^{\alpha} f \right\|_{\infty,[a,x]} (x-a)^{\alpha} \right\}.$$

Next when $k = \lfloor nx\rfloor + 1, \ldots, \lfloor nb\rfloor$ we consider

$$\gamma_{2k} := \left| \int_x^{\frac{k}{n}} \left(\frac{k}{n} - J\right)^{\alpha-1} \left(D_{*x}^{\alpha} f(J) - D_{*x}^{\alpha} f(x)\right) dJ \right| \leq$$

$$\int_x^{\frac{k}{n}} \left(\frac{k}{n} - J\right)^{\alpha-1} \left|D_{*x}^{\alpha} f(J) - D_{*x}^{\alpha} f(x)\right| dJ =$$

$$\int_x^{\frac{k}{n}} \left(\frac{k}{n} - J\right)^{\alpha-1} \left|D_{*x}^{\alpha} f(J)\right| dJ \leq \left\|D_{*x}^{\alpha} f\right\|_{\infty,[x,b]} \frac{\left(\frac{k}{n} - x\right)^{\alpha}}{\alpha} \leq \tag{17.109}$$

$$\left\|D_{*x}^{\alpha} f\right\|_{\infty,[x,b]} \frac{(b-x)^{\alpha}}{\alpha}. \tag{17.110}$$

Therefore when $k = \lfloor nx \rfloor + 1, \ldots, \lfloor nb \rfloor$ we get that

$$\gamma_{2k} \leq \left\|D_{*x}^{\alpha} f\right\|_{\infty,[x,b]} \frac{(b-x)^{\alpha}}{\alpha}. \tag{17.111}$$

In case of $\left|\frac{k}{n} - x\right| \leq \frac{1}{n^{\beta}}$, we get

$$\gamma_{2k} \leq \int_x^{\frac{k}{n}} \left(\frac{k}{n} - J\right)^{\alpha-1} \omega_1\left(D_{*x}^{\alpha} f, |J - x|\right)_{[x,b]} dJ \leq \tag{17.112}$$

$$\omega_1\left(D_{*x}^{\alpha} f, \left|\frac{k}{n} - x\right|\right)_{[x,b]} \int_x^{\frac{k}{n}} \left(\frac{k}{n} - J\right)^{\alpha-1} dJ \leq$$

$$\omega_1\left(D_{*x}^{\alpha} f, \frac{1}{n^{\beta}}\right)_{[x,b]} \frac{\left(\frac{k}{n} - x\right)^{\alpha}}{\alpha} \leq \omega_1\left(D_{*x}^{\alpha} f, \frac{1}{n^{\beta}}\right)_{[x,b]} \frac{1}{\alpha n^{\alpha\beta}}. \tag{17.113}$$

So when $\left|\frac{k}{n} - x\right| \leq \frac{1}{n^{\beta}}$ we derived that

$$\gamma_{2k} \leq \frac{\omega_1\left(D_{*x}^{\alpha} f, \frac{1}{n^{\beta}}\right)_{[x,b]}}{\alpha n^{\alpha\beta}}. \tag{17.114}$$

Similarly we have that

$$|\theta_{2n}(x)| \leq \frac{1}{\Gamma(\alpha)} \left(\sum_{k=\lfloor nx \rfloor+1}^{\lfloor nb \rfloor} \chi(nx - k)\gamma_{2k}\right) = \tag{17.115}$$

$$\frac{1}{\Gamma(\alpha)} \left\{ \sum_{\substack{k = \lfloor nx \rfloor + 1 \\ : \left|\frac{k}{n} - x\right| \leq \frac{1}{n^{\beta}}}}^{\lfloor nb \rfloor} \chi(nx - k)\gamma_{2k} + \sum_{\substack{k = \lfloor nx \rfloor + 1 \\ : \left|\frac{k}{n} - x\right| > \frac{1}{n^{\beta}}}}^{\lfloor nb \rfloor} \chi(nx - k)\gamma_{2k} \right\} \leq$$

$$\frac{1}{\Gamma(\alpha)}\left\{\left(\sum_{\substack{k=\lfloor nx\rfloor+1\\ :\left|\frac{k}{n}-x\right|\le\frac{1}{n^{\beta}}}}^{\lfloor nb\rfloor}\chi(nx-k)\right)\frac{\omega_1\left(D_{*x}^{\alpha}f,\frac{1}{n^{\beta}}\right)_{[x,b]}}{\alpha n^{\alpha\beta}}+\right.$$

$$\left.\left(\sum_{\substack{k=\lfloor nx\rfloor+1\\ :\left|\frac{k}{n}-x\right|>\frac{1}{n^{\beta}}}}^{\lfloor nb\rfloor}\chi(nx-k)\right)\left\|D_{*x}^{\alpha}f\right\|_{\infty,[x,b]}\frac{(b-x)^{\alpha}}{\alpha}\right\}\le \tag{17.116}$$

$$\frac{1}{\Gamma(\alpha+1)}\left\{\frac{\omega_1\left(D_{*x}^{\alpha}f,\frac{1}{n^{\beta}}\right)_{[x,b]}}{n^{\alpha\beta}}+\right.$$

$$\left.\left(\sum_{\substack{k=-\infty\\ :\left|\frac{k}{n}-x\right|>\frac{1}{n^{\beta}}}}^{\infty}\chi(nx-k)\right)\left\|D_{*x}^{\alpha}f\right\|_{\infty,[x,b]}(b-x)^{\alpha}\right\}\overset{(17.18)}{\le} \tag{17.117}$$

$$\frac{1}{\Gamma(\alpha+1)}\left\{\frac{\omega_1\left(D_{*x}^{\alpha}f,\frac{1}{n^{\beta}}\right)_{[x,b]}}{n^{\alpha\beta}}+\right.$$

$$\left.\frac{1}{2\sqrt{\pi}\left(n^{1-\beta}-2\right)e^{\left(n^{1-\beta}-2\right)^{2}}}\left\|D_{*x}^{\alpha}f\right\|_{\infty,[x,b]}(b-x)^{\alpha}\right\}.$$

So we have proved that

$$|\theta_{2n}(x)|\le\frac{1}{\Gamma(\alpha+1)}\left\{\frac{\omega_1\left(D_{*x}^{\alpha}f,\frac{1}{n^{\beta}}\right)_{[x,b]}}{n^{\alpha\beta}}+\right. \tag{17.118}$$

$$\left.\frac{1}{2\sqrt{\pi}\left(n^{1-\beta}-2\right)e^{\left(n^{1-\beta}-2\right)^2}}\left\|D_{*x}^{\alpha}f\right\|_{\infty,[x,b]}(b-x)^{\alpha}\right\}.$$

Therefore

$$|\theta_n(x)| \leq |\theta_{1n}(x)| + |\theta_{2n}(x)| \leq \tag{17.119}$$

$$\frac{1}{\Gamma(\alpha+1)}\left\{\frac{\omega_1\left(D_{x-}^{\alpha}f, \frac{1}{n^{\beta}}\right)_{[a,x]} + \omega_1\left(D_{*x}^{\alpha}f, \frac{1}{n^{\beta}}\right)_{[x,b]}}{n^{\alpha\beta}} + \right. \tag{17.120}$$

$$\left.\frac{1}{2\sqrt{\pi}\left(n^{1-\beta}-2\right)e^{\left(n^{1-\beta}-2\right)^2}}\left(\left\|D_{x-}^{\alpha}f\right\|_{\infty,[a,x]}(x-a)^{\alpha} + \left\|D_{*x}^{\alpha}f\right\|_{\infty,[x,b]}(b-x)^{\alpha}\right)\right\}.$$

As in (17.69) we get that

$$\left|A_n^*\left((\cdot-x)^j\right)(x)\right| \leq \frac{1}{n^{\beta j}} + (b-a)^j\frac{1}{2\sqrt{\pi}\left(n^{1-\beta}-2\right)e^{\left(n^{1-\beta}-2\right)^2}}, \tag{17.121}$$

for $j = 1, \ldots, N-1$, $\forall\, x \in [a, b]$.

Putting things together, we have established

$$\left|A_n^*(f,x) - f(x)\left(\sum_{k=\lceil na\rceil}^{\lfloor nb\rfloor}\chi(nx-k)\right)\right| \leq \tag{17.122}$$

$$\sum_{j=1}^{N-1}\frac{\left|f^{(j)}(x)\right|}{j!}\left[\frac{1}{n^{\beta j}} + (b-a)^j\frac{1}{2\sqrt{\pi}\left(n^{1-\beta}-2\right)e^{\left(n^{1-\beta}-2\right)^2}}\right] +$$

$$\frac{1}{\Gamma(\alpha+1)}\left\{\frac{\left(\omega_1\left(D_{x-}^{\alpha}f, \frac{1}{n^{\beta}}\right)_{[a,x]} + \omega_1\left(D_{*x}^{\alpha}f, \frac{1}{n^{\beta}}\right)_{[x,b]}\right)}{n^{\alpha\beta}} + \right.$$

$$\frac{1}{2\sqrt{\pi}\left(n^{1-\beta}-2\right)e^{\left(n^{1-\beta}-2\right)^2}}\cdot$$

$$\left.\left(\left\|D_{x-}^{\alpha}f\right\|_{\infty,[a,x]}(x-a)^{\alpha} + \left\|D_{*x}^{\alpha}f\right\|_{\infty,[x,b]}(b-x)^{\alpha}\right)\right\} =: T_n(x). \tag{17.123}$$

As a result, see (17.39), we derive

$$|A_n(f,x) - f(x)| \leq (4.019)\, T_n(x), \tag{17.124}$$

$\forall\, x \in [a,b]$.

We further have that

$$\|T_n\|_\infty \leq \sum_{j=1}^{N-1} \frac{\|f^{(j)}\|_\infty}{j!} \left[\frac{1}{n^{\beta j}} + (b-a)^j \frac{1}{2\sqrt{\pi}\left(n^{1-\beta}-2\right) e^{\left(n^{1-\beta}-2\right)^2}} \right] + \tag{17.125}$$

$$\frac{1}{\Gamma(\alpha+1)} \left\{ \frac{\left\{ \sup_{x\in[a,b]} \left(\omega_1\left(D_{x-}^{\alpha} f, \frac{1}{n^{\beta}}\right)_{[a,x]} \right) + \sup_{x\in[a,b]} \left(\omega_1\left(D_{*x}^{\alpha} f, \frac{1}{n^{\beta}}\right)_{[x,b]} \right) \right\}}{n^{\alpha\beta}} \right.$$

$$+ \frac{1}{2\sqrt{\pi}\left(n^{1-\beta}-2\right) e^{\left(n^{1-\beta}-2\right)^2}} (b-a)^{\alpha} \cdot$$

$$\left. \left\{ \left(\sup_{x\in[a,b]} \left(\|D_{x-}^{\alpha} f\|_{\infty,[a,x]} \right) + \sup_{x\in[a,b]} \left(\|D_{*x}^{\alpha} f\|_{\infty,[x,b]} \right) \right) \right\} \right\} =: E_n.$$

Hence it holds

$$\|A_n f - f\|_\infty \leq (4.019)\, E_n. \tag{17.126}$$

Since $f \in AC^N([a,b])$, $N = \lceil \alpha \rceil$, $\alpha > 0$, $\alpha \notin \mathbb{N}$, $f^{(N)} \in L_\infty([a,b])$, $x \in [a,b]$, then we get that $f \in AC^N([a,x])$, $f^{(N)} \in L_\infty([a,x])$ and $f \in AC^N([x,b])$, $f^{(N)} \in L_\infty([x,b])$.

We have

$$\left(D_{x-}^{\alpha} f\right)(y) = \frac{(-1)^N}{\Gamma(N-\alpha)} \int_y^x (J-y)^{N-\alpha-1} f^{(N)}(J)\, dJ, \tag{17.127}$$

$\forall\, y \in [a,x]$ and

$$\left|\left(D_{x-}^{\alpha} f\right)(y)\right| \leq \frac{1}{\Gamma(N-\alpha)} \left(\int_y^x (J-y)^{N-\alpha-1} dJ \right) \left\| f^{(N)} \right\|_\infty \tag{17.128}$$

$$= \frac{1}{\Gamma(N-\alpha)} \frac{(x-y)^{N-\alpha}}{(N-\alpha)} \left\| f^{(N)} \right\|_\infty =$$

$$\frac{(x-y)^{N-\alpha}}{\Gamma(N-\alpha+1)} \left\| f^{(N)} \right\|_\infty \leq \frac{(b-a)^{N-\alpha}}{\Gamma(N-\alpha+1)} \left\| f^{(N)} \right\|_\infty.$$

That is

$$\left\| D_{x-}^{\alpha} f \right\|_{\infty,[a,x]} \leq \frac{(b-a)^{N-\alpha}}{\Gamma(N-\alpha+1)} \left\| f^{(N)} \right\|_{\infty}, \tag{17.129}$$

and

$$\sup_{x\in[a,b]} \left\| D_{x-}^{\alpha} f \right\|_{\infty,[a,x]} \leq \frac{(b-a)^{N-\alpha}}{\Gamma(N-\alpha+1)} \left\| f^{(N)} \right\|_{\infty}. \tag{17.130}$$

Similarly we have

$$\left(D_{*x}^{\alpha} f\right)(y) = \frac{1}{\Gamma(N-\alpha)} \int_x^y (y-t)^{N-\alpha-1} f^{(N)}(t)\, dt, \tag{17.131}$$

$\forall\, y \in [x, b]$.

Thus we get

$$\left|\left(D_{*x}^{\alpha} f\right)(y)\right| \leq \frac{1}{\Gamma(N-\alpha)} \left(\int_x^y (y-t)^{N-\alpha-1} dt\right) \left\| f^{(N)} \right\|_{\infty} \leq$$

$$\frac{1}{\Gamma(N-\alpha)} \frac{(y-x)^{N-\alpha}}{(N-\alpha)} \left\| f^{(N)} \right\|_{\infty} \leq \frac{(b-a)^{N-\alpha}}{\Gamma(N-\alpha+1)} \left\| f^{(N)} \right\|_{\infty}.$$

Hence

$$\left\| D_{*x}^{\alpha} f \right\|_{\infty,[x,b]} \leq \frac{(b-a)^{N-\alpha}}{\Gamma(N-\alpha+1)} \left\| f^{(N)} \right\|_{\infty}, \tag{17.132}$$

and

$$\sup_{x\in[a,b]} \left\| D_{*x}^{\alpha} f \right\|_{\infty,[x,b]} \leq \frac{(b-a)^{N-\alpha}}{\Gamma(N-\alpha+1)} \left\| f^{(N)} \right\|_{\infty}. \tag{17.133}$$

From (17.82) and (17.83) we get

$$\sup_{x\in[a,b]} \omega_1 \left(D_{x-}^{\alpha} f, \frac{1}{n^{\beta}}\right)_{[a,x]} \leq \frac{2 \left\| f^{(N)} \right\|_{\infty}}{\Gamma(N-\alpha+1)} (b-a)^{N-\alpha}, \tag{17.134}$$

and

$$\sup_{x\in[a,b]} \omega_1 \left(D_{*x}^{\alpha} f, \frac{1}{n^{\beta}}\right)_{[x,b]} \leq \frac{2 \left\| f^{(N)} \right\|_{\infty}}{\Gamma(N-\alpha+1)} (b-a)^{N-\alpha}. \tag{17.135}$$

So that $E_n < \infty$.

We finally notice that

$$A_n(f,x) - \sum_{j=1}^{N-1} \frac{f^{(j)}(x)}{j!} A_n\left((\cdot - x)^j\right)(x) - f(x) = \frac{A_n^*(f,x)}{\left(\sum_{k=\lceil na \rceil}^{\lfloor nb \rfloor} \chi(nx-k)\right)}$$

$$-\frac{1}{\left(\sum_{k=\lceil na\rceil}^{\lfloor nb\rfloor}\chi(nx-k)\right)}\left(\sum_{j=1}^{N-1}\frac{f^{(j)}(x)}{j!}A_n^*\left((\cdot-x)^j\right)(x)\right)-f(x)$$

$$=\frac{1}{\left(\sum_{k=\lceil na\rceil}^{\lfloor nb\rfloor}\chi(nx-k)\right)}\cdot \tag{17.136}$$

$$\left[A_n^*(f,x)-\left(\sum_{j=1}^{N-1}\frac{f^{(j)}(x)}{j!}A_n^*\left((\cdot-x)^j\right)(x)\right)-\left(\sum_{k=\lceil na\rceil}^{\lfloor nb\rfloor}\chi(nx-k)\right)f(x)\right].$$

Therefore we get

$$\left|A_n(f,x)-\sum_{j=1}^{N-1}\frac{f^{(j)}(x)}{j!}A_n\left((\cdot-x)^j\right)(x)-f(x)\right|\leq(4.019)\cdot$$

$$\left|A_n^*(f,x)-\left(\sum_{j=1}^{N-1}\frac{f^{(j)}(x)}{j!}A_n^*\left((\cdot-x)^j\right)(x)\right)-\left(\sum_{k=\lceil na\rceil}^{\lfloor nb\rfloor}\chi(nx-k)\right)f(x)\right|, \tag{17.137}$$

$\forall\, x\in[a,b]$.

The proof of the theorem is now complete. ■

Next we apply Theorem 17.30 for $N=1$.

Corollary 17.31 *Let $0<\alpha,\beta<1$, $n^{1-\beta}\geq 3$, $f\in AC([a,b])$, $f'\in L_\infty([a,b])$, $n\in\mathbb{N}$. Then*

$$\|A_nf-f\|_\infty\leq\frac{(4.019)}{\Gamma(\alpha+1)}\cdot \tag{17.138}$$

$$\left\{\frac{\left(\sup_{x\in[a,b]}\omega_1\left(D_{x-}^\alpha f,\frac{1}{n^\beta}\right)_{[a,x]}+\sup_{x\in[a,b]}\omega_1\left(D_{*x}^\alpha f,\frac{1}{n^\beta}\right)_{[x,b]}\right)}{n^{\alpha\beta}}+\right.$$

$$\frac{1}{2\sqrt{\pi}\left(n^{1-\beta}-2\right)e^{\left(n^{1-\beta}-2\right)^2}}(b-a)^\alpha\cdot$$

$$\left.\left(\sup_{x\in[a,b]}\left\|D_{x-}^\alpha f\right\|_{\infty,[a,x]}+\sup_{x\in[a,b]}\left\|D_{*x}^\alpha f\right\|_{\infty,[x,b]}\right)\right\}.$$

Remark 17.32 Let $0 < \alpha < 1$, then by (17.130), we get

$$\sup_{x\in[a,b]} \left\| D_{x-}^{\alpha} f \right\|_{\infty,[a,x]} \leq \frac{(b-a)^{1-\alpha}}{\Gamma(2-\alpha)} \left\| f' \right\|_{\infty}, \tag{17.139}$$

and by (17.133), we obtain

$$\sup_{x\in[a,b]} \left\| D_{*x}^{\alpha} f \right\|_{\infty,[x,b]} \leq \frac{(b-a)^{1-\alpha}}{\Gamma(2-\alpha)} \left\| f' \right\|_{\infty}, \tag{17.140}$$

given that $f \in AC([a,b])$ and $f' \in L_\infty([a,b])$.

Next we specialize to $\alpha = \frac{1}{2}$.

Corollary 17.33 *Let* $0 < \beta < 1$, $n^{1-\beta} \geq 3$, $f \in AC([a,b])$, $f' \in L_\infty([a,b])$, $n \in \mathbb{N}$. *Then*

$$\|A_n f - f\|_\infty \leq \frac{(8.038)}{\sqrt{\pi}}\cdot$$

$$\left\{ \frac{\left(\sup_{x\in[a,b]} \omega_1\left(D_{x-}^{\frac{1}{2}} f, \frac{1}{n^\beta}\right)_{[a,x]} + \sup_{x\in[a,b]} \omega_1\left(D_{*x}^{\frac{1}{2}} f, \frac{1}{n^\beta}\right)_{[x,b]} \right)}{n^{\frac{\beta}{2}}} + \right.$$

$$\frac{1}{2\sqrt{\pi}\left(n^{1-\beta}-2\right) e^{\left(n^{1-\beta}-2\right)^2}} \sqrt{b-a}\cdot$$

$$\left. \left(\sup_{x\in[a,b]} \left\| D_{x-}^{\frac{1}{2}} f \right\|_{\infty,[a,x]} + \sup_{x\in[a,b]} \left\| D_{*x}^{\frac{1}{2}} f \right\|_{\infty,[x,b]} \right) \right\}, \tag{17.141}$$

Remark 17.34 (*to Corollary* 17.33) Assume that

$$\omega_1\left(D_{x-}^{\frac{1}{2}} f, \frac{1}{n^\beta}\right)_{[a,x]} \leq \frac{K_1}{n^\beta}, \tag{17.142}$$

and

$$\omega_1\left(D_{*x}^{\frac{1}{2}} f, \frac{1}{n_\beta}\right)_{[x,b]} \leq \frac{K_2}{n^\beta}, \tag{17.143}$$

$\forall\, x \in [a,b]$, $\forall\, n \in \mathbb{N}$, where $K_1, K_2 > 0$.

Then for large enough $n \in \mathbb{N}$, by (17.141), we obtain

$$\|A_n f - f\|_\infty \le \frac{M}{n^{\frac{3}{2}\beta}}, \tag{17.144}$$

for some $M > 0$.

The speed of convergence in (17.144) is much higher than the corresponding speeds achieved in (17.40), which were there $\frac{1}{n^\beta}$.

17.5 Complex Neural Network Approximations

We make

Remark 17.35 Let $X := [a, b]$, $\mathbb{R}$ and $f : X \to \mathbb{C}$ with real and imaginary parts $f_1, f_2 : f = f_1 + i f_2$, $i = \sqrt{-1}$. Clearly f is continuous iff f_1 and f_2 are continuous.

Also it holds

$$f^{(j)}(x) = f_1^{(j)}(x) + i f_2^{(j)}(x), \tag{17.145}$$

for all $j = 1, \ldots, N$, given that $f_1, f_2 \in C^N(X)$, $N \in \mathbb{N}$.

We denote by $C_B(\mathbb{R}, \mathbb{C})$ the space of continuous and bounded functions $f : \mathbb{R} \to \mathbb{C}$. Clearly f is bounded, iff both f_1, f_2 are bounded from $\mathbb{R}$ into $\mathbb{R}$, where $f = f_1 + i f_2$.

Here we define

$$A_n(f, x) := A_n(f_1, x) + i A_n(f_2, x), \tag{17.146}$$

and

$$B_n(f, x) := B_n(f_1, x) + i B_n(f_2, x). \tag{17.147}$$

We observe here that

$$|A_n(f, x) - f(x)| \le |A_n(f_1, x) - f_1(x)| + |A_n(f_2, x) - f_2(x)|, \tag{17.148}$$

and

$$\|A_n(f) - f\|_\infty \le \|A_n(f_1) - f_1\|_\infty + \|A_n(f_2) - f_2\|_\infty. \tag{17.149}$$

Similarly we get

$$|B_n(f, x) - f(x)| \le |B_n(f_1, x) - f_1(x)| + |B_n(f_2, x) - f_2(x)|, \tag{17.150}$$

and

$$\|B_n(f) - f\|_\infty \le \|B_n(f_1) - f_1\|_\infty + \|B_n(f_2) - f_2\|_\infty. \tag{17.151}$$

We present

Theorem 17.36 *Let* $f \in C([a,b], \mathbb{C})$, $f = f_1 + i f_2$, $0 < \alpha < 1$, $n \in \mathbb{N}$, $n^{1-\alpha} \geq 3$, $x \in [a,b]$. *Then*

(i)

$$|A_n(f,x) - f(x)| \leq (4.019) \cdot \tag{17.152}$$

$$\left[\left(\omega_1\left(f_1, \frac{1}{n^\alpha}\right) + \omega_1\left(f_2, \frac{1}{n^\alpha}\right)\right) + (\|f_1\|_\infty + \|f_2\|_\infty) \frac{1}{\sqrt{\pi}\left(n^{1-\alpha} - 2\right) e^{\left(n^{1-\alpha}-2\right)^2}}\right]$$

$$=: \psi_1,$$

and

(ii)

$$\|A_n(f) - f\|_\infty \leq \psi_1. \tag{17.153}$$

Proof Based on Remark 17.35 and Theorem 17.12. ■

We give

Theorem 17.37 *Let* $f \in C_B(\mathbb{R}, \mathbb{C})$, $f = f_1 + i f_2$, $0 < \alpha < 1$, $n \in \mathbb{N}$, $n^{1-\alpha} \geq 3$, $x \in \mathbb{R}$. *Then*

(i)

$$|B_n(f,x) - f(x)| \leq \left(\omega_1\left(f_1, \frac{1}{n^\alpha}\right) + \omega_1\left(f_2, \frac{1}{n^\alpha}\right)\right) + \tag{17.154}$$

$$(\|f_1\|_\infty + \|f_2\|_\infty) \frac{1}{\sqrt{\pi}\left(n^{1-\alpha} - 2\right) e^{\left(n^{1-\alpha}-2\right)^2}} =: \psi_2,$$

(ii)

$$\|B_n(f) - f\|_\infty \leq \psi_2. \tag{17.155}$$

Proof Based on Remark 17.35 and Theorem 17.13. ■

Next we present a result of high order complex neural network approximation.

Theorem 17.38 *Let* $f : [a,b] \to \mathbb{C}$, $[a,b] \subset \mathbb{R}$, *such that* $f = f_1 + i f_2$. *Assume* $f_1, f_2 \in C^N([a,b])$, $n, N \in \mathbb{N}$, $n^{1-\alpha} \geq 3$, $0 < \alpha < 1$, $x \in [a,b]$. *Then*

(i)

$$|A_n(f,x) - f(x)| \leq (4.019) \cdot \tag{17.156}$$

$$\left\{ \sum_{j=1}^{N} \frac{\left(\left| f_1^{(j)}(x) \right| + \left| f_2^{(j)}(x) \right| \right)}{j!} \left[\frac{1}{n^{\alpha j}} + \frac{(b-a)^j}{2\sqrt{\pi}\left(n^{1-\alpha}-2\right) e^{\left(n^{1-\alpha}-2\right)^2}} \right] + \right.$$

$$\left[\frac{\omega_1\left(f_1^{(N)}, \frac{1}{n^\alpha}\right) + \omega_1\left(f_2^{(N)}, \frac{1}{n^\alpha}\right)}{n^{\alpha N} N!} + \right.$$

$$\left. \left. \left(\frac{\left(\left\| f_1^{(N)} \right\|_\infty + \left\| f_2^{(N)} \right\|_\infty \right) (b-a)^N}{N! \sqrt{\pi} \left(n^{1-\alpha}-2\right) e^{\left(n^{1-\alpha}-2\right)^2}} \right) \right] \right\},$$

(ii) assume further $f_1^{(j)}(x_0) = f_2^{(j)}(x_0) = 0$, $j = 1, \ldots, N$, *for some* $x_0 \in [a, b]$, *it holds*

$$|A_n(f, x_0) - f(x_0)| \leq (4.019) \cdot \tag{17.157}$$

$$\left[\frac{\omega_1\left(f_1^{(N)}, \frac{1}{n^\alpha}\right) + \omega_1\left(f_2^{(N)}, \frac{1}{n^\alpha}\right)}{n^{\alpha N} N!} + \right.$$

$$\left. \left(\frac{\left(\left\| f_1^{(N)} \right\|_\infty + \left\| f_2^{(N)} \right\|_\infty \right) (b-a)^N}{N! \sqrt{\pi} \left(n^{1-\alpha}-2\right) e^{\left(n^{1-\alpha}-2\right)^2}} \right) \right],$$

notice here the extremely high rate of convergence at $n^{-(N+1)\alpha}$,

(iii)

$$\|A_n(f) - f\|_\infty \leq (4.019) \cdot \tag{17.158}$$

$$\left\{ \sum_{j=1}^{N} \frac{\left(\left\| f_1^{(j)} \right\|_\infty + \left\| f_2^{(j)} \right\|_\infty \right)}{j!} \left[\frac{1}{n^{\alpha j}} + \frac{(b-a)^j}{2\sqrt{\pi}\left(n^{1-\alpha}-2\right) e^{\left(n^{1-\alpha}-2\right)^2}} \right] + \right.$$

$$\left[\frac{\left(\omega_1\left(f_1^{(N)}, \frac{1}{n^\alpha}\right) + \omega_1\left(f_2^{(N)}, \frac{1}{n^\alpha}\right)\right)}{n^{\alpha N} N!} + \right.$$

$$\left. \left. \frac{\left(\left\| f_1^{(N)} \right\|_\infty + \left\| f_2^{(N)} \right\|_\infty \right) (b-a)^N}{N! \sqrt{\pi} \left(n^{1-\alpha}-2\right) e^{\left(n^{1-\alpha}-2\right)^2}} \right] \right\}.$$

Proof Based on Remark 17.35 and Theorem 17.16. ■

We continue with high order complex fractional neural network approximation.

Theorem 17.39 *Let* $f:[a,b]\to\mathbb{C}$, $[a,b]\subset\mathbb{R}$, *such that* $f=f_1+if_2$; $\alpha>0$, $N=\lceil\alpha\rceil$, $\alpha\notin\mathbb{N}$, $0<\beta<1$, $x\in[a,b]$, $n\in\mathbb{N}$, $n^{1-\beta}\geq 3$. *Assume* $f_1,f_2\in AC^N([a,b])$, *with* $f_1^{(N)},f_2^{(N)}\in L_\infty([a,b])$. *Then*

(i) *assume further* $f_1^{(j)}(x)=f_2^{(j)}(x)=0$, $j=1,\ldots,N-1$, *we have*

$$|A_n(f,x)-f(x)|\leq\frac{(4.019)}{\Gamma(\alpha+1)}\cdot$$

$$\left\{\frac{1}{n^{\alpha\beta}}\left[\left(\omega_1\left(D_{x-}^{\alpha}f_1,\frac{1}{n^\beta}\right)_{[a,x]}+\omega_1\left(D_{*x}^{\alpha}f_1,\frac{1}{n^\beta}\right)_{[x,b]}\right)+\right.\right.$$

$$\left.\left(\omega_1\left(D_{x-}^{\alpha}f_2,\frac{1}{n^\beta}\right)_{[a,x]}+\omega_1\left(D_{*x}^{\alpha}f_2,\frac{1}{n^\beta}\right)_{[x,b]}\right)\right]+$$

$$\frac{1}{2\sqrt{\pi}\left(n^{1-\beta}-2\right)e^{\left(n^{1-\beta}-2\right)^2}}\cdot$$

$$\left[\left(\left\|D_{x-}^{\alpha}f_1\right\|_{\infty,[a,x]}(x-a)^\alpha+\left\|D_{*x}^{\alpha}f_1\right\|_{\infty,[x,b]}(b-x)^\alpha\right)+\right.$$

$$\left.\left.\left(\left\|D_{x-}^{\alpha}f_2\right\|_{\infty,[a,x]}(x-a)^\alpha+\left\|D_{*x}^{\alpha}f_2\right\|_{\infty,[x,b]}(b-x)^\alpha\right)\right]\right\}, \tag{17.159}$$

when $\alpha>1$ *notice here the extremely high rate of convergence at* $n^{-(\alpha+1)\beta}$,

(ii)

$$|A_n(f,x)-f(x)|\leq(4.019)\cdot\left\{\sum_{j=1}^{N-1}\frac{\left(\left|f_1^{(j)}(x)\right|+\left|f_2^{(j)}(x)\right|\right)}{j!}\right.$$

$$\left\{\frac{1}{n^{\beta j}}+\frac{(b-a)^j}{2\sqrt{\pi}\left(n^{1-\beta}-2\right)e^{\left(n^{1-\beta}-2\right)^2}}\right\}+$$

$$\frac{1}{\Gamma(\alpha+1)}\left\{\frac{1}{n^{\alpha\beta}}\left[\left(\omega_1\left(D_{x-}^{\alpha}f_1,\frac{1}{n^\beta}\right)_{[a,x]}+\omega_1\left(D_{*x}^{\alpha}f_1,\frac{1}{n^\beta}\right)_{[x,b]}\right)+\right.\right.$$

$$\left.\left(\omega_1\left(D_{x-}^{\alpha}f_2,\frac{1}{n^\beta}\right)_{[a,x]}+\omega_1\left(D_{*x}^{\alpha}f_2,\frac{1}{n^\beta}\right)_{[x,b]}\right)\right]+$$

$$\frac{1}{2\sqrt{\pi}\left(n^{1-\beta}-2\right)e^{\left(n^{1-\beta}-2\right)^2}}\cdot$$

$$\left[\left(\left\|D_{x-}^{\alpha}f_1\right\|_{\infty,[a,x]}(x-a)^{\alpha}+\left\|D_{*x}^{\alpha}f_1\right\|_{\infty,[x,b]}(b-x)^{\alpha}\right)+\right.$$

$$\left.\left.\left(\left\|D_{x-}^{\alpha}f_2\right\|_{\infty,[a,x]}(x-a)^{\alpha}+\left\|D_{*x}^{\alpha}f_2\right\|_{\infty,[x,b]}(b-x)^{\alpha}\right)\right]\right\}, \quad (17.160)$$

and

(iii)

$$\left\|A_n(f)-f\right\|_{\infty}\leq(4.019)\cdot$$

$$\left\{\sum_{j=1}^{N-1}\frac{\left(\left\|f_1^{(j)}\right\|_{\infty}+\left\|f_2^{(j)}\right\|_{\infty}\right)}{j!}\left\{\frac{1}{n^{\beta j}}+\frac{(b-a)^j}{2\sqrt{\pi}\left(n^{1-\beta}-2\right)e^{\left(n^{1-\beta}-2\right)^2}}\right\}+\right.$$

$$\frac{1}{\Gamma(\alpha+1)}\left\{\frac{1}{n^{\alpha\beta}}\left\{\left[\sup_{x\in[a,b]}\omega_1\left(D_{x-}^{\alpha}f_1,\frac{1}{n^{\beta}}\right)_{[a,x]}+\sup_{x\in[a,b]}\omega_1\left(D_{*x}^{\alpha}f_1,\frac{1}{n^{\beta}}\right)_{[x,b]}+\right.\right.\right.$$

$$\left.\left.\sup_{x\in[a,b]}\omega_1\left(D_{x-}^{\alpha}f_2,\frac{1}{n^{\beta}}\right)_{[a,x]}+\sup_{x\in[a,b]}\omega_1\left(D_{*x}^{\alpha}f_2,\frac{1}{n^{\beta}}\right)_{[x,b]}\right]\right\}+$$

$$\frac{(b-a)^{\alpha}}{2\sqrt{\pi}\left(n^{1-\beta}-2\right)e^{\left(n^{1-\beta}-2\right)^2}}\cdot$$

$$\left[\left(\sup_{x\in[a,b]}\left\|D_{x-}^{\alpha}f_1\right\|_{\infty,[a,x]}+\sup_{x\in[a,b]}\left\|D_{*x}^{\alpha}f_1\right\|_{\infty,[x,b]}\right)+\right.$$

$$\left.\left.\left.\left(\sup_{x\in[a,b]}\left\|D_{x-}^{\alpha}f_2\right\|_{\infty,[a,x]}+\sup_{x\in[a,b]}\left\|D_{*x}^{\alpha}f_2\right\|_{\infty,[x,b]}\right)\right]\right\}. \quad (17.161)$$

Above, when $N=1$ the sum $\sum_{j=1}^{N-1}\cdot=0$.
As we see here we obtain fractionally type pointwise and uniform convergence with rates of complex $A_n\to I$ the unit operator, as $n\to\infty$.

Proof Using Theorem 17.30 and Remark 17.35. ■

We need

Definition 17.40 Let $f\in C_B(\mathbb{R},\mathbb{C})$, with $f=f_1+if_2$. We define

$$C_n(f,x) := C_n(f_1,x) + iC_n(f_2,x), \quad (17.162)$$
$$D_n(f,x) := D_n(f_1,x) + iD_n(f_2,x), \quad \forall\, x \in \mathbb{R}, n \in \mathbb{N}.$$

We finish with

Theorem 17.41 *Let $f \in C_B(\mathbb{R}, \mathbb{C})$, $f = f_1 + if_2$, $0 < \alpha < 1$, $n \in \mathbb{N}$, $n^{1-\alpha} \geq 3$, $x \in \mathbb{R}$. Then*

(i)

$$\left\{ \begin{array}{l} |C_n(f,x) - f(x)| \\ |D_n(f,x) - f(x)| \end{array} \right. \leq \left(\omega_1 \left(f_1, \frac{1}{n} + \frac{1}{n^\alpha} \right) + \omega_1 \left(f_2, \frac{1}{n} + \frac{1}{n^\alpha} \right) \right)$$

$$+ \frac{\left(\|f_1\|_\infty + \|f_2\|_\infty \right)}{\sqrt{\pi}\left(n^{1-\alpha} - 2\right) e^{\left(n^{1-\alpha}-2\right)^2}} =: \mu_{3n}(f_1, f_2), \quad (17.163)$$

and

(ii)

$$\left\{ \begin{array}{l} \|C_n(f) - f\|_\infty \\ \|D_n(f) - f\|_\infty \end{array} \right. \leq \mu_{3n}(f_1, f_2). \quad (17.164)$$

Proof By Theorems 17.14, 17.15, also see (17.162). ∎

References

1. M. Abramowitz, I.A. Stegun (eds.), *Handbook of Mathematical Functions with Formulas, Graphs, and Mathematical Tables* (Dover Publications, New York, 1972)
2. G.A. Anastassiou, Rate of convergence of some neural network operators to the unit-univariate case. J. Math. Anal. Appli. **212**, 237–262 (1997)
3. G.A. Anastassiou, *Quantitative Approximations* (Chapman & Hall/CRC, Boca Raton, New York, 2001)
4. G.A. Anastassiou, On right fractional calculus. Chaos, Solitons Fractals **42**, 365–376 (2009)
5. G.A. Anastassiou, *Fractional Differentiation Inequalities* (Springer, New York, 2009)
6. G.A. Anastassiou, Fractional Korovkin theory. Chaos, Solitons Fractals **42**(4), 2080–2094 (2009)
7. G.A. Anastassiou, *Inteligent Systems: Approximation by Artificial Neural Networks, Intelligent Systems Reference Library*, vol. 19 (Springer, Heidelberg, 2011)
8. G.A. Anastassiou, Fractional representation formulae and right fractional inequalities. Math. Comput. Model. **54**(11–12), 3098–3115 (2011)
9. G.A. Anastassiou, Univariate hyperbolic tangent neural network approximation. Math. Comput. Model. **53**, 1111–1132 (2011)
10. G.A. Anastassiou, Multivariate hyperbolic tangent neural network approximation. Comput. Math. **61**, 809–821 (2011)
11. G.A. Anastassiou, Multivariate sigmoidal neural network approximation. Neural Networks **24**, 378–386 (2011)
12. G.A. Anastassiou, Univariate sigmoidal neural network approximation. J. Comput. Anal. Appl. **14**(4), 659–690 (2012)

13. G.A. Anastassiou, Fractional neural network approximation. Comput. Math. Appl. **64**, 1655–1676 (2012)
14. G.A. Anastassiou, Univariate error function based neural network approximation. Indian J. Math. (2014)
15. L.C. Andrews, *Special Functions of Mathematics for Engineers*, 2nd edn. (Mc Graw-Hill, New York, 1992)
16. Z. Chen, F. Cao, The approximation operators with sigmoidal functions. Comput. Math. Appl. **58**, 758–765 (2009)
17. K. Diethelm, *The Analysis of Fractional Differential Equations*, vol. 2004, Lecture Notes in Mathematics (Springer, Berlin, 2010)
18. A.M.A. El-Sayed, M. Gaber, On the finite caputo and finite riesz derivatives. Electron. J. Theor. Phys. **3**(12), 81–95 (2006)
19. G.S. Frederico, D.F.M. Torres, Fractional optimal control in the sense of caputo and the fractional Noether's theorem. Int. Math. forum **3**(10), 479–493 (2008)
20. S. Haykin, *Neural Networks: a Comprehensive Foundation*, 2nd edn. (Prentice Hall, New York, 1998)
21. W. McCulloch, W. Pitts, A logical calculus of the ideas immanent in nervous activity. Bull. Math. Biophys. **7**, 115–133 (1943)
22. T.M. Mitchell, *Machine Learning* (WCB-McGraw-Hill, New York, 1997)
23. D.S. Mitrinovic, *Analytical Inequalities* (Springer, New York, 1970)
24. S.G. Samko, A.A. Kilbas and O.I. Marichev, *Fractional Integrals and Derivatives, Theory and Applications*, (Gordon and Breach, Amsterdam, 1993) [English translation from the Russian, Integrals and Derivatives of Fractional Order and Some of Their Applications (Nauka i Tekhnika, Minsk, 1987)]

Chapter 18
Multivariate Error Function Based Neural Network Operators Approximation

Here we present multivariate quantitative approximations of real and complex valued continuous multivariate functions on a box or $\mathbb{R}^N$, $N \in \mathbb{N}$, by the multivariate quasi-interpolation, Baskakov type and quadrature type neural network operators. We treat also the case of approximation by iterated operators of the last three types. These approximations are derived by establishing multidimensional Jackson type inequalities involving the multivariate modulus of continuity of the engaged function or its high order partial derivatives. Our multivariate operators are defined by using a multidimensional density function induced by the Gaussian error special function. The approximations are pointwise and uniform. The related feed-forward neural network is with one hidden layer. It follows [10].

18.1 Introduction

The author in [2, 3], see Chaps. 2–5, was the first to establish neural network approximations to continuous functions with rates by very specifically defined neural network operators of Cardaliagnet-Euvrard and "Squashing" types, by employing the modulus of continuity of the engaged function or its high order derivative, and producing very tight Jackson type inequalities. He treats there both the univariate and multivariate cases. The defining these operators "bell-shaped" and "squashing" functions are assumed to be of compact support. Also in [3] he gives the Nth order asymptotic expansion for the error of weak approximation of these two operators to a special natural class of smooth functions, see Chaps. 4–5 there.

For this chapter the author is motivated by the article [13] of Chen and Cao, also by [4–9, 11, 14, 15].

The author here performs multivariate error function based neural network approximations to continuous functions over boxes or over the whole $\mathbb{R}^N$, $N \in \mathbb{N}$, then he extends his results to complex valued multivariate functions. Also he does iterated approximation. All convergences here are with rates expressed via the multivariate

G.A. Anastassiou, *Intelligent Systems II: Complete Approximation by Neural Network Operators*, Studies in Computational Intelligence 608,
DOI 10.1007/978-3-319-20505-2_18

modulus of continuity of the involved function or its high order partial derivative and given by very tight multidimensional Jackson type inequalities.

The author here comes up with the “right” precisely defined multivariate quasi-interpolation neural network operators related to boxes or $\mathbb{R}^N$, as well as Baskakov type and quadrature type related operators on $\mathbb{R}^N$. Our boxes are not necessarily symmetric to the origin. In preparation to prove our results we establish important properties of the basic multivariate density function induced by error function and defining our operators.

Feed-forward neural networks (FNNs) with one hidden layer, the only type of networks we deal with in this chapter, are mathematically expressed as

$$N_n(x) = \sum_{j=0}^{n} c_j \sigma\left(\langle a_j \cdot x \rangle + b_j\right), \quad x \in \mathbb{R}^s, \quad s \in \mathbb{N},$$

where for $0 \leq j \leq n$, $b_j \in \mathbb{R}$ are the thresholds, $a_j \in \mathbb{R}^s$ are the connection weights, $c_j \in \mathbb{R}$ are the coefficients, $\langle a_j \cdot x \rangle$ is the inner product of a_j and x, and σ is the activation function of the network. In many fundamental network models, the activation function is the error function. About neural networks read [16–18].

18.2 Basics

We consider here the (Gauss) error special function [1, 12]

$$\operatorname{erf}(x) = \frac{2}{\sqrt{\pi}} \int_0^x e^{-t^2} dt, \quad x \in \mathbb{R}, \tag{18.1}$$

which is a sigmoidal type function and is a strictly increasing function.

It has the basic properties

$$\operatorname{erf}(0) = 0, \ \operatorname{erf}(-x) = -\operatorname{erf}(x), \ \operatorname{erf}(+\infty) = 1, \ \operatorname{erf}(-\infty) = -1.$$

We consider the activation function [11]

$$\chi(x) = \frac{1}{4}\left(\operatorname{erf}(x+1) - \operatorname{erf}(x-1)\right) > 0, \ \text{any}\, x \in \mathbb{R}, \tag{18.2}$$

which is an even function.

Next we follow [11] on χ. We got there $\chi(0) \simeq 0.4215$, and that χ is strictly decreasing on $[0, \infty)$ and strictly increasing on $(-\infty, 0]$, and the x-axis is the horizontal asymptote on χ, i.e. χ is a bell symmetric function.

Theorem 18.1 ([11]) *We have that*

$$\sum_{i=-\infty}^{\infty} \chi(x-i) = 1, \text{ all } x \in \mathbb{R}, \tag{18.3}$$

$$\sum_{i=-\infty}^{\infty} \chi(nx-i) = 1, \text{ all } x \in \mathbb{R}, n \in \mathbb{N}, \tag{18.4}$$

and

$$\int_{-\infty}^{\infty} \chi(x)\,dx = 1, \tag{18.5}$$

that is $\chi(x)$ *is a density function on* $\mathbb{R}$.

We need the important

Theorem 18.2 ([11]) *Let* $0 < \alpha < 1$, *and* $n \in \mathbb{N}$ *with* $n^{1-\alpha} \geq 3$. *It holds*

$$\sum_{\substack{k=-\infty \\ : |nx-k| \geq n^{1-\alpha}}}^{\infty} \chi(nx-k) < \frac{1}{2\sqrt{\pi}\left(n^{1-\alpha}-2\right)e^{\left(n^{1-\alpha}-2\right)^2}}. \tag{18.6}$$

Denote by $\lfloor \cdot \rfloor$ the integral part of the number and by $\lceil \cdot \rceil$ the ceiling of the number.

Theorem 18.3 ([11]) *Let* $x \in [a,b] \subset \mathbb{R}$ *and* $n \in \mathbb{N}$ *so that* $\lceil na \rceil \leq \lfloor nb \rfloor$. *It holds*

$$\frac{1}{\sum_{k=\lceil na \rceil}^{\lfloor nb \rfloor} \chi(nx-k)} < \frac{1}{\chi(1)} \simeq 4.019, \ \forall x \in [a,b]. \tag{18.7}$$

Also from [11] we get

$$\lim_{n\to\infty} \sum_{k=\lceil na \rceil}^{\lfloor nb \rfloor} \chi(nx-k) \neq 1, \tag{18.8}$$

at least for some $x \in [a,b]$.

For large enough n we always obtain $\lceil na \rceil \leq \lfloor nb \rfloor$. Also $a \leq \frac{k}{n} \leq b$, iff $\lceil na \rceil \leq k \leq \lfloor nb \rfloor$. In general it holds by (18.4) that

$$\sum_{k=\lceil na \rceil}^{\lfloor nb \rfloor} \chi(nx-k) \leq 1. \tag{18.9}$$

We introduce

$$Z(x_1,\ldots,x_N) := Z(x) := \prod_{i=1}^{N} \chi(x_i),\ x = (x_1,\ldots,x_N) \in \mathbb{R}^N,\ N \in \mathbb{N}. \quad (18.10)$$

It has the properties:

(i) $Z(x) > 0,\ \forall\, x \in \mathbb{R}^N$,

(ii)

$$\sum_{k=-\infty}^{\infty} Z(x-k) := \sum_{k_1=-\infty}^{\infty}\sum_{k_2=-\infty}^{\infty}\cdots\sum_{k_N=-\infty}^{\infty} Z(x_1-k_1,\ldots,x_N-k_N) = 1, \quad (18.11)$$

where $k := (k_1,\ldots,k_n) \in \mathbb{Z}^N, \forall\, x \in \mathbb{R}^N$,

hence

(iii)

$$\sum_{k=-\infty}^{\infty} Z(nx-k) =$$

$$\sum_{k_1=-\infty}^{\infty}\sum_{k_2=-\infty}^{\infty}\cdots\sum_{k_N=-\infty}^{\infty} Z(nx_1-k_1,\ldots,nx_N-k_N) = 1, \quad (18.12)$$

$\forall\, x \in \mathbb{R}^N;\ n \in \mathbb{N}$,

and

(iv)

$$\int_{\mathbb{R}^N} Z(x)\,dx = 1, \quad (18.13)$$

that is Z is a multivariate density function.

Here $\|x\|_\infty := \max\{|x_1|,\ldots,|x_N|\}$, $x \in \mathbb{R}^N$, also set $\infty := (\infty,\ldots,\infty)$, $-\infty := (-\infty,\ldots,-\infty)$ upon the multivariate context, and

$$\lceil na \rceil := (\lceil na_1 \rceil,\ldots,\lceil na_N \rceil), \quad (18.14)$$
$$\lfloor nb \rfloor := (\lfloor nb_1 \rfloor,\ldots,\lfloor nb_N \rfloor),$$

where $a := (a_1,\ldots,a_N)$, $b := (b_1,\ldots,b_N)$.

We obviously see that

$$\sum_{k=\lceil na \rceil}^{\lfloor nb \rfloor} Z(nx-k) = \sum_{k=\lceil na \rceil}^{\lfloor nb \rfloor} \left(\prod_{i=1}^{N} \chi(nx_i-k_i)\right) =$$

$$\sum_{k_1=\lceil na_1\rceil}^{\lfloor nb_1\rfloor}\cdots\sum_{k_N=\lceil na_N\rceil}^{\lfloor nb_N\rfloor}\left(\prod_{i=1}^{N}\chi\left(nx_i-k_i\right)\right)=\prod_{i=1}^{N}\left(\sum_{k_i=\lceil na_i\rceil}^{\lfloor nb_i\rfloor}\chi\left(nx_i-k_i\right)\right). \quad (18.15)$$

For $0<\beta<1$ and $n\in\mathbb{N}$, a fixed $x\in\mathbb{R}^N$, we have that

$$\sum_{k=\lceil na\rceil}^{\lfloor nb\rfloor}\chi\left(nx-k\right)=$$

$$\sum_{\substack{k=\lceil na\rceil \\ \left\|\frac{k}{n}-x\right\|_\infty\leq\frac{1}{n^\beta}}}^{\lfloor nb\rfloor}\chi\left(nx-k\right)+\sum_{\substack{k=\lceil na\rceil \\ \left\|\frac{k}{n}-x\right\|_\infty>\frac{1}{n^\beta}}}^{\lfloor nb\rfloor}\chi\left(nx-k\right). \quad (18.16)$$

In the last two sums the counting is over disjoint vector sets of k's, because the condition $\left\|\frac{k}{n}-x\right\|_\infty>\frac{1}{n^\beta}$ implies that there exists at least one $\left|\frac{k_r}{n}-x_r\right|>\frac{1}{n^\beta}$, where $r\in\{1,\ldots,N\}$.

We treat

$$\sum_{\substack{k=\lceil na\rceil \\ \left\|\frac{k}{n}-x\right\|_\infty>\frac{1}{n^\beta}}}^{\lfloor nb\rfloor}Z\left(nx-k\right)=\prod_{i=1}^{N}\left(\sum_{\substack{k_i=\lceil na_i\rceil \\ \left\|\frac{k}{n}-x\right\|_\infty>\frac{1}{n^\beta}}}^{\lfloor nb_i\rfloor}\chi\left(nx_i-k_i\right)\right)$$

$$\leq\left(\prod_{\substack{i=1\\i\neq r}}^{N}\left(\sum_{k_i=-\infty}^{\infty}\chi\left(nx_i-k_i\right)\right)\right)\cdot\left(\sum_{\substack{k_r=\lceil na_r\rceil \\ \left|\frac{k_r}{n}-x_r\right|>\frac{1}{n^\beta}}}^{\lfloor nb_r\rfloor}\chi\left(nx_r-k_r\right)\right)= \quad (18.17)$$

$$\left(\sum_{\substack{k_r=\lceil na_r\rceil \\ \left|\frac{k_r}{n}-x_r\right|>\frac{1}{n^\beta}}}^{\lfloor nb_r\rfloor}\chi\left(nx_r-k_r\right)\right)\leq\sum_{\substack{k_r=-\infty \\ \left|\frac{k_r}{n}-x_r\right|>\frac{1}{n^\beta}}}^{\infty}\chi\left(nx_r-k_r\right)$$

$$= \sum_{\substack{k_r = -\infty \\ : |nx_r - k_r| > n^{1-\beta}}}^{\infty} \chi(nx_r - k_r) \overset{(18.6)}{\leq} \frac{1}{2\sqrt{\pi}\left(n^{1-\beta} - 2\right) e^{\left(n^{1-\beta}-2\right)^2}}, \quad (18.18)$$

when $n^{1-\beta} \geq 3$.

We have proved that

(v)

$$\sum_{\substack{k = \lceil na \rceil \\ \left\| \frac{k}{n} - x \right\|_\infty > \frac{1}{n^\beta}}}^{\lfloor nb \rfloor} Z(nx - k) \leq \frac{1}{2\sqrt{\pi}\left(n^{1-\beta} - 2\right) e^{\left(n^{1-\beta}-2\right)^2}}, \quad (18.19)$$

$0 < \beta < 1, n \in \mathbb{N}; n^{1-\beta} \geq 3, x \in \left(\prod_{i=1}^{N} [a_i, b_i]\right)$.

By Theorem 18.3 clearly we obtain

$$0 < \frac{1}{\sum_{k=\lceil na \rceil}^{\lfloor nb \rfloor} Z(nx - k)} = \frac{1}{\prod_{i=1}^{N} \left(\sum_{k_i=\lceil na_i \rceil}^{\lfloor nb_i \rfloor} \chi(nx_i - k_i)\right)} \quad (18.20)$$

$$< \frac{1}{(\chi(1))^N} \simeq (4.019)^N.$$

That is,

(vi) it holds

$$0 < \frac{1}{\sum_{k=\lceil na \rceil}^{\lfloor nb \rfloor} Z(nx - k)} < \frac{1}{(\chi(1))^N} \simeq (4.019)^N, \quad (18.21)$$

$\forall\, x \in \left(\prod_{i=1}^{N} [a_i, b_i]\right)$, $n \in \mathbb{N}$.

It is also clear that

(vii)

$$\sum_{\substack{k = -\infty \\ \left\| \frac{k}{n} - x \right\|_\infty > \frac{1}{n^\beta}}}^{\infty} Z(nx - k) \leq \frac{1}{2\sqrt{\pi}\left(n^{1-\beta} - 2\right) e^{\left(n^{1-\beta}-2\right)^2}}, \quad (18.22)$$

$0 < \beta < 1, n \in \mathbb{N} : n^{1-\beta} \geq 3, x \in \left(\prod_{i=1}^{N} [a_i, b_i]\right)$.

Also we get that

$$\lim_{n\to\infty} \sum_{k=\lceil na\rceil}^{\lfloor nb\rfloor} Z(nx-k) \neq 1, \tag{18.23}$$

for at least some $x \in \left(\prod_{i=1}^{N}[a_i,b_i]\right)$.

Let $f \in C\left(\prod_{i=1}^{N}[a_i,b_i]\right)$ and $n \in \mathbb{N}$ such that $\lceil na_i\rceil \leq \lfloor nb_i\rfloor$, $i=1,\dots,N$.

We introduce and define the multivariate positive linear neural network operator $(x := (x_1,\dots,x_N) \in \left(\prod_{i=1}^{N}[a_i,b_i]\right))$

$$A_n(f,x_1,\dots,x_N) := A_n(f,x) := \frac{\sum_{k=\lceil na\rceil}^{\lfloor nb\rfloor} f\left(\frac{k}{n}\right) Z(nx-k)}{\sum_{k=\lceil na\rceil}^{\lfloor nb\rfloor} Z(nx-k)} \tag{18.24}$$

$$:= \frac{\sum_{k_1=\lceil na_1\rceil}^{\lfloor nb_1\rfloor}\sum_{k_2=\lceil na_2\rceil}^{\lfloor nb_2\rfloor}\cdots\sum_{k_N=\lceil na_N\rceil}^{\lfloor nb_N\rfloor} f\left(\frac{k_1}{n},\dots,\frac{k_N}{n}\right)\left(\prod_{i=1}^{N}\chi(nx_i-k_i)\right)}{\prod_{i=1}^{N}\left(\sum_{k_i=\lceil na_i\rceil}^{\lfloor nb_i\rfloor}\chi(nx_i-k_i)\right)}.$$

For large enough n we always obtain $\lceil na_i\rceil \leq \lfloor nb_i\rfloor$, $i=1,\dots,N$. Also $a_i \leq \frac{k_i}{n} \leq b_i$, iff $\lceil na_i\rceil \leq k_i \leq \lfloor nb_i\rfloor$, $i=1,\dots,N$.

For convinience we call

$$A_n^*(f,x) := \sum_{k=\lceil na\rceil}^{\lfloor nb\rfloor} f\left(\frac{k}{n}\right) Z(nx-k) \tag{18.25}$$

$$:= \sum_{k_1=\lceil na_1\rceil}^{\lfloor nb_1\rfloor}\sum_{k_2=\lceil na_2\rceil}^{\lfloor nb_2\rfloor}\cdots\sum_{k_N=\lceil na_N\rceil}^{\lfloor nb_N\rfloor} f\left(\frac{k_1}{n},\dots,\frac{k_N}{n}\right)\left(\prod_{i=1}^{N}\chi(nx_i-k_i)\right),$$

$\forall\, x \in \left(\prod_{i=1}^{N}[a_i,b_i]\right)$.

That is

$$A_n(f,x) := \frac{A_n^*(f,x)}{\sum_{k=\lceil na\rceil}^{\lfloor nb\rfloor} Z(nx-k)}, \tag{18.26}$$

$\forall\, x \in \left(\prod_{i=1}^{N}[a_i,b_i]\right)$, $n \in \mathbb{N}$.

Hence

$$A_n(f,x) - f(x) = \frac{A_n^*(f,x) - f(x)\left(\sum_{k=\lceil na\rceil}^{\lfloor nb\rfloor} Z(nx-k)\right)}{\sum_{k=\lceil na\rceil}^{\lfloor nb\rfloor} Z(nx-k)}. \tag{18.27}$$

Consequently we derive

$$|A_n(f,x) - f(x)| \leq (4.019)^N \left| A_n^*(f,x) - f(x) \sum_{k=\lceil na \rceil}^{\lfloor nb \rfloor} Z(nx-k) \right|, \quad (18.28)$$

$\forall\, x \in \left(\prod_{i=1}^N [a_i, b_i]\right)$.

We will estimate the right hand side of (18.28).

For the last we need, for $f \in C\left(\prod_{i=1}^N [a_i, b_i]\right)$ the first multivariate modulus of continuity

$$\omega_1(f,h) := \sup_{\substack{x,\, y \in \prod_{i=1}^N [a_i, b_i] \\ \|x-y\|_\infty \leq h}} |f(x) - f(y)|, h > 0. \quad (18.29)$$

It holds that

$$\lim_{h \to 0} \omega_1(f,h) = 0. \quad (18.30)$$

Similarly it is defined for $f \in C_B\left(\mathbb{R}^N\right)$ (continuous and bounded functions on $\mathbb{R}^N$) the $\omega_1(f,h)$, and it has the property (18.30), given that $f \in C_U\left(\mathbb{R}^N\right)$ (uniformly continuous functions on $\mathbb{R}^N$).

When $f \in C_B\left(\mathbb{R}^N\right)$ we define,

$$B_n(f,x) := B_n(f, x_1, \ldots, x_N) := \sum_{k=-\infty}^{\infty} f\left(\frac{k}{n}\right) Z(nx-k) := \quad (18.31)$$

$$\sum_{k_1=-\infty}^{\infty} \sum_{k_2=-\infty}^{\infty} \cdots \sum_{k_N=-\infty}^{\infty} f\left(\frac{k_1}{n}, \frac{k_2}{n}, \ldots, \frac{k_N}{n}\right) \left(\prod_{i=1}^N \chi(nx_i - k_i)\right),$$

$n \in \mathbb{N}$, $\forall\, x \in \mathbb{R}^N$, $N \in \mathbb{N}$, the multivariate quasi-interpolation neural network operator.

Also for $f \in C_B\left(\mathbb{R}^N\right)$ we define the multivariate Kantorovich type neural network operator

$$C_n(f,x) := C_n(f, x_1, \ldots, x_N) := \sum_{k=-\infty}^{\infty} \left(n^N \int_{\frac{k}{n}}^{\frac{k+1}{n}} f(t)\, dt\right) Z(nx-k) := \quad (18.32)$$

$$\sum_{k_1=-\infty}^{\infty} \sum_{k_2=-\infty}^{\infty} \cdots \sum_{k_N=-\infty}^{\infty} \left(n^N \int_{\frac{k_1}{n}}^{\frac{k_1+1}{n}} \int_{\frac{k_2}{n}}^{\frac{k_2+1}{n}} \cdots \int_{\frac{k_N}{n}}^{\frac{k_N+1}{n}} f(t_1, \ldots, t_N)\, dt_1 \ldots dt_N\right)$$

$$\cdot \left(\prod_{i=1}^{N} \chi (nx_i - k_i) \right),$$

$n \in \mathbb{N}$, $\forall\, x \in \mathbb{R}^N$.

Again for $f \in C_B\left(\mathbb{R}^N\right)$, $N \in \mathbb{N}$, we define the multivariate neural network operator of quadrature type $D_n(f, x)$, $n \in \mathbb{N}$, as follows. Let $\theta = (\theta_1, \ldots, \theta_N) \in \mathbb{N}^N$, $r = (r_1, \ldots, r_N) \in \mathbb{Z}_+^N$, $w_r = w_{r_1, r_2, \ldots r_N} \geq 0$, such that $\sum_{r=0}^{\theta} w_r = \sum_{r_1=0}^{\theta_1} \sum_{r_2=0}^{\theta_2} \cdots \sum_{r_N=0}^{\theta_N} w_{r_1, r_2, \ldots r_N} = 1$; $k \in \mathbb{Z}^N$ and

$$\delta_{nk}(f) := \delta_{n, k_1, k_2, \ldots, k_N}(f) := \sum_{r=0}^{\theta} w_r f\left(\frac{k}{n} + \frac{r}{n\theta}\right) :=$$

$$\sum_{r_1=0}^{\theta_1} \sum_{r_2=0}^{\theta_2} \cdots \sum_{r_N=0}^{\theta_N} w_{r_1, r_2, \ldots r_N} f\left(\frac{k_1}{n} + \frac{r_1}{n\theta_1}, \frac{k_2}{n} + \frac{r_2}{n\theta_2}, \ldots, \frac{k_N}{n} + \frac{r_N}{n\theta_N}\right), \quad (18.33)$$

where $\frac{r}{\theta} := \left(\frac{r_1}{\theta_1}, \frac{r_2}{\theta_2}, \ldots, \frac{r_N}{\theta_N}\right)$.

We put

$$D_n(f, x) := D_n(f, x_1, \ldots, x_N) := \sum_{k=-\infty}^{\infty} \delta_{nk}(f)\, Z(nx - k) := \quad (18.34)$$

$$\sum_{k_1=-\infty}^{\infty} \sum_{k_2=-\infty}^{\infty} \cdots \sum_{k_N=-\infty}^{\infty} \delta_{n, k_1, k_2, \ldots, k_N}(f) \left(\prod_{i=1}^{N} \chi (nx_i - k_i) \right),$$

$\forall\, x \in \mathbb{R}^N$.

Let fixed $j \in \mathbb{N}$, $0 < \beta < 1$, and $A, B > 0$. For large enough $n \in \mathbb{N}$: $n^{1-\beta} \geq 3$, in the linear combination $\left(\frac{A}{n^{\beta j}} + \frac{B}{\left(n^{1-\beta}-2\right) e^{\left(n^{1-\beta}-2\right)^2}} \right)$, the dominant rate of convergence, as $n \to \infty$, is $n^{-\beta j}$. The closer β is to 1 we get faster and better rate of convergence to zero.

Let $f \in C^m\left(\prod_{i=1}^N [a_i, b_i]\right)$, $m, N \in \mathbb{N}$. Here f_α denotes a partial derivative of f, $\alpha := (\alpha_1, \ldots, \alpha_N)$, $\alpha_i \in \mathbb{Z}_+$, $i = 1, \ldots, N$, and $|\alpha| := \sum_{i=1}^N \alpha_i = l$, where $l = 0, 1, \ldots, m$. We write also $f_\alpha := \frac{\partial^\alpha f}{\partial x^\alpha}$ and we say it is of order l.

We denote

$$\omega_{1,m}^{\max}(f_\alpha, h) := \max_{\alpha : |\alpha| = m} \omega_1(f_\alpha, h). \quad (18.35)$$

Call also

$$\|f_{\alpha}\|_{\infty,m}^{\max} := \max_{|\alpha|=m} \{\|f_{\alpha}\|_{\infty}\}, \tag{18.36}$$

$\|\cdot\|_{\infty}$ is the supremum norm.

In this chapter we study the basic approximation properties of A_n, B_n, C_n, D_n neural network operators and as well of their iterates. That is, the quantitative pointwise and uniform convergence of these operators to the unit operator I. We study also the complex functions related approximation.

18.3 Multidimensional Real Neural Network Approximations

Here we present a series of neural network approximations to a function given with rates.

We give

Theorem 18.4 *Let* $f \in C\left(\prod_{i=1}^{N}[a_i, b_i]\right)$, $0 < \beta < 1$, $x \in \left(\prod_{i=1}^{N}[a_i, b_i]\right)$, $N, n \in \mathbb{N}$ *with* $n^{1-\beta} \geq 3$. *Then*

(1)

$$|A_n(f,x) - f(x)| \leq (4.019)^N \left[\omega_1\left(f, \frac{1}{n^{\beta}}\right) + \frac{\|f\|_{\infty}}{\sqrt{\pi}\left(n^{1-\beta}-2\right)e^{\left(n^{1-\beta}-2\right)^2}}\right] =: \lambda_1, \tag{18.37}$$

and

(2)

$$\|A_n(f) - f\|_{\infty} \leq \lambda_1. \tag{18.38}$$

We notice that $\lim_{n\to\infty} A_n(f) = f$, *pointwise and uniformly.*

Proof We observe that

$$\Delta(x) := A_n^*(f,x) - f(x) \sum_{k=\lceil na\rceil}^{\lfloor nb\rfloor} Z(nx-k) =$$

$$\sum_{k=\lceil na\rceil}^{\lfloor nb\rfloor} f\left(\frac{k}{n}\right) Z(nx-k) - \sum_{k=\lceil na\rceil}^{\lfloor nb\rfloor} f(x) Z(nx-k) =$$

$$\sum_{k=\lceil na\rceil}^{\lfloor nb\rfloor} \left(f\left(\frac{k}{n}\right) - f(x)\right) Z(nx-k). \tag{18.39}$$

Thus

$$|\Delta(x)| \leq \sum_{k=\lceil na \rceil}^{\lfloor nb \rfloor} \left| f\left(\frac{k}{n}\right) - f(x) \right| Z(nx-k) =$$

$$\sum_{\begin{cases} k = \lceil na \rceil \\ \left\| \frac{k}{n} - x \right\|_\infty \leq \frac{1}{n^\beta} \end{cases}}^{\lfloor nb \rfloor} \left| f\left(\frac{k}{n}\right) - f(x) \right| Z(nx-k) +$$

$$\sum_{\begin{cases} k = \lceil na \rceil \\ \left\| \frac{k}{n} - x \right\|_\infty > \frac{1}{n^\beta} \end{cases}}^{\lfloor nb \rfloor} \left| f\left(\frac{k}{n}\right) - f(x) \right| Z(nx-k) \overset{\text{(by (18.12))}}{\leq}$$

$$\omega_1\left(f, \frac{1}{n^\beta}\right) + 2\|f\|_\infty \sum_{\begin{cases} k = \lceil na \rceil \\ \left\| \frac{k}{n} - x \right\|_\infty > \frac{1}{n^\beta} \end{cases}}^{\lfloor nb \rfloor} Z(nx-k) \overset{\text{(by (18.19))}}{\leq}$$

$$\omega_1\left(f, \frac{1}{n^\beta}\right) + \frac{\|f\|_\infty}{\sqrt{\pi}\left(n^{1-\beta}-2\right) e^{\left(n^{1-\beta}-2\right)^2}}. \tag{18.40}$$

So that

$$|\Delta| \leq \omega_1\left(f, \frac{1}{n^\beta}\right) + \frac{\|f\|_\infty}{\sqrt{\pi}\left(n^{1-\beta}-2\right) e^{\left(n^{1-\beta}-2\right)^2}}.$$

Now using (18.28) we finish proof. ■

We continue with

Theorem 18.5 *Let $f \in C_B\left(\mathbb{R}^N\right)$, $0 < \beta < 1$, $x \in \mathbb{R}^N$, $N, n \in \mathbb{N}$ with $n^{1-\beta} \geq 3$. Then*

(1)

$$|B_n(f,x) - f(x)| \leq \omega_1\left(f, \frac{1}{n^\beta}\right) + \frac{\|f\|_\infty}{\sqrt{\pi}\left(n^{1-\beta}-2\right) e^{\left(n^{1-\beta}-2\right)^2}} =: \lambda_2, \tag{18.41}$$

(2)

$$\|B_n(f) - f\|_\infty \leq \lambda_2. \tag{18.42}$$

Given that $f \in \left(C_U\left(\mathbb{R}^N\right) \cap C_B\left(\mathbb{R}^N\right)\right)$, we obtain $\lim_{n\to\infty} B_n(f) = f$, uniformly.

Proof We have that

$$B_n(f,x)-f(x)\overset{(18.12)}{=}\sum_{k=-\infty}^{\infty}f\left(\frac{k}{n}\right)Z(nx-k)-f(x)\sum_{k=-\infty}^{\infty}Z(nx-k)= \tag{18.43}$$

$$\sum_{k=-\infty}^{\infty}\left(f\left(\frac{k}{n}\right)-f(x)\right)Z(nx-k).$$

Hence

$$|B_n(f,x)-f(x)|\leq\sum_{k=-\infty}^{\infty}\left|f\left(\frac{k}{n}\right)-f(x)\right|Z(nx-k)=$$

$$\sum_{\substack{k=-\infty \\ \left\|\frac{k}{n}-x\right\|_\infty\leq\frac{1}{n^\beta}}}^{\infty}\left|f\left(\frac{k}{n}\right)-f(x)\right|Z(nx-k)+$$

$$\sum_{\substack{k=-\infty \\ \left\|\frac{k}{n}-x\right\|_\infty>\frac{1}{n^\beta}}}^{\infty}\left|f\left(\frac{k}{n}\right)-f(x)\right|Z(nx-k)\overset{(18.12)}{\leq}$$

$$\omega_1\left(f,\frac{1}{n^\beta}\right)+2\|f\|_\infty\sum_{\substack{k=-\infty \\ \left\|\frac{k}{n}-x\right\|_\infty>\frac{1}{n^\beta}}}^{\infty}Z(nx-k)\overset{(18.19)}{\leq}$$

$$\omega_1\left(f,\frac{1}{n^\beta}\right)+\frac{\|f\|_\infty}{\sqrt{\pi}\left(n^{1-\beta}-2\right)e^{\left(n^{1-\beta}-2\right)^2}}, \tag{18.44}$$

proving the claim. ∎

We give

Theorem 18.6 *Let $f\in C_B\left(\mathbb{R}^N\right)$, $0<\beta<1$, $x\in\mathbb{R}^N$, $N,n\in\mathbb{N}$ with $n^{1-\beta}\geq 3$. Then*

(1)

$$|C_n(f,x)-f(x)|\leq\omega_1\left(f,\frac{1}{n}+\frac{1}{n^\beta}\right)+\frac{\|f\|_\infty}{\sqrt{\pi}\left(n^{1-\beta}-2\right)e^{\left(n^{1-\beta}-2\right)^2}}=:\lambda_3, \tag{18.45}$$

(2)

$$\|C_n(f) - f\|_\infty \leq \lambda_3. \tag{18.46}$$

Given that $f \in \left(C_U\left(\mathbb{R}^N\right) \cap C_B\left(\mathbb{R}^N\right)\right)$, *we obtain* $\lim_{n\to\infty} C_n(f) = f$, *uniformly.*

Proof We notice that

$$\int_{\frac{k}{n}}^{\frac{k+1}{n}} f(t)\,dt = \int_{\frac{k_1}{n}}^{\frac{k_1+1}{n}} \int_{\frac{k_2}{n}}^{\frac{k_2+1}{n}} \cdots \int_{\frac{k_N}{n}}^{\frac{k_N+1}{n}} f(t_1, t_2, \ldots, t_N)\,dt_1 dt_2 \ldots dt_N =$$

$$\int_0^{\frac{1}{n}} \int_0^{\frac{1}{n}} \cdots \int_0^{\frac{1}{n}} f\left(t_1 + \frac{k_1}{n}, t_2 + \frac{k_2}{n}, \ldots, t_N + \frac{k_N}{n}\right) dt_1 \ldots dt_N = \int_0^{\frac{1}{n}} f\left(t + \frac{k}{n}\right) dt. \tag{18.47}$$

Thus it holds

$$C_n(f, x) = \sum_{k=-\infty}^{\infty} \left(n^N \int_0^{\frac{1}{n}} f\left(t + \frac{k}{n}\right) dt\right) Z(nx - k). \tag{18.48}$$

We observe that

$$|C_n(f, x) - f(x)| =$$

$$\left|\sum_{k=-\infty}^{\infty} \left(n^N \int_0^{\frac{1}{n}} f\left(t + \frac{k}{n}\right) dt\right) Z(nx - k) - \sum_{k=-\infty}^{\infty} f(x)\, Z(nx - k)\right| =$$

$$\left|\sum_{k=-\infty}^{\infty} \left(\left(n^N \int_0^{\frac{1}{n}} f\left(t + \frac{k}{n}\right) dt\right) - f(x)\right) Z(nx - k)\right| =$$

$$\left|\sum_{k=-\infty}^{\infty} \left(n^N \int_0^{\frac{1}{n}} \left(f\left(t + \frac{k}{n}\right) - f(x)\right) dt\right) Z(nx - k)\right| \leq \tag{18.49}$$

$$\sum_{k=-\infty}^{\infty} \left(n^N \int_0^{\frac{1}{n}} \left|f\left(t + \frac{k}{n}\right) - f(x)\right| dt\right) Z(nx - k) =$$

$$\sum_{\substack{k = -\infty \\ \left\|\frac{k}{n} - x\right\|_\infty \leq \frac{1}{n^\beta}}}^{\infty} \left(n^N \int_0^{\frac{1}{n}} \left|f\left(t + \frac{k}{n}\right) - f(x)\right| dt\right) Z(nx - k) +$$

$$\sum_{\begin{cases} k=-\infty \\ \left\|\frac{k}{n}-x\right\|_{\infty} > \frac{1}{n^{\beta}} \end{cases}}^{\infty} \left(n^N \int_0^{\frac{1}{n}} \left| f\left(t+\frac{k}{n}\right) - f(x) \right| dt \right) Z(nx-k) \leq$$

$$\sum_{\begin{cases} k=-\infty \\ \left\|\frac{k}{n}-x\right\|_{\infty} \leq \frac{1}{n^{\beta}} \end{cases}}^{\infty} \left(n^N \int_0^{\frac{1}{n}} \omega_1 \left(f, \|t\|_{\infty} + \left\|\frac{k}{n}-x\right\|_{\infty} \right) dt \right) Z(nx-k) +$$

$$2\|f\|_{\infty} \left(\sum_{\begin{cases} k=-\infty \\ \left\|\frac{k}{n}-x\right\|_{\infty} > \frac{1}{n^{\beta}} \end{cases}}^{\infty} Z(|nx-k|) \right) \leq$$

$$\omega_1\left(f, \frac{1}{n} + \frac{1}{n^{\beta}}\right) + \frac{\|f\|_{\infty}}{\sqrt{\pi}\left(n^{1-\beta}-2\right) e^{\left(n^{1-\beta}-2\right)^2}}, \tag{18.50}$$

proving the claim. ∎

We also present

Theorem 18.7 *Let $f \in C_B\left(\mathbb{R}^N\right)$, $0 < \beta < 1$, $x \in \mathbb{R}^N$, $N, n \in \mathbb{N}$ with $n^{1-\beta} \geq 3$. Then*

(1)

$$|D_n(f,x) - f(x)| \leq \omega_1\left(f, \frac{1}{n} + \frac{1}{n^{\beta}}\right) + \frac{\|f\|_{\infty}}{\sqrt{\pi}\left(n^{1-\beta}-2\right) e^{\left(n^{1-\beta}-2\right)^2}} = \lambda_3, \tag{18.51}$$

(2)

$$\|D_n(f) - f\|_{\infty} \leq \lambda_3. \tag{18.52}$$

Given that $f \in \left(C_U\left(\mathbb{R}^N\right) \cap C_B\left(\mathbb{R}^N\right)\right)$, we obtain $\lim_{n\to\infty} D_n(f) = f$, uniformly.

Proof We have that

$$|D_n(f,x) - f(x)| = \left| \sum_{k=-\infty}^{\infty} \delta_{nk}(f) Z(nx-k) - \sum_{k=-\infty}^{\infty} f(x) Z(nx-k) \right| = \tag{18.53}$$

$$\left| \sum_{k=-\infty}^{\infty} \left(\delta_{nk}(f) - f(x)\right) Z(nx-k) \right| =$$

$$\left| \sum_{k=-\infty}^{\infty} \left(\sum_{r=0}^{\theta} w_r \left(f \left(\frac{k}{n} + \frac{r}{n\theta} \right) - f(x) \right) \right) Z(nx-k) \right| \leq$$

$$\sum_{k=-\infty}^{\infty} \left(\sum_{r=0}^{\theta} w_r \left| f \left(\frac{k}{n} + \frac{r}{n\theta} \right) - f(x) \right| \right) Z(nx-k) =$$

$$\sum_{\substack{k=-\infty \\ \left\| \frac{k}{n} - x \right\|_\infty \leq \frac{1}{n^\beta}}}^{\infty} \left(\sum_{r=0}^{\theta} w_r \left| f \left(\frac{k}{n} + \frac{r}{n\theta} \right) - f(x) \right| \right) Z(nx-k) +$$

$$\sum_{\substack{k=-\infty \\ \left\| \frac{k}{n} - x \right\|_\infty > \frac{1}{n^\beta}}}^{\infty} \left(\sum_{r=0}^{\theta} w_r \left| f \left(\frac{k}{n} + \frac{r}{n\theta} \right) - f(x) \right| \right) Z(nx-k) \leq$$

$$\sum_{\substack{k=-\infty \\ \left\| \frac{k}{n} - x \right\|_\infty \leq \frac{1}{n^\beta}}}^{\infty} \left(\sum_{r=0}^{\theta} w_r \omega_1 \left(f, \left\| \frac{k}{n} - x \right\|_\infty + \left\| \frac{r}{n\theta} \right\|_\infty \right) \right) Z(nx-k) +$$

$$2 \|f\|_\infty \left(\sum_{\substack{k=-\infty \\ \left\| \frac{k}{n} - x \right\|_\infty > \frac{1}{n^\beta}}}^{\infty} Z(nx-k) \right) \leq$$

$$\omega_1 \left(f, \frac{1}{n} + \frac{1}{n^\beta} \right) + \frac{\|f\|_\infty}{\sqrt{\pi} \left(n^{1-\beta} - 2 \right) e^{\left(n^{1-\beta} - 2 \right)^2}}, \tag{18.54}$$

proving the claim. ■

In the next we discuss high order of approximation by using the smoothness of f.

We give

Theorem 18.8 *Let* $f \in C^m \left(\prod_{i=1}^N [a_i, b_i] \right)$, $0 < \beta < 1$, $n, m, N \in \mathbb{N}$, $n^{1-\beta} \geq 3$, $x \in \left(\prod_{i=1}^N [a_i, b_i] \right)$. *Then*

(i)

$$\left| A_n(f,x) - f(x) - \sum_{j=1}^{m} \left(\sum_{|\alpha|=j} \left(\frac{f_\alpha(x)}{\prod_{i=1}^{N} \alpha_i!} \right) A_n \left(\prod_{i=1}^{N} (\cdot - x_i)^{\alpha_i}, x \right) \right) \right| \leq \tag{18.55}$$

$$(4.019)^N \cdot \left\{ \frac{N^m}{m! n^{m\beta}} \omega_{1,m}^{\max} \left(f_\alpha, \frac{1}{n^\beta} \right) + \right.$$

$$\left. \left(\frac{\|b-a\|_\infty^m \|f_\alpha\|_{\infty,m}^{\max} N^m}{m!} \right) \frac{1}{\sqrt{\pi}\left(n^{1-\beta}-2\right) e^{\left(n^{1-\beta}-2\right)^2}} \right\},$$

(ii)

$$|A_n(f,x) - f(x)| \leq (4.019)^N \cdot \tag{18.56}$$

$$\left\{ \sum_{j=1}^{m} \left(\sum_{|\alpha|=j} \left(\frac{|f_\alpha(x)|}{\prod_{i=1}^{N} \alpha_i!} \right) \left[\frac{1}{n^{\beta j}} + \left(\prod_{i=1}^{N} (b_i - a_i)^{\alpha_i} \right) \cdot \right. \right. \right.$$

$$\left. \left. \frac{1}{2\sqrt{\pi}\left(n^{1-\beta}-2\right) e^{\left(n^{1-\beta}-2\right)^2}} \right] \right) + \frac{N^m}{m! n^{m\beta}} \omega_{1,m}^{\max} \left(f_\alpha, \frac{1}{n^\beta} \right)$$

$$\left. + \left(\frac{\|b-a\|_\infty^m \|f_\alpha\|_{\infty,m}^{\max} N^m}{m!} \right) \frac{1}{\sqrt{\pi}\left(n^{1-\beta}-2\right) e^{\left(n^{1-\beta}-2\right)^2}} \right\},$$

(iii)

$$\|A_n(f) - f\|_\infty \leq (4.019)^N \cdot \tag{18.57}$$

$$\left\{ \sum_{j=1}^{m} \left(\sum_{|\alpha|=j} \left(\frac{\|f_\alpha\|_\infty}{\prod_{i=1}^{N} \alpha_i!} \right) \left[\frac{1}{n^{\beta j}} + \left(\prod_{i=1}^{N} (b_i - a_i)^{\alpha_i} \right) \cdot \right. \right. \right.$$

$$\left. \left. \frac{1}{2\sqrt{\pi}\left(n^{1-\beta}-2\right) e^{\left(n^{1-\beta}-2\right)^2}} \right] \right) + \frac{N^m}{m! n^{m\beta}} \omega_{1,m}^{\max} \left(f_\alpha, \frac{1}{n^\beta} \right)$$

$$\left. + \left(\frac{\|b-a\|_\infty^m \|f_\alpha\|_{\infty,m}^{\max} N^m}{m!} \right) \frac{1}{\sqrt{\pi}\left(n^{1-\beta}-2\right) e^{\left(n^{1-\beta}-2\right)^2}} \right\} =: K_n,$$

(iv) Assume $f_{\alpha}(x_0) = 0$, *for all* $\alpha : |\alpha| = 1, \ldots, m$; $x_0 \in \left(\prod_{i=1}^{N}[a_i, b_i]\right)$. *Then*

$$|A_n(f, x_0) - f(x_0)| \leq (4.019)^N \left\{ \frac{N^m}{m! n^{m\beta}} \omega_1^{\max}\left(f_{\alpha}, \frac{1}{n^{\beta}}\right) + \right. \tag{18.58}$$

$$\left. \left(\frac{\|b-a\|_{\infty}^m \|f_{\alpha}\|_{\infty,m}^{\max} N^m}{m!} \right) \frac{1}{\sqrt{\pi}\left(n^{1-\beta}-2\right) e^{\left(n^{1-\beta}-2\right)^2}} \right\},$$

notice in the last the extremely high rate of convergence at $n^{-\beta(m+1)}$.

Proof Consider $g_z(t) := f(x_0 + t(z - x_0))$, $t \geq 0$; $x_0, z \in \prod_{i=1}^{N}[a_i, b_i]$.
Then

$$g_z^{(j)}(t) = \left[\left(\sum_{i=1}^{N}(z_i - x_{0i})\frac{\partial}{\partial x_i}\right)^j f\right](x_{01} + t(z_1 - x_{01}), \ldots, x_{0N} + t(z_N - x_{0N})), \tag{18.59}$$

for all $j = 0, 1, \ldots, m$.

We have the multivariate Taylor's formula

$$f(z_1, \ldots, z_N) = g_z(1) =$$

$$\sum_{j=0}^{m} \frac{g_z^{(j)}(0)}{j!} + \frac{1}{(m-1)!}\int_0^1 (1-\theta)^{m-1}\left(g_z^{(m)}(\theta) - g_z^{(m)}(0)\right) d\theta. \tag{18.60}$$

Notice $g_z(0) = f(x_0)$. Also for $j = 0, 1, \ldots, m$, we have

$$g_z^{(j)}(0) = \sum_{\substack{\alpha:=(\alpha_1,\ldots,\alpha_N),\alpha_i\in\mathbb{Z}^+,\\ i=1,\ldots,N,|\alpha|:=\sum_{i=1}^{N}\alpha_i=j}} \left(\frac{j!}{\prod_{i=1}^{N}\alpha_i!}\right)\left(\prod_{i=1}^{N}(z_i - x_{0i})^{\alpha_i}\right) f_{\alpha}(x_0). \tag{18.61}$$

Furthermore

$$g_z^{(m)}(\theta) =$$

$$\sum_{\substack{\alpha:=(\alpha_1,\ldots,\alpha_N),\alpha_i\in\mathbb{Z}^+,\\ i=1,\ldots,N,|\alpha|:=\sum_{i=1}^{N}\alpha_i=m}} \left(\frac{m!}{\prod_{i=1}^{N}\alpha_i!}\right)\left(\prod_{i=1}^{N}(z_i - x_{0i})^{\alpha_i}\right) f_{\alpha}(x_0 + \theta(z - x_0)), \tag{18.62}$$

$0 \leq \theta \leq 1$.

So we treat $f \in C^m\left(\prod_{i=1}^{N}[a_i, b_i]\right)$.

Thus, we have for $\frac{k}{n}, x \in \left(\prod_{i=1}^{N} [a_i, b_i]\right)$ that

$$f\left(\frac{k_1}{n}, \dots, \frac{k_N}{n}\right) - f(x) =$$

$$\sum_{j=1}^{m} \sum_{\substack{\alpha:=(\alpha_1,\dots,\alpha_N),\alpha_i\in\mathbb{Z}^+,\\ i=1,\dots,N,|\alpha|:=\sum_{i=1}^{N}\alpha_i=j}} \left(\frac{1}{\prod_{i=1}^{N}\alpha_i!}\right)\left(\prod_{i=1}^{N}\left(\frac{k_i}{n}-x_i\right)^{\alpha_i}\right) f_\alpha(x) + R, \quad (18.63)$$

where

$$R := m\int_0^1 (1-\theta)^{m-1} \sum_{\substack{\alpha:=(\alpha_1,\dots,\alpha_N),\alpha_i\in\mathbb{Z}^+,\\ i=1,\dots,N,|\alpha|:=\sum_{i=1}^{N}\alpha_i=m}} \left(\frac{1}{\prod_{i=1}^{N}\alpha_i!}\right)\left(\prod_{i=1}^{N}\left(\frac{k_i}{n}-x_i\right)^{\alpha_i}\right)$$

$$\cdot\left[f_\alpha\left(x+\theta\left(\frac{k}{n}-x\right)\right) - f_\alpha(x)\right]d\theta. \quad (18.64)$$

We see that

$$|R| \le m\int_0^1 (1-\theta)^{m-1} \sum_{|\alpha|=m}\left(\frac{1}{\prod_{i=1}^{N}\alpha_i!}\right)\left(\prod_{i=1}^{N}\left|\frac{k_i}{n}-x_i\right|^{\alpha_i}\right)\cdot$$

$$\left|f_\alpha\left(x+\theta\left(\frac{k}{n}-x\right)\right) - f_\alpha(x)\right| d\theta \le m\int_0^1 (1-\theta)^{m-1}\cdot \quad (18.65)$$

$$\left(\sum_{|\alpha|=m}\left(\frac{1}{\prod_{i=1}^{N}\alpha_i!}\right)\left(\prod_{i=1}^{N}\left|\frac{k_i}{n}-x_i\right|^{\alpha_i}\right)\omega_1\left(f_\alpha, \theta\left\|\frac{k}{n}-x\right\|_\infty\right)\right)d\theta \le (*).$$

Notice here that

$$\left\|\frac{k}{n}-x\right\|_\infty \le \frac{1}{n^\beta} \Leftrightarrow \left|\frac{k_i}{n}-x_i\right| \le \frac{1}{n^\beta}, i = 1, \dots, N. \quad (18.66)$$

We further see that

$$(*) \le m\cdot\omega_{1,m}^{\max}\left(f_\alpha, \frac{1}{n^\beta}\right)\int_0^1 (1-\theta)^{m-1}\left(\sum_{|\alpha|=m}\left(\frac{1}{\prod_{i=1}^{N}\alpha_i!}\right)\left(\prod_{i=1}^{N}\left(\frac{1}{n^\beta}\right)^{\alpha_i}\right)\right)d\theta =$$

$$\left(\frac{\omega_{1,m}^{\max}\left(f_\alpha, \frac{1}{n^\beta}\right)}{(m!)\,n^{m\beta}}\right)\left(\sum_{|\alpha|=m}\frac{m!}{\prod_{i=1}^{N}\alpha_i!}\right) = \left(\frac{\omega_{1,m}^{\max}\left(f_\alpha, \frac{1}{n^\beta}\right)}{(m!)\,n^{m\beta}}\right)N^m. \quad (18.67)$$

Conclusion: When $\left\| \frac{k}{n} - x \right\|_\infty \leq \frac{1}{n^\beta}$, we proved that

$$|R| \leq \left(\frac{N^m}{m! n^{m\beta}} \right) \omega_{1,m}^{\max} \left(f_\alpha, \frac{1}{n^\beta} \right). \tag{18.68}$$

In general we notice that

$$|R| \leq m \int_0^1 (1-\theta)^{m-1} \left(\sum_{|\alpha|=m} \left(\frac{1}{\prod_{i=1}^N \alpha_i!} \right) \left(\prod_{i=1}^N (b_i - a_i)^{\alpha_i} \right) 2 \| f_\alpha \|_\infty \right) d\theta =$$

$$2 \sum_{|\alpha|=m} \frac{1}{\prod_{i=1}^N \alpha_i!} \left(\prod_{i=1}^N (b_i - a_i)^{\alpha_i} \right) \| f_\alpha \|_\infty \leq$$

$$\left(\frac{2 \|b-a\|_\infty^m \|f_\alpha\|_{\infty,m}^{\max}}{m!} \right) \left(\sum_{|\alpha|=m} \frac{m!}{\prod_{i=1}^N \alpha_i!} \right) = \frac{2 \|b-a\|_\infty^m \|f_\alpha\|_{\infty,m}^{\max} N^m}{m!}. \tag{18.69}$$

We proved in general that

$$|R| \leq \frac{2 \|b-a\|_\infty^m \|f_\alpha\|_{\infty,m}^{\max} N^m}{m!} := \rho. \tag{18.70}$$

Next we see that

$$U_n := \sum_{k=\lceil na \rceil}^{\lfloor nb \rfloor} Z(nx-k) R =$$

$$\sum_{\substack{k=\lceil na \rceil \\ : \left\| \frac{k}{n} - x \right\|_\infty \leq \frac{1}{n^\beta}}}^{\lfloor nb \rfloor} Z(nx-k) R + \sum_{\substack{k=\lceil na \rceil \\ : \left\| \frac{k}{n} - x \right\|_\infty > \frac{1}{n^\beta}}}^{\lfloor nb \rfloor} Z(nx-k) R.$$

Consequently

$$|U_n| \leq \left(\sum_{\substack{k=\lceil na \rceil \\ : \left\| \frac{k}{n} - x \right\|_\infty \leq \frac{1}{n^\beta}}}^{\lfloor nb \rfloor} Z(nx-k) \right) \frac{N^m}{m! n^{m\beta}} \omega_{1,m}^{\max} \left(f_\alpha, \frac{1}{n^\beta} \right)$$

$$+\rho \frac{1}{2\sqrt{\pi}\left(n^{1-\beta}-2\right)e^{\left(n^{1-\beta}-2\right)^2}}$$

$$\leq \frac{N^m}{m!n^{m\beta}}\omega_{1,m}^{\max}\left(f_\alpha, \frac{1}{n^\beta}\right)+\rho \frac{1}{2\sqrt{\pi}\left(n^{1-\beta}-2\right)e^{\left(n^{1-\beta}-2\right)^2}}. \tag{18.71}$$

We have established that

$$|U_n| \leq \frac{N^m}{m!n^{m\beta}}\omega_{1,m}^{\max}\left(f_\alpha, \frac{1}{n^\beta}\right)+$$

$$\left(\frac{\|b-a\|_\infty^m \|f_\alpha\|_{\infty,m}^{\max} N^m}{m!}\right)\frac{1}{\sqrt{\pi}\left(n^{1-\beta}-2\right)e^{\left(n^{1-\beta}-2\right)^2}}. \tag{18.72}$$

We observe that

$$\sum_{k=\lceil na\rceil}^{\lfloor nb\rfloor} f\left(\frac{k}{n}\right)Z(nx-k) - f(x)\sum_{k=\lceil na\rceil}^{\lfloor nb\rfloor} Z(nx-k) =$$

$$\sum_{j=1}^{m}\left(\sum_{|\alpha|=j}\left(\frac{f_\alpha(x)}{\prod_{i=1}^{N}\alpha_i!}\right)\left(\sum_{k=\lceil na\rceil}^{\lfloor nb\rfloor} Z(nx-k)\left(\prod_{i=1}^{N}\left(\frac{k_i}{n}-x_i\right)^{\alpha_i}\right)\right)\right)$$

$$+\sum_{k=\lceil na\rceil}^{\lfloor nb\rfloor} Z(nx-k)R. \tag{18.73}$$

The last says

$$A_n^*(f,x) - f(x)\left(\sum_{k=\lceil na\rceil}^{\lfloor nb\rfloor} Z(nx-k)\right) -$$

$$\sum_{j=1}^{m}\left(\sum_{|\alpha|=j}\left(\frac{f_\alpha(x)}{\prod_{i=1}^{N}\alpha_i!}\right)A_n^*\left(\prod_{i=1}^{N}(\cdot - x_i)^{\alpha_i}, x\right)\right) = U_n. \tag{18.74}$$

Clearly A_n^* is a positive linear operator.

Thus (here $\alpha_i \in \mathbb{Z}^+ : |\alpha| = \sum_{i=1}^{N}\alpha_i = j$)

$$\left|A_n^*\left(\prod_{i=1}^N(\cdot - x_i)^{\alpha_i}, x\right)\right| \leq A_n^*\left(\prod_{i=1}^N|\cdot - x_i|^{\alpha_i}, x\right) =$$

$$\sum_{k=\lceil na\rceil}^{\lfloor nb\rfloor}\left(\prod_{i=1}^N\left|\frac{k_i}{n} - x_i\right|^{\alpha_i}\right)Z(nx - k) =$$

$$\sum_{\substack{k=\lceil na\rceil \\ :\left\|\frac{k}{n}-x\right\|_\infty \leq \frac{1}{n^\beta}}}^{\lfloor nb\rfloor}\left(\prod_{i=1}^N\left|\frac{k_i}{n} - x_i\right|^{\alpha_i}\right)Z(nx - k) +$$

$$\sum_{\substack{k=\lceil na\rceil \\ :\left\|\frac{k}{n}-x\right\|_\infty > \frac{1}{n^\beta}}}^{\lfloor nb\rfloor}\left(\prod_{i=1}^N\left|\frac{k_i}{n} - x_i\right|^{\alpha_i}\right)Z(nx - k) \leq$$

$$\frac{1}{n^{\beta j}} + \prod_{i=1}^N(b_i - a_i)^{\alpha_i}\left(\sum_{\substack{k=\lceil na\rceil \\ :\left\|\frac{k}{n}-x\right\|_\infty > \frac{1}{n^\beta}}}^{\lfloor nb\rfloor} Z(nx - k)\right) \leq$$

$$\frac{1}{n^{\beta j}} + \left(\prod_{i=1}^N(b_i - a_i)^{\alpha_i}\right)\frac{1}{2\sqrt{\pi}\left(n^{1-\beta} - 2\right)e^{\left(n^{1-\beta}-2\right)^2}}. \tag{18.75}$$

So we have proved that

$$\left|A_n^*\left(\prod_{i=1}^N(\cdot - x_i)^{\alpha_i}, x\right)\right| \leq \frac{1}{n^{\beta j}} + \left(\prod_{i=1}^N(b_i - a_i)^{\alpha_i}\right)\frac{1}{2\sqrt{\pi}\left(n^{1-\beta} - 2\right)e^{\left(n^{1-\beta}-2\right)^2}}, \tag{18.76}$$

for all $j = 1, \ldots, m$.

At last we observe

$$\left|A_n(f, x) - f(x) - \sum_{j=1}^m\left(\sum_{|\alpha|=j}\left(\frac{f_\alpha(x)}{\prod_{i=1}^N \alpha_i!}\right)A_n\left(\prod_{i=1}^N(\cdot - x_i)^{\alpha_i}, x\right)\right)\right| \leq$$

$$(4.019)^N \cdot \left|A_n^*(f, x) - f(x)\sum_{k=\lceil na\rceil}^{\lfloor nb\rfloor} Z(nx - k) -\right.$$

$$\sum_{j=1}^{m}\left(\sum_{|\alpha|=j}\left(\frac{f_{\alpha}(x)}{\prod_{i=1}^{N}\alpha_i!}\right)A_n^*\left(\prod_{i=1}^{N}(\cdot-x_i)^{\alpha_i},x\right)\right)\Bigg|. \tag{18.77}$$

Putting all of the above together we prove theorem. ■

We make

Definition 18.9 Let $f\in C_B\left(\mathbb{R}^N\right)$, $N\in\mathbb{N}$. We define the general neural network operator

$$F_n(f,x):=\sum_{k=-\infty}^{\infty}l_{nk}(f)\,Z(nx-k)=$$

$$\begin{cases}B_n(f,x), \text{ if } l_{nk}(f)=f\left(\frac{k}{n}\right),\\ C_n(f,x), \text{ if } l_{nk}(f)=n^N\int_{\frac{k}{n}}^{\frac{k+1}{n}}f(t)\,dt,\\ D_n(f,x), \text{ if } l_{nk}(f)=\delta_{nk}(f).\end{cases} \tag{18.78}$$

Clearly $l_{nk}(f)$ is a positive linear functional such that $|l_{nk}(f)|\leq\|f\|_{\infty}$.

Hence $F_n(f)$ is a positive linear operator with $\|F_n(f)\|_{\infty}\leq\|f\|_{\infty}$, a continuous bounded linear operator.

We need

Theorem 18.10 *Let $f\in C_B\left(\mathbb{R}^N\right)$, $N\geq 1$. Then $F_n(f)\in C_B\left(\mathbb{R}^N\right)$.*

Proof Clearly $F_n(f)$ is a bounded function.

Next we prove the continuity of $F_n(f)$. Notice for $N=1$, $Z=\chi$ by (18.10).

We will use the Weierstrass M test: If a sequence of positive constants M_1, M_2, $M_3,\ldots$, can be found such that in some interval

(a) $|u_n(x)|\leq M_n$, $n=1,2,3,\ldots$

(b) $\sum M_n$ converges,

then $\sum u_n(x)$ is uniformly and absolutely convergent in the interval.

Also we will use:

If $\{u_n(x)\}$, $n=1,2,3,\ldots$ are continuous in $[a,b]$ and if $\sum u_n(x)$ converges uniformly to the sum $S(x)$ in $[a,b]$, then $S(x)$ is continuous in $[a,b]$. I.e. a uniformly convergent series of continuous functions is a continuous function. First we prove claim for $N=1$.

We will prove that $\sum_{k=-\infty}^{\infty}l_{nk}(f)\,\chi(nx-k)$ is continuous in $x\in\mathbb{R}$.

There always exists $\lambda\in\mathbb{N}$ such that $nx\in[-\lambda,\lambda]$.

Since $nx\leq\lambda$, then $-nx\geq-\lambda$ and $k-nx\geq k-\lambda\geq 0$, when $k\geq\lambda$. Therefore

$$\sum_{k=\lambda}^{\infty}\chi(nx-k)=\sum_{k=\lambda}^{\infty}\chi(k-nx)\leq\sum_{k=\lambda}^{\infty}\chi(k-\lambda)=\sum_{k'=0}^{\infty}\chi\left(k'\right)\leq 1. \tag{18.79}$$

So for $k \geq \lambda$ we get

$$|l_{nk}(f)| \chi(nx-k) \leq \|f\|_\infty \chi(k-\lambda),$$

and

$$\|f\|_\infty \sum_{k=\lambda}^{\infty} \chi(k-\lambda) \leq \|f\|_\infty.$$

Hence by Weierstrass M test we obtain that $\sum_{k=\lambda}^{\infty} l_{nk}(f) \chi(nx-k)$ is uniformly and absolutely convergent on $\left[-\frac{\lambda}{n}, \frac{\lambda}{n}\right]$.

Since $l_{nk}(f) \chi(nx-k)$ is continuous in x, then $\sum_{k=\lambda}^{\infty} l_{nk}(f) \chi(nx-k)$ is continuous on $\left[-\frac{\lambda}{n}, \frac{\lambda}{n}\right]$.

Because $nx \geq -\lambda$, then $-nx \leq \lambda$, and $k - nx \leq k + \lambda \leq 0$, when $k \leq -\lambda$. Therefore

$$\sum_{k=-\infty}^{-\lambda} \chi(nx-k) = \sum_{k=-\infty}^{-\lambda} \chi(k-nx) \leq \sum_{k=-\infty}^{-\lambda} \chi(k+\lambda) = \sum_{k'=-\infty}^{0} \chi(k') \leq 1.$$

So for $k \leq -\lambda$ we get

$$|l_{nk}(f)| \chi(nx-k) \leq \|f\|_\infty \chi(k+\lambda), \tag{18.80}$$

and

$$\|f\|_\infty \sum_{k=-\infty}^{-\lambda} \chi(k+\lambda) \leq \|f\|_\infty.$$

Hence by Weierstrass M test we obtain that $\sum_{k=-\infty}^{-\lambda} l_{nk}(f) \chi(nx-k)$ is uniformly and absolutely convergent on $\left[-\frac{\lambda}{n}, \frac{\lambda}{n}\right]$.

Since $l_{nk}(f) \chi(nx-k)$ is continuous in x, then $\sum_{k=-\infty}^{-\lambda} l_{nk}(f) \chi(nx-k)$ is continuous on $\left[-\frac{\lambda}{n}, \frac{\lambda}{n}\right]$.

So we proved that $\sum_{k=\lambda}^{\infty} l_{nk}(f) \chi(nx-k)$ and $\sum_{k=-\infty}^{-\lambda} l_{nk}(f) \chi(nx-k)$ are continuous on $\mathbb{R}$. Since $\sum_{k=-\lambda+1}^{\lambda-1} l_{nk}(f) \chi(nx-k)$ is a finite sum of continuous functions on $\mathbb{R}$, it is also a continuous function on $\mathbb{R}$.

Writing

$$\sum_{k=-\infty}^{\infty} l_{nk}(f) \chi(nx-k) = \sum_{k=-\infty}^{-\lambda} l_{nk}(f) \chi(nx-k) +$$

$$\sum_{k=-\lambda+1}^{\lambda-1} l_{nk}(f) \chi(nx-k) + \sum_{k=\lambda}^{\infty} l_{nk}(f) \chi(nx-k) \tag{18.81}$$

we have it as a continuous function on $\mathbb{R}$. Therefore $F_n(f)$, when $N = 1$, is a continuous function on $\mathbb{R}$.

When $N = 2$ we have

$$F_n(f, x_1, x_2) = \sum_{k_1=-\infty}^{\infty} \sum_{k_2=-\infty}^{\infty} l_{nk}(f) \chi(nx_1 - k_1) \chi(nx_2 - k_2) =$$

$$\sum_{k_1=-\infty}^{\infty} \chi(nx_1 - k_1) \left(\sum_{k_2=-\infty}^{\infty} l_{nk}(f) \chi(nx_2 - k_2) \right)$$

(there always exist $\lambda_1, \lambda_2 \in \mathbb{N}$ such that $nx_1 \in [-\lambda_1, \lambda_1]$ and $nx_2 \in [-\lambda_2, \lambda_2]$)

$$= \sum_{k_1=-\infty}^{\infty} \chi(nx_1 - k_1) \left[\sum_{k_2=-\infty}^{-\lambda_2} l_{nk}(f) \chi(nx_2 - k_2) + \right.$$

$$\left. \sum_{k_2=-\lambda_2+1}^{\lambda_2-1} l_{nk}(f) \chi(nx_2 - k_2) + \sum_{k_2=\lambda_2}^{\infty} l_{nk}(f) \chi(nx_2 - k_2) \right] =$$

$$= \sum_{k_1=-\infty}^{\infty} \sum_{k_2=-\infty}^{-\lambda_2} l_{nk}(f) \chi(nx_1 - k_1) \chi(nx_2 - k_2) +$$

$$\sum_{k_1=-\infty}^{\infty} \sum_{k_2=-\lambda_2+1}^{\lambda_2-1} l_{nk}(f) \chi(nx_1 - k_1) \chi(nx_2 - k_2) +$$

$$\sum_{k_1=-\infty}^{\infty} \sum_{k_2=\lambda_2}^{\infty} l_{nk}(f) \chi(nx_1 - k_1) \chi(nx_2 - k_2) =: (*) .$$

(For convenience call

$$F(k_1, k_2, x_1, x_2) := l_{nk}(f) \chi(nx_1 - k_1) \chi(nx_2 - k_2) .)$$

Thus

$$(*) = \sum_{k_1=-\infty}^{-\lambda_1} \sum_{k_2=-\infty}^{-\lambda_2} F(k_1, k_2, x_1, x_2) + \sum_{k_1=-\lambda_1+1}^{\lambda_1-1} \sum_{k_2=-\infty}^{-\lambda_2} F(k_1, k_2, x_1, x_2) +$$

$$\sum_{k_1=\lambda_1}^{\infty} \sum_{k_2=-\infty}^{-\lambda_2} F(k_1, k_2, x_1, x_2) + \sum_{k_1=-\infty}^{-\lambda_1} \sum_{k_2=-\lambda_2+1}^{\lambda_2-1} F(k_1, k_2, x_1, x_2) +$$

$$\sum_{k_1=-\lambda_1+1}^{\lambda_1-1}\sum_{k_2=-\lambda_2+1}^{\lambda_2-1} F(k_1,k_2,x_1,x_2)+\sum_{k_1=\lambda_1}^{\infty}\sum_{k_2=-\lambda_2+1}^{\lambda_2-1} F(k_1,k_2,x_1,x_2)+$$

$$\sum_{k_1=-\infty}^{-\lambda_1}\sum_{k_2=\lambda_2}^{\infty} F(k_1,k_2,x_1,x_2)+\sum_{k_1=-\lambda_1+1}^{\lambda_1-1}\sum_{k_2=\lambda_2}^{\infty} F(k_1,k_2,x_1,x_2)+ \tag{18.82}$$

$$\sum_{k_1=\lambda_1}^{\infty}\sum_{k_2=\lambda_2}^{\infty} F(k_1,k_2,x_1,x_2).$$

Notice that the finite sum of continuous functions $F(k_1,k_2,x_1,x_2)$, $\sum_{k_1=-\lambda_1+1}^{\lambda_1-1}\sum_{k_2=-\lambda_2+1}^{\lambda_2-1} F(k_1,k_2,x_1,x_2)$ is a continuous function.

The rest of the summands of $F_n(f,x_1,x_2)$ are treated all the same way and similarly to the case of $N=1$. The method is demonstrated as follows.

We will prove that $\sum_{k_1=\lambda_1}^{\infty}\sum_{k_2=-\infty}^{-\lambda_2} l_{nk}(f)\,\chi(nx_1-k_1)\,\chi(nx_2-k_2)$ is continuous in $(x_1,x_2)\in\mathbb{R}^2$.

The continuous function

$$|l_{nk}(f)|\,\chi(nx_1-k_1)\,\chi(nx_2-k_2)\le\|f\|_\infty\,\chi(k_1-\lambda_1)\,\chi(k_2+\lambda_2),$$

and

$$\|f\|_\infty\sum_{k_1=\lambda_1}^{\infty}\sum_{k_2=-\infty}^{-\lambda_2}\chi(k_1-\lambda_1)\,\chi(k_2+\lambda_2)=$$

$$\|f\|_\infty\left(\sum_{k_1=\lambda_1}^{\infty}\chi(k_1-\lambda_1)\right)\left(\sum_{k_2=-\infty}^{-\lambda_2}\chi(k_2+\lambda_2)\right)\le$$

$$\|f\|_\infty\left(\sum_{k_1'=0}^{\infty}\chi(k_1')\right)\left(\sum_{k_2'=-\infty}^{0}\chi(k_2')\right)\le\|f\|_\infty.$$

So by the Weierstrass M test we get that $\sum_{k_1=\lambda_1}^{\infty}\sum_{k_2=-\infty}^{-\lambda_2} l_{nk}(f)\,\chi(nx_1-k_1)\,\chi(nx_2-k_2)$ is uniformly and absolutely convergent. Therefore it is continuous on $\mathbb{R}^2$.

Next we prove continuity on $\mathbb{R}^2$ of $\sum_{k_1=-\lambda_1+1}^{\lambda_1-1}\sum_{k_2=-\infty}^{-\lambda_2} l_{nk}(f)\,\chi(nx_1-k_1)\,\chi(nx_2-k_2)$.

Notice here that

$$|l_{nk}(f)|\,\chi(nx_1-k_1)\,\chi(nx_2-k_2)\le\|f\|_\infty\,\chi(nx_1-k_1)\,\chi(k_2+\lambda_2)$$

$$\le\|f\|_\infty\,\chi(0)\,\chi(k_2+\lambda_2)=0.4215\cdot\|f\|_\infty\,\chi(k_2+\lambda_2),$$

and

$$0.4215 \cdot \|f\|_{\infty} \left(\sum_{k_1=-\lambda_1+1}^{\lambda_1-1} 1 \right) \left(\sum_{k_2=-\infty}^{-\lambda_2} \chi(k_2+\lambda_2) \right) =$$

$$0.4215 \cdot \|f\|_{\infty} (2\lambda_1 - 1) \left(\sum_{k_2'=-\infty}^{0} \chi(k_2') \right) \leq 0.4215 \cdot (2\lambda_1 - 1) \|f\|_{\infty}. \quad (18.83)$$

So the double series under consideration is uniformly convergent and continuous. Clearly $F_n(f, x_1, x_2)$ is proved to be continuous on $\mathbb{R}^2$.

Similarly reasoning one can prove easily now, but with more tedious work, that $F_n(f, x_1, \ldots, x_N)$ is continuous on $\mathbb{R}^N$, for any $N \geq 1$. We choose to omit this similar extra work. ■

Remark 18.11 By (18.24) it is obvious that $\|A_n(f)\|_{\infty} \leq \|f\|_{\infty} < \infty$, and $A_n(f) \in C\left(\prod_{i=1}^{N} [a_i, b_i]\right)$, given that $f \in C\left(\prod_{i=1}^{N} [a_i, b_i]\right)$.

Call L_n any of the operators A_n, B_n, C_n, D_n.

Clearly then

$$\left\| L_n^2(f) \right\|_{\infty} = \|L_n(L_n(f))\|_{\infty} \leq \|L_n(f)\|_{\infty} \leq \|f\|_{\infty}, \quad (18.84)$$

etc.

Therefore we get

$$\left\| L_n^k(f) \right\|_{\infty} \leq \|f\|_{\infty}, \forall k \in \mathbb{N}, \quad (18.85)$$

the contraction property.

Also we see that

$$\left\| L_n^k(f) \right\|_{\infty} \leq \left\| L_n^{k-1}(f) \right\|_{\infty} \leq \cdots \leq \|L_n(f)\|_{\infty} \leq \|f\|_{\infty}. \quad (18.86)$$

Also $L_n(1) = 1$, $L_n^k(1) = 1, \forall\, k \in \mathbb{N}$.

Here L_n^k are positive linear operators.

Notation 18.12 *Here $N \in \mathbb{N}$, $0 < \beta < 1$. Denote by*

$$c_N := \begin{cases} (4.019)^N, \text{ if } L_n = A_n, \\ 1, \text{ if } L_n = B_n, C_n, D_n, \end{cases} \quad (18.87)$$

$$\varphi(n) := \begin{cases} \frac{1}{n^{\beta}}, \text{ if } L_n = A_n, B_n, \\ \frac{1}{n} + \frac{1}{n^{\beta}}, \text{ if } L_n = C_n, D_n, \end{cases} \quad (18.88)$$

$$\Omega := \begin{cases} C\left(\prod_{i=1}^{N} [a_i, b_i]\right), \text{ if } L_n = A_n, \\ C_B\left(\mathbb{R}^N\right), \text{ if } L_n = B_n, C_n, D_n, \end{cases} \tag{18.89}$$

and

$$Y := \begin{cases} \prod_{i=1}^{N} [a_i, b_i], \text{ if } L_n = A_n, \\ \mathbb{R}^N, \text{ if } L_n = B_n, C_n, D_n. \end{cases} \tag{18.90}$$

We give the condensed

Theorem 18.13 *Let $f \in \Omega$, $0 < \beta < 1$, $x \in Y$; $n, N \in \mathbb{N}$ with $n^{1-\beta} \geq 3$. Then*

(i)

$$|L_n(f, x) - f(x)| \leq c_N \left[\omega_1(f, \varphi(n)) + \frac{\|f\|_\infty}{\sqrt{\pi}\left(n^{1-\beta} - 2\right) e^{\left(n^{1-\beta}-2\right)^2}} \right] =: \tau, \tag{18.91}$$

(ii)

$$\|L_n(f) - f\|_\infty \leq \tau. \tag{18.92}$$

For f uniformly continuous and in Ω we obtain

$$\lim_{n \to \infty} L_n(f) = f,$$

pointwise and uniformly.

Proof By Theorems 18.4–18.7. ■

Next we do iterated neural network approximation (see also [9]).
We make

Remark 18.14 Let $r \in \mathbb{N}$ and L_n as above. We observe that

$$L_n^r f - f = \left(L_n^r f - L_n^{r-1} f\right) + \left(L_n^{r-1} f - L_n^{r-2} f\right) +$$

$$\left(L_n^{r-2} f - L_n^{r-3} f\right) + \cdots + \left(L_n^2 f - L_n f\right) + \left(L_n f - f\right).$$

Then

$$\|L_n^r f - f\|_\infty \leq \left\|L_n^r f - L_n^{r-1} f\right\|_\infty + \left\|L_n^{r-1} f - L_n^{r-2} f\right\|_\infty +$$

$$\left\|L_n^{r-2} f - L_n^{r-3} f\right\|_\infty + \cdots + \left\|L_n^2 f - L_n f\right\|_\infty + \|L_n f - f\|_\infty =$$

$$\left\|L_n^{r-1}(L_n f - f)\right\|_\infty + \left\|L_n^{r-2}(L_n f - f)\right\|_\infty + \left\|L_n^{r-3}(L_n f - f)\right\|_\infty$$

$$+ \cdots + \|L_n(L_n f - f)\|_\infty + \|L_n f - f\|_\infty \leq r\,\|L_n f - f\|_\infty. \quad (18.93)$$

That is

$$\left\|L_n^r f - f\right\|_\infty \leq r\,\|L_n f - f\|_\infty. \quad (18.94)$$

We give

Theorem 18.15 *All here as in Theorem 18.13 and $r \in \mathbb{N}$, τ as in (18.91). Then*

$$\left\|L_n^r f - f\right\|_\infty \leq r\tau. \quad (18.95)$$

So that the speed of convergence to the unit operator of L_n^r is not worse than of L_n.

Proof By (18.94) and (18.92). ■

We make

Remark 18.16 Let $m_1, \ldots, m_r \in \mathbb{N} : m_1 \leq m_2 \leq \cdots \leq m_r$, $0 < \beta < 1$, $f \in \Omega$. Then $\varphi(m_1) \geq \varphi(m_2) \geq \cdots \geq \varphi(m_r)$, φ as in (18.88).

Therefore

$$\omega_1(f, \varphi(m_1)) \geq \omega_1(f, \varphi(m_2)) \geq \cdots \geq \omega_1(f, \varphi(m_r)). \quad (18.96)$$

Assume further that $m_i^{1-\beta} \geq 3$, $i = 1, \ldots, r$. Then

$$\frac{1}{\left(m_1^{1-\beta} - 2\right) e^{\left(m_1^{1-\beta}-2\right)^2}} \geq \frac{1}{\left(m_2^{1-\beta} - 2\right) e^{\left(m_2^{1-\beta}-2\right)^2}}$$

$$\geq \cdots \geq \frac{1}{\left(m_r^{1-\beta} - 2\right) e^{\left(m_r^{1-\beta}-2\right)^2}}. \quad (18.97)$$

Let L_{m_i} as above, $i = 1, \ldots, r$, all of the same kind.

We write

$$L_{m_r}\left(L_{m_{r-1}}\left(\ldots L_{m_2}\left(L_{m_1} f\right)\right)\right) - f =$$

$$L_{m_r}\left(L_{m_{r-1}}\left(\ldots L_{m_2}\left(L_{m_1} f\right)\right)\right) - L_{m_r}\left(L_{m_{r-1}}\left(\ldots L_{m_2} f\right)\right) +$$

$$L_{m_r}\left(L_{m_{r-1}}\left(\ldots L_{m_2} f\right)\right) - L_{m_r}\left(L_{m_{r-1}}\left(\ldots L_{m_3} f\right)\right) +$$

$$L_{m_r}\left(L_{m_{r-1}}\left(\ldots L_{m_3} f\right)\right) - L_{m_r}\left(L_{m_{r-1}}\left(\ldots L_{m_4} f\right)\right) + \cdots + \quad (18.98)$$

$$L_{m_r}\left(L_{m_{r-1}}f\right) - L_{m_r}f + L_{m_r}f - f =$$

$$L_{m_r}\left(L_{m_{r-1}}\left(\ldots L_{m_2}\right)\right)\left(L_{m_1}f - f\right) + L_{m_r}\left(L_{m_{r-1}}\left(\ldots L_{m_3}\right)\right)\left(L_{m_2}f - f\right) +$$

$$L_{m_r}\left(L_{m_{r-1}}\left(\ldots L_{m_4}\right)\right)\left(L_{m_3}f - f\right) + \cdots + L_{m_r}\left(L_{m_{r-1}}f - f\right) + L_{m_r}f - f.$$

Hence by the triangle inequality property of $\|\cdot\|_\infty$ we get

$$\left\|L_{m_r}\left(L_{m_{r-1}}\left(\ldots L_{m_2}\left(L_{m_1}f\right)\right)\right) - f\right\|_\infty \leq$$

$$\left\|L_{m_r}\left(L_{m_{r-1}}\left(\ldots L_{m_2}\right)\right)\left(L_{m_1}f - f\right)\right\|_\infty + \left\|L_{m_r}\left(L_{m_{r-1}}\left(\ldots L_{m_3}\right)\right)\left(L_{m_2}f - f\right)\right\|_\infty +$$

$$\left\|L_{m_r}\left(L_{m_{r-1}}\left(\ldots L_{m_4}\right)\right)\left(L_{m_3}f - f\right)\right\|_\infty + \cdots +$$

$$\left\|L_{m_r}\left(L_{m_{r-1}}f - f\right)\right\|_\infty + \left\|L_{m_r}f - f\right\|_\infty$$

(repeatedly applying (18.84))

$$\leq \left\|L_{m_1}f - f\right\|_\infty + \left\|L_{m_2}f - f\right\|_\infty + \left\|L_{m_3}f - f\right\|_\infty + \cdots +$$

$$\left\|L_{m_{r-1}}f - f\right\|_\infty + \left\|L_{m_r}f - f\right\|_\infty = \sum_{i=1}^{r}\left\|L_{m_i}f - f\right\|_\infty. \tag{18.99}$$

That is, we proved

$$\left\|L_{m_r}\left(L_{m_{r-1}}\left(\ldots L_{m_2}\left(L_{m_1}f\right)\right)\right) - f\right\|_\infty \leq \sum_{i=1}^{r}\left\|L_{m_i}f - f\right\|_\infty. \tag{18.100}$$

We give

Theorem 18.17 *Let $f \in \Omega$; $N, m_1, m_2, \ldots, m_r \in \mathbb{N} : m_1 \leq m_2 \leq \cdots \leq m_r$, $0 < \beta < 1$; $m_i^{1-\beta} \geq 3$, $i = 1, \ldots, r$, $x \in Y$, and let $\left(L_{m_1}, \ldots, L_{m_r}\right)$ as $\left(A_{m_1}, \ldots, A_{m_r}\right)$ or $\left(B_{m_1}, \ldots, B_{m_r}\right)$ or $\left(C_{m_1}, \ldots, C_{m_r}\right)$ or $\left(D_{m_1}, \ldots, D_{m_r}\right)$. Then*

$$\left|L_{m_r}\left(L_{m_{r-1}}\left(\ldots L_{m_2}\left(L_{m_1}f\right)\right)\right)(x) - f(x)\right| \leq$$

$$\left\|L_{m_r}\left(L_{m_{r-1}}\left(\ldots L_{m_2}\left(L_{m_1}f\right)\right)\right) - f\right\|_\infty \leq$$

$$\sum_{i=1}^{r}\left\|L_{m_i}f - f\right\|_\infty \leq$$

$$c_N \sum_{i=1}^{r} \left[\omega_1 \left(f, \varphi \left(m_i \right) \right) + \frac{\| f \|_\infty}{\sqrt{\pi} \left(m_i^{1-\beta} - 2 \right) e^{\left(m_i^{1-\beta} - 2 \right)^2}} \right] \leq$$

$$r c_N \left[\omega_1 \left(f, \varphi \left(m_1 \right) \right) + \frac{\| f \|_\infty}{\sqrt{\pi} \left(m_1^{1-\beta} - 2 \right) e^{\left(m_1^{1-\beta} - 2 \right)^2}} \right]. \quad (18.101)$$

Clearly, we notice that the speed of convergence to the unit operator of the multiply iterated operator is not worse than the speed of L_{m_1}.

Proof Using (18.100), (18.96), (18.97) and (18.91), (18.92). ■

We continue with

Theorem 18.18 *Let all as in Theorem 18.8, and* $r \in \mathbb{N}$. *Here* K_n *is as in (18.57). Then*

$$\left\| A_n^r f - f \right\|_\infty \leq r \left\| A_n f - f \right\|_\infty \leq r K_n. \quad (18.102)$$

Proof By (18.94) and (18.57). ■

18.4 Complex Multivariate Neural Network Approximations

We make

Remark 18.19 Let $Y = \prod_{i=1}^{n} [a_i, b_i]$ or $\mathbb{R}^N$, and $f : Y \to \mathbb{C}$ with real and imaginary parts $f_1, f_2 : f = f_1 + i f_2$, $i = \sqrt{-1}$. Clearly f is continuous iff f_1 and f_2 are continuous.

Given that $f_1, f_2 \in C^m(Y)$, $m \in \mathbb{N}$, it holds

$$f_\alpha(x) = f_{1,\alpha}(x) + i f_{2,\alpha}(x), \quad (18.103)$$

where α indicates a partial derivative of any order and arrangement.

We denote by $C_B\left(\mathbb{R}^N, \mathbb{C}\right)$ the space of continuous and bounded functions $f : \mathbb{R}^N \to \mathbb{C}$. Clearly f is bounded, iff both f_1, f_2 are bounded from $\mathbb{R}^N$ into $\mathbb{R}$, where $f = f_1 + i f_2$.

Here L_n is any of A_n, B_n, C_n, D_n, $n \in \mathbb{N}$.

We define

$$L_n(f, x) := L_n(f_1, x) + i L_n(f_2, x), \ \forall x \in Y. \quad (18.104)$$

We observe that

$$|L_n(f,x)-f(x)| \leq |L_n(f_1,x)-f_1(x)|+|L_n(f_2,x)-f_2(x)|, \quad (18.105)$$

and

$$\|L_n(f)-f\|_\infty \leq \|L_n(f_1)-f_1\|_\infty+\|L_n(f_2)-f_2\|_\infty. \quad (18.106)$$

We present

Theorem 18.20 *Let $f \in C(Y,\mathbb{C})$ which is bounded, $f=f_1+if_2$, $0<\beta<1$, $n,N\in\mathbb{N}: n^{1-\beta}\geq 3$, $x\in Y$. Then*

(i)

$$|L_n(f,x)-f(x)| \leq c_N \cdot$$

$$\left[\omega_1(f_1,\varphi(n))+\omega_1(f_2,\varphi(n_2))+\frac{(\|f_1\|_\infty+\|f_2\|_\infty)}{\sqrt{\pi}\left(n^{1-\beta}-2\right)e^{\left(n^{1-\beta}-2\right)^2}}\right]=:\varepsilon, \quad (18.107)$$

(ii)

$$\|L_n(f)-f\|_\infty \leq \varepsilon. \quad (18.108)$$

Proof Use of (18.91). ∎

In the next we discuss high order of complex approximation by using the smoothness of f.

We give

Theorem 18.21 *Let $f:\prod_{i=1}^{n}[a_i,b_i]\to\mathbb{C}$, such that $f=f_1+if_2$. Assume $f_1,f_2\in C^m\left(\prod_{i=1}^{n}[a_i,b_i]\right)$, $0<\beta<1$, $n,m,N\in\mathbb{N}$, $n^{1-\beta}\geq 3$, $x\in\left(\prod_{i=1}^{n}[a_i,b_i]\right)$. Then*

(i)

$$\left|A_n(f,x)-f(x)-\sum_{j=1}^{m}\left(\sum_{|\alpha|=j}\left(\frac{f_\alpha(x)}{\prod_{i=1}^{N}\alpha_i!}\right)A_n\left(\prod_{i=1}^{N}(\cdot-x_i)^{\alpha_i},x\right)\right)\right| \leq \quad (18.109)$$

$$(4.019)^N\cdot\left\{\frac{N^m}{m!n^{m\beta}}\left(\omega_{1,m}^{\max}\left(f_{1,\alpha},\frac{1}{n^\beta}\right)+\omega_{1,m}^{\max}\left(f_{2,\alpha},\frac{1}{n^\beta}\right)\right)+\right.$$

$$\left.\left(\frac{\|b-a\|_\infty^m\left(\|f_{1,\alpha}\|_{\infty,m}^{\max}+\|f_{2,\alpha}\|_{\infty,m}^{\max}\right)N^m}{m!}\right)\frac{1}{\sqrt{\pi}\left(n^{1-\beta}-2\right)e^{\left(n^{1-\beta}-2\right)^2}}\right\},$$

(ii)

$$|A_n(f,x) - f(x)| \leq (4.019)^N \cdot \tag{18.110}$$

$$\left\{ \sum_{j=1}^{m} \left(\sum_{|\alpha|=j} \left(\frac{|f_{1,\alpha}(x)| + |f_{2,\alpha}(x)|}{\prod_{i=1}^{N} \alpha_i!} \right) \left[\frac{1}{n^{\beta j}} + \right.\right.\right.$$

$$\left.\left.\left(\prod_{i=1}^{N} (b_i - a_i)^{\alpha_i} \right) \cdot \frac{1}{2\sqrt{\pi}\left(n^{1-\beta} - 2\right) e^{\left(n^{1-\beta}-2\right)^2}} \right] \right) +$$

$$\frac{N^m}{m! n^{m\beta}} \left(\omega_{1,m}^{\max}\left(f_{1,\alpha}, \frac{1}{n^\beta}\right) + \omega_{1,m}^{\max}\left(f_{2,\alpha}, \frac{1}{n^\beta}\right) \right) +$$

$$\left.\left(\frac{\|b-a\|_\infty^m \left(\|f_{1,\alpha}\|_{\infty,m}^{\max} + \|f_{2,\alpha}\|_{\infty,m}^{\max} \right) N^m}{m!} \right) \frac{1}{\sqrt{\pi}\left(n^{1-\beta} - 2\right) e^{\left(n^{1-\beta}-2\right)^2}} \right\},$$

(iii)

$$\|A_n(f) - f\|_\infty \leq (4.019)^N \cdot \tag{18.111}$$

$$\left\{ \sum_{j=1}^{m} \left(\sum_{|\alpha|=j} \left(\frac{\|f_{1,\alpha}\|_\infty + \|f_{2,\alpha}\|_\infty}{\prod_{i=1}^{N} \alpha_i!} \right) \left[\frac{1}{n^{\beta j}} + \right.\right.\right.$$

$$\left.\left.\left(\prod_{i=1}^{N} (b_i - a_i)^{\alpha_i} \right) \cdot \frac{1}{2\sqrt{\pi}\left(n^{1-\beta} - 2\right) e^{\left(n^{1-\beta}-2\right)^2}} \right] \right) +$$

$$\frac{N^m}{m! n^{m\beta}} \left(\omega_{1,m}^{\max}\left(f_{1,\alpha}, \frac{1}{n^\beta}\right) + \omega_{1,m}^{\max}\left(f_{2,\alpha}, \frac{1}{n^\beta}\right) \right) +$$

$$\left. + \left(\frac{\|b-a\|_\infty^m \left(\|f_{1,\alpha}\|_{\infty,m}^{\max} + \|f_{2,\alpha}\|_{\infty,m}^{\max} \right) N^m}{m!} \right) \frac{1}{\sqrt{\pi}\left(n^{1-\beta} - 2\right) e^{\left(n^{1-\beta}-2\right)^2}} \right\},$$

(iv) Assume $f_\alpha(x_0) = 0$*, for all* $\alpha : |\alpha| = 1, \ldots, m;$ $x_0 \in \left(\prod_{i=1}^{N} [a_i, b_i] \right)$*. Then*

$$|A_n(f, x_0) - f(x_0)| \leq (4.019)^N \cdot \tag{18.112}$$

$$\left\{ \frac{N^m}{m! n^{m\beta}} \left(\omega_{1,m}^{\max}\left(f_{1,\alpha}, \frac{1}{n^\beta}\right) + \omega_{1,m}^{\max}\left(f_{2,\alpha}, \frac{1}{n^\beta}\right) \right) + \right.$$

$$\left(\frac{\|b-a\|_\infty^m \left(\left\| f_{1,\alpha} \right\|_{\infty,m}^{\max} + \left\| f_{2,\alpha} \right\|_{\infty,m}^{\max} \right) N^m}{m!} \right) \frac{1}{\sqrt{\pi} \left(n^{1-\beta} - 2 \right) e^{\left(n^{1-\beta} - 2 \right)^2}} \Bigg\},$$

notice in the last the extremely high rate of convergence at $n^{-\beta(m+1)}$.

Proof By Theorem 18.8 and Remark 18.19. ∎

References

1. M. Abramowitz, I.A. Stegun (eds.), *Handbook of Mathematical Functions with Formulas, Graphs, and Mathematical Tables* (Dover Publications, New York, 1972)
2. G.A. Anastassiou, Rate of convergence of some neural network operators to the unit-univariate case. J. Math. Anal. Appli. **212**, 237–262 (1997)
3. G.A. Anastassiou, *Quantitative Approximations* (Chapman & Hall/CRC, Boca Raton, New York, 2001)
4. G.A. Anastassiou, *Inteligent Systems: Approximation by Artificial Neural Networks, Intelligent Systems Reference Library*, vol. 19 (Springer, Heidelberg, 2011)
5. G.A. Anastassiou, Univariate hyperbolic tangent neural network approximation. Math. Comput. Modell. **53**, 1111–1132 (2011)
6. G.A. Anastassiou, Multivariate hyperbolic tangent neural network approximation. Comput. Math. **61**, 809–821 (2011)
7. G.A. Anastassiou, Multivariate sigmoidal neural network approximation. Neural Netw. **24**, 378–386 (2011)
8. G.A. Anastassiou, Univariate sigmoidal neural network approximation. J. Computat. Anal. Appl. **14**(4), 659–690 (2012)
9. G.A. Anastassiou, *Approximation by Neural Networks Iterates, Advancesin Applied Mathematics and Approximation Theory*, ed. by G. Anastassiou, O. Duman. Springer Proceedings in Mathematics and Statistics (Springer, New York, 2013), pp. 1–20
10. G.A. Anastassiou, Multivariate error function based neural network approximations. Rev. Anal. Numer. Theor. Approx., Romania. Accepted 2014
11. G.A. Anastassiou, Univariate error function based neural network approximation. Indian J. Math. Accepted 2014
12. L.C. Andrews, *Special Functions of Mathematics for Engineers*, 2nd edn. (Mc Graw-Hill, New York, 1992)
13. Z. Chen, F. Cao, The approximation operators with sigmoidal functions. Comput. Math. Appl. **58**, 758–765 (2009)
14. D. Costarelli, R. Spigler, Approximation results for neural network operators activated by sigmoidal functions. Neural Netw. **44**, 101–106 (2013)
15. D. Costarelli, R. Spigler, Multivariate neural network operators with sigmoidal activation functions. Neural Netw. **48**, 72–77 (2013)
16. S. Haykin, *Neural Networks: A Comprehensive Foundation*, 2nd edn. (Prentice Hall, New York, 1998)
17. W. McCulloch, W. Pitts, A logical calculus of the ideas immanent in nervous activity. Bull. Math. Biophys. **7**, 115–133 (1943)
18. T.M. Mitchell, *Machine Learning* (WCB-McGraw-Hill, New York, 1997)

Chapter 19
Voronovskaya Type Asymptotic Expansions for Error Function Based Quasi-interpolation Neural Networks

Here we examine further the quasi-interpolation error function based neural network operators of one hidden layer. Based on fractional calculus theory we derive a fractional Voronovskaya type asymptotic expansion for the error of approximation of these operators the unit operator, as we are studying the univariate case. We treat also analogously the multivariate case. It follows [15].

19.1 Background

We consider here the (Gauss) error special function [1, 18]

$$\operatorname{erf}(x)=\frac{2}{\sqrt{\pi}}\int_0^x e^{-t^2}dt,\ x\in\mathbb{R}, \tag{19.1}$$

which is a sigmoidal type function and a strictly increasing function.

It has the basic properties

$$\operatorname{erf}(0)=0, \operatorname{erf}(-x)=-\operatorname{erf}(x),\ \operatorname{erf}(+\infty)=1, \operatorname{erf}(-\infty)=-1.$$

We consider the activation function [16]

$$\chi(x)=\frac{1}{4}(\operatorname{erf}(x+1)-\operatorname{erf}(x-1)),\ \text{any } x\in\mathbb{R}, \tag{19.2}$$

which is an even positive function.

Next we follow [16] on χ. We got there $\chi(0)\simeq 0.4215$, and that χ is strictly decreasing on $[0,\infty)$ and strictly increasing on $(-\infty,0]$, and the x-axis is the horizontal asymptote on χ, i.e. χ is a bell symmetric function.

G.A. Anastassiou, *Intelligent Systems II: Complete Approximation by Neural Network Operators*, Studies in Computational Intelligence 608,
DOI 10.1007/978-3-319-20505-2_19

Theorem 19.1 ([16]) *We have that*

$$\sum_{i=-\infty}^{\infty} \chi(x-i) = 1, \text{ all } x \in \mathbb{R}, \tag{19.3}$$

$$\sum_{i=-\infty}^{\infty} \chi(nx-i) = 1, \text{ all } x \in \mathbb{R},\ n \in \mathbb{N}, \tag{19.4}$$

and

$$\int_{-\infty}^{\infty} \chi(x)\,dx = 1, \tag{19.5}$$

that is $\chi(x)$ *is a density function on* $\mathbb{R}$.

We need the important

Theorem 19.2 ([16]) *Let* $0 < \alpha < 1$, *and* $n \in \mathbb{N}$ *with* $n^{1-\alpha} \geq 3$. *It holds*

$$\sum_{\substack{k=-\infty \\ : |nx-k| \geq n^{1-\alpha}}}^{\infty} \chi(nx-k) < \frac{1}{2\sqrt{\pi}\left(n^{1-\alpha}-2\right) e^{\left(n^{1-\alpha}-2\right)^2}}. \tag{19.6}$$

Denote by $\lfloor \cdot \rfloor$ the integral part of the number and by $\lceil \cdot \rceil$ the ceiling of the number.

Theorem 19.3 ([16]) *Let* $x \in [a,b] \subset \mathbb{R}$ *and* $n \in \mathbb{N}$ *so that* $\lceil na \rceil \leq \lfloor nb \rfloor$. *It holds*

$$\frac{1}{\sum_{k=\lceil na \rceil}^{\lfloor nb \rfloor} \chi(nx-k)} < \frac{1}{\chi(1)} \simeq 4.019, \ \forall\, x \in [a,b]. \tag{19.7}$$

Also from [16] we get

$$\lim_{n\to\infty} \sum_{k=\lceil na \rceil}^{\lfloor nb \rfloor} \chi(nx-k) \neq 1, \tag{19.8}$$

at least for some $x \in [a,b]$.

For large enough n we always obtain $\lceil na \rceil \leq \lfloor nb \rfloor$. Also $a \leq \frac{k}{n} \leq b$, iff $\lceil na \rceil \leq k \leq \lfloor nb \rfloor$. In general it holds by (19.4) that

$$\sum_{k=\lceil na \rceil}^{\lfloor nb \rfloor} \chi(nx-k) \leq 1. \tag{19.9}$$

We need the univariate neural network operator

Definition 19.4 ([16]) Let $f \in C([a,b])$, $n \in \mathbb{N}$. We set

$$A_n(f,x) := \frac{\sum_{k=\lceil na \rceil}^{\lfloor nb \rfloor} f\left(\frac{k}{n}\right) \chi(nx-k)}{\sum_{k=\lceil na \rceil}^{\lfloor nb \rfloor} \chi(nx-k)}, \forall x \in [a,b], \tag{19.10}$$

A_n is a univariate neural network operator.

We mention from [17] the following:
We define

$$Z(x_1,\ldots,x_N) := Z(x) := \prod_{i=1}^{N} \chi(x_i), x = (x_1,\ldots,x_N) \in \mathbb{R}^N, \ N \in \mathbb{N}. \tag{19.11}$$

It has the properties:

(i) $Z(x) > 0$, $\forall x \in \mathbb{R}^N$,

(ii)

$$\sum_{k=-\infty}^{\infty} Z(x-k) := \sum_{k_1=-\infty}^{\infty} \sum_{k_2=-\infty}^{\infty} \cdots \sum_{k_N=-\infty}^{\infty} Z(x_1-k_1,\ldots,x_N-k_N) = 1, \tag{19.12}$$

where $k := (k_1,\ldots,k_n) \in \mathbb{Z}^N$, $\forall x \in \mathbb{R}^N$,

hence

(iii)

$$\sum_{k=-\infty}^{\infty} Z(nx-k) :=$$

$$\sum_{k_1=-\infty}^{\infty} \sum_{k_2=-\infty}^{\infty} \cdots \sum_{k_N=-\infty}^{\infty} Z(nx_1-k_1,\ldots,nx_N-k_N) = 1, \tag{19.13}$$

$\forall x \in \mathbb{R}^N$; $n \in \mathbb{N}$,

and

(iv)

$$\int_{\mathbb{R}^N} Z(x)\,dx = 1, \tag{19.14}$$

that is Z is a multivariate density function.

Here $\|x\|_\infty := \max\{|x_1|, \ldots, |x_N|\}$, $x \in \mathbb{R}^N$, also set $\infty := (\infty, \ldots, \infty)$, $-\infty := (-\infty, \ldots, -\infty)$ upon the multivariate context, and

$$\lceil na \rceil : = (\lceil na_1 \rceil, \ldots, \lceil na_N \rceil), \quad (19.15)$$
$$\lfloor nb \rfloor : = (\lfloor nb_1 \rfloor, \ldots, \lfloor nb_N \rfloor),$$

where $a := (a_1, \ldots, a_N)$, $b := (b_1, \ldots, b_N)$.

We obviously see that

$$\sum_{k=\lceil na \rceil}^{\lfloor nb \rfloor} Z(nx - k) = \sum_{k=\lceil na \rceil}^{\lfloor nb \rfloor} \left(\prod_{i=1}^{N} \chi(nx_i - k_i) \right) =$$

$$\sum_{k_1=\lceil na_1 \rceil}^{\lfloor nb_1 \rfloor} \cdots \sum_{k_N=\lceil na_N \rceil}^{\lfloor nb_N \rfloor} \left(\prod_{i=1}^{N} \chi(nx_i - k_i) \right) = \prod_{i=1}^{N} \left(\sum_{k_i=\lceil na_i \rceil}^{\lfloor nb_i \rfloor} \chi(nx_i - k_i) \right). \quad (19.16)$$

For $0 < \beta < 1$ and $n \in \mathbb{N}$, a fixed $x \in \mathbb{R}^N$, we have that

$$\sum_{k=\lceil na \rceil}^{\lfloor nb \rfloor} \chi(nx - k) =$$

$$\sum_{\substack{k = \lceil na \rceil \\ \left\| \frac{k}{n} - x \right\|_\infty \le \frac{1}{n^\beta}}}^{\lfloor nb \rfloor} \chi(nx - k) + \sum_{\substack{k = \lceil na \rceil \\ \left\| \frac{k}{n} - x \right\|_\infty > \frac{1}{n^\beta}}}^{\lfloor nb \rfloor} \chi(nx - k). \quad (19.17)$$

In the last two sums the counting is over disjoint vector sets of k's, because the condition $\left\| \frac{k}{n} - x \right\|_\infty > \frac{1}{n^\beta}$ implies that there exists at least one $\left| \frac{k_r}{n} - x_r \right| > \frac{1}{n^\beta}$, where $r \in \{1, \ldots, N\}$.

From [17] we need

(v)

$$\sum_{\substack{k = \lceil na \rceil \\ \left\| \frac{k}{n} - x \right\|_\infty > \frac{1}{n^\beta}}}^{\lfloor nb \rfloor} Z(nx - k) \le \frac{1}{2\sqrt{\pi}\left(n^{1-\beta} - 2\right) e^{\left(n^{1-\beta}-2\right)^2}}, \quad (19.18)$$

$0 < \beta < 1$, $n \in \mathbb{N}$; $n^{1-\beta} \ge 3$, $x \in \left(\prod_{i=1}^{N} [a_i, b_i] \right)$,

(vi)

$$0 < \frac{1}{\sum_{k=\lceil na \rceil}^{\lfloor nb \rfloor} Z(nx-k)} < \frac{1}{(\chi(1))^N} \simeq (4.019)^N, \tag{19.19}$$

$\forall\, x \in \left(\prod_{i=1}^N [a_i, b_i]\right)$, $n \in \mathbb{N}$,

and

(vii)

$$\sum_{\substack{k=-\infty \\ \left\| \frac{k}{n} - x \right\|_\infty > \frac{1}{n^\beta}}}^{\infty} Z(nx-k) \leq \frac{1}{2\sqrt{\pi}\left(n^{1-\beta}-2\right) e^{\left(n^{1-\beta}-2\right)^2}}, \tag{19.20}$$

$0 < \beta < 1$, $n \in \mathbb{N} : n^{1-\beta} \geq 3$, $x \in \left(\prod_{i=1}^N [a_i, b_i]\right)$.

Also we get that

$$\lim_{n\to\infty} \sum_{k=\lceil na \rceil}^{\lfloor nb \rfloor} Z(nx-k) \neq 1, \tag{19.21}$$

for at least some $x \in \left(\prod_{i=1}^N [a_i, b_i]\right)$.

Let $f \in C\left(\prod_{i=1}^N [a_i, b_i]\right)$ and $n \in \mathbb{N}$ such that $\lceil na_i \rceil \leq \lfloor nb_i \rfloor$, $i = 1, \ldots, N$.

We mention from [17] the multivariate positive linear neural network operator ($x := (x_1, \ldots, x_N) \in \left(\prod_{i=1}^N [a_i, b_i]\right)$)

$$H_n(f, x) := H_n(f, x_1, \ldots, x_N) := \frac{\sum_{k=\lceil na \rceil}^{\lfloor nb \rfloor} f\left(\frac{k}{n}\right) Z(nx-k)}{\sum_{k=\lceil na \rceil}^{\lfloor nb \rfloor} Z(nx-k)} \tag{19.22}$$

$$:= \frac{\sum_{k_1=\lceil na_1 \rceil}^{\lfloor nb_1 \rfloor} \sum_{k_2=\lceil na_2 \rceil}^{\lfloor nb_2 \rfloor} \cdots \sum_{k_N=\lceil na_N \rceil}^{\lfloor nb_N \rfloor} f\left(\frac{k_1}{n}, \ldots, \frac{k_N}{n}\right) \left(\prod_{i=1}^N \chi(nx_i - k_i)\right)}{\prod_{i=1}^N \left(\sum_{k_i=\lceil na_i \rceil}^{\lfloor nb_i \rfloor} \chi(nx_i - k_i)\right)}.$$

For large enough n we always obtain $\lceil na_i \rceil \leq \lfloor nb_i \rfloor$, $i = 1, \ldots, N$. Also $a_i \leq \frac{k_i}{n} \leq b_i$, iff $\lceil na_i \rceil \leq k_i \leq \lfloor nb_i \rfloor$, $i = 1, \ldots, N$.

By $AC^m\left(\prod_{i=1}^N [a_i, b_i]\right)$, $m, N \in \mathbb{N}$, we denote the space of functions such that all partial derivatives of order $(m-1)$ of f are coordinatewise absolutely continuous functions, also $f \in C^{m-1}\left(\prod_{i=1}^N [a_i, b_i]\right)$.

Let $f \in AC^m\left(\prod_{i=1}^{N}[a_i, b_i]\right)$, $m, N \in \mathbb{N}$. Here f_α denotes a partial derivative of f, $\alpha := (\alpha_1, \ldots, \alpha_N)$, $\alpha_i \in \mathbb{Z}^+$, $i = 1, \ldots, N$, and $|\alpha| := \sum_{i=1}^{N} \alpha_i = l$, were $l = 0, 1, \ldots, m$. We write also $f_\alpha := \frac{\partial^\alpha f}{\partial x^\alpha}$ and we say it is order l.

We denote

$$\|f_\alpha\|_{\infty,m}^{\max} := \max_{|\alpha|=m} \left\{\|f_\alpha\|_\infty\right\}, \tag{19.23}$$

where $\|\cdot\|_\infty$ is the supremum norm.

We assume here that $\|f_\alpha\|_{\infty,m}^{\max} < \infty$.

We need

Definition 19.5 Let $\nu > 0$, $n = \lceil \nu \rceil$ ($\lceil \cdot \rceil$ is the ceiling of the number), $f \in AC^n([a, b])$ (space of functions f with $f^{(n-1)} \in AC([a, b])$, absolutely continuous functions). We call left Caputo fractional derivative (see [22], pp. 49–52) the function

$$D_{*a}^{\nu} f(x) = \frac{1}{\Gamma(n-\nu)} \int_a^x (x-t)^{n-\nu-1} f^{(n)}(t)\, dt, \tag{19.24}$$

$\forall\, x \in [a, b]$, where Γ is the gamma function $\Gamma(\nu) = \int_0^\infty e^{-t} t^{\nu-1} dt$, $\nu > 0$. Notice $D_{*a}^{\nu} f \in L_1([a, b])$ and $D_{*a}^{\nu} f$ exists a.e. on $[a, b]$.

We set $D_{*a}^{0} f(x) = f(x)$, $\forall\, x \in [a, b]$.

Definition 19.6 (see also [4, 23, 24]). Let $f \in AC^m([a, b])$, $m = \lceil \alpha \rceil$, $\alpha > 0$. The right Caputo fractional derivative of order $\alpha > 0$ is given by

$$D_{b-}^{\alpha} f(x) = \frac{(-1)^m}{\Gamma(m-\alpha)} \int_x^b (\zeta - x)^{m-\alpha-1} f^{(m)}(\zeta)\, d\zeta, \tag{19.25}$$

$\forall\, x \in [a, b]$. We set $D_{b-}^{0} f(x) = f(x)$. Notice $D_{b-}^{\alpha} f \in L_1([a, b])$ and $D_{b-}^{\alpha} f$ exists a.e. on $[a, b]$.

Convention 19.7 *We assume that*

$$D_{*x_0}^{\alpha} f(x) = 0, \text{ for } x < x_0, \tag{19.26}$$

and

$$D_{x_0-}^{\alpha} f(x) = 0, \text{ for } x > x_0, \tag{19.27}$$

for all $x, x_0 \in (a, b]$.

We mention

Proposition 19.8 (By [6]) *Let* $f \in C^n([a, b])$, $n = \lceil \nu \rceil$, $\nu > 0$. *Then* $D_{*a}^{\nu} f(x)$ *is continuous in* $x \in [a, b]$.

Also we have

Proposition 19.9 (By [6]) *Let* $f \in C^m([a,b])$, $m = \lceil \alpha \rceil$, $\alpha > 0$. *Then* $D_{b-}^{\alpha} f(x)$ *is continuous in* $x \in [a,b]$.

Theorem 19.10 ([6]) *Let* $f \in C^m([a,b])$, $m = \lceil \alpha \rceil$, $\alpha > 0$, $x, x_0 \in [a,b]$. *Then* $D_{*x_0}^{\alpha} f(x)$, $D_{x_0-}^{\alpha} f(x)$ *are jointly continuous functions in* (x, x_0) *from* $[a,b]^2$ *into* $\mathbb{R}$.

We mention the left Caputo fractional Taylor formula with integral remainder.

Theorem 19.11 ([22], p. 54) *Let* $f \in AC^m([a,b])$, $[a,b] \subset \mathbb{R}$, $m = \lceil \alpha \rceil$, $\alpha > 0$. *Then*

$$f(x) = \sum_{k=0}^{m-1} \frac{f^{(k)}(x_0)}{k!}(x - x_0)^k + \frac{1}{\Gamma(\alpha)} \int_{x_0}^{x} (x - J)^{\alpha-1} D_{*x_0}^{\alpha} f(J)\, dJ, \quad (19.28)$$

$\forall\, x \geq x_0;\ x, x_0 \in [a,b]$.

Also we mention the right Caputo fractional Taylor formula.

Theorem 19.12 ([4]) *Let* $f \in AC^m([a,b])$, $[a,b] \subset \mathbb{R}$, $m = \lceil \alpha \rceil$, $\alpha > 0$. *Then*

$$f(x) = \sum_{j=0}^{m-1} \frac{f^{(k)}(x_0)}{k!}(x - x_0)^k + \frac{1}{\Gamma(\alpha)} \int_{x}^{x_0} (J - x)^{\alpha-1} D_{x_0-}^{\alpha} f(J)\, dJ, \quad (19.29)$$

$\forall\, x \leq x_0;\ x, x_0 \in [a,b]$.

For more on fractional calculus related to this work see [3, 5, 8].
Next we follow [9], pp. 284–286.

About Taylor Formula-Multivariate Case and Estimates

Let Q be a compact convex subset of $\mathbb{R}^N$; $N \geq 2$; $z := (z_1, \ldots, z_N)$, $x_0 := (x_{01}, \ldots, x_{0N}) \in Q$.

Let $f : Q \to \mathbb{R}$ be such that all partial derivatives of order $(m-1)$ are coordinatewise absolutely continuous functions, $m \in \mathbb{N}$. Also $f \in C^{m-1}(Q)$. That is $f \in AC^m(Q)$. Each mth order partial derivative is denoted by $f_\alpha := \frac{\partial^\alpha f}{\partial x^\alpha}$, where $\alpha := (\alpha_1, \ldots, \alpha_N)$, $\alpha_i \in \mathbb{Z}^+$, $i = 1, \ldots, N$ and $|\alpha| := \sum_{i=1}^{N} \alpha_i = m$. Consider $g_z(t) := f(x_0 + t(z - x_0))$, $t \geq 0$. Then

$$g_z^{(j)}(t) = \left[\left(\sum_{i=1}^{N} (z_i - x_{0i}) \frac{\partial}{\partial x_i} \right)^j f \right] (x_{01} + t(z_1 - x_{01}), \ldots, x_{0N} + t(z_N - x_{0N})), \quad (19.30)$$

for all $j = 0, 1, 2, \ldots, m$.

Example 19.13 Let $m = N = 2$. Then

$$g_z(t) = f(x_{01} + t(z_1 - x_{01}), x_{02} + t(z_2 - x_{02})),\ t \in \mathbb{R},$$

and

$$g_z'(t) = (z_1 - x_{01}) \frac{\partial f}{\partial x_1}(x_0 + t(z - x_0)) + (z_2 - x_{02}) \frac{\partial f}{\partial x_2}(x_0 + t(z - x_0)). \tag{19.31}$$

Setting

$$(*) = (x_{01} + t(z_1 - x_{01}), x_{02} + t(z_2 - x_{02})) = (x_0 + t(z - x_0)),$$

we get

$$g_z''(t) = (z_1 - x_{01})^2 \frac{\partial f^2}{\partial x_1^2}(*) + (z_1 - x_{01})(z_2 - x_{02}) \frac{\partial f^2}{\partial x_2 \partial x_1}(*) +$$

$$(z_1 - x_{01})(z_2 - x_{02}) \frac{\partial f^2}{\partial x_1 \partial x_2}(*) + (z_2 - x_{02})^2 \frac{\partial f^2}{\partial x_2^2}(*). \tag{19.32}$$

Similarly, we have the general case of $m, N \in \mathbb{N}$ for $g_z^{(m)}(t)$.

We mention the following multivariate Taylor theorem.

Theorem 19.14 ([9]) *Under the above assumptions we have*

$$f(z_1, \ldots, z_N) = g_z(1) = \sum_{j=0}^{m-1} \frac{g_z^{(j)}(0)}{j!} + R_m(z, 0), \tag{19.33}$$

where

$$R_m(z, 0) := \int_0^1 \left(\int_0^{t_1} \cdots \left(\int_0^{t_{m-1}} g_z^{(m)}(t_m)\, dt_m \right) \cdots \right) dt_1, \tag{19.34}$$

or

$$R_m(z, 0) = \frac{1}{(m-1)!} \int_0^1 (1 - \theta)^{m-1} g_z^{(m)}(\theta)\, d\theta. \tag{19.35}$$

Notice that $g_z(0) = f(x_0)$.

We make

Remark 19.15 Assume here that

$$\|f_\alpha\|_{\infty,Q,m}^{\max} := \max_{|\alpha|=m} \|f_\alpha\|_{\infty,Q} < \infty.$$

Then

$$\left\| g_z^{(m)} \right\|_{\infty,[0,1]} = \left\| \left[\left(\sum_{i=1}^{N} (z_i - x_{0i}) \frac{\partial}{\partial x_i} \right)^m f \right] (x_0 + t(z - x_0)) \right\|_{\infty,[0,1]} \leq \tag{19.36}$$

$$\left(\sum_{i=1}^{N} |z_i - x_{0i}| \right)^m \|f_\alpha\|_{\infty,Q,m}^{\max},$$

that is

$$\left\| g_z^{(m)} \right\|_{\infty,[0,1]} \leq \left(\|z - x_0\|_{l_1} \right)^m \|f_\alpha\|_{\infty,Q,m}^{\max} < \infty. \tag{19.37}$$

Hence we get by (19.35) that

$$|R_m(z, 0)| \leq \frac{\left\| g_z^{(m)} \right\|_{\infty,[0,1]}}{m!} < \infty. \tag{19.38}$$

And it holds

$$|R_m(z, 0)| \leq \frac{\left(\|z - x_0\|_{l_1} \right)^m}{m!} \|f_\alpha\|_{\infty,Q,m}^{\max}, \tag{19.39}$$

$\forall\, z, x_0 \in Q$.

Inequality (19.39) will be an important tool in proving our multivariate main result.

In this chapter first we find fractional Voronskaya type asymptotic expansion for $A_n(f, x)$, $x \in [a, b]$, then we find multivariate Voronskaya type asymptotic expansion for $H_n(f, x)$, $x \in \left(\prod_{i=1}^{N} [a_i, b_i] \right)$; $n \in \mathbb{N}$.

Our considered neural networks here are of one hidden layer.

For other neural networks related work, see [2, 7, 10–14, 19–21]. For neural networks in general, read [25–27].

19.2 Main Results

We present our first univariate main result

Theorem 19.16 *Let* $\alpha > 0$, $N \in \mathbb{N}$, $N = \lceil \alpha \rceil$, $f \in AC^N([a,b])$, $0 < \beta < 1$, $x \in [a,b]$, $n \in \mathbb{N}$ *large enough and* $n^{1-\beta} \geq 3$. *Assume that* $\left\| D_{x-}^{\alpha} f \right\|_{\infty,[a,x]}$, $\left\| D_{*x}^{\alpha} f \right\|_{\infty,[x,b]} \leq M$, $M > 0$. *Then*

$$A_n(f,x) - f(x) = \sum_{j=1}^{N-1} \frac{f^{(j)}(x)}{j!} A_n((\cdot - x)^j, x) + o\left(\frac{1}{n^{\beta(\alpha-\varepsilon)}}\right), \tag{19.40}$$

where $0 < \varepsilon \leq \alpha$.

If $N = 1$, *the sum in (19.40) collapses.*

The last (19.40) implies that

$$n^{\beta(\alpha-\varepsilon)} \left[A_n(f,x) - f(x) - \sum_{j=1}^{N-1} \frac{f^{(j)}(x)}{j!} A_n((\cdot - x)^j, x) \right] \to 0, \tag{19.41}$$

as $n \to \infty$, $0 < \varepsilon \leq \alpha$.

When $N = 1$, *or* $f^{(j)}(x) = 0$, $j = 1, \ldots, N-1$, *then we derive that*

$$n^{\beta(\alpha-\varepsilon)} \left[A_n(f,x) - f(x) \right] \to 0$$

as $n \to \infty$, $0 < \varepsilon \leq \alpha$. *Of great interest is the case of* $\alpha = \frac{1}{2}$.

Proof From [22], p. 54; (28), we get by the left Caputo fractional Taylor formula that

$$f\left(\frac{k}{n}\right) = \sum_{j=0}^{N-1} \frac{f^{(j)}(x)}{j!} \left(\frac{k}{n} - x\right)^j + \frac{1}{\Gamma(\alpha)} \int_x^{\frac{k}{n}} \left(\frac{k}{n} - J\right)^{\alpha-1} D_{*x}^{\alpha} f(J) \, dJ, \tag{19.42}$$

for all $x \leq \frac{k}{n} \leq b$.

Also from [4]; (29), using the right Caputo fractional Taylor formula we get

$$f\left(\frac{k}{n}\right) = \sum_{j=0}^{N-1} \frac{f^{(j)}(x)}{j!} \left(\frac{k}{n} - x\right)^j + \frac{1}{\Gamma(\alpha)} \int_{\frac{k}{n}}^{x} \left(J - \frac{k}{n}\right)^{\alpha-1} D_{x-}^{\alpha} f(J) \, dJ, \tag{19.43}$$

for all $a \leq \frac{k}{n} \leq x$.

We call

$$W(x) := \sum_{k=\lceil na \rceil}^{\lfloor nb \rfloor} \chi(nx-k). \tag{19.44}$$

Hence we have

$$\frac{f\left(\frac{k}{n}\right)\chi(nx-k)}{W(x)} = \sum_{j=0}^{N-1} \frac{f^{(j)}(x)}{j!} \frac{\chi(nx-k)}{W(x)} \left(\frac{k}{n}-x\right)^j + \tag{19.45}$$

$$\frac{\chi(nx-k)}{W(x)\Gamma(\alpha)} \int_x^{\frac{k}{n}} \left(\frac{k}{n}-J\right)^{\alpha-1} D_{*x}^{\alpha} f(J)\, dJ,$$

all $x \leq \frac{k}{n} \leq b$, iff $\lceil nx \rceil \leq k \leq \lfloor nb \rfloor$, and

$$\frac{f\left(\frac{k}{n}\right)\chi(nx-k)}{W(x)} = \sum_{j=0}^{N-1} \frac{f^{(j)}(x)}{j!} \frac{\chi(nx-k)}{W(x)} \left(\frac{k}{n}-x\right)^j + \tag{19.46}$$

$$\frac{\chi(nx-k)}{W(x)\Gamma(\alpha)} \int_{\frac{k}{n}}^{x} \left(J-\frac{k}{n}\right)^{\alpha-1} D_{x-}^{\alpha} f(J)\, dJ,$$

for all $a \leq \frac{k}{n} \leq x$, iff $\lceil na \rceil \leq k \leq \lfloor nx \rfloor$.

We have that $\lceil nx \rceil \leq \lfloor nx \rfloor + 1$.

Therefore it holds

$$\sum_{k=\lfloor nx \rfloor+1}^{\lfloor nb \rfloor} \frac{f\left(\frac{k}{n}\right)\chi(nx-k)}{W(x)} = \sum_{j=0}^{N-1} \frac{f^{(j)}(x)}{j!} \sum_{k=\lfloor nx \rfloor+1}^{\lfloor nb \rfloor} \frac{\chi(nx-k)\left(\frac{k}{n}-x\right)^j}{W(x)} + \tag{19.47}$$

$$\frac{1}{\Gamma(\alpha)} \left(\frac{\sum_{k=\lfloor nx \rfloor+1}^{\lfloor nb \rfloor} \chi(nx-k)}{W(x)} \int_x^{\frac{k}{n}} \left(\frac{k}{n}-J\right)^{\alpha-1} D_{*x}^{\alpha} f(J)\, dJ \right),$$

and

$$\sum_{k=\lceil na \rceil}^{\lfloor nx \rfloor} f\left(\frac{k}{n}\right) \frac{\chi(nx-k)}{W(x)} = \sum_{j=0}^{N-1} \frac{f^{(j)}(x)}{j!} \sum_{k=\lceil na \rceil}^{\lfloor nx \rfloor} \frac{\chi(nx-k)}{W(x)} \left(\frac{k}{n}-x\right)^j + \tag{19.48}$$

$$\frac{1}{\Gamma(\alpha)} \left(\sum_{k=\lceil na \rceil}^{\lfloor nx \rfloor} \frac{\chi(nx-k)}{W(x)} \int_{\frac{k}{n}}^{x} \left(J-\frac{k}{n}\right)^{\alpha-1} D_{x-}^{\alpha} f(J)\, dJ \right).$$

Adding the last two equalities (19.47) and (19.48) we obtain

$$A_n(f,x) = \sum_{k=\lceil na \rceil}^{\lfloor nb \rfloor} f\left(\frac{k}{n}\right) \frac{\chi(nx-k)}{W(x)} = \tag{19.49}$$

$$\sum_{j=0}^{N-1} \frac{f^{(j)}(x)}{j!} \sum_{k=\lceil na \rceil}^{\lfloor nb \rfloor} \frac{\chi(nx-k)}{W(x)} \left(\frac{k}{n} - x\right)^j +$$

$$\frac{1}{\Gamma(\alpha) W(x)} \left\{ \sum_{k=\lceil na \rceil}^{\lfloor nx \rfloor} \chi(nx-k) \int_{\frac{k}{n}}^{x} \left(J - \frac{k}{n}\right)^{\alpha-1} D_{x-}^{\alpha} f(J)\, dJ + \right.$$

$$\left. \sum_{k=\lfloor nx \rfloor + 1}^{\lfloor nb \rfloor} \chi(nx-k) \int_{x}^{\frac{k}{n}} \left(\frac{k}{n} - J\right)^{\alpha-1} \left(D_{*x}^{\alpha} f(J)\right) dJ \right\}.$$

So we have derived

$$\theta(x) := A_n(f,x) - f(x) - \sum_{j=1}^{N-1} \frac{f^{(j)}(x)}{j!} A_n\left((\cdot - x)^j\right)(x) = \theta_n^*(x), \tag{19.50}$$

where

$$\theta_n^*(x) := \frac{1}{\Gamma(\alpha) W(x)} \left\{ \sum_{k=\lceil na \rceil}^{\lfloor nx \rfloor} \chi(nx-k) \int_{\frac{k}{n}}^{x} \left(J - \frac{k}{n}\right)^{\alpha-1} D_{x-}^{\alpha} f(J)\, dJ \right.$$

$$\left. + \sum_{k=\lfloor nx \rfloor + 1}^{\lfloor nb \rfloor} \chi(nx-k) \int_{x}^{\frac{k}{n}} \left(\frac{k}{n} - J\right)^{\alpha-1} D_{*x}^{\alpha} f(J)\, dJ \right\}. \tag{19.51}$$

We set

$$\theta_{1n}^*(x) := \frac{1}{\Gamma(\alpha)} \left(\frac{\sum_{k=\lceil na \rceil}^{\lfloor nx \rfloor} \chi(nx-k)}{W(x)} \int_{\frac{k}{n}}^{x} \left(J - \frac{k}{n}\right)^{\alpha-1} D_{x-}^{\alpha} f(J)\, dJ \right), \tag{19.52}$$

and

$$\theta_{2n}^* := \frac{1}{\Gamma(\alpha)} \left(\frac{\sum_{k=\lfloor nx \rfloor + 1}^{\lfloor nb \rfloor} \chi(nx-k)}{W(x)} \int_{x}^{\frac{k}{n}} \left(\frac{k}{n} - J\right)^{\alpha-1} D_{*x}^{\alpha} f(J)\, dJ \right), \tag{19.53}$$

i.e.

$$\theta_n^*(x) = \theta_{1n}^*(x) + \theta_{2n}^*(x). \tag{19.54}$$

We assume $b - a > \frac{1}{n^\beta}$, $0 < \beta < 1$, which is always the case for large enough $n \in \mathbb{N}$, that is when $n > \left\lceil (b-a)^{-\frac{1}{\beta}} \right\rceil$. It is always true that either $\left|\frac{k}{n} - x\right| \leq \frac{1}{n^\beta}$ or $\left|\frac{k}{n} - x\right| > \frac{1}{n^\beta}$.

For $k = \lceil na \rceil, \ldots, \lfloor nx \rfloor$, we consider

$$\gamma_{1k} := \left| \int_{\frac{k}{n}}^{x} \left(J - \frac{k}{n}\right)^{\alpha-1} D_{x-}^{\alpha} f(J)\, dJ \right| \leq \tag{19.55}$$

$$\int_{\frac{k}{n}}^{x} \left(J - \frac{k}{n}\right)^{\alpha-1} \left| D_{x-}^{\alpha} f(J) \right| dJ$$

$$\leq \left\| D_{x-}^{\alpha} f \right\|_{\infty,[a,x]} \frac{\left(x - \frac{\kappa}{n}\right)^\alpha}{\alpha} \leq \left\| D_{x-}^{\alpha} f \right\|_{\infty,[a,x]} \frac{(x-a)^\alpha}{\alpha}. \tag{19.56}$$

That is

$$\gamma_{1k} \leq \left\| D_{x-}^{\alpha} f \right\|_{\infty,[a,x]} \frac{(x-a)^\alpha}{\alpha}, \tag{19.57}$$

for $k = \lceil na \rceil, \ldots, \lfloor nx \rfloor$.

Also we have in case of $\left|\frac{k}{n} - x\right| \leq \frac{1}{n^\beta}$ that

$$\gamma_{1k} \leq \int_{\frac{k}{n}}^{x} \left(J - \frac{k}{n}\right)^{\alpha-1} \left| D_{x-}^{\alpha} f(J) \right| dJ \tag{19.58}$$

$$\leq \left\| D_{x-}^{\alpha} f \right\|_{\infty,[a,x]} \frac{\left(x - \frac{\kappa}{n}\right)^\alpha}{\alpha} \leq \left\| D_{x-}^{\alpha} f \right\|_{\infty,[a,x]} \frac{1}{n^{\alpha\beta}\alpha}.$$

So that, when $\left(x - \frac{k}{n}\right) \leq \frac{1}{n^\beta}$, we get

$$\gamma_{1k} \leq \left\| D_{x-}^{\alpha} f \right\|_{\infty,[a,x]} \frac{1}{\alpha n^{a\beta}}. \tag{19.59}$$

Therefore

$$\left| \theta_{1n}^*(x) \right| \leq \frac{1}{\Gamma(\alpha)} \left(\frac{\sum_{k=\lceil na \rceil}^{\lfloor nx \rfloor} \chi(nx-k)}{W(x)} \gamma_{1k} \right) = \frac{1}{\Gamma(\alpha)}.$$

$$\left\{ \frac{\sum_{\substack{k=\lceil na\rceil \\ : \left|\frac{k}{n}-x\right| \leq \frac{1}{n^{\beta}}}}^{\lfloor nx\rfloor} \chi(nx-k)}{W(x)} \gamma_{1k} + \frac{\sum_{\substack{k=\lceil na\rceil \\ : \left|\frac{k}{n}-x\right| > \frac{1}{n^{\beta}}}}^{\lfloor nx\rfloor} \chi(nx-k)}{W(x)} \gamma_{1k} \right\}$$

$$\leq \frac{1}{\Gamma(\alpha)} \left\{ \left(\frac{\sum_{\substack{k=\lceil na\rceil \\ : \left|\frac{k}{n}-x\right| \leq \frac{1}{n^{\beta}}}}^{\lfloor nx\rfloor} \chi(nx-k)}{W(x)} \right) \left\| D_{x-}^{\alpha} f \right\|_{\infty,[a,x]} \frac{1}{\alpha n^{\alpha\beta}} + \right.$$

$$\left. \frac{1}{W(x)} \left(\sum_{\substack{k=\lceil na\rceil \\ : \left|\frac{k}{n}-x\right| > \frac{1}{n^{\beta}}}}^{\lfloor nx\rfloor} \chi(nx-k) \right) \left\| D_{x-}^{\alpha} f \right\|_{\infty,[a,x]} \frac{(x-a)^{\alpha}}{\alpha} \right\} \overset{\text{(by (19.6), (19.7))}}{\leq}$$

$$\frac{\left\| D_{x-}^{\alpha} f \right\|_{\infty,[a,x]}}{\Gamma(\alpha+1)} \left\{ \frac{1}{n^{\alpha\beta}} + (4.019) \frac{1}{2\sqrt{\pi}\left(n^{1-\beta}-2\right) e^{\left(n^{1-\beta}-2\right)^{2}}} (x-a)^{\alpha} \right\}. \tag{19.60}$$

Therefore we proved

$$\left|\theta_{1n}^{*}(x)\right| \leq \frac{\left\| D_{x-}^{\alpha} f \right\|_{\infty,[a,x]}}{\Gamma(\alpha+1)} \left\{ \frac{1}{n^{\alpha\beta}} + \frac{2.0095}{\sqrt{\pi}\left(n^{1-\beta}-2\right) e^{\left(n^{1-\beta}-2\right)^{2}}} (x-a)^{\alpha} \right\}. \tag{19.61}$$

But for large enough $n \in \mathbb{N}$ we get

$$\left|\theta_{1n}^{*}(x)\right| \leq \frac{2\left\| D_{x-}^{\alpha} f \right\|_{\infty,[a,x]}}{\Gamma(\alpha+1) n^{\alpha\beta}}. \tag{19.62}$$

Similarly we have

$$\gamma_{2k} := \left| \int_{x}^{\frac{k}{n}} \left(\frac{k}{n} - J\right)^{\alpha-1} D_{*x}^{\alpha} f(J)\, dJ \right| \leq$$

$$\int_{x}^{\frac{k}{n}} \left(\frac{k}{n} - J\right)^{\alpha-1} \left| D_{*x}^{\alpha} f(J) \right| dJ \leq$$

$$\left\| D_{*x}^{\alpha} f \right\|_{\infty,[x,b]} \frac{\left(\frac{k}{n} - x\right)^{\alpha}}{\alpha} \leq \left\| D_{*x}^{\alpha} f \right\|_{\infty,[x,b]} \frac{(b-x)^{\alpha}}{\alpha}. \tag{19.63}$$

That is

$$\gamma_{2k} \leq \left\| D_{*x}^{\alpha} f \right\|_{\infty,[x,b]} \frac{(b-x)^{\alpha}}{\alpha}, \tag{19.64}$$

for $k = \lfloor nx \rfloor + 1, \ldots, \lfloor nb \rfloor$.

Also we have in case of $\left| \frac{k}{n} - x \right| \leq \frac{1}{n^{\beta}}$ that

$$\gamma_{2k} \leq \frac{\left\| D_{*x}^{\alpha} f \right\|_{\infty,[x,b]}}{\alpha n^{\alpha\beta}}. \tag{19.65}$$

Consequently it holds

$$\left| \theta_{2n}^{*}(x) \right| \leq \frac{1}{\Gamma(\alpha)} \left(\frac{\sum_{k=\lfloor nx \rfloor+1}^{\lfloor nb \rfloor} \chi(nx-k)}{W(x)} \gamma_{2k} \right) =$$

$$\frac{1}{\Gamma(\alpha)} \left\{ \left(\frac{\sum_{\left\{ \begin{array}{c} k = \lfloor nx \rfloor + 1 \\ : \left| \frac{k}{n} - x \right| \leq \frac{1}{n^{\beta}} \end{array} \right.}^{\lfloor nb \rfloor} \chi(nx-k)}{W(x)} \right) \frac{\left\| D_{*x}^{\alpha} f \right\|_{\infty,[x,b]}}{\alpha n^{\alpha\beta}} + \right.$$

$$\left. \frac{1}{W(x)} \left(\sum_{\left\{ \begin{array}{c} k = \lfloor nx \rfloor + 1 \\ : \left| \frac{k}{n} - x \right| > \frac{1}{n^{\beta}} \end{array} \right.}^{\lfloor nb \rfloor} \chi(nx-k) \right) \left\| D_{*x}^{\alpha} f \right\|_{\infty,[x,b]} \frac{(b-x)^{\alpha}}{\alpha} \right\} \leq$$

$$\frac{\left\| D_{*x}^{\alpha} f \right\|_{\infty,[x,b]}}{\Gamma(\alpha+1)} \left\{ \frac{1}{n^{\alpha\beta}} + \frac{2.0095}{\sqrt{\pi}\left(n^{1-\beta}-2\right) e^{\left(n^{1-\beta}-2\right)^{2}}} (b-x)^{\alpha} \right\}. \tag{19.66}$$

That is

$$\left| \theta_{2n}^{*}(x) \right| \leq \frac{\left\| D_{*x}^{\alpha} f \right\|_{\infty,[x,b]}}{\Gamma(\alpha+1)} \left\{ \frac{1}{n^{\alpha\beta}} + \frac{2.0095}{\sqrt{\pi}\left(n^{1-\beta}-2\right) e^{\left(n^{1-\beta}-2\right)^{2}}} (b-x)^{\alpha} \right\}. \tag{19.67}$$

But for large enough $n \in \mathbb{N}$ we get

$$\left|\theta_{2n}^{*}(x)\right| \leq \frac{2\left\|D_{*x}^{\alpha}f\right\|_{\infty,[x,b]}}{\Gamma(\alpha+1)\,n^{\alpha\beta}}. \tag{19.68}$$

Since $\left\|D_{x-}^{\alpha}f\right\|_{\infty,[a,x]}, \left\|D_{*x}^{\alpha}f\right\|_{\infty,[x,b]} \leq M$, $M > 0$, we derive

$$\left|\theta_{n}^{*}(x)\right| \leq \left|\theta_{1n}^{*}(x)\right| + \left|\theta_{2n}^{*}(x)\right| \overset{\text{(by (19.62), (19.68))}}{\leq} \frac{4M}{\Gamma(\alpha+1)\,n^{\alpha\beta}}. \tag{19.69}$$

That is for large enough $n \in \mathbb{N}$ we get

$$|\theta(x)| = \left|\theta_{n}^{*}(x)\right| \leq \left(\frac{4M}{\Gamma(\alpha+1)}\right)\left(\frac{1}{n^{\alpha\beta}}\right), \tag{19.70}$$

resulting to

$$|\theta(x)| = O\left(\frac{1}{n^{\alpha\beta}}\right), \tag{19.71}$$

and

$$|\theta(x)| = o(1). \tag{19.72}$$

And, letting $0 < \varepsilon \leq \alpha$, we derive

$$\frac{|\theta(x)|}{\left(\frac{1}{n^{\beta(\alpha-\varepsilon)}}\right)} \leq \left(\frac{4M}{\Gamma(\alpha+1)}\right)\left(\frac{1}{n^{\beta\varepsilon}}\right) \to 0, \tag{19.73}$$

as $n \to \infty$.

I.e.

$$|\theta(x)| = o\left(\frac{1}{n^{\beta(\alpha-\varepsilon)}}\right), \tag{19.74}$$

proving the claim. ■

We present our second main result which is a multivariate one.

Theorem 19.17 *Let* $0 < \beta < 1$, $x \in \prod_{i=1}^{N}[a_i, b_i]$, $n \in \mathbb{N}$ *large enough and* $n^{1-\beta} \geq 3$, $f \in AC^{m}\left(\prod_{i=1}^{N}[a_i, b_i]\right)$, $m, N \in \mathbb{N}$. *Assume further that* $\|f_{\alpha}\|_{\infty,m}^{\max} < \infty$. *Then*

$$H_n(f,x) - f(x)$$

$$= \sum_{j=1}^{m-1} \left(\sum_{|\alpha|=j} \left(\frac{f_\alpha(x)}{\prod_{i=1}^{N} \alpha_i!} \right) H_n \left(\prod_{i=1}^{N} (\cdot - x_i)^{\alpha_i}, x \right) \right) + o\left(\frac{1}{n^{\beta(m-\varepsilon)}} \right), \quad (19.75)$$

where $0 < \varepsilon \leq m$.

If $m = 1$, *the sum in (19.75) collapses.*

The last (19.75) implies that

$$n^{\beta(m-\varepsilon)} \left[H_n(f,x) - f(x) - \sum_{j=1}^{m-1} \left(\sum_{|\alpha|=j} \left(\frac{f_\alpha(x)}{\prod_{i=1}^{N} \alpha_i!} \right) H_n \left(\prod_{i=1}^{N} (\cdot - x_i)^{\alpha_i}, x \right) \right) \right] \quad (19.76)$$

$$\to 0, \text{ as } n \to \infty, \ 0 < \varepsilon \leq m.$$

When $m = 1$, *or* $f_\alpha(x) = 0$, *for* $|\alpha| = j$, $j = 1, \ldots, m-1$, *then we derive that*

$$n^{\beta(m-\varepsilon)} [H_n(f,x) - f(x)] \to 0,$$

as $n \to \infty$, $0 < \varepsilon \leq m$.

Proof Consider $g_z(t) := f(x_0 + t(z - x_0))$, $t \geq 0$; $x_0, z \in \prod_{i=1}^{N} [a_i, b_i]$. Then

$$g_z^{(j)}(t) = \left[\left(\sum_{i=1}^{N} (z_i - x_{0i}) \frac{\partial}{\partial x_i} \right)^j f \right] (x_{01} + t(z_1 - x_{01}), \ldots, x_{0N} + t(z_N - x_{0N})), \quad (19.77)$$

for all $j = 0, 1, \ldots, m$.

By (19.33) we have the multivariate Taylor's formula

$$f(z_1, \ldots, z_N) = g_z(1) = \sum_{j=0}^{m-1} \frac{g_z^{(j)}(0)}{j!} + \frac{1}{(m-1)!} \int_0^1 (1-\theta)^{m-1} g_z^{(m)}(\theta) \, d\theta. \quad (19.78)$$

Notice $g_z(0) = f(x_0)$. Also for $j = 0, 1, \ldots, m-1$, we have

$$g_z^{(j)}(0) = \sum_{\substack{\alpha := (\alpha_1, \ldots, \alpha_N),\ \alpha_i \in \mathbb{Z}^+, \\ i=1,\ldots,N,\ |\alpha| := \sum_{i=1}^{N} \alpha_i = j}} \left(\frac{j!}{\prod_{i=1}^{N} \alpha_i!} \right) \left(\prod_{i=1}^{N} (z_i - x_{0i})^{\alpha_i} \right) f_\alpha(x_0). \quad (19.79)$$

Furthermore

$$g_z^{(m)}(\theta)$$

$$= \sum_{\substack{\alpha:=(\alpha_1,\ldots,\alpha_N),\ \alpha_i\in\mathbb{Z}^+,\\ i=1,\ldots,N,\ |\alpha|:=\sum_{i=1}^N \alpha_i=m}} \left(\frac{m!}{\prod_{i=1}^N \alpha_i!}\right)\left(\prod_{i=1}^N (z_i - x_{0i})^{\alpha_i}\right) f_\alpha\left(x_0 + \theta\left(z - x_0\right)\right), \tag{19.80}$$

$0 \leq \theta \leq 1$.

So we treat $f \in AC^m\left(\prod_{i=1}^N [a_i, b_i]\right)$ with $\|f_\alpha\|_{\infty,m}^{\max} < \infty$.

Thus, by (19.78) we have for $\frac{k}{n}, x \in \left(\prod_{i=1}^N [a_i, b_i]\right)$ that

$$f\left(\frac{k_1}{n}, \ldots, \frac{k_N}{n}\right) - f(x) =$$

$$\sum_{j=1}^{m-1} \sum_{\substack{\alpha:=(\alpha_1,\ldots,\alpha_N),\ \alpha_i\in\mathbb{Z}^+,\\ i=1,\ldots,N,\ |\alpha|:=\sum_{i=1}^N \alpha_i=j}} \left(\frac{1}{\prod_{i=1}^N \alpha_i!}\right)\left(\prod_{i=1}^N \left(\frac{k_i}{n} - x_i\right)^{\alpha_i}\right) f_\alpha(x) + R, \tag{19.81}$$

where

$$R := m\int_0^1 (1-\theta)^{m-1} \sum_{\substack{\alpha:=(\alpha_1,\ldots,\alpha_N),\ \alpha_i\in\mathbb{Z}^+,\\ i=1,\ldots,N,\ |\alpha|:=\sum_{i=1}^N \alpha_i=m}} \left(\frac{1}{\prod_{i=1}^N \alpha_i!}\right) \cdot$$

$$\left(\prod_{i=1}^N \left(\frac{k_i}{n} - x_i\right)^{\alpha_i}\right) f_\alpha\left(x + \theta\left(\frac{k}{n} - x\right)\right) d\theta. \tag{19.82}$$

By (19.39) we obtain

$$|R| \leq \frac{\left(\left\|x - \frac{k}{n}\right\|_{l_1}\right)^m}{m!} \|f_\alpha\|_{\infty,m}^{\max}. \tag{19.83}$$

Notice here that

$$\left\|\frac{k}{n} - x\right\|_\infty \leq \frac{1}{n^\beta} \Leftrightarrow \left|\frac{k_i}{n} - x_i\right| \leq \frac{1}{n^\beta},\ i = 1, \ldots, N. \tag{19.84}$$

So, if $\left\| \frac{k}{n} - x \right\|_{\infty} \leq \frac{1}{n^{\beta}}$ we get that $\left\| x - \frac{k}{n} \right\|_{l_1} \leq \frac{N}{n^{\beta}}$, and

$$|R| \leq \frac{N^m}{n^{m\beta} m!} \| f_{\alpha} \|_{\infty,m}^{\max} . \tag{19.85}$$

Also we see that

$$\left\| x - \frac{k}{n} \right\|_{l_1} = \sum_{i=1}^{N} \left| x_i - \frac{k_i}{n} \right| \leq \sum_{i=1}^{N} (b_i - a_i) = \| b - a \|_{l_1} ,$$

therefore in general it holds

$$|R| \leq \frac{\left(\| b - a \|_{l_1} \right)^m}{m!} \| f_{\alpha} \|_{\infty,m}^{\max} . \tag{19.86}$$

Call

$$V(x) := \sum_{k=\lceil na \rceil}^{\lfloor nb \rfloor} Z(nx - k) .$$

Hence we have

$$U_n(x) := \frac{\sum_{k=\lceil na \rceil}^{\lfloor nb \rfloor} Z(nx - k) R}{V(x)} \tag{19.87}$$

$$= \frac{\sum_{\begin{cases} k = \lceil na \rceil \\ : \left\| \frac{k}{n} - x \right\|_{\infty} \leq \frac{1}{n^{\beta}} \end{cases}}^{\lfloor nb \rfloor} Z(nx - k) R}{V(x)} + \frac{\sum_{\begin{cases} k = \lceil na \rceil \\ : \left\| \frac{k}{n} - x \right\|_{\infty} > \frac{1}{n^{\beta}} \end{cases}}^{\lfloor nb \rfloor} Z(nx - k) R}{V(x)} .$$

Consequently we obtain

$$|U_n(x)| \leq \left(\frac{\sum_{\begin{cases} k = \lceil na \rceil \\ : \left\| \frac{k}{n} - x \right\|_{\infty} \leq \frac{1}{n^{\beta}} \end{cases}}^{\lfloor nb \rfloor} Z(nx - k)}{V(x)} \right) \left(\frac{N^m}{n^{m\beta} m!} \| f_{\alpha} \|_{\infty,m}^{\max} \right) +$$

$$\frac{1}{V(x)} \left(\sum_{\begin{cases} k = \lceil na \rceil \\ : \left\| \frac{k}{n} - x \right\|_{\infty} > \frac{1}{n^{\beta}} \end{cases}}^{\lfloor nb \rfloor} Z(nx - k) \right) \frac{\left(\| b - a \|_{l_1} \right)^m}{m!} \| f_{\alpha} \|_{\infty,m}^{\max} \overset{\text{(by (19.19), (19.18))}}{\leq}$$

$$\frac{N^m}{n^{m\beta}m!}\|f_\alpha\|_{\infty,m}^{\max}+(4.019)^N\frac{1}{2\sqrt{\pi}\left(n^{1-\beta}-2\right)e^{\left(n^{1-\beta}-2\right)^2}}\frac{\left(\|b-a\|_{l_1}\right)^m}{m!}\|f_\alpha\|_{\infty,m}^{\max}. \tag{19.88}$$

Therefore we have found

$$|U_n(x)|\leq\frac{\|f_\alpha\|_{\infty,m}^{\max}}{m!}\left\{\frac{N^m}{n^{m\beta}}+(4.019)^N\frac{1}{2\sqrt{\pi}\left(n^{1-\beta}-2\right)e^{\left(n^{1-\beta}-2\right)^2}}\left(\|b-a\|_{l_1}\right)^m\right\}. \tag{19.89}$$

For large enough $n\in\mathbb{N}$ we get

$$|U_n(x)|\leq\left(\frac{2\|f_\alpha\|_{\infty,m}^{\max}N^m}{m!}\right)\left(\frac{1}{n^{m\beta}}\right). \tag{19.90}$$

That is

$$|U_n(x)|=O\left(\frac{1}{n^{m\beta}}\right), \tag{19.91}$$

and

$$|U_n(x)|=o(1). \tag{19.92}$$

And, letting $0<\varepsilon\leq m$, we derive

$$\frac{|U_n(x)|}{\left(\frac{1}{n^{\beta(m-\varepsilon)}}\right)}\leq\left(\frac{2\|f_\alpha\|_{\infty,m}^{\max}N^m}{m!}\right)\frac{1}{n^{\beta\varepsilon}}\to 0, \tag{19.93}$$

as $n\to\infty$.

I.e.

$$|U_n(x)|=o\left(\frac{1}{n^{\beta(m-\varepsilon)}}\right). \tag{19.94}$$

By (19.81) we observe that

$$\frac{\sum_{k=\lceil na\rceil}^{\lfloor nb\rfloor}f\left(\frac{k}{n}\right)Z(nx-k)}{V(x)}-f(x)=$$

$$\sum_{j=1}^{m-1}\left(\sum_{|\alpha|=j}\left(\frac{f_\alpha(x)}{\prod_{i=1}^N\alpha_i!}\right)\right)\frac{\left(\sum_{k=\lceil na\rceil}^{\lfloor nb\rfloor}Z(nx-k)\left(\prod_{i=1}^N\left(\frac{k_i}{n}-x_i\right)^{\alpha_i}\right)\right)}{V(x)}+ \tag{19.95}$$

$$\frac{\sum_{k=\lceil na\rceil}^{\lfloor nb\rfloor}Z(nx-k)R}{V(x)}.$$

The last says

$$H_n(f,x) - f(x) - \sum_{j=1}^{m-1} \left(\sum_{|\alpha|=j} \left(\frac{f_\alpha(x)}{\prod_{i=1}^N \alpha_i!} \right) H_n \left(\prod_{i=1}^N (\cdot - x_i)^{\alpha_i}, x \right) \right) = U_n(x). \tag{19.96}$$

The proof of the theorem is complete. ■

References

1. M. Abramowitz, I.A. Stegun (eds.), *Handbook of Mathematical Functions with Formulas, Graphs, and Mathematical Tables* (Dover Publications, New York, 1972)
2. G.A. Anastassiou, Rate of convergence of some neural network operators to the unit-univariate case. J. Math. Anal. Appli. **212**, 237–262 (1997)
3. G.A. Anastassiou, *Quantitative Approximations* (Chapman & Hall/CRC, Boca Raton, New York, 2001)
4. G.A. Anastassiou, On right fractional calculus. Chaos, Solitons Fractals **42**, 365–376 (2009)
5. G.A. Anastassiou, *Fractional Differentiation Inequalities* (Springer, New York, 2009)
6. G.A. Anastassiou, Fractional Korovkin theory. Chaos, Solitons Fractals **42**(4), 2080–2094 (2009)
7. G.A. Anastassiou, *Inteligent Systems: Approximation by Artificial Neural Networks, Intelligent Systems Reference Library*, vol. 19 (Springer, Heidelberg, 2011)
8. G.A. Anastassiou, Fractional representation formulae and right fractional inequalities. Math. Comput. Modell. **54**(11–12), 3098–3115 (2011)
9. G.A. Anastassiou, *Advanced Inequalities* (World Scientific Publishing Company, Singapore, New Jersey, 2011)
10. G.A. Anastassiou, Univariate hyperbolic tangent neural network approximation. Math. Comput. Modell. **53**, 1111–1132 (2011)
11. G.A. Anastassiou, Multivariate hyperbolic tangent neural network approximation. Comput. Math. **61**, 809–821 (2011)
12. G.A. Anastassiou, Multivariate sigmoidal neural network approximation. Neural Netw. **24**, 378–386 (2011)
13. G.A. Anastassiou, Univariate sigmoidal neural network approximation. J. Comput. Anal. Appl. **14**(4), 659–690 (2012)
14. G.A. Anastassiou, Fractional neural network approximation. Comput. Math. Appl. **64**, 1655–1676 (2012)
15. G.A. Anastassiou, Voronovskaya type asymptotic expansions for error function based quasi-interpolation neural network operators. Revista Colombiana De Matematicas. Accepted 2014
16. G.A. Anastassiou, Univariate error function based neural network approximation, Indian J. Math. Accepted 2014
17. G.A. Anastassiou, Multivariate error function based neural network approximations. Rev. Anal. Numer. Theor. Approx. Accepted 2014
18. L.C. Andrews, *Special Functions of Mathematics for Engineers*, 2nd edn. (Mc Graw-Hill, New York, 1992)
19. Z. Chen, F. Cao, The approximation operators with sigmoidal functions. Comput. Math. Appl. **58**, 758–765 (2009)
20. D. Costarelli, R. Spigler, Approximation results for neural network operators activated by sigmoidal functions. Neural Netw. **44**, 101–106 (2013)
21. D. Costarelli, R. Spigler, Multivariate neural network operators with sigmoidal activation functions. Neural Netw. **48**, 72–77 (2013)

22. K. Diethelm, *The Analysis of Fractional Differential Equations*, vol. 2004, Lecture Notes in Mathematics (Springer, Berlin, Heidelberg, 2010)
23. A.M.A. El-Sayed, M. Gaber, On the finite Caputo and finite Riesz derivatives. Electronic J. Theor. Phys. **3**(12), 81–95 (2006)
24. G.S. Frederico, D.F.M. Torres, Fractional optimal control in the sense of Caputo and the fractional Noether's theorem. Int. Math. Forum **3**(10), 479–493 (2008)
25. S. Haykin, *Neural Networks: A Comprehensive Foundation*, 2nd edn. (Prentice Hall, New York, 1998)
26. W. McCulloch, W. Pitts, A logical calculus of the ideas immanent in nervous activity. Bull. Math. Biophys. **7**, 115–133 (1943)
27. T.M. Mitchell, *Machine Learning* (WCB-McGraw-Hill, New York, 1997)

Chapter 20
Fuzzy Fractional Error Function Relied Neural Network Approximations

Here we treat the univariate fuzzy fractional quantitative approximation of fuzzy real valued functions on a compact interval by quasi-interpolation error function based fuzzy neural network operators. These approximations are derived by establishing fuzzy Jackson type inequalities involving the fuzzy moduli of continuity of the right and left Caputo fuzzy fractional derivatives of the engaged function. The approximations are fuzzy pointwise and fuzzy uniform. The related feed-forward fuzzy neural networks are with one hidden layer. We study also the fuzzy integer derivative and just fuzzy continuous cases. Our fuzzy fractional approximation result using higher order fuzzy differentiation converges better than in the fuzzy just continuous case. It follows [24].

20.1 Introduction

The author in [2, 3], see Chaps. 2–5, was the first to derive neural network approximations to continuous functions with rates by very specifically defined neural network operators of Cardaliaguet-Euvrard and "Squashing" types, by employing the modulus of continuity of the engaged function or its high order derivative, and producing very tight Jackson type inequalities. He treats there both the univariate and multivariate cases. The defining these operators "bell-shaped" and "squashing" function are assumed to be of compact support.

The author inspired by [27], continued his studies on neural networks approximation by introducing and using the proper quasi-interpolation operators of sigmoidal and hyperbolic tangent type which resulted into [13, 15–19], by treating both the univariate and multivariate cases.

Continuation of the author's works [19, 22, 23] is this chapter where fuzzy neural network approximation based on error function is taken at the fractional and ordinary levels resulting into higher rates of approximation. We involve the fuzzy ordinary derivatives and the right and left Caputo fuzzy fractional derivatives of the fuzzy func-

G.A. Anastassiou, *Intelligent Systems II: Complete Approximation by Neural Network Operators*, Studies in Computational Intelligence 608,
DOI 10.1007/978-3-319-20505-2_20

tion under approximation and we establish tight fuzzy Jackson type inequalities. An extensive background is given on fuzzyness, fractional calculus and neural networks, all needed to present our work.

Our fuzzy feed-forward neural networks (FFNNs) are with one hidden layer. About neural networks in general study [33, 36, 37].

20.2 Fuzzy Fractional Mathematical Analysis Background

We need the following basic background

Definition 20.1 (*see* [40]) Let $\mu : \mathbb{R} \to [0, 1]$ with the following properties:

(i) is normal, i.e., $\exists\, x_0 \in \mathbb{R}$; $\mu(x_0) = 1$.
(ii) $\mu(\lambda x + (1-\lambda) y) \geq \min\{\mu(x), \mu(y)\}$, $\forall\, x, y \in \mathbb{R}$, $\forall\, \lambda \in [0, 1]$ (μ is called a convex fuzzy subset).
(iii) μ is upper semicontinuous on $\mathbb{R}$, i.e. $\forall\, x_0 \in \mathbb{R}$ and $\forall\, \varepsilon > 0$, $\exists$ neighborhood $V(x_0) : \mu(x) \leq \mu(x_0) + \varepsilon$, $\forall\, x \in V(x_0)$.
(iv) The set $\overline{\text{supp}(\mu)}$ is compact in $\mathbb{R}$ (where $\text{supp}(\mu) := \{x \in \mathbb{R} : \mu(x) > 0\}$).

We call μ a fuzzy real number. Denote the set of all μ with $\mathbb{R}_{\mathcal{F}}$.

E.g. $\chi_{\{x_0\}} \in \mathbb{R}_{\mathcal{F}}$, for any $x_0 \in \mathbb{R}$, where $\chi_{\{x_0\}}$ is the characteristic function at x_0.

For $0 < r \leq 1$ and $\mu \in \mathbb{R}_{\mathcal{F}}$ define

$$[\mu]^r := \{x \in \mathbb{R} : \mu(x) \geq r\}$$

and

$$[\mu]^0 := \overline{\{x \in \mathbb{R} : \mu(x) \geq 0\}}.$$

Then it is well known that for each $r \in [0, 1]$, $[\mu]^r$ is a closed and bounded interval on $\mathbb{R}$ [32].

For $u, v \in \mathbb{R}_{\mathcal{F}}$ and $\lambda \in \mathbb{R}$, we define uniquely the sum $u \oplus v$ and the product $\lambda \odot u$ by

$$[u \oplus v]^r = [u]^r + [v]^r, \quad [\lambda \odot u]^r = \lambda [u]^r, \ \forall\, r \in [0, 1],$$

where

$[u]^r + [v]^r$ means the usual addition of two intervals (as substes of $\mathbb{R}$) and $\lambda [u]^r$ means the usual product between a scalar and a subset of $\mathbb{R}$ (see, e.g. [40]).

Notice $1 \odot u = u$ and it holds

$$u \oplus v = v \oplus u, \lambda \odot u = u \odot \lambda.$$

If $0 \leq r_1 \leq r_2 \leq 1$ then

$$[u]^{r_2} \subseteq [u]^{r_1}.$$

Actually $[u]^r = \left[u_-^{(r)}, u_+^{(r)}\right]$, where $u_-^{(r)} \leq u_+^{(r)}, u_-^{(r)}, u_+^{(r)} \in \mathbb{R}, \forall\, r \in [0,1]$.

For $\lambda > 0$ one has $\lambda u_\pm^{(r)} = (\lambda \odot u)_\pm^{(r)}$, respectively.

Define $D : \mathbb{R}_\mathcal{F} \times \mathbb{R}_\mathcal{F} \to \mathbb{R}_\mathcal{F}$ by

$$D(u,v) := \sup_{r\in[0,1]} \max\left\{\left|u_-^{(r)} - v_-^{(r)}\right|, \left|u_+^{(r)} - v_+^{(r)}\right|\right\},$$

where

$$[v]^r = \left[v_-^{(r)}, v_+^{(r)}\right];\ u, v \in \mathbb{R}_\mathcal{F}.$$

We have that D is a metric on $\mathbb{R}_\mathcal{F}$.

Then $(\mathbb{R}_\mathcal{F}, D)$ is a complete metric space, see [40, 41].

Here $\sum^*$ stands for fuzzy summation and $\widetilde{o} := \chi_{\{0\}} \in \mathbb{R}_\mathcal{F}$ is the neural element with respect to $\oplus$, i.e.,

$$u \oplus \widetilde{0} = \widetilde{0} \oplus u = u,\ \forall\, u \in \mathbb{R}_\mathcal{F}.$$

Denote

$$D^*(f,g) = \sup_{x\in X\subseteq\mathbb{R}} D(f,g),$$

where $f, g : X \to \mathbb{R}_\mathcal{F}$.

We mention

Definition 20.2 Let $f : X \subseteq \mathbb{R} \to \mathbb{R}_\mathcal{F}$, X interval, we define the (first) fuzzy modulus of continuity of f by

$$\omega_1^{(\mathcal{F})}(f,\delta)_X = \sup_{x,y\in X,\ |x-y|\leq\delta} D(f(x), f(y)),\ \delta > 0.$$

When $g : X \subseteq \mathbb{R} \to \mathbb{R}$, we define

$$\omega_1(g,\delta)_X = \sup_{x,y\in X,\ |x-y|\leq\delta} |g(x) - g(y)|.$$

We define by $C_\mathcal{F}^U(\mathbb{R})$ the space of fuzzy uniformly continuous functions from $\mathbb{R} \to \mathbb{R}_\mathcal{F}$, also $C_\mathcal{F}(\mathbb{R})$ is the space of fuzzy continuous functions on $\mathbb{R}$, and $C_b(\mathbb{R}, \mathbb{R}_\mathcal{F})$ is the fuzzy continuous and bounded functions.

We mention

Proposition 20.3 ([6]) *Let $f \in C_\mathcal{F}^U(X)$. Then $\omega_1^{(\mathcal{F})}(f,\delta)_X < \infty$, for any $\delta > 0$.*

By [10], p. 129 we have that $C_\mathcal{F}^U([a,b]) = C_\mathcal{F}([a,b])$, fuzzy continuous functions on $[a,b] \subset \mathbb{R}$.

Proposition 20.4 ([6]) *It holds*

$$\lim_{\delta \to 0} \omega_1^{(\mathcal{F})}(f, \delta)_X = \omega_1^{(\mathcal{F})}(f, 0)_X = 0,$$

iff $f \in C_{\mathcal{F}}^U(X)$.

Proposition 20.5 ([6]) *Here* $[f]^r = \left[f_-^{(r)}, f_+^{(r)}\right]$, $r \in [0,1]$. *Let* $f \in C_{\mathcal{F}}(\mathbb{R})$. *Then* $f_\pm^{(r)}$ *are equicontinuous with respect to* $r \in [0,1]$ *over* $\mathbb{R}$, *respectively in* $\pm$.

Note 20.6 It is clear by Propositions 20.4, 20.5, that if $f \in C_{\mathcal{F}}^U(\mathbb{R})$, then $f_\pm^{(r)} \in C_U(\mathbb{R})$ (uniformly continuous on $\mathbb{R}$). Also if $f \in C_b(\mathbb{R}, \mathbb{R}_{\mathcal{F}})$ implies $f_\pm^{(r)} \in C_b(\mathbb{R})$ (continuous and bounded functions on $\mathbb{R}$).

Proposition 20.7 *Let* $f : \mathbb{R} \to \mathbb{R}_{\mathcal{F}}$. *Assume that* $\omega_1^{\mathcal{F}}(f, \delta)_X$, $\omega_1\left(f_-^{(r)}, \delta\right)_X$, $\omega_1\left(f_+^{(r)}, \delta\right)_X$ *are finite for any* $\delta > 0$, $r \in [0,1]$, *where* X *any interval of* $\mathbb{R}$.
Then

$$\omega_1^{(\mathcal{F})}(f, \delta)_X = \sup_{r \in [0,1]} \max\left\{\omega_1\left(f_-^{(r)}, \delta\right)_X, \omega_1\left(f_+^{(r)}, \delta\right)_X\right\}.$$

Proof Similar to Proposition 14.15, p. 246 of [10]. ■

We need

Remark 20.8 ([4]). Here $r \in [0,1]$, $x_i^{(r)}, y_i^{(r)} \in \mathbb{R}$, $i = 1, \ldots, m \in \mathbb{N}$. Suppose that

$$\sup_{r \in [0,1]} \max\left(x_i^{(r)}, y_i^{(r)}\right) \in \mathbb{R}, \text{ for } i = 1, \ldots, m.$$

Then one sees easily that

$$\sup_{r \in [0,1]} \max\left(\sum_{i=1}^m x_i^{(r)}, \sum_{i=1}^m y_i^{(r)}\right) \leq \sum_{i=1}^m \sup_{r \in [0,1]} \max\left(x_i^{(r)}, y_i^{(r)}\right). \tag{20.1}$$

We need

Definition 20.9 Let $x, y \in \mathbb{R}_{\mathcal{F}}$. If there exists $z \in \mathbb{R}_{\mathcal{F}} : x = y \oplus z$, then we call z the H-difference on x and y, denoted $x - y$.

Definition 20.10 ([39]) Let $T := [x_0, x_0 + \beta] \subset \mathbb{R}$, with $\beta > 0$. A function $f : T \to \mathbb{R}_{\mathcal{F}}$ is H-difference at $x \in T$ if there exists an $f'(x) \in \mathbb{R}_{\mathcal{F}}$ such that the limits (with respect to D)

$$\lim_{h\to 0+}\frac{f(x+h)-f(x)}{h}, \quad \lim_{h\to 0+}\frac{f(x)-f(x-h)}{h} \tag{20.2}$$

exist and are equal to $f'(x)$.

We call f' the H-derivative or fuzzy derivative of f at x.

Above is assumed that the H-differences $f(x+h)-f(x)$, $f(x)-f(x-h)$ exists in $\mathbb{R}_{\mathcal{F}}$ in a neighborhood of x.

Higher order H-fuzzy derivatives are defined the obvious way, like in the real case.

We denote by $C^N_{\mathcal{F}}(\mathbb{R})$, $N \geq 1$, the space of all N-times continuously H-fuzzy differentiable functions from $\mathbb{R}$ into $\mathbb{R}_{\mathcal{F}}$, similarly is defined $C^N_{\mathcal{F}}([a,b])$, $[a,b] \subset \mathbb{R}$.

We mention

Theorem 20.11 ([34]) *Let $f:\mathbb{R}\to\mathbb{R}_{\mathcal{F}}$ be H-fuzzy differentiable. Let $t\in\mathbb{R}$, $0\leq r\leq 1$. Clearly*

$$[f(t)]^r = \left[f(t)^{(r)}_-, f(t)^{(r)}_+\right] \subseteq \mathbb{R}.$$

Then $(f(t))^{(r)}_{\pm}$ are differentiable and

$$[f'(t)]^r = \left[\left(f(t)^{(r)}_-\right)', \left(f(t)^{(r)}_+\right)'\right].$$

I.e.

$$(f')^{(r)}_{\pm} = \left(f^{(r)}_{\pm}\right)', \quad \forall\, r\in[0,1].$$

Remark 20.12 ([5]) Let $f\in C^N_{\mathcal{F}}(\mathbb{R})$, $N\geq 1$. Then by Theorem 20.11 we obtain

$$\left[f^{(i)}(t)\right]^r = \left[\left(f(t)^{(r)}_-\right)^{(i)}, \left(f(t)^{(r)}_+\right)^{(i)}\right],$$

for $i=0,1,2,\ldots,N$, and in particular we have that

$$\left(f^{(i)}\right)^{(r)}_{\pm} = \left(f^{(r)}_{\pm}\right)^{(i)},$$

for any $r\in[0,1]$, all $i=0,1,2,\ldots,N$.

Note 20.13 ([5]) Let $f\in C^N_{\mathcal{F}}(\mathbb{R})$, $N\geq 1$. Then by Theorem 20.11 we have $f^{(r)}_{\pm}\in C^N(\mathbb{R})$, for any $r\in[0,1]$.

Items (20.11)–(20.13) are valid also on $[a,b]$.

By [10], p. 131, if $f\in C_{\mathcal{F}}([a,b])$, then f is a fuzzy bounded function.

We need also a particular case of the Fuzzy Henstock integral ($\delta(x)=\frac{\delta}{2}$), see [40].

Definition 20.14 ([31], *p. 644*) Let $f : [a, b] \to \mathbb{R}_{\mathcal{F}}$. We say that f is Fuzzy-Riemann integrable to $I \in \mathbb{R}_{\mathcal{F}}$ if for any $\varepsilon > 0$, there exists $\delta > 0$ such that for any division $P = \{[u, v]; \xi\}$ of $[a, b]$ with the norms $\Delta(P) < \delta$, we have

$$D\left(\sum_P^{*} (v - u) \odot f(\xi), I\right) < \varepsilon.$$

We write

$$I := (FR)\int_a^b f(x)\,dx. \tag{20.3}$$

We mention

Theorem 20.15 ([32]) *Let $f : [a, b] \to \mathbb{R}_{\mathcal{F}}$ be fuzzy continuous. Then*

$$(FR)\int_a^b f(x)\,dx$$

exists and belongs to $\mathbb{R}_{\mathcal{F}}$, furthermore it holds

$$\left[(FR)\int_a^b f(x)\,dx\right]^r = \left[\int_a^b (f)_-^{(r)}(x)\,dx, \int_a^b (f)_+^{(r)}(x)\,dx\right],$$

$\forall\, r \in [0, 1]$.

For the definition of general fuzzy integral we follow [35] next.

Definition 20.16 Let (Ω, Σ, μ) be a complete σ-finite measure space. We call $F : \Omega \to R_{\mathcal{F}}$ measurable iff $\forall$ closed $B \subseteq \mathbb{R}$ the function $F^{-1}(B) : \Omega \to [0, 1]$ defined by

$$F^{-1}(B)(w) := \sup_{x \in B} F(w)(x), \text{ all } w \in \Omega$$

is measurable, see [35].

Theorem 20.17 ([35]) *For $F : \Omega \to \mathbb{R}_{\mathcal{F}}$,*

$$F(w) = \left\{\left(F_-^{(r)}(w), F_+^{(r)}(w)\right) | 0 \le r \le 1\right\},$$

the following are equivalent

(1) F is measurable,
(2) $\forall\, r \in [0, 1]$, $F_-^{(r)}$, $F_+^{(r)}$ are measurable.

Following [35], given that for each $r \in [0, 1]$, $F_-^{(r)}$, $F_+^{(r)}$ are integrable we have that the parametrized representation

$$\left\{ \left(\int_A F_-^{(r)} d\mu, \int_A F_+^{(r)} d\mu \right) | 0 \leq r \leq 1 \right\} \tag{20.4}$$

is a fuzzy real number for each $A \in \Sigma$.

The last fact leads to

Definition 20.18 ([35]) A measurable function $F : \Omega \to \mathbb{R}_{\mathcal{F}}$,

$$F(w) = \left\{ \left(F_-^{(r)}(w), F_+^{(r)}(w) \right) | 0 \leq r \leq 1 \right\}$$

is integrable if for each $r \in [0, 1]$, $F_\pm^{(r)}$ are integrable, or equivalently, if $F_\pm^{(0)}$ are integrable.

In this case, the fuzzy integral of F over $A \in \Sigma$ is defined by

$$\int_A F d\mu := \left\{ \left(\int_A F_-^{(r)} d\mu, \int_A F_+^{(r)} d\mu \right) | 0 \leq r \leq 1 \right\}.$$

By [35], F is integrable iff $w \to \|F(w)\|_{\mathcal{F}}$ is real-valued integrable.

Here denote

$$\|u\|_{\mathcal{F}} := D\left(u, \widetilde{0}\right), \ \forall\, u \in \mathbb{R}_{\mathcal{F}}.$$

We need also

Theorem 20.19 ([35]) *Let $F, G : \Omega \to \mathbb{R}_{\mathcal{F}}$ be integrable. Then*

(1) Let $a, b \in \mathbb{R}$, then $aF + bG$ is integrable and for each $A \in \Sigma$,

$$\int_A (aF + bG)\, d\mu = a \int_A F d\mu + b \int_A G d\mu;$$

(2) $D(F, G)$ is a real-valued integrable function and for each $A \in \Sigma$,

$$D\left(\int_A F d\mu, \int_A G d\mu \right) \leq \int_A D(F, G)\, d\mu.$$

In particular,

$$\left\| \int_A F d\mu \right\|_{\mathcal{F}} \leq \int_A \|F\|_{\mathcal{F}}\, d\mu.$$

Above μ could be the Lebesgue measure, with all the basic properties valid here too.

Basically here we have

$$\left[\int_A F d\mu\right]^r = \left[\int_A F_-^{(r)} d\mu, \int_A F_+^{(r)} d\mu\right], \tag{20.5}$$

I.e.

$$\left(\int_A F d\mu\right)_\pm^{(r)} = \int_A F_\pm^{(r)} d\mu, \ \forall\, r \in [0, 1].$$

We need

Definition 20.20 Let $\nu \geq 0$, $n = \lceil \nu \rceil$ ($\lceil \cdot \rceil$ is the ceiling of the number), $f \in AC^n([a, b])$ (space of functions f with $f^{(n-1)} \in AC([a, b])$, absolutely continuous functions). We call left Caputo fractional derivative (see [28], pp. 49–52, [30, 38]) the function

$$D_{*a}^{\nu} f(x) = \frac{1}{\Gamma(n - \nu)} \int_a^x (x - t)^{n-\nu-1} f^{(n)}(t)\, dt, \tag{20.6}$$

$\forall\, x \in [a, b]$, where Γ is the gamma function $\Gamma(\nu) = \int_0^\infty e^{-t} t^{\nu-1} dt$, $\nu > 0$.
Notice $D_{*a}^{\nu} f \in L_1([a, b])$ and $D_{*a}^{\nu} f$ exists a.e. on $[a, b]$.
We set $D_{*a}^{0} f(x) = f(x)$, $\forall\, x \in [a, b]$.

Lemma 20.21 ([9]) *Let $\nu > 0$, $\nu \notin \mathbb{N}$, $n = \lceil \nu \rceil$, $f \in C^{n-1}([a, b])$ and $f^{(n)} \in L_\infty([a, b])$. Then $D_{*a}^{\nu} f(a) = 0$.*

Definition 20.22 (*see also* [7, 29, 30]) Let $f \in AC^m([a, b])$, $m = \lceil \beta \rceil$, $\beta > 0$. The right Caputo fractional derivative of order $\beta > 0$ is given by

$$D_{b-}^{\beta} f(x) = \frac{(-1)^m}{\Gamma(m - \beta)} \int_x^b (\zeta - x)^{m-\beta-1} f^{(m)}(\zeta)\, d\zeta, \tag{20.7}$$

$\forall\, x \in [a, b]$. We set $D_{b-}^{0} f(x) = f(x)$. Notice that $D_{b-}^{\beta} f \in L_1([a, b])$ and $D_{b-}^{\beta} f$ exists a.e. on $[a, b]$.

Lemma 20.23 ([9]) *Let $f \in C^{m-1}([a, b])$, $f^{(m)} \in L_\infty([a, b])$, $m = \lceil \beta \rceil$, $\beta > 0$, $\beta \notin \mathbb{N}$. Then $D_{b-}^{\beta} f(b) = 0$.*

Convention 20.24 *We assume that*

$$D_{*x_0}^{\beta} f(x) = 0, \text{ for } x < x_0, \tag{20.8}$$

and

$$D_{x_0-}^{\beta} f(x) = 0, \text{ for } x > x_0, \tag{20.9}$$

for all $x, x_0 \in (a, b]$.

We mention

Proposition 20.25 ([9]) *Let $f \in C^n([a,b])$, $n = \lceil \nu \rceil$, $\nu > 0$. Then $D_{*a}^{\nu} f(x)$ is continuous in $x \in [a,b]$.*

Also we have

Proposition 20.26 ([9]) *Let $f \in C^m([a,b])$, $m = \lceil \beta \rceil$, $\beta > 0$. Then $D_{b-}^{\beta} f(x)$ is continuous in $x \in [a,b]$.*

We further mention

Proposition 20.27 ([9]) *Let $f \in C^{m-1}([a,b])$, $f^{(m)} \in L_\infty([a,b])$, $m = \lceil \beta \rceil$, $\beta > 0$ and*

$$D_{*x_0}^{\beta} f(x) = \frac{1}{\Gamma(m-\beta)} \int_{x_0}^{x} (x-t)^{m-\beta-1} f^{(m)}(t)\, dt, \tag{20.10}$$

for all $x, x_0 \in [a,b] : x \geq x_0$.

Then $D_{*x_0}^{\beta} f(x)$ is continuous in x_0.

Proposition 20.28 ([9]) *Let $f \in C^{m-1}([a,b])$, $f^{(m)} \in L_\infty([a,b])$, $m = \lceil \beta \rceil$, $\beta > 0$ and*

$$D_{x_0-}^{\beta} f(x) = \frac{(-1)^m}{\Gamma(m-\beta)} \int_{x}^{x_0} (\zeta - x)^{m-\beta-1} f^{(m)}(\zeta)\, d\zeta, \tag{20.11}$$

for all $x, x_0 \in [a,b] : x \leq x_0$.

Then $D_{x_0-}^{\beta} f(x)$ is continuous in x_0.

We need

Proposition 20.29 ([9]) *Let $g \in C([a,b])$, $0 < c < 1$, $x, x_0 \in [a,b]$. Define*

$$L(x, x_0) = \int_{x_0}^{x} (x-t)^{c-1} g(t)\, dt, \text{ for } x \geq x_0, \tag{20.12}$$

and $L(x, x_0) = 0$, for $x < x_0$.

Then L is jointly continuous in (x, x_0) on $[a,b]^2$.

We mention

Proposition 20.30 ([9]) *Let $g \in C([a,b])$, $0 < c < 1$, $x, x_0 \in [a,b]$. Define*

$$K(x, x_0) = \int_{x_0}^{x} (\zeta - x)^{c-1} g(\zeta)\, d\zeta, \text{ for } x \leq x_0, \tag{20.13}$$

and $K(x, x_0) = 0$, for $x > x_0$.

Then $K(x, x_0)$ is jointly continuous from $[a,b]^2$ into $\mathbb{R}$.

Based on Propositions 20.29, 20.30 we derive

Corollary 20.31 ([9]) *Let $f \in C^m([a,b])$, $m = \lceil \beta \rceil$, $\beta > 0$, $\beta \notin \mathbb{N}$, $x, x_0 \in [a,b]$. Then $D^{\beta}_{*x_0} f(x)$, $D^{\beta}_{x_0-} f(x)$ are jointly continuous functions in (x, x_0) from $[a,b]^2$ into $\mathbb{R}$.*

We need

Theorem 20.32 ([9]) *Let $f : [a,b]^2 \to \mathbb{R}$ be jointly continuous. Consider*

$$G(x) = \omega_1 (f(\cdot, x), \delta)_{[x,b]}, \tag{20.14}$$

$\delta > 0$, $x \in [a,b]$.

Then G is continuous in $x \in [a,b]$.

Also it holds

Theorem 20.33 ([9]) *Let $f : [a,b]^2 \to \mathbb{R}$ be jointly continuous. Then*

$$H(x) = \omega_1 (f(\cdot, x), \delta)_{[a,x]}, \tag{20.15}$$

$x \in [a,b]$, is continuous in $x \in [a,b]$, $\delta > 0$.

So that for $f \in C^m([a,b])$, $m = \lceil \beta \rceil$, $\beta > 0$, $\beta \notin \mathbb{N}$, $x, x_0 \in [a,b]$, we have that $\omega_1 \left(D^{\beta}_{*x} f, h \right)_{[x,b]}$, $\omega_1 \left(D^{\beta}_{x-} f, h \right)_{[a,x]}$ are continuous functions in $x \in [a,b]$, $h > 0$ is fixed.

We make

Remark 20.34 ([9]) Let $f \in C^{n-1}([a,b])$, $f^{(n)} \in L_\infty([a,b])$, $n = \lceil \nu \rceil$, $\nu > 0$, $\nu \notin \mathbb{N}$. Then we have

$$\left| D^{\nu}_{*a} f(x) \right| \leq \frac{\left\| f^{(n)} \right\|_\infty}{\Gamma(n - \nu + 1)} (x - a)^{n-\nu}, \ \forall\, x \in [a,b]. \tag{20.16}$$

Thus we observe

$$\omega_1 \left(D^{\nu}_{*a} f, \delta \right) = \sup_{\substack{x,y \in [a,b] \\ |x-y| \leq \delta}} \left| D^{\nu}_{*a} f(x) - D^{\nu}_{*a} f(y) \right| \tag{20.17}$$

$$\leq \sup_{\substack{x,y \in [a,b] \\ |x-y| \leq \delta}} \left(\frac{\left\| f^{(n)} \right\|_\infty}{\Gamma(n - \nu + 1)} (x - a)^{n-\nu} + \frac{\left\| f^{(n)} \right\|_\infty}{\Gamma(n - \nu + 1)} (y - a)^{n-\nu} \right)$$

$$\leq \frac{2 \left\| f^{(n)} \right\|_\infty}{\Gamma(n - \nu + 1)} (b - a)^{n-\nu}. \tag{20.18}$$

Consequently

$$\omega_1\left(D_{*a}^{\nu} f, \delta\right) \leq \frac{2\left\|f^{(n)}\right\|_{\infty}}{\Gamma(n-\nu+1)}(b-a)^{n-\nu}. \tag{20.19}$$

Similarly, let $f \in C^{m-1}([a, b])$, $f^{(m)} \in L_{\infty}([a, b])$, $m = \lceil \beta \rceil$, $\beta > 0$, $\beta \notin \mathbb{N}$, then

$$\omega_1\left(D_{b-}^{\beta} f, \delta\right) \leq \frac{2\left\|f^{(m)}\right\|_{\infty}}{\Gamma(m-\beta+1)}(b-a)^{m-\beta}. \tag{20.20}$$

So for $f \in C^{m-1}([a, b])$, $f^{(m)} \in L_{\infty}([a, b])$, $m = \lceil \beta \rceil$, $\beta > 0$, $\beta \notin \mathbb{N}$, we find

$$\sup_{x_0 \in [a,b]} \omega_1\left(D_{*x_0}^{\beta} f, \delta\right)_{[x_0,b]} \leq \frac{2\left\|f^{(m)}\right\|_{\infty}}{\Gamma(m-\beta+1)}(b-a)^{m-\beta}, \tag{20.21}$$

and

$$\sup_{x_0 \in [a,b]} \omega_1\left(D_{x_0-}^{\beta} f, \delta\right)_{[a,x_0]} \leq \frac{2\left\|f^{(m)}\right\|_{\infty}}{\Gamma(m-\beta+1)}(b-a)^{m-\beta}. \tag{20.22}$$

By Proposition 15.114, p. 388 of [8], we get here that $D_{*x_0}^{\beta} f \in C([x_0, b])$, and by [14] we obtain that $D_{x_0-}^{\beta} f \in C([a, x_0])$.

We need

Definition 20.35 ([12]) Let $f \in C_{\mathcal{F}}([a, b])$ (fuzzy continuous on $[a, b] \subset \mathbb{R}$), $\nu > 0$.

We define the Fuzzy Fractional left Riemann-Liouville operator as

$$J_a^{\nu} f(x) := \frac{1}{\Gamma(\nu)} \odot \int_a^x (x-t)^{\nu-1} \odot f(t)\, dt, \quad x \in [a, b], \tag{20.23}$$

$$J_a^0 f := f.$$

Also, we define the Fuzzy Fractional right Riemann-Liouville operator as

$$I_{b-}^{\nu} f(x) := \frac{1}{\Gamma(\nu)} \odot \int_x^b (t-x)^{\nu-1} \odot f(t)\, dt, \quad x \in [a, b], \tag{20.24}$$

$$I_{b-}^0 f := f.$$

We mention

Definition 20.36 ([12]) Let $f : [a, b] \to \mathbb{R}_{\mathcal{F}}$ is called fuzzy absolutely continuous iff $\forall\ \epsilon > 0$, $\exists\ \delta > 0$ for every finite, pairwise disjoint, family

$$(c_k, d_k)_{k=1}^{n} \subseteq (a, b) \text{ with } \sum_{k=1}^{n} (d_k - c_k) < \delta$$

we get

$$\sum_{k=1}^{n} D(f(d_k), f(c_k)) < \epsilon. \tag{20.25}$$

We denote the related space of functions by $AC_{\mathcal{F}}([a, b])$.

If $f \in AC_{\mathcal{F}}([a, b])$, then $f \in C_{\mathcal{F}}([a, b])$.

It holds

Proposition 20.37 ([12]) $f \in AC_{\mathcal{F}}([a, b]) \Leftrightarrow f_{\pm}^{(r)} \in AEC([a, b])$, $\forall\, r \in [0, 1]$ *(absolutely equicontinuous).*

We need

Definition 20.38 ([12]) We define the Fuzzy Fractional left Caputo derivative, $x \in [a, b]$.

Let $f \in C_{\mathcal{F}}^{n}([a, b])$, $n = \lceil \nu \rceil$, $\nu > 0$ ($\lceil \cdot \rceil$ denotes the ceiling). We define

$$D_{*a}^{\nu\mathcal{F}} f(x) := \frac{1}{\Gamma(n-\nu)} \odot \int_a^x (x-t)^{n-\nu-1} \odot f^{(n)}(t)\, dt \tag{20.26}$$

$$= \left\{ \left(\frac{1}{\Gamma(n-\nu)} \int_a^x (x-t)^{n-\nu-1} \left(f^{(n)}\right)_{-}^{(r)}(t)\, dt, \right. \right.$$

$$\left. \left. \frac{1}{\Gamma(n-\nu)} \int_a^x (x-t)^{n-\nu-1} \left(f^{(n)}\right)_{+}^{(r)}(t)\, dt \right) | 0 \leq r \leq 1 \right\}$$

$$= \left\{ \left(\frac{1}{\Gamma(n-\nu)} \int_a^x (x-t)^{n-\nu-1} \left(f_{-}^{(r)}\right)^{(n)}(t)\, dt, \right. \right.$$

$$\left. \left. \frac{1}{\Gamma(n-\nu)} \int_a^x (x-t)^{n-\nu-1} \left(f_{+}^{(r)}\right)^{(n)}(t)\, dt \right) | 0 \leq r \leq 1 \right\}. \tag{20.27}$$

So, we get

$$\left[D_{*a}^{\nu\mathcal{F}} f(x) \right]^r = \left[\left(\frac{1}{\Gamma(n-\nu)} \int_a^x (x-t)^{n-\nu-1} \left(f_{-}^{(r)}\right)^{(n)}(t)\, dt, \right. \right.$$

$$\left. \left. \frac{1}{\Gamma(n-\nu)} \int_a^x (x-t)^{n-\nu-1} \left(f_{+}^{(r)}\right)^{(n)}(t)\, dt \right) \right], \quad 0 \leq r \leq 1. \tag{20.28}$$

That is

$$\left(D_{*a}^{\nu\mathcal{F}} f(x)\right)_{\pm}^{(r)} = \frac{1}{\Gamma(n-\nu)} \int_a^x (x-t)^{n-\nu-1} \left(f_{\pm}^{(r)}\right)^{(n)} (t)\, dt = \left(D_{*a}^{\nu}\left(f_{\pm}^{(r)}\right)\right)(x),$$

see [8, 28].

I.e. we get that

$$\left(D_{*a}^{\nu\mathcal{F}} f(x)\right)_{\pm}^{(r)} = \left(D_{*a}^{\nu}\left(f_{\pm}^{(r)}\right)\right)(x), \tag{20.29}$$

$\forall\, x \in [a,b]$, in short

$$\left(D_{*a}^{\nu\mathcal{F}} f\right)_{\pm}^{(r)} = D_{*a}^{\nu}\left(f_{\pm}^{(r)}\right), \quad \forall\, r \in [0,1]. \tag{20.30}$$

We need

Lemma 20.39 ([12]) *$D_{*a}^{\nu\mathcal{F}} f(x)$ is fuzzy continuous in $x \in [a,b]$.*

We need

Definition 20.40 ([12]) We define the Fuzzy Fractional right Caputo derivative, $x \in [a,b]$.

Let $f \in C_{\mathcal{F}}^{n}([a,b])$, $n = \lceil \nu \rceil$, $\nu > 0$. We define

$$D_{b-}^{\nu\mathcal{F}} f(x) := \frac{(-1)^n}{\Gamma(n-\nu)} \odot \int_x^b (t-x)^{n-\nu-1} \odot f^{(n)}(t)\, dt$$

$$= \left\{ \left(\frac{(-1)^n}{\Gamma(n-\nu)} \int_x^b (t-x)^{n-\nu-1} \left(f^{(n)}\right)_{-}^{(r)} (t)\, dt, \right.\right.$$

$$\left.\left. \frac{(-1)^n}{\Gamma(n-\nu)} \int_x^b (t-x)^{n-\nu-1} \left(f^{(n)}\right)_{+}^{(r)} (t)\, dt \right) \mid 0 \le r \le 1 \right\} \tag{20.31}$$

$$= \left\{ \left(\frac{(-1)^n}{\Gamma(n-\nu)} \int_x^b (t-x)^{n-\nu-1} \left(f_{-}^{(r)}\right)^{(n)} (t)\, dt, \right.\right.$$

$$\left.\left. \frac{(-1)^n}{\Gamma(n-\nu)} \int_x^b (t-x)^{n-\nu-1} \left(f_{+}^{(r)}\right)^{(n)} (t)\, dt \right) \mid 0 \le r \le 1 \right\}.$$

We get

$$\left[D_{b-}^{\nu\mathcal{F}} f(x)\right]^r = \left[\left(\frac{(-1)^n}{\Gamma(n-\nu)} \int_x^b (t-x)^{n-\nu-1} \left(f_{-}^{(r)}\right)^{(n)} (t)\, dt, \right.\right.$$

$$\frac{(-1)^n}{\Gamma(n-\nu)}\int_x^b (t-x)^{n-\nu-1}\left(f_+^{(r)}\right)^{(n)}(t)\,dt\Big)\Big],\quad 0\le r\le 1.$$

That is

$$\left(D_{b-}^{\nu\mathcal{F}} f(x)\right)_\pm^{(r)} = \frac{(-1)^n}{\Gamma(n-\nu)}\int_x^b (t-x)^{n-\nu-1}\left(f_\pm^{(r)}\right)^{(n)}(t)\,dt = \left(D_{b-}^{\nu}\left(f_\pm^{(r)}\right)\right)(x),$$

see [7].

I.e. we get that

$$\left(D_{b-}^{\nu\mathcal{F}} f(x)\right)_\pm^{(r)} = \left(D_{b-}^{\nu}\left(f_\pm^{(r)}\right)\right)(x), \tag{20.32}$$

$\forall\, x \in [a,b]$, in short

$$\left(D_{b-}^{\nu\mathcal{F}} f\right)_\pm^{(r)} = D_{b-}^{\nu}\left(f_\pm^{(r)}\right),\quad \forall\, r\in[0,1]. \tag{20.33}$$

Clearly,

$$D_{b-}^{\nu}\left(f_-^{(r)}\right) \le D_{b-}^{\nu}\left(f_+^{(r)}\right),\quad \forall\, r\in[0,1].$$

We need

Lemma 20.41 ([12]) *$D_{b-}^{\nu\mathcal{F}} f(x)$ is fuzzy continuous in $x\in[a,b]$.*

20.3 Real Neural Network Approximation Basics

We consider here the (Gauss) error special function [1, 25]

$$\operatorname{erf}(x) = \frac{2}{\sqrt{\pi}}\int_0^x e^{-t^2}dt,\quad x\in\mathbb{R}, \tag{20.34}$$

which is a sigmoidal type function and a strictly increasing function.

It has the basic properties

$$\operatorname{erf}(0)=0,\quad \operatorname{erf}(-x) = -\operatorname{erf}(x),\quad \operatorname{erf}(+\infty)=1,\quad \operatorname{erf}(-\infty)=-1, \tag{20.35}$$

and

$$(\operatorname{erf}(x))' = \frac{2}{\sqrt{\pi}}e^{-x^2},\quad x\in\mathbb{R}, \tag{20.36}$$

$$\int \operatorname{erf}(x)\,dx = x\operatorname{erf}(x) + \frac{e^{-x^2}}{\sqrt{\pi}} + C, \tag{20.37}$$

where C is a constant.

The error function is related to the cumulative probability distribution function of the standard normal distribution

$$\Phi(x) = \frac{1}{2} + \frac{1}{2}\operatorname{erf}\left(\frac{x}{\sqrt{2}}\right).$$

We consider the activation function

$$\chi(x) = \frac{1}{4}\left(\operatorname{erf}(x+1) - \operatorname{erf}(x-1)\right),\ x \in \mathbb{R}, \tag{20.38}$$

and we notice that

$$\chi(-x) = \chi(x), \tag{20.39}$$

thus χ is an even function.

Since $x + 1 > x - 1$, then $\operatorname{erf}(x+1) > \operatorname{erf}(x-1)$, and $\chi(x) > 0$, all $x \in \mathbb{R}$.

We see that

$$\chi(0) = \frac{\operatorname{erf}(1)}{2} \simeq 0.4215. \tag{20.40}$$

Let $x > 0$, we have

$$\chi'(x) = \frac{1}{2\sqrt{\pi}}\left(\frac{e^{(x-1)^2} - e^{(x+1)^2}}{e^{(x+1)^2}e^{(x-1)^2}}\right) < 0, \tag{20.41}$$

proving $\chi'(x) < 0$, for $x > 0$.

That is χ is strictly decreasing on $[0, \infty)$ and is strictly increasing on $(-\infty, 0]$, and $\chi'(0) = 0$.

Clearly the x-axis is the horizontal asymptote on χ.

Conclusion, χ is a bell symmetric function with maximum $\chi(0) \simeq 0.4215$.

We further mention

Theorem 20.42 ([23]) *We have that*

$$\sum_{i=-\infty}^{\infty} \chi(x-i) = 1,\ all\ x \in \mathbb{R}. \tag{20.42}$$

Thus

$$\sum_{i=-\infty}^{\infty} \chi(nx-i) = 1, \quad \forall\, n \in \mathbb{N},\ \forall\, x \in \mathbb{R}. \tag{20.43}$$

Furthermore we get:

Since χ is even it holds $\sum_{i=-\infty}^{\infty} \chi(i-x) = 1$, for any $x \in \mathbb{R}$.

Hence $\sum_{i=-\infty}^{\infty} \chi(i+x) = 1,\ \forall\, x \in \mathbb{R}$, and $\sum_{i=-\infty}^{\infty} \chi(x+i) = 1\,,\ \forall\, x \in \mathbb{R}$.

Theorem 20.43 ([23]) *It holds*

$$\int_{-\infty}^{\infty} \chi(x)\, dx = 1. \tag{20.44}$$

So $\chi(x)$ is a density function on $\mathbb{R}$.

Theorem 20.44 ([23]) *Let* $0 < \alpha < 1$, *and* $n \in \mathbb{N}$ *with* $n^{1-\alpha} \geq 3$. *It holds*

$$\sum_{\substack{k=-\infty \\ : |nx-k| \geq n^{1-\alpha}}}^{\infty} \chi(nx-k) < \frac{1}{2\sqrt{\pi}\left(n^{1-\alpha}-2\right) e^{\left(n^{1-\alpha}-2\right)^2}}. \tag{20.45}$$

Denote by $\lfloor \cdot \rfloor$ the integral part of the number and by $\lceil \cdot \rceil$ the ceiling of the number.

Theorem 20.45 ([23]) *Let* $x \in [a,b] \subset \mathbb{R}$ *and* $n \in \mathbb{N}$ *so that* $\lceil na \rceil \leq \lfloor nb \rfloor$. *It holds*

$$\frac{1}{\sum_{k=\lceil na \rceil}^{\lfloor nb \rfloor} \chi(nx-k)} < \frac{1}{\chi(1)} \simeq 4.019, \ \forall\, x \in [a,b]. \tag{20.46}$$

Also we have that [23]

$$\lim_{n\to\infty} \sum_{k=\lceil na \rceil}^{\lfloor nb \rfloor} \chi(nx-k) \neq 1, \tag{20.47}$$

for at least some $x \in [a,b]$.

Note 20.46 For large enough n we always obtain $\lceil na \rceil \leq \lfloor nb \rfloor$. Also $a \leq \frac{k}{n} \leq b$, iff $\lceil na \rceil \leq k \leq \lfloor nb \rfloor$. In general it holds (by (20.43)) that

$$\sum_{k=\lceil na \rceil}^{\lfloor nb \rfloor} \chi(nx-k) \leq 1. \tag{20.48}$$

We mention

Definition 20.47 ([23]) Let $f \in C([a,b])$ $n \in \mathbb{N}$. We set

$$A_n(f,x) = \frac{\sum_{k=\lceil na \rceil}^{\lfloor nb \rfloor} f\left(\frac{k}{n}\right) \chi(nx-k)}{\sum_{k=\lceil na \rceil}^{\lfloor nb \rfloor} \chi(nx-k)}, \ \forall\, x \in [a.b], \tag{20.49}$$

A_n is a neural network operator.

We need

Definition 20.48 For $f \in C([a,b])$, the first modulus of continuity is given by

$$\omega_1(f,\delta) := \sup_{\substack{x,y\in[a,b]\\ |x-y|\le\delta}} |f(x)-f(y)|,\ \delta>0. \tag{20.50}$$

We have that $\lim_{\delta\to 0}\omega_1(f,\delta)=0$.

We make

Remark 20.49 We notice the following, that

$$A_n(f,x)-f(x) \overset{(20.49)}{=} \frac{\sum_{k=\lceil na\rceil}^{\lfloor nb\rfloor} f\left(\frac{k}{n}\right)\chi(nx-k) - f(x)\sum_{k=\lceil na\rceil}^{\lfloor nb\rfloor}\chi(nx-k)}{\sum_{k=\lceil na\rceil}^{\lfloor nb\rfloor}\chi(nx-k)}, \tag{20.51}$$

using (20.46) we get,

$$|A_n(f,x)-f(x)| \le (4.019)\left|\sum_{k=\lceil na\rceil}^{\lfloor nb\rfloor} f\left(\frac{k}{n}\right)\chi(nx-k) - f(x)\sum_{k=\lceil na\rceil}^{\lfloor nb\rfloor}\chi(nx-k)\right|.$$

We need

Theorem 20.50 ([23]) *Let $f\in C([a,b])$, $0<\alpha<1$, $x\in[a,b]$, $n\in\mathbb{N}$ with $n^{1-\alpha}\ge 3$, $\|\cdot\|_\infty$ is the supremum norm. Then*

(1)

$$|A_n(f,x)-f(x)| \le (4.019)\left[\omega_1\left(f,\frac{1}{n^\alpha}\right)+\frac{\|f\|_\infty}{\sqrt{\pi}\left(n^{1-\alpha}-2\right)e^{\left(n^{1-\alpha}-2\right)^2}}\right] =: \mu_{1n}, \tag{20.52}$$

(2)

$$\|A_n(f)-f\|_\infty \le \mu_{1n}. \tag{20.53}$$

We notice that $\lim_{n\to\infty} A_n(f)=f$, pointwise and uniformly.

In the next we mention high order of approximation by using the smoothness of f.

Theorem 20.51 ([23]) *Let $f\in C^N([a,b])$, $n,N\in\mathbb{N}$, $n^{1-\alpha}\ge 3$, $0<\alpha<1$, $x\in[a,b]$. Then*

(i)

$$|A_n(f,x)-f(x)| \le (4,019)\cdot \tag{20.54}$$

$$\left\{\sum_{j=1}^{N}\frac{\left|f^{(j)}(x)\right|}{j!}\left[\frac{1}{n^{\alpha j}}+\frac{(b-a)^{j}}{2\sqrt{\pi}\left(n^{1-\alpha}-2\right)e^{\left(n^{1-\alpha}-2\right)^{2}}}\right]+\right.$$

$$\left.\left[\omega_{1}\left(f^{(N)},\frac{1}{n^{\alpha}}\right)\frac{1}{n^{\alpha N}N!}+\frac{\left\|f^{(N)}\right\|_{\infty}(b-a)^{N}}{N!\sqrt{\pi}\left(n^{1-\alpha}-2\right)e^{\left(n^{1-\alpha}-2\right)^{2}}}\right]\right\},$$

(ii) assume further $f^{(j)}(x_0)=0$, $j=1,\ldots,N$, *for some* $x_0\in[a,b]$, *it holds*

$$|A_n(f,x_0)-f(x_0)|\leq(4.019)\cdot \tag{20.55}$$

$$\left[\omega_{1}\left(f^{(N)},\frac{1}{n^{\alpha}}\right)\frac{1}{n^{\alpha N}N!}+\frac{\left\|f^{(N)}\right\|_{\infty}(b-a)^{N}}{N!\sqrt{\pi}\left(n^{1-\alpha}-2\right)e^{\left(n^{1-\alpha}-2\right)^{2}}}\right],$$

notice here the extremely high rate of convergence at $n^{-(N+1)\alpha}$,

(iii)

$$\|A_n(f)-f\|_{\infty}\leq(4.019)\cdot \tag{20.56}$$

$$\left\{\sum_{j=1}^{N}\frac{\left\|f^{(j)}\right\|_{\infty}}{j!}\left[\frac{1}{n^{\alpha j}}+\frac{(b-a)^{j}}{2\sqrt{\pi}\left(n^{1-\alpha}-2\right)e^{\left(n^{1-\alpha}-2\right)^{2}}}\right]+\right.$$

$$\left.\left[\omega_{1}\left(f^{(N)},\frac{1}{n^{\alpha}}\right)\frac{1}{n^{\alpha N}N!}+\frac{\left\|f^{(N)}\right\|_{\infty}(b-a)^{N}}{N!\sqrt{\pi}\left(n^{1-\alpha}-2\right)e^{\left(n^{1-\alpha}-2\right)^{2}}}\right]\right\}.$$

Here comes the related fractional result

Theorem 20.52 ([23]) *Let* $\alpha>0$, $N=\lceil\alpha\rceil$, $\alpha\notin\mathbb{N}$, $f\in AC^{N}([a,b])$, *with* $f^{(N)}\in L_{\infty}([a,b])$, $0<\beta<1$, $x\in[a,b]$, $n\in\mathbb{N}$, $n^{1-\beta}\geq3$. *Then*

(i)

$$\left|A_n(f,x)-\sum_{j=1}^{N-1}\frac{f^{(j)}(x)}{j!}A_n\left((\cdot-x)^{j}\right)(x)-f(x)\right|\leq \tag{20.57}$$

$$\frac{(4.019)}{\Gamma(\alpha+1)}\cdot\left\{\frac{\left(\omega_{1}\left(D_{x-}^{\alpha}f,\frac{1}{n^{\beta}}\right)_{[a,x]}+\omega_{1}\left(D_{*x}^{\alpha}f,\frac{1}{n^{\beta}}\right)_{[x,b]}\right)}{n^{\alpha\beta}}+\right.$$

$$\frac{1}{2\sqrt{\pi}\left(n^{1-\beta}-2\right)e^{\left(n^{1-\beta}-2\right)^{2}}}\cdot$$

$$\left.\left(\left\|D_{x-}^{\alpha}f\right\|_{\infty,[a,x]}(x-a)^{\alpha}+\left\|D_{*x}^{\alpha}f\right\|_{\infty,[x,b]}(b-x)^{\alpha}\right)\right\},$$

(ii) if $f^{(j)}(x)=0$, for $j=1,\dots,N-1$, we have

$$|A_n(f,x)-f(x)|\leq\frac{(4.019)}{\Gamma(\alpha+1)}\cdot\tag{20.58}$$

$$\left\{\frac{\left(\omega_1\left(D_{x-}^{\alpha}f,\frac{1}{n^{\beta}}\right)_{[a,x]}+\omega_1\left(D_{*x}^{\alpha}f,\frac{1}{n^{\beta}}\right)_{[x,b]}\right)}{n^{\alpha\beta}}+\right.$$

$$\frac{1}{2\sqrt{\pi}\left(n^{1-\beta}-2\right)e^{\left(n^{1-\beta}-2\right)^2}}\cdot$$

$$\left.\left(\left\|D_{x-}^{\alpha}f\right\|_{\infty,[a,x]}(x-a)^{\alpha}+\left\|D_{*x}^{\alpha}f\right\|_{\infty,[x,b]}(b-x)^{\alpha}\right)\right\},$$

when $\alpha>1$ notice here the extremely high rate of convergence at $n^{-(\alpha+1)\beta}$,

(iii)

$$|A_n(f,x)-f(x)|\leq(4.019)\cdot\tag{20.59}$$

$$\left\{\sum_{j=1}^{N-1}\frac{\left|f^{(j)}(x)\right|}{j!}\left\{\frac{1}{n^{\beta j}}+(b-a)^j\frac{1}{2\sqrt{\pi}\left(n^{1-\beta}-2\right)e^{\left(n^{1-\beta}-2\right)^2}}\right\}+\right.$$

$$\frac{1}{\Gamma(\alpha+1)}\left\{\frac{\left(\omega_1\left(D_{x-}^{\alpha}f,\frac{1}{n^{\beta}}\right)_{[a,x]}+\omega_1\left(D_{*x}^{\alpha}f,\frac{1}{n^{\beta}}\right)_{[x,b]}\right)}{n^{\alpha\beta}}+\right.$$

$$\frac{1}{2\sqrt{\pi}\left(n^{1-\beta}-2\right)e^{\left(n^{1-\beta}-2\right)^2}}\cdot$$

$$\left.\left.\left(\left\|D_{x-}^{\alpha}f\right\|_{\infty,[a,x]}(x-a)^{\alpha}+\left\|D_{*x}^{\alpha}f\right\|_{\infty,[x,b]}(b-x)^{\alpha}\right)\right\}\right\},$$

$\forall\, x \in [a, b]$,
and

(iv)

$$\|A_n f - f\|_\infty \leq (4.019) \cdot$$

$$\left\{ \sum_{j=1}^{N-1} \frac{\|f^{(j)}\|_\infty}{j!} \left\{ \frac{1}{n^{\beta j}} + (b-a)^j \frac{1}{2\sqrt{\pi}\left(n^{1-\beta}-2\right) e^{\left(n^{1-\beta}-2\right)^2}} \right\} + \right.$$

$$\frac{1}{\Gamma(\alpha+1)} \left\{ \frac{\left(\sup_{x\in[a,b]} \omega_1 \left(D_{x-}^{\alpha} f, \frac{1}{n^\beta}\right)_{[a,x]} + \sup_{x\in[a,b]} \omega_1 \left(D_{*x}^{\alpha} f, \frac{1}{n^\beta}\right)_{[x,b]} \right)}{n^{\alpha\beta}} + \right. \tag{20.60}$$

$$\frac{1}{2\sqrt{\pi}\left(n^{1-\beta}-2\right) e^{\left(n^{1-\beta}-2\right)^2}} \cdot$$

$$\left. \left. (b-a)^\alpha \left(\sup_{x\in[a,b]} \left\| D_{x-}^{\alpha} f \right\|_{\infty,[a,x]} + \sup_{x\in[a,b]} \left\| D_{*x}^{\alpha} f \right\|_{\infty,[x,b]} \right) \right\} \right\}.$$

Above, when $N = 1$ the sum $\sum_{j=1}^{N-1} \cdot = 0$.
As we see here we obtain fractional type pointwise and uniform convergence with rates of $A_n \to I$ the unit operator, as $n \to \infty$.

Next we apply Theorem 20.52 for $N = 1$.

Corollary 20.53 ([23]) *Let $0 < \alpha, \beta < 1$, $n^{1-\beta} \geq 3$, $f \in AC([a, b])$, $f' \in L_\infty([a, b])$, $n \in \mathbb{N}$. Then*

$$\|A_n f - f\|_\infty \leq \frac{(4.019)}{\Gamma(\alpha+1)} \cdot \tag{20.61}$$

$$\left\{ \frac{\left(\sup_{x\in[a,b]} \omega_1 \left(D_{x-}^{\alpha} f, \frac{1}{n^\beta}\right)_{[a,x]} + \sup_{x\in[a,b]} \omega_1 \left(D_{*x}^{\alpha} f, \frac{1}{n^\beta}\right)_{[x,b]} \right)}{n^{\alpha\beta}} + \right.$$

$$\frac{1}{2\sqrt{\pi}\left(n^{1-\beta}-2\right) e^{\left(n^{1-\beta}-2\right)^2}} (b-a)^\alpha \cdot$$

$$\left.\left(\sup_{x\in[a,b]}\left\|D_{x-}^{\alpha}f\right\|_{\infty,[a,x]}+\sup_{x\in[a,b]}\left\|D_{*x}^{\alpha}f\right\|_{\infty,[x,b]}\right)\right\}.$$

Finally we specialize to $\alpha=\frac{1}{2}$.

Corollary 20.54 ([23]) *Let* $0<\beta<1$, $n^{1-\beta}\geq 3$, $f\in AC([a,b])$, $f'\in L_{\infty}([a,b])$, $n\in\mathbb{N}$. *Then*

$$\|A_n f-f\|_{\infty}\leq\frac{(8.038)}{\sqrt{\pi}}\cdot$$

$$\left\{\frac{\left(\sup_{x\in[a,b]}\omega_1\left(D_{x-}^{\frac{1}{2}}f,\frac{1}{n^{\beta}}\right)_{[a,x]}+\sup_{x\in[a,b]}\omega_1\left(D_{*x}^{\frac{1}{2}}f,\frac{1}{n^{\beta}}\right)_{[x,b]}\right)}{n^{\frac{\beta}{2}}}+\right.$$

$$\frac{1}{2\sqrt{\pi}\left(n^{1-\beta}-2\right)e^{\left(n^{1-\beta}-2\right)^2}}\sqrt{b-a}\cdot \tag{20.62}$$

$$\left.\left(\sup_{x\in[a,b]}\left\|D_{x-}^{\frac{1}{2}}f\right\|_{\infty,[a,x]}+\sup_{x\in[a,b]}\left\|D_{*x}^{\frac{1}{2}}f\right\|_{\infty,[x,b]}\right)\right\}.$$

We next extend items (20.50)–(20.54) to the fuzzy level.

20.4 Main Results: Approximation by Fuzzy Quasi-interpolation Neural Network Operators with Respect to Error Function

Let $f\in C_{\mathcal{F}}([a,b])$ (fuzzy continuous functions on $[a,b]\subset\mathbb{R}$), $n\in\mathbb{N}$. We define the following Fuzzy Quasi-Interpolation Neural Network operators

$$A_n^{\mathcal{F}}(f,x)=\sum_{k=\lceil na\rceil}^{\lfloor nb\rfloor *}f\left(\frac{k}{n}\right)\odot\frac{\chi(nx-k)}{\sum_{k=\lceil na\rceil}^{\lfloor nb\rfloor}\chi(nx-k)}, \tag{20.63}$$

$\forall\, x\in[a,b]$, see also (20.49).

The fuzzy sums in (20.63) are finite.

Let $r \in [0,1]$, we observe that

$$\left[A_n^{\mathcal{F}}(f,x)\right]^r = \sum_{k=\lceil na \rceil}^{\lfloor nb \rfloor} \left[f\left(\frac{k}{n}\right)\right]^r \left(\frac{\chi(nx-k)}{\sum_{k=\lceil na \rceil}^{\lfloor nb \rfloor} \chi(nx-k)}\right) =$$

$$\sum_{k=\lceil na \rceil}^{\lfloor nb \rfloor} \left[f_-^{(r)}\left(\frac{k}{n}\right), f_+^{(r)}\left(\frac{k}{n}\right)\right] \left(\frac{\chi(nx-k)}{\sum_{k=\lceil na \rceil}^{\lfloor nb \rfloor} \chi(nx-k)}\right) =$$

$$\left[\sum_{k=\lceil na \rceil}^{\lfloor nb \rfloor} f_-^{(r)}\left(\frac{k}{n}\right) \left(\frac{\chi(nx-k)}{\sum_{k=\lceil na \rceil}^{\lfloor nb \rfloor} \chi(nx-k)}\right), \sum_{k=\lceil na \rceil}^{\lfloor nb \rfloor} f_+^{(r)}\left(\frac{k}{n}\right) \left(\frac{\chi(nx-k)}{\sum_{k=\lceil na \rceil}^{\lfloor nb \rfloor} \chi(nx-k)}\right)\right] \tag{20.64}$$

$$= \left[A_n\left(f_-^{(r)}, x\right), A_n\left(f_+^{(r)}, x\right)\right]. \tag{20.65}$$

We have proved that

$$\left(A_n^{\mathcal{F}}(f,x)\right)_{\pm}^{(r)} = A_n\left(f_{\pm}^{(r)}, x\right), \tag{20.66}$$

respectively, $\forall\, r \in [0,1]$, $\forall\, x \in [a,b]$.

Therefore we get

$$D\left(A_n^{\mathcal{F}}(f,x), f(x)\right) =$$

$$\sup_{r\in[0,1]} \max\left\{\left|A_n\left(f_-^{(r)}, x\right) - f_-^{(r)}(x)\right|, \left|A_n\left(f_+^{(r)}, x\right) - f_+^{(r)}(x)\right|\right\}, \tag{20.67}$$

$\forall\, x \in [a,b]$.

We present

Theorem 20.55 *Let $f \in C_{\mathcal{F}}([a,b])$, $0<\alpha<1$, $x \in [a,b]$, $n \in \mathbb{N}$ with $n^{1-\alpha} \geq 3$. Then*

(1)

$$D\left(A_n^{\mathcal{F}}(f,x), f(x)\right) \leq \tag{20.68}$$

$$(4.019)\left[\omega_1^{(\mathcal{F})}\left(f, \frac{1}{n^{\alpha}}\right) + \frac{D^*\left(f, \widetilde{o}\right)}{\sqrt{\pi}\left(n^{1-\alpha}-2\right) e^{\left(n^{1-\alpha}-2\right)^2}}\right] =: \lambda_{1n},$$

(2)

$$D^{*}\left(A_{n}^{\mathcal{F}}(f), f\right) \leq \lambda_{1n}. \tag{20.69}$$

We notice that $\lim_{n\to\infty}\left(A_{n}^{\mathcal{F}}(f)\right)(x) \overset{D}{\rightarrow} f(x)$, $\lim_{n\to\infty} A_{n}^{\mathcal{F}}(f) \overset{D^{*}}{\rightarrow} f$, *pointwise and uniformly.*

Proof We have that $f_{\pm}^{(r)} \in C([a,b])$, $\forall\, r \in [0,1]$. Hence by (20.52) we obtain

$$\left|A_{n}\left(f_{\pm}^{(r)}, x\right) - f_{\pm}^{(r)}(x)\right|$$

$$\leq (4.019)\left[\omega_{1}\left(f_{\pm}^{(r)}, \frac{1}{n^{\alpha}}\right) + \frac{\left\|f_{\pm}^{(r)}\right\|_{\infty}}{\sqrt{\pi}\left(n^{1-\alpha}-2\right)e^{\left(n^{1-\alpha}-2\right)^{2}}}\right] \tag{20.70}$$

(where $\|\cdot\|_{\infty}$ is the supremum norm, and by Proposition 20.7 and $\left\|f_{\pm}^{(r)}\right\|_{\infty} \leq D^{*}(f, \widetilde{o})$)

$$\leq (4.019)\left[\omega_{1}^{(\mathcal{F})}\left(f, \frac{1}{n^{\alpha}}\right) + \frac{D^{*}(f, \widetilde{o})}{\sqrt{\pi}\left(n^{1-\alpha}-2\right)e^{\left(n^{1-\alpha}-2\right)^{2}}}\right].$$

Taking into account (20.67) the proof of the claim is completed. ■

We also give

Theorem 20.56 *Let* $f \in C_{\mathcal{F}}^{N}([a,b])$, $n, N \in \mathbb{N}$, $0 < \alpha < 1$, $x \in [a,b]$ *with* $n^{1-\alpha} \geq 3$. *Then*

(1)

$$D\left(A_{n}^{\mathcal{F}}(f, x), f(x)\right) \leq (4,019)\cdot \tag{20.71}$$

$$\left\{\sum_{j=1}^{N} \frac{D\left(f^{(j)}(x), \widetilde{o}\right)}{j!}\left[\frac{1}{n^{\alpha j}} + \frac{(b-a)^{j}}{2\sqrt{\pi}\left(n^{1-\alpha}-2\right)e^{\left(n^{1-\alpha}-2\right)^{2}}}\right] + \right.$$

$$\left.\left[\omega_{1}^{(\mathcal{F})}\left(f^{(N)}, \frac{1}{n^{\alpha}}\right)\frac{1}{n^{\alpha N}N!} + \frac{D^{*}\left(f^{(N)}, \widetilde{o}\right)(b-a)^{N}}{N!\sqrt{\pi}\left(n^{1-\alpha}-2\right)e^{\left(n^{1-\alpha}-2\right)^{2}}}\right]\right\},$$

(2) assume further $f^{(j)}(x_0) = \widetilde{o}$, $j = 1, \ldots, N$, *for some* $x_0 \in [a,b]$, *it holds*

$$D\left(A_{n}^{\mathcal{F}}(f, x_0), f(x_0)\right) \leq (4.019)\cdot \tag{20.72}$$

$$\left[\omega_1^{(\mathcal{F})}\left(f^{(N)}, \frac{1}{n^\alpha}\right)\frac{1}{n^{\alpha N} N!} + \frac{D^*\left(f^{(N)}, \widetilde{o}\right)(b-a)^N}{N!\sqrt{\pi}\left(n^{1-\alpha}-2\right)e^{\left(n^{1-\alpha}-2\right)^2}}\right],$$

notice here the extremely high rate of convergence at $n^{-(N+1)\alpha}$,

(3)

$$D^*\left(A_n^{\mathcal{F}}(f), f\right) \leq (4.019)\cdot \tag{20.73}$$

$$\left\{\sum_{j=1}^{N}\frac{D^*\left(f^{(j)}, \widetilde{o}\right)}{j!}\left[\frac{1}{n^{\alpha j}} + \frac{(b-a)^j}{2\sqrt{\pi}\left(n^{1-\alpha}-2\right)e^{\left(n^{1-\alpha}-2\right)^2}}\right] + \right.$$

$$\left.\left[\omega_1^{(\mathcal{F})}\left(f^{(N)}, \frac{1}{n^\alpha}\right)\frac{1}{n^{\alpha N} N!} + \frac{D^*\left(f^{(N)}, \widetilde{o}\right)(b-a)^N}{N!\sqrt{\pi}\left(n^{1-\alpha}-2\right)e^{\left(n^{1-\alpha}-2\right)^2}}\right]\right\}.$$

Proof Since $f \in C_{\mathcal{F}}^N([a,b])$, $N \geq 1$, we have that $f_\pm^{(r)} \in C^N([a,b])$, $\forall\, r \in [0,1]$. Using (20.54) we get

$$\left|A_n\left(f_\pm^{(r)}, x\right) - f_\pm^{(r)}(x)\right| \leq (4.019)\cdot \tag{20.74}$$

$$\left\{\sum_{j=1}^{N}\frac{\left|\left(f_\pm^{(r)}\right)^{(j)}(x)\right|}{j!}\left[\frac{1}{n^{\alpha j}} + \frac{(b-a)^j}{2\sqrt{\pi}\left(n^{1-\alpha}-2\right)e^{\left(n^{1-\alpha}-2\right)^2}}\right] + \right.$$

$$\left.\left[\omega_1\left(\left(f_\pm^{(r)}\right)^{(N)}, \frac{1}{n^\alpha}\right)\frac{1}{n^{\alpha N} N!} + \frac{\left\|\left(f_\pm^{(r)}\right)^{(N)}\right\|_\infty (b-a)^N}{N!\sqrt{\pi}\left(n^{1-\alpha}-2\right)e^{\left(n^{1-\alpha}-2\right)^2}}\right]\right\} \overset{\text{(by Remark 20.12)}}{=}$$

$$(4.019)\left\{\sum_{j=1}^{N}\frac{\left|\left(f^{(j)}\right)_\pm^{(r)}(x)\right|}{j!}\left[\frac{1}{n^{\alpha j}} + \frac{(b-a)^j}{2\sqrt{\pi}\left(n^{1-\alpha}-2\right)e^{\left(n^{1-\alpha}-2\right)^2}}\right] + \right. \tag{20.75}$$

$$\left.\left[\omega_1\left(\left(f^{(N)}\right)_\pm^{(r)}, \frac{1}{n^\alpha}\right)\frac{1}{n^{\alpha N} N!} + \frac{\left\|\left(f^{(N)}\right)_\pm^{(r)}\right\|_\infty (b-a)^N}{N!\sqrt{\pi}\left(n^{1-\alpha}-2\right)e^{\left(n^{1-\alpha}-2\right)^2}}\right]\right\} \leq$$

$$(4.019)\left\{\sum_{j=1}^{N}\frac{D\left(f^{(j)}(x), \widetilde{o}\right)}{j!}\left[\frac{1}{n^{\alpha j}} + \frac{(b-a)^j}{2\sqrt{\pi}\left(n^{1-\alpha}-2\right)e^{\left(n^{1-\alpha}-2\right)^2}}\right] + \right.$$

$$\left[\omega_1^{(\mathcal{F})}\left(f^{(N)}, \frac{1}{n^\alpha}\right)\frac{1}{n^{\alpha N} N!} + \frac{D^*\left(f^{(N)}, \widetilde{o}\right)(b-a)^N}{N!\sqrt{\pi}\left(n^{1-\alpha}-2\right)e^{\left(n^{1-\alpha}-2\right)^2}}\right]\Bigg\},$$

by Proposition 20.7, $\left\|\left(f^{(N)}\right)_{\pm}^{(r)}\right\|_\infty \leq D^*\left(f^{(N)}, \widetilde{o}\right)$ and (20.67).

The theorem is proved. ∎

Next we present

Theorem 20.57 *Let* $\alpha > 0$, $N = \lceil \alpha \rceil$, $\alpha \notin \mathbb{N}$, $f \in C_{\mathcal{F}}^N([a,b])$, $0 < \beta < 1$, $x \in [a,b]$, $n \in \mathbb{N}$, $n^{1-\beta} \geq 3$. *Then*

(i)

$$D\left(A_n^{\mathcal{F}}(f,x), f(x)\right) \leq$$

$$(4.019)\left\{\sum_{j=1}^{N-1}\frac{D\left(f^{(j)}(x), \widetilde{o}\right)}{j!}\left[\frac{1}{n^{\beta j}} + \frac{(b-a)^j}{2\sqrt{\pi}\left(n^{1-\beta}-2\right)e^{\left(n^{1-\beta}-2\right)^2}}\right] + \right.$$

$$\frac{1}{\Gamma(\alpha+1)}\left\{\frac{\left[\omega_1^{(\mathcal{F})}\left(\left(D_{x-}^{\alpha\mathcal{F}} f\right), \frac{1}{n^\beta}\right)_{[a,x]} + \omega_1^{(\mathcal{F})}\left(\left(D_{*x}^{\alpha\mathcal{F}} f\right), \frac{1}{n^\beta}\right)_{[x,b]}\right]}{n^{\alpha\beta}} + \right. \tag{20.77}$$

$$\frac{1}{2\sqrt{\pi}\left(n^{1-\beta}-2\right)e^{\left(n^{1-\beta}-2\right)^2}}\cdot$$

$$\left.\left.\left[D^*\left(\left(D_{x-}^{\alpha\mathcal{F}} f\right), \widetilde{o}\right)_{[a,x]}(x-a)^\alpha + D^*\left(\left(D_{*x}^{\alpha\mathcal{F}} f\right), \widetilde{o}\right)_{[x,b]}(b-x)^\alpha\right]\right\}\right\},$$

(ii) if $f^{(j)}(x_0) = 0$, $j = 1, \ldots, N-1$, *for some* $x_0 \in [a,b]$, *we have*

$$D\left(A_n^{\mathcal{F}}(f,x_0), f(x_0)\right) \leq$$

$$\frac{(4.019)}{\Gamma(\alpha+1)}\left\{\frac{\left[\omega_1^{(\mathcal{F})}\left(\left(D_{x_0-}^{\alpha\mathcal{F}} f\right), \frac{1}{n^\beta}\right)_{[a,x_0]} + \omega_1^{(\mathcal{F})}\left(\left(D_{*x_0}^{\alpha\mathcal{F}} f\right), \frac{1}{n^\beta}\right)_{[x_0,b]}\right]}{n^{\alpha\beta}} + \right. \tag{20.78}$$

$$\frac{1}{2\sqrt{\pi}\left(n^{1-\beta}-2\right)e^{\left(n^{1-\beta}-2\right)^2}}\cdot$$

$$\left.\left[D^*\left(\left(D_{x_0-}^{\alpha\mathcal{F}}f\right),\widetilde{o}\right)_{[a,x_0]}(x_0-a)^\alpha+D^*\left(\left(D_{*x_0}^{\alpha\mathcal{F}}f\right),\widetilde{o}\right)_{[x_0,b]}(b-x_0)^\alpha\right]\right\},$$

when $\alpha > 1$ *notice here the extremely high rate of convergence at* $n^{-(\alpha+1)\beta}$, *and*

(iii)

$$D^*\left(A_n^{\mathcal{F}}(f),f\right)\leq$$

$$(4.019)\left\{\sum_{j=1}^{N-1}\frac{D^*\left(f^{(j)},\widetilde{o}\right)}{j!}\left[\frac{1}{n^{\beta j}}+\frac{(b-a)^j}{2\sqrt{\pi}\left(n^{1-\beta}-2\right)e^{\left(n^{1-\beta}-2\right)^2}}\right]+\right.$$

$$\frac{1}{\Gamma(\alpha+1)}\left\{\frac{\left[\sup_{x\in[a,b]}\omega_1^{(\mathcal{F})}\left(\left(D_{x-}^{\alpha\mathcal{F}}f\right),\frac{1}{n^\beta}\right)_{[a,x]}+\sup_{x\in[a,b]}\omega_1^{(\mathcal{F})}\left(\left(D_{*x}^{\alpha\mathcal{F}}f\right),\frac{1}{n^\beta}\right)_{[x,b]}\right]}{n^{\alpha\beta}}+\right. \tag{20.79}$$

$$\frac{(b-a)^\alpha}{2\sqrt{\pi}\left(n^{1-\beta}-2\right)e^{\left(n^{1-\beta}-2\right)^2}}\cdot$$

$$\left.\left.\left[\sup_{x\in[a,b]}D^*\left(\left(D_{x-}^{\alpha\mathcal{F}}f\right),\widetilde{o}\right)_{[a,x]}+\sup_{x\in[a,b]}D^*\left(\left(D_{*x}^{\alpha\mathcal{F}}f\right),\widetilde{o}\right)_{[x,b]}\right]\right\}\right\},$$

above, when $N = 1$ *the sum* $\sum_{j=1}^{N-1}\cdot = 0$.

As we see here we obtain fractionally the fuzzy pointwise and uniform convergence with rates of $A_n^{\mathcal{F}} \to I$ *the unit operator, as* $n \to \infty$.

Proof Here $f_\pm^{(r)} \in C^N([a,b])$, $\forall\, r \in [0,1]$, and $D_{x-}^{\alpha\mathcal{F}}f$, $D_{*x}^{\alpha\mathcal{F}}f$ are fuzzy continuous over $[a,b]$, $\forall\, x \in [a,b]$, so that $\left(D_{x-}^{\alpha\mathcal{F}}f\right)_\pm^{(r)}$, $\left(D_{*x}^{\alpha\mathcal{F}}f\right)_\pm^{(r)} \in C([a,b])$, $\forall\, r \in [0,1]$, $\forall\, x \in [a,b]$.

We observe by (20.59), $\forall\, x \in [a,b]$, that (respectively in $\pm$)

$$\left|A_n\left(f_\pm^{(r)},x\right)-f_\pm^{(r)}(x)\right|\leq(4.019)\cdot$$

$$\left\{ \sum_{j=1}^{N-1} \frac{\left| \left(f_{\pm}^{(r)}\right)^{(j)}(x) \right|}{j!} \left\{ \frac{1}{n^{\beta j}} + \frac{(b-a)^j}{2\sqrt{\pi}\left(n^{1-\beta}-2\right) e^{\left(n^{1-\beta}-2\right)^2}} \right\} + \right. \tag{20.80}$$

$$\frac{1}{\Gamma(\alpha+1)} \left\{ \frac{\left(\omega_1 \left(D_{x-}^{\alpha}\left(f_{\pm}^{(r)}\right), \frac{1}{n^{\beta}} \right)_{[a,x]} + \omega_1 \left(D_{*x}^{\alpha}\left(f_{\pm}^{(r)}\right), \frac{1}{n^{\beta}} \right)_{[x,b]} \right)}{n^{\alpha\beta}} + \right.$$

$$\frac{1}{2\sqrt{\pi}\left(n^{1-\beta}-2\right) e^{\left(n^{1-\beta}-2\right)^2}} \cdot$$

$$\left. \left. \left(\left\| D_{x-}^{\alpha}\left(f_{\pm}^{(r)}\right) \right\|_{\infty,[a,x]} (x-a)^{\alpha} + \left\| D_{*x}^{\alpha}\left(f_{\pm}^{(r)}\right) \right\|_{\infty,[x,b]} (b-x)^{\alpha} \right) \right\} \right\} =$$

(by Remark 20.12, (20.30), (20.33))

$$\overset{(4.019)}{} \left\{ \sum_{j=1}^{N-1} \frac{\left| \left(f^{(j)}(x)\right)_{\pm}^{(r)} \right|}{j!} \left\{ \frac{1}{n^{\beta j}} + \frac{(b-a)^j}{2\sqrt{\pi}\left(n^{1-\beta}-2\right) e^{\left(n^{1-\beta}-2\right)^2}} \right\} + \right.$$

$$\frac{1}{\Gamma(\alpha+1)} \left\{ \frac{\left(\omega_1 \left(\left(D_{x-}^{\alpha\mathcal{F}} f\right)_{\pm}^{(r)}, \frac{1}{n^{\beta}} \right)_{[a,x]} + \omega_1 \left(\left(D_{*x}^{\alpha\mathcal{F}} f\right)_{\pm}^{(r)}, \frac{1}{n^{\beta}} \right)_{[x,b]} \right)}{n^{\alpha\beta}} + \right. \tag{20.81}$$

$$\frac{1}{2\sqrt{\pi}\left(n^{1-\beta}-2\right) e^{\left(n^{1-\beta}-2\right)^2}} \cdot$$

$$\left. \left. \left(\left\| \left(D_{x-}^{\alpha\mathcal{F}} f\right)_{\pm}^{(r)} \right\|_{\infty,[a,x]} (x-a)^{\alpha} + \left\| \left(D_{*x}^{\alpha\mathcal{F}} f\right)_{\pm}^{(r)} \right\|_{\infty,[x,b]} (b-x)^{\alpha} \right) \right\} \right\} \leq$$

$$(4.019)\left\{\sum_{j=1}^{N-1}\frac{D\left(f^{(j)}(x),\widetilde{o}\right)}{j!}\left\{\frac{1}{n^{\beta j}}+\frac{(b-a)^{j}}{2\sqrt{\pi}\left(n^{1-\beta}-2\right)e^{\left(n^{1-\beta}-2\right)^{2}}}\right\}+\right.$$

$$\frac{1}{\Gamma(\alpha+1)}\left\{\frac{\left[\omega_{1}^{(\mathcal{F})}\left(\left(D_{x-}^{\alpha\mathcal{F}}f\right),\frac{1}{n^{\beta}}\right)_{[a,x]}+\omega_{1}^{(\mathcal{F})}\left(\left(D_{*x}^{\alpha\mathcal{F}}f\right),\frac{1}{n^{\beta}}\right)_{[x,b]}\right]}{n^{\alpha\beta}}+\right. \quad (20.82)$$

$$\frac{1}{2\sqrt{\pi}\left(n^{1-\beta}-2\right)e^{\left(n^{1-\beta}-2\right)^{2}}}\cdot$$

$$\left.\left.\left[D^{*}\left(\left(D_{x-}^{\alpha\mathcal{F}}f\right),\widetilde{o}\right)_{[a,x]}(x-a)^{\alpha}+D^{*}\left(\left(D_{*x}^{\alpha\mathcal{F}}f\right),\widetilde{o}\right)_{[x,b]}(b-x)^{\alpha}\right]\right\}\right\},$$

along with (20.67) proving all inequalities of theorem.

Here we notice that

$$\left(D_{x-}^{\alpha\mathcal{F}}f\right)_{\pm}^{(r)}(t)=\left(D_{x-}^{\alpha}\left(f_{\pm}^{(r)}\right)\right)(t)$$

$$=\frac{(-1)^{N}}{\Gamma(N-\alpha)}\int_{t}^{x}(s-t)^{N-\alpha-1}\left(f_{\pm}^{(r)}\right)^{(N)}(s)\,ds,$$

where $a\leq t\leq x$.

Hence

$$\left|\left(D_{x-}^{\alpha\mathcal{F}}f\right)_{\pm}^{(r)}(t)\right|\leq\frac{1}{\Gamma(N-\alpha)}\int_{t}^{x}(s-t)^{N-\alpha-1}\left|\left(f_{\pm}^{(r)}\right)^{(N)}(s)\right|ds$$

$$\leq\frac{\left\|\left(f^{(N)}\right)_{\pm}^{(r)}\right\|_{\infty}}{\Gamma(N-\alpha+1)}(b-a)^{N-\alpha}\leq\frac{D^{*}\left(f^{(N)},\widetilde{o}\right)}{\Gamma(N-\alpha+1)}(b-a)^{N-\alpha}.$$

So we have

$$\left|\left(D_{x-}^{\alpha\mathcal{F}}f\right)_{\pm}^{(r)}(t)\right|\leq\frac{D^{*}\left(f^{(N)},\widetilde{o}\right)}{\Gamma(N-\alpha+1)}(b-a)^{N-\alpha},$$

all $a\leq t\leq x$.

And it holds

$$\left\| \left(D_{x-}^{\alpha\mathcal{F}} f \right)_{\pm}^{(r)} \right\|_{\infty,[a,x]} \leq \frac{D^* \left(f^{(N)}, \widetilde{o} \right)}{\Gamma(N-\alpha+1)} (b-a)^{N-\alpha}, \tag{20.83}$$

that is

$$D^* \left(\left(D_{x-}^{\alpha\mathcal{F}} f \right), \widetilde{o} \right)_{[a,x]} \leq \frac{D^* \left(f^{(N)}, \widetilde{o} \right)}{\Gamma(N-\alpha+1)} (b-a)^{N-\alpha},$$

and

$$\sup_{x\in[a,b]} D^* \left(\left(D_{x-}^{\alpha\mathcal{F}} f \right), \widetilde{o} \right)_{[a,x]} \leq \frac{D^* \left(f^{(N)}, \widetilde{o} \right)}{\Gamma(N-\alpha+1)} (b-a)^{N-\alpha} < \infty. \tag{20.84}$$

Similarly we have

$$\left(D_{*x}^{\alpha\mathcal{F}} f \right)_{\pm}^{(r)} (t) = \left(D_{*x}^{\alpha} \left(f_{\pm}^{(r)} \right) \right) (t)$$

$$= \frac{1}{\Gamma(N-\alpha)} \int_x^t (t-s)^{N-\alpha-1} \left(f_{\pm}^{(r)} \right)^{(N)} (s)\, ds,$$

where $x \leq t \leq b$.

Hence

$$\left| \left(D_{*x}^{\alpha\mathcal{F}} f \right)_{\pm}^{(r)} (t) \right| \leq \frac{1}{\Gamma(N-\alpha)} \int_x^t (t-s)^{N-\alpha-1} \left| \left(f^{(N)} \right)_{\pm}^{(r)} (s) \right| ds \leq$$

$$\frac{\left\| \left(f^{(N)} \right)_{\pm}^{(r)} \right\|_{\infty}}{\Gamma(N-\alpha+1)} (b-a)^{N-\alpha} \leq \frac{D^* \left(f^{(N)}, \widetilde{o} \right)}{\Gamma(N-\alpha+1)} (b-a)^{N-\alpha},$$

$x \leq t \leq b$.

So we have

$$\left\| \left(D_{*x}^{\alpha\mathcal{F}} f \right)_{\pm}^{(r)} \right\|_{\infty,[x,b]} \leq \frac{D^* \left(f^{(N)}, \widetilde{o} \right)}{\Gamma(N-\alpha+1)} (b-a)^{N-\alpha}, \tag{20.85}$$

that is

$$D^* \left(\left(D_{*x}^{\alpha\mathcal{F}} f \right), \widetilde{o} \right)_{[x,b]} \leq \frac{D^* \left(f^{(N)}, \widetilde{o} \right)}{\Gamma(N-\alpha+1)} (b-a)^{N-\alpha},$$

and

$$\sup_{x\in[a,b]} D^* \left(\left(D_{*x}^{\alpha\mathcal{F}} f \right), \widetilde{o} \right)_{[x,b]} \leq \frac{D^* \left(f^{(N)}, \widetilde{o} \right)}{\Gamma(N-\alpha+1)} (b-a)^{N-\alpha} < +\infty. \tag{20.86}$$

Furthermore we notice

$$\omega_1^{(\mathcal{F})}\left(\left(D_{x-}^{\alpha\mathcal{F}}f\right),\frac{1}{n^\beta}\right)_{[a,x]}=\sup_{\substack{s,t\in[a,x]\\|s-t|\leq\frac{1}{n^\beta}}}D\left(\left(D_{x-}^{\alpha\mathcal{F}}f\right)(s),\left(D_{x-}^{\alpha\mathcal{F}}f\right)(t)\right)\leq$$

$$\sup_{\substack{s,t\in[a,x]\\|s-t|\leq\frac{1}{n^\beta}}}\left\{D\left(\left(D_{x-}^{\alpha\mathcal{F}}f\right)(s),\widetilde{o}\right)+D\left(\left(D_{x-}^{\alpha\mathcal{F}}f\right)(t),\widetilde{o}\right)\right\}\leq 2D^*\left(\left(D_{x-}^{\alpha\mathcal{F}}f\right),\widetilde{o}\right)_{[a,x]}$$

$$\leq\frac{2D^*\left(f^{(N)},\widetilde{o}\right)}{\Gamma(N-\alpha+1)}(b-a)^{N-\alpha}.$$

Therefore it holds

$$\sup_{x\in[a,b]}\omega_1^{(\mathcal{F})}\left(\left(D_{x-}^{\alpha\mathcal{F}}f\right),\frac{1}{n^\beta}\right)_{[a,x]}\leq\frac{2D^*\left(f^{(N)},\widetilde{o}\right)}{\Gamma(N-\alpha+1)}(b-a)^{N-\alpha}<+\infty. \tag{20.87}$$

Similarly we observe

$$\omega_1^{(\mathcal{F})}\left(\left(D_{*x}^{\alpha\mathcal{F}}f\right),\frac{1}{n^\beta}\right)_{[x,b]}=\sup_{\substack{s,t\in[x,b]\\|s-t|\leq\frac{1}{n^\beta}}}D\left(\left(D_{*x}^{\alpha\mathcal{F}}f\right)(s),\left(D_{*x}^{\alpha\mathcal{F}}f\right)(t)\right)\leq$$

$$2D^*\left(\left(D_{*x}^{\alpha\mathcal{F}}f\right),\widetilde{o}\right)_{[x,b]}\leq\frac{2D^*\left(f^{(N)},\widetilde{o}\right)}{\Gamma(N-\alpha+1)}(b-a)^{N-\alpha}.$$

Consequently it holds

$$\sup_{x\in[a,b]}\omega_1^{(\mathcal{F})}\left(\left(D_{*x}^{\alpha\mathcal{F}}f\right),\frac{1}{n^\beta}\right)_{[x,b]}\leq\frac{2D^*\left(f^{(N)},\widetilde{o}\right)}{\Gamma(N-\alpha+1)}(b-a)^{N-\alpha}<+\infty. \tag{20.88}$$

So everything in the statements of the theorem makes sense.

The proof of the theorem is now completed. ■

Corollary 20.58 (To Theorem 20.57, $N=1$ case) *Let* $0<\alpha,\beta<1$, $f\in C_{\mathcal{F}}^1([a,b])$, $n\in\mathbb{N}$, $n^{1-\beta}\geq 3$. *Then*

$$D^*\left(A_n^{\mathcal{F}}(f),f\right)\leq$$

$$\frac{4.019}{\Gamma(\alpha+1)}\left\{\frac{\left[\sup_{x\in[a,b]}\omega_1^{(\mathcal{F})}\left(\left(D_{x-}^{\alpha\mathcal{F}}f\right),\frac{1}{n^\beta}\right)_{[a,x]}+\sup_{x\in[a,b]}\omega_1^{(\mathcal{F})}\left(\left(D_{*x}^{\alpha\mathcal{F}}f\right),\frac{1}{n^\beta}\right)_{[x,b]}\right]}{n^{\alpha\beta}}+\right.$$

$$\frac{(b-a)^\alpha}{2\sqrt{\pi}\left(n^{1-\beta}-2\right)e^{\left(n^{1-\beta}-2\right)^2}}\cdot \tag{20.89}$$

$$\left.\left[\sup_{x\in[a,b]}D^*\left(\left(D_{x-}^{\alpha\mathcal{F}}f\right),\widetilde{o}\right)_{[a,x]}+\sup_{x\in[a,b]}D^*\left(\left(D_{*x}^{\alpha\mathcal{F}}f\right),\widetilde{o}\right)_{[x,b]}\right]\right\}.$$

Proof By (20.79). ■

Finally we specialize to $\alpha=\frac{1}{2}$.

Corollary 20.59 (To Theorem 20.57) *Let* $0<\beta<1$, $f\in C_{\mathcal{F}}^1([a,b])$, $n\in\mathbb{N}$, $n^{1-\beta}\geq 3$. *Then*

$$D^*\left(A_n^{\mathcal{F}}(f),f\right)\leq$$

$$\frac{8.038}{\sqrt{\pi}}\left\{\frac{\left[\sup_{x\in[a,b]}\omega_1^{(\mathcal{F})}\left(\left(D_{x-}^{\frac{1}{2}\mathcal{F}}f\right),\frac{1}{n^\beta}\right)_{[a,x]}+\sup_{x\in[a,b]}\omega_1^{(\mathcal{F})}\left(\left(D_{*x}^{\frac{1}{2}\mathcal{F}}f\right),\frac{1}{n^\beta}\right)_{[x,b]}\right]}{n^{\frac{\beta}{2}}}+\right.$$

$$\frac{\sqrt{b-a}}{2\sqrt{\pi}\left(n^{1-\beta}-2\right)e^{\left(n^{1-\beta}-2\right)^2}}\cdot \tag{20.90}$$

$$\left.\left[\sup_{x\in[a,b]}D^*\left(\left(D_{x-}^{\frac{1}{2}\mathcal{F}}f\right),\widetilde{o}\right)_{[a,x]}+\sup_{x\in[a,b]}D^*\left(\left(D_{*x}^{\frac{1}{2}\mathcal{F}}f\right),\widetilde{o}\right)_{[x,b]}\right]\right\}.$$

Proof By (20.89). ■

Conclusion 20.60 *We have extended to the fuzzy setting all of Theorems 20.50, 20.51, 20.52, and Corollaries 20.53, 20.54.*

References

1. M. Abramowitz, I.A. Stegun (eds.), *Handbook of Mathematical Functions with Formulas, Graphs, and Mathematical Tables* (Dover Publications, New York, 1972)
2. G.A. Anastassiou, Rate of convergence of some neural network operators to the unit-univariate case. J. Math. Anal. Appl. **212**, 237–262 (1997)
3. G.A. Anastassiou, *Quantitative Approximation* (Chapmann and Hall/CRC, Boca Raton, 2001)
4. G.A. Anastassiou, Fuzzy Approximation by Fuzzy Convolution type Operators. Computers and Mathematics **48**, 1369–1386 (2004)
5. G.A. Anastassiou, Higher order Fuzzy Korovkin theory via inequalities. Commun. Appl. Anal. **10**(2), 359–392 (2006)
6. G.A. Anastassiou, Fuzzy Korovkin theorems and inequalities. J. Fuzzy Math. **15**(1), 169–205 (2007)
7. G.A. Anastassiou, On right fractional calculus. Chaos, Solitons Fractals **42**, 365–376 (2009)
8. G.A. Anastassiou, *Fractional Differentiation Inequalities* (Springer, New York, 2009)
9. G.A. Anastassiou, Fractional Korovkin theory. Chaos, Solitons Fractals **42**(4), 2080–2094 (2009)
10. G.A. Anastassiou, *Fuzzy Mathematics: Approximation Theory* (Springer, Heildelberg, 2010)
11. G.A. Anastassiou, Quantitative approximation by fractional smooth Picard singular operators. Math. Eng. Sci. Aerosp. **2**(1), 71–87 (2011)
12. G.A. Anastassiou, Fuzzy fractional Calculus and Ostrowski inequality. J. Fuzzy Math. **19**(3), 577–590 (2011)
13. G.A. Anastassiou, *Intelligent Systems: Approximation by Artificial Neural Networks*, Intelligent Systems Reference Library, vol. 19 (Springer, Heidelberg, 2011)
14. G.A. Anastassiou, Fractional representation formulae and right fractional inequalities. Math. Comput. Model. **54**(11–12), 3098–3115 (2011)
15. G.A. Anastassiou, Univariate hyperbolic tangent neural network approximation. Math. Comput. Model. **53**, 1111–1132 (2011)
16. G.A. Anastassiou, Multivariate hyperbolic tangent neural network approximation. Comput. Math. **61**, 809–821 (2011)
17. G.A. Anastassiou, Multivariate sigmoidal neural network approximation. Neural Netw. **24**, 378–386 (2011)
18. G.A. Anastassiou, Univariate sigmoidal neural network approximation. J. Comput. Anal. Appl. **14**(4), 659–690 (2012)
19. G.A. Anastassiou, Fractional neural network approximation. Comput. Math. Appl. **64**(6), 1655–1676 (2012)
20. G.A. Anastassiou, Rate of convergence of some neural network operators to the unit-univariate case, revisited. Matematicki Vesnik **65**(4), 511–518 (2013)
21. G.A. Anastassiou, Fractional approximation by normalized Bell and Squashing type neural network operators. New Math. Nat. Comput. **9**(1), 43–63 (2013)
22. G.A. Anastassiou, Fuzzy fractional neural network approximation by fuzzy quasi-interpolation operators. J. Appl. Nonlinear Dyn. **2**(3), 235–259 (2013)
23. G.A. Anastassiou, Univariate error function based neural network approximation, Indian J. Math. Accepted 2014
24. G.A. Anastassiou, Fuzzy fractional error function based neural network approximation. Submitted 2014
25. L.C. Andrews, *Special Functions of Mathematics for Engineers*, 2nd edn. (Mc Graw-Hill, New York, 1992)
26. P. Cardaliaguet, G. Euvrard, Approximation of a function and its derivative with a neural network. Neural Netw. **5**, 207–220 (1992)
27. Z. Chen, F. Cao, The approximation operators with sigmoidal functions. Comput. Math. Appl. **58**, 758–765 (2009)
28. K. Diethelm, *The Analysis of Fractional Differential Equations*, Lecture Notes in Mathematics, 2004 (Springer, Berlin, 2010)

29. A.M.A. El-Sayed, M. Gaber, On the finite Caputo and finite Riesz derivatives. Electron. J. Theor. Phys. **3**(12), 81–95 (2006)
30. G.S. Frederico, D.F.M. Torres, Fractional optimal control in the sense of Caputo and the fractional Noether's theorem. Int. Math. Forum **3**(10), 479–493 (2008)
31. S. Gal, in *Approximation Theory in Fuzzy Setting, Chapter 13*, Handbook of Analytic-Computational Methods in Applied Mathematics, ed. by G. Anastassiou (Chapman & Hall/CRC, Boca Raton, 2000), pp. 617–666
32. R. Goetschel Jr, W. Voxman, Elementary fuzzy calculus. Fuzzy Sets Syst. **18**, 31–43 (1986)
33. S. Haykin, *Neural Networks: A Comprehensive Foundation*, 2nd edn. (Prentice Hall, New York, 1998)
34. O. Kaleva, Fuzzy differential equations. Fuzzy Sets Syst. **24**, 301–317 (1987)
35. Y.K. Kim, B.M. Ghil, Integrals of fuzzy-number-valued functions. Fuzzy Sets Syst. **86**, 213–222 (1997)
36. W. McCulloch, W. Pitts, A logical calculus of the ideas immanent in nervous activity. Bull. Math. Biophys. **7**, 115–133 (1943)
37. T.M. Mitchell, *Machine Learning* (WCB-McGraw-Hill, New York, 1997)
38. S.G. Samko, A.A. Kilbas, O.I. Marichev, *Fractional Integrals and Derivatives, Theory and Applications*, (Gordon and Breach, Amsterdam, 1993) [English translation from the Russian, Integrals and Derivatives of Fractional Order and Some of Their Applications (Nauka i Tekhnika, Minsk, 1987)]
39. Wu Congxin, Gong Zengtai, On Henstock integrals of interval-valued functions and fuzzy valued functions. Fuzzy Sets Syst. **115**(3), 377–391 (2000)
40. C. Wu, Z. Gong, On Henstock integral of fuzzy-number-valued functions (I). Fuzzy Sets Syst. **120**(3), 523–532 (2001)
41. C. Wu, M. Ma, On embedding problem of fuzzy numer spaces: part 1. Fuzzy Sets Syst. **44**, 33–38 (1991)

Chapter 21
High Degree Multivariate Fuzzy Approximation by Neural Network Operators Using the Error Function

Here we study in terms of multivariate fuzzy high approximation to the multivariate unit sequences of multivariate fuzzy error function based neural network operators. These operators are multivariate fuzzy analogs of earlier considered multivariate real ones. The derived results generalize earlier real ones into the fuzzy level. Here the high order multivariate fuzzy pointwise and uniform convergences with rates to the multivariate fuzzy unit operator are given through multivariate fuzzy Jackson type inequalities involving the multivariate fuzzy moduli of continuity of the mth order ($m \geq 0$) H-fuzzy partial derivatives, of the involved multivariate fuzzy number valued function. The treated operators are of quasi-interpolation, Kantorovich and quadrature types at the multivariate fuzzy setting. It follows [19].

21.1 Introduction

The author in [2, 3] and [4], see Chaps. 2–5, was the first to derive neural network approximations to continuous functions with rates by very specifically defined neural network operators of Cardaliaguet-Euvrard and "Squashing" types, by employing the modulus of continuity of the engaged function or its high order derivative, and deriving very tight Jackson type inequalities. He treats there both the univariate and multivariate cases. The defining these operators "bell-shaped" and "squashing" function are assumed to be of compact support.

The author motivated by [22], continued his studies on neural networks approximation by introducing and using the proper quasi-interpolation operators of sigmoidal and hyperbolic tangent type which resulted into [9–13, 16], by treating both the univariate and multivariate cases.

Continuation of the author's works [14, 15, 17, 18] is this chapter where multivariate fuzzy neural network approximation based on error function among others results into higher rates of approximation. We involve the fuzzy partial derivatives of the multivariate fuzzy function under approximation or itself, and we establish

G.A. Anastassiou, *Intelligent Systems II: Complete Approximation by Neural Network Operators*, Studies in Computational Intelligence 608,
DOI 10.1007/978-3-319-20505-2_21

tight multivariate fuzzy Jackson type inequalities. An extensive background is given on fuzzy multivariate analysis and real neural network approximation, all needed to present our results.

Our fuzzy multivariate feed-forward neural networks (FFNNs) are with one hidden layer. For neural networks in general you may read [25, 28, 29].

21.2 Fuzzy Real Analysis Basics

We need the following background

Definition 21.1 (*see* [30]) Let $\mu : \mathbb{R} \to [0, 1]$ with the following properties

(i) is normal, i.e., $\exists\, x_0 \in \mathbb{R}$; $\mu(x_0) = 1$.
(ii) $\mu(\lambda x + (1-\lambda) y) \geq \min\{\mu(x), \mu(y)\}$, $\forall\, x, y \in \mathbb{R}$, $\forall\, \lambda \in [0, 1]$ (μ is called a convex fuzzy subset).
(iii) μ is upper semicontinuous on $\mathbb{R}$, i.e. $\forall\, x_0 \in \mathbb{R}$ and $\forall\, \varepsilon > 0$, $\exists$ neighborhood $V(x_0) : \mu(x) \leq \mu(x_0) + \varepsilon$, $\forall\, x \in V(x_0)$.
(iv) The set $\overline{\sup p(\mu)}$ is compact in $\mathbb{R}$, (where $\sup p(\mu) := \{x \in \mathbb{R} : \mu(x) > 0\}$).

We call μ a fuzzy real number. Denote the set of all μ with $\mathbb{R}_{\mathcal{F}}$.

E.g. $\chi_{\{x_0\}} \in \mathbb{R}_{\mathcal{F}}$, for any $x_0 \in \mathbb{R}$, where $\chi_{\{x_0\}}$ is the characteristic function at x_0.

For $0 < r \leq 1$ and $\mu \in \mathbb{R}_{\mathcal{F}}$ define

$$[\mu]^r := \{x \in \mathbb{R} : \mu(x) \geq r\} \tag{21.1}$$

and

$$[\mu]^0 := \overline{\{x \in \mathbb{R} : \mu(x) \geq 0\}}.$$

Then it is well known that for each $r \in [0, 1]$, $[\mu]^r$ is a closed and bounded interval on $\mathbb{R}$ [24].

For $u, v \in \mathbb{R}_{\mathcal{F}}$ and $\lambda \in \mathbb{R}$, we define uniquely the sum $u \oplus v$ and the product $\lambda \odot u$ by

$$[u \oplus v]^r = [u]^r + [v]^r, \quad [\lambda \odot u]^r = \lambda [u]^r, \quad \forall\, r \in [0, 1],$$

where $[u]^r + [v]^r$ means the usual addition of two intervals (as substes of $\mathbb{R}$) and $\lambda [u]^r$ means the usual product between a scalar and a subset of $\mathbb{R}$ (see, e.g. [30]).

Notice $1 \odot u = u$ and it holds

$$u \oplus v = v \oplus u, \lambda \odot u = u \odot \lambda.$$

If $0 \leq r_1 \leq r_2 \leq 1$ then

$$[u]^{r_2} \subseteq [u]^{r_1}.$$

Actually $[u]^r = \left[u_-^{(r)}, u_+^{(r)}\right]$, where $u_-^{(r)} \leq u_+^{(r)}$, $u_-^{(r)}, u_+^{(r)} \in \mathbb{R}$, $\forall\, r \in [0,1]$.

For $\lambda > 0$ one has $\lambda u_\pm^{(r)} = (\lambda \odot u)_\pm^{(r)}$, respectively.

Define $D : \mathbb{R}_{\mathcal{F}} \times \mathbb{R}_{\mathcal{F}} \to \mathbb{R}_{\mathcal{F}}$ by

$$D(u,v) := \sup_{r\in[0,1]} \max\left\{\left|u_-^{(r)} - v_-^{(r)}\right|, \left|u_+^{(r)} - v_+^{(r)}\right|\right\}, \tag{21.2}$$

where

$$[v]^r = \left[v_-^{(r)}, v_+^{(r)}\right];\; u, v \in \mathbb{R}_{\mathcal{F}}.$$

We have that D is a metric on $\mathbb{R}_{\mathcal{F}}$.

Then $(\mathbb{R}_{\mathcal{F}}, D)$ is a complete metric space, see [30, 31].

Let $f, g : W \subseteq \mathbb{R}^m \to \mathbb{R}_{\mathcal{F}}$. We define the distance

$$D^*(f,g) = \sup_{x\in W} D(f(x), g(x)).$$

Remark 21.2 We determine and use

$$D^*(f,\widetilde{o}) = \sup_{x\in W} D(f(x),\widetilde{o}) =$$

$$\sup_{x\in W}\sup_{r\in[0,1]} \max\left\{\left|f_-^{(r)}(x)\right|, \left|f_+^{(r)}(x)\right|\right\}.$$

By the principle of iterated suprema we find that

$$D^*(f,\widetilde{o}) = \sup_{r\in[0,1]} \max\left\{\left\|f_-^{(r)}\right\|_\infty, \left\|f_+^{(r)}\right\|_\infty\right\}, \tag{21.3}$$

under the assumption $D^*(f,\widetilde{o}) < \infty$, that is f is a fuzzy bounded function.

Above $\|\cdot\|_\infty$ is the supremum norm of the function over $W \subseteq \mathbb{R}^m$.

Here Σ^* stands for fuzzy summation and $\widetilde{0} := \chi_{\{0\}} \in \mathbb{R}_{\mathcal{F}}$ is the neutral element with respect to $\oplus$, i.e.,

$$u \oplus \widetilde{0} = \widetilde{0} \oplus u = u,\; \forall\, u \in \mathbb{R}_{\mathcal{F}}.$$

We need

Remark 21.3 ([6]) Here $r \in [0,1]$, $x_i^{(r)}, y_i^{(r)} \in \mathbb{R}$, $i = 1, \ldots, m \in \mathbb{N}$. Suppose that

$$\sup_{r\in[0,1]} \max\left(x_i^{(r)}, y_i^{(r)}\right) \in \mathbb{R}, \text{ for } i = 1, \ldots, m.$$

Then one sees easily that

$$\sup_{r\in[0,1]} \max\left(\sum_{i=1}^{m} x_i^{(r)}, \sum_{i=1}^{m} y_i^{(r)}\right) \leq \sum_{i=1}^{m} \sup_{r\in[0,1]} \max\left(x_i^{(r)}, y_i^{(r)}\right). \tag{21.4}$$

Definition 21.4 Let $f \in C(W)$, $W \subseteq \mathbb{R}^m$, $m \in \mathbb{N}$, which is bounded or uniformly continuous, we define ($h > 0$)

$$\omega_1(f,h) := \sup_{x,y\in W,\ \|x-y\|_\infty \leq h} |f(x) - f(y)|, \tag{21.5}$$

where $x = (x_1, \ldots, x_m)$, $y = (y_1, \ldots, y_m)$.

Definition 21.5 Let $f : W \to \mathbb{R}_{\mathcal{F}}$, $W \subseteq \mathbb{R}^m$, we define the fuzzy modulus of continuity of f by

$$\omega_1^{(\mathcal{F})}(f,h) = \sup_{x,y\in W,\ \|x-y\|_\infty \leq h} D(f(x), f(y)),\ h > 0. \tag{21.6}$$

where $x = (x_1, \ldots, x_m)$, $y = (y_1, \ldots, y_m)$.

For $f : W \to \mathbb{R}_{\mathcal{F}}$, $W \subseteq \mathbb{R}^m$, we use

$$[f]^r = \left[f_-^{(r)}, f_+^{(r)}\right], \tag{21.7}$$

where $f_\pm^{(r)} : W \to \mathbb{R}$, $\forall\, r \in [0,1]$.

We need

Proposition 21.6 ([6]) *Let $f : W \to \mathbb{R}_{\mathcal{F}}$. Assume that $\omega_1^{\mathcal{F}}(f,\delta)$, $\omega_1\left(f_-^{(r)}, \delta\right)$, $\omega_1\left(f_+^{(r)}, \delta\right)$ are finite for any $\delta > 0$, $r \in [0,1]$.*

Then

$$\omega_1^{(\mathcal{F})}(f,\delta) = \sup_{r\in[0,1]} \max\left\{\omega_1\left(f_-^{(r)}, \delta\right), \omega_1\left(f_+^{(r)}, \delta\right)\right\}. \tag{21.8}$$

We denote by $C_{\mathcal{F}}^U(W)$ the space of fuzzy uniformly continuous functions from $W \to \mathbb{R}_{\mathcal{F}}$, also $C_{\mathcal{F}}(W)$ is the space of fuzzy continuous functions on $W \subseteq \mathbb{R}^m$, and $C_B(W, \mathbb{R}_{\mathcal{F}})$ is the fuzzy continuous and bounded functions.

We mention

Proposition 21.7 ([8]) *Let $f \in C_{\mathcal{F}}^U(W)$, where $W \subseteq \mathbb{R}^m$ is convex. Then $\omega_1^{(\mathcal{F})}(f,\delta) < \infty$, for any $\delta > 0$.*

Proposition 21.8 ([8]) *It holds*

$$\lim_{\delta \to 0} \omega_1^{(\mathcal{F})}(f, \delta) = \omega_1^{(\mathcal{F})}(f, 0) = 0, \tag{21.9}$$

iff $f \in C_{\mathcal{F}}^U(W)$, $W \subseteq \mathbb{R}^m$, *where* W *is open or compact.*

Proposition 21.9 ([8]) *Let* $f \in C_{\mathcal{F}}(W)$, $W \subseteq \mathbb{R}^m$ *open or compact. Then* $f_{\pm}^{(r)}$ *are equicontinuous with respect to* $r \in [0, 1]$ *over* W, *respectively in* $\pm$.

Note 21.10 It is clear by Propositions 21.6, 21.8, that if $f \in C_{\mathcal{F}}^U(W)$, then $f_{\pm}^{(r)} \in C_U(W)$ (uniformly continuous on W). Also if $f \in C_B(W, \mathbb{R}_{\mathcal{F}})$, it implies by (21.3) that $f_{\pm}^{(r)} \in C_B(W)$ (continuous and bounded functions on W).

We need

Definition 21.11 Let $x, y \in \mathbb{R}_{\mathcal{F}}$. If there exists $z \in \mathbb{R}_{\mathcal{F}} : x = y \oplus z$, then we call z the H-difference on x and y, denoted $x - y$.

Definition 21.12 ([30]) Let $T := [x_0, x_0 + \beta] \subset \mathbb{R}$, with $\beta > 0$. A function $f : T \to \mathbb{R}_{\mathcal{F}}$ is H-difference at $x \in T$ if there exists an $f'(x) \in \mathbb{R}_{\mathcal{F}}$ such that the limits (with respect to D)

$$\lim_{h \to 0+} \frac{f(x+h) - f(x)}{h}, \quad \lim_{h \to 0+} \frac{f(x) - f(x-h)}{h} \tag{21.10}$$

exist and are equal to $f'(x)$.

We call f' the H-derivative or fuzzy derivative of f at x.

Above is assumed that the H-differences $f(x+h) - f(x)$, $f(x) - f(x-h)$ exists in $\mathbb{R}_{\mathcal{F}}$ in a neighborhood of x.

Definition 21.13 We denote by $C_{\mathcal{F}}^{N^*}(W)$, $N^* \in \mathbb{N}$, the space of all N^*-times fuzzy continuously differentiable functions from W into $\mathbb{R}_{\mathcal{F}}$, $W \subseteq \mathbb{R}^m$ open or compact which is convex.

Here fuzzy partial derivatives are defined via Definition 21.12 in the obvious way as in the ordinary real case.

We mention

Theorem 21.14 ([26]) *Let* $f : [a, b] \subseteq \mathbb{R} \to \mathbb{R}_{\mathcal{F}}$ *be* H*-fuzzy differentiable. Let* $t \in [a, b]$, $0 \leq r \leq 1$. *Clearly*

$$[f(t)]^r = \left[f(t)_-^{(r)}, f(t)_+^{(r)} \right] \subseteq \mathbb{R}.$$

Then $(f(t))_{\pm}^{(r)}$ *are differentiable and*

$$\left[f'(t)\right]^r = \left[\left(f(t)_-^{(r)}\right)', \left(f(t)_+^{(r)}\right)'\right].$$

I.e.

$$\left(f'\right)_{\pm}^{(r)} = \left(f_{\pm}^{(r)}\right)', \quad \forall\, r \in [0,1]. \tag{21.11}$$

Remark 21.15 (see also [7]) Let $f \in C^{N^*}([a,b], \mathbb{R}_{\mathcal{F}})$, $N^* \geq 1$. Then by Theorem 21.14 we obtain $f_{\pm}^{(r)} \in C^{N^*}([a,b])$ and

$$\left[f^{(i)}(t)\right]^r = \left[\left(f(t)_-^{(r)}\right)^{(i)}, \left(f(t)_+^{(r)}\right)^{(i)}\right],$$

for $i = 0, 1, 2, \ldots, N^*$, and in particular we have

$$\left(f^{(i)}\right)_{\pm}^{(r)} = \left(f_{\pm}^{(r)}\right)^{(i)}, \tag{21.12}$$

for any $r \in [0,1]$.

Let $f \in C_{\mathcal{F}}^{N^*}(W)$, $W \subseteq \mathbb{R}^m$, open or compact, which is convex, denote $f_{\widetilde{\alpha}} := \frac{\partial^{\widetilde{\alpha}} f}{\partial x^{\widetilde{\alpha}}}$, where $\widetilde{\alpha} := (\widetilde{\alpha_1}, \ldots, \widetilde{\alpha_m})$, $\widetilde{\alpha_i} \in \mathbb{Z}^+$, $i = 1, \ldots, m$ and

$$0 < |\widetilde{\alpha}| := \sum_{i=1}^{m} \widetilde{\alpha_i} \leq N^*, \quad N^* > 1.$$

Then by Theorem 21.14 we get that

$$\left(f_{\pm}^{(r)}\right)_{\widetilde{\alpha}} = (f_{\widetilde{\alpha}})_{\pm}^{(r)}, \quad \forall\, r \in [0,1], \tag{21.13}$$

and any $\widetilde{\alpha} : |\widetilde{\alpha}| \leq N^*$. Here $f_{\pm}^{(r)} \in C^{N^*}(W)$.

Notation 21.16 *We denote*

$$\left(\sum_{i=1}^{2} D\left(\frac{\partial}{\partial x_i}, \widetilde{o}\right)\right)^2 f(\overrightarrow{x}) := \tag{21.14}$$

$$D\left(\frac{\partial^2 f(x_1,x_2)}{\partial x_1^2}, \widetilde{o}\right) + D\left(\frac{\partial^2 f(x_1,x_2)}{\partial x_2^2}, \widetilde{o}\right) + 2D\left(\frac{\partial^2 f(x_1,x_2)}{\partial x_1 \partial x_2}, \widetilde{o}\right).$$

In general we denote ($j = 1, \ldots, N^*$)

$$\left(\sum_{i=1}^{m} D\left(\frac{\partial}{\partial x_i}, \tilde{o} \right) \right)^j f(\vec{x}) := \tag{21.15}$$

$$\sum_{(j_1,\ldots,j_m)\in\mathbb{Z}_+^m : \sum_{i=1}^m j_i = j} \frac{j!}{j_1! j_2! \ldots j_m!} D\left(\frac{\partial^j f(x_1,\ldots,x_m)}{\partial x_1^{j_1} \partial x_2^{j_2} \ldots \partial x_m^{j_m}}, \tilde{o} \right).$$

We mention also a particular case of the Fuzzy Henstock integral ($\delta(x) = \frac{\delta}{2}$), see [30].

Definition 21.17 ([23], p. 644) Let $f : [a, b] \to \mathbb{R}_{\mathcal{F}}$. We say that f is Fuzzy-Riemann integrable to $I \in \mathbb{R}_{\mathcal{F}}$ if for any $\varepsilon > 0$, there exists $\delta > 0$ such that for any division $P = \{[u, v]; \xi\}$ of $[a, b]$ with the norms $\Delta(P) < \delta$, we have

$$D\left(\sum_{P}^{*} (v - u) \odot f(\xi), I \right) < \varepsilon.$$

We write

$$I := (FR) \int_a^b f(x)\,dx. \tag{21.16}$$

We mention

Theorem 21.18 ([24]) *Let* $f : [a, b] \to \mathbb{R}_{\mathcal{F}}$ *be fuzzy continuous. Then*

$$(FR) \int_a^b f(x)\,dx$$

exists and belongs to $\mathbb{R}_{\mathcal{F}}$, *furthermore it holds*

$$\left[(FR) \int_a^b f(x)\,dx \right]^r = \left[\int_a^b (f)_-^{(r)}(x)\,dx, \int_a^b (f)_+^{(r)}(x)\,dx \right],$$

$\forall r \in [0, 1]$.

For the definition of general fuzzy integral we follow [27] next.

Definition 21.19 Let (Ω, Σ, μ) be a complete σ-finite measure space. We call $F : \Omega \to R_{\mathcal{F}}$ measurable iff $\forall$ closed $B \subseteq \mathbb{R}$ the function $F^{-1}(B) : \Omega \to [0, 1]$ defined by

$$F^{-1}(B)(w) := \sup_{x \in B} F(w)(x), \text{ all } w \in \Omega$$

is measurable, see [27].

Theorem 21.20 ([27]) *For* $F : \Omega \to \mathbb{R}_{\mathcal{F}}$,

$$F(w) = \left\{ \left(F_{-}^{(r)}(w), F_{+}^{(r)}(w) \right) | 0 \le r \le 1 \right\},$$

the following are equivalent

(1) F is measurable,
(2) $\forall\, r \in [0,1]$, $F_{-}^{(r)}$, $F_{+}^{(r)}$ *are measurable.*

Following [27], given that for each $r \in [0,1]$, $F_{-}^{(r)}$, $F_{+}^{(r)}$ are integrable we have that the parametrized representation

$$\left\{ \left(\int_A F_{-}^{(r)} d\mu, \int_A F_{+}^{(r)} d\mu \right) | 0 \le r \le 1 \right\} \tag{21.17}$$

is a fuzzy real number for each $A \in \Sigma$.

The last fact leads to

Definition 21.21 ([27]) A measurable function $F : \Omega \to \mathbb{R}_{\mathcal{F}}$,

$$F(w) = \left\{ \left(F_{-}^{(r)}(w), F_{+}^{(r)}(w) \right) | 0 \le r \le 1 \right\}$$

is integrable if for each $r \in [0,1]$, $F_{\pm}^{(r)}$ are integrable, or equivalently, if $F_{\pm}^{(0)}$ are integrable.

In this case, the fuzzy integral of F over $A \in \Sigma$ is defined by

$$\int_A F d\mu := \left\{ \left(\int_A F_{-}^{(r)} d\mu, \int_A F_{+}^{(r)} d\mu \right) | 0 \le r \le 1 \right\}. \tag{21.18}$$

By [27], F is integrable iff $w \to \|F(w)\|_{\mathcal{F}}$ is real-valued integrable.

Here denote

$$\|u\|_{\mathcal{F}} := D\left(u, \widetilde{0}\right), \ \forall\, u \in \mathbb{R}_{\mathcal{F}}.$$

We need also

Theorem 21.22 ([27]) *Let* $F, G : \Omega \to \mathbb{R}_{\mathcal{F}}$ *be integrable. Then*

(1) Let $a, b \in \mathbb{R}$, *then* $aF + bG$ *is integrable and for each* $A \in \Sigma$,

$$\int_A (aF + bG)\, d\mu = a \int_A F d\mu + b \int_A G d\mu;$$

(2) $D(F, G)$ *is a real- valued integrable function and for each* $A \in \Sigma$,

$$D\left(\int_A F d\mu, \int_A G d\mu \right) \le \int_A D(F, G)\, d\mu.$$

In particular,

$$\left\| \int_A F d\mu \right\|_{\mathcal{F}} \leq \int_A \|F\|_{\mathcal{F}} d\mu.$$

Above μ could be the multivariate Lebesgue measure, which we use in this chapter, with all the basic properties valid here too. Notice by [27], Fubini's theorem is valid for fuzzy integral (21.18).

Basically here we have that

$$\left[\int_A F d\mu\right]^r = \left[\int_A F_-^{(r)} d\mu, \int_A F_+^{(r)} d\mu\right], \tag{21.19}$$

i.e.

$$\left(\int_A F d\mu\right)_{\pm}^{(r)} = \int_A F_{\pm}^{(r)} d\mu, \ \forall \, r \in [0,1]. \tag{21.20}$$

21.3 Basics on Multivariate Real Neural Network Operators Approximation Relied on the Error Function (see [18])

We consider here the (Gauss) error special function [1, 20]

$$\text{erf}\,(x) = \frac{2}{\sqrt{\pi}} \int_0^x e^{-t^2} dt, \quad x \in \mathbb{R}, \tag{21.21}$$

which is a sigmoidal type function and a strictly increasing function.

It has the basic properties

$$\text{erf}\,(0) = 0, \ \text{erf}\,(-x) = -\,\text{erf}\,(x), \quad \text{erf}\,(+\infty) = 1, \quad \text{erf}\,(-\infty) = -1.$$

We consider also the activation function [18]

$$\chi\,(x) = \frac{1}{4}\left(\text{erf}\,(x+1) - \text{erf}\,(x-1)\right), \quad \text{any } x \in \mathbb{R}, \tag{21.22}$$

which is an even function.

Next we follow [18] on χ. We got there $\chi\,(0) \simeq 0.4215$, and that χ is strictly decreasing on $[0, \infty)$ and strictly increasing on $(-\infty, 0]$, and the x-axis is the horizontal asymptote on χ, i.e. χ is a bell symmetric function.

Theorem 21.23 ([18]) *We have that*

$$\sum_{i=-\infty}^{\infty} \chi(x-i) = 1, \text{ all } x \in \mathbb{R}, \tag{21.23}$$

$$\sum_{i=-\infty}^{\infty} \chi(nx-i) = 1, \text{ all } x \in \mathbb{R},\ n \in \mathbb{N}, \tag{21.24}$$

and

$$\int_{-\infty}^{\infty} \chi(x)\, dx = 1, \tag{21.25}$$

that is $\chi(x)$ is a density function on $\mathbb{R}$.

We need the important

Theorem 21.24 ([18]) *Let $0 < \alpha < 1$, and $n \in \mathbb{N}$ with $n^{1-\alpha} \geq 3$. It holds*

$$\sum_{\substack{k=-\infty \\ : |nx-k| \geq n^{1-\alpha}}}^{\infty} \chi(nx-k) < \frac{1}{2\sqrt{\pi}\left(n^{1-\alpha}-2\right) e^{\left(n^{1-\alpha}-2\right)^2}}. \tag{21.26}$$

Denote by $\lfloor \cdot \rfloor$ the integral part of the number and by $\lceil \cdot \rceil$ the ceiling of the number.

Theorem 21.25 ([18]) *Let $x \in [a,b] \subset \mathbb{R}$ and $n \in \mathbb{N}$ so that $\lceil na \rceil \leq \lfloor nb \rfloor$. It holds*

$$\frac{1}{\sum_{k=\lceil na \rceil}^{\lfloor nb \rfloor} \chi(nx-k)} < \frac{1}{\chi(1)} \simeq 4.019, \quad \forall\, x \in [a,b]. \tag{21.27}$$

Also from [18] we get

$$\lim_{n\to\infty} \sum_{k=\lceil na \rceil}^{\lfloor nb \rfloor} \chi(nx-k) \neq 1, \tag{21.28}$$

at least for some $x \in [a,b]$.

For large enough n we always obtain $\lceil na \rceil \leq \lfloor nb \rfloor$. Also $a \leq \frac{k}{n} \leq b$, iff $\lceil na \rceil \leq k \leq \lfloor nb \rfloor$. In general it holds by (21.24) that

$$\sum_{k=\lceil na \rceil}^{\lfloor nb \rfloor} \chi(nx-k) \leq 1. \tag{21.29}$$

We introduce (see [18])

$$Z(x_1, \ldots, x_N) := Z(x) := \prod_{i=1}^{N} \chi(x_i), \quad x = (x_1, \ldots, x_N) \in \mathbb{R}^N, \quad N \in \mathbb{N}. \tag{21.30}$$

It has the properties:

(i) $Z(x) > 0, \ \forall\, x \in \mathbb{R}^N$,

(ii)

$$\sum_{k=-\infty}^{\infty} Z(x-k) := \sum_{k_1=-\infty}^{\infty} \sum_{k_2=-\infty}^{\infty} \cdots \sum_{k_N=-\infty}^{\infty} Z(x_1 - k_1, \ldots, x_N - k_N) = 1, \tag{21.31}$$

where $k := (k_1, \ldots, k_n) \in \mathbb{Z}^N, \forall\, x \in \mathbb{R}^N$,
hence

(iii)

$$\sum_{k=-\infty}^{\infty} Z(nx-k) :=$$

$$\sum_{k_1=-\infty}^{\infty} \sum_{k_2=-\infty}^{\infty} \cdots \sum_{k_N=-\infty}^{\infty} Z(nx_1 - k_1, \ldots, nx_N - k_N) = 1, \tag{21.32}$$

$\forall\, x \in \mathbb{R}^N; \ n \in \mathbb{N}$,
and

(iv)

$$\int_{\mathbb{R}^N} Z(x)\, dx = 1, \tag{21.33}$$

that is Z is a multivariate density function.

Here $\|x\|_\infty := \max\{|x_1|, \ldots, |x_N|\}$, $x \in \mathbb{R}^N$, also set $\infty := (\infty, \ldots, \infty)$, $-\infty := (-\infty, \ldots, -\infty)$ upon the multivariate context, and

$$\begin{aligned} \lceil na \rceil &:= (\lceil na_1 \rceil, \ldots, \lceil na_N \rceil), \\ \lfloor nb \rfloor &:= (\lfloor nb_1 \rfloor, \ldots, \lfloor nb_N \rfloor), \end{aligned} \tag{21.34}$$

where $a := (a_1, \ldots, a_N)$, $b := (b_1, \ldots, b_N)$.

We obviously see that

$$\sum_{k=\lceil na \rceil}^{\lfloor nb \rfloor} Z(nx-k) = \sum_{k=\lceil na \rceil}^{\lfloor nb \rfloor} \left(\prod_{i=1}^{N} \chi(nx_i - k_i) \right) =$$

$$\sum_{k_1=\lceil na_1\rceil}^{\lfloor nb_1\rfloor}\cdots\sum_{k_N=\lceil na_N\rceil}^{\lfloor nb_N\rfloor}\left(\prod_{i=1}^{N}\chi\left(nx_i-k_i\right)\right)=\prod_{i=1}^{N}\left(\sum_{k_i=\lceil na_i\rceil}^{\lfloor nb_i\rfloor}\chi\left(nx_i-k_i\right)\right). \quad (21.35)$$

For $0<\beta<1$ and $n\in\mathbb{N}$, a fixed $x\in\mathbb{R}^N$, we have that

$$\sum_{k=\lceil na\rceil}^{\lfloor nb\rfloor}\chi\left(nx-k\right)=$$

$$\sum_{\begin{cases}k=\lceil na\rceil\\ \left\|\frac{k}{n}-x\right\|_\infty\leq\frac{1}{n^\beta}\end{cases}}^{\lfloor nb\rfloor}\chi\left(nx-k\right)+\sum_{\begin{cases}k=\lceil na\rceil\\ \left\|\frac{k}{n}-x\right\|_\infty>\frac{1}{n^\beta}\end{cases}}^{\lfloor nb\rfloor}\chi\left(nx-k\right). \quad (21.36)$$

In the last two sums the counting is over disjoint vector sets of k's, because the condition $\left\|\frac{k}{n}-x\right\|_\infty>\frac{1}{n^\beta}$ implies that there exists at least one $\left|\frac{k_{\overline{r}}}{n}-x_{\overline{r}}\right|>\frac{1}{n^\beta}$, where $\overline{r}\in\{1,\ldots,N\}$.

We have that

$$\sum_{\begin{cases}k=\lceil na\rceil\\ \left\|\frac{k}{n}-x\right\|_\infty>\frac{1}{n^\beta}\end{cases}}^{\lfloor nb\rfloor}Z\left(nx-k\right)=\prod_{i=1}^{N}\left(\sum_{\begin{cases}k_i=\lceil na_i\rceil\\ \left\|\frac{k}{n}-x\right\|_\infty>\frac{1}{n^\beta}\end{cases}}^{\lfloor nb_i\rfloor}\chi\left(nx_i-k_i\right)\right)$$

$$\leq\sum_{\substack{k_{\overline{r}}=-\infty\\ :\,|nx_{\overline{r}}-k_{\overline{r}}|>n^{1-\beta}}}^{\infty}\chi\left(nx_{\overline{r}}-k_{\overline{r}}\right)\overset{(21.26)}{\leq}\frac{1}{2\sqrt{\pi}\left(n^{1-\beta}-2\right)e^{\left(n^{1-\beta}-2\right)^2}}, \quad (21.37)$$

when $n^{1-\beta}\geq 3$.

We have proved (see also [18] for details) that

(v)

$$\sum_{\begin{cases}k=\lceil na\rceil\\ \left\|\frac{k}{n}-x\right\|_\infty>\frac{1}{n^\beta}\end{cases}}^{\lfloor nb\rfloor}Z\left(nx-k\right)\leq\frac{1}{2\sqrt{\pi}\left(n^{1-\beta}-2\right)e^{\left(n^{1-\beta}-2\right)^2}}, \quad (21.38)$$

$0<\beta<1$, $n\in\mathbb{N}$; $n^{1-\beta}\geq 3$, $x\in\left(\prod_{i=1}^{N}\left[a_i,b_i\right]\right)$.

By Theorem 21.25 clearly we obtain

$$0 < \frac{1}{\sum_{k=\lceil na \rceil}^{\lfloor nb \rfloor} Z(nx-k)} = \frac{1}{\prod_{i=1}^{N} \left(\sum_{k_i=\lceil na_i \rceil}^{\lfloor nb_i \rfloor} \chi(nx_i - k_i) \right)} \tag{21.39}$$

$$< \frac{1}{(\chi(1))^N} \simeq (4.019)^N .$$

That is,

(vi) it holds

$$0 < \frac{1}{\sum_{k=\lceil na \rceil}^{\lfloor nb \rfloor} Z(nx-k)} < \frac{1}{(\chi(1))^N} \simeq (4.019)^N , \tag{21.40}$$

$\forall\, x \in \left(\prod_{i=1}^{N} [a_i, b_i] \right)$, $n \in \mathbb{N}$.

It is also clear that

(vii)

$$\sum_{\substack{k=-\infty \\ \left\| \frac{k}{n} - x \right\|_\infty > \frac{1}{n^\beta}}}^{\infty} Z(nx-k) \leq \frac{1}{2\sqrt{\pi}\left(n^{1-\beta}-2\right) e^{\left(n^{1-\beta}-2\right)^2}}, \tag{21.41}$$

$0 < \beta < 1, n \in \mathbb{N} : n^{1-\beta} \geq 3, x \in \left(\prod_{i=1}^{N} [a_i, b_i] \right)$.

Also we get

$$\lim_{n\to\infty} \sum_{k=\lceil na \rceil}^{\lfloor nb \rfloor} Z(nx-k) \neq 1, \tag{21.42}$$

for at least some $x \in \left(\prod_{i=1}^{N} [a_i, b_i] \right)$.

Let $f \in C\left(\prod_{i=1}^{N} [a_i, b_i] \right)$ and $n \in \mathbb{N}$ such that $\lceil na_i \rceil \leq \lfloor nb_i \rfloor$, $i = 1, \ldots, N$.

We introduce and define (see [18]) the multivariate positive linear neural network operator ($x := (x_1, \ldots, x_N) \in \left(\prod_{i=1}^{N} [a_i, b_i] \right)$)

$$A_n(f, x_1, \ldots, x_N) := A_n(f, x) := \frac{\sum_{k=\lceil na \rceil}^{\lfloor nb \rfloor} f\left(\frac{k}{n}\right) Z(nx-k)}{\sum_{k=\lceil na \rceil}^{\lfloor nb \rfloor} Z(nx-k)} \tag{21.43}$$

$$:= \frac{\sum_{k_1=\lceil na_1\rceil}^{\lfloor nb_1\rfloor} \sum_{k_2=\lceil na_2\rceil}^{\lfloor nb_2\rfloor} \cdots \sum_{k_N=\lceil na_N\rceil}^{\lfloor nb_N\rfloor} f\left(\frac{k_1}{n}, \ldots, \frac{k_N}{n}\right)\left(\prod_{i=1}^{N} \chi(nx_i - k_i)\right)}{\prod_{i=1}^{N}\left(\sum_{k_i=\lceil na_i\rceil}^{\lfloor nb_i\rfloor} \chi(nx_i - k_i)\right)}.$$

For large enough n we always obtain $\lceil na_i\rceil \leq \lfloor nb_i\rfloor$, $i = 1, \ldots, N$. Also $a_i \leq \frac{k_i}{n} \leq b_i$, iff $\lceil na_i\rceil \leq k_i \leq \lfloor nb_i\rfloor$, $i = 1, \ldots, N$.

We need, for $f \in C\left(\prod_{i=1}^{N} [a_i, b_i]\right)$ the first multivariate modulus of continuity

$$\omega_1(f, h) := \sup_{\substack{x, y \in \prod_{i=1}^{N} [a_i, b_i] \\ \|x - y\|_\infty \leq h}} |f(x) - f(y)|, \quad h > 0. \tag{21.44}$$

It holds that

$$\lim_{h \to 0} \omega_1(f, h) = 0. \tag{21.45}$$

Similarly it is defined for $f \in C_B\left(\mathbb{R}^N\right)$ (continuous and bounded functions on $\mathbb{R}^N$) the $\omega_1(f, h)$, and it has the property (21.45), given that $f \in C_U\left(\mathbb{R}^N\right)$ (uniformly continuous functions on $\mathbb{R}^N$).

When $f \in C_B\left(\mathbb{R}^N\right)$ we define ([18]),

$$B_n(f, x) := B_n(f, x_1, \ldots, x_N) := \sum_{k=-\infty}^{\infty} f\left(\frac{k}{n}\right) Z(nx - k) := \tag{21.46}$$

$$\sum_{k_1=-\infty}^{\infty} \sum_{k_2=-\infty}^{\infty} \cdots \sum_{k_N=-\infty}^{\infty} f\left(\frac{k_1}{n}, \frac{k_2}{n}, \ldots, \frac{k_N}{n}\right)\left(\prod_{i=1}^{N} \chi(nx_i - k_i)\right),$$

$n \in \mathbb{N}$, $\forall\, x \in \mathbb{R}^N$, $N \in \mathbb{N}$, the multivariate quasi-interpolation neural network operator.

Also for $f \in C_B\left(\mathbb{R}^N\right)$ we define [18] the multivariate Kantorovich type neural network operator

$$C_n(f, x) := C_n(f, x_1, \ldots, x_N) := \sum_{k=-\infty}^{\infty}\left(n^N \int_{\frac{k}{n}}^{\frac{k+1}{n}} f(t)\, dt\right) Z(nx - k) := \tag{21.47}$$

$$\sum_{k_1=-\infty}^{\infty} \sum_{k_2=-\infty}^{\infty} \cdots \sum_{k_N=-\infty}^{\infty}\left(n^N \int_{\frac{k_1}{n}}^{\frac{k_1+1}{n}} \int_{\frac{k_2}{n}}^{\frac{k_2+1}{n}} \cdots \int_{\frac{k_N}{n}}^{\frac{k_N+1}{n}} f(t_1, \ldots, t_N)\, dt_1 \ldots dt_N\right)$$

$$\cdot \left(\prod_{i=1}^{N} \chi \left(nx_i - k_i \right) \right),$$

$n \in \mathbb{N}$, $\forall \, x \in \mathbb{R}^N$.

Again for $f \in C_B\left(\mathbb{R}^N\right)$, $N \in \mathbb{N}$, we define [18] the multivariate neural network operator of quadrature type $D_n(f, x)$, $n \in \mathbb{N}$, as follows. Let $\theta = (\theta_1, \ldots, \theta_N) \in \mathbb{N}^N$, $\overline{r} = (r_1, \ldots, r_N) \in \mathbb{Z}_+^N$, $w_{\overline{r}} = w_{r_1, r_2, \ldots r_N} \geq 0$, such that $\sum_{\overline{r}=0}^{\theta} w_{\overline{r}} = \sum_{r_1=0}^{\theta_1} \sum_{r_2=0}^{\theta_2} \cdots \sum_{r_N=0}^{\theta_N} w_{r_1, r_2, \ldots r_N} = 1$; $k \in \mathbb{Z}^N$ and

$$\delta_{nk}(f) := \delta_{n, k_1, k_2, \ldots, k_N}(f) := \sum_{\overline{r}=0}^{\theta} w_{\overline{r}} f\left(\frac{k}{n} + \frac{\overline{r}}{n\theta}\right) :=$$

$$\sum_{r_1=0}^{\theta_1} \sum_{r_2=0}^{\theta_2} \cdots \sum_{r_N=0}^{\theta_N} w_{r_1, r_2, \ldots r_N} f\left(\frac{k_1}{n} + \frac{r_1}{n\theta_1}, \frac{k_2}{n} + \frac{r_2}{n\theta_2}, \ldots, \frac{k_N}{n} + \frac{r_N}{n\theta_N}\right), \quad (21.48)$$

where $\frac{\overline{r}}{\theta} := \left(\frac{r_1}{\theta_1}, \frac{r_2}{\theta_2}, \ldots, \frac{r_N}{\theta_N}\right)$.

We put (see [18])

$$D_n(f, x) := D_n(f, x_1, \ldots, x_N) := \sum_{k=-\infty}^{\infty} \delta_{nk}(f) \, Z(nx - k) := \quad (21.49)$$

$$\sum_{k_1=-\infty}^{\infty} \sum_{k_2=-\infty}^{\infty} \cdots \sum_{k_N=-\infty}^{\infty} \delta_{n, k_1, k_2, \ldots, k_N}(f) \left(\prod_{i=1}^{N} \chi \left(nx_i - k_i \right) \right),$$

$\forall \, x \in \mathbb{R}^N$.

Let $f \in C^m\left(\prod_{i=1}^N [a_i, b_i]\right)$, $m, N \in \mathbb{N}$. Here f_α denotes a partial derivative of f, $\alpha := (\alpha_1, \ldots, \alpha_N)$, $\alpha_i \in \mathbb{Z}_+$, $i = 1, \ldots, N$, and $|\alpha| := \sum_{i=1}^N \alpha_i = l$, where $l = 0, 1, \ldots, m$. We write also $f_\alpha := \frac{\partial^\alpha f}{\partial x^\alpha}$ and we say it is of order l.

We denote

$$\omega_{1,m}^{\max}(f_\alpha, h) := \max_{\alpha : |\alpha| = m} \omega_1(f_\alpha, h). \quad (21.50)$$

Call also

$$\|f_\alpha\|_{\infty, m}^{\max} := \max_{\alpha : |\alpha| = m} \left\{ \|f_\alpha\|_\infty \right\}, \quad (21.51)$$

where $\|\cdot\|_\infty$ is the supremum norm.

In [18] we studied the basic approximation properties of A_n, B_n, C_n, D_n neural network operators and as well of their iterates. That is, the quantitative pointwise and uniform convergence of these operators to the unit operator I. We studied also there the complex functions related approximation.

We need

Theorem 21.26 (see [18]) *Let* $f \in C\left(\prod_{i=1}^{N}[a_i, b_i]\right)$, $0 < \beta < 1$, $x \in \left(\prod_{i=1}^{N}[a_i, b_i]\right)$, $N, n \in \mathbb{N}$ *with* $n^{1-\beta} \geq 3$. *Then*

(1)

$$|A_n(f, x) - f(x)| \leq (4.019)^N \left[\omega_1\left(f, \frac{1}{n^\beta}\right) + \frac{\|f\|_\infty}{\sqrt{\pi}\left(n^{1-\beta} - 2\right) e^{\left(n^{1-\beta}-2\right)^2}}\right] =: \lambda_1, \tag{21.52}$$

and

(2)

$$\|A_n(f) - f\|_\infty \leq \lambda_1. \tag{21.53}$$

We notice that $\lim_{n\to\infty} A_n(f) = f$, *pointwise and uniformly.*

We need

Theorem 21.27 (see [18]) *Let* $f \in C_B\left(\mathbb{R}^N\right)$, $0 < \beta < 1$, $x \in \mathbb{R}^N$, $N, n \in \mathbb{N}$ *with* $n^{1-\beta} \geq 3$. *Then*

(1)

$$|B_n(f, x) - f(x)| \leq \omega_1\left(f, \frac{1}{n^\beta}\right) + \frac{\|f\|_\infty}{\sqrt{\pi}\left(n^{1-\beta} - 2\right) e^{\left(n^{1-\beta}-2\right)^2}} =: \lambda_2, \tag{21.54}$$

(2)

$$\|B_n(f) - f\|_\infty \leq \lambda_2. \tag{21.55}$$

Given that $f \in \left(C_U\left(\mathbb{R}^N\right) \cap C_B\left(\mathbb{R}^N\right)\right)$, *we obtain* $\lim_{n\to\infty} B_n(f) = f$, *uniformly.*

We also need

Theorem 21.28 (see [18]) *Let* $f \in C_B\left(\mathbb{R}^N\right)$, $0 < \beta < 1$, $x \in \mathbb{R}^N$, $N, n \in \mathbb{N}$ *with* $n^{1-\beta} \geq 3$. *Then*

(1)

$$|C_n(f, x) - f(x)| \leq \omega_1\left(f, \frac{1}{n} + \frac{1}{n^\beta}\right) + \frac{\|f\|_\infty}{\sqrt{\pi}\left(n^{1-\beta} - 2\right) e^{\left(n^{1-\beta}-2\right)^2}} =: \lambda_3, \tag{21.56}$$

(2)

$$\|C_n(f) - f\|_\infty \le \lambda_3. \tag{21.57}$$

Given that $f \in \left(C_U\left(\mathbb{R}^N\right) \cap C_B\left(\mathbb{R}^N\right)\right)$, *we obtain* $\lim_{n\to\infty} C_n(f) = f$, *uniformly.*

We also need

Theorem 21.29 (see [18]) *Let* $f \in C_B\left(\mathbb{R}^N\right)$, $0 < \beta < 1$, $x \in \mathbb{R}^N$, $N, n \in \mathbb{N}$ *with* $n^{1-\beta} \ge 3$. *Then*

(1)

$$|D_n(f,x) - f(x)| \le \omega_1\left(f, \frac{1}{n} + \frac{1}{n^\beta}\right) + \frac{\|f\|_\infty}{\sqrt{\pi}\left(n^{1-\beta} - 2\right) e^{\left(n^{1-\beta}-2\right)^2}} = \lambda_3, \tag{21.58}$$

(2)

$$\|D_n(f) - f\|_\infty \le \lambda_3. \tag{21.59}$$

Given that $f \in \left(C_U\left(\mathbb{R}^N\right) \cap C_B\left(\mathbb{R}^N\right)\right)$, *we obtain* $\lim_{n\to\infty} D_n(f) = f$, *uniformly.*

We finally mention

Theorem 21.30 (see [18]) *Let* $f \in C^m\left(\prod_{i=1}^N [a_i, b_i]\right)$, $0 < \beta < 1$, $n, m, N \in \mathbb{N}$, $n^{1-\beta} \ge 3$, $x \in \left(\prod_{i=1}^N [a_i, b_i]\right)$. *Then*

(i)

$$\left| A_n(f,x) - f(x) - \sum_{j=1}^m \left(\sum_{|\alpha|=j} \left(\frac{f_\alpha(x)}{\prod_{i=1}^N \alpha_i!} \right) A_n\left(\prod_{i=1}^N (\cdot - x_i)^{\alpha_i}, x \right) \right) \right| \le \tag{21.60}$$

$$(4.019)^N \cdot \left\{ \frac{N^m}{m! n^{m\beta}} \omega_{1,m}^{\max}\left(f_\alpha, \frac{1}{n^\beta}\right) + \right.$$

$$\left. \left(\frac{\|b-a\|_\infty^m \|f_\alpha\|_{\infty,m}^{\max} N^m}{m!} \right) \frac{1}{\sqrt{\pi}\left(n^{1-\beta}-2\right) e^{\left(n^{1-\beta}-2\right)^2}} \right\},$$

(ii)

$$|A_n(f,x) - f(x)| \le (4.019)^N \cdot \tag{21.61}$$

$$\left\{ \sum_{j=1}^m \left(\sum_{|\alpha|=j} \left(\frac{|f_\alpha(x)|}{\prod_{i=1}^N \alpha_i!} \right) \left[\frac{1}{n^{\beta j}} + \left(\prod_{i=1}^N (b_i - a_i)^{\alpha_i} \right) \cdot \right. \right. \right.$$

$$\left.\left.\frac{1}{2\sqrt{\pi}\left(n^{1-\beta}-2\right)e^{\left(n^{1-\beta}-2\right)^2}}\right]\right)+\frac{N^m}{m!n^{m\beta}}\omega_{1,m}^{\max}\left(f_\alpha,\frac{1}{n^\beta}\right)$$

$$\left.+\left(\frac{\|b-a\|_\infty^m \|f_\alpha\|_{\infty,m}^{\max} N^m}{m!}\right)\frac{1}{\sqrt{\pi}\left(n^{1-\beta}-2\right)e^{\left(n^{1-\beta}-2\right)^2}}\right\},$$

(iii)

$$\|A_n(f)-f\|_\infty \leq (4.019)^N \cdot \tag{21.62}$$

$$\left\{\sum_{j=1}^m \left(\sum_{|\alpha|=j}\left(\frac{\|f_\alpha\|_\infty}{\prod_{i=1}^N \alpha_i!}\right)\left[\frac{1}{n^{\beta j}}+\left(\prod_{i=1}^N (b_i-a_i)^{\alpha_i}\right)\cdot\right.\right.\right.$$

$$\left.\left.\frac{1}{2\sqrt{\pi}\left(n^{1-\beta}-2\right)e^{\left(n^{1-\beta}-2\right)^2}}\right]\right)+\frac{N^m}{m!n^{m\beta}}\omega_{1,m}^{\max}\left(f_\alpha,\frac{1}{n^\beta}\right)$$

$$\left.+\left(\frac{\|b-a\|_\infty^m \|f_\alpha\|_{\infty,m}^{\max} N^m}{m!}\right)\frac{1}{\sqrt{\pi}\left(n^{1-\beta}-2\right)e^{\left(n^{1-\beta}-2\right)^2}}\right\},$$

(iv) additionally assume $f_\alpha(x_0)=0$, *for all* $\alpha : |\alpha| = 1, \ldots, m$; $x_0 \in \left(\prod_{i=1}^N [a_i,b_i]\right)$, *then*

$$|A_n(f,x_0)-f(x_0)| \leq (4.019)^N \left\{\frac{N^m}{m!n^{m\beta}}\omega_{1,m}^{\max}\left(f_\alpha,\frac{1}{n^\beta}\right)+\right. \tag{21.63}$$

$$\left.\left(\frac{\|b-a\|_\infty^m \|f_\alpha\|_{\infty,m}^{\max} N^m}{m!}\right)\frac{1}{\sqrt{\pi}\left(n^{1-\beta}-2\right)e^{\left(n^{1-\beta}-2\right)^2}}\right\},$$

notice in the last the extremely high rate of convergence at $n^{-\beta(m+1)}$.

21.4 Main Results: Fuzzy Multivariate Neural Network Approximation Based on the Error Function

We define the following Fuzzy multivariate Neural Network Operators $A_n^{\mathcal{F}}$, $B_n^{\mathcal{F}}$, $C_n^{\mathcal{F}}$, $D_n^{\mathcal{F}}$ based on the error function. These are analogs of the real A_n, B_n, C_n, D_n, see (21.43), (21.46), (21.47) and (21.49), respectively.

Let $f \in C_{\mathcal{F}}\left(\prod_{i=1}^N [a_i,b_i]\right)$, $N \in \mathbb{N}$, we set

$$A_n^{\mathcal{F}}(f, x_1, \ldots, x_N) := A_n^{\mathcal{F}}(f, x) := \frac{\sum_{k=\lceil na \rceil}^{\lfloor nb \rfloor *} f\left(\frac{k}{n}\right) \odot Z(nx - k)}{\sum_{k=\lceil na \rceil}^{\lfloor nb \rfloor} Z(nx - k)}$$

$$= \frac{\sum_{k_1=\lceil na_1 \rceil}^{\lfloor nb_1 \rfloor *} \cdots \sum_{k_N=\lceil na_N \rceil}^{\lfloor nb_N \rfloor *} f\left(\frac{k_1}{n}, \ldots, \frac{k_N}{n}\right) \odot \left(\prod_{i=1}^{N} \chi(nx_i - k_i)\right)}{\prod_{i=1}^{N} \left(\sum_{k_i=\lceil na_i \rceil}^{\lfloor nb_i \rfloor} \chi(nx_i - k_i)\right)}, \quad (21.64)$$

$x \in \prod_{i=1}^{N} [a_i, b_i]$, $n \in \mathbb{N}$.

Let $f \in C_B(\mathbb{R}^N, \mathbb{R}_{\mathcal{F}})$, we put

$$B_n^{\mathcal{F}}(f, x) := B_n^{\mathcal{F}}(f, x_1, \ldots, x_N) := \sum_{k=-\infty}^{\infty *} f\left(\frac{k}{n}\right) \odot Z(nx - k)$$

$$:= \sum_{k_1=-\infty}^{\infty *} \cdots \sum_{k_N=-\infty}^{\infty *} f\left(\frac{k_1}{n}, \ldots, \frac{k_N}{n}\right) \odot \left(\prod_{i=1}^{N} \chi(nx_i - k_i)\right), \quad (21.65)$$

$x \in \mathbb{R}^N$, $n \in \mathbb{N}$.

Let $f \in C_B(\mathbb{R}^N, \mathbb{R}_{\mathcal{F}})$, we define the multivariate fuzzy Kantorovich type neural network operator,

$$C_n^{\mathcal{F}}(f, x) := C_n^{\mathcal{F}}(f, x_1, \ldots, x_N) := \sum_{k=-\infty}^{\infty *} \left(n^N \odot \int_{\frac{k}{n}}^{\frac{k+1}{n}} f(t)\, dt\right) \odot Z(nx - k) :=$$

$$\sum_{k_1=-\infty}^{\infty *} \cdots \sum_{k_N=-\infty}^{\infty *} \left(n^N \odot \int_{\frac{k_1}{n}}^{\frac{k_1+1}{n}} \cdots \int_{\frac{k_N}{n}}^{\frac{k_N+1}{n}} f(t_1, \ldots, t_N)\, dt_1 \ldots dt_N\right)$$

$$\odot \left(\prod_{i=1}^{N} \chi(nx_i - k_i)\right), \quad (21.66)$$

$x \in \mathbb{R}^N$, $n \in \mathbb{N}$.

Let $f \in C_B(\mathbb{R}^N, \mathbb{R}_{\mathcal{F}})$, we define the multivariate fuzzy quadrature type neural network operator. Let here

$$\delta_{nk}^{\mathcal{F}}(f) := \delta_{n,k_1,\ldots,k_N}^{\mathcal{F}}(f) := \sum_{\overline{r}=0}^{\theta *} w_{\overline{r}} \odot f\left(\frac{k}{n} + \frac{\overline{r}}{n\theta}\right) := \quad (21.67)$$

$$\sum_{r_1=0}^{\theta_1 *} \cdots \sum_{r_N=0}^{\theta_N *} w_{r_1,\dots r_N} \odot f\left(\frac{k_1}{n}+\frac{r_1}{n\theta_1},\dots,\frac{k_N}{n}+\frac{r_N}{n\theta_N}\right).$$

We put

$$D_n^{\mathcal{F}}(f,x) := D_n^{\mathcal{F}}(f,x_1,\dots,x_N) := \sum_{k=-\infty}^{\infty *} \delta_{nk}^{\mathcal{F}}(f) \odot Z(nx-k) :=$$

$$\sum_{k_1=-\infty}^{\infty *} \cdots \sum_{k_N=-\infty}^{\infty *} \delta_{n,k_1,\dots,k_N}^{\mathcal{F}}(f) \odot \left(\prod_{i=1}^{N} \chi(nx_i-k_i)\right), \tag{21.68}$$

$x \in \mathbb{R}^N$, $n \in \mathbb{N}$.

We can put together all $B_n^{\mathcal{F}}$, $C_n^{\mathcal{F}}$, $D_n^{\mathcal{F}}$ fuzzy operators as follows:

$$L_n^{\mathcal{F}}(f,x) := \sum_{k=-\infty}^{\infty *} l_{nk}^{\mathcal{F}}(f) \odot Z(nx-k), \tag{21.69}$$

where

$$l_{nk}^{\mathcal{F}}(f) := \begin{cases} f\left(\frac{k}{n}\right), \text{ if } L_n^{\mathcal{F}} = B_n^{\mathcal{F}}, \\ n^N \odot \int_{\frac{k}{n}}^{\frac{k+1}{n}} f(t)\,dt, \text{ if } L_n^{\mathcal{F}} = C_n^{\mathcal{F}}, \\ \delta_{nk}^{\mathcal{F}}(f), \text{ if } L_n^{\mathcal{F}} = D_n^{\mathcal{F}}, \end{cases} \tag{21.70}$$

$x \in \mathbb{R}^N$, $n \in \mathbb{N}$.

Similarly we can put together all B_n, C_n, D_n real operators as

$$L_n(f,x) := \sum_{k=-\infty}^{\infty} l_{nk}(f) Z(nx-k), \tag{21.71}$$

where

$$l_{nk}(f) := \begin{cases} f\left(\frac{k}{n}\right), \text{ if } L_n = B_n, \\ n^N \int_{\frac{k}{n}}^{\frac{k+1}{n}} f(t)\,dt, \text{ if } L_n = C_n, \\ \delta_{nk}(f), \text{ if } L_n = D_n, \end{cases} \tag{21.72}$$

$x \in \mathbb{R}^N$, $n \in \mathbb{N}$.

Let $r \in [0,1]$, we observe that

$$\left[A_n^{\mathcal{F}}(f,x)\right]^r = \sum_{k=\lceil na \rceil}^{\lfloor nb \rfloor} \left[f\left(\frac{k}{n}\right)\right]^r \left(\frac{Z(nx-k)}{\sum_{k=\lceil na \rceil}^{\lfloor nb \rfloor} Z(nx-k)} \right) =$$

$$\sum_{k=\lceil na \rceil}^{\lfloor nb \rfloor} \left[f_-^{(r)}\left(\frac{k}{n}\right), f_+^{(r)}\left(\frac{k}{n}\right)\right] \left(\frac{Z(nx-k)}{\sum_{k=\lceil na \rceil}^{\lfloor nb \rfloor} Z(nx-k)} \right) = \tag{21.73}$$

$$\left[\sum_{k=\lceil na \rceil}^{\lfloor nb \rfloor} f_-^{(r)}\left(\frac{k}{n}\right) \left(\frac{Z(nx-k)}{\sum_{k=\lceil na \rceil}^{\lfloor nb \rfloor} Z(nx-k)} \right), \sum_{k=\lceil na \rceil}^{\lfloor nb \rfloor} f_+^{(r)}\left(\frac{k}{n}\right) \left(\frac{Z(nx-k)}{\sum_{k=\lceil na \rceil}^{\lfloor nb \rfloor} Z(nx-k)} \right) \right]$$

$$= \left[A_n\left(f_-^{(r)}, x\right), A_n\left(f_+^{(r)}, x\right)\right]. \tag{21.74}$$

We have proved that

$$\left(A_n^{\mathcal{F}}(f,x)\right)_{\pm}^{(r)} = A_n\left(f_{\pm}^{(r)}, x\right), \tag{21.75}$$

respectively, $\forall\, r \in [0,1]$, $\forall\, x \in \left(\prod_{i=1}^{N} [a_i, b_i]\right)$.

We will prove also that

$$\left(L_n^{\mathcal{F}}(f,x)\right)_{\pm}^{(r)} = L_n\left(f_{\pm}^{(r)}, x\right), \tag{21.76}$$

respectively, $\forall\, r \in [0,1]$, $\forall\, x \in \mathbb{R}^N$.

The sum in (21.76) is doubly infinite and the proof is more complicated and follows.

We need

Remark 21.31 (see also [27] and [17]) (1) Here $k = (k_1, k_2) \in \mathbb{Z}^2$, $m = (m_1, m_2) \in \mathbb{Z}_-^2$, $n = (n_1, n_2) \in \mathbb{Z}_+^2$, $\infty = (\infty, \infty)$, $-\infty = (-\infty, -\infty)$, $\sum$ is a double sum.

Let $(u_k)_{k\in\mathbb{Z}^2} \in \mathbb{R}_{\mathcal{F}}$. We denote the fuzzy double infinite series by $\sum_{k=-\infty}^{\infty *} u_k$ and we say that it converges to $u \in \mathbb{R}_{\mathcal{F}}$ iff $\lim_{m\to\infty}\lim_{n\to\infty} D\left(\sum_{k=m}^{n*} u_k, u\right) = 0$. We denote the last by $\sum_{k=-\infty}^{\infty *} u_k = u$.

Let $(u_k)_{k\in\mathbb{Z}^2}, (v_k)_{k\in\mathbb{Z}^2}, u, v \in \mathbb{R}_{\mathcal{F}}$ such that $\sum_{k=-\infty}^{\infty *} u_k = u$, $\sum_{k=-\infty}^{\infty *} v_k = v$. Then (by [17])

$$\sum_{k=-\infty}^{\infty *} (u_k \oplus v_k) = u \oplus v = \sum_{k=-\infty}^{\infty *} u_k \oplus \sum_{k=-\infty}^{\infty *} v_k. \tag{21.77}$$

Let $\sum_{k=-\infty}^{\infty *} u_k = u \in \mathbb{R}_{\mathcal{F}}$, then one has that (see [17])

$$\sum_{k=-\infty}^{\infty} (u_k)_-^{(r)} = u_-^{(r)} = \left(\sum_{k=-\infty}^{\infty *} u_k\right)_-^{(r)}, \tag{21.78}$$

and

$$\sum_{k=-\infty}^{\infty} (u_k)_+^{(r)} = u_+^{(r)} = \left(\sum_{k=-\infty}^{\infty *} u_k\right)_+^{(r)}, \tag{21.79}$$

$\forall\, r \in [0, 1]$.

Also we need: let $(u_k)_{k\in\mathbb{Z}^2} \in \mathbb{R}_{\mathcal{F}}$ with $\sum_{k=-\infty}^{\infty *} u_k = u \in \mathbb{R}_{\mathcal{F}}$, then clearly one [17] has for any $\lambda \in \mathbb{R}$ that $\sum_{k=-\infty}^{\infty *} \lambda u_k = \lambda u$.

Clearly also here $\sum_{k=-\infty}^{\infty *} u_k = u \in \mathbb{R}_{\mathcal{F}}$, where $(u_k)_{k\in\mathbb{Z}^2} \in \mathbb{R}_{\mathcal{F}}$, iff $\sum_{k=-\infty}^{\infty} (u_k)_-^{(r)} = (u)_-^{(r)}$ and $\sum_{k=-\infty}^{\infty} (u_k)_+^{(r)} = (u)_+^{(r)}$, uniformly in $r \in [0, 1]$, see also [17, 27].

(2) By [17, 27] we see: Let $\left(k \in \mathbb{Z}^2\right)$ $u_k := \left\{\left((u_k)_-^{(r)}, (u_k)_+^{(r)}\right) | 0 \le r \le 1\right\} \in \mathbb{R}_{\mathcal{F}}$ such that $\sum_{k=-\infty}^{\infty} (u_k)_-^{(r)} =: (u)_-^{(r)}$ and $\sum_{k=-\infty}^{\infty} (u_k)_+^{(r)} =: (u)_+^{(r)}$ converge uniformly in $r \in [0,1]$, then $u := \left\{\left((u)_-^{(r)}, (u)_+^{(r)}\right) | 0 \le r \le 1\right\} \in \mathbb{R}_{\mathcal{F}}$ and $u = \sum_{k=-\infty}^{\infty *} u_k$.

I.e. we have

$$\sum_{k=-\infty}^{\infty *} \left\{\left((u_k)_-^{(r)}, (u_k)_+^{(r)}\right) | 0 \le r \le 1\right\} = \tag{21.80}$$

$$\left\{\left(\sum_{k=-\infty}^{\infty *} (u_k)_-^{(r)}, \sum_{k=-\infty}^{\infty *} (u_k)_+^{(r)}\right) | 0 \le r \le 1\right\}.$$

All the content of Remark 21.31 goes through and is valid for $k \in Z^N$, $N \ge 2$.

Proof (of (21.76))

Let $f \in C_B\left(\mathbb{R}^N, \mathbb{R}_{\mathcal{F}}\right)$, then $f_{\pm}^{(r)} \in C_B\left(\mathbb{R}^N\right)$, $\forall\, r \in [0,1]$. We have

$$(f(x))_-^{(0)} \le (f(x))_-^{(r)} \le (f(x))_-^{(1)} \le (f(x))_+^{(1)}$$

$$\le (f(x))_+^{(r)} \le (f(x))_+^{(0)}, \ \forall\, x \in \mathbb{R}^N. \tag{21.81}$$

We get that

$$\left|(f(x))_-^{(r)}\right| \le \max\left\{\left|(f(x))_-^{(0)}\right|, \left|(f(x))_-^{(1)}\right|\right\} =$$

$$\frac{1}{2}\left\{\left|(f(x))_-^{(0)}\right| + \left|(f(x))_-^{(1)}\right| - \left|\left|(f(x))_-^{(0)}\right| - \left|(f(x))_-^{(1)}\right|\right|\right\}$$

$$=: A_-^{(0,1)}(x) \in C_B\left(\mathbb{R}^N\right). \tag{21.82}$$

Also it holds

$$\left|(f(x))_+^{(r)}\right| \le \max\left\{\left|(f(x))_+^{(0)}\right|, \left|(f(x))_+^{(1)}\right|\right\} =$$

$$\frac{1}{2}\left\{\left|(f(x))_+^{(0)}\right| + \left|(f(x))_+^{(1)}\right| - \left|\left|(f(x))_+^{(0)}\right| - \left|(f(x))_+^{(1)}\right|\right|\right\}$$

$$=: A_+^{(0,1)}(x) \in C_B\left(\mathbb{R}^N\right). \tag{21.83}$$

I.e. we have obtained that

$$0 \leq \left| (f(x))_{-}^{(r)} \right| \leq A_{-}^{(0,1)}(x), \tag{21.84}$$

$$0 \leq \left| (f(x))_{+}^{(r)} \right| \leq A_{+}^{(0,1)}(x), \ \forall\, r \in [0,1], \forall\, x \in \mathbb{R}^N.$$

One easily derives that

$$\left(l_{nk}^{\mathcal{F}}(f) \right)_{\pm}^{(r)} = l_{nk}\left(f_{\pm}^{(r)} \right), \tag{21.85}$$

$\forall\, r \in [0,1]$, respectively in $\pm$.

Notice here that l_{nk} is a positive linear functional. Hence it holds

$$0 \leq \left| l_{nk}\left(f_{\pm}^{(r)} \right) \right| \leq l_{nk}\left(\left| f_{\pm}^{(r)} \right| \right) \leq l_{nk}\left(A_{\pm}^{(0,1)} \right), \tag{21.86}$$

respectively in $\pm$, $\forall\, r \in [0,1]$.

In details one has

$$\left| l_{nk}\left(f_{\pm}^{(r)} \right) \right| Z(nx-k) \leq l_{nk}\left(\left| f_{\pm}^{(r)} \right| \right) Z(nx-k) \leq l_{nk}\left(A_{\pm}^{(0,1)} \right) Z(nx-k), \tag{21.87}$$

$\forall\, r \in [0,1]$, $k \in \mathbb{Z}^N$, $n \in \mathbb{N}$ fixed, any $x \in \mathbb{R}^N$ fixed, respectively in $\pm$.

We notice that

$$\sum_{k=-\infty}^{\infty} l_{nk}\left(A_{\pm}^{(0,1)} \right) Z(nx-k) \leq \left\| A_{\pm}^{(0,1)} \right\|_{\infty} < \infty, \tag{21.88}$$

i.e. $\sum_{k=-\infty}^{\infty} l_{nk}\left(A_{\pm}^{(0,1)} \right) Z(nx-k)$ converges, respectively in $\pm$.

Thus by Weierstrass M-test we obtain that $L_n\left(f_{\pm}^{(r)}, x \right)$ as a series converges uniformly in $r \in [0,1]$, respectively in $\pm$, $\forall\, n \in \mathbb{N}$, for any $x \in \mathbb{R}^N$.

Here for $k \in \mathbb{Z}^N$,

$$l_{nk}^{\mathcal{F}}(f) \overset{(21.85)}{=} \left\{ \left(l_{nk}\left(f_{-}^{(r)} \right), l_{nk}\left(f_{+}^{(r)} \right) \right) | 0 \leq r \leq 1 \right\} \in \mathbb{R}_{\mathcal{F}}, \tag{21.89}$$

and also

$$u_k := \left\{ \left(l_{nk}\left(f_{-}^{(r)} \right) Z(nx-k), l_{nk}\left(f_{+}^{(r)} \right) Z(nx-k) \right) | 0 \leq r \leq 1 \right\}$$

$$= l_{nk}^{\mathcal{F}}(f) \odot Z(nx-k) \in \mathbb{R}_{\mathcal{F}}. \tag{21.90}$$

That is

$$(u_k)_{\pm}^{(r)} = l_{nk}\left(f_{\pm}^{(r)}\right) Z(nx - k),$$

respectively in $\pm$.

But we proved that

$$\sum_{k=-\infty}^{\infty} (u_k)_{\pm}^{(r)} = L_n\left(f_{\pm}^{(r)}, x\right) =: (u)_{\pm}^{(r)}, \tag{21.91}$$

converges uniformly in $r \in [0, 1]$, respectively in $\pm$.

Then by Remark 21.31, part (2) we get that

$$u := \left\{\left((u)_{-}^{(r)}, (u)_{+}^{(r)}\right) | 0 \le r \le 1\right\} =$$

$$\left\{\left(L_n\left(f_{-}^{(r)}, x\right), L_n\left(f_{+}^{(r)}, x\right)\right) | 0 \le r \le 1\right\} \in \mathbb{R}_{\mathcal{F}}, \tag{21.92}$$

and

$$u = \sum_{k=-\infty}^{\infty *} u_k = \tag{21.93}$$

$$\sum_{k=-\infty}^{\infty *} \left\{\left(l_{nk}\left(f_{-}^{(r)}\right) Z(nx - k), l_{nk}\left(f_{+}^{(r)}\right) Z(nx - k)\right) | 0 \le r \le 1\right\} \overset{(21.85)}{=}$$

$$\sum_{k=-\infty}^{\infty *} \left\{\left(\left(l_{nk}^{\mathcal{F}}(f)\right)_{-}^{(r)}, \left(l_{nk}^{\mathcal{F}}(f)\right)_{+}^{(r)}\right) | 0 \le r \le 1\right\} \odot Z(nx - k) =$$

$$\sum_{k=-\infty}^{\infty *} l_k^{\mathcal{F}}(f) \odot Z(nx - k) = L_n^{\mathcal{F}}(f, x). \tag{21.94}$$

So we have proved (21.76). ■

Based on (21.75) and (21.76) now one may write

$$D\left(A_n^{\mathcal{F}}(f, x), f(x)\right) = \tag{21.95}$$

$$\sup_{r \in [0,1]} \max\left\{\left|\left(A_n\left(f_{-}^{(r)}, x\right)\right) - f_{-}^{(r)}(x)\right|, \left|A_n\left(f_{+}^{(r)}, x\right) - f_{+}^{(r)}(x)\right|\right\},$$

and

$$D\left(L_n^{\mathcal{F}}(f,x), f(x)\right) =$$

$$\sup_{r\in[0,1]} \max\left\{\left|\left(L_n\left(f_-^{(r)}, x\right)\right) - f_-^{(r)}(x)\right|, \left|L_n\left(f_+^{(r)}, x\right) - f_+^{(r)}(x)\right|\right\}. \quad (21.96)$$

We present

Theorem 21.32 *Let* $f \in C_{\mathcal{F}}\left(\prod_{i=1}^N [a_i, b_i]\right)$, $0 < \beta < 1$, $x \in \left(\prod_{i=1}^N [a_i, b_i]\right)$, $N, n \in \mathbb{N}$ *with* $n^{1-\beta} \geq 3$. *Then*

(1)

$$D\left(A_n^{\mathcal{F}}(f,x), f(x)\right) \leq$$

$$(4.019)^N \left[\omega_1^{(\mathcal{F})}\left(f, \frac{1}{n^\beta}\right) + \frac{D^*(f, \widetilde{o})}{\sqrt{\pi}\left(n^{1-\beta}-2\right)e^{\left(n^{1-\beta}-2\right)^2}}\right] =: \rho_1, \quad (21.97)$$

and

(2)

$$D^*\left(A_n^{\mathcal{F}}(f), f\right) \leq \rho_1. \quad (21.98)$$

We notice that $A_n^{\mathcal{F}}(f,x) \overset{D}{\to} f(x)$, *and* $A_n^{\mathcal{F}}(f) \overset{D^*}{\to} f$, *as* $n \to \infty$, *quantitatively with rates.*

Proof Since $f \in C_{\mathcal{F}}\left(\prod_{i=1}^N [a_i, b_i]\right)$ we have that $f_\pm^{(r)} \in C\left(\prod_{i=1}^N [a_i, b_i]\right)$, $\forall\, r \in [0,1]$. Hence by (21.52) we obtain

$$\left|A_n\left(f_\pm^{(r)}, x\right) - f_\pm^{(r)}(x)\right| \leq (4.019)^N \left[\omega_1\left(f_\pm^{(r)}, \frac{1}{n^\beta}\right) + \frac{\left\|f_\pm^{(r)}\right\|_\infty}{\sqrt{\pi}\left(n^{1-\beta}-2\right)e^{\left(n^{1-\beta}-2\right)^2}}\right]$$

$$\overset{\text{(by (21.8), (21.3))}}{\leq} (4.019)^N \left[\omega_1^{(\mathcal{F})}\left(f, \frac{1}{n^\beta}\right) + \frac{D^*(f, \widetilde{o})}{\sqrt{\pi}\left(n^{1-\beta}-2\right)e^{\left(n^{1-\beta}-2\right)^2}}\right]. \quad (21.99)$$

By (21.95) now we are proving the claim. ■

We give

Theorem 21.33 *Let* $f \in C_B\left(\mathbb{R}^N, \mathbb{R}_{\mathcal{F}}\right)$, $0 < \beta < 1$, $x \in \mathbb{R}^N$, $N, n \in \mathbb{N}$, *with* $n^{1-\beta} \geq 3$. *Then*

(1)

$$D\left(B_n^{\mathcal{F}}(f,x), f(x)\right) \leq \tag{21.100}$$

$$\omega_1^{(\mathcal{F})}\left(f, \frac{1}{n^{\beta}}\right) + \frac{D^*(f,\widetilde{o})}{\sqrt{\pi}\left(n^{1-\beta}-2\right)e^{\left(n^{1-\beta}-2\right)^2}} =: \rho_2,$$

and

(2)

$$D^*\left(B_n^{\mathcal{F}}(f), f\right) \leq \rho_2. \tag{21.101}$$

Proof Similar to Theorem 21.32. We use (21.54) and (21.96), along with (21.3) and (21.8). ■

We also present

Theorem 21.34 *All as in Theorem 21.33. Then*

(1)

$$D\left(C_n^{\mathcal{F}}(f,x), f(x)\right) \leq$$

$$\omega_1^{(\mathcal{F})}\left(f, \frac{1}{n} + \frac{1}{n^{\beta}}\right) + \frac{D^*(f,\widetilde{o})}{\sqrt{\pi}\left(n^{1-\beta}-2\right)e^{\left(n^{1-\beta}-2\right)^2}} =: \rho_3, \tag{21.102}$$

and

(2)

$$D^*\left(C_n^{\mathcal{F}}(f), f\right) \leq \rho_3. \tag{21.103}$$

Proof Similar to Theorem 21.32. We use (21.56) and (21.96), along with (21.3) and (21.8). ■

We also give

Theorem 21.35 *All as in Theorem 21.33. Then*

(1)

$$D\left(D_n^{\mathcal{F}}(f,x), f(x)\right) \leq$$

$$\omega_1^{(\mathcal{F})}\left(f, \frac{1}{n} + \frac{1}{n^{\beta}}\right) + \frac{D^*(f,\widetilde{o})}{\sqrt{\pi}\left(n^{1-\beta}-2\right)e^{\left(n^{1-\beta}-2\right)^2}} = \rho_3, \tag{21.104}$$

and

(2)

$$D^*\left(D_n^{\mathcal{F}}(f), f\right) \leq \rho_3. \tag{21.105}$$

Proof Similar to Theorem 21.32. We use (21.58) and (21.96), along with (21.3) and (21.8). ■

Note 21.36 By Theorems 21.33, 21.34, 21.35 for $f \in \left(C_B\left(\mathbb{R}^N, \mathbb{R}_{\mathcal{F}}\right) \cap C_{\mathcal{F}}^U\left(\mathbb{R}^N\right)\right)$, we obtain $\lim_{n\to\infty} D\left(L_n^{\mathcal{F}}(f,x), f(x)\right) = 0$, and $\lim_{n\to\infty} D^*\left(L_n^{\mathcal{F}}(f), f\right) = 0$, quantitatively with rates, where $L_n^{\mathcal{F}}$ is as in (21.69) and (21.70).

Notation 21.37 *Let* $f \in C_{\mathcal{F}}^m\left(\prod_{i=1}^N [a_i, b_i]\right)$, $m, N \in \mathbb{N}$. *Here* f_α *denotes a fuzzy partial derivative with all related notation similar to the real case, see also Remark 21.15 and Notation 21.16. We denote*

$$\omega_{1,m}^{(\mathcal{F})\max}(f_\alpha, h) := \max_{\alpha:|\alpha|=m} \omega_1^{(\mathcal{F})}(f_\alpha, h), \quad h > 0. \tag{21.106}$$

Call also

$$D_m^{*\max}(f_\alpha, \widetilde{o}) := \max_{\alpha:|\alpha|=m} \left\{D^*(f_\alpha, \widetilde{o})\right\}. \tag{21.107}$$

We finally present

Theorem 21.38 *Let* $f \in C_{\mathcal{F}}^m\left(\prod_{i=1}^N [a_i, b_i]\right)$, $0 < \beta < 1$, $n, m, N \in \mathbb{N}$ *with* $n^{1-\beta} \geq 3$, *and* $x \in \left(\prod_{i=1}^N [a_i, b_i]\right)$. *Then*

(1)

$$D\left(A_n^{\mathcal{F}}(f,x), f(x)\right) \leq (4.019)^N \cdot$$

$$\left\{\sum_{j=1}^m \left(\sum_{|\alpha|=j} \frac{D(f_\alpha(x), \widetilde{o})}{\prod_{i=1}^N \alpha_i!}\left[\frac{1}{n^{\beta j}} + \left(\prod_{i=1}^N (b_i - a_i)^{\alpha_i}\right)\frac{1}{2\sqrt{\pi}\left(n^{1-\beta}-2\right)e^{\left(n^{1-\beta}-2\right)^2}}\right]\right)\right.$$

$$+ \frac{N^m}{m! n^{m\beta}} \omega_{1,m}^{(\mathcal{F})\max}\left(f_\alpha, \frac{1}{n^\beta}\right) +$$

$$\left.\left(\frac{\|b-a\|_\infty^m D_m^{*\max}(f_\alpha, \widetilde{o}) N^m}{m!}\right)\frac{1}{\sqrt{\pi}\left(n^{1-\beta}-2\right)e^{\left(n^{1-\beta}-2\right)^2}}\right\}, \tag{21.108}$$

(2)

$$D^{*}\left(A_{n}^{\mathcal{F}}(f), f\right) \leq (4.019)^{N} \cdot$$

$$\left\{\sum_{j=1}^{m}\left(\sum_{|\alpha|=j} \frac{D^{*}\left(f_{\alpha}, \widetilde{o}\right)}{\prod_{i=1}^{N} \alpha_{i}!}\left[\frac{1}{n^{\beta j}}+\left(\prod_{i=1}^{N}\left(b_{i}-a_{i}\right)^{\alpha_{i}}\right) \frac{1}{2 \sqrt{\pi}\left(n^{1-\beta}-2\right) e^{\left(n^{1-\beta}-2\right)^{2}}}\right]\right)\right.$$

$$+\frac{N^{m}}{m! n^{m \beta}} \omega_{1, m}^{(\mathcal{F}) \max }\left(f_{\alpha}, \frac{1}{n^{\beta}}\right)+$$

$$\left.\left(\frac{\|b-a\|_{\infty}^{m} D_{m}^{* \max }\left(f_{\alpha}, \widetilde{o}\right) N^{m}}{m!}\right) \frac{1}{\sqrt{\pi}\left(n^{1-\beta}-2\right) e^{\left(n^{1-\beta}-2\right)^{2}}}\right\}, \quad (21.109)$$

(3) additionally assume that $f_{\alpha}\left(x_{0}\right)=\widetilde{o}$, *for all* $\alpha : |\alpha|=1, \ldots, m$; $x_{0} \in \left(\prod_{i=1}^{N}\left[a_{i}, b_{i}\right]\right)$, *then*

$$D\left(A_{n}^{\mathcal{F}}\left(f, x_{0}\right), f\left(x_{0}\right)\right) \leq (4.019)^{N}\left\{\frac{N^{m}}{m! n^{m \beta}} \omega_{1, m}^{(\mathcal{F}) \max }\left(f_{\alpha}, \frac{1}{n^{\beta}}\right)+\right.$$

$$\left.\left(\frac{\|b-a\|_{\infty}^{m} D_{m}^{* \max }\left(f_{\alpha}, \widetilde{o}\right) N^{m}}{m!}\right) \frac{1}{\sqrt{\pi}\left(n^{1-\beta}-2\right) e^{\left(n^{1-\beta}-2\right)^{2}}}\right\}, \quad (21.110)$$

notice in the last the extremely high rate of convergence at $n^{-\beta(m+1)}$.
Above we derive quantitatively with rates the high speed approximation of $D\left(A_{n}^{\mathcal{F}}(f, x), f(x)\right) \rightarrow 0$, *as* $n \rightarrow \infty$.
Also we establish with rates that $D^{*}\left(A_{n}^{\mathcal{F}}(f), f\right) \rightarrow 0$, *as* $n \rightarrow \infty$, *involving the fuzzy smoothness of* f.

Proof Here $f_{\pm}^{(r)} \in C^{m}\left(\prod_{i=1}^{N}\left[a_{i}, b_{i}\right]\right)$. We observe that

$$\left|A_{n}\left(f_{\pm}^{(r)}, x\right)-f_{\pm}^{(r)}(x)\right| \overset{(21.61)}{\leq} (4.019)^{N} \cdot$$

$$\left\{\sum_{j=1}^{m}\left(\sum_{|\alpha|=j}\left(\frac{\left|\left(f_{\pm}^{(r)}\right)_{\alpha}(x)\right|}{\prod_{i=1}^{N} \alpha_{i}!}\right)\left[\frac{1}{n^{\beta j}}+\left(\prod_{i=1}^{N}\left(b_{i}-a_{i}\right)^{\alpha_{i}}\right)\right.\right.\right.$$

$$\left.\left.\cdot \frac{1}{2\sqrt{\pi}\left(n^{1-\beta}-2\right)e^{\left(n^{1-\beta}-2\right)^2}}\right]\right) + \frac{N^m}{m!n^{m\beta}}\omega_{1,m}^{\max}\left(\left(f_{\pm}^{(r)}\right)_{\alpha}, \frac{1}{n^{\beta}}\right) +$$

$$\left.\left(\frac{\|b-a\|_{\infty}^{m}\left\|\left(f_{\pm}^{(r)}\right)_{\alpha}\right\|_{\infty,m}^{\max} N^m}{m!}\right)\frac{1}{\sqrt{\pi}\left(n^{1-\beta}-2\right)e^{\left(n^{1-\beta}-2\right)^2}}\right\} \overset{(21.13)}{=} \quad (21.111)$$

$$(4.019)^N \left\{\sum_{j=1}^{m}\left(\sum_{|\alpha|=j}\left(\frac{\left|(f_\alpha)_{\pm}^{(r)}(x)\right|}{\prod_{i=1}^{N}\alpha_i!}\right)\left[\frac{1}{n^{\beta j}}+\left(\prod_{i=1}^{N}(b_i-a_i)^{\alpha_i}\right)\right.\right.\right.$$

$$\left.\left.\cdot \frac{1}{2\sqrt{\pi}\left(n^{1-\beta}-2\right)e^{\left(n^{1-\beta}-2\right)^2}}\right]\right) + \frac{N^m}{m!n^{m\beta}}\omega_{1,m}^{\max}\left((f_\alpha)_{\pm}^{(r)}, \frac{1}{n^{\beta}}\right) +$$

$$\left.\left(\frac{\|b-a\|_{\infty}^{m}\left\|(f_\alpha)_{\pm}^{(r)}\right\|_{\infty,m}^{\max} N^m}{m!}\right)\frac{1}{\sqrt{\pi}\left(n^{1-\beta}-2\right)e^{\left(n^{1-\beta}-2\right)^2}}\right\} \overset{(\text{by } (21.3),\ (21.8))}{\leq}$$

$$(21.112)$$

$$(4.019)^N \left\{\sum_{j=1}^{m}\left(\sum_{|\alpha|=j}\left(\frac{D\left(f_\alpha(x),\widetilde{o}\right)}{\prod_{i=1}^{N}\alpha_i!}\right)\left[\frac{1}{n^{\beta j}}+\left(\prod_{i=1}^{N}(b_i-a_i)^{\alpha_i}\right)\right.\right.\right.$$

$$\left.\left.\cdot \frac{1}{2\sqrt{\pi}\left(n^{1-\beta}-2\right)e^{\left(n^{1-\beta}-2\right)^2}}\right]\right) + \frac{N^m}{m!n^{m\beta}}\omega_{1,m}^{(\mathcal{F})\max}\left(f_\alpha, \frac{1}{n^{\beta}}\right) +$$

$$\left.\left(\frac{\|b-a\|_{\infty}^{m} D_m^{*\max}\left(f_\alpha,\widetilde{o}\right)N^m}{m!}\right)\frac{1}{\sqrt{\pi}\left(n^{1-\beta}-2\right)e^{\left(n^{1-\beta}-2\right)^2}}\right\} =: T, \quad (21.113)$$

respectively in $\pm$.

We have proved that

$$\left| A_n \left(f_{\pm}^{(r)}, x \right) - f_{\pm}^{(r)} (x) \right| \leq T, \tag{21.114}$$

$\forall \, r \in [0, 1]$, respectively in $\pm$.

Using (21.95) we obtain

$$D \left(A_n^{\mathcal{F}} (f, x), f (x) \right) \leq T, \tag{21.115}$$

proving the theorem. ∎

References

1. M. Abramowitz, I.A. Stegun (eds.), *Handbook of Mathematical Functions with Formulas, Graphs, and Mathematical Tables* (Dover Publications, New York, 1972)
2. G.A. Anastassiou, Rate of convergence of some neural network operators to the unit-univariate case. J. Math. Anal. Appl. **212**, 237–262 (1997)
3. G.A. Anastassiou, Rate of convergence of some multivariate neural network operators to the unit. Comput. Math. **40**, 1–19 (2000)
4. G.A. Anastassiou, *Quantitative Approximations* (Chapman & Hall/CRC, Boca Raton, New York, 2001)
5. G.A. Anastassiou, Higher order fuzzy approximation by fuzzy wavelet type and neural network operators. Comput. Math. **48**, 1387–1401 (2004)
6. G.A. Anastassiou, Fuzzy approximation by fuzzy convolution type operators. Comput. Math. **48**, 1369–1386 (2004)
7. G.A. Anastassiou, Higher order fuzzy korovkin theory via inequalities. Commun. Appl. Anal. **10**(2), 359–392 (2006)
8. G.A. Anastassiou, Fuzzy korovkin theorems and inequalities. J. Fuzzy Math. **15**(1), 169–205 (2007)
9. G.A. Anastassiou, *Inteligent Systems: Approximation by Artificial Neural Networks* (Springer, Heidelberg, 2011)
10. G.A. Anastassiou, Univariate hyperbolic tangent neural network approximation. Math. Comput. Modell. **53**, 1111–1132 (2011)
11. G.A. Anastassiou, Multivariate hyperbolic tangent neural network approximation. Comput. Math. **61**, 809–821 (2011)
12. G.A. Anastassiou, Multivariate sigmoidal neural network approximation. Neural Netw. **24**, 378–386 (2011)
13. G.A. Anastassiou, Univariate sigmoidal neural network approximation. J. Comput. Anal. Appl. **14**(4), 659–690 (2012)
14. G.A. Anastassiou, Higher order multivariate fuzzy approximation by multivariate fuzzy wavelet type and neural network operators. J. Fuzzy Math. **19**(3), 601–618 (2011)
15. G.A. Anastassiou, Rate of convergence of some multivariate neural network operators to the unit, revisited. J. Comput. Anal. Appl. **15**(7), 1300–1309 (2013)
16. G.A. Anastassiou, in *Approximation by Neural Network Iterates*, ed. by G. Anastassiou, O. Duman, Advances in Applied Mathematics and Approximation Theory: Contributions from AMAT 2012 (Springer, New York, 2013), pp. 1–20
17. G.A. Anastassiou, High degree multivariate fuzzy approximation by quasi-interpolation neural network operators. Discontinuity, Nonlinearity Complex. **2**(2), 125–146 (2013)

18. G.A. Anastassiou, Multivariate error function based neural network approximations. Rev. Anal. Numer. Theor. Approx. Accepted (2014)
19. G.A. Anastassiou, *High Order Multivariate Fuzzy Approximation by Neural Network Operators Based on the Error Function*. Submitted (2014)
20. L.C. Andrews, *Special Functions of Mathematics for Engineers*, 2nd edn. (Mc Graw-Hill, New York, 1992)
21. P. Cardaliaguet, G. Euvrard, Approximation of a function and its derivative with a neural network. Neural Netw. **5**, 207–220 (1992)
22. Z. Chen, F. Cao, The approximation operators with sigmoidal functions. Comput. Math. Appl. **58**, 758–765 (2009)
23. S. Gal, in *Approximation Theory in Fuzzy Setting*, ed. by G. Anastassiou. Handbook of Analytic-Computational Methods in Applied Mathematics, Chap.13 (Chapman & Hall/CRC, Boca Raton, New York, 2000), pp. 617–666
24. R. Goetschel Jr, W. Voxman, Elementary fuzzy calculus. Fuzzy Sets Syst. **18**, 31–43 (1986)
25. S. Haykin, *Neural Networks: A Comprehensive Foundation*, 2nd edn. (Prentice Hall, New York, 1998)
26. O. Kaleva, Fuzzy differential equations. Fuzzy Sets Syst. **24**, 301–317 (1987)
27. Y.K. Kim, B.M. Ghil, Integrals of fuzzy-number-valued functions. Fuzzy Sets Syst. **86**, 213–222 (1997)
28. T.M. Mitchell, *Machine Learning* (WCB-McGraw-Hill, New York, 1997)
29. W. McCulloch, W. Pitts, A logical calculus of the ideas immanent in nervous activity. Bull. Math. Biophys. **7**, 115–133 (1943)
30. C. Wu, Z. Gong, On Henstock integral of fuzzy-number-valued functions (I). Fuzzy Sets Syst. **120**(3), 523–532 (2001)
31. C. Wu, M. Ma, On embedding problem of fuzzy number space: Part 1. Fuzzy Sets Syst. **44**, 33–38 (1991)

Chapter 22
Multivariate Fuzzy-Random Error Function Relied Neural Network Approximations

In this chapter we deal with the rate of multivariate pointwise and uniform convergence in the q-mean to the Fuzzy-Random unit operator multivariate Fuzzy-Random Quasi-Interpolation error function based neural network operators. These multivariate Fuzzy-Random operators arise in a natural way among multivariate Fuzzy-Random neural networks. The rates are given through multivariate Probabilistic-Jackson type inequalities involving the multivariate Fuzzy-Random modulus of continuity of the engaged multivariate Fuzzy-Random function. The plain stochastic extreme analog of this theory is also treated for the stochastic analogous to the above operators, the stochastic Kantorovich type operator and the stochastic quadrature type operator. It follows [18].

22.1 Fuzzy-Random Functions Basics

We start with

Definition 22.1 (*see* [20]) Let $\mu : \mathbb{R} \to [0, 1]$ with the following properties:

(i) is normal, i.e., $\exists\, x_0 \in \mathbb{R} : \mu(x_0) = 1$.
(ii) $\mu(\lambda x + (1-\lambda) y) \geq \min\{\mu(x), \mu(y)\}, \forall\, x, y \in \mathbb{R}, \forall\, \lambda \in [0, 1]$ (μ is called a convex fuzzy subset).
(iii) μ is upper semicontinuous on $\mathbb{R}$, i.e., $\forall\, x_0 \in \mathbb{R}$ and $\forall\, \varepsilon > 0$, $\exists$ neighborhood $V(x_0) : \mu(x) \leq \mu(x_0) + \varepsilon, \forall\, x \in V(x_0)$.
(iv) the set $\overline{\operatorname{supp}(\mu)}$ is compact in $\mathbb{R}$ (where $\operatorname{supp}(\mu) := \{x \in \mathbb{R}; \mu(x) > 0\}$).

We call μ a fuzzy real number. Denote the set of all μ with $\mathbb{R}_{\mathcal{F}}$.

E.g., $\chi_{\{x_0\}} \in \mathbb{R}_{\mathcal{F}}$, for any $x_0 \in \mathbb{R}$, where $\chi_{\{x_0\}}$ is the characteristic function at x_0.

For $0 < r \leq 1$ and $\mu \in \mathbb{R}_{\mathcal{F}}$ define $[\mu]^r := \{x \in \mathbb{R} : \mu(x) \geq r\}$ and $[\mu]^0 := \overline{\{x \in \mathbb{R} : \mu(x) > 0\}}$.

G.A. Anastassiou, *Intelligent Systems II: Complete Approximation by Neural Network Operators*, Studies in Computational Intelligence 608,
DOI 10.1007/978-3-319-20505-2_22

Then it is well known that for each $r \in [0, 1]$, $[\mu]^r$ is a closed and bounded interval of $\mathbb{R}$. For $u, v \in \mathbb{R}_{\mathcal{F}}$ and $\lambda \in \mathbb{R}$, we define uniquely the sum $u \oplus v$ and the product $\lambda \odot u$ by

$$[u \oplus v]^r = [u]^r + [v]^r, \; [\lambda \odot u]^r = \lambda [u]^r, \; \forall r \in [0, 1],$$

where $[u]^r + [v]^r$ means the usual addition of two intervals (as subsets of $\mathbb{R}$) and $\lambda [u]^r$ means the usual product between a scalar and a subset of $\mathbb{R}$ (see, e.g., [20]). Notice $1 \odot u = u$ and it holds $u \oplus v = v \oplus u$, $\lambda \odot u = u \odot \lambda$. If $0 \le r_1 \le r_2 \le 1$ then $[u]^{r_2} \subseteq [u]^{r_1}$. Actually $[u]^r = \left[u_-^{(r)}, u_+^{(r)}\right]$, where $u_-^{(r)} < u_+^{(r)}$, $u_-^{(r)}, u_+^{(r)} \in \mathbb{R}$, $\forall$ $r \in [0, 1]$.

Define

$$D : \mathbb{R}_{\mathcal{F}} \times \mathbb{R}_{\mathcal{F}} \to \mathbb{R}_+ \cup \{0\}$$

by

$$D(u, v) := \sup_{r \in [0,1]} \max \left\{ \left| u_-^{(r)} - v_-^{(r)} \right|, \left| u_+^{(r)} - v_+^{(r)} \right| \right\},$$

where $[v]^r = \left[v_-^{(r)}, v_+^{(r)}\right]$; $u, v \in \mathbb{R}_{\mathcal{F}}$. We have that D is a metric on $\mathbb{R}_{\mathcal{F}}$. Then $(\mathbb{R}_{\mathcal{F}}, D)$ is a complete metric space, see [20], with the properties

$$\begin{aligned} D(u \oplus w, v \oplus w) &= D(u, v), \;\; \forall u, v, w \in \mathbb{R}_{\mathcal{F}}, \\ D(k \odot u, k \odot v) &= |k| \, D(u, v), \;\; \forall u, v \in \mathbb{R}_{\mathcal{F}}, \forall k \in \mathbb{R}, \\ D(u \oplus v, w \oplus e) &\le D(u, w) + D(v, e), \;\; \forall u, v, w, e \in \mathbb{R}_{\mathcal{F}}. \end{aligned} \tag{22.1}$$

Let (M, d) metric space and $f, g : M \to \mathbb{R}_{\mathcal{F}}$ be fuzzy real number valued functions. The distance between f, g is defined by

$$D^*(f, g) := \sup_{x \in M} D(f(x), g(x)).$$

On $\mathbb{R}_{\mathcal{F}}$ we define a partial order by "$\le$": $u, v \in \mathbb{R}_{\mathcal{F}}$, $u \le v$ iff $u_-^{(r)} \le v_-^{(r)}$ and $u_+^{(r)} \le v_+^{(r)}$, $\forall$ $r \in [0, 1]$.

$\sum^*$ denotes the fuzzy summation, $\widetilde{o} := \chi_{\{0\}} \in \mathbb{R}_{\mathcal{F}}$ the neutral element with respect to $\oplus$. For more see also [26, 27].

We need

Definition 22.2 (*see also* [22], *Definition 13.16, p. 654*) Let $(X, \mathcal{B}, P)$ be a probability space. A fuzzy-random variable is a $\mathcal{B}$-measurable mapping $g : X \to \mathbb{R}_{\mathcal{F}}$ (i.e., for any open set $U \subseteq \mathbb{R}_{\mathcal{F}}$, in the topology of $\mathbb{R}_{\mathcal{F}}$ generated by the metric D, we have

$$g^{-1}(U) = \{s \in X; g(s) \in U\} \in \mathcal{B}). \tag{22.2}$$

The set of all fuzzy-random variables is denoted by $\mathcal{L}_{\mathcal{F}}(X, \mathcal{B}, P)$. Let $g_n, g \in \mathcal{L}_{\mathcal{F}}(X, \mathcal{B}, P)$, $n \in \mathbb{N}$ and $0 < q < +\infty$. We say $g_n(s) \overset{\text{“}q-mean\text{”}}{\underset{n \to +\infty}{\to}} g(s)$ if

$$\lim_{n \to +\infty} \int_X D(g_n(s), g(s))^q P(ds) = 0. \tag{22.3}$$

Remark 22.3 (see [22], p. 654) If $f, g \in \mathcal{L}_{\mathcal{F}}(X, \mathcal{B}, P)$, let us denote $F : X \to \mathbb{R}_+ \cup \{0\}$ by $F(s) = D(f(s), g(s))$, $s \in X$. Here, F is $\mathcal{B}$-measurable, because $F = G \circ H$, where $G(u, v) = D(u, v)$ is continuous on $\mathbb{R}_{\mathcal{F}} \times \mathbb{R}_{\mathcal{F}}$, and $H : X \to \mathbb{R}_{\mathcal{F}} \times \mathbb{R}_{\mathcal{F}}$, $H(s) = (f(s), g(s))$, $s \in X$, is $\mathcal{B}$-measurable. This shows that the above convergence in q -mean makes sense.

Definition 22.4 (*see* [22], *p. 654, Definition 13.17*) Let $(T, \mathcal{T})$ be a topological space. A mapping $f : T \to \mathcal{L}_{\mathcal{F}}(X, \mathcal{B}, P)$ will be called fuzzy-random function (or fuzzy-stochastic process) on T. We denote $f(t)(s) = f(t, s)$, $t \in T$, $s \in X$.

Remark 22.5 (see [22], p. 655) Any usual fuzzy real function $f : T \to \mathbb{R}_{\mathcal{F}}$ can be identified with the degenerate fuzzy-random function $f(t, s) = f(t)$, $\forall\, t \in T$, $s \in X$.

Remark 22.6 (see [22], p. 655) Fuzzy-random functions that coincide with probability one for each $t \in T$ will be consider equivalent.

Remark 22.7 (see [22], p. 655) Let $f, g : T \to \mathcal{L}_{\mathcal{F}}(X, \mathcal{B}, P)$. Then $f \oplus g$ and $k \odot f$ are defined pointwise, i.e.,

$$(f \oplus g)(t, s) = f(t, s) \oplus g(t, s),$$
$$(k \odot f)(t, s) = k \odot f(t, s), \; t \in T, s \in X.$$

Definition 22.8 (*see also Definition 13.18, pp. 655–656,* [22]) For a fuzzy-random function $f : W \subseteq \mathbb{R}^N \to \mathcal{L}_{\mathcal{F}}(X, \mathcal{B}, P)$, $N \in \mathbb{N}$, we define the (first) fuzzy-random modulus of continuity

$$\Omega_1^{(\mathcal{F})}(f, \delta)_{L^q} =$$
$$\sup\left\{\left(\int_X D^q(f(x, s), f(y, s)) P(ds)\right)^{\frac{1}{q}} : x, y \in W, \|x - y\|_\infty \le \delta\right\},$$

$0 < \delta, 1 \le q < \infty$.

Definition 22.9 ([16]) Here $1 \le q < +\infty$. Let $f : W \subseteq \mathbb{R}^N \to \mathcal{L}_{\mathcal{F}}(X, \mathcal{B}, P)$, $N \in \mathbb{N}$, be a fuzzy random function. We call f a (q-mean) uniformly continuous fuzzy random function over W, iff $\forall\, \varepsilon > 0\; \exists\, \delta > 0$: whenever $\|x - y\|_\infty \le \delta$, $x, y \in W$, implies that

$$\int_X (D(f(x,s), f(y,s)))^q P(ds) \leq \varepsilon.$$

We denote it as $f \in C_{FR}^{U_q}(W)$.

Proposition 22.10 ([16]) *Let $f \in C_{FR}^{U_q}(W)$, where $W \subseteq \mathbb{R}^N$ is convex.*
Then $\Omega_1^{(\mathcal{F})}(f,\delta)_{L^q} < \infty$, any $\delta > 0$.

Proposition 22.11 ([16]) *Let $f, g : W \subseteq \mathbb{R}^N \to \mathcal{L}_{\mathcal{F}}(X, \mathcal{B}, P)$, $N \in \mathbb{N}$, be fuzzy random functions. It holds*

(i) $\Omega_1^{(\mathcal{F})}(f,\delta)_{L^q}$ is nonnegative and nondecreasing in $\delta > 0$.
(ii) $\lim_{\delta \downarrow 0} \Omega_1^{(\mathcal{F})}(f,\delta)_{L^q} = \Omega_1^{(\mathcal{F})}(f,0)_{L^q} = 0$, iff $f \in C_{FR}^{U_q}(W)$.

We mention

Definition 22.12 (*see also* [6]) Let $f(t,s)$ be a random function (stochastic process) from $W \times (X, \mathcal{B}, P)$, $W \subseteq \mathbb{R}^N$, into $\mathbb{R}$, where $(X, \mathcal{B}, P)$ is a probability space. We define the q-mean multivariate first modulus of continuity of f by

$$\Omega_1(f,\delta)_{L^q} :=$$

$$\sup\left\{\left(\int_X |f(x,s) - f(y,s)|^q P(ds)\right)^{\frac{1}{q}} : x, y \in W, \|x-y\|_\infty \leq \delta\right\},$$

$$\delta > 0, 1 \leq q < \infty. \tag{22.4}$$

The concept of f being (q-mean) uniformly continuous random function is defined the same way as in Definition 22.9, just replace D by $|\cdot|$, etc. We denote it as $f \in C_{\mathbb{R}}^{U_q}(W)$.

Similar properties as in Propositions 22.10 and 22.11 are valid for $\Omega_1(f,\delta)_{L^q}$.

Also we have

Proposition 22.13 ([3]) *Let $Y(t,\omega)$ be a real valued stochastic process such that Y is continuous in $t \in [a,b]$. Then Y is jointly measurable in (t,ω).*

According to [21], p. 94 we have the following.

Definition 22.14 Let $(Y, \mathcal{T})$ be a topological space, with its σ-algebra of Borel sets $\mathcal{B} := \mathcal{B}(Y, \mathcal{T})$ generated by $\mathcal{T}$. If $(X, \mathcal{S})$ is a measurable space, a function $f : X \to Y$ is called measurable iff $f^{-1}(B) \in \mathcal{S}$ for all $B \in \mathcal{B}$.

By Theorem 4.1.6 of [21], p. 89 f as above is measurable iff

$$f^{-1}(C) \in \mathcal{S} \text{ for all } C \in \mathcal{T}.$$

We mention

Theorem 22.15 (see [21], p. 95) *Let $(X, \mathcal{S})$ be a measurable space and (Y, d) be a metric space. Let f_n be measurable functions from X into Y such that for all $x \in X$, $f_n(x) \to f(x)$ in Y. Then f is measurable. I.e., $\lim_{n\to\infty} f_n = f$ is measurable.*

We need also

Proposition 22.16 ([16]) *Let f, g be fuzzy random variables from $\mathcal{S}$ into $\mathbb{R}_{\mathcal{F}}$. Then*

(i) Let $c \in \mathbb{R}$, then $c \odot f$ is a fuzzy random variable.
(ii) $f \oplus g$ is a fuzzy random variable.

Proposition 22.17 *Let $Y\left(\overrightarrow{t}, \omega\right)$ be a real valued multivariate random function (stochastic process) such that Y is continuous in $\overrightarrow{t} \in \prod_{i=1}^{N}[a_i, b_i]$. Then Y is jointly measurable in $\left(\overrightarrow{t}, \omega\right)$ and $\int_{\prod_{i=1}^{N}[a_i, b_i]} Y\left(\overrightarrow{t}, \omega\right) d\overrightarrow{t}$ is a real valued random variable.*

Proof Similar to Proposition 18.14, p. 353 of [7]. ■

22.2 Real Error Function Based Neural Network Approximation Basics [17])

We consider here the (Gauss) error special function

$$\operatorname{erf}(x) = \frac{2}{\sqrt{\pi}} \int_0^x e^{-t^2} dt, \quad x \in \mathbb{R}, \tag{22.5}$$

which is a sigmoidal type function and is a strictly increasing function.

It has the basic properties

$$\operatorname{erf}(0) = 0, \ \operatorname{erf}(-x) = -\operatorname{erf}(x), \ \operatorname{erf}(+\infty) = 1, \ \operatorname{erf}(-\infty) = -1.$$

We also consider the activation function [17]

$$\chi(x) = \frac{1}{4}\left(\operatorname{erf}(x+1) - \operatorname{erf}(x-1)\right) > 0, \ \text{any } x \in \mathbb{R}, \tag{22.6}$$

which is an even function.

Next we follow [17] on χ. We got there $\chi(0) \simeq 0.4215$, and that χ is strictly decreasing on $[0, \infty)$ and strictly increasing on $(-\infty, 0]$, and the x-axis is the horizontal asymptote on χ, i.e. χ is a bell symmetric function.

Theorem 22.18 ([17]) *We have that*

$$\sum_{i=-\infty}^{\infty} \chi(x-i) = 1, \quad all\ x \in \mathbb{R}, \tag{22.7}$$

$$\sum_{i=-\infty}^{\infty} \chi(nx-i) = 1, \quad all\ x \in \mathbb{R}, n \in \mathbb{N}, \tag{22.8}$$

and

$$\int_{-\infty}^{\infty} \chi(x)\, dx = 1, \tag{22.9}$$

that is $\chi(x)$ is a density function on $\mathbb{R}$.

We mention the important

Theorem 22.19 ([17]) *Let $0 < \alpha < 1$, and $n \in \mathbb{N}$ with $n^{1-\alpha} \geq 3$. It holds*

$$\sum_{\substack{k=-\infty \\ : |nx-k| \geq n^{1-\alpha}}}^{\infty} \chi(nx-k) < \frac{1}{2\sqrt{\pi}\left(n^{1-\alpha}-2\right) e^{\left(n^{1-\alpha}-2\right)^2}}. \tag{22.10}$$

Denote by $\lfloor \cdot \rfloor$ the integral part of the number and by $\lceil \cdot \rceil$ the ceiling of the number.

Theorem 22.20 ([17]) *Let $x \in [a,b] \subset \mathbb{R}$ and $n \in \mathbb{N}$ so that $\lceil na \rceil \leq \lfloor nb \rfloor$. It holds*

$$\frac{1}{\sum_{k=\lceil na \rceil}^{\lfloor nb \rfloor} \chi(nx-k)} < \frac{1}{\chi(1)} \simeq 4.019, \ \forall x \in [a,b]. \tag{22.11}$$

Also from [17] we get

$$\lim_{n\to\infty} \sum_{k=\lceil na \rceil}^{\lfloor nb \rfloor} \chi(nx-k) \neq 1, \tag{22.12}$$

at least for some $x \in [a,b]$.

For large enough n we always obtain $\lceil na \rceil \leq \lfloor nb \rfloor$. Also $a \leq \frac{k}{n} \leq b$, iff $\lceil na \rceil \leq k \leq \lfloor nb \rfloor$. In general it holds by (22.8) that

$$\sum_{k=\lceil na \rceil}^{\lfloor nb \rfloor} \chi(nx-k) \leq 1. \tag{22.13}$$

We introduce

$$Z(x_1,\dots,x_N) := Z(x) := \prod_{i=1}^{N} \chi(x_i), \quad x=(x_1,\dots,x_N)\in\mathbb{R}^N,\ N\in\mathbb{N}. \tag{22.14}$$

It has the properties:

(i) $Z(x)>0,\ \forall\, x\in\mathbb{R}^N$,

(ii)

$$\sum_{k=-\infty}^{\infty} Z(x-k) := \sum_{k_1=-\infty}^{\infty}\sum_{k_2=-\infty}^{\infty}\dots\sum_{k_N=-\infty}^{\infty} Z(x_1-k_1,\dots,x_N-k_N)=1, \tag{22.15}$$

where $k:=(k_1,\dots,k_n)\in\mathbb{Z}^N,\ \forall\, x\in\mathbb{R}^N$,

hence

(iii)

$$\sum_{k=-\infty}^{\infty} Z(nx-k) =$$

$$\sum_{k_1=-\infty}^{\infty}\sum_{k_2=-\infty}^{\infty}\dots\sum_{k_N=-\infty}^{\infty} Z(nx_1-k_1,\dots,nx_N-k_N)=1, \tag{22.16}$$

$\forall\, x\in\mathbb{R}^N;\ n\in\mathbb{N}$,

and

(iv)

$$\int_{\mathbb{R}^N} Z(x)\,dx = 1, \tag{22.17}$$

that is Z is a multivariate density function.

Here $\|x\|_\infty := \max\{|x_1|,\dots,|x_N|\}$, $x\in\mathbb{R}^N$, also set $\infty := (\infty,\dots,\infty)$, $-\infty := (-\infty,\dots,-\infty)$ upon the multivariate context, and

$$\begin{aligned}\lceil na\rceil &:= (\lceil na_1\rceil,\dots,\lceil na_N\rceil),\\ \lfloor nb\rfloor &:= (\lfloor nb_1\rfloor,\dots,\lfloor nb_N\rfloor),\end{aligned} \tag{22.18}$$

where $a:=(a_1,\dots,a_N)$, $b:=(b_1,\dots,b_N)$, $k:=(k_1,\dots,k_N)$.

We obviously see that

$$\sum_{k=\lceil na\rceil}^{\lfloor nb\rfloor} Z(nx-k) = \sum_{k=\lceil na\rceil}^{\lfloor nb\rfloor}\left(\prod_{i=1}^{N}\chi(nx_i-k_i)\right) =$$

$$\sum_{k_1=\lceil na_1 \rceil}^{\lfloor nb_1 \rfloor} \cdots \sum_{k_N=\lceil na_N \rceil}^{\lfloor nb_N \rfloor} \left(\prod_{i=1}^{N} \chi(nx_i - k_i) \right) = \prod_{i=1}^{N} \left(\sum_{k_i=\lceil na_i \rceil}^{\lfloor nb_i \rfloor} \chi(nx_i - k_i) \right). \quad (22.19)$$

(v) It holds [17]

$$\sum_{\begin{cases} k = \lceil na \rceil \\ \left\| \frac{k}{n} - x \right\|_{\infty} > \frac{1}{n^{\beta}} \end{cases}}^{\lfloor nb \rfloor} Z(nx - k) \le \frac{1}{2\sqrt{\pi}\left(n^{1-\beta} - 2\right) e^{\left(n^{1-\beta}-2\right)^2}}, \quad (22.20)$$

$0 < \beta < 1$, $n \in \mathbb{N}$; $n^{1-\beta} \ge 3$, $x \in \left(\prod_{i=1}^{N} [a_i, b_i]\right)$.

(vi) it also holds [17]

$$0 < \frac{1}{\sum_{k=\lceil na \rceil}^{\lfloor nb \rfloor} Z(nx - k)} < \frac{1}{(\chi(1))^N} \simeq (4.019)^N, \quad (22.21)$$

$\forall\, x \in \left(\prod_{i=1}^{N} [a_i, b_i]\right)$, $n \in \mathbb{N}$.

It is clear that

(vii)

$$\sum_{\begin{cases} k = -\infty \\ \left\| \frac{k}{n} - x \right\|_{\infty} > \frac{1}{n^{\beta}} \end{cases}}^{\infty} Z(nx - k) \le \frac{1}{2\sqrt{\pi}\left(n^{1-\beta} - 2\right) e^{\left(n^{1-\beta}-2\right)^2}}, \quad (22.22)$$

$0 < \beta < 1$, $n \in \mathbb{N} : n^{1-\beta} \ge 3$, $x \in \left(\prod_{i=1}^{N} [a_i, b_i]\right)$.

Also we get

$$\lim_{n \to \infty} \sum_{k=\lceil na \rceil}^{\lfloor nb \rfloor} Z(nx - k) \ne 1, \quad (22.23)$$

for at least some $x \in \left(\prod_{i=1}^{N} [a_i, b_i]\right)$.

Let $f \in C\left(\prod_{i=1}^{N} [a_i, b_i]\right)$ and $n \in \mathbb{N}$ such that $\lceil na_i \rceil \le \lfloor nb_i \rfloor$, $i = 1, \ldots, N$.

We introduce and define [17] the multivariate positive linear neural network operator ($x := (x_1, \ldots, x_N) \in \left(\prod_{i=1}^{N} [a_i, b_i]\right)$)

$$A_n(f, x_1, \ldots, x_N) := A_n(f, x) := \frac{\sum_{k=\lceil na \rceil}^{\lfloor nb \rfloor} f\left(\frac{k}{n}\right) Z(nx - k)}{\sum_{k=\lceil na \rceil}^{\lfloor nb \rfloor} Z(nx - k)} \quad (22.24)$$

$$:= \frac{\sum_{k_1=\lceil na_1 \rceil}^{\lfloor nb_1 \rfloor} \sum_{k_2=\lceil na_2 \rceil}^{\lfloor nb_2 \rfloor} \cdots \sum_{k_N=\lceil na_N \rceil}^{\lfloor nb_N \rfloor} f\left(\frac{k_1}{n}, \ldots, \frac{k_N}{n}\right)\left(\prod_{i=1}^{N} \chi(nx_i - k_i)\right)}{\prod_{i=1}^{N}\left(\sum_{k_i=\lceil na_i \rceil}^{\lfloor nb_i \rfloor} \chi(nx_i - k_i)\right)}.$$

For large enough n we always obtain $\lceil na_i \rceil \leq \lfloor nb_i \rfloor$, $i = 1, \ldots, N$. Also $a_i \leq \frac{k_i}{n} \leq b_i$, iff $\lceil na_i \rceil \leq k_i \leq \lfloor nb_i \rfloor$, $i = 1, \ldots, N$.

For the last we need, for $f \in C\left(\prod_{i=1}^{N} [a_i, b_i]\right)$ the first multivariate modulus of continuity

$$\omega_1(f, h) := \sup_{\substack{x, y \in \prod_{i=1}^{N} [a_i, b_i] \\ \|x - y\|_\infty \leq h}} |f(x) - f(y)|, \; h > 0. \quad (22.25)$$

It holds that

$$\lim_{h \to 0} \omega_1(f, h) = 0. \quad (22.26)$$

Similarly it is defined for $f \in C_B\left(\mathbb{R}^N\right)$ (continuous and bounded functions on $\mathbb{R}^N$) the $\omega_1(f, h)$, and it has the property (22.26), given that $f \in C_U\left(\mathbb{R}^N\right)$ (uniformly continuous functions on $\mathbb{R}^N$).

When $f \in C_B\left(\mathbb{R}^N\right)$ we define [17],

$$B_n(f, x) := B_n(f, x_1, \ldots, x_N) := \sum_{k=-\infty}^{\infty} f\left(\frac{k}{n}\right) Z(nx - k) := \quad (22.27)$$

$$\sum_{k_1=-\infty}^{\infty} \sum_{k_2=-\infty}^{\infty} \cdots \sum_{k_N=-\infty}^{\infty} f\left(\frac{k_1}{n}, \frac{k_2}{n}, \ldots, \frac{k_N}{n}\right)\left(\prod_{i=1}^{N} \chi(nx_i - k_i)\right),$$

$n \in \mathbb{N}$, $\forall\, x \in \mathbb{R}^N$, $N \in \mathbb{N}$, the multivariate quasi-interpolation neural network operator.

Also for $f \in C_B\left(\mathbb{R}^N\right)$ we define [15] the multivariate Kantorovich type neural network operator

$$C_n(f, x) := C_n(f, x_1, \ldots, x_N) := \sum_{k=-\infty}^{\infty}\left(n^N \int_{\frac{k}{n}}^{\frac{k+1}{n}} f(t)\, dt\right) Z(nx - k) :=$$

$$\sum_{k_1=-\infty}^{\infty}\sum_{k_2=-\infty}^{\infty}\cdots\sum_{k_N=-\infty}^{\infty}\left(n^N\int_{\frac{k_1}{n}}^{\frac{k_1+1}{n}}\int_{\frac{k_2}{n}}^{\frac{k_2+1}{n}}\cdots\int_{\frac{k_N}{n}}^{\frac{k_N+1}{n}} f(t_1,\ldots,t_N)\,dt_1\ldots dt_N\right)$$

$$\cdot\left(\prod_{i=1}^{N}\chi(nx_i-k_i)\right), \tag{22.28}$$

$n \in \mathbb{N}$, $\forall\, x \in \mathbb{R}^N$.

Again for $f \in C_B(\mathbb{R}^N)$, $N \in \mathbb{N}$, we define [17] the multivariate neural network operator of quadrature type $D_n(f,x)$, $n \in \mathbb{N}$, as follows. Let $\theta = (\theta_1,\ldots,\theta_N) \in \mathbb{N}^N$, $\overline{r} = (r_1,\ldots,r_N) \in \mathbb{Z}_+^N$, $w_{\overline{r}} = w_{r_1,r_2,\ldots,r_N} \geq 0$, such that $\sum_{\overline{r}=0}^{\theta} w_{\overline{r}} = \sum_{r_1=0}^{\theta_1}\sum_{r_2=0}^{\theta_2}\ldots,\sum_{r_N=0}^{\theta_N} w_{r_1,r_2,\ldots,r_N} = 1$; $k \in \mathbb{Z}^N$ and

$$\delta_{nk}(f) := \delta_{n,k_1,k_2,\ldots,k_N}(f) := \sum_{\overline{r}=0}^{\theta} w_{\overline{r}} f\left(\frac{k}{n}+\frac{\overline{r}}{n\theta}\right)$$

$$:= \sum_{r_1=0}^{\theta_1}\sum_{r_2=0}^{\theta_2}\cdots\sum_{r_N=0}^{\theta_N} w_{r_1,r_2,\ldots,r_N} f\left(\frac{k_1}{n}+\frac{r_1}{n\theta_1},\frac{k_2}{n}+\frac{r_2}{n\theta_2},\ldots,\frac{k_N}{n}+\frac{r_N}{n\theta_N}\right), \tag{22.29}$$

where $\frac{\overline{r}}{\theta} := \left(\frac{r_1}{\theta_1},\frac{r_2}{\theta_2},\ldots,\frac{r_N}{\theta_N}\right)$.

We put

$$D_n(f,x) := D_n(f,x_1,\ldots,x_N) := \sum_{k=-\infty}^{\infty}\delta_{nk}(f)\,Z(nx-k) \tag{22.30}$$

$$:= \sum_{k_1=-\infty}^{\infty}\sum_{k_2=-\infty}^{\infty}\cdots\sum_{k_N=-\infty}^{\infty}\delta_{n,k_1,k_2,\ldots,k_N}(f)\left(\prod_{i=1}^{N}\chi(nx_i-k_i)\right),$$

$\forall\, x \in \mathbb{R}^N$.

We mention

Theorem 22.21 ([17]) *Let $f \in C\left(\prod_{i=1}^{N}[a_i,b_i]\right)$, $0<\beta<1$, $x \in \left(\prod_{i=1}^{N}[a_i,b_i]\right)$, $N, n \in \mathbb{N}$ with $n^{1-\beta} \geq 3$. Then*

(1)

$$|A_n(f,x)-f(x)| \leq (4.019)^N\left[\omega_1\left(f,\frac{1}{n^\beta}\right)+\frac{\|f\|_\infty}{\sqrt{\pi}\left(n^{1-\beta}-2\right)e^{\left(n^{1-\beta}-2\right)^2}}\right] =: \lambda_1, \tag{22.31}$$

and

(2)

$$\|A_n(f) - f\|_\infty \leq \lambda_1. \tag{22.32}$$

We notice that $\lim_{n\to\infty} A_n(f) = f$, *pointwise and uniformly.*

In this chapter we extend Theorem 22.21 to the fuzzy-random level.

Theorem 22.22 ([17]) *Let* $f \in C_B(\mathbb{R}^N)$, $0 < \beta < 1$, $x \in \mathbb{R}^N$, $N, n \in \mathbb{N}$ *with* $n^{1-\beta} \geq 3$. *Then*

(1)

$$|B_n(f,x) - f(x)| \leq \omega_1\left(f, \frac{1}{n^\beta}\right) + \frac{\|f\|_\infty}{\sqrt{\pi}\left(n^{1-\beta} - 2\right) e^{\left(n^{1-\beta}-2\right)^2}} =: \lambda_2, \tag{22.33}$$

(2)

$$\|B_n(f) - f\|_\infty \leq \lambda_2. \tag{22.34}$$

Given that $f \in \left(C_U\left(\mathbb{R}^N\right) \cap C_B\left(\mathbb{R}^N\right)\right)$, *we obtain* $\lim_{n\to\infty} B_n(f) = f$, *uniformly.*

Theorem 22.23 ([17]) *Let* $f \in C_B(\mathbb{R}^N)$, $0 < \beta < 1$, $x \in \mathbb{R}^N$, $N, n \in \mathbb{N}$ *with* $n^{1-\beta} \geq 3$. *Then*

(1)

$$|C_n(f,x) - f(x)| \leq \omega_1\left(f, \frac{1}{n} + \frac{1}{n^\beta}\right) + \frac{\|f\|_\infty}{\sqrt{\pi}\left(n^{1-\beta} - 2\right) e^{\left(n^{1-\beta}-2\right)^2}} =: \lambda_3, \tag{22.35}$$

(2)

$$\|C_n(f) - f\|_\infty \leq \lambda_3. \tag{22.36}$$

Given that $f \in \left(C_U\left(\mathbb{R}^N\right) \cap C_B\left(\mathbb{R}^N\right)\right)$, *we obtain* $\lim_{n\to\infty} C_n(f) = f$, *uniformly.*

Theorem 22.24 ([17]) *Let* $f \in C_B(\mathbb{R}^N)$, $0 < \beta < 1$, $x \in \mathbb{R}^N$, $N, n \in \mathbb{N}$ *with* $n^{1-\beta} \geq 3$. *Then*

(1)

$$|D_n(f,x) - f(x)| \leq \omega_1\left(f, \frac{1}{n} + \frac{1}{n^\beta}\right) + \frac{\|f\|_\infty}{\sqrt{\pi}\left(n^{1-\beta} - 2\right) e^{\left(n^{1-\beta}-2\right)^2}} = \lambda_3, \tag{22.37}$$

(2)

$$\|D_n(f) - f\|_\infty \leq \lambda_3. \tag{22.38}$$

Given that $f \in \left(C_U\left(\mathbb{R}^N\right) \cap C_B\left(\mathbb{R}^N\right)\right)$, *we obtain* $\lim_{n\to\infty} D_n(f) = f$, *uniformly.*

In this chapter we extend Theorems 22.22–22.24 to the random level. We are also motivated by [1, 2, 4–7, 12, 14–16]. For general knowledge on neural networks we recommend [23–25].

22.3 Main Results

(I) q-mean Approximation by Fuzzy-Random Error Function Based Quasi-interpolation Neural Network Operators

All terms and assumptions here as in Sects. 22.1 and 22.2.

Let $f \in C^{U_q}_{\mathcal{FR}}\left(\prod_{i=1}^{N}[a_i,b_i]\right)$, $1 \le q < +\infty$, $n, N \in \mathbb{N}$, $0 < \beta < 1$, $\overrightarrow{x} \in \left(\prod_{i=1}^{N}[a_i,b_i]\right)$, $(X, \mathcal{B}, P)$ probability space, $s \in X$.

We define the following multivariate fuzzy random error function based quasi-interpolation linear neural network operator

$$\left(A_n^{\mathcal{FR}}(f)\right)\left(\overrightarrow{x}, s\right) := \sum_{\overrightarrow{k}=\lceil na \rceil}^{\lfloor nb \rfloor *} f\left(\frac{\overrightarrow{k}}{n}, s\right) \odot \frac{Z\left(n\overrightarrow{x} - \overrightarrow{k}\right)}{\sum_{\overrightarrow{k}=\lceil na \rceil}^{\lfloor nb \rfloor} Z\left(n\overrightarrow{x} - \overrightarrow{k}\right)}, \qquad (22.39)$$

(see also (22.24)).

We present

Theorem 22.25 *Let $f \in C^{U_q}_{\mathcal{FR}}\left(\prod_{i=1}^{N}[a_i,b_i]\right)$, $0 < \beta < 1$, $\overrightarrow{x} \in \left(\prod_{i=1}^{N}[a_i,b_i]\right)$, $n, N \in \mathbb{N}$, with $n^{1-\beta} \ge 3$, $1 \le q < +\infty$. Assume that $\int_X \left(D^*\left(f(\cdot, s), \widetilde{o}\right)\right)^q P(ds) < \infty$. Then*

(1)

$$\left(\int_X D^q\left(\left(A_n^{\mathcal{FR}}(f)\right)\left(\overrightarrow{x}, s\right), f\left(\overrightarrow{x}, s\right)\right) P(ds)\right)^{\frac{1}{q}} \le (4.019)^N \cdot \qquad (22.40)$$

$$\left\{ \Omega_1\left(f, \frac{1}{n^\beta}\right)_{L^q} + \frac{\left(\int_X \left(D^*\left(f(\cdot,s), \widetilde{o}\right)\right)^q P(ds)\right)^{\frac{1}{q}}}{\sqrt{\pi}\left(n^{1-\beta}-2\right) e^{\left(n^{1-\beta}-2\right)^2}} \right\} =: \lambda_1^{(\mathcal{FR})},$$

(2)

$$\left\| \left(\int_X D^q \left(\left(A_n^{\mathcal{FR}}(f) \right)(\overrightarrow{x}, s), f(\overrightarrow{x}, s) \right) P(ds) \right)^{\frac{1}{q}} \right\|_{\infty, \left(\prod_{i=1}^{N} [a_i, b_i] \right)} \leq \lambda_1^{(\mathcal{FR})}. \tag{22.41}$$

Proof We notice that

$$D\left(f\left(\frac{\overrightarrow{k}}{n}, s \right), f(\overrightarrow{x}, s) \right) \leq D\left(f\left(\frac{\overrightarrow{k}}{n}, s \right), \widetilde{o} \right) + D\left(f(\overrightarrow{x}, s), \widetilde{o} \right) \tag{22.42}$$

$$\leq 2D^* \left(f(\cdot, s), \widetilde{o} \right).$$

Hence

$$D^q \left(f\left(\frac{\overrightarrow{k}}{n}, s \right), f(\overrightarrow{x}, s) \right) \leq 2^q D^{*q} \left(f(\cdot, s), \widetilde{o} \right), \tag{22.43}$$

and

$$\left(\int_X D^q \left(f\left(\frac{\overrightarrow{k}}{n}, s \right), f(\overrightarrow{x}, s) \right) P(ds) \right)^{\frac{1}{q}} \leq 2 \left(\int_X \left(D^* \left(f(\cdot, s), \widetilde{o} \right) \right)^q P(ds) \right)^{\frac{1}{q}}. \tag{22.44}$$

We observe that

$$D\left(\left(A_n^{\mathcal{FR}}(f) \right)(\overrightarrow{x}, s), f(\overrightarrow{x}, s) \right) = \tag{22.45}$$

$$D\left(\sum_{\overrightarrow{k} = \lceil na \rceil}^{\lfloor nb \rfloor *} f\left(\frac{\overrightarrow{k}}{n}, s \right) \odot \frac{Z(nx - k)}{\sum_{\overrightarrow{k} = \lceil na \rceil}^{\lfloor nb \rfloor} Z(nx - k)}, f(\overrightarrow{x}, s) \odot 1 \right) =$$

$$D\left(\sum_{\overrightarrow{k} = \lceil na \rceil}^{\lfloor nb \rfloor *} f\left(\frac{\overrightarrow{k}}{n}, s \right) \odot \frac{Z(nx - k)}{\sum_{\overrightarrow{k} = \lceil na \rceil}^{\lfloor nb \rfloor} Z(nx - k)}, f(\overrightarrow{x}, s) \odot \frac{\sum_{\overrightarrow{k} = \lceil na \rceil}^{\lfloor nb \rfloor} Z(nx - k)}{\sum_{\overrightarrow{k} = \lceil na \rceil}^{\lfloor nb \rfloor} Z(nx - k)} \right) = \tag{22.46}$$

$$D\left(\sum_{\vec{k}=\lceil na\rceil}^{\lfloor nb\rfloor *} f\left(\frac{\vec{k}}{n}, s\right) \odot \frac{Z(nx-k)}{\sum_{\vec{k}=\lceil na\rceil}^{\lfloor nb\rfloor} Z(nx-k)}, \sum_{\vec{k}=\lceil na\rceil}^{\lfloor nb\rfloor *} f(\vec{x}, s) \odot \frac{Z(nx-k)}{\sum_{\vec{k}=\lceil na\rceil}^{\lfloor nb\rfloor} Z(nx-k)}\right)$$

$$\leq \sum_{\vec{k}=\lceil na\rceil}^{\lfloor nb\rfloor} \left(\frac{Z(nx-k)}{\sum_{\vec{k}=\lceil na\rceil}^{\lfloor nb\rfloor} Z(nx-k)}\right) D\left(f\left(\frac{\vec{k}}{n}, s\right), f(\vec{x}, s)\right). \tag{22.47}$$

So that

$$D\left(\left(A_n^{\mathcal{FR}}(f)\right)(\vec{x}, s), f(\vec{x}, s)\right) \leq$$

$$\sum_{\vec{k}=\lceil na\rceil}^{\lfloor nb\rfloor} \left(\frac{Z(nx-k)}{\sum_{\vec{k}=\lceil na\rceil}^{\lfloor nb\rfloor} Z(nx-k)}\right) D\left(f\left(\frac{\vec{k}}{n}, s\right), f(\vec{x}, s)\right) = \tag{22.48}$$

$$\sum_{\substack{\vec{k}=\lceil na\rceil \\ \left\|\frac{\vec{k}}{n}-\vec{x}\right\|_\infty \leq \frac{1}{n^\beta}}}^{\lfloor nb\rfloor} \left(\frac{Z(nx-k)}{\sum_{\vec{k}=\lceil na\rceil}^{\lfloor nb\rfloor} Z(nx-k)}\right) D\left(f\left(\frac{\vec{k}}{n}, s\right), f(\vec{x}, s)\right) +$$

$$\sum_{\substack{\vec{k}=\lceil na\rceil \\ \left\|\frac{\vec{k}}{n}-\vec{x}\right\|_\infty > \frac{1}{n^\beta}}}^{\lfloor nb\rfloor} \left(\frac{Z(nx-k)}{\sum_{\vec{k}=\lceil na\rceil}^{\lfloor nb\rfloor} Z(nx-k)}\right) D\left(f\left(\frac{\vec{k}}{n}, s\right), f(\vec{x}, s)\right).$$

Hence it holds

$$\left(\int_X D^q\left(\left(A_n^{\mathcal{FR}}(f)\right)(\vec{x},s), f(\vec{x},s)\right) P(ds)\right)^{\frac{1}{q}} \leq \tag{22.49}$$

$$\sum_{\substack{\vec{k}=\lceil na\rceil \\ \left\|\frac{\vec{k}}{n}-\vec{x}\right\|_\infty \leq \frac{1}{n^\beta}}}^{\lfloor nb\rfloor} \left(\frac{Z(nx-k)}{\sum_{\vec{k}=\lceil na\rceil}^{\lfloor nb\rfloor} Z(nx-k)}\right)\left(\int_X D^q\left(f\left(\frac{\vec{k}}{n},s\right), f(\vec{x},s)\right) P(ds)\right)^{\frac{1}{q}} +$$

$$\sum_{\substack{\vec{k}=\lceil na\rceil \\ \left\|\frac{\vec{k}}{n}-\vec{x}\right\|_\infty > \frac{1}{n^\beta}}}^{\lfloor nb\rfloor} \left(\frac{Z(nx-k)}{\sum_{\vec{k}=\lceil na\rceil}^{\lfloor nb\rfloor} Z(nx-k)}\right)\left(\int_X D^q\left(f\left(\frac{\vec{k}}{n},s\right), f(\vec{x},s)\right) P(ds)\right)^{\frac{1}{q}} \leq$$

$$\left(\frac{1}{\sum_{\vec{k}=\lceil na\rceil}^{\lfloor nb\rfloor} Z(nx-k)}\right) \cdot \left\{\Omega_1^{(\mathcal{F})}\left(f, \frac{1}{n^\beta}\right)_{L^q} + \right. \tag{22.50}$$

$$\left. 2\left(\int_X \left(D^*\left(f(\cdot,s), \widetilde{o}\right)\right)^q P(ds)\right)^{\frac{1}{q}} \left(\sum_{\substack{\vec{k}=\lceil na\rceil \\ \left\|\frac{\vec{k}}{n}-\vec{x}\right\|_\infty > \frac{1}{n^\beta}}}^{\lfloor nb\rfloor} Z(nx-k)\right)\right\}$$

$$\overset{\text{(by (22.20), (22.21))}}{\leq} (4.019)^N \cdot$$

$$\left\{\Omega_1^{(\mathcal{F})}\left(f, \frac{1}{n^\beta}\right)_{L^q} + \frac{\left(\int_X \left(D^*\left(f(\cdot,s), \widetilde{o}\right)\right)^q P(ds)\right)^{\frac{1}{q}}}{\sqrt{\pi}\left(n^{1-\beta}-2\right) e^{\left(n^{1-\beta}-2\right)^2}}\right\}. \tag{22.51}$$

We have proved claim. ■

Conclusion 22.26 *By Theorem 22.25 we obtain the pointwise and uniform convergences with rates in the q-mean and D-metric of the operator* $A_n^{\mathcal{FR}}$ *to the unit operator for* $f \in C_{\mathcal{FR}}^{U_q}\left(\prod_{i=1}^{N}[a_i, b_i]\right)$.

(II) 1-mean Approximation by Random Error Function Based Quasi-interpolation Neural Network Operators

Let $g \in C_{\mathcal{R}}^{U_1}(\mathbb{R}^N)$, $0 < \beta < 1$, $\overrightarrow{x} \in \mathbb{R}^N$, $n, N \in \mathbb{N}$, with $\|g\|_{\infty,\mathbb{R}^N,X} < \infty$, $(X, \mathcal{B}, P)$ probability space, $s \in X$.

We define

$$B_n^{(\mathcal{R})}(g)\left(\overrightarrow{x}, s\right) := \sum_{\overrightarrow{k}=-\infty}^{\infty} g\left(\frac{\overrightarrow{k}}{n}, s\right) Z\left(n\overrightarrow{x} - \overrightarrow{k}\right), \tag{22.52}$$

(see also (22.27)).

We give

Theorem 22.27 *Let* $g \in C_{\mathcal{R}}^{U_1}(\mathbb{R}^N)$, $0 < \beta < 1$, $\overrightarrow{x} \in \mathbb{R}^N$, $n, N \in \mathbb{N}$, *with* $n^{1-\beta} \geq 3$, $\|g\|_{\infty,\mathbb{R}^N,X} < \infty$. *Then*

(1)

$$\int_X \left|\left(B_n^{(\mathcal{R})}(g)\right)\left(\overrightarrow{x}, s\right) - g\left(\overrightarrow{x}, s\right)\right| P(ds) \tag{22.53}$$

$$\leq \left\{\Omega_1\left(g, \frac{1}{n^{\beta}}\right)_{L^1} + \frac{\|g\|_{\infty,\mathbb{R}^N,X}}{\sqrt{\pi}\left(n^{1-\beta} - 2\right) e^{\left(n^{1-\beta}-2\right)^2}}\right\} =: \mu_1^{(\mathcal{R})},$$

(2)

$$\left\|\int_X \left|\left(B_n^{(\mathcal{R})}(g)\right)\left(\overrightarrow{x}, s\right) - g\left(\overrightarrow{x}, s\right)\right| P(ds)\right\|_{\infty,\mathbb{R}^N} \leq \mu_1^{(\mathcal{R})}. \tag{22.54}$$

Proof Since $\|g\|_{\infty,\mathbb{R}^N,X} < \infty$, then

$$\left|g\left(\frac{\overrightarrow{k}}{n}, s\right) - g\left(\overrightarrow{x}, s\right)\right| \leq 2\|g\|_{\infty,\mathbb{R}^N,X} < \infty. \tag{22.55}$$

Hence

$$\int_X \left|g\left(\frac{\overrightarrow{k}}{n}, s\right) - g\left(\overrightarrow{x}, s\right)\right| P(ds) \leq 2\|g\|_{\infty,\mathbb{R}^N,X} < \infty. \tag{22.56}$$

We observe that

$$\left(B_n^{(\mathcal{R})}(g)\right)\left(\overrightarrow{x},s\right)-g\left(\overrightarrow{x},s\right)=$$

$$\sum_{\overrightarrow{k}=-\infty}^{\infty} g\left(\frac{\overrightarrow{k}}{n},s\right)Z\left(nx-k\right)-g\left(\overrightarrow{x},s\right)\sum_{\overrightarrow{k}=-\infty}^{\infty} Z\left(nx-k\right)= \tag{22.57}$$

$$\left(\sum_{\overrightarrow{k}=-\infty}^{\infty} g\left(\frac{\overrightarrow{k}}{n},s\right)-g\left(\overrightarrow{x},s\right)\right)Z\left(nx-k\right).$$

However it holds

$$\sum_{\overrightarrow{k}=-\infty}^{\infty}\left|g\left(\frac{\overrightarrow{k}}{n},s\right)-g\left(\overrightarrow{x},s\right)\right|Z\left(nx-k\right)\leq 2\left\|g\right\|_{\infty,\mathbb{R}^N,X}<\infty. \tag{22.58}$$

Hence

$$\left|\left(B_n^{(\mathcal{R})}(g)\right)\left(\overrightarrow{x},s\right)-g\left(\overrightarrow{x},s\right)\right|$$

$$\leq\sum_{\overrightarrow{k}=-\infty}^{\infty}\left|g\left(\frac{\overrightarrow{k}}{n},s\right)-g\left(\overrightarrow{x},s\right)\right|Z\left(nx-k\right)= \tag{22.59}$$

$$\sum_{\substack{\overrightarrow{k}=-\infty\\ \left\|\frac{\overrightarrow{k}}{n}-\overrightarrow{x}\right\|_{\infty}\leq\frac{1}{n^{\beta}}}}^{\infty}\left|g\left(\frac{\overrightarrow{k}}{n},s\right)-g\left(\overrightarrow{x},s\right)\right|Z\left(nx-k\right)+$$

$$\sum_{\substack{\overrightarrow{k}=-\infty\\ \left\|\frac{\overrightarrow{k}}{n}-\overrightarrow{x}\right\|_{\infty}>\frac{1}{n^{\beta}}}}^{\infty}\left|g\left(\frac{\overrightarrow{k}}{n},s\right)-g\left(\overrightarrow{x},s\right)\right|Z\left(nx-k\right).$$

Furthermore it holds

$$\left(\int_X\left|\left(B_n^{(\mathcal{R})}(g)\right)\left(\overrightarrow{x},s\right)-g\left(\overrightarrow{x},s\right)\right|P\left(ds\right)\right)\leq$$

$$\sum_{\substack{\vec{k}=-\infty \\ \left\| \frac{\vec{k}}{n} - \vec{x} \right\|_\infty \leq \frac{1}{n^\beta}}}^{\infty} \left(\int_X \left| g\left(\frac{\vec{k}}{n}, s \right) - g\left(\vec{x}, s \right) \right| P(ds) \right) Z(nx - k) + \tag{22.60}$$

$$\sum_{\substack{\vec{k}=-\infty \\ \left\| \frac{\vec{k}}{n} - \vec{x} \right\|_\infty > \frac{1}{n^\beta}}}^{\infty} \left(\int_X \left| g\left(\frac{\vec{k}}{n}, s \right) - g\left(\vec{x}, s \right) \right| P(ds) \right) Z(nx - k) \leq$$

$$\Omega_1 \left(g, \frac{1}{n^\beta} \right)_{L^1} + 2 \|g\|_{\infty, \mathbb{R}^N, X} \sum_{\substack{\vec{k}=-\infty \\ \left\| \frac{\vec{k}}{n} - \vec{x} \right\|_\infty > \frac{1}{n^\beta}}}^{\infty} Z(nx - k) \overset{(22.22)}{\leq}$$

$$\Omega_1 \left(g, \frac{1}{n^\beta} \right)_{L^1} + \frac{\|g\|_{\infty, \mathbb{R}^N, X}}{\sqrt{\pi} \left(n^{1-\beta} - 2 \right) e^{\left(n^{1-\beta} - 2 \right)^2}},$$

proving the claim. ∎

Conclusion 22.28 *By Theorem 22.27 we obtain pointwise and uniform convergences with rates in the* 1*-mean of random operators* $B_n^{(\mathcal{R})}$ *to the unit operator for* $g \in C_{\mathcal{R}}^{U_1}\left(\mathbb{R}^N\right)$.

(III) 1-mean Approximation by Random Error Function Based Multivariate Kantorovich Type Neural Network Operator

Let $g \in C_{\mathcal{R}}^{U_1}\left(\mathbb{R}^N\right)$, $0 < \beta < 1$, $\vec{x} \in \mathbb{R}^N$, $n, N \in \mathbb{N}$, with $\|g\|_{\infty, \mathbb{R}^N, X} < \infty$, $(X, \mathcal{B}, P)$ probability space, $s \in X$.

We define

$$C_n^{(\mathcal{R})}(g)\left(\vec{x}, s\right) := \sum_{\vec{k}=-\infty}^{\infty} \left(n^N \int_{\frac{\vec{k}}{n}}^{\frac{\vec{k}+1}{n}} g\left(\vec{t}, s \right) d\vec{t} \right) Z\left(n\vec{x} - \vec{k} \right), \tag{22.61}$$

(see also (22.28)).

We present

Theorem 22.29 *Let* $g \in C_{\mathcal{R}}^{U_1}\left(\mathbb{R}^N\right)$, $0 < \beta < 1$, $\vec{x} \in \mathbb{R}^N$, $n, N \in \mathbb{N}$, *with* $n^{1-\beta} \geq 3$, $\|g\|_{\infty, \mathbb{R}^N, X} < \infty$. *Then*

(1)

$$\int_X \left| \left(C_n^{(\mathcal{R})}(g) \right) \left(\vec{x}, s \right) - g\left(\vec{x}, s \right) \right| P(ds) \leq$$

$$\left[\Omega_1\left(g, \frac{1}{n}+\frac{1}{n^{\beta}}\right)_{L^1}+\frac{\|g\|_{\infty,\mathbb{R}^N,X}}{\sqrt{\pi}\left(n^{1-\beta}-2\right)e^{\left(n^{1-\beta}-2\right)^2}}\right]=\gamma_1^{(\mathcal{R})}, \tag{22.62}$$

(2)

$$\left\|\int_X\left|\left(C_n^{(\mathcal{R})}(g)\right)(\vec{x},s)-g(\vec{x},s)\right|P(ds)\right\|_{\infty,\mathbb{R}^N}\leq\gamma_1^{(\mathcal{R})}. \tag{22.63}$$

Proof Since $\|g\|_{\infty,\mathbb{R}^N,X}<\infty$, then

$$\left|n^N\int_{\frac{\vec{k}}{n}}^{\frac{\vec{k}+1}{n}}g\left(\vec{t},s\right)d\vec{t}-g(\vec{x},s)\right|=\left|n^N\int_{\frac{\vec{k}}{n}}^{\frac{\vec{k}+1}{n}}\left(g\left(\vec{t},s\right)-g(\vec{x},s)\right)d\vec{t}\right|\leq$$

$$n^N\int_{\frac{\vec{k}}{n}}^{\frac{\vec{k}+1}{n}}\left|g\left(\vec{t},s\right)-g(\vec{x},s)\right|d\vec{t}\leq 2\|g\|_{\infty,\mathbb{R}^N,X}<\infty. \tag{22.64}$$

Hence

$$\int_X\left|n^N\int_{\frac{\vec{k}}{n}}^{\frac{\vec{k}+1}{n}}g\left(\vec{t},s\right)d\vec{t}-g(\vec{x},s)\right|P(ds)\leq 2\|g\|_{\infty,\mathbb{R}^N,X}<\infty. \tag{22.65}$$

We observe that

$$\left(C_n^{(\mathcal{R})}(g)\right)(\vec{x},s)-g(\vec{x},s)=$$

$$\sum_{\vec{k}=-\infty}^{\infty}\left(n^N\int_{\frac{\vec{k}}{n}}^{\frac{\vec{k}+1}{n}}g\left(\vec{t},s\right)d\vec{t}\right)Z\left(n\vec{x}-\vec{k}\right)-g(\vec{x},s)=$$

$$\sum_{\vec{k}=-\infty}^{\infty}\left(n^N\int_{\frac{\vec{k}}{n}}^{\frac{\vec{k}+1}{n}}g\left(\vec{t},s\right)d\vec{t}\right)Z\left(n\vec{x}-\vec{k}\right)-g(\vec{x},s)\sum_{\vec{k}=-\infty}^{\infty}Z\left(n\vec{x}-\vec{k}\right)= \tag{22.66}$$

$$\sum_{\vec{k}=-\infty}^{\infty}\left[\left(n^N\int_{\frac{\vec{k}}{n}}^{\frac{\vec{k}+1}{n}}g\left(\vec{t},s\right)d\vec{t}\right)-g(\vec{x},s)\right]Z\left(n\vec{x}-\vec{k}\right)=$$

$$\sum_{\vec{k}=-\infty}^{\infty}\left[n^N\int_{\frac{\vec{k}}{n}}^{\frac{\vec{k}+1}{n}}\left(g\left(\vec{t},s\right)-g(\vec{x},s)\right)d\vec{t}\right]Z\left(n\vec{x}-\vec{k}\right).$$

However it holds

$$\sum_{\vec{k}=-\infty}^{\infty}\left[n^N\int_{\frac{\vec{k}}{n}}^{\frac{\vec{k}+1}{n}}\left|g\left(\vec{t},s\right)-g\left(\vec{x},s\right)\right|d\vec{t}\right]Z\left(n\vec{x}-\vec{k}\right)\leq 2\|g\|_{\infty,\mathbb{R}^N,X}<\infty. \tag{22.67}$$

Hence

$$\left|\left(C_n^{(\mathcal{R})}(g)\right)\left(\vec{x},s\right)-g\left(\vec{x},s\right)\right|\leq$$

$$\sum_{\vec{k}=-\infty}^{\infty}\left[n^N\int_{\frac{\vec{k}}{n}}^{\frac{\vec{k}+1}{n}}\left|g\left(\vec{t},s\right)-g\left(\vec{x},s\right)\right|d\vec{t}\right]Z\left(n\vec{x}-\vec{k}\right)= \tag{22.68}$$

$$\sum_{\substack{\vec{k}=-\infty\\ \left\|\frac{\vec{k}}{n}-\vec{x}\right\|_\infty\leq\frac{1}{n^\beta}}}^{\infty}\left[n^N\int_{\frac{\vec{k}}{n}}^{\frac{\vec{k}+1}{n}}\left|g\left(\vec{t},s\right)-g\left(\vec{x},s\right)\right|d\vec{t}\right]Z\left(n\vec{x}-\vec{k}\right)+$$

$$\sum_{\substack{\vec{k}=-\infty\\ \left\|\frac{\vec{k}}{n}-\vec{x}\right\|_\infty>\frac{1}{n^\beta}}}^{\infty}\left[n^N\int_{\frac{\vec{k}}{n}}^{\frac{\vec{k}+1}{n}}\left|g\left(\vec{t},s\right)-g\left(\vec{x},s\right)\right|d\vec{t}\right]Z\left(n\vec{x}-\vec{k}\right)= \tag{22.69}$$

$$\sum_{\substack{\vec{k}=-\infty\\ \left\|\frac{\vec{k}}{n}-\vec{x}\right\|_\infty\leq\frac{1}{n^\beta}}}^{\infty}\left[n^N\int_{0}^{\frac{1}{n}}\left|g\left(\vec{t}+\frac{\vec{k}}{n},s\right)-g\left(\vec{x},s\right)\right|d\vec{t}\right]Z\left(n\vec{x}-\vec{k}\right)+$$

$$\sum_{\substack{\vec{k}=-\infty\\ \left\|\frac{\vec{k}}{n}-\vec{x}\right\|_\infty>\frac{1}{n^\beta}}}^{\infty}\left[n^N\int_{0}^{\frac{1}{n}}\left|g\left(\vec{t}+\frac{\vec{k}}{n},s\right)-g\left(\vec{x},s\right)\right|d\vec{t}\right]Z\left(n\vec{x}-\vec{k}\right). \tag{22.70}$$

Furthermore it holds

$$\left(\int_X\left|\left(C_n^{(\mathcal{R})}(g)\right)\left(\vec{x},s\right)-g\left(\vec{x},s\right)\right|P(ds)\right)\underset{\text{(by Fubini's theorem)}}{\leq}$$

$$\sum_{\substack{\vec{k}=-\infty \\ \left\|\frac{\vec{k}}{n}-\vec{x}\right\|_{\infty}\leq\frac{1}{n^{\beta}}}}^{\infty}\left[n^{N}\int_{0}^{\frac{1}{n}}\left(\int_{X}\left|g\left(\vec{t}+\frac{\vec{k}}{n},s\right)-g\left(\vec{x},s\right)\right|P(ds)\right)d\vec{t}\right]Z\left(n\vec{x}-\vec{k}\right)+$$

$$\sum_{\substack{\vec{k}=-\infty \\ \left\|\frac{\vec{k}}{n}-\vec{x}\right\|_{\infty}>\frac{1}{n^{\beta}}}}^{\infty}\left[n^{N}\int_{0}^{\frac{1}{n}}\left(\int_{X}\left|g\left(\vec{t}+\frac{\vec{k}}{n},s\right)-g\left(\vec{x},s\right)\right|P(ds)\right)d\vec{t}\right]Z\left(n\vec{x}-\vec{k}\right)\leq \tag{22.71}$$

$$\Omega_1\left(g,\frac{1}{n}+\frac{1}{n^{\beta}}\right)_{L^1}+2\|g\|_{\infty,\mathbb{R}^N,X}\sum_{\substack{\vec{k}=-\infty \\ \left\|\frac{\vec{k}}{n}-\vec{x}\right\|_{\infty}>\frac{1}{n^{\beta}}}}^{\infty}Z\left(n\vec{x}-\vec{k}\right)\overset{(22.22)}{\leq}$$

$$\Omega_1\left(g,\frac{1}{n}+\frac{1}{n^{\beta}}\right)_{L^1}+\frac{\|g\|_{\infty,\mathbb{R}^N,X}}{\sqrt{\pi}\left(n^{1-\beta}-2\right)e^{\left(n^{1-\beta}-2\right)^2}}, \tag{22.72}$$

proving the claim. ■

Conclusion 22.30 *By Theorem 22.29 we obtain pointwise and uniform convergences with rates in the 1-mean of random operators $C_n^{(\mathcal{R})}$ to the unit operator for $g \in C_{\mathcal{R}}^{U_1}(\mathbb{R}^N)$.*

(IV) 1-mean Approximation by Random error function based multivariate quadrature type neural network operator

Let $g \in C_{\mathcal{R}}^{U_1}(\mathbb{R}^N)$, $0<\beta<1$, $\vec{x}\in\mathbb{R}^N$, $n,N\in\mathbb{N}$, with $\|g\|_{\infty,\mathbb{R}^N,X}<\infty$, $(X,\mathcal{B},P)$ probability space, $s\in X$.

We define

$$D_n^{(\mathcal{R})}(g)\left(\vec{x},s\right):=\sum_{\vec{k}=-\infty}^{\infty}\left(\delta_{n\vec{k}}(g)\right)(s)\,Z\left(n\vec{x}-\vec{k}\right), \tag{22.73}$$

where

$$\left(\delta_{n\vec{k}}(g)\right)(s):=\sum_{\vec{r}=0}^{\vec{\theta}}w_{\vec{r}}\,g\left(\frac{\vec{k}}{n}+\frac{\vec{r}}{n\vec{\theta}},s\right), \tag{22.74}$$

(see also (22.29), (22.30)).

We finally give

Theorem 22.31 *Let $g \in C_{\mathcal{R}}^{U_1}(\mathbb{R}^N)$, $0<\beta<1$, $\vec{x}\in\mathbb{R}^N$, $n,N\in\mathbb{N}$, with $n^{1-\beta}\geq 3$, $\|g\|_{\infty,\mathbb{R}^N,X}<\infty$. Then*

(1)

$$\int_X \left| \left(D_n^{(\mathcal{R})}(g) \right) \left(\overrightarrow{x}, s \right) - g \left(\overrightarrow{x}, s \right) \right| P(ds) \leq$$

$$\left\{ \Omega_1 \left(g, \frac{1}{n} + \frac{1}{n^\beta} \right)_{L^1} + \frac{\|g\|_{\infty, \mathbb{R}^N, X}}{\sqrt{\pi} \left(n^{1-\beta} - 2 \right) e^{\left(n^{1-\beta} - 2 \right)^2}} \right\} =: \gamma_1^{(\mathcal{R})}, \quad (22.75)$$

(2)

$$\left\| \int_X \left| \left(D_n^{(\mathcal{R})}(g) \right) \left(\overrightarrow{x}, s \right) - g \left(\overrightarrow{x}, s \right) \right| P(ds) \right\|_{\infty, \mathbb{R}^N} \leq \gamma_1^{(\mathcal{R})}. \quad (22.76)$$

Proof Notice that

$$\left| \left(\delta_{n \overrightarrow{k}}(g) \right)(s) - g \left(\overrightarrow{x}, s \right) \right| =$$

$$\left| \sum_{\overrightarrow{r}=0}^{\overrightarrow{\theta}} w_{\overrightarrow{r}} \left(g \left(\frac{\overrightarrow{k}}{n} + \frac{\overrightarrow{r}}{n \overrightarrow{\theta}}, s \right) - g \left(\overrightarrow{x}, s \right) \right) \right| \leq$$

$$\sum_{\overrightarrow{r}=0}^{\overrightarrow{\theta}} w_{\overrightarrow{r}} \left| g \left(\frac{\overrightarrow{k}}{n} + \frac{\overrightarrow{r}}{n \overrightarrow{\theta}}, s \right) - g \left(\overrightarrow{x}, s \right) \right| \leq 2 \|g\|_{\infty, \mathbb{R}^N, X} < \infty. \quad (22.77)$$

Hence

$$\int_X \left| \left(\delta_{n \overrightarrow{k}}(g) \right)(s) - g \left(\overrightarrow{x}, s \right) \right| P(ds) \leq 2 \|g\|_{\infty, \mathbb{R}^N, X} < \infty. \quad (22.78)$$

We observe that

$$\left(D_n^{(\mathcal{R})}(g) \right) \left(\overrightarrow{x}, s \right) - g \left(\overrightarrow{x}, s \right) =$$

$$\sum_{\overrightarrow{k}=-\infty}^{\infty} \left(\delta_{n \overrightarrow{k}}(g) \right)(s) \, Z \left(n \overrightarrow{x} - \overrightarrow{k} \right) - g \left(\overrightarrow{x}, s \right) =$$

$$\sum_{\overrightarrow{k}=-\infty}^{\infty} \left(\left(\delta_{n \overrightarrow{k}}(g) \right)(s) - g \left(\overrightarrow{x}, s \right) \right) Z \left(n \overrightarrow{x} - \overrightarrow{k} \right). \quad (22.79)$$

Thus

$$\left|D_n^{(\mathcal{R})}(g)\left(\overrightarrow{x}, s\right) - g\left(\overrightarrow{x}, s\right)\right| \leq$$

$$\sum_{\overrightarrow{k}=-\infty}^{\infty} \left|\left(\delta_{n\overrightarrow{k}}(g)\right)(s) - g\left(\overrightarrow{x}, s\right)\right| Z\left(n\overrightarrow{x} - \overrightarrow{k}\right) \leq 2\|g\|_{\infty,\mathbb{R}^N,X} < \infty. \tag{22.80}$$

Hence it holds

$$\left|\left(D_n^{(\mathcal{R})}(g)\right)\left(\overrightarrow{x}, s\right) - g\left(\overrightarrow{x}, s\right)\right| \leq$$

$$\sum_{\overrightarrow{k}=-\infty}^{\infty} \left|\left(\delta_{n\overrightarrow{k}}(g)\right)(s) - g\left(\overrightarrow{x}, s\right)\right| Z\left(n\overrightarrow{x} - \overrightarrow{k}\right) =$$

$$\sum_{\substack{\overrightarrow{k}=-\infty \\ \left\|\frac{\overrightarrow{k}}{n} - \overrightarrow{x}\right\|_{\infty} \leq \frac{1}{n^{\beta}}}}^{\infty} \left|\left(\delta_{n\overrightarrow{k}}(g)\right)(s) - g\left(\overrightarrow{x}, s\right)\right| Z\left(n\overrightarrow{x} - \overrightarrow{k}\right) +$$

$$\sum_{\substack{\overrightarrow{k}=-\infty \\ \left\|\frac{\overrightarrow{k}}{n} - \overrightarrow{x}\right\|_{\infty} > \frac{1}{n^{\beta}}}}^{\infty} \left|\left(\delta_{n\overrightarrow{k}}(g)\right)(s) - g\left(\overrightarrow{x}, s\right)\right| Z\left(n\overrightarrow{x} - \overrightarrow{k}\right). \tag{22.81}$$

Furthermore we derive

$$\left(\int_X \left|\left(D_n^{(\mathcal{R})}(g)\right)\left(\overrightarrow{x}, s\right) - g\left(\overrightarrow{x}, s\right)\right| P(ds)\right) \leq$$

$$\sum_{\substack{\overrightarrow{k}=-\infty \\ \left\|\frac{\overrightarrow{k}}{n} - \overrightarrow{x}\right\|_{\infty} \leq \frac{1}{n^{\beta}}}}^{\infty} \sum_{\overrightarrow{r}=0}^{\overrightarrow{\theta}} w_{\overrightarrow{r}} \left(\int_X \left|g\left(\frac{\overrightarrow{k}}{n} + \frac{\overrightarrow{r}}{n\overrightarrow{\theta}}, s\right) - g\left(\overrightarrow{x}, s\right)\right| P(ds)\right) Z\left(n\overrightarrow{x} - \overrightarrow{k}\right) \tag{22.82}$$

$$+\left(\sum_{\substack{\vec{k}=-\infty \\ \left\|\frac{\vec{k}}{n}-\vec{x}\right\|_{\infty}>\frac{1}{n^{\beta}}}}^{\infty} Z\left(n\vec{x}-\vec{k}\right)\right) 2\|g\|_{\infty,\mathbb{R}^N,X} \overset{(22.22)}{\leq}$$

$$\Omega_1\left(g, \frac{1}{n}+\frac{1}{n^{\beta}}\right)_{L^1} + \frac{\|g\|_{\infty,\mathbb{R}^N,X}}{\sqrt{\pi}\left(n^{1-\beta}-2\right)e^{\left(n^{1-\beta}-2\right)^2}}, \tag{22.83}$$

proving the claim. ■

Conclusion 22.32 *From Theorem 22.31 we obtain pointwise and uniform convergences with rates in the* 1*-mean of random operators* $D_n^{(\mathcal{R})}$ *to the unit operator for* $g \in C_{\mathcal{R}}^{U_1}\left(\mathbb{R}^N\right)$.

References

1. G.A. Anastassiou, Rate of convergence of Fuzzy neural network operators, univariate case. J. Fuzzy Math. **10**(3), 755–780 (2002)
2. G.A. Anastassiou, Higher order fuzzy approximation by fuzzy wavelet type and neural network operators. Comput. Math. **48**, 1387–1401 (2004)
3. G.A. Anastassiou, Univariate fuzzy-random neural network approximation operators, in *Computers and Mathematics with Applications*, ed. by G. Anastassiou. Special Issue/ Proceedings of Special Session "Computational Methods in Analysis", AMS meeting in Orlando, vol. 48, pp. 1263–1283. Florida, Nov. 2002
4. G.A. Anastassiou, Higher order fuzzy korovkin theory via inequalities. Commun. Appl. Anal. **10**(2), 359–392 (2006)
5. G.A. Anastassiou, Fuzzy korovkin theorems and inequalities. J. Fuzzy Math. **15**(1), 169–205 (2007)
6. G.A. Anastassiou, Multivariate stochastic korovkin theory given quantitatively. Math. Comput. Model. **48**, 558–580 (2008)
7. G.A. Anastassiou, *Fuzzy Mathematics: Approximation Theory* (Springer, Heidelberg, New York, 2010)
8. G.A. Anastassiou, *Inteligent Systems: Approximation by Artificial Neural Networks* (Springer, Heidelberg, 2011)
9. G.A. Anastassiou, Univariate hyperbolic tangent neural network approximation. Math. Comput. Model. **53**, 1111–1132 (2011)
10. G.A. Anastassiou, Multivariate hyperbolic tangent neural network approximation. Comput. Math. **61**, 809–821 (2011)
11. G.A. Anastassiou, Multivariate sigmoidal neural network approximation. Neural Netw. **24**, 378–386 (2011)
12. G.A. Anastassiou, Higher order multivariate fuzzy approximation by multivariate fuzzy wavelet type and neural network operators. J. Fuzzy Math. **19**(3), 601–618 (2011)
13. G.A. Anastassiou, Univariate sigmoidal neural network approximation. J. Comput. Anal. Appl. **14**(4), 659–690 (2012)

14. G.A. Anastassiou, Rate of convergence of some multivariate neural network operators to the unit, revisited. J. Comput. Anal. Appl. **15**(7), 1300–1309 (2013)
15. G.A. Anastassiou, Higher order multivariate fuzzy approximation by basic neural network operators. Cubo **16**(03), 21–35 (2014)
16. G.A. Anastassiou, Multivariate fuzzy-random quasi-interpolation neural network approximation operators. J. Fuzzy Math. **22**(1), 167–184 (2014)
17. G.A. Anastassiou, Multivariate error function based neural network approximations. Rev. Anal. Numer. Theor. Approx., accepted 2014
18. G.A. Anastassiou, Multivariate fuzzy-random error function based neural network approximation, submitted 2014
19. Z. Chen, F. Cao, The approximation operators with sigmoidal functions. Comput. Math. Appl. **58**, 758–765 (2009)
20. W. Congxin, G. Zengtai, On henstock integral of interval-valued functions and fuzzy valued functions. Fuzzy Sets Syst. **115**(3), 377–391 (2000)
21. R.M. Dudley, *Real Analysis and Probability* (Wadsworth and Brooks/Cole Mathematics Series, Pacific Grove, 1989)
22. S. Gal, in *Approximation Theory in Fuzzy Setting*, ed. by G. Anastassiou. Handbook of Analytic-Computational Methods in Applied Mathematics, Chap. 13, (Chapman and Hall/CRC, Boca Raton, 2000) pp. 617–666
23. S. Haykin, *Neural Networks: A Comprehensive Foundation*, 2nd edn. (Prentice Hall, New York, 1998)
24. W. McCulloch, W. Pitts, A logical calculus of the ideas immanent in nervous activity. Bull. Math. Biophys. **7**, 115–133 (1943)
25. T.M. Mitchell, *Machine Learning* (WCB-McGraw-Hill, New York, 1997)
26. C. Wu, Z. Gong, On henstock integral of fuzzy-number-valued functions. Fuzzy Sets Syst. **120**(3), 523–532 (2001)
27. C. Wu, M. Ma, On embedding problem of fuzzy number space: part 1. Fuzzy Sets Syst. **44**, 33–38 (1991)

Chapter 23
Approximation by Perturbed Neural Networks

This chapter deals with the determination of the rate of convergence to the unit of each of three newly introduced here perturbed normalized neural network operators of one hidden layer. These are given through the modulus of continuity of the involved function or its high order derivative and that appears in the right-hand side of the associated Jackson type inequalities. The activation function is very general, especially it can derive from any sigmoid or bell-shaped function. The right hand sides of our convergence inequalities do not depend on the activation function. The sample functionals are of Stancu, Kantorovich and Quadrature types. We give applications for the first derivative of the involved function. It follows [6].

23.1 Introduction

Feed-forward neural networks (FNNs) with one hidden layer, the only type of networks we deal with in this chapter, are mathematically expressed as

$$N_n(x) = \sum_{j=0}^{n} c_j \sigma(\langle a_j \cdot x \rangle + b_j), \quad x \in \mathbb{R}^s, s \in \mathbb{N},$$

where for $0 \leq j \leq n$, $b_j \in \mathbb{R}$ are the thresholds, $a_j \in \mathbb{R}^s$ are the connection weights, $c_j \in \mathbb{R}$ are the coefficients, $\langle a_j \cdot x \rangle$ is the inner product of a_j and x, and σ is the activation function of the network. In many fundamental network models, the activation function is the sigmoidal function of logistic type or other sigmoidal function or bell-shaped function.

It is well known that FNNs are universal approximators. Theoretically, any continuous function defined on a compact set can be approximated to any desired degree of accuracy by increasing the number of hidden neurons. It was proved by Cybenko [15] and Funahashi [17], that any continuous function can be approximated on a compact

G.A. Anastassiou, *Intelligent Systems II: Complete Approximation by Neural Network Operators*, Studies in Computational Intelligence 608,
DOI 10.1007/978-3-319-20505-2_23

set with uniform topology by a network of the form $N_n(x)$, using any continuous, sigmoidal activation function. Hornik et al. in [20], have shown that any measurable function can be approached with such a network. Furthermore, these authors proved in [21], that any function of the Sobolev spaces can be approached with all derivatives. A variety of density results on FNN approximations to multivariate functions were later established by many authors using different methods, for more or less general situations: [22] by Leshno et al. [26] by Mhaskar and Micchelli, [12] by Chui and Li, [11] by Chen and Chen, [18] by Hahm and Hong, etc.

Usually these results only give theorems about the existence of an approximation. A related and important problem is that of complexity: determining the number of neurons required to guarantee that all functions belonging to a space can be approximated to the prescribed degree of accuracy ϵ.

Barron [7] shows that if the function is supposed to satisfy certain conditions expressed in terms of its Fourier transform, and if each of the neurons evaluates a sigmoidal activation function, then at most $O\left(\epsilon^{-2}\right)$ neurons are needed to achieve the order of approximation ϵ. Some other authors have published similar results on the complexity of FNN approximations: Mhaskar and Micchelli [27], Suzuki [30], Maiorov and Meir [23], Makovoz [24], Ferrari and Stengel [16], Xu and Cao [31], Cao et al. [8], etc.

Cardaliaguet and Euvrard were the first, see [9], to describe precisely and study neural network approximation operators to the unit operator. Namely they proved: be given $f : \mathbb{R} \to \mathbb{R}$ a continuous bounded function and b a centered bell-shaped function, then the functions

$$F_n(x) = \sum_{k=-n^2}^{n^2} \frac{f\left(\frac{k}{n}\right)}{In^{\alpha}} b\left(n^{1-\alpha}\left(x - \frac{k}{n}\right)\right),$$

where $I := \int_{-\infty}^{\infty} b(t)\, dt$, $0 < \alpha < 1$, converge uniformly on compacta to f.

You see above that the weights $\frac{f\left(\frac{k}{n}\right)}{In^{\alpha}}$ are explicitly known, for the first time shown in [9].

Still the work [9] is qualitative and not quantitative.

The author in [1–3], see Chaps. 2–5, was the first to establish neural network approximations to continuous functions with rates, that is quantitative works, by very specifically defined neural network operators of Cardaliagnet-Euvrard and "Squashing" types, by employing the modulus of continuity of the engaged function or its high order derivative, and producing very tight Jackson type inequalities. He treats there both the univariate and multivariate cases. The defining these operators "bell-shaped" and "squashing" function are assumed to be of compact support. Also in [3] he gives the Nth order asymptotic expansion for the error of weak approximation of these two operators to a special natural class of smooth functions, see Chaps. 4 and 5 there.

Though the work in [1–3], was quantitative, the rate of convergence was not precisely determined.

Finally the author in [4, 5], by normalizing his operators he achieved to determine the exact rates of convergence.

In this chapter the author continuous and completes his related work, by introducing three new perturbed neural network operators of Cardaliaguet-Euvrard type.

The sample coefficients $f\left(\frac{k}{n}\right)$ are replaced by three suitable natural perturbations, what is actually happens in reality of a neural network operation.

The calculation of $f\left(\frac{k}{n}\right)$ at the neurons many times are not calculated as such, but rather in a distored way.

Next we justify why we take here the activation function to be of compact support, of course it helps us to conduct our study.

The activation function, same as transfer function or learning rule, is connected and associated to firing of neurons. Firing, which sends electric pulses or an output signal to other neurons, occurs when the activation level is above the threshold level set by the learning rule.

Each Neural Network firing is essentially of finite time duration. Essentially the firing in time decays, but in practice sends positive energy over a finite time interval.

Thus by using an activation function of compact support, in practice we do not alter much of the good results of our approximation.

To be more precise, we may take the compact support to be a large symmetric to the origin interval. This is what happens in real time with the firing of neurons.

For more about neural networks in general we refer to [10, 13, 14, 19, 25, 28].

23.2 Basics

Here the activation function $b : \mathbb{R} \to \mathbb{R}_+$ is of compact support $[-T, T]$, $T > 0$. That is $b(x) > 0$ for any $x \in [-T, T]$, and clearly b may have jump discontinuities. Also the shape of the graph of b could be anything. Typically in neural networks approximation we take b as a sigmoidal function or bell-shaped function, of course here of compact support $[-T, T]$, $T > 0$.

Example 23.1 (i) b can be the characteristic function on $[-1, 1]$,

(ii) b can be that hat function over $[-1, 1]$, i.e.,

$$b(x) = \begin{cases} 1 + x, & -1 \leq x \leq 0, \\ 1 - x, & 0 < x \leq 1, \\ 0, & elsewhere, \end{cases}$$

(iii) the truncated sigmoidals

$$b(x) = \begin{cases} \frac{1}{1+e^{-x}} \text{ or } \tanh x \text{ or } \operatorname{erf}(x), \text{ for } x \in [-T, T], \text{ with large } T > 0, \\ 0, \; x \in \mathbb{R} - [-T, T], \end{cases}$$

(iv) the truncated Gompertz function

$$b(x) = \begin{cases} e^{-\alpha e^{-\beta x}}, x \in [-T, T]; \ \alpha, \beta > 0; \ \text{large } T > 0, \\ 0, x \in \mathbb{R} - [-T, T], \end{cases}$$

The Gompertz functions are also sigmoidal functions, with wide applications to many applied fields, e.g. demography and tumor growth modeling, etc.

So the general function b we will be using here covers all kinds of activation functions in neural network approximations.

Here we consider functions $f : \mathbb{R} \to \mathbb{R}$ that are either continuous and bounded, or uniformly continuous.

Let here the parameters $\mu, \nu \geq 0$; $\mu_i, \nu_i \geq 0$, $i = 1, \ldots, r \in \mathbb{N}$; $w_i \geq 0$: $\sum_{i=1}^{r} w_i = 1$; $0 < \alpha < 1$, $x \in \mathbb{R}$, $n \in \mathbb{N}$.

We use here the first modulus of continuity

$$\omega_1(f, \delta) := \sup_{\substack{x, y \in \mathbb{R} \\ |x-y| \leq \delta}} |f(x) - f(y)|,$$

and given that f is uniformly continuous we get $\lim_{\delta \to 0} \omega_1(f, \delta) = 0$.

In this chapter mainly we study the pointwise convergence with rates over $\mathbb{R}$, to the unit operator, of the following one hidden layer normalized neural network perturbed operators,

(i)

$$\left(H_n^*(f)\right)(x) = \frac{\sum_{k=-n^2}^{n^2} \left(\sum_{i=1}^{r} w_i f\left(\frac{k+\mu_i}{n+\nu_i}\right)\right) b\left(n^{1-\alpha}\left(x - \frac{k}{n}\right)\right)}{\sum_{k=-n^2}^{n^2} b\left(n^{1-\alpha}\left(x - \frac{k}{n}\right)\right)}, \quad (23.1)$$

(ii) the Kantorovich type

$$\left(K_n^*(f)\right)(x) = \frac{\sum_{k=-n^2}^{n^2} \left(\sum_{i=1}^{r} w_i (n+\nu_i) \int_{\frac{k+\mu_i}{n+\nu_i}}^{\frac{k+\mu_i+1}{n+\nu_i}} f(t)\, dt\right) b\left(n^{1-\alpha}\left(x - \frac{k}{n}\right)\right)}{\sum_{k=-n^2}^{n^2} b\left(n^{1-\alpha}\left(x - \frac{k}{n}\right)\right)}, \quad (23.2)$$

and

(iii) the quadrature type

$$\left(M_n^*(f)\right)(x) = \frac{\sum_{k=-n^2}^{n^2} \left(\sum_{i=1}^{r} w_i f\left(\frac{k}{n} + \frac{i}{nr}\right)\right) b\left(n^{1-\alpha}\left(x - \frac{k}{n}\right)\right)}{\sum_{k=-n^2}^{n^2} b\left(n^{1-\alpha}\left(x - \frac{k}{n}\right)\right)}. \quad (23.3)$$

Similar operators defined for bell-shaped functions and sample coefficients $f\left(\frac{k}{n}\right)$ were studied initially in [1–5, 9], etc.

Here we study the generalized perturbed cases of these operators.

Operator K_n^* in the corresponding Signal Processing context, represents the natural called "time-jitter" error, where the sample information is calculated in a perturbed neighborhood of $\frac{k+\mu}{n+\nu}$ rather than exactly at the node $\frac{k}{n}$.

The perturbed sample coefficients $f\left(\frac{k+\mu}{n+\nu}\right)$ with $0 \le \mu \le \nu$, were first used by Stancu [29], in a totally different context, generalizing Bernstein operators approximation on $C([0,1])$.

The terms in the ratio of sums (23.1)–(23.3) are nonzero, iff

$$\left| n^{1-\alpha}\left(x - \frac{k}{n}\right)\right| \le T \text{ , i.e. } \left| x - \frac{k}{n}\right| \le \frac{T}{n^{1-\alpha}} \tag{23.4}$$

iff

$$nx - Tn^{\alpha} \le k \le nx + Tn^{\alpha}. \tag{23.5}$$

In order to have the desired order of the numbers

$$-n^2 \le nx - Tn^{\alpha} \le nx + Tn^{\alpha} \le n^2, \tag{23.6}$$

it is sufficiently enough to assume that

$$n \ge T + |x|. \tag{23.7}$$

When $x \in [-T, T]$ it is enough to assume $n \ge 2T$, which implies (23.7).

Proposition 23.2 ([1]) *Let $a \le b$, $a, b \in \mathbb{R}$. Let $card(k)$ (≥ 0) be the maximum number of integers contained in $[a, b]$. Then*

$$\max(0, (b-a)-1) \le card(k) \le (b-a)+1. \tag{23.8}$$

Note 23.3 *We would like to establish a lower bound on $card(k)$ over the interval $[nx - Tn^{\alpha}, nx + Tn^{\alpha}]$. From Proposition 23.2 we get that*

$$card(k) \ge \max\left(2Tn^{\alpha} - 1, 0\right). \tag{23.9}$$

We obtain $card(k) \ge 1$, if

$$2Tn^{\alpha} - 1 \ge 1 \text{ iff } n \ge T^{-\frac{1}{\alpha}}. \tag{23.10}$$

So to have the desired order (23.6) and $card(k) \ge 1$ over $[nx - Tn^{\alpha}, nx + Tn^{\alpha}]$, we need to consider

$$n \ge \max\left(T + |x|, T^{-\frac{1}{\alpha}}\right). \tag{23.11}$$

Also notice that card $(k) \to +\infty$, as $n \to +\infty$.

Denote by $[\cdot]$ the integral part of a number and by $\lceil \cdot \rceil$ its ceiling.

So under assumption (23.11), the operators H_n^*, K_n^*, M_n^*, collapse to

(i)

$$\left(H_n^*(f)\right)(x) = \frac{\sum_{k=\lceil nx-Tn^{\alpha}\rceil}^{[nx+Tn^{\alpha}]} \left(\sum_{i=1}^{r} w_i f\left(\frac{k+\mu_i}{n+\nu_i}\right)\right) b\left(n^{1-\alpha}\left(x-\frac{k}{n}\right)\right)}{\sum_{k=\lceil nx-Tn^{\alpha}\rceil}^{[nx+Tn^{\alpha}]} b\left(n^{1-\alpha}\left(x-\frac{k}{n}\right)\right)}, \tag{23.12}$$

(ii)

$$\left(K_n^*(f)\right)(x) = \frac{\sum_{k=\lceil nx-Tn^{\alpha}\rceil}^{[nx+Tn^{\alpha}]} \left(\sum_{i=1}^{r} w_i (n+\nu_i) \int_{\frac{k+\mu_i}{n+\nu_i}}^{\frac{k+\mu_i+1}{n+\nu_i}} f(t)\,dt\right) b\left(n^{1-\alpha}\left(x-\frac{k}{n}\right)\right)}{\sum_{k=\lceil nx-Tn^{\alpha}\rceil}^{[nx+Tn^{\alpha}]} b\left(n^{1-\alpha}\left(x-\frac{k}{n}\right)\right)}, \tag{23.13}$$

and

(iii)

$$\left(M_n^*(f)\right)(x) = \frac{\sum_{k=\lceil nx-Tn^{\alpha}\rceil}^{[nx+Tn^{\alpha}]} \left(\sum_{i=1}^{r} w_i f\left(\frac{k}{n}+\frac{i}{nr}\right)\right) b\left(n^{1-\alpha}\left(x-\frac{k}{n}\right)\right)}{\sum_{k=\lceil nx-Tn^{\alpha}\rceil}^{[nx+Tn^{\alpha}]} b\left(n^{1-\alpha}\left(x-\frac{k}{n}\right)\right)}. \tag{23.14}$$

We make

Remark 23.4 Let k as in (23.5). We observe that

$$\left|\frac{k+\mu}{n+\nu} - x\right| \leq \left|\frac{k}{n+\nu} - x\right| + \frac{\mu}{n+\nu}. \tag{23.15}$$

Next we see

$$\left|\frac{k}{n+\nu} - x\right| \leq \left|\frac{k}{n+\nu} - \frac{k}{n}\right| + \left|\frac{k}{n} - x\right| \overset{(23.4)}{\leq} \frac{\nu|k|}{n(n+\nu)} + \frac{T}{n^{1-\alpha}}$$

(by $|k| \leq \max\left(|nx - Tn^{\alpha}|, |nx + Tn^{\alpha}|\right) \leq n|x| + Tn^{\alpha}$)

$$\leq \left(\frac{\nu}{n+\nu}\right)\left(|x| + \frac{T}{n^{1-\alpha}}\right) + \frac{T}{n^{1-\alpha}}. \tag{23.16}$$

Consequently it holds

$$\left|\frac{k+\mu}{n+\nu} - x\right| \leq \left(\frac{\nu}{n+\nu}\right)\left(|x| + \frac{T}{n^{1-\alpha}}\right) + \frac{T}{n^{1-\alpha}} + \frac{\mu}{n+\nu} = \tag{23.17}$$

$$\left(\frac{\nu |x| + \mu}{n+\nu}\right) + \left(1 + \frac{\nu}{n+\nu}\right)\frac{T}{n^{1-\alpha}}.$$

Hence we obtain

$$\omega_1\left(f, \left|\frac{k+\mu}{n+\nu} - x\right|\right) \overset{(23.17)}{\leq} \omega_1\left(f, \left(\frac{\nu |x| + \mu}{n+\nu}\right) + \left(1 + \frac{\nu}{n+\nu}\right)\frac{T}{n^{1-\alpha}}\right), \tag{23.18}$$

where $\mu, \nu \geq 0$, $0 < \alpha < 1$, so that the dominant speed above is $\frac{1}{n^{1-\alpha}}$.

Also, by change of variable method, the operator K_n^* could be written conveniently as follows:

(ii)'

$$\left(K_n^*(f)\right)(x) =$$

$$\frac{\sum_{k=\lceil nx-Tn^{\alpha}\rceil}^{[nx+Tn^{\alpha}]} \left(\sum_{i=1}^{r} w_i (n+\nu_i) \int_0^{\frac{1}{n+\nu_i}} f\left(t + \frac{k+\mu_i}{n+\nu_i}\right) dt\right) b\left(n^{1-\alpha}\left(x - \frac{k}{n}\right)\right)}{\sum_{k=\lceil nx-Tn^{\alpha}\rceil}^{[nx+Tn^{\alpha}]} b\left(n^{1-\alpha}\left(x - \frac{k}{n}\right)\right)}. \tag{23.19}$$

23.3 Main Results

We present our first approximation result

Theorem 23.5 *Let $x \in \mathbb{R}$, $T > 0$ and $n \in \mathbb{N}$ such that $n \geq \max\left(T + |x|, T^{-\frac{1}{\alpha}}\right)$. Then*

$$\left|\left(H_n^*(f)\right)(x) - f(x)\right| \leq \sum_{i=1}^{r} w_i \omega_1\left(f, \left(\frac{\nu_i |x| + \mu_i}{n+\nu_i}\right) + \left(1 + \frac{\nu_i}{n+\nu_i}\right)\frac{T}{n^{1-\alpha}}\right) \tag{23.20}$$

$$\leq \max_{i\in\{1,\dots,r\}} \left\{\omega_1\left(f, \left(\frac{\nu_i |x| + \mu_i}{n+\nu_i}\right) + \left(1 + \frac{\nu_i}{n+\nu_i}\right)\frac{T}{n^{1-\alpha}}\right)\right\}.$$

Proof We notice that

$$\left(H_n^*(f)\right)(x) - f(x) = \frac{\sum_{k=\lceil nx-Tn^{\alpha}\rceil}^{[nx+Tn^{\alpha}]} \left(\sum_{i=1}^{r} w_i f\left(\frac{k+\mu_i}{n+\nu_i}\right)\right) b\left(n^{1-\alpha}\left(x - \frac{k}{n}\right)\right)}{\sum_{k=\lceil nx-Tn^{\alpha}\rceil}^{[nx+Tn^{\alpha}]} b\left(n^{1-\alpha}\left(x - \frac{k}{n}\right)\right)} - f(x) = \tag{23.21}$$

$$\frac{\sum_{k=\lceil nx-Tn^{\alpha}\rceil}^{[nx+Tn^{\alpha}]}\left(\sum_{i=1}^{r} w_i f\left(\frac{k+\mu_i}{n+\nu_i}\right)\right) b\left(n^{1-\alpha}\left(x-\frac{k}{n}\right)\right) - f(x)\sum_{k=\lceil nx-Tn^{\alpha}\rceil}^{[nx+Tn^{\alpha}]} b\left(n^{1-\alpha}\left(x-\frac{k}{n}\right)\right)}{\sum_{k=\lceil nx-Tn^{\alpha}\rceil}^{[nx+Tn^{\alpha}]} b\left(n^{1-\alpha}\left(x-\frac{k}{n}\right)\right)} =$$

$$\frac{\sum_{k=\lceil nx-Tn^{\alpha}\rceil}^{[nx+Tn^{\alpha}]}\left(\left(\sum_{i=1}^{r} w_i f\left(\frac{k+\mu_i}{n+\nu_i}\right)\right) - f(x)\right) b\left(n^{1-\alpha}\left(x-\frac{k}{n}\right)\right)}{\sum_{k=\lceil nx-Tn^{\alpha}\rceil}^{[nx+Tn^{\alpha}]} b\left(n^{1-\alpha}\left(x-\frac{k}{n}\right)\right)} = \tag{23.22}$$

$$\frac{\sum_{k=\lceil nx-Tn^{\alpha}\rceil}^{[nx+Tn^{\alpha}]}\left(\sum_{i=1}^{r} w_i \left(f\left(\frac{k+\mu_i}{n+\nu_i}\right) - f(x)\right)\right) b\left(n^{1-\alpha}\left(x-\frac{k}{n}\right)\right)}{\sum_{k=\lceil nx-Tn^{\alpha}\rceil}^{[nx+Tn^{\alpha}]} b\left(n^{1-\alpha}\left(x-\frac{k}{n}\right)\right)}.$$

Hence it holds

$$\left|\left(H_n^*(f)\right)(x) - f(x)\right| \le$$

$$\frac{\sum_{k=\lceil nx-Tn^{\alpha}\rceil}^{[nx+Tn^{\alpha}]}\left(\sum_{i=1}^{r} w_i \left|f\left(\frac{k+\mu_i}{n+\nu_i}\right) - f(x)\right|\right) b\left(n^{1-\alpha}\left(x-\frac{k}{n}\right)\right)}{\sum_{k=\lceil nx-Tn^{\alpha}\rceil}^{[nx+Tn^{\alpha}]} b\left(n^{1-\alpha}\left(x-\frac{k}{n}\right)\right)} \le \tag{23.23}$$

$$\frac{\sum_{k=\lceil nx-Tn^{\alpha}\rceil}^{[nx+Tn^{\alpha}]}\left(\sum_{i=1}^{r} w_i \omega_1\left(f, \left|\frac{k+\mu_i}{n+\nu_i} - x\right|\right)\right) b\left(n^{1-\alpha}\left(x-\frac{k}{n}\right)\right)}{\sum_{k=\lceil nx-Tn^{\alpha}\rceil}^{[nx+Tn^{\alpha}]} b\left(n^{1-\alpha}\left(x-\frac{k}{n}\right)\right)} \overset{(23.18)}{\le}$$

$$\frac{\sum_{k=\lceil nx-Tn^{\alpha}\rceil}^{[nx+Tn^{\alpha}]}\left(\sum_{i=1}^{r} w_i \omega_1\left(f, \left(\frac{\nu_i|x|+\mu_i}{n+\nu_i}\right) + \left(1+\frac{\nu_i}{n+\nu_i}\right)\frac{T}{n^{1-\alpha}}\right)\right) b\left(n^{1-\alpha}\left(x-\frac{k}{n}\right)\right)}{\sum_{k=\lceil nx-Tn^{\alpha}\rceil}^{[nx+Tn^{\alpha}]} b\left(n^{1-\alpha}\left(x-\frac{k}{n}\right)\right)} = \tag{23.24}$$

$$\frac{\left[\sum_{i=1}^{r} w_i \omega_1\left(f, \left(\frac{\nu_i|x|+\mu_i}{n+\nu_i}\right) + \left(1+\frac{\nu_i}{n+\nu_i}\right)\frac{T}{n^{1-\alpha}}\right)\right]\left(\sum_{k=\lceil nx-Tn^{\alpha}\rceil}^{[nx+Tn^{\alpha}]} b\left(n^{1-\alpha}\left(x-\frac{k}{n}\right)\right)\right)}{\sum_{k=\lceil nx-Tn^{\alpha}\rceil}^{[nx+Tn^{\alpha}]} b\left(n^{1-\alpha}\left(x-\frac{k}{n}\right)\right)}$$

$$= \sum_{i=1}^{r} w_i \omega_1\left(f, \left(\frac{\nu_i|x|+\mu_i}{n+\nu_i}\right) + \left(1+\frac{\nu_i}{n+\nu_i}\right)\frac{T}{n^{1-\alpha}}\right) \le$$

$$\max_{i\in\{1,\ldots,r\}}\left\{\omega_1\left(f, \left(\frac{\nu_i|x|+\mu_i}{n+\nu_i}\right) + \left(1+\frac{\nu_i}{n+\nu_i}\right)\frac{T}{n^{1-\alpha}}\right)\right\}, \tag{23.25}$$

proving the claim. ■

Corollary 23.6 *Let* $x \in \left[-T^*, T^*\right]$, $T^* > 0$, $n \in \mathbb{N} : n \geq \max\left(T + T^*, T^{-\frac{1}{\alpha}}\right)$, $T > 0$. *Then*

$$\left\| H_n^*(f) - f \right\|_{\infty,[-T^*,T^*]} \leq \sum_{i=1}^{r} w_i \omega_1 \left(f, \left(\frac{\nu_i T^* + \mu_i}{n + \nu_i} \right) + \left(1 + \frac{\nu_i}{n + \nu_i} \right) \frac{T}{n^{1-\alpha}} \right) \tag{23.26}$$

$$\leq \max_{i \in \{1,\ldots,r\}} \left\{ \omega_1 \left(f, \left(\frac{\nu_i T^* + \mu_i}{n + \nu_i} \right) + \left(1 + \frac{\nu_i}{n + \nu_i} \right) \frac{T}{n^{1-\alpha}} \right) \right\}.$$

Proof By (23.20). ■

We continue with

Theorem 23.7 *Let* $x \in \mathbb{R}$, $T > 0$ *and* $n \in \mathbb{N}$ *such that* $n \geq \max\left(T + |x|, T^{-\frac{1}{\alpha}}\right)$. *Then*

$$\left| \left(K_n^*(f) \right)(x) - f(x) \right| \leq \tag{23.27}$$

$$\max_{i \in \{1,\ldots,r\}} \left\{ \omega_1 \left(f, \left(\frac{\nu_i |x| + \mu_i + 1}{n + \nu_i} \right) + \left(1 + \frac{\nu_i}{n + \nu_i} \right) \frac{T}{n^{1-\alpha}} \right) \right\}.$$

Proof Call

$$\delta_{n,k}(f) = \sum_{i=1}^{r} w_i (n + \nu_i) \int_0^{\frac{1}{n+\nu_i}} f\left(t + \frac{k + \mu_i}{n + \nu_i} \right) dt. \tag{23.28}$$

We observe that

$$\delta_{n,k}(f) - f(x) = \sum_{i=1}^{r} w_i (n + \nu_i) \int_0^{\frac{1}{n+\nu_i}} f\left(t + \frac{k + \mu_i}{n + \nu_i} \right) dt - f(x) =$$

$$\sum_{i=1}^{r} w_i (n + \nu_i) \int_0^{\frac{1}{n+\nu_i}} \left(f\left(t + \frac{k + \mu_i}{n + \nu_i} \right) - f(x) \right) dt. \tag{23.29}$$

Hence it holds

$$\left| \delta_{n,k}(f) - f(x) \right| \leq \sum_{i=1}^{r} w_i (n + \nu_i) \int_0^{\frac{1}{n+\nu_i}} \left| f\left(t + \frac{k + \mu_i}{n + \nu_i} \right) - f(x) \right| dt \leq$$

$$\sum_{i=1}^{r} w_i (n + \nu_i) \int_0^{\frac{1}{n+\nu_i}} \omega_1 \left(f, |t| + \left| \frac{k + \mu_i}{n + \nu_i} - x \right| \right) dt \leq \tag{23.30}$$

$$\sum_{i=1}^{r} w_i \omega_1 \left(f, \frac{1}{n+\nu_i} + \left| \frac{k+\mu_i}{n+\nu_i} - x \right| \right) \overset{(23.17)}{\leq}$$

$$\sum_{i=1}^{r} w_i \omega_1 \left(f, \left(\frac{\nu_i |x| + \mu_i + 1}{n+\nu_i} \right) + \left(1 + \frac{\nu_i}{n+\nu_i} \right) \frac{T}{n^{1-\alpha}} \right) \leq \tag{23.31}$$

$$\max_{i \in \{1,\ldots,r\}} \left\{ \omega_1 \left(f, \left(\frac{\nu_i |x| + \mu_i + 1}{n+\nu_i} \right) + \left(1 + \frac{\nu_i}{n+\nu_i} \right) \frac{T}{n^{1-\alpha}} \right) \right\}.$$

We proved that

$$\left| \delta_{n,k}(f) - f(x) \right| \leq$$

$$\max_{i \in \{1,\ldots,r\}} \left\{ \omega_1 \left(f, \left(\frac{\nu_i |x| + \mu_i + 1}{n+\nu_i} \right) + \left(1 + \frac{\nu_i}{n+\nu_i} \right) \frac{T}{n^{1-\alpha}} \right) \right\}. \tag{23.32}$$

Therefore by (23.19) and (23.28) we get

$$\left(K_n^*(f) \right)(x) - f(x) = \frac{\sum_{k=\lceil nx-Tn^{\alpha} \rceil}^{[nx+Tn^{\alpha}]} \delta_{n,k}(f)\, b\left(n^{1-\alpha}\left(x - \frac{k}{n} \right) \right)}{\sum_{k=\lceil nx-Tn^{\alpha} \rceil}^{[nx+Tn^{\alpha}]} b\left(n^{1-\alpha}\left(x - \frac{k}{n} \right) \right)} - f(x) =$$

$$\frac{\sum_{k=\lceil nx-Tn^{\alpha} \rceil}^{[nx+Tn^{\alpha}]} \delta_{n,k}(f)\, b\left(n^{1-\alpha}\left(x - \frac{k}{n} \right) \right) - f(x) \sum_{k=\lceil nx-Tn^{\alpha} \rceil}^{[nx+Tn^{\alpha}]} b\left(n^{1-\alpha}\left(x - \frac{k}{n} \right) \right)}{\sum_{k=\lceil nx-Tn^{\alpha} \rceil}^{[nx+Tn^{\alpha}]} b\left(n^{1-\alpha}\left(x - \frac{k}{n} \right) \right)}$$

$$= \frac{\sum_{k=\lceil nx-Tn^{\alpha} \rceil}^{[nx+Tn^{\alpha}]} \left(\delta_{n,k}(f) - f(x) \right) b\left(n^{1-\alpha}\left(x - \frac{k}{n} \right) \right)}{\sum_{k=\lceil nx-Tn^{\alpha} \rceil}^{[nx+Tn^{\alpha}]} b\left(n^{1-\alpha}\left(x - \frac{k}{n} \right) \right)}. \tag{23.33}$$

Consequently we obtain

$$\left| \left(K_n^*(f) \right)(x) - f(x) \right| \leq \frac{\sum_{k=\lceil nx-Tn^{\alpha} \rceil}^{[nx+Tn^{\alpha}]} \left| \delta_{n,k}(f) - f(x) \right| b\left(n^{1-\alpha}\left(x - \frac{k}{n} \right) \right)}{\sum_{k=\lceil nx-Tn^{\alpha} \rceil}^{[nx+Tn^{\alpha}]} b\left(n^{1-\alpha}\left(x - \frac{k}{n} \right) \right)} \tag{23.34}$$

$$\overset{(23.32)}{\leq} \left(\sum_{k=\lceil nx-Tn^{\alpha} \rceil}^{[nx+Tn^{\alpha}]} b\left(n^{1-\alpha}\left(x - \frac{k}{n} \right) \right) \right) \cdot$$

$$\frac{\max_{i \in \{1,\ldots,r\}} \left\{ \omega_1 \left(f, \left(\frac{\nu_i |x| + \mu_i + 1}{n+\nu_i} \right) + \left(1 + \frac{\nu_i}{n+\nu_i} \right) \frac{T}{n^{1-\alpha}} \right) \right\}}{\sum_{k=\lceil nx-Tn^{\alpha} \rceil}^{[nx+Tn^{\alpha}]} b\left(n^{1-\alpha}\left(x - \frac{k}{n} \right) \right)} =$$

$$\max_{i\in\{1,\ldots,r\}}\left\{\omega_1\left(f,\left(\frac{\nu_i|x|+\mu_i+1}{n+\nu_i}\right)+\left(1+\frac{\nu_i}{n+\nu_i}\right)\frac{T}{n^{1-\alpha}}\right)\right\}, \tag{23.35}$$

proving the claim. ■

Corollary 23.8 *Let* $x\in[-T^*,T^*]$, $T^*>0$, $n\in\mathbb{N}: n\geq\max\left(T+T^*,T^{-\frac{1}{\alpha}}\right)$, $T>0$. *Then*

$$\left\|K_n^*(f)-f\right\|_{\infty,[-T^*,T^*]}\leq \tag{23.36}$$

$$\max_{i\in\{1,\ldots,r\}}\left\{\omega_1\left(f,\left(\frac{\nu_iT^*+\mu_i+1}{n+\nu_i}\right)+\left(1+\frac{\nu_i}{n+\nu_i}\right)\frac{T}{n^{1-\alpha}}\right)\right\}.$$

Proof By (23.27). ■

We also give

Theorem 23.9 *Let* $x\in\mathbb{R}$, $T>0$ *and* $n\in\mathbb{N}$ *such that* $n\geq\max\left(T+|x|,T^{-\frac{1}{\alpha}}\right)$. *Then*

$$\left|M_n^*(f)(x)-f(x)\right|\leq\omega_1\left(f,\frac{1}{n}+\frac{T}{n^{1-\alpha}}\right). \tag{23.37}$$

Proof Let k as in (23.5). Set

$$\lambda_{nk}(f)=\sum_{i=1}^{r}w_if\left(\frac{k}{n}+\frac{i}{nr}\right),$$

then

$$\lambda_{nk}(f)-f(x)=\sum_{i=1}^{r}w_i\left(f\left(\frac{k}{n}+\frac{i}{nr}\right)-f(x)\right). \tag{23.38}$$

Then

$$|\lambda_{nk}(f)-f(x)|\leq\sum_{i=1}^{r}w_i\left|f\left(\frac{k}{n}+\frac{i}{nr}\right)-f(x)\right|\leq$$

$$\sum_{i=1}^{r}w_i\omega_1\left(f,\left|\frac{k}{n}-x\right|+\frac{i}{nr}\right)\leq \tag{23.39}$$

$$\sum_{i=1}^{r}w_i\omega_1\left(f,\frac{T}{n^{1-\alpha}}+\frac{1}{n}\right)=\omega_1\left(f,\frac{1}{n}+\frac{T}{n^{1-\alpha}}\right).$$

Hence it holds

$$|\lambda_{nk}(f)-f(x)|\leq\omega_1\left(f,\frac{1}{n}+\frac{T}{n^{1-\alpha}}\right). \tag{23.40}$$

By (23.14) we can write and use next

$$\left(M_n^*(f)\right)(x) = \frac{\sum_{k=\lceil nx-Tn^{\alpha}\rceil}^{[nx+Tn^{\alpha}]} \lambda_{nk}(f)\, b\left(n^{1-\alpha}\left(x-\frac{k}{n}\right)\right)}{\sum_{k=\lceil nx-Tn^{\alpha}\rceil}^{[nx+Tn^{\alpha}]} b\left(n^{1-\alpha}\left(x-\frac{k}{n}\right)\right)}. \tag{23.41}$$

That is we have

$$M_n^*(f)(x) - f(x) = \frac{\sum_{k=\lceil nx-Tn^{\alpha}\rceil}^{[nx+Tn^{\alpha}]} \left(\lambda_{nk}(f) - f(x)\right) b\left(n^{1-\alpha}\left(x-\frac{k}{n}\right)\right)}{\sum_{k=\lceil nx-Tn^{\alpha}\rceil}^{[nx+Tn^{\alpha}]} b\left(n^{1-\alpha}\left(x-\frac{k}{n}\right)\right)}. \tag{23.42}$$

Hence we easily derive by (23.40), as before, that

$$\left|M_n^*(f)(x) - f(x)\right| \le \omega_1\left(f, \frac{1}{n} + \frac{T}{n^{1-\alpha}}\right), \tag{23.43}$$

proving the claim. ■

Corollary 23.10 *Let* $x \in \left[-T^*, T^*\right]$, $T^* > 0$, $n \in \mathbb{N} : n \ge \max\left(T + T^*, T^{-\frac{1}{\alpha}}\right)$, $T > 0$. *Then*

$$\left\|M_n^*(f) - f\right\|_{\infty,[-T^*,T^*]} \le \omega_1\left(f, \frac{1}{n} + \frac{T}{n^{1-\alpha}}\right). \tag{23.44}$$

Proof By (23.37). ■

Theorems 23.5, 23.7 and 23.9 and Corollaries 23.6, 23.8 and 23.10 given that f is uniformly continuous, produce the pointwise and uniform convergences with rates, at speed $\frac{1}{n^{1-\alpha}}$, of neural network operators H_n^*, K_n^*, M_n^* to the unit operator. Notice that the right hand sides of inequalities (23.20), (23.26), (23.27), (23.36), (23.37) and (23.44) do not depend on b.

We proceed to the following results where we use the smoothness of a derivative of f.

Theorem 23.11 *Let* $x \in \mathbb{R}$, $T > 0$ *and* $n \in \mathbb{N}$ *such that* $n \ge \max(T + |x|, T^{-\frac{1}{\alpha}})$. *Let* $f \in C^N(\mathbb{R})$, $N \in \mathbb{N}$, *such that* $f^{(N)}$ *is uniformly continuous or is continuous and bounded. Then*

$$\left|\left(H_n^*(f)\right)(x) - f(x)\right| \le$$

$$\sum_{j=1}^{N} \frac{\left|f^{(j)}(x)\right|}{j!} \left\{\sum_{i=1}^{r} w_i \left[\left(\frac{\nu_i|x| + \mu_i}{n+\nu_i}\right) + \left(1 + \frac{\nu_i}{n+\nu_i}\right)\frac{T}{n^{1-\alpha}}\right]^j\right\} + \tag{23.45}$$

$$\sum_{i=1}^{r} w_i \omega_1\left(f^{(N)}, \left(\frac{\nu_i|x| + \mu_i}{n+\nu_i}\right) + \left(1 + \frac{\nu_i}{n+\nu_i}\right)\frac{T}{n^{1-\alpha}}\right).$$

$$\frac{\left(\left(\frac{\nu_i|x|+\mu_i}{n+\nu_i}\right)+\left(1+\frac{\nu_i}{n+\nu_i}\right)\frac{T}{n^{1-\alpha}}\right)^N}{N!}.$$

Inequality (23.45) implies the pointwise convergence with rates of $\left(H_n^*(f)\right)(x)$ *to* $f(x)$, *as* $n\to\infty$, *at speed* $\frac{1}{n^{1-\alpha}}$.

Proof Let k as in (23.5). We observe that

$$w_i f\left(\frac{k+\mu_i}{n+\nu_i}\right)=\sum_{j=0}^{N}\frac{f^{(j)}(x)}{j!}w_i\left(\frac{k+\mu_i}{n+\nu_i}-x\right)^j+ \tag{23.46}$$

$$w_i\int_x^{\frac{k+\mu_i}{n+\nu_i}}\left(f^{(N)}(t)-f^{(N)}(x)\right)\frac{\left(\frac{k+\mu_i}{n+\nu_i}-t\right)^{N-1}}{(N-1)!}dt,\ i=1,\dots,r.$$

Call

$$V(x)=\sum_{k=\lceil nx-Tn^{\alpha}\rceil}^{[nx+Tn^{\alpha}]}b\left(n^{1-\alpha}\left(x-\frac{k}{n}\right)\right). \tag{23.47}$$

Hence

$$\frac{\left(\sum_{i=1}^{r}w_i f\left(\frac{k+\mu_i}{n+\nu_i}\right)\right)b\left(n^{1-\alpha}\left(x-\frac{k}{n}\right)\right)}{V(x)}=$$

$$\sum_{j=0}^{N}\frac{f^{(j)}(x)}{j!}\left(\sum_{i=1}^{r}w_i\left(\frac{k+\mu_i}{n+\nu_i}-x\right)^j\right)\frac{b\left(n^{1-\alpha}\left(x-\frac{k}{n}\right)\right)}{V(x)}+$$

$$\frac{b\left(n^{1-\alpha}\left(x-\frac{k}{n}\right)\right)}{V(x)}\left(\sum_{i=1}^{r}w_i\int_x^{\frac{k+\mu_i}{n+\nu_i}}\left(f^{(N)}(t)-f^{(N)}(x)\right)\frac{\left(\frac{k+\mu_i}{n+\nu_i}-t\right)^{N-1}}{(N-1)!}dt\right). \tag{23.48}$$

Therefore it holds (see (23.12))

$$\left(H_n^*(f)\right)(x)-f(x)= \tag{23.49}$$

$$\sum_{j=1}^{N}\frac{f^{(j)}(x)}{j!}\left(\sum_{k=\lceil nx-Tn^{\alpha}\rceil}^{[nx+Tn^{\alpha}]}\left\{\left(\sum_{i=1}^{r}w_i\left(\frac{k+\mu_i}{n+\nu_i}-x\right)^j\right)\frac{b\left(n^{1-\alpha}\left(x-\frac{k}{n}\right)\right)}{V(x)}\right\}\right)+R,$$

where

$$R=\frac{\sum_{k=\lceil nx-Tn^{\alpha}\rceil}^{[nx+Tn^{\alpha}]}b\left(n^{1-\alpha}\left(x-\frac{k}{n}\right)\right)}{V(x)}. \tag{23.50}$$

$$\left(\sum_{i=1}^{r} w_i \int_x^{\frac{k+\mu_i}{n+\nu_i}} \left(f^{(N)}(t) - f^{(N)}(x) \right) \frac{\left(\frac{k+\mu_i}{n+\nu_i} - t \right)^{N-1}}{(N-1)!} dt \right).$$

So that

$$\left| \left(H_n^*(f) \right)(x) - f(x) \right| \leq \tag{23.51}$$

$$\sum_{j=1}^{N} \frac{\left| f^{(j)}(x) \right|}{j!} \left(\sum_{k=\lceil nx - Tn^{\alpha} \rceil}^{[nx+Tn^{\alpha}]} \left\{ \left(\sum_{i=1}^{r} w_i \left| \frac{k+\mu_i}{n+\nu_i} - x \right|^j \right) \frac{b\left(n^{1-\alpha}\left(x - \frac{k}{n} \right) \right)}{V(x)} \right\} \right) + |R| \leq$$

$$\sum_{j=1}^{N} \frac{\left| f^{(j)}(x) \right|}{j!} \left(\sum_{k=\lceil nx - Tn^{\alpha} \rceil}^{[nx+Tn^{\alpha}]} \left\{ \left(\sum_{i=1}^{r} w_i \left[\left(\frac{\nu_i |x| + \mu_i}{n+\nu_i} \right) + \left(1 + \frac{\nu_i}{n+\nu_i} \right) \frac{T}{n^{1-\alpha}} \right]^j \right) \right. \right. \tag{23.52}$$

$$\left. \left. \frac{b\left(n^{1-\alpha}\left(x - \frac{k}{n} \right) \right)}{V(x)} \right\} \right) + |R| =$$

$$\sum_{j=0}^{N} \frac{\left| f^{(j)}(x) \right|}{j!} \left\{ \sum_{i=1}^{r} w_i \left[\left(\frac{\nu_i |x| + \mu_i}{n+\nu_i} \right) + \left(1 + \frac{\nu_i}{n+\nu_i} \right) \frac{T}{n^{1-\alpha}} \right]^j \right\} + |R|. \tag{23.53}$$

So that thus far we have

$$\left| \left(H_n^*(f) \right)(x) - f(x) \right| \leq$$

$$\sum_{j=0}^{N} \frac{\left| f^{(j)}(x) \right|}{j!} \left\{ \sum_{i=1}^{r} w_i \left[\left(\frac{\nu_i |x| + \mu_i}{n+\nu_i} \right) + \left(1 + \frac{\nu_i}{n+\nu_i} \right) \frac{T}{n^{1-\alpha}} \right]^j \right\} + |R|. \tag{23.54}$$

Furthermore we see

$$|R| \leq \frac{\sum_{k=\lceil nx - Tn^{\alpha} \rceil}^{[nx+Tn^{\alpha}]} b\left(n^{1-\alpha}\left(x - \frac{k}{n} \right) \right)}{V(x)}. \tag{23.55}$$

$$\left(\sum_{i=1}^{r} w_i \left| \int_x^{\frac{k+\mu_i}{n+\nu_i}} \left(f^{(N)}(t) - f^{(N)}(x) \right) \frac{\left(\frac{k+\mu_i}{n+\nu_i} - t \right)^{N-1}}{(N-1)!} dt \right| \right) \leq$$

$$\frac{\sum_{k=\lceil nx - Tn^{\alpha} \rceil}^{[nx+Tn^{\alpha}]} b\left(n^{1-\alpha}\left(x - \frac{k}{n} \right) \right)}{V(x)} \gamma, \tag{23.56}$$

where

$$\gamma := \sum_{i=1}^{r} w_i \left| \int_x^{\frac{k+\mu_i}{n+\nu_i}} \left| f^{(N)}(t) - f^{(N)}(x) \right| \frac{\left| \frac{k+\mu_i}{n+\nu_i} - t \right|^{N-1}}{(N-1)!} dt \right|. \tag{23.57}$$

(i) Let $x \leq \frac{k+\mu_i}{n+\nu_i}$, then

$$\varepsilon_i := \int_x^{\frac{k+\mu_i}{n+\nu_i}} \left| f^{(N)}(t) - f^{(N)}(x) \right| \frac{\left| \frac{k+\mu_i}{n+\nu_i} - t \right|^{N-1}}{(N-1)!} dt \leq \tag{23.58}$$

$$\omega_1 \left(f^{(N)}, \left| \frac{k+\mu_i}{n+\nu_i} - x \right| \right) \int_x^{\frac{k+\mu_i}{n+\nu_i}} \frac{\left(\frac{k+\mu_i}{n+\nu_i} - t \right)^{N-1}}{(N-1)!} dt =$$

$$\omega_1 \left(f^{(N)}, \left| \frac{k+\mu_i}{n+\nu_i} - x \right| \right) \frac{\left| \left(\frac{k+\mu_i}{n+\nu_i} - t \right) \right|^{N}}{N!} \overset{(23.17)}{\leq}$$

$$\omega_1 \left(f^{(N)}, \left(\frac{\nu_i |x| + \mu_i}{n+\nu_i} \right) + \left(1 + \frac{\nu_i}{n+\nu_i} \right) \frac{T}{n^{1-\alpha}} \right) \cdot$$

$$\frac{\left[\left(\frac{\nu_i |x| + \mu_i}{n+\nu_i} \right) + \left(1 + \frac{\nu_i}{n+\nu_i} \right) \frac{T}{n^{1-\alpha}} \right]^N}{N!}. \tag{23.59}$$

So when $x \leq \frac{k+\mu_i}{n+\nu_i}$, we got

$$\varepsilon_i \leq \omega_1 \left(f^{(N)}, \left(\frac{\nu_i |x| + \mu_i}{n+\nu_i} \right) + \left(1 + \frac{\nu_i}{n+\nu_i} \right) \frac{T}{n^{1-\alpha}} \right) \cdot$$

$$\frac{\left[\left(\frac{\nu_i |x| + \mu_i}{n+\nu_i} \right) + \left(1 + \frac{\nu_i}{n+\nu_i} \right) \frac{T}{n^{1-\alpha}} \right]^N}{N!}. \tag{23.60}$$

(ii) Let $x > \frac{k+\mu_i}{n+\nu_i}$, then

$$\rho_i := \int_{\frac{k+\mu_i}{n+\nu_i}}^{x} \left| f^{(N)}(t) - f^{(N)}(x) \right| \frac{\left(t - \frac{k+\mu_i}{n+\nu_i} \right)^{N-1}}{(N-1)!} dt \leq$$

$$\omega_1 \left(f^{(N)}, \left| \frac{k+\mu_i}{n+\nu_i} - x \right| \right) \frac{\left(x - \frac{k+\mu_i}{n+\nu_i} \right)^{N}}{N!} = \tag{23.61}$$

$$\omega_1\left(f^{(N)},\left|\frac{k+\mu_i}{n+\nu_i}-x\right|\right)\frac{\left|\frac{k+\mu_i}{n+\nu_i}-x\right|^N}{N!}\leq$$

$$\omega_1\left(f^{(N)},\left(\frac{\nu_i\left|x\right|+\mu_i}{n+\nu_i}\right)+\left(1+\frac{\nu_i}{n+\nu_i}\right)\frac{T}{n^{1-\alpha}}\right)\cdot$$

$$\frac{\left(\left(\frac{\nu_i|x|+\mu_i}{n+\nu_i}\right)+\left(1+\frac{\nu_i}{n+\nu_i}\right)\frac{T}{n^{1-\alpha}}\right)^N}{N!}. \tag{23.62}$$

Hence when $x>\frac{k+\mu_i}{n+\nu_i}$, then

$$\rho_i\leq\omega_1\left(f^{(N)},\left(\frac{\nu_i\left|x\right|+\mu_i}{n+\nu_i}\right)+\left(1+\frac{\nu_i}{n+\nu_i}\right)\frac{T}{n^{1-\alpha}}\right)\cdot$$

$$\frac{\left(\left(\frac{\nu_i|x|+\mu_i}{n+\nu_i}\right)+\left(1+\frac{\nu_i}{n+\nu_i}\right)\frac{T}{n^{1-\alpha}}\right)^N}{N!}. \tag{23.63}$$

Notice in (23.60) and (23.63) we obtained the same upper bound. Hence it holds

$$\gamma\leq\sum_{i=1}^{r}w_i\omega_1\left(f^{(N)},\left(\frac{\nu_i\left|x\right|+\mu_i}{n+\nu_i}\right)+\left(1+\frac{\nu_i}{n+\nu_i}\right)\frac{T}{n^{1-\alpha}}\right)\cdot$$

$$\frac{\left(\left(\frac{\nu_i|x|+\mu_i}{n+\nu_i}\right)+\left(1+\frac{\nu_i}{n+\nu_i}\right)\frac{T}{n^{1-\alpha}}\right)^N}{N!}=:E. \tag{23.64}$$

Thus

$$|R|\leq E, \tag{23.65}$$

proving the claim. ■

Corollary 23.12 *All as in Theorem 23.11, plus* $f^{(j)}(x)=0$, $j=1,\ldots,N$. *Then*

$$\left|\left(H_n^*(f)\right)(x)-f(x)\right|\leq$$

$$\sum_{i=1}^{r}w_i\omega_1\left(f^{(N)},\left(\frac{\nu_i\left|x\right|+\mu_i}{n+\nu_i}\right)+\left(1+\frac{\nu_i}{n+\nu_i}\right)\frac{T}{n^{1-\alpha}}\right)\cdot$$

$$\frac{\left(\left(\frac{\nu_i|x|+\mu_i}{n+\nu_i}\right)+\left(1+\frac{\nu_i}{n+\nu_i}\right)\frac{T}{n^{1-\alpha}}\right)^N}{N!}. \tag{23.66}$$

Proof By (23.49), (23.50), (23.64) and (23.65). ■

In (23.66) notice the extremely high speed of convergence $\frac{1}{n^{(1-\alpha)(N+1)}}$.
The uniform convergence with rates follows from

Corollary 23.13 *Let $x \in [-T^*, T^*]$, $T^* > 0$; $T > 0$ and $n \in \mathbb{N}$ such that $n \geq \max\left(T + T^*, T^{-\frac{1}{\alpha}}\right)$. Let $f \in C^N(\mathbb{R})$, $N \in \mathbb{N}$, such that $f^{(N)}$ is uniformly continuous or is continuous and bounded. Then*

$$\left\| H_n^*(f) - f \right\|_{\infty,[-T^*,T^*]} \leq \tag{23.67}$$

$$\sum_{j=1}^{N} \frac{\left\| f^{(j)} \right\|_{\infty,[-T^*,T^*]}}{j!} \left\{ \sum_{i=1}^{r} w_i \left[\left(\frac{\nu_i T^* + \mu_i}{n + \nu_i} \right) + \left(1 + \frac{\nu_i}{n + \nu_i} \right) \frac{T}{n^{1-\alpha}} \right]^j \right\} +$$

$$\sum_{i=1}^{r} w_i \omega_1 \left(f^{(N)}, \left(\frac{\nu_i T^* + \mu_i}{n + \nu_i} \right) + \left(1 + \frac{\nu_i}{n + \nu_i} \right) \frac{T}{n^{1-\alpha}} \right) \cdot$$

$$\frac{\left(\left(\frac{\nu_i T^* + \mu_i}{n + \nu_i} \right) + \left(1 + \frac{\nu_i}{n + \nu_i} \right) \frac{T}{n^{1-\alpha}} \right)^N}{N!}.$$

Proof By (23.45). ■

Corollary 23.14 *All as in Theorem 23.11, case of $N = 1$. It holds*

$$\left| \left(H_n^*(f) \right)(x) - f(x) \right| \leq$$

$$\left| f'(x) \right| \left\{ \sum_{i=1}^{r} w_i \left[\left(\frac{\nu_i |x| + \mu_i}{n + \nu_i} \right) + \left(1 + \frac{\nu_i}{n + \nu_i} \right) \frac{T}{n^{1-\alpha}} \right] \right\} + \tag{23.68}$$

$$\sum_{i=1}^{r} w_i \omega_1 \left(f', \left(\frac{\nu_i |x| + \mu_i}{n + \nu_i} \right) + \left(1 + \frac{\nu_i}{n + \nu_i} \right) \frac{T}{n^{1-\alpha}} \right) \cdot$$

$$\left(\left(\frac{\nu_i |x| + \mu_i}{n + \nu_i} \right) + \left(1 + \frac{\nu_i}{n + \nu_i} \right) \frac{T}{n^{1-\alpha}} \right).$$

We continue with

Theorem 23.15 *Same assumptions as in Theorem 23.11, with $0 < \alpha < 1$. Then*

$$\left| \left(K_n^*(f) \right)(x) - f(x) \right| \leq \sum_{j=1}^{N} \frac{\left| f^{(j)}(x) \right|}{j!} \cdot \tag{23.69}$$

$$\left(\sum_{i=1}^{r} w_i \left(\left(\frac{\nu_i |x| + \mu_i + 1}{n + \nu_i}\right) + \left(1 + \frac{\nu_i}{n + \nu_i}\right)\frac{T}{n^{1-\alpha}}\right)^j\right) +$$

$$\sum_{i=1}^{r} w_i \left\{ \omega_1 \left(f^{(N)}, \left(\frac{\nu_i |x| + \mu_i + 1}{n + \nu_i}\right) + \left(1 + \frac{\nu_i}{n + \nu_i}\right)\frac{T}{n^{1-\alpha}}\right) \cdot \right.$$

$$\left. \frac{\left(\left(\frac{\nu_i|x|+\mu_i+1}{n+\nu_i}\right) + \left(1 + \frac{\nu_i}{n+\nu_i}\right)\frac{T}{n^{1-\alpha}}\right)^N}{N!} \right\}.$$

Inequality (23.69) implies the pointwise convergence with rates of $\left(K_n^* (f)\right)(x)$ *to* $f(x)$*, as* $n \to \infty$*, at the speed* $\frac{1}{n^{1-\alpha}}$*.*

Proof Let k as in (23.5). We observe that

$$f\left(t + \frac{k + \mu_i}{n + \nu_i}\right) = \sum_{j=0}^{N} \frac{f^{(j)}(x)}{j!} \left(t + \frac{k + \mu_i}{n + \nu_i} - x\right)^j +$$

$$\int_x^{t+\frac{k+\mu_i}{n+\nu_i}} \left(f^{(N)}(z) - f^{(N)}(x)\right) \frac{\left(t + \frac{k+\mu_i}{n+\nu_i} - z\right)^{N-1}}{(N-1)!} dz, \tag{23.70}$$

$i = 1, \ldots, r$.

That is

$$\int_0^{\frac{1}{n+\nu_i}} f\left(t + \frac{k + \mu_i}{n + \nu_i}\right) dt =$$

$$\sum_{j=0}^{N} \frac{f^{(j)}(x)}{j!} \int_0^{\frac{1}{n+\nu_i}} \left(t + \frac{k + \mu_i}{n + \nu_i} - x\right)^j dt +$$

$$\int_0^{\frac{1}{n+\nu_i}} \left(\int_x^{t+\frac{k+\mu_i}{n+\nu_i}} \left(f^{(N)}(z) - f^{(N)}(x)\right) \frac{\left(t + \frac{k+\mu_i}{n+\nu_i} - z\right)^{N-1}}{(N-1)!} dz\right) dt, \tag{23.71}$$

$i = 1, \ldots, r$.

Furthermore we have

$$\sum_{i=1}^{r} w_i (n + \nu_i) \int_0^{\frac{1}{n+\nu_i}} f\left(t + \frac{k + \mu_i}{n + \nu_i}\right) dt =$$

$$\sum_{j=0}^{N} \frac{f^{(j)}(x)}{j!} \left(\sum_{i=1}^{r} w_i (n+\nu_i) \int_0^{\frac{1}{n+\nu_i}} \left(t + \frac{k+\mu_i}{n+\nu_i} - x \right)^j dt \right) +$$

$$\sum_{i=1}^{r} w_i (n+\nu_i) \int_0^{\frac{1}{n+\nu_i}} \left(\int_x^{t+\frac{k+\mu_i}{n+\nu_i}} \left(f^{(N)}(z) - f^{(N)}(x) \right) \frac{\left(t + \frac{k+\mu_i}{n+\nu_i} - z \right)^{N-1}}{(N-1)!} dz \right) dt. \tag{23.72}$$

Call

$$V(x) = \sum_{k=\lceil nx - Tn^{\alpha} \rceil}^{[nx+Tn^{\alpha}]} b\left(n^{1-\alpha} \left(x - \frac{k}{n} \right) \right). \tag{23.73}$$

Consequently we get

$$\left(K_n^*(f) \right)(x) =$$

$$\frac{\sum_{k=\lceil nx-Tn^{\alpha} \rceil}^{[nx+Tn^{\alpha}]} \left(\sum_{i=1}^{r} w_i (n+\nu_i) \int_0^{\frac{1}{n+\nu_i}} f\left(t + \frac{k+\mu_i}{n+\nu_i} \right) dt \right) b\left(n^{1-\alpha} \left(x - \frac{k}{n} \right) \right)}{V(x)}$$

$$= \sum_{j=0}^{N} \frac{f^{(j)}(x)}{j!} \frac{\sum_{k=\lceil nx-Tn^{\alpha} \rceil}^{[nx+Tn^{\alpha}]} b\left(n^{1-\alpha} \left(x - \frac{k}{n} \right) \right)}{V(x)} \cdot$$

$$\left(\sum_{i=1}^{r} w_i (n+\nu_i) \int_0^{\frac{1}{n+\nu_i}} \left(t + \frac{k+\mu_i}{n+\nu_i} - x \right)^j dt \right) +$$

$$\frac{\sum_{k=\lceil nx-Tn^{\alpha} \rceil}^{[nx+Tn^{\alpha}]} b\left(n^{1-\alpha} \left(x - \frac{k}{n} \right) \right)}{V(x)} \left(\sum_{i=1}^{r} w_i (n+\nu_i) \right.$$

$$\left. \int_0^{\frac{1}{n+\nu_i}} \left(\int_x^{t+\frac{k+\mu_i}{n+\nu_i}} \left(f^{(N)}(z) - f^{(N)}(x) \right) \frac{\left(t + \frac{k+\mu_i}{n+\nu_i} - z \right)^{N-1}}{(N-1)!} dz \right) dt \right). \tag{23.74}$$

Therefore it holds

$$\left(K_n^*(f) \right)(x) - f(x) = \sum_{j=1}^{N} \frac{f^{(j)}(x)}{j!} \frac{\sum_{k=\lceil nx-Tn^{\alpha} \rceil}^{[nx+Tn^{\alpha}]} b\left(n^{1-\alpha} \left(x - \frac{k}{n} \right) \right)}{V(x)} \cdot$$

$$\left(\sum_{i=1}^{r} w_i (n+\nu_i) \int_0^{\frac{1}{n+\nu_i}} \left(t + \frac{k+\mu_i}{n+\nu_i} - x \right)^j dt \right) + R, \tag{23.75}$$

where

$$R=\frac{\sum_{k=\lceil nx-Tn^{\alpha}\rceil}^{[nx+Tn^{\alpha}]} b\left(n^{1-\alpha}\left(x-\frac{k}{n}\right)\right)}{V(x)}\left(\sum_{i=1}^{r} w_i\left(n+\nu_i\right)\right.$$

$$\left.\int_0^{\frac{1}{n+\nu_i}}\left(\int_x^{t+\frac{k+\mu_i}{n+\nu_i}}\left(f^{(N)}(z)-f^{(N)}(x)\right)\frac{\left(t+\frac{k+\mu_i}{n+\nu_i}-z\right)^{N-1}}{(N-1)!}dz\right)dt\right). \quad (23.76)$$

We derive that

$$\left|\left(K_n^*(f)\right)(x)-f(x)\right|\leq\sum_{j=1}^{N}\frac{f^{(j)}(x)}{j!}\frac{\sum_{k=\lceil nx-Tn^{\alpha}\rceil}^{[nx+Tn^{\alpha}]} b\left(n^{1-\alpha}\left(x-\frac{k}{n}\right)\right)}{V(x)}\cdot \quad (23.77)$$

$$\left(\sum_{i=1}^{r} w_i\left(n+\nu_i\right)\int_0^{\frac{1}{n+\nu_i}}|t+\frac{k+\mu_i}{n+\nu_i}-x|^j dt\right)+|R| \overset{(23.17)}{\leq}$$

$$\sum_{j=1}^{N}\frac{\left|f^{(j)}(x)\right|}{j!}\frac{\sum_{k=\lceil nx-Tn^{\alpha}\rceil}^{[nx+Tn^{\alpha}]} b\left(n^{1-\alpha}\left(x-\frac{k}{n}\right)\right)}{V(x)}\cdot$$

$$\sum_{i=1}^{r} w_i[\left(\frac{\nu_i|x|+\mu_i+1}{n+\nu_i}\right)+\left(1+\frac{\nu_i}{n+\nu_i}\right)\frac{T}{n^{1-\alpha}}]^j$$

$$+|R|. \quad (23.78)$$

Above we used

$$|t+\frac{k+\mu_i}{n+\nu_i}-x|$$

$$\leq[\left(\frac{\nu_i|x|+\mu_i+1}{n+\nu_i}\right)+\left(1+\frac{\nu_i}{n+\nu_i}\right)\frac{T}{n^{1-\alpha}}]. \quad (23.79)$$

We have found that

$$\left|\left(K_n^*(f)\right)(x)-f(x)\right|\leq$$

$$\sum_{j=1}^{N}\frac{\left|f^{(j)}(x)\right|}{j!}\sum_{i=1}^{r} w_i[\left(\frac{\nu_i|x|+\mu_i+1}{n+\nu_i}\right)+\left(1+\frac{\nu_i}{n+\nu_i}\right)\frac{T}{n^{1-\alpha}}]^j$$

$$+|R|. \quad (23.80)$$

Notice that

$$\left[\left(\frac{\nu_i\,|x|+\mu_i+1}{n+\nu_i}\right)+\left(1+\frac{\nu_i}{n+\nu_i}\right)\frac{T}{n^{1-\alpha}}\right]\to 0,\text{ as } n\to\infty,\ 0<\alpha<1. \tag{23.81}$$

We observe that

$$|R|\overset{(23.76)}{\leq}\frac{\sum_{k=\lceil nx-Tn^{\alpha}\rceil}^{[nx+Tn^{\alpha}]} b\left(n^{1-\alpha}\left(x-\frac{k}{n}\right)\right)}{V(x)}\left(\sum_{i=1}^{r} w_i\,(n+\nu_i)\right. \tag{23.82}$$

$$\left.\int_0^{\frac{1}{n+\nu_i}}\left|\int_x^{t+\frac{k+\mu_i}{n+\nu_i}}\left|f^{(N)}(z)-f^{(N)}(x)\right|\frac{\left|t+\frac{k+\mu_i}{n+\nu_i}-z\right|^{N-1}}{(N-1)!}dz\right|dt\right)=:(\xi).$$

We distinguish the cases:

(i) if $t+\frac{k+\mu_i}{n+\nu_i}\geq x$, then

$$\theta_i:=\left|\int_x^{t+\frac{k+\mu_i}{n+\nu_i}}\left|f^{(N)}(t)-f^{(N)}(x)\right|\frac{\left|t+\frac{k+\mu_i}{n+\nu_i}-z\right|^{N-1}}{(N-1)!}dz\right|= \tag{23.83}$$

$$\int_x^{t+\frac{k+\mu_i}{n+\nu_i}}\left|f^{(N)}(t)-f^{(N)}(x)\right|\frac{\left(t+\frac{k+\mu_i}{n+\nu_i}-z\right)^{N-1}}{(N-1)!}dz\leq$$

$$\omega_1\left(f^{(N)},|t|+\left|\frac{k+\mu_i}{n+\nu_i}-x\right|\right)\frac{\left(|t|+\left|\frac{k+\mu_i}{n+\nu_i}-x\right|\right)^N}{N!}\overset{(23.17)}{\leq}$$

$$\omega_1\left(f^{(N)},\left(\frac{\nu_i\,|x|+\mu_i+1}{n+\nu_i}\right)+\left(1+\frac{\nu_i}{n+\nu_i}\right)\frac{T}{n^{1-\alpha}}\right)\cdot$$

$$\frac{\left(\left(\frac{\nu_i|x|+\mu_i+1}{n+\nu_i}\right)+\left(1+\frac{\nu_i}{n+\nu_i}\right)\frac{T}{n^{1-\alpha}}\right)^N}{N!}. \tag{23.84}$$

That is, if $t+\frac{k+\mu_i}{n+\nu_i}\geq x$, we proved that

$$\theta_i\leq\omega_1\left(f^{(N)},\left(\frac{\nu_i\,|x|+\mu_i+1}{n+\nu_i}\right)+\left(1+\frac{\nu_i}{n+\nu_i}\right)\frac{T}{n^{1-\alpha}}\right)\cdot$$

$$\frac{\left(\left(\frac{\nu_i|x|+\mu_i+1}{n+\nu_i}\right)+\left(1+\frac{\nu_i}{n+\nu_i}\right)\frac{T}{n^{1-\alpha}}\right)^N}{N!}. \tag{23.85}$$

(ii) if $t + \frac{k+\mu_i}{n+\nu_i} < x$, then

$$\theta_i := \int_{t+\frac{k+\mu_i}{n+\nu_i}}^{x} \left| f^{(N)}(z) - f^{(N)}(x) \right| \frac{\left(z - \left(t + \frac{k+\mu_i}{n+\nu_i}\right)\right)^{N-1}}{(N-1)!} dz \le$$

$$\omega_1 \left(f^{(N)}, |t| + \left| \frac{k+\mu_i}{n+\nu_i} - x \right| \right) \frac{\left(|t| + \left| \frac{k+\mu_i}{n+\nu_i} - x \right| \right)^N}{N!} \overset{(23.17)}{\le}$$

$$\omega_1 \left(f^{(N)}, \left(\frac{\nu_i |x| + \mu_i + 1}{n+\nu_i} \right) + \left(1 + \frac{\nu_i}{n+\nu_i} \right) \frac{T}{n^{1-\alpha}} \right) \cdot$$

$$\frac{\left(\left(\frac{\nu_i |x| + \mu_i + 1}{n+\nu_i} \right) + \left(1 + \frac{\nu_i}{n+\nu_i} \right) \frac{T}{n^{1-\alpha}} \right)^N}{N!}, \tag{23.86}$$

same estimate as in (23.85).

Therefore we derive (see (23.82))

$$(\xi) \le \sum_{k=\lceil nx - Tn^{\alpha} \rceil}^{[nx+Tn^{\alpha}]} \frac{b\left(n^{1-\alpha}\left(x - \frac{k}{n}\right)\right)}{V(x)} \cdot$$

$$\left(\sum_{i=1}^{r} w_i \left\{ \omega_1 \left(f^{(N)}, \left(\frac{\nu_i |x| + \mu_i + 1}{n+\nu_i} \right) + \left(1 + \frac{\nu_i}{n+\nu_i} \right) \frac{T}{n^{1-\alpha}} \right) \cdot \right. \right. \tag{23.87}$$

$$\left. \left. \frac{\left(\left(\frac{\nu_i |x| + \mu_i + 1}{n+\nu_i} \right) + \left(1 + \frac{\nu_i}{n+\nu_i} \right) \frac{T}{n^{1-\alpha}} \right)^N}{N!} \right\} \right).$$

Clearly we have found the estimate

$$|R| \le \left(\sum_{i=1}^{r} w_i \left\{ \omega_1 \left(f^{(N)}, \left(\frac{\nu_i |x| + \mu_i + 1}{n+\nu_i} \right) + \left(1 + \frac{\nu_i}{n+\nu_i} \right) \frac{T}{n^{1-\alpha}} \right) \cdot \right. \right. \tag{23.88}$$

$$\left. \left. \frac{\left(\left(\frac{\nu_i |x| + \mu_i + 1}{n+\nu_i} \right) + \left(1 + \frac{\nu_i}{n+\nu_i} \right) \frac{T}{n^{1-\alpha}} \right)^N}{N!} \right\} \right).$$

Based on (23.80) and (23.88) we derive (23.69). ■

Corollary 23.16 *All as in Theorem 23.15, plus* $f^{(j)}(x) = 0$, $j = 1, \ldots, N$; $0 < \alpha < 1$. *Then*

$$\left|\left(K_n^*(f)\right)(x) - f(x)\right| \leq$$

$$\sum_{i=1}^{r} w_i \left\{ \omega_1 \left(f^{(N)}, \left(\frac{\nu_i |x| + \mu_i + 1}{n + \nu_i} \right) + \left(1 + \frac{\nu_i}{n + \nu_i} \right) \frac{T}{n^{1-\alpha}} \right) \cdot \right. \tag{23.89}$$

$$\left. \frac{\left(\left(\frac{\nu_i |x| + \mu_i + 1}{n + \nu_i} \right) + \left(1 + \frac{\nu_i}{n + \nu_i} \right) \frac{T}{n^{1-\alpha}} \right)^N}{N!} \right\}.$$

Proof By (23.75), (23.76) and (23.88). ■

In (23.89) notice the extremely high speed of convergence $\frac{1}{n^{(1-\alpha)(N+1)}}$.
The uniform convergence with rates follows from

Corollary 23.17 *Let* $x \in \left[-T^*, T^*\right]$, $T^* > 0$; $T > 0$ *and* $n \in \mathbb{N}$ *such that* $n \geq \max\left(T + T^*, T^{-\frac{1}{\alpha}}\right)$, $0 < \alpha < 1$. *Let* $f \in C^N(\mathbb{R})$, $N \in \mathbb{N}$, *such that* $f^{(N)}$ *is uniformly continuous or is continuous and bounded. Then*

$$\left\| K_n^*(f) - f \right\|_{\infty,[-T^*,T^*]} \leq \sum_{j=1}^{N} \frac{\left\| f^{(j)} \right\|_{\infty,[-T^*,T^*]}}{j!} \cdot \tag{23.90}$$

$$\left(\sum_{i=1}^{r} w_i \left[\left(\frac{\nu_i T^* + \mu_i + 1}{n + \nu_i} \right) + \left(1 + \frac{\nu_i}{n + \nu_i} \right) \frac{T}{n^{1-\alpha}} \right]^j \right) +$$

$$\sum_{i=1}^{r} w_i \left\{ \omega_1 \left(f^{(N)}, \left(\frac{\nu_i T^* + \mu_i + 1}{n + \nu_i} \right) + \left(1 + \frac{\nu_i}{n + \nu_i} \right) \frac{T}{n^{1-\alpha}} \right) \cdot \right.$$

$$\left. \frac{\left(\left(\frac{\nu_i T^* + \mu_i + 1}{n + \nu_i} \right) + \left(1 + \frac{\nu_i}{n + \nu_i} \right) \frac{T}{n^{1-\alpha}} \right)^N}{N!} \right\}.$$

Proof By (23.69). ■

Corollary 23.18 *All as in Theorem 23.15, case of* $N = 1$. *It holds*

$$\left|\left(K_n^*(f)\right)(x) - f(x)\right| \leq$$

$$\left|f'(x)\right|\left(\sum_{i=1}^{r} w_i \left(\left(\frac{\nu_i |x| + \mu_i + 1}{n+\nu_i}\right) + \left(1 + \frac{\nu_i}{n+\nu_i}\right)\frac{T}{n^{1-\alpha}}\right)\right) + \tag{23.91}$$

$$\sum_{i=1}^{r} w_i \left\{\omega_1\left(f', \left(\frac{\nu_i |x| + \mu_i + 1}{n+\nu_i}\right) + \left(1 + \frac{\nu_i}{n+\nu_i}\right)\frac{T}{n^{1-\alpha}}\right) \cdot \right.$$

$$\left.\left(\left(\frac{\nu_i |x| + \mu_i + 1}{n+\nu_i}\right) + \left(1 + \frac{\nu_i}{n+\nu_i}\right)\frac{T}{n^{1-\alpha}}\right)\right\}.$$

Proof By (23.69). ■

We also present

Theorem 23.19 *Let all as in Theorem 23.11. Then*

$$\left|\left(M_n^*(f)\right)(x) - f(x)\right| \leq \sum_{j=1}^{N} \frac{\left|f^{(j)}(x)\right|}{j!}\left[\frac{T}{n^{1-\alpha}} + \frac{1}{n}\right]^j + \tag{23.92}$$

$$\omega_1\left(f^{(N)}, \frac{T}{n^{1-\alpha}} + \frac{1}{n}\right)\frac{\left(\frac{T}{n^{1-\alpha}} + \frac{1}{n}\right)^N}{N!}.$$

Inequality (23.92) implies the pointwise convergence with rates of $\left(M_n^*(f)\right)(x)$ *to* $f(x)$*, as* $n \to \infty$*, at the speed* $\frac{1}{n^{1-\alpha}}$.

Proof Let k as in (23.5). Again by Taylor's formula we have that

$$\sum_{i=1}^{r} w_i f\left(\frac{k}{n} + \frac{i}{nr}\right) = \sum_{j=0}^{N} \frac{f^{(j)}(x)}{j!} \sum_{i=1}^{r} w_i \left(\frac{k}{n} + \frac{i}{nr} - x\right)^j + \tag{23.93}$$

$$\sum_{i=1}^{r} w_i \int_x^{\frac{k}{n}+\frac{i}{nr}} \left(f^{(N)}(t) - f^{(N)}(x)\right)\frac{\left(\frac{k}{n} + \frac{i}{nr} - t\right)^{N-1}}{(N-1)!}dt.$$

Call

$$V(x) = \sum_{k=\lceil nx - Tn^{\alpha}\rceil}^{[nx+Tn^{\alpha}]} b\left(n^{1-\alpha}\left(x - \frac{k}{n}\right)\right). \tag{23.94}$$

Then

$$\left(M_n^*(f)\right)(x) = \frac{\sum_{k=\lceil nx-Tn^{\alpha}\rceil}^{[nx+Tn^{\alpha}]}\left(\sum_{i=1}^{r} w_i f\left(\frac{k}{n} + \frac{i}{nr}\right)\right) b\left(n^{1-\alpha}\left(x - \frac{k}{n}\right)\right)}{V(x)}$$

$$= \sum_{j=0}^{N} \frac{f^{(j)}(x)}{j!} \frac{\sum_{k=\lceil nx-Tn^{\alpha}\rceil}^{[nx+Tn^{\alpha}]} b\left(n^{1-\alpha}\left(x-\frac{k}{n}\right)\right)}{V(x)} \cdot$$

$$\sum_{i=1}^{r} w_i \left(\frac{k}{n}+\frac{i}{nr}-x\right)^j + \frac{\sum_{k=\lceil nx-Tn^{\alpha}\rceil}^{[nx+Tn^{\alpha}]} b\left(n^{1-\alpha}\left(x-\frac{k}{n}\right)\right)}{V(x)} \cdot$$

$$\sum_{i=1}^{r} w_i \int_{x}^{\frac{k}{n}+\frac{i}{nr}} \left(f^{(N)}(t)-f^{(N)}(x)\right) \frac{\left(\frac{k}{n}+\frac{i}{nr}-t\right)^{N-1}}{(N-1)!} dt. \tag{23.95}$$

Therefore we get

$$\left(M_n^*(f)\right)(x) - f(x) = \sum_{j=1}^{N} \frac{f^{(j)}(x)}{j!} \frac{\sum_{k=\lceil nx-Tn^{\alpha}\rceil}^{[nx+Tn^{\alpha}]} b\left(n^{1-\alpha}\left(x-\frac{k}{n}\right)\right)}{V(x)} \cdot$$

$$\sum_{i=1}^{r} w_i \left(\frac{k}{n}+\frac{i}{nr}-x\right)^j + R, \tag{23.96}$$

where

$$R = \frac{\sum_{k=\lceil nx-Tn^{\alpha}\rceil}^{[nx+Tn^{\alpha}]} b\left(n^{1-\alpha}\left(x-\frac{k}{n}\right)\right)}{V(x)} \cdot$$

$$\sum_{i=1}^{r} w_i \int_{x}^{\frac{k}{n}+\frac{i}{nr}} \left(f^{(N)}(t)-f^{(N)}(x)\right) \frac{\left(\frac{k}{n}+\frac{i}{nr}-t\right)^{N-1}}{(N-1)!} dt. \tag{23.97}$$

Hence it holds

$$\left|\left(M_n^*(f)\right)(x) - f(x)\right| \leq \sum_{j=1}^{N} \frac{f^{(j)}(x)}{j!} \frac{\sum_{k=\lceil nx-Tn^{\alpha}\rceil}^{[nx+Tn^{\alpha}]} b\left(n^{1-\alpha}\left(x-\frac{k}{n}\right)\right)}{V(x)} \cdot \tag{23.98}$$

$$\sum_{i=1}^{r} w_i \left[\left|\frac{k}{n}-x\right| + \frac{i}{nr}\right]^j + |R| \leq$$

$$\sum_{j=1}^{N} \frac{\left|f^{(j)}(x)\right|}{j!} \frac{\sum_{k=\lceil nx-Tn^{\alpha}\rceil}^{[nx+Tn^{\alpha}]} b\left(n^{1-\alpha}\left(x-\frac{k}{n}\right)\right)}{V(x)} \cdot$$

$$\sum_{i=1}^{r} w_i \left[\frac{T}{n^{1-\alpha}}+\frac{1}{n}\right]^j + |R| = \sum_{j=1}^{N} \frac{\left|f^{(j)}(x)\right|}{j!} \left[\frac{T}{n^{1-\alpha}}+\frac{1}{n}\right]^j + |R|. \tag{23.99}$$

We have proved that

$$\left|\left(M_n^*(f)\right)(x) - f(x)\right| \leq \sum_{j=1}^{N} \frac{\left|f^{(j)}(x)\right|}{j!}\left[\frac{T}{n^{1-\alpha}} + \frac{1}{n}\right]^j + |R|. \tag{23.100}$$

Next we observe it holds

$$|R| \leq \frac{\sum_{k=\lceil nx-Tn^{\alpha}\rceil}^{[nx+Tn^{\alpha}]} b\left(n^{1-\alpha}\left(x-\frac{k}{n}\right)\right)}{V(x)}. \tag{23.101}$$

$$\sum_{i=1}^{r} w_i \left|\int_x^{\frac{k}{n}+\frac{i}{nr}} \left|f^{(N)}(t) - f^{(N)}(x)\right| \frac{\left|\frac{k}{n}+\frac{i}{nr}-t\right|^{N-1}}{(N-1)!} dt\right|.$$

Call

$$\varepsilon_i := \left|\int_x^{\frac{k}{n}+\frac{i}{nr}} \left|f^{(N)}(t) - f^{(N)}(x)\right| \frac{\left|\frac{k}{n}+\frac{i}{nr}-t\right|^{N-1}}{(N-1)!} dt\right|. \tag{23.102}$$

We distinguish the cases:

(i) if $\frac{k}{n} + \frac{i}{nr} \geq x$, then

$$\varepsilon_i := \int_x^{\frac{k}{n}+\frac{i}{nr}} \left|f^{(N)}(t) - f^{(N)}(x)\right| \frac{\left(\frac{k}{n}+\frac{i}{nr}-t\right)^{N-1}}{(N-1)!} dt \leq$$

$$\omega_1\left(f^{(N)}, \left|\frac{k}{n} - x\right| + \frac{1}{n}\right) \frac{\left(\frac{k}{n}+\frac{i}{nr}-x\right)^N}{N!} \leq \tag{23.103}$$

$$\omega_1\left(f^{(N)}, \frac{T}{n^{1-\alpha}} + \frac{1}{n}\right) \frac{\left(\frac{T}{n^{1-\alpha}}+\frac{1}{n}\right)^N}{N!}. \tag{23.104}$$

Thus

$$\varepsilon_i \leq \omega_1\left(f^{(N)}, \frac{T}{n^{1-\alpha}} + \frac{1}{n}\right) \frac{\left(\frac{T}{n^{1-\alpha}}+\frac{1}{n}\right)^N}{N!}. \tag{23.105}$$

(ii) if $\frac{k}{n} + \frac{i}{nr} < x$, then

$$\varepsilon_i := \int_{\frac{k}{n}+\frac{i}{nr}}^{x} \left|f^{(N)}(t) - f^{(N)}(x)\right| \frac{\left(t-\left(\frac{k}{n}+\frac{i}{nr}\right)\right)^{N-1}}{(N-1)!} dt \leq$$

$$\omega_1\left(f^{(N)},\left(x-\left(\frac{k}{n}+\frac{i}{nr}\right)\right)\right)\frac{\left(x-\left(\frac{k}{n}+\frac{i}{nr}\right)\right)^N}{N!}\leq \tag{23.106}$$

$$\omega_1\left(f^{(N)},\frac{T}{n^{1-\alpha}}+\frac{1}{n}\right)\frac{\left(\frac{T}{n^{1-\alpha}}+\frac{1}{n}\right)^N}{N!}.$$

So we obtain again (23.105).

Clearly now by (23.101) we derive that

$$|R|\leq\omega_1\left(f^{(N)},\frac{T}{n^{1-\alpha}}+\frac{1}{n}\right)\frac{\left(\frac{T}{n^{1-\alpha}}+\frac{1}{n}\right)^N}{N!}, \tag{23.107}$$

proving the claim. ■

Corollary 23.20 *All as in Theorem 23.19, plus $f^{(j)}(x)=0$, $j=1,\ldots,N$. Then*

$$\left|\left(M_n^*(f)\right)(x)-f(x)\right|\leq\omega_1\left(f^{(N)},\frac{T}{n^{1-\alpha}}+\frac{1}{n}\right)\frac{\left(\frac{T}{n^{1-\alpha}}+\frac{1}{n}\right)^N}{N!}. \tag{23.108}$$

Proof By (23.92). ■

In (23.108) notice the extremely high speed of convergence $\frac{1}{n^{(1-\alpha)(N+1)}}$. Uniform convergence estimate follows

Corollary 23.21 *All here as in Corollary 23.13. Then*

$$\left\|M_n^*(f)-f\right\|_{\infty,[-T^*,T^*]}\leq\sum_{j=1}^{N}\frac{\left\|f^{(j)}\right\|_{\infty,[-T^*,T^*]}}{j!}\left(\frac{T}{n^{1-\alpha}}+\frac{1}{n}\right)^j \tag{23.109}$$

$$+\omega_1\left(f^{(N)},\frac{T}{n^{1-\alpha}}+\frac{1}{n}\right)\frac{\left(\frac{T}{n^{1-\alpha}}+\frac{1}{n}\right)^N}{N!}.$$

Proof By (23.92). ■

Corollary 23.22 *All as in Theorem 23.19, $N=1$ case. It holds*

$$\left|\left(M_n^*(f)\right)(x)-f(x)\right|\leq\left[\left|f'(x)\right|\right.$$

$$\left.+\omega_1\left(f',\frac{T}{n^{1-\alpha}}+\frac{1}{n}\right)\right]\left(\frac{T}{n^{1-\alpha}}+\frac{1}{n}\right). \tag{23.110}$$

Proof By (23.92). ■

Note 23.23 *We also observe that all the right hand sides of convergence inequalities (23.45), (23.66)–(23.69), (23.89)–(23.92), (23.108)–(23.110), are independent of b.*

Note 23.24 *We observe that*

$$H_n^* (1) = K_n^* (1) = M_n^* (1) = 1, \tag{23.111}$$

thus unitary operators.

Also, given that f is bounded, we get

$$\left\| H_n^* (f) \right\|_{\infty,\mathbb{R}} \leq \| f \|_{\infty,\mathbb{R}} , \tag{23.112}$$

$$\left\| K_n^* (f) \right\|_{\infty,\mathbb{R}} \leq \| f \|_{\infty,\mathbb{R}} , \tag{23.113}$$

and

$$\left\| M_n^* (f) \right\|_{\infty,\mathbb{R}} \leq \| f \|_{\infty,\mathbb{R}} . \tag{23.114}$$

Operators H_n^*, K_n^*, M_n^* are positive linear operators, and of course bounded operators directly by (23.112)–(23.114).

References

1. G.A. Anastassiou, Rate of convergence of some neural network operators to the unit-univariate case. J. Math. Anal. Appli. **212**, 237–262 (1997)
2. G.A. Anastassiou, Rate of convergence of some multivariate neural network operators to the unit. J. Comp. Math. with Appl. **40**, 1–19 (2000)
3. G.A. Anastassiou, *Quantitative Approximations* (Chapman and Hall/CRC, Boca Raton, 2001)
4. G.A. Anastassiou, Rate of convergence of some neural network operators to the unit-univariate case, revisited. Vesnik **65**(4), 511–518 (2013)
5. G.A. Anastassiou, Rate of convergence of some multivariate neural network operators to the unit, revisited. J. Comput. Anal. Appl. **15**(7), 1300–1309 (2013)
6. G.A. Anastassiou, Approximation by perturbed neural network operators. Submitted 2014
7. A.R. Barron, Universal approximation bounds for superpositions of a sigmoidal function. IEEE Trans. Inf. Theory **39**, 930–945 (1993)
8. F.L. Cao, T.F. Xie, Z.B. Xu, The estimate for approximation error of neural networks: a constructive approach. Neurocomputing **71**, 626–630 (2008)
9. P. Cardaliaguet, G. Euvrard, Approximation of a function and its derivative with a neural network. Neural Netw. **5**, 207–220 (1992)
10. Z. Chen, F. Cao, The approximation operators with sigmoidal functions. Comput. Math. Appl. **58**, 758–765 (2009)
11. T.P. Chen, H. Chen, Universal approximation to nonlinear operators by neural networks with arbitrary activation functions and its applications to a dynamic system. IEEE Trans. Neural Netw. **6**, 911–917 (1995)
12. C.K. Chui, X. Li, Approximation by ridge functions and neural networks with one hidden layer. J. Approx. Theor. **70**, 131–141 (1992)
13. D. Costarelli, R. Spigler, Approximation results for neural network operators activated by sigmoidal functions. Neural Netw. **44**, 101–106 (2013)

14. D. Costarelli, R. Spigler, Multivariate neural network operators with sigmoidal activation functions. Neural Netw. **48**, 72–77 (2013)
15. G. Cybenko, Approximation by superpositions of sigmoidal function. Math. Contr. Signals Syst. **2**, 303–314 (1989)
16. S. Ferrari, R.F. Stengel, Smooth function approximation using neural networks. IEEE Trans. Neural Netw. **16**, 24–38 (2005)
17. K.I. Funahashi, On the approximate realization of continuous mappings by neural networks. Neural Netw. **2**, 183–192 (1989)
18. N. Hahm, B.I. Hong, An approximation by neural networks with a fixed weight. Comput. Math. Appl. **47**, 1897–1903 (2004)
19. S. Haykin, *Neural Networks: A Comprehensive Foundation*, 2nd edn. (Prentice Hall, New York, 1998)
20. K. Hornik, M. Stinchombe, H. White, Multilayer feedforward networks are universal approximations. Neural Netw. **2**, 359–366 (1989)
21. K. Hornik, M. Stinchombe, H. White, Universal approximation of an unknown mapping and its derivatives using multilayer feedforward networks. Neural Netw. **3**, 551–560 (1990)
22. M. Leshno, V.Y. Lin, A. Pinks, S. Schocken, Multilayer feedforward networks with a nonpolynomial activation function can approximate any function. Neural Netw. **6**, 861–867 (1993)
23. V. Maiorov, R.S. Meir, Approximation bounds for smooth functions in $C\left(R^d\right)$ by neural and mixture networks. IEEE Trans. Neural Netw. **9**, 969–978 (1998)
24. Y. Makovoz, Uniform approximation by neural networks. J. Approx. Theor. **95**, 215–228 (1998)
25. W. McCulloch, W. Pitts, A logical calculus of the ideas immanent in nervous activity. Bull. Math. Biophys. **7**, 115–133 (1943)
26. H.N. Mhaskar, C.A. Micchelli, Approximation by superposition of a sigmoidal function. Adv. Appl. Math. **13**, 350–373 (1992)
27. H.N. Mhaskar, C.A. Micchelli, Degree of approximation by neural networks with a single hidden layer. Adv. Appl. Math. **16**, 151–183 (1995)
28. T.M. Mitchell, *Machanical Learning* (WCB-McGraw-Hill, New York, 1997)
29. D.D. Stancu, On a generalization of the Bernstein polynomials. Studia Universitatis Babeş-Bolyai, Series Mathematica-Physica **14**, 31–45 (1969)
30. S. Suzuki, Constructive function approximation by three-layer artificial neural networks. Neural Netw. **11**, 1049–1058 (1998)
31. Z.B. Xu, F.L. Cao, The essential order of approximation for neural networks, Sci. China (Ser. F) **47**, 97–112 (2004)

Chapter 24
Approximations by Multivariate Perturbed Neural Networks

This chapter deals with the determination of the rate of convergence to the unit of each of three newly introduced here multivariate perturbed normalized neural network operators of one hidden layer. These are given through the multivariate modulus of continuity of the involved multivariate function or its high order partial derivatives and that appears in the right-hand side of the associated multivariate Jackson type inequalities. The multivariate activation function is very general, especially it can derive from any multivariate sigmoid or multivariate bell-shaped function. The right hand sides of our convergence inequalities do not depend on the activation function. The sample functionals are of multivariate Stancu, Kantorovich and Quadrature types. We give applications for the first partial derivatives of the involved function. It follows [7].

24.1 Introduction

Feed-forward neural networks (FNNs) with one hidden layer, the only type of networks we deal with in this chapter, are mathematically expressed as

$$N_n(x) = \sum_{j=0}^{n} c_j \sigma\left(\left\langle a_j \cdot x\right\rangle + b_j\right), \quad x \in \mathbb{R}^s, \ s \in \mathbb{N},$$

where for $0 \le j \le n$, $b_j \in \mathbb{R}$ are the thresholds, $a_j \in \mathbb{R}^s$ are the connection weights, $c_j \in \mathbb{R}$ are the coefficients, $\left\langle a_j \cdot x\right\rangle$ is the inner product of a_j and x, and σ is the activation function of the network. In many fundamental network models, the activation function is the sigmoid function of logistic type or other sigmoid function or bell-shaped function.

It is well known that FNNs are universal approximators. Theoretically, any continuous function defined on a compact set can be approximated to any desired degree of

G.A. Anastassiou, *Intelligent Systems II: Complete Approximation by Neural Network Operators*, Studies in Computational Intelligence 608,
DOI 10.1007/978-3-319-20505-2_24

accuracy by increasing the number of hidden neurons. It was proved by Cybenko [16] and Funahashi [18], that any continuous function can be approximated on a compact set with uniform topology by a network of the form $N_n(x)$, using any continuous, sigmoid activation function. Hornik et al. in [21], have shown that any measurable function can be approached with such a network. Furthermore, these authors proved in [22], that any function of the Sobolev spaces can be approached with all derivatives. A variety of density results on FNN approximations to multivariate functions were later established by many authors using different methods, for more or less general situations: [23] by Leshno et al. [27] by Mhaskar and Micchelli, [13] by Chui and Li, [12] by Chen and Chen, [19] by Hahm and Hong, etc.

Usually these results only give theorems about the existence of an approximation. A related and important problem is that of complexity: determining the number of neurons required to guarantee that all functions belonging to a space can be approximated to the prescribed degree of accuracy ϵ.

Barron [8] shows that if the function is supposed to satisfy certain conditions expressed in terms of its Fourier transform, and if each of the neurons evaluates a sigmoid activation function, then at most $O\left(\epsilon^{-2}\right)$ neurons are needed to achieve the order of approximation ϵ. Some other authors have published similar results on the complexity of FNN approximations: Mhaskar and Micchelli [28], Suzuki [31], Maiorov and Meir [24], Makovoz [25], Ferrari and Stengel [17], Xu and Cao [32], Cao et al. [9], etc.

Cardaliaguet and Euvrard were the first, see [10], to describe precisely and study neural network approximation operators to the unit operator. Namely they proved: be given $f : \mathbb{R} \to \mathbb{R}$ a continuous bounded function and b a centered bell-shaped function, then the functions

$$F_n(x) = \sum_{k=-n^2}^{n^2} \frac{f\left(\frac{k}{n}\right)}{In^{\alpha}} b\left(n^{1-\alpha}\left(x - \frac{k}{n}\right)\right),$$

where $I := \int_{-\infty}^{\infty} b(t)\,dt$, $0 < \alpha < 1$, converge uniformly on compacta to f.

You see above that the weights $\frac{f\left(\frac{k}{n}\right)}{In^{\alpha}}$ are explicitly known, for the first time shown in [10].

Furthermore the authors in [10] proved that: let $f : \mathbb{R}^p \to \mathbb{R}$, $p \in \mathbb{N}$, be a continuous bounded function and b a p-dimensional bell-shaped function. Then the functions

$$G_n(x) =$$

$$\sum_{k_1=-n^2}^{n^2} \cdots \sum_{k_p=-n^2}^{n^2} \frac{f\left(\frac{k_1}{n}, \ldots, \frac{k_p}{n}\right)}{In^{\alpha p}} b\left(n^{1-\alpha}\left(x_1 - \frac{k_1}{n}\right), \ldots, n^{1-\alpha}\left(x_p - \frac{k_p}{n}\right)\right),$$

where I is the integral of b on $\mathbb{R}^p$ and $0 < \alpha < 1$, converge uniformly on compacta to f.

Still the work [10] is qualitative and not quantitative.

The author in [1–3], see Chaps. 2–5, was the first to establish neural network approximations to continuous functions with rates, that is quantitative works, by very specifically defined neural network operators of Cardaliagnet-Euvrard and "Squashing" types, by employing the modulus of continuity of the engaged function or its high order derivative or partial derivatives, and producing very tight Jackson type inequalities. He treats there both the univariate and multivariate cases. The defining these operators "bell-shaped" and "squashing" functions are assumed to be of compact support. Also in [3] he gives the Nth order asymptotic expansion for the error of weak approximation of these two operators to a special natural class of smooth functions, see Chaps. 4 and 5 there.

Though the work in [1–3], was quantitative, the rate of convergence was not precisely determined.

Finally the author in [4, 5], by normalizing his operators he achieved to determine the exact rates of convergence.

In this chapter the author continuous and completes his related work, by introducing three new multivariate perturbed neural network operators of Cardaliaguet-Euvrard type. This started with the univariate treatment in [6].

The sample coefficients $f\left(\frac{k}{n}\right)$ are replaced by three suitable natural perturbations, what is actuality happens in reality of neural network operations.

The calculuation of $f\left(\frac{k}{n}\right)$ at the neurons many times are not calculated as such, but rather in a distored way.

Next we justify why we take here the multivariate activation function to be of compact support, of course it helps us to conduct our study.

The multivariate activation function, same as transfer function or learning rule, is connected and associated to firing of neurons. Firing, which sends electric pulses or an output signal to other neurons, occurs when the activation level is above the threshold level set by the learning rule.

Each Neural Network firing is essentially of finite time duration. Essentially the firing in time decays, but in practice sends positive energy over a finite time interval.

Thus by using an activation function of compact support, in practice we do not alter much of the good results of our approximation.

To be more precise, we may take the compact support to be a large symmetric to the origin box. This is what happens in real time with the firing of neurons.

For more about neural networks in general we refer to [11, 14, 15, 20, 26, 29].

24.2 Basics

Here the activation function $b : \mathbb{R}^d \to \mathbb{R}_+$, $d \in \mathbb{N}$, is of compact support $B := \prod_{j=1}^{d} [-T_j, T_j]$, $T_j > 0$, $j = 1, \ldots, d$. That is $b(x) > 0$ for any $x \in B$, and clearly b may have jump discontinuities. Also the shape of the graph of b is immaterial.

Typically in neural networks approximation we take b to be a d-dimensional bell-shaped function (i.e. per coordinate is a centered bell-shaped function), or a product of univariate centered bell-shaped functions, or a product of sigmoid functions, in our case all of them are of compact support B.

Example 24.1 Take $b(x) = \beta(x_1)\beta(x_2)\ldots\beta(x_d)$, where β is any of the following functions, $j = 1, \ldots, d$:

(i) $\beta(x_j)$ is the characteristic function on $[-1, 1]$,
(ii) $\beta(x_j)$ is the hat function over $[-1, 1]$, that is,

$$\beta(x_j) = \begin{cases} 1 + x_j, & -1 \le x_j \le 0, \\ 1 - x_j, & 0 < x_j \le 1, \\ 0, & elsewhere, \end{cases}$$

(iii) the truncated sigmoids

$$\beta(x_j) = \begin{cases} \frac{1}{1+e^{-x_j}} \text{ or } \tanh x_j \text{ or } \operatorname{erf}(x_j), \text{ for } x_j \in [-T_j, T_j], \text{ with large } T_j > 0, \\ 0, \ x_j \in \mathbb{R} - [-T_j, T_j], \end{cases}$$

(iv) the truncated Gompertz function

$$\beta(x_j) = \begin{cases} e^{-\alpha e^{-\beta x_j}}, \ x_j \in [-T_j, T_j]; \ \alpha, \beta > 0; \text{ large } T_j > 0, \\ 0, \ x \in \mathbb{R} - [-T_j, T_j], \end{cases}$$

The Gompertz functions are also sigmoid functions, with wide applications to many applied fields, e.g. demography and tumor growth modeling, etc.

Thus the general activation function b we will be using here includes all kinds of activation functions in neural network approximations.

Here we consider functions $f : \mathbb{R}^d \to \mathbb{R}$ that either continuous and bounded, or uniformly continuous.

Let here the parameters: $0 < \alpha < 1$, $x = (x_1, \ldots, x_d) \in \mathbb{R}^d$, $n \in \mathbb{N}$; $r = (r_1, \ldots, r_d) \in \mathbb{N}^d$, $i = (i_1, \ldots, i_d) \in \mathbb{N}^d$, with $i_j = 1, 2, \ldots, r_j$, $j = 1, \ldots, d$; also let $w_i = w_{i_1,\ldots,i_d} \ge 0$, such that $\sum_{i_1=1}^{r_1} \sum_{i_2=1}^{r_2} \ldots \sum_{i_d=1}^{r_d} w_{i_1,\ldots,i_d} = 1$, in brief written as $\sum_{i=1}^{r} w_i = 1$. We further consider the parameters $k = (k_1, \ldots, k_d) \in \mathbb{Z}^d$; $\mu_i = (\mu_{i_1}, \ldots, \mu_{i_d}) \in \mathbb{R}^d_+$, $\nu_i = (\nu_{i_1}, \ldots, \nu_{i_d}) \in \mathbb{R}^d_+$; and $\lambda_i = \lambda_{i_1,\ldots,i_d}$, $\rho_i = \rho_{i_1,\ldots,i_d} \ge 0$; $\mu, \nu \ge 0$. Call $\nu_i^{\min} = \min\{\nu_{i_1}, \ldots, \nu_{i_d}\}$.

We use here the first modulus of continuity, with $\delta > 0$,

$$\omega_1(f, \delta) := \sup_{\substack{x, y \in \mathbb{R}^d \\ \|x - y\|_\infty \le \delta}} |f(x) - f(y)|,$$

where $\|x\|_\infty = \max(|x_1|, \ldots, |x_d|)$.

Given that f is uniformly continuous we get $\lim_{\delta \to 0} \omega_1(f, \delta) = 0$.

This chapter is a continuation of [6] at the multivariate level.

So in this chapter mainly we study the pointwise convergence with rates over $\mathbb{R}^d$, to the unit operator, of the following one hidden layer multivariate normalized neural network perturbed operators,

(i) the Stancu type (see [30])

$$\left(H_n^*(f)\right)(x) = \left(H_n^*(f)\right)(x_1, \ldots, x_d) = \tag{24.1}$$

$$\frac{\sum_{k=-n^2}^{n^2} \left(\sum_{i=1}^{r} w_i f\left(\frac{k+\mu_i}{n+\nu_i}\right)\right) b\left(n^{1-\alpha}\left(x - \frac{k}{n}\right)\right)}{\sum_{k=-n^2}^{n^2} b\left(n^{1-\alpha}\left(x - \frac{k}{n}\right)\right)} =$$

$$\frac{\sum_{k_1=-n^2}^{n^2} \cdots \sum_{k_d=-n^2}^{n^2} \left(\sum_{i_1=1}^{r} \cdots \sum_{i_d=1}^{r} w_{i_1,\ldots,i_d} f\left(\frac{k_1+\mu_{i_1}}{n+\nu_{i_1}}, \ldots, \frac{k_d+\mu_{i_d}}{n+\nu_{i_d}}\right)\right) \cdot}{\sum_{k_1=-n^2}^{n^2} \cdots \sum_{k_d=-n^2}^{n^2} b\left(n^{1-\alpha}\left(x_1 - \frac{k_1}{n}\right), \ldots, n^{1-\alpha}\left(x_d - \frac{k_d}{n}\right)\right)}$$

$$b\left(n^{1-\alpha}\left(x_1 - \frac{k_1}{n}\right), \ldots, n^{1-\alpha}\left(x_d - \frac{k_d}{n}\right)\right),$$

(ii) the Kantorovich type

$$\left(K_n^*(f)\right)(x) = \tag{24.2}$$

$$\frac{\sum_{k=-n^2}^{n^2} \left(\sum_{i=1}^{r} w_i (n+\rho_i)^d \int_0^{\frac{1}{n+\rho_i}} f\left(t + \frac{k+\lambda_i}{n+\rho_i}\right) dt\right) b\left(n^{1-\alpha}\left(x - \frac{k}{n}\right)\right)}{\sum_{k=-n^2}^{n^2} b\left(n^{1-\alpha}\left(x - \frac{k}{n}\right)\right)} =$$

$$\sum_{k_1=-n^2}^{n^2} \cdots \sum_{k_d=-n^2}^{n^2} \left(\sum_{i_1=1}^{r} \cdots \sum_{i_d=1}^{r} w_{i_1,\ldots,i_d} \left(n + \rho_{i_1,\ldots,i_d}\right)^d \cdot\right.$$

$$\frac{\left.\int \ldots \int \ldots \int_0^{\frac{1}{n+\rho_{i_1,\ldots,i_d}}} f\left(t_1 + \frac{k_1+\lambda_{i_1,\ldots,i_d}}{n+\rho_{i_1,\ldots,i_d}}, \ldots, t_d + \frac{k_d+\lambda_{i_1,\ldots,i_d}}{n+\rho_{i_1,\ldots,i_d}}\right) dt_1 \ldots dt_d\right) \cdot}{\sum_{k_1=-n^2}^{n^2} \cdots \sum_{k_d=-n^2}^{n^2} b\left(n^{1-\alpha}\left(x_1 - \frac{k_1}{n}\right), \ldots, n^{1-\alpha}\left(x_d - \frac{k_d}{n}\right)\right)} \tag{24.3}$$

$$b\left(n^{1-\alpha}\left(x_1 - \frac{k_1}{n}\right), \ldots, n^{1-\alpha}\left(x_d - \frac{k_d}{n}\right)\right),$$

and

(iii) the quadrature type

$$\left(M_n^*(f)\right)(x) = \frac{\sum_{k=-n^2}^{n^2}\left(\sum_{i=1}^{r} w_i f\left(\frac{k}{n}+\frac{i}{nr}\right)\right) b\left(n^{1-\alpha}\left(x-\frac{k}{n}\right)\right)}{\sum_{k=-n^2}^{n^2} b\left(n^{1-\alpha}\left(x-\frac{k}{n}\right)\right)} = \tag{24.4}$$

$$\frac{\sum_{k_1=-n^2}^{n^2}\cdots\sum_{k_d=-n^2}^{n^2}\left(\sum_{i_1=1}^{r_1}\cdots\sum_{i_d=1}^{r_d} w_{i_1,\ldots,i_d} f\left(\frac{k_1}{n}+\frac{i_1}{nr_1},\ldots,\frac{k_d}{n}+\frac{i_d}{nr_d}\right)\right)\cdot}{\sum_{k_1=-n^2}^{n^2}\cdots\sum_{k_d=-n^2}^{n^2} b\left(n^{1-\alpha}\left(x_1-\frac{k_1}{n}\right),\ldots,n^{1-\alpha}\left(x_d-\frac{k_d}{n}\right)\right)}$$

$$b\left(n^{1-\alpha}\left(x_1-\frac{k_1}{n}\right),\ldots,n^{1-\alpha}\left(x_d-\frac{k_d}{n}\right)\right).$$

Similar operators defined for d-dimensional bell-shaped activation functions and sample coefficients $f\left(\frac{k}{n}\right) = f\left(\frac{k_1}{n},\ldots,\frac{k_d}{n}\right)$ were studied initially in [1–5, 10], etc.

Here we study the multivariate generalized perturbed cases of these operators.

Operator K_n^* in the corresponding Signal Processing context, represents the natural so called "time-jitter" error, where the sample information is calculated in a perturbed neighborhood of $\frac{k+\mu}{n+\nu}$ rather than exactly at the node $\frac{k}{n}$.

The perturbed sample coefficients $f\left(\frac{k+\mu}{n+\nu}\right)$ with $0 \leq \mu \leq \nu$, were first used by Stancu [30], in a totally different context, generalizing Bernstein operators approximation on $C([0, 1])$.

The terms in the ratio of sums (24.1)–(24.4) can be nonzero, iff simultaneously

$$\left|n^{1-\alpha}\left(x_j-\frac{k_j}{n}\right)\right| \leq T_j \text{ , all } j = 1,\ldots,d, \tag{24.5}$$

i.e. $\left|x_j - \frac{k_j}{n}\right| \leq \frac{T_j}{n^{1-\alpha}}$, all $j = 1,\ldots,d$, iff

$$nx_j - T_j n^{\alpha} \leq k_j \leq nx_j + T_j n^{\alpha}, \text{ all } j = 1,\ldots,d. \tag{24.6}$$

To have the order

$$-n^2 \leq nx_j - T_j n^{\alpha} \leq k_j \leq nx_j + T_j n^{\alpha} \leq n^2, \tag{24.7}$$

we need $n \geq T_j + |x_j|$, all $j = 1,\ldots,d$. So (24.7) is true when we take

$$n \geq \max_{j\in\{1,\ldots,d\}}\left(T_j + |x_j|\right). \tag{24.8}$$

When $x \in B$ in order to have (24.7) it is enough to assume that $n \geq 2T^*$, where $T^* := \max\{T_1, \ldots, T_d\} > 0$. Consider

$$\widetilde{I}_j := \left[nx_j - T_j n^\alpha, nx_j + T_j n^\alpha\right], \ j = 1, \ldots, d, \ n \in \mathbb{N}.$$

The length of $\widetilde{I}_j$ is $2T_j n^\alpha$. By Proposition 1 of [1], we get that the cardinality of $k_j \in \mathbb{Z}$ that belong to $\widetilde{I}_j := card\left(k_j\right) \geq \max\left(2T_j n^\alpha - 1, 0\right)$, any $j \in \{1, \ldots, d\}$. In order to have $card\left(k_j\right) \geq 1$, we need $2T_j n^\alpha - 1 \geq 1$ iff $n \geq T_j^{-\frac{1}{\alpha}}$, any $j \in \{1, \ldots, d\}$.

Therefore, a sufficient condition in order to obtain the order (24.7) along with the interval $\widetilde{I}_j$ to contain at least one integer for all $j = 1, \ldots, d$ is that

$$n \geq \max_{j \in \{1, \ldots, d\}} \left\{T_j + \left|x_j\right|, T_j^{-\frac{1}{\alpha}}\right\}. \tag{24.9}$$

Clearly as $n \to +\infty$ we get that $card\left(k_j\right) \to +\infty$, all $j = 1, \ldots, d$. Also notice that $card\left(k_j\right)$ equals to the cardinality of integers in $\left[\lceil nx_j - T_j n^\alpha\rceil, \left[nx_j + T_j n^\alpha\right]\right]$ for all $j = 1, \ldots, d$. Here $[\cdot]$ denotes the integral part of the number while $\lceil\cdot\rceil$ denotes its ceiling.

From now on, in this chapter we assume (24.9).

We denote by $T = (T_1, \ldots, T_d)$, $[nx + Tn^\alpha] = ([nx_1 + T_1 n^\alpha], \ldots, [nx_d + T_d n^\alpha])$, and $\lceil nx - Tn^\alpha\rceil = (\lceil nx_1 - T_1 n^\alpha\rceil, \ldots, \lceil nx_d - T_d n^\alpha\rceil)$. Furthermore it holds

(i)

$$\left(H_n^*(f)\right)(x) = \left(H_n^*(f)\right)(x_1, \ldots, x_d) = \tag{24.10}$$

$$\frac{\sum_{k=\lceil nx - Tn^\alpha\rceil}^{[nx+Tn^\alpha]} \left(\sum_{i=1}^{r} w_i f\left(\frac{k+\mu_i}{n+\nu_i}\right)\right) b\left(n^{1-\alpha}\left(x - \frac{k}{n}\right)\right)}{\sum_{k=\lceil nx - Tn^\alpha\rceil}^{[nx+Tn^\alpha]} b\left(n^{1-\alpha}\left(x - \frac{k}{n}\right)\right)} =$$

$$\frac{\sum_{k_1=\lceil nx_1 - T_1 n^\alpha\rceil}^{[nx_1+T_1 n^\alpha]} \cdots \sum_{k_d=\lceil nx_d - T_d n^\alpha\rceil}^{[nx_d+T_d n^\alpha]} \left(\sum_{i_1=1}^{r_1} \cdots \sum_{i_d=1}^{r_d} w_{i_1,\ldots,i_d} f\left(\frac{k_1+\mu_{i_1}}{n+\nu_{i_1}}, \ldots, \frac{k_d+\mu_{i_d}}{n+\nu_{i_d}}\right)\right) \cdot}{\sum_{k_1=\lceil nx_1 - T_1 n^\alpha\rceil}^{[nx_1+T_1 n^\alpha]} \cdots \sum_{k_d=\lceil nx_d - T_d n^\alpha\rceil}^{[nx_d+T_d n^\alpha]} b\left(n^{1-\alpha}\left(x_1 - \frac{k_1}{n}\right), \ldots, n^{1-\alpha}\left(x_d - \frac{k_d}{n}\right)\right)}$$

$$b\left(n^{1-\alpha}\left(x_1 - \frac{k_1}{n}\right), \ldots, n^{1-\alpha}\left(x_d - \frac{k_d}{n}\right)\right),$$

(ii)

$$\left(K_n^*(f)\right)(x) = \tag{24.11}$$

$$\frac{\sum_{k=\lceil nx - Tn^\alpha\rceil}^{[nx+Tn^\alpha]} \left(\sum_{i=1}^{r} w_i \left(n + \rho_i\right) \int_0^{\frac{1}{n+\rho_i}} f\left(t + \frac{k+\lambda_i}{n+\rho_i}\right) dt\right) b\left(n^{1-\alpha}\left(x - \frac{k}{n}\right)\right)}{\sum_{k=\lceil nx - Tn^\alpha\rceil}^{[nx+Tn^\alpha]} b\left(n^{1-\alpha}\left(x - \frac{k}{n}\right)\right)} =$$

$$\sum_{k_1=\lceil nx_1-T_1n^{\alpha}\rceil}^{[nx_1+T_1n^{\alpha}]} \cdots \sum_{k_d=\lceil nx_d-T_dn^{\alpha}\rceil}^{[nx_d+T_dn^{\alpha}]} \left(\sum_{i_1=1}^{r_1} \cdots \sum_{i_d=1}^{r_d} w_{i_1,\ldots,i_d} \left(n+\rho_{i_1,\ldots,i_d}\right)^d \cdot \right.$$

$$\frac{\left. \int \ldots \int \ldots \int_0^{\frac{1}{n+\rho_{i_1,\ldots,i_d}}} f\left(t_1 + \frac{k_1+\lambda_{i_1,\ldots,i_d}}{n+\rho_{i_1,\ldots,i_d}}, \ldots, t_d + \frac{k_d+\lambda_{i_1,\ldots,i_d}}{n+\rho_{i_1,\ldots,i_d}} \right) dt_1 \ldots dt_d \right) \cdot}{\sum_{k_1=\lceil nx_1-T_1n^{\alpha}\rceil}^{[nx_1+T_1n^{\alpha}]} \cdots \sum_{k_d=\lceil nx_d-T_dn^{\alpha}\rceil}^{[nx_d+T_dn^{\alpha}]} b\left(n^{1-\alpha}\left(x_1-\frac{k_1}{n}\right), \ldots, n^{1-\alpha}\left(x_d-\frac{k_d}{n}\right)\right)} \tag{24.12}$$

$$b\left(n^{1-\alpha}\left(x_1-\frac{k_1}{n}\right), \ldots, n^{1-\alpha}\left(x_d-\frac{k_d}{n}\right)\right),$$

and

(iii)

$$\left(M_n^*(f)\right)(x) = \frac{\sum_{k=\lceil nx-Tn^{\alpha}\rceil}^{[nx+Tn^{\alpha}]} \left(\sum_{i=1}^{r} w_i f\left(\frac{k}{n}+\frac{i}{nr}\right)\right) b\left(n^{1-\alpha}\left(x-\frac{k}{n}\right)\right)}{\sum_{k=\lceil nx-Tn^{\alpha}\rceil}^{[nx+Tn^{\alpha}]} b\left(n^{1-\alpha}\left(x-\frac{k}{n}\right)\right)} = \tag{24.13}$$

$$\frac{\sum_{k_1=\lceil nx_1-T_1n^{\alpha}\rceil}^{[nx_1+T_1n^{\alpha}]} \cdots \sum_{k_d=\lceil nx_d-T_dn^{\alpha}\rceil}^{[nx_d+T_dn^{\alpha}]} \left(\sum_{i_1=1}^{r_1} \cdots \sum_{i_d=1}^{r_d} w_{i_1,\ldots,i_d} f\left(\frac{k_1}{n}+\frac{i_1}{nr_1}, \ldots, \frac{k_d}{n}+\frac{i_d}{nr_d}\right)\right) \cdot}{\sum_{k_1=\lceil nx_1-T_1n^{\alpha}\rceil}^{[nx_1+T_1n^{\alpha}]} \cdots \sum_{k_d=\lceil nx_d-T_dn^{\alpha}\rceil}^{[nx_d+T_dn^{\alpha}]} b\left(n^{1-\alpha}\left(x_1-\frac{k_1}{n}\right), \ldots, n^{1-\alpha}\left(x_d-\frac{k_d}{n}\right)\right)}$$

$$b\left(n^{1-\alpha}\left(x_1-\frac{k_1}{n}\right), \ldots, n^{1-\alpha}\left(x_d-\frac{k_d}{n}\right)\right).$$

So if $\left| n^{1-\alpha}\left(x_j-\frac{k_j}{n}\right)\right| \leq T_j$, all $j=1,\ldots,d$, we get that

$$\left\| x-\frac{k}{n}\right\|_{\infty} \leq \frac{T^*}{n^{1-\alpha}}. \tag{24.14}$$

For convenience we call

$$V(x) = \sum_{k=\lceil nx-Tn^{\alpha}\rceil}^{[nx+Tn^{\alpha}]} b\left(n^{1-\alpha}\left(x-\frac{k}{n}\right)\right) =$$

$$\sum_{k_1=\lceil nx_1-T_1n^{\alpha}\rceil}^{[nx_1+T_1n^{\alpha}]} \cdots \sum_{k_d=\lceil nx_d-T_dn^{\alpha}\rceil}^{[nx_d+T_dn^{\alpha}]} b\left(n^{1-\alpha}\left(x_1-\frac{k_1}{n}\right),\ldots,n^{1-\alpha}\left(x_d-\frac{k_d}{n}\right)\right). \tag{24.15}$$

We make

Remark 24.2 Here always k is as in (24.7).

(I) We observe that

$$\left\|\frac{k+\mu_i}{n+\nu_i}-x\right\|_{\infty} \le \left\|\frac{k}{n+\nu_i}-x\right\|_{\infty} + \left\|\frac{\mu_i}{n+\nu_i}\right\|_{\infty} \tag{24.16}$$

$$\le \left\|\frac{k}{n+\nu_i}-x\right\|_{\infty} + \frac{\|\mu_i\|_{\infty}}{n+\nu_i^{\min}}.$$

Next we see

$$\left\|\frac{k}{n+\nu_i}-x\right\|_{\infty} \le \left\|\frac{k}{n+\nu_i}-\frac{k}{n}\right\|_{\infty} + \left\|\frac{k}{n}-x\right\|_{\infty} \overset{(24.14)}{\le} \left\|\frac{\nu_i k}{n(n+\nu_i)}\right\|_{\infty} + \frac{T^*}{n^{1-\alpha}} \tag{24.17}$$

$$\le \|k\|_{\infty}\frac{\|\nu_i\|_{\infty}}{n\left(n+\nu_i^{\min}\right)} + \frac{T^*}{n^{1-\alpha}} =: (*).$$

We notice for $j=1,\ldots,d$ we get that

$$\left|k_j\right| \le n\left|x_j\right| + T_j n^{\alpha}.$$

Therefore

$$\|k\|_{\infty} \le \left\|n|x| + Tn^{\alpha}\right\|_{\infty} \le n\|x\|_{\infty} + T^*n^{\alpha}, \tag{24.18}$$

where $|x| = (|x_1|,\ldots,|x|_d)$.
Thus

$$(*) \le \left(n\|x\|_{\infty} + T^*n^{\alpha}\right)\frac{\|\nu_i\|_{\infty}}{n\left(n+\nu_i^{\min}\right)} + \frac{T^*}{n^{1-\alpha}} = \tag{24.19}$$

$$\left(\|x\|_{\infty} + T^*n^{\alpha-1}\right)\frac{\|\nu_i\|_{\infty}}{\left(n+\nu_i^{\min}\right)} + \frac{T^*}{n^{1-\alpha}}.$$

So we get

$$\left\|\frac{k}{n+\nu_i}-x\right\|_{\infty} \le \left(\|x\|_{\infty} + \frac{T^*}{n^{1-\alpha}}\right)\frac{\|\nu_i\|_{\infty}}{\left(n+\nu_i^{\min}\right)} + \frac{T^*}{n^{1-\alpha}}. \tag{24.20}$$

Consequently we obtain

$$\left\| \frac{k+\mu_i}{n+\nu_i} - x \right\|_\infty \leq \left(\frac{\|\nu_i\|_\infty \|x\|_\infty + \|\mu_i\|_\infty}{n+\nu_i^{\min}} \right) + \left(1 + \frac{\|\nu_i\|_\infty}{\left(n+\nu_i^{\min}\right)} \right) \frac{T^*}{n^{1-\alpha}}. \tag{24.21}$$

Hence we derive

$$\omega_1 \left(f, \left\| \frac{k+\mu_i}{n+\nu_i} - x \right\|_\infty \right) \leq$$

$$\omega_1 \left(f, \left(\frac{\|\nu_i\|_\infty \|x\|_\infty + \|\mu_i\|_\infty}{n+\nu_i^{\min}} \right) + \left(1 + \frac{\|\nu_i\|_\infty}{\left(n+\nu_i^{\min}\right)} \right) \frac{T^*}{n^{1-\alpha}} \right), \tag{24.22}$$

with dominant speed of convergence $\frac{1}{n^{1-\alpha}}$.

(II) We also have for

$$0 \leq t_j \leq \frac{1}{n+\rho_{i_1,\ldots,i_d}}, \quad j = 1, \ldots, d, \tag{24.23}$$

that

$$\left\| t + \frac{k+\lambda_{i_1,\ldots,i_d}}{n+\rho_{i_1,\ldots,i_d}} - x \right\|_\infty \leq \frac{1}{n+\rho_{i_1,\ldots,i_d}} + \left\| \frac{k+\lambda_{i_1,\ldots,i_d}}{n+\rho_{i_1,\ldots,i_d}} - x \right\|_\infty \leq \tag{24.24}$$

$$\frac{1+\lambda_{i_1,\ldots,i_d}}{n+\rho_{i_1,\ldots,i_d}} + \left\| \frac{k}{n+\rho_{i_1,\ldots,i_d}} - x \right\|_\infty \leq$$

$$\frac{1+\lambda_{i_1,\ldots,i_d}}{n+\rho_{i_1,\ldots,i_d}} + \left\| \frac{k}{n+\rho_{i_1,\ldots,i_d}} - \frac{k}{n} \right\|_\infty + \left\| \frac{k}{n} - x \right\|_\infty \leq$$

$$\frac{1+\lambda_{i_1,\ldots,i_d}}{n+\rho_{i_1,\ldots,i_d}} + \frac{T^*}{n^{1-\alpha}} + \frac{\rho_{i_1,\ldots,i_d} \|k\|_\infty}{\left(n+\rho_{i_1,\ldots,i_d}\right) n} \leq$$

$$\frac{1+\lambda_{i_1,\ldots,i_d}}{n+\rho_{i_1,\ldots,i_d}} + \frac{T^*}{n^{1-\alpha}} + \frac{\rho_{i_1,\ldots,i_d}}{n\left(n+\rho_{i_1,\ldots,i_d}\right)} \left(n \|x\|_\infty + T^* n^\alpha\right) = \tag{24.25}$$

$$\frac{1+\lambda_{i_1,\ldots,i_d}}{n+\rho_{i_1,\ldots,i_d}} + \frac{T^*}{n^{1-\alpha}} + \left(\frac{\rho_{i_1,\ldots,i_d}}{n+\rho_{i_1,\ldots,i_d}} \right) \left(\|x\|_\infty + \frac{T^*}{n^{1-\alpha}} \right). \tag{24.26}$$

We have found that

$$\left\| t + \frac{k+\lambda_{i_1,\ldots,i_d}}{n+\rho_{i_1,\ldots,i_d}} - x \right\|_\infty \leq \tag{24.27}$$

$$\left(\frac{\rho_{i_1,\ldots,i_d} \left\| x \right\|_\infty + \lambda_{i_1,\ldots,i_d} + 1}{n + \rho_{i_1,\ldots,i_d}} \right) + \left(1 + \frac{\rho_{i_1,\ldots,i_d}}{n + \rho_{i_1,\ldots,i_d}} \right) \frac{T^*}{n^{1-\alpha}}.$$

So when $0 \leq t_j \leq \frac{1}{n+\rho_{i_1,\ldots,i_d}}$, $j = 1, \ldots, d$, we get that

$$\omega_1 \left(f, \left\| t + \frac{k + \lambda_{i_1,\ldots,i_d}}{n + \rho_{i_1,\ldots,i_d}} - x \right\|_\infty \right) \leq \tag{24.28}$$

$$\omega_1 \left(f, \left(\frac{\rho_{i_1,\ldots,i_d} \left\| x \right\|_\infty + \lambda_{i_1,\ldots,i_d} + 1}{n + \rho_{i_1,\ldots,i_d}} \right) + \left(1 + \frac{\rho_{i_1,\ldots,i_d}}{n + \rho_{i_1,\ldots,i_d}} \right) \frac{T^*}{n^{1-\alpha}} \right),$$

with dominant speed $\frac{1}{n^{1-\alpha}}$.

(III) We observe that

$$\left\| \frac{k}{n} + \frac{i}{nr} - x \right\|_\infty \leq \left\| \frac{k}{n} - x \right\|_\infty + \frac{1}{n} \left\| \frac{i}{r} \right\|_\infty \leq \frac{T^*}{n^{1-\alpha}} + \frac{1}{n}. \tag{24.29}$$

Hence

$$\omega_1 \left(f, \left\| \frac{k}{n} + \frac{i}{nr} - x \right\|_\infty \right) \leq \omega_1 \left(f, \frac{T^*}{n^{1-\alpha}} + \frac{1}{n} \right), \tag{24.30}$$

with dominant speed $\frac{1}{n^{1-\alpha}}$.

24.3 Main Results

We present our first approximation result

Theorem 24.3 *Let $x \in \mathbb{R}^d$ and $n \in \mathbb{N}$ such that $n \geq \max_{j \in \{1,\ldots,d\}} \left(T_j + \left| x_j \right|, T_j^{-\frac{1}{\alpha}} \right)$, $T_j > 0$, $0 < \alpha < 1$. Then*

$$\left| \left(H_n^* (f) \right) (x) - f (x) \right| \leq$$

$$\sum_{i=1}^{r} w_i \omega_1 \left(f, \left(\frac{\left\| \nu_i \right\|_\infty \left\| x \right\|_\infty + \left\| \mu_i \right\|_\infty}{n + \nu_i^{\min}} \right) + \left(1 + \frac{\left\| \nu_i \right\|_\infty}{n + \nu_i^{\min}} \right) \frac{T^*}{n^{1-\alpha}} \right) = \tag{24.31}$$

$$\sum_{i_1=1}^{r_1} \cdots \sum_{i_d=1}^{r_d} w_{i_1,\ldots,i_d} \omega_1 \left(f, \left(\frac{\left\| \nu_i \right\|_\infty \left\| x \right\|_\infty + \left\| \mu_i \right\|_\infty}{n + \nu_i^{\min}} \right) + \left(1 + \frac{\left\| \nu_i \right\|_\infty}{n + \nu_i^{\min}} \right) \frac{T^*}{n^{1-\alpha}} \right),$$

where $i = (i_1, \ldots, i_d)$.

Proof We notice that

$$\left(H_n^*(f)\right)(x) - f(x) = \frac{\sum_{k=\lceil nx-Tn^{\alpha}\rceil}^{[nx+Tn^{\alpha}]}\left(\sum_{i=1}^{r} w_i f\left(\frac{k+\mu_i}{n+\nu_i}\right)\right) b\left(n^{1-\alpha}\left(x-\frac{k}{n}\right)\right)}{\sum_{k=\lceil nx-Tn^{\alpha}\rceil}^{[nx+Tn^{\alpha}]} b\left(n^{1-\alpha}\left(x-\frac{k}{n}\right)\right)} - f(x) \tag{24.32}$$

$$= \frac{\sum_{k=\lceil nx-Tn^{\alpha}\rceil}^{[nx+Tn^{\alpha}]}\left(\sum_{i=1}^{r} w_i f\left(\frac{k+\mu_i}{n+\nu_i}\right)\right) b\left(n^{1-\alpha}\left(x-\frac{k}{n}\right)\right) - f(x)\sum_{k=\lceil nx-Tn^{\alpha}\rceil}^{[nx+Tn^{\alpha}]} b\left(n^{1-\alpha}\left(x-\frac{k}{n}\right)\right)}{\sum_{k=\lceil nx-Tn^{\alpha}\rceil}^{[nx+Tn^{\alpha}]} b\left(n^{1-\alpha}\left(x-\frac{k}{n}\right)\right)} =$$

$$\frac{\sum_{k=\lceil nx-Tn^{\alpha}\rceil}^{[nx+Tn^{\alpha}]}\left(\left(\sum_{i=1}^{r} w_i f\left(\frac{k+\mu_i}{n+\nu_i}\right)\right) - f(x)\right) b\left(n^{1-\alpha}\left(x-\frac{k}{n}\right)\right)}{\sum_{k=\lceil nx-Tn^{\alpha}\rceil}^{[nx+Tn^{\alpha}]} b\left(n^{1-\alpha}\left(x-\frac{k}{n}\right)\right)} = \tag{24.33}$$

$$\frac{\sum_{k=\lceil nx-Tn^{\alpha}\rceil}^{[nx+Tn^{\alpha}]}\left(\sum_{i=1}^{r} w_i \left(f\left(\frac{k+\mu_i}{n+\nu_i}\right) - f(x)\right)\right) b\left(n^{1-\alpha}\left(x-\frac{k}{n}\right)\right)}{\sum_{k=\lceil nx-Tn^{\alpha}\rceil}^{[nx+Tn^{\alpha}]} b\left(n^{1-\alpha}\left(x-\frac{k}{n}\right)\right)}.$$

Hence it holds

$$\left|\left(H_n^*(f)\right)(x) - f(x)\right| \leq$$

$$\frac{\sum_{k=\lceil nx-Tn^{\alpha}\rceil}^{[nx+Tn^{\alpha}]}\left(\sum_{i=1}^{r} w_i \left|f\left(\frac{k+\mu_i}{n+\nu_i}\right) - f(x)\right|\right) b\left(n^{1-\alpha}\left(x-\frac{k}{n}\right)\right)}{\sum_{k=\lceil nx-Tn^{\alpha}\rceil}^{[nx+Tn^{\alpha}]} b\left(n^{1-\alpha}\left(x-\frac{k}{n}\right)\right)} \leq \tag{24.34}$$

$$\frac{\sum_{k=\lceil nx-Tn^{\alpha}\rceil}^{[nx+Tn^{\alpha}]}\left(\sum_{i=1}^{r} w_i \omega_1\left(f, \left\|\frac{k+\mu_i}{n+\nu_i} - x\right\|_{\infty}\right)\right) b\left(n^{1-\alpha}\left(x-\frac{k}{n}\right)\right)}{\sum_{k=\lceil nx-Tn^{\alpha}\rceil}^{[nx+Tn^{\alpha}]} b\left(n^{1-\alpha}\left(x-\frac{k}{n}\right)\right)} \overset{(24.22)}{\leq}$$

$$\frac{\sum_{k=\lceil nx-Tn^{\alpha}\rceil}^{[nx+Tn^{\alpha}]}\left(\sum_{i=1}^{r} w_i \omega_1\left(f, \left(\frac{\|\nu_i\|_{\infty}\|x\|_{\infty}+\|\mu_i\|_{\infty}}{n+\nu_i^{\min}}\right) + \left(1+\frac{\|\nu_i\|_{\infty}}{n+\nu_i^{\min}}\right)\frac{T^*}{n^{1-\alpha}}\right)\right)\cdot}{\sum_{k=\lceil nx-Tn^{\alpha}\rceil}^{[nx+Tn^{\alpha}]} b\left(n^{1-\alpha}\left(x-\frac{k}{n}\right)\right)} \tag{24.35}$$

$$b\left(n^{1-\alpha}\left(x-\frac{k}{n}\right)\right) =$$

$$\frac{\left[\sum_{i=1}^{r} w_i \omega_1\left(f, \left(\frac{\|\nu_i\|_{\infty}\|x\|_{\infty}+\|\mu_i\|_{\infty}}{n+\nu_i^{\min}}\right) + \left(1+\frac{\|\nu_i\|_{\infty}}{n+\nu_i^{\min}}\right)\frac{T^*}{n^{1-\alpha}}\right)\right]\cdot}{\sum_{k=\lceil nx-Tn^{\alpha}\rceil}^{[nx+Tn^{\alpha}]} b\left(n^{1-\alpha}\left(x-\frac{k}{n}\right)\right)}$$

$$\left(\sum_{k=\lceil nx-Tn^{\alpha}\rceil}^{[nx+Tn^{\alpha}]} b\left(n^{1-\alpha}\left(x-\frac{k}{n}\right)\right)\right)$$

$$=\sum_{i=1}^{r} w_i \omega_1 \left(f, \left(\frac{\|\nu_i\|_{\infty} \|x\|_{\infty} + \|\mu_i\|_{\infty}}{n+\nu_i^{\min}} \right) + \left(1 + \frac{\|\nu_i\|_{\infty}}{n+\nu_i^{\min}} \right) \frac{T^*}{n^{1-\alpha}} \right), \tag{24.36}$$

proving the claim. ∎

Corollary 24.4 (to Theorem 24.3) *Let* $x \in \prod_{j=1}^{d} \left[-\gamma_j, \gamma_j\right] \subset \mathbb{R}^d$, $\gamma_j > 0$, $\gamma^* = \max\{\gamma_1, \ldots, \gamma_d\}$ *and* $n \in \mathbb{N}$ *such that* $n \geq \max_{j\in\{1,\ldots,d\}} \{T_j + \gamma_j, T_j^{-\frac{1}{\alpha}}\}$. *Then*

$$\left\| H_n^* (f) - f \right\|_{\infty, \prod_{j=1}^{d} [-\gamma_j, \gamma_j]} \leq$$

$$\sum_{i_1=1}^{r_1} \cdots \sum_{i_d=1}^{r_d} w_{i_1,\ldots,i_d} \omega_1 \left(f, \left(\frac{\|\nu_i\|_{\infty} \gamma^* + \|\mu_i\|_{\infty}}{n+\nu_i^{\min}} \right) + \left(1 + \frac{\|\nu_i\|_{\infty}}{n+\nu_i^{\min}} \right) \frac{T^*}{n^{1-\alpha}} \right). \tag{24.37}$$

Proof By (24.31). ∎

We continue with

Theorem 24.5 *All assumptions as in Theorem 24.3. Then*

$$\left| \left(K_n^* (f)\right)(x) - f(x) \right| \leq$$

$$\sum_{i=1}^{r} w_i \omega_1 \left(f, \left(\frac{\rho_i \|x\|_{\infty} + \lambda_i + 1}{n+\rho_i} \right) + \left(1 + \frac{\rho_i}{n+\rho_i} \right) \frac{T^*}{n^{1-\alpha}} \right) = \tag{24.38}$$

$$\sum_{i_1=1}^{r_1} \cdots \sum_{i_d=1}^{r_d} w_{i_1,\ldots,i_d} \omega_1 \left(f, \left(\frac{\rho_{i_1,\ldots,i_d} \|x\|_{\infty} + \lambda_{i_1,\ldots,i_d} + 1}{n+\rho_{i_1,\ldots,i_d}} \right) + \left(1 + \frac{\rho_{i_1,\ldots,i_d}}{n+\rho_{i_1,\ldots,i_d}} \right) \frac{T^*}{n^{1-\alpha}} \right).$$

Proof We observe the following

$$\left| \left(K_n^* (f)\right)(x) - f(x) \right| =$$

$$\left| \frac{\sum_{k=\lceil nx-Tn^{\alpha}\rceil}^{[nx+Tn^{\alpha}]} \left(\sum_{i=1}^{r} w_i (n+\rho_i)^d \int_0^{\frac{1}{n+\rho_i}} f\left(t + \frac{k+\lambda_i}{n+\rho_i}\right) dt \right) b\left(n^{1-\alpha}\left(x-\frac{k}{n}\right)\right)}{V(x)} - f(x) \right| = \tag{24.39}$$

$$\left|\frac{\sum_{k=\lceil nx-Tn^{\alpha}\rceil}^{[nx+Tn^{\alpha}]}\left(\sum_{i=1}^{r} w_i (n+\rho_i)^d \int_0^{\frac{1}{n+\rho_i}} f\left(t+\frac{k+\lambda_i}{n+\rho_i}\right)dt\right)b\left(n^{1-\alpha}\left(x-\frac{k}{n}\right)\right)-f(x)V(x)}{V(x)}\right| =$$

$$\left|\frac{\sum_{k=\lceil nx-Tn^{\alpha}\rceil}^{[nx+Tn^{\alpha}]}\left[\left(\sum_{i=1}^{r} w_i (n+\rho_i)^d \int_0^{\frac{1}{n+\rho_i}} f\left(t+\frac{k+\lambda_i}{n+\rho_i}\right)dt\right)-f(x)\right]b\left(n^{1-\alpha}\left(x-\frac{k}{n}\right)\right)}{V(x)}\right| =$$

$$\left|\frac{\sum_{k=\lceil nx-Tn^{\alpha}\rceil}^{[nx+Tn^{\alpha}]}\left[\sum_{i=1}^{r} w_i \left[(n+\rho_i)^d \int_0^{\frac{1}{n+\rho_i}} f\left(t+\frac{k+\lambda_i}{n+\rho_i}\right)dt - f(x)\right]\right]b\left(n^{1-\alpha}\left(x-\frac{k}{n}\right)\right)}{V(x)}\right| \tag{24.40}$$

$$= \frac{\left|\sum_{k=\lceil nx-Tn^{\alpha}\rceil}^{[nx+Tn^{\alpha}]}\left[\sum_{i=1}^{r} w_i \left[(n+\rho_i)^d \int_0^{\frac{1}{n+\rho_i}} \left[f\left(t+\frac{k+\lambda_i}{n+\rho_i}\right)-f(x)\right]dt\right]\right]b\left(n^{1-\alpha}\left(x-\frac{k}{n}\right)\right)\right|}{V(x)} \le$$

$$\frac{\sum_{k=\lceil nx-Tn^{\alpha}\rceil}^{[nx+Tn^{\alpha}]}\left[\sum_{i=1}^{r} w_i \left[(n+\rho_i)^d \int_0^{\frac{1}{n+\rho_i}} \left|f\left(t+\frac{k+\lambda_i}{n+\rho_i}\right)-f(x)\right|dt\right]\right]b\left(n^{1-\alpha}\left(x-\frac{k}{n}\right)\right)}{V(x)} \le$$

$$\frac{\sum_{k=\lceil nx-Tn^{\alpha}\rceil}^{[nx+Tn^{\alpha}]}\left[\sum_{i=1}^{r} w_i \left[(n+\rho_i)^d \int_0^{\frac{1}{n+\rho_i}} \omega_1\left(f,\left\|t+\frac{k+\lambda_i}{n+\rho_i}-x\right\|_{\infty}\right)dt\right]\right]b\left(n^{1-\alpha}\left(x-\frac{k}{n}\right)\right)}{V(x)} \overset{\text{(by (24.28))}}{\le}$$

$$\frac{\sum_{k=\lceil nx-Tn^{\alpha}\rceil}^{[nx+Tn^{\alpha}]}\left[\sum_{i=1}^{r} w_i \left[(n+\rho_i)^d \int_0^{\frac{1}{n+\rho_i}} \omega_1\left(f,\left(\frac{\rho_i\|x\|_{\infty}+\lambda_i+1}{n+\rho_i}\right)+\left(1+\frac{\rho_i}{n+\rho_i}\right)\frac{T^*}{n^{1-\alpha}}\right)dt\right]\right]b\left(n^{1-\alpha}\left(x-\frac{k}{n}\right)\right)}{V(x)} =$$

$$\frac{\sum_{k=\lceil nx-Tn^{\alpha}\rceil}^{[nx+Tn^{\alpha}]}\left[\sum_{i=1}^{r} w_i \omega_1\left(f,\left(\frac{\rho_i\|x\|_{\infty}+\lambda_i+1}{n+\rho_i}\right)+\left(1+\frac{\rho_i}{n+\rho_i}\right)\frac{T^*}{n^{1-\alpha}}\right)\right]b\left(n^{1-\alpha}\left(x-\frac{k}{n}\right)\right)}{V(x)} \tag{24.41}$$

$$= \sum_{i=1}^{r} w_i \omega_1\left(f,\left(\frac{\rho_i\|x\|_{\infty}+\lambda_i+1}{n+\rho_i}\right)+\left(1+\frac{\rho_i}{n+\rho_i}\right)\frac{T^*}{n^{1-\alpha}}\right), \tag{24.42}$$

proving the claim. ■

Corollary 24.6 (to Theorem 24.5) *All here as in Corollary 24.4. It holds*

$$\left\|K_n^*(f)-f\right\|_{\infty,\prod_{j=1}^{d}[-\gamma_j,\gamma_j]} \le$$

$$\sum_{i=1}^{r} w_i \omega_1\left(f,\left(\frac{\rho_i\gamma^*+\lambda_i+1}{n+\rho_i}\right)+\left(1+\frac{\rho_i}{n+\rho_i}\right)\frac{T^*}{n^{1-\alpha}}\right) = \tag{24.43}$$

$$\sum_{i_s 1=1}^{r_1} \cdots \sum_{i_d=1}^{r_d} w_{i_1,\ldots,i_d} \omega_1 \left(f, \left(\frac{\rho_{i_1,\ldots,i_d} \gamma^* + \lambda_{i_1,\ldots,i_d} + 1}{n + \rho_{i_1,\ldots,i_d}} \right) + \left(1 + \frac{\rho_{i_1,\ldots,i_d}}{n + \rho_{i_1,\ldots,i_d}} \right) \frac{T^*}{n^{1-\alpha}} \right).$$

Proof By (24.38). ■

We also present

Theorem 24.7 *All here as in Theorem 24.3. Then*

$$\left| M_n^* (f) (x) - f (x) \right| \leq \omega_1 \left(f, \frac{T^*}{n^{1-\alpha}} + \frac{1}{n} \right). \tag{24.44}$$

Proof We observe that

$$\left| M_n^* (f) (x) - f (x) \right| =$$

$$\left| \frac{\sum_{k=\lceil nx - Tn^\alpha \rceil}^{[nx+Tn^\alpha]} \left(\sum_{i=1}^{r} w_i f \left(\frac{k}{n} + \frac{i}{nr} \right) \right) b \left(n^{1-\alpha} \left(x - \frac{k}{n} \right) \right)}{V (x)} - f (x) \right| = \tag{24.45}$$

$$\left| \frac{\sum_{k=\lceil nx - Tn^\alpha \rceil}^{[nx+Tn^\alpha]} \left(\sum_{i=1}^{r} w_i f \left(\frac{k}{n} + \frac{i}{nr} \right) \right) b \left(n^{1-\alpha} \left(x - \frac{k}{n} \right) \right) - f (x) V (x)}{V (x)} \right| =$$

$$\frac{\left| \sum_{k=\lceil nx - Tn^\alpha \rceil}^{[nx+Tn^\alpha]} \left[\left(\sum_{i=1}^{r} w_i f \left(\frac{k}{n} + \frac{i}{nr} \right) \right) - f (x) \right] b \left(n^{1-\alpha} \left(x - \frac{k}{n} \right) \right) \right|}{V (x)} = \tag{24.46}$$

$$\frac{\left| \sum_{k=\lceil nx - Tn^\alpha \rceil}^{[nx+Tn^\alpha]} \left[\sum_{i=1}^{r} w_i \left(f \left(\frac{k}{n} + \frac{i}{nr} \right) - f (x) \right) \right] b \left(n^{1-\alpha} \left(x - \frac{k}{n} \right) \right) \right|}{V (x)} \leq$$

$$\frac{\sum_{k=\lceil nx - Tn^\alpha \rceil}^{[nx+Tn^\alpha]} \left[\sum_{i=1}^{r} w_i \left| f \left(\frac{k}{n} + \frac{i}{nr} \right) - f (x) \right| \right] b \left(n^{1-\alpha} \left(x - \frac{k}{n} \right) \right)}{V (x)} \leq$$

$$\frac{\sum_{k=\lceil nx - Tn^\alpha \rceil}^{[nx+Tn^\alpha]} \left[\sum_{i=1}^{r} w_i \omega_1 \left(f, \left\| \frac{k}{n} + \frac{i}{nr} - x \right\|_\infty \right) \right] b \left(n^{1-\alpha} \left(x - \frac{k}{n} \right) \right)}{V (x)} \overset{\text{(by (24.30))}}{\leq} \tag{24.47}$$

$$\frac{\sum_{k=\lceil nx - Tn^\alpha \rceil}^{[nx+Tn^\alpha]} \left[\sum_{i=1}^{r} w_i \omega_1 \left(f, \frac{T^*}{n^{1-\alpha}} + \frac{1}{n} \right) \right] b \left(n^{1-\alpha} \left(x - \frac{k}{n} \right) \right)}{V (x)}$$

$$= \omega_1 \left(f, \frac{T^*}{n^{1-\alpha}} + \frac{1}{n} \right), \tag{24.48}$$

proving the claim. ■

Corollary 24.8 (to Theorem 24.7) *All here as in Corollary 24.4. It holds*

$$\left\| M_n^* (f) - f \right\|_{\infty, \prod_{j=1}^{d} [-\gamma_j, \gamma_j]} \leq \omega_1 \left(f, \frac{T^*}{n^{1-\alpha}} + \frac{1}{n} \right). \tag{24.49}$$

Proof By (24.44). ■

Note 24.9 *Theorems 24.3, 24.5 and 24.7 and Corollaries 24.4, 24.6 and 24.8 given that f is uniformly continuous, produce the pointwise and uniform convergences with rates, at speed $\frac{1}{n^{1-\alpha}}$, of multivariate neural network operators H_n^*, K_n^*, M_n^* to the unit operator. Notice that the right hand sides of inequalities (24.31), (24.37), (24.38), (24.43), (24.44) and (24.49) do not depend on b.*

Next we present higher order of approximation results based on the high order differentiability of the approximated function.

Theorem 24.10 *Let $x \in \mathbb{R}^d$ and $n \in \mathbb{N}$ such that $n \geq \max_{j \in \{1,\ldots,d\}} \left(T_j + \left| x_j \right|, T_j^{-\frac{1}{\alpha}} \right)$, $T_j > 0$, $0 < \alpha < 1$. Let also $f \in C^N \left(\mathbb{R}^d \right)$, $N \in \mathbb{N}$, such that all of its partial derivatives $f_{\widetilde{\alpha}}$ of order N, $\widetilde{\alpha} : |\widetilde{\alpha}| = \sum_{j=1}^{d} \alpha_j = N$, are uniformly continuous or continuous and bounded. Then*

$$\left| \left(H_n^* (f) \right) (x) - f (x) \right| \leq \sum_{l=1}^{N} \frac{1}{l!} \left(\left(\sum_{j=1}^{d} \left| \frac{\partial}{\partial x_j} \right| \right)^l f (x) \right) \cdot$$

$$\left[\sum_{i=1}^{r} w_i \left[\left(\frac{\|\nu_i\|_\infty \|x\|_\infty + \|\mu_i\|_\infty}{n + \nu_i^{\min}} \right) + \left(1 + \frac{\|\nu_i\|_\infty}{n + \nu_i^{\min}} \right) \frac{T^*}{n^{1-\alpha}} \right] \right]^l +$$

$$\frac{d^N}{N!} \sum_{i=1}^{r} w_i \left[\left(\frac{\|\nu_i\|_\infty \|x\|_\infty + \|\mu_i\|_\infty}{n + \nu_i^{\min}} \right) + \left(1 + \frac{\|\nu_i\|_\infty}{n + \nu_i^{\min}} \right) \frac{T^*}{n^{1-\alpha}} \right]^N \cdot$$

$$\max_{\widetilde{\alpha} : |\widetilde{\alpha}| = N} \omega_1 \left(f_{\widetilde{\alpha}}, \left[\left(\frac{\|\nu_i\|_\infty \|x\|_\infty + \|\mu_i\|_\infty}{n + \nu_i^{\min}} \right) + \left(1 + \frac{\|\nu_i\|_\infty}{n + \nu_i^{\min}} \right) \frac{T^*}{n^{1-\alpha}} \right] \right). \tag{24.50}$$

Inequality (24.50) implies the pointwise convergence with rates on $\left(H_n^* (f) \right) (x)$ to $f (x)$, as $n \to \infty$, at speed $\frac{1}{n^{1-\alpha}}$.

Proof Set

$$g_{\frac{k+\mu_i}{n+\nu_i}}(t) = f\left(x + t\left(\frac{k+\mu_i}{n+\nu_i} - x\right)\right), \quad 0 \le t \le 1. \tag{24.51}$$

Then we have

$$g^{(l)}_{\frac{k+\mu_i}{n+\nu_i}}(t) = \left[\left(\sum_{j=1}^{d}\left(\frac{k_j+\mu_{i_j}}{n+\nu_{i_j}} - x_j\right)\frac{\partial}{\partial x_j}\right)^l f\right] \tag{24.52}$$

$$\left(x_1 + t\left(\frac{k_1+\mu_{i_1}}{n+\nu_{i_1}} - x_1\right), \ldots, x_d + t\left(\frac{k_d+\mu_{i_d}}{n+\nu_{i_d}} - x_d\right)\right),$$

and

$$g_{\frac{k+\mu_i}{n+\nu_i}}(0) = f(x).$$

By Taylor's formula, we get

$$f\left(\frac{k_1+\mu_{i_1}}{n+\nu_{i_1}}, \ldots, \frac{k_d+\mu_{i_d}}{n+\nu_{i_d}}\right) = g_{\frac{k+\mu_i}{n+\nu_i}}(1) = \tag{24.53}$$

$$\sum_{l=0}^{N}\frac{g^{(l)}_{\frac{k+\mu_i}{n+\nu_i}}(0)}{l!} + R_N\left(\frac{k+\mu_i}{n+\nu_i}, 0\right),$$

where

$$R_N\left(\frac{k+\mu_i}{n+\nu_i}, 0\right) = \int_0^1\left(\int_0^{t_1}\ldots\left(\int_0^{t_{N-1}}\left(g^{(N)}_{\frac{k+\mu_i}{n+\nu_i}}(t_N) - g^{(N)}_{\frac{k+\mu_i}{n+\nu_i}}(0)\right)dt_N\right)\ldots\right)dt_1. \tag{24.54}$$

Here we denote by

$$f_{\widetilde{\alpha}} := \frac{\partial^{\widetilde{\alpha}} f}{\partial x^{\widetilde{\alpha}}}, \quad \widetilde{\alpha} := (\alpha_1, \ldots, \alpha_d), \ \alpha_j \in \mathbb{Z}^+, \ j = 1, \ldots, d,$$

such that $|\widetilde{\alpha}| = \sum_{j=1}^{d} \alpha_j = N$. Thus

$$\sum_{i=1}^{r} w_i f\left(\frac{k+\mu_i}{n+\nu_i}\right) = \sum_{l=0}^{N}\sum_{i=1}^{r} w_i \frac{g^{(l)}_{\frac{k+\mu_i}{n+\nu_i}}(0)}{l!} + \sum_{i=1}^{r} w_i R_N\left(\frac{k+\mu_i}{n+\nu_i}, 0\right), \tag{24.55}$$

and

$$\left(H_n^*(f)\right)(x) = \frac{\sum_{k=\lceil nx-Tn^\alpha\rceil}^{[nx+Tn^\alpha]}\left(\sum_{i=1}^{r} w_i f\left(\frac{k+\mu_i}{n+\nu_i}\right)\right) b\left(n^{1-\alpha}\left(x-\frac{k}{n}\right)\right)}{V(x)} =$$

$$\sum_{l=0}^{N} \frac{\sum_{k=\lceil nx-Tn^\alpha\rceil}^{[nx+Tn^\alpha]}\left(\sum_{i=1}^{r} w_i \frac{g^{(l)}_{\frac{k+\mu_i}{n+\nu_i}}(0)}{l!}\right) b\left(n^{1-\alpha}\left(x-\frac{k}{n}\right)\right)}{V(x)} + \tag{24.56}$$

$$\frac{\sum_{k=\lceil nx-Tn^\alpha\rceil}^{[nx+Tn^\alpha]} b\left(n^{1-\alpha}\left(x-\frac{k}{n}\right)\right)}{V(x)}\left(\sum_{i=1}^{r} w_i R_N\left(\frac{k+\mu_i}{n+\nu_i}, 0\right)\right).$$

Therefore it holds

$$\left(H_n^*(f)\right)(x) - f(x) = \tag{24.57}$$

$$\sum_{l=1}^{N} \frac{\sum_{k=\lceil nx-Tn^\alpha\rceil}^{[nx+Tn^\alpha]}\left(\sum_{i=1}^{r} w_i \frac{g^{(l)}_{\frac{k+\mu_i}{n+\nu_i}}(0)}{l!}\right) b\left(n^{1-\alpha}\left(x-\frac{k}{n}\right)\right)}{V(x)} + R^*,$$

where

$$R^* = \sum_{k=\lceil nx-Tn^\alpha\rceil}^{[nx+Tn^\alpha]} \frac{b\left(n^{1-\alpha}\left(x-\frac{k}{n}\right)\right)}{V(x)}\left(\sum_{i=1}^{r} w_i R_N\left(\frac{k+\mu_i}{n+\nu_i}, 0\right)\right). \tag{24.58}$$

Consequently, we obtain

$$\left|\left(H_n^*(f)\right)(x) - f(x)\right| \leq \tag{24.51}$$

$$\sum_{l=1}^{N} \frac{\sum_{k=\lceil nx-Tn^\alpha\rceil}^{[nx+Tn^\alpha]}\left(\sum_{i=1}^{r} w_i \frac{\left|g^{(l)}_{\frac{k+\mu_i}{n+\nu_i}}(0)\right|}{l!}\right) b\left(n^{1-\alpha}\left(x-\frac{k}{n}\right)\right)}{V(x)} + \left|R^*\right|. \tag{24.59}$$

Notice that

$$\left|g^{(l)}_{\frac{k+\mu_i}{n+\nu_i}}(0)\right| \overset{(24.21)}{\leq} \left(\left(\sum_{j=1}^{d}\left|\frac{\partial}{\partial x_j}\right|\right)^{l} f(x)\right)$$

$$\left[\left(\frac{\|\nu_i\|_\infty \|x\|_\infty + \|\mu_i\|_\infty}{n+\nu_i^{\min}}\right) + \left(1 + \frac{\|\nu_i\|_\infty}{n+\nu_i^{\min}}\right)\frac{T^*}{n^{1-\alpha}}\right]^l, \tag{24.60}$$

and so far we have

$$\left|\left(H_n^*(f)\right)(x) - f(x)\right| \le \sum_{l=1}^{N} \frac{1}{l!}\left(\left(\sum_{j=1}^{d}\left|\frac{\partial}{\partial x_j}\right|\right)^l f(x)\right)$$

$$\left[\sum_{i=1}^{r} w_i \left[\left(\frac{\|\nu_i\|_\infty \|x\|_\infty + \|\mu_i\|_\infty}{n+\nu_i^{\min}}\right) + \left(1 + \frac{\|\nu_i\|_\infty}{n+\nu_i^{\min}}\right)\frac{T^*}{n^{1-\alpha}}\right]^l\right] + |R^*|. \tag{24.61}$$

Next, we need to estimate $|R^*|$. For that, we observe $(0 \le t_N \le 1)$

$$\left|g^{(N)}_{\frac{k+\mu_i}{n+\nu_i}}(t_N) - g^{(N)}_{\frac{k+\mu_i}{n+\nu_i}}(0)\right| =$$

$$\left|\left[\left(\sum_{j=1}^{d}\left(\frac{k_j+\mu_{i_j}}{n+\nu_{i_j}} - x_j\right)\frac{\partial}{\partial x_j}\right)^N f\right]\left(x + t_N\left(\frac{k+\mu_i}{n+\nu_i} - x\right)\right) - \right. \tag{24.62}$$

$$\left.\left[\left(\sum_{j=1}^{d}\left(\frac{k_j+\mu_{i_j}}{n+\nu_{i_j}} - x_j\right)\frac{\partial}{\partial x_j}\right)^N f\right](x)\right| \overset{\text{(by (24.21), (24.22))}}{\le}$$

$$d^N\left[\left(\frac{\|\nu_i\|_\infty \|x\|_\infty + \|\mu_i\|_\infty}{n+\nu_i^{\min}}\right) + \left(1 + \frac{\|\nu_i\|_\infty}{n+\nu_i^{\min}}\right)\frac{T^*}{n^{1-\alpha}}\right]^N \cdot$$

$$\max_{\widetilde{\alpha}:|\widetilde{\alpha}|=N} \omega_1\left(f_{\widetilde{\alpha}}, \left[\left(\frac{\|\nu_i\|_\infty \|x\|_\infty + \|\mu_i\|_\infty}{n+\nu_i^{\min}}\right) + \left(1 + \frac{\|\nu_i\|_\infty}{n+\nu_i^{\min}}\right)\frac{T^*}{n^{1-\alpha}}\right]\right). \tag{24.63}$$

Thus we find

$$\left|R_N\left(\frac{k+\mu_i}{n+\nu_i}, 0\right)\right| \le$$

$$\int_0^1\left(\int_0^{t_1} \cdots \left(\int_0^{t_{N-1}} \left|g^{(N)}_{\frac{k+\mu_i}{n+\nu_i}}(t_N) - g^{(N)}_{\frac{k+\mu_i}{n+\nu_i}}(0)\right| dt_N\right) \cdots\right) dt_1 \le$$

$$\frac{d^N}{N!}\left[\left(\frac{\|\nu_i\|_\infty \|x\|_\infty + \|\mu_i\|_\infty}{n+\nu_i^{\min}}\right)+\left(1+\frac{\|\nu_i\|_\infty}{n+\nu_i^{\min}}\right)\frac{T^*}{n^{1-\alpha}}\right]^N \cdot$$

$$\max_{\widetilde{\alpha}:|\widetilde{\alpha}|=N}\omega_1\left(f_{\widetilde{\alpha}},\left[\left(\frac{\|\nu_i\|_\infty \|x\|_\infty + \|\mu_i\|_\infty}{n+\nu_i^{\min}}\right)+\left(1+\frac{\|\nu_i\|_\infty}{n+\nu_i^{\min}}\right)\frac{T^*}{n^{1-\alpha}}\right]\right). \tag{24.64}$$

Finally we obtain

$$\left|R^*\right| \leq \sum_{k=\lceil nx-Tn^\alpha\rceil}^{[nx+Tn^\alpha]} \frac{b\left(n^{1-\alpha}\left(x-\frac{k}{n}\right)\right)}{V(x)}\left(\sum_{i=1}^r w_i \left|R_N\left(\frac{k+\mu_i}{n+\nu_i},0\right)\right|\right) \leq$$

$$\sum_{i=1}^r w_i \frac{d^N}{N!}\left[\left(\frac{\|\nu_i\|_\infty \|x\|_\infty + \|\mu_i\|_\infty}{n+\nu_i^{\min}}\right)+\left(1+\frac{\|\nu_i\|_\infty}{n+\nu_i^{\min}}\right)\frac{T^*}{n^{1-\alpha}}\right]^N \cdot$$

$$\max_{\widetilde{\alpha}:|\widetilde{\alpha}|=N}\omega_1\left(f_{\widetilde{\alpha}},\left[\left(\frac{\|\nu_i\|_\infty \|x\|_\infty + \|\mu_i\|_\infty}{n+\nu_i^{\min}}\right)+\left(1+\frac{\|\nu_i\|_\infty}{n+\nu_i^{\min}}\right)\frac{T^*}{n^{1-\alpha}}\right]\right). \tag{24.65}$$

Using (24.61) and (24.65) we derive (24.50). ■

Corollary 24.11 (to Theorem 24.10) *Let all as in Theorem 24.10. Additionally assume that all* $f_{\widetilde{\alpha}}(x) = 0$, $\widetilde{\alpha} : |\widetilde{\alpha}| = \rho$, $1 \leq \rho \leq N$. *Then*

$$\left|\left(H_n^*(f)\right)(x) - f(x)\right| \leq$$

$$\frac{d^N}{N!}\sum_{i=1}^r w_i \left[\left(\frac{\|\nu_i\|_\infty \|x\|_\infty + \|\mu_i\|_\infty}{n+\nu_i^{\min}}\right)+\left(1+\frac{\|\nu_i\|_\infty}{n+\nu_i^{\min}}\right)\frac{T^*}{n^{1-\alpha}}\right]^N \cdot$$

$$\max_{\widetilde{\alpha}:|\widetilde{\alpha}|=N}\omega_1\left(f_{\widetilde{\alpha}},\left[\left(\frac{\|\nu_i\|_\infty \|x\|_\infty + \|\mu_i\|_\infty}{n+\nu_i^{\min}}\right)+\left(1+\frac{\|\nu_i\|_\infty}{n+\nu_i^{\min}}\right)\frac{T^*}{n^{1-\alpha}}\right]\right). \tag{24.66}$$

Inequality (24.66) implies the pointwise convergence with rates of $\left(H_n^*(f)\right)(x)$ *to* $f(x)$, *as* $n \to \infty$, *at the high speed* $\frac{1}{n^{(1-\alpha)(N+1)}}$.

Proof By (24.50). ■

The uniform convergence with rates follows from

Corollary 24.12 (to Theorem 24.10) *All as in Theorem 24.10, but now* $x \in G = \prod_{j=1}^d \left[-\gamma_j, \gamma_j\right] \subset \mathbb{R}^d$, $\gamma_j > 0$, $\gamma^* = \max\{\gamma_1, \ldots, \gamma_d\}$ *and* $n \in \mathbb{N}$:

$n \geq \max_{j \in \{1,\ldots,d\}} \left(T_j + \gamma_j, T_j^{-\frac{1}{\alpha}} \right)$. *Then*

$$\left\| H_n^* (f) - f \right\|_{\infty, \prod_{j=1}^{d} [-\gamma_j, \gamma_j]} \leq \sum_{l=1}^{N} \frac{1}{l!} \left(\left(\sum_{j=1}^{d} \left| \frac{\partial}{\partial x_j} \right| \right)^l \| f \|_{\infty, \prod_{j=1}^{d} \left[-\gamma_j, \gamma_j\right]} \right)$$

$$\left[\sum_{i=1}^{r} w_i \left[\left(\frac{\| \nu_i \|_\infty \gamma^* + \| \mu_i \|_\infty}{n + \nu_i^{\min}} \right) + \left(1 + \frac{\| \nu_i \|_\infty}{n + \nu_i^{\min}} \right) \frac{T^*}{n^{1-\alpha}} \right]^l \right] +$$

$$\frac{d^N}{N!} \sum_{i=1}^{r} w_i \left[\left(\frac{\| \nu_i \|_\infty \gamma^* + \| \mu_i \|_\infty}{n + \nu_i^{\min}} \right) + \left(1 + \frac{\| \nu_i \|_\infty}{n + \nu_i^{\min}} \right) \frac{T^*}{n^{1-\alpha}} \right]^N \cdot$$

$$\max_{\widetilde{\alpha}:|\widetilde{\alpha}|=N} \omega_1 \left(f_{\widetilde{\alpha}}, \left[\left(\frac{\| \nu_i \|_\infty \gamma^* + \| \mu_i \|_\infty}{n + \nu_i^{\min}} \right) + \left(1 + \frac{\| \nu_i \|_\infty}{n + \nu_i^{\min}} \right) \frac{T^*}{n^{1-\alpha}} \right] \right). \quad (24.67)$$

Inequality (24.67) implies the uniform convergence with rates of $H_n^* (f)$ to f on G, as $n \to \infty$, at speed $\frac{1}{n^{1-\alpha}}$.

Proof By (24.50). ∎

Corollary 24.13 (to Theorem 24.10) *All as in Theorem 24.10 with $N = 1$. Then*

$$\left| \left(H_n^* (f) \right) (x) - f (x) \right| \leq \left(\sum_{j=1}^{d} \left| \frac{\partial f (x)}{\partial x_j} \right| \right)$$

$$\left[\sum_{i=1}^{r} w_i \left[\left(\frac{\| \nu_i \|_\infty \| x \|_\infty + \| \mu_i \|_\infty}{n + \nu_i^{\min}} \right) + \left(1 + \frac{\| \nu_i \|_\infty}{n + \nu_i^{\min}} \right) \frac{T^*}{n^{1-\alpha}} \right] \right] +$$

$$d \sum_{i=1}^{r} w_i \left[\left(\frac{\| \nu_i \|_\infty \| x \|_\infty + \| \mu_i \|_\infty}{n + \nu_i^{\min}} \right) + \left(1 + \frac{\| \nu_i \|_\infty}{n + \nu_i^{\min}} \right) \frac{T^*}{n^{1-\alpha}} \right] \cdot$$

$$\max_{\widetilde{\alpha}:|\widetilde{\alpha}|=N} \omega_1 \left(f_{\widetilde{\alpha}}, \left[\left(\frac{\| \nu_i \|_\infty \| x \|_\infty + \| \mu_i \|_\infty}{n + \nu_i^{\min}} \right) + \left(1 + \frac{\| \nu_i \|_\infty}{n + \nu_i^{\min}} \right) \frac{T^*}{n^{1-\alpha}} \right] \right). \quad (24.68)$$

Inequality (24.68) implies the pointwise convergence with rates of $\left(H_n^* (f) \right) (x)$ to $f (x)$, as $n \to \infty$, at speed $\frac{1}{n^{1-\alpha}}$.

Proof By (24.50). ∎

We continue with

Theorem 24.14 *All here as in Theorem 24.10. Then*

$$\left|\left(K_n^*(f)\right)(x) - f(x)\right| \le$$

$$\sum_{l=1}^{N} \frac{1}{l!} \left(\sum_{i=1}^{r} w_i \left[\left(\frac{\rho_i \|x\|_\infty + \lambda_i + 1}{n + \rho_i} \right) + \left(1 + \frac{\rho_i}{n + \rho_i} \right) \frac{T^*}{n^{1-\alpha}} \right]^l \right) \cdot$$

$$\left(\left(\sum_{j=1}^{d} \left| \frac{\partial}{\partial x_j} \right| \right)^l f(x) \right) + \tag{24.69}$$

$$\frac{d^N}{N!} \sum_{i=1}^{r} w_i \left[\left(\frac{\rho_i \|x\|_\infty + \lambda_i + 1}{n + \rho_i} \right) + \left(1 + \frac{\rho_i}{n + \rho_i} \right) \frac{T^*}{n^{1-\alpha}} \right]^N \cdot$$

$$\max_{\widetilde{\alpha}:|\widetilde{\alpha}|=N} \omega_1 \left(f_{\widetilde{\alpha}}, \left[\left(\frac{\rho_i \|x\|_\infty + \lambda_i + 1}{n + \rho_i} \right) + \left(1 + \frac{\rho_i}{n + \rho_i} \right) \frac{T^*}{n^{1-\alpha}} \right] \right).$$

Inequality (24.69) implies the pointwise convergence with rates of $\left(K_n^*(f)\right)(x)$ *to* $f(x)$*, as* $n \to \infty$*, at speed* $\frac{1}{n^{1-\alpha}}$.

Proof Set

$$g_{t+\frac{k+\lambda_i}{n+\rho_i}}(\lambda^*) = f\left(x + \lambda^* \left(t + \frac{k + \lambda_i}{n + \rho_i} - x \right) \right), \quad 0 \le \lambda^* \le 1. \tag{24.70}$$

Then we have

$$g^{(l)}_{t+\frac{k+\lambda_i}{n+\rho_i}}(\lambda^*) = \tag{24.71}$$

$$\left[\left(\sum_{j=1}^{d} \left(t_j + \frac{k_j + \lambda_i}{n + \rho_i} - x_j \right) \frac{\partial}{\partial x_j} \right)^l f \right] \left(x + \lambda^* \left(t + \frac{k + \lambda_i}{n + \rho_i} - x \right) \right),$$

and

$$g_{t+\frac{k+\lambda_i}{n+\rho_i}}(0) = f(x). \tag{24.72}$$

By Taylor's formula, we get

$$f\left(t + \frac{k + \lambda_i}{n + \rho_i} \right) = g_{t+\frac{k+\lambda_i}{n+\rho_i}}(1) = \tag{24.73}$$

$$\sum_{l=0}^{N} \frac{g^{(l)}_{t+\frac{k+\lambda_i}{n+\rho_i}}(0)}{l!} + R_N\left(t + \frac{k+\lambda_i}{n+\rho_i}, 0\right),$$

where

$$R_N\left(t + \frac{k+\lambda_i}{n+\rho_i}, 0\right) =$$

$$\int_0^1 \left(\int_0^{\lambda_1^*} \ldots \left(\int_0^{\lambda_{N-1}^*} \left(g^{(N)}_{t+\frac{k+\lambda_i}{n+\rho_i}}\left(\lambda_N^*\right) - g^{(N)}_{t+\frac{k+\lambda_i}{n+\rho_i}}(0)\right) d\lambda_N^*\right) \ldots\right) d\lambda_1^*. \quad (24.74)$$

Here we denote by

$$f_{\widetilde{\alpha}} := \frac{\partial^{\widetilde{\alpha}} f}{\partial x^{\widetilde{\alpha}}}, \ \widetilde{\alpha} := (\alpha_1, \ldots, \alpha_d), \alpha_j \in \mathbb{Z}^+, \ j = 1, \ldots, d,$$

such that $|\widetilde{\alpha}| = \sum_{j=1}^{d} \alpha_j = N$. Thus

$$\sum_{i=1}^{r} w_i (n+\rho_i)^d \int_0^{\frac{1}{n+\rho_i}} f\left(t + \frac{k+\lambda_i}{n+\rho_i}\right) dt = \quad (24.75)$$

$$\sum_{l=0}^{N} \frac{\sum_{i=1}^{r} w_i (n+\rho_i)^d \int_0^{\frac{1}{n+\rho_i}} g^{(l)}_{t+\frac{k+\lambda_i}{n+\rho_i}}(0)\, dt}{l!} +$$

$$\sum_{i=1}^{r} w_i (n+\rho_i)^d \int_0^{\frac{1}{n+\rho_i}} R_N\left(t + \frac{k+\lambda_i}{n+\rho_i}, 0\right) dt.$$

Hence it holds

$$\left(K_n^*(f)\right)(x) =$$

$$\frac{\sum_{k=\lceil nx-Tn^{\alpha}\rceil}^{[nx+Tn^{\alpha}]} \left(\sum_{i=1}^{r} w_i (n+\rho_i)^d \int_0^{\frac{1}{n+\rho_i}} f\left(t + \frac{k+\lambda_i}{n+\rho_i}\right) dt\right) b\left(n^{1-\alpha}\left(x - \frac{k}{n}\right)\right)}{V(x)} =$$

$$\sum_{l=0}^{N} \frac{1}{l!} \sum_{k=\lceil nx-Tn^{\alpha}\rceil}^{[nx+Tn^{\alpha}]} \frac{b\left(n^{1-\alpha}\left(x - \frac{k}{n}\right)\right)}{V(x)} \left(\sum_{i=1}^{r} w_i (n+\rho_i)^d \int_0^{\frac{1}{n+\rho_i}} g^{(l)}_{t+\frac{k+\lambda_i}{n+\rho_i}}(0)\, dt\right) + \quad (24.76)$$

$$\sum_{k=\lceil nx-Tn^{\alpha}\rceil}^{[nx+Tn^{\alpha}]}\frac{b\left(n^{1-\alpha}\left(x-\frac{k}{n}\right)\right)}{V(x)}\left(\sum_{i=1}^{r}w_i\left(n+\rho_i\right)^d\int_0^{\frac{1}{n+\rho_i}}R_N\left(t+\frac{k+\lambda_i}{n+\rho_i},0\right)dt\right).$$

So we see that

$$\left(K_n^*(f)\right)(x)-f(x)= \tag{24.77}$$

$$\sum_{l=1}^{N}\frac{1}{l!}\sum_{k=\lceil nx-Tn^{\alpha}\rceil}^{[nx+Tn^{\alpha}]}\frac{b\left(n^{1-\alpha}\left(x-\frac{k}{n}\right)\right)}{V(x)}\left(\sum_{i=1}^{r}w_i\left(n+\rho_i\right)^d\int_0^{\frac{1}{n+\rho_i}}g^{(l)}_{t+\frac{k+\lambda_i}{n+\rho_i}}(0)\,dt\right)+R^*,$$

where

$$R^*=$$

$$\sum_{k=\lceil nx-Tn^{\alpha}\rceil}^{[nx+Tn^{\alpha}]}\frac{b\left(n^{1-\alpha}\left(x-\frac{k}{n}\right)\right)}{V(x)}\left(\sum_{i=1}^{r}w_i\left(n+\rho_i\right)^d\int_0^{\frac{1}{n+\rho_i}}R_N\left(t+\frac{k+\lambda_i}{n+\rho_i},0\right)dt\right). \tag{24.78}$$

Consequently, we obtain

$$\left|\left(K_n^*(f)\right)(x)-f(x)\right|\leq$$

$$\sum_{l=1}^{N}\frac{1}{l!}\sum_{k=\lceil nx-Tn^{\alpha}\rceil}^{[nx+Tn^{\alpha}]}\frac{b\left(n^{1-\alpha}\left(x-\frac{k}{n}\right)\right)}{V(x)}\left(\sum_{i=1}^{r}w_i\left(n+\rho_i\right)^d\int_0^{\frac{1}{n+\rho_i}}\left|g^{(l)}_{t+\frac{k+\lambda_i}{n+\rho_i}}(0)\right|dt\right) \tag{24.79}$$

$$+\left|R^*\right|.$$

Notice that

$$\left|g^{(l)}_{t+\frac{k+\lambda_i}{n+\rho_i}}(0)\right|\overset{(24.27)}{\leq}\left(\left(\sum_{j=1}^{d}\left|\frac{\partial}{\partial x_j}\right|\right)^l f(x)\right)\cdot$$

$$\left[\left(\frac{\rho_i\left\|x\right\|_\infty+\lambda_i+1}{n+\rho_i}\right)+\left(1+\frac{\rho_i}{n+\rho_i}\right)\frac{T^*}{n^{1-\alpha}}\right]^l, \tag{24.80}$$

and so far we have

$$\left|\left(K_n^*(f)\right)(x)-f(x)\right|\leq$$

$$\sum_{l=1}^{N}\frac{1}{l!}\left(\sum_{i=1}^{r}w_i\left[\left(\frac{\rho_i\left\|x\right\|_\infty+\lambda_i+1}{n+\rho_i}\right)+\left(1+\frac{\rho_i}{n+\rho_i}\right)\frac{T^*}{n^{1-\alpha}}\right]^l\right) \tag{24.81}$$

$$\cdot \left(\left(\sum_{j=1}^{d} \left| \frac{\partial}{\partial x_j} \right| \right)^{l} f(x) \right) + |R^*| .$$

Next, we need to estimate $|R^*|$. For that, we observe $(0 \leq \lambda_N^* \leq 1)$

$$\left| g_{t+\frac{k+\lambda_i}{n+\rho_i}}^{(N)} \left(\lambda_N^* \right) - g_{t+\frac{k+\lambda_i}{n+\rho_i}}^{(N)} (0) \right| =$$

$$| \left[\left(\sum_{j=1}^{d} \left(t_j + \frac{k_j + \lambda_i}{n + \rho_i} - x_j \right) \frac{\partial}{\partial x_j} \right)^{N} f \right] \left(x + \lambda_N^* \left(t + \frac{k + \lambda_i}{n + \rho_i} - x \right) \right) -$$

$$\left[\left(\sum_{j=1}^{d} \left(t_j + \frac{k_j + \lambda_i}{n + \rho_i} - x_j \right) \frac{\partial}{\partial x_j} \right)^{N} f \right] (x) | \overset{\text{(by (24.27), (24.28))}}{\leq} \tag{24.82}$$

$$d^N \left[\left(\frac{\rho_i \|x\|_\infty + \lambda_i + 1}{n + \rho_i} \right) + \left(1 + \frac{\rho_i}{n + \rho_i} \right) \frac{T^*}{n^{1-\alpha}} \right]^N \cdot$$

$$\max_{\widetilde{\alpha}:|\widetilde{\alpha}|=N} \omega_1 \left(f_{\widetilde{\alpha}}, \left[\left(\frac{\rho_i \|x\|_\infty + \lambda_i + 1}{n + \rho_i} \right) + \left(1 + \frac{\rho_i}{n + \rho_i} \right) \frac{T^*}{n^{1-\alpha}} \right] \right) . \tag{24.83}$$

Thus we find

$$\left| R_N \left(t + \frac{k + \lambda_i}{n + \rho_i}, 0 \right) \right| \leq$$

$$\int_0^1 \left(\int_0^{\lambda_1^*} \ldots \left(\int_0^{\lambda_{N-1}^*} \left| g_{t+\frac{k+\lambda_i}{n+\rho_i}}^{(N)} \left(\lambda_N^* \right) - g_{t+\frac{k+\lambda_i}{n+\rho_i}}^{(N)} (0) \right| d\lambda_N^* \right) \ldots \right) d\lambda_1^* \leq \tag{24.84}$$

$$\frac{d^N}{N!} \left[\left(\frac{\rho_i \|x\|_\infty + \lambda_i + 1}{n + \rho_i} \right) + \left(1 + \frac{\rho_i}{n + \rho_i} \right) \frac{T^*}{n^{1-\alpha}} \right]^N \cdot \tag{24.85}$$

$$\max_{\widetilde{\alpha}:|\widetilde{\alpha}|=N} \omega_1 \left(f_{\widetilde{\alpha}}, \left[\left(\frac{\rho_i \|x\|_\infty + \lambda_i + 1}{n + \rho_i} \right) + \left(1 + \frac{\rho_i}{n + \rho_i} \right) \frac{T^*}{n^{1-\alpha}} \right] \right) .$$

Finally we obtain

$$|R^*| \leq \sum_{k=\lceil nx - Tn^{\alpha} \rceil}^{[nx+Tn^{\alpha}]} \frac{b \left(n^{1-\alpha} \left(x - \frac{k}{n} \right) \right)}{V(x)} \cdot$$

$$\left(\sum_{i=1}^{r} w_i (n+\rho_i)^d \int_0^{\frac{1}{n+\rho_i}} \left| R_N \left(t + \frac{k+\lambda_i}{n+\rho_i}, 0 \right) \right| dt \right) \leq$$

$$\sum_{i=1}^{r} w_i \frac{d^N}{N!} \left[\left(\frac{\rho_i \|x\|_\infty + \lambda_i + 1}{n+\rho_i} \right) + \left(1 + \frac{\rho_i}{n+\rho_i} \right) \frac{T^*}{n^{1-\alpha}} \right]^N \cdot$$

$$\max_{\widetilde{\alpha}:|\widetilde{\alpha}|=N} \omega_1 \left(f_{\widetilde{\alpha}}, \left[\left(\frac{\rho_i \|x\|_\infty + \lambda_i + 1}{n+\rho_i} \right) + \left(1 + \frac{\rho_i}{n+\rho_i} \right) \frac{T^*}{n^{1-\alpha}} \right] \right). \tag{24.86}$$

Using (24.81) and (24.86) we derive (24.69). ■

Corollary 24.15 (to Theorem 24.14) *Let all as in Corollary 24.11. Then*

$$\left| \left(K_n^* (f) \right) (x) - f (x) \right| \leq$$

$$\frac{d^N}{N!} \sum_{i=1}^{r} w_i \left[\left(\frac{\rho_i \|x\|_\infty + \lambda_i + 1}{n+\rho_i} \right) + \left(1 + \frac{\rho_i}{n+\rho_i} \right) \frac{T^*}{n^{1-\alpha}} \right]^N \cdot \tag{24.87}$$

$$\max_{\widetilde{\alpha}:|\widetilde{\alpha}|=N} \omega_1 \left(f_{\widetilde{\alpha}}, \left[\left(\frac{\rho_i \|x\|_\infty + \lambda_i + 1}{n+\rho_i} \right) + \left(1 + \frac{\rho_i}{n+\rho_i} \right) \frac{T^*}{n^{1-\alpha}} \right] \right).$$

Inequality (24.87) implies the pointwise convergence with rates of $\left(K_n^* (f) \right) (x)$ to $f (x)$, as $n \to \infty$, at the high speed $\frac{1}{n^{(1-\alpha)(N+1)}}$.

Proof By (24.69). ■

The uniform convergence with rates follows from

Corollary 24.16 (to Theorem 24.14) *Let all as in Corollary 24.12. Then*

$$\left\| K_n^* (f) - f \right\|_{\infty, \prod_{j=1}^{d} [-\gamma_j, \gamma_j]} \leq \sum_{l=1}^{N} \frac{1}{l!} \left(\left(\sum_{j=1}^{d} \left| \frac{\partial}{\partial x_j} \right| \right)^l \|f\|_{\infty, \prod_{j=1}^{d} [-\gamma_j, \gamma_j]} \right)$$

$$\left[\sum_{i=1}^{r} w_i \left[\left(\frac{\rho_i \gamma^* + \lambda_i + 1}{n+\rho_i} \right) + \left(1 + \frac{\rho_i}{n+\rho_i} \right) \frac{T^*}{n^{1-\alpha}} \right]^l \right] + \tag{24.88}$$

$$\frac{d^N}{N!} \sum_{i=1}^{r} w_i \left[\left(\frac{\rho_i \gamma^* + \lambda_i + 1}{n+\rho_i} \right) + \left(1 + \frac{\rho_i}{n+\rho_i} \right) \frac{T^*}{n^{1-\alpha}} \right]^N \cdot$$

$$\max_{\widetilde{\alpha}:|\widetilde{\alpha}|=N} \omega_1 \left(f_{\widetilde{\alpha}}, \left[\left(\frac{\rho_i \gamma^* + \lambda_i + 1}{n+\rho_i} \right) + \left(1 + \frac{\rho_i}{n+\rho_i} \right) \frac{T^*}{n^{1-\alpha}} \right] \right).$$

Inequality (24.88) implies the uniform convergence with rates of $K_n^*(f)$ to f on G, as $n \to \infty$, at speed $\frac{1}{n^{1-\alpha}}$.

Proof By (24.69). ∎

Corollary 24.17 (to Theorem 24.14) *All as in Theorem 24.14 with $N = 1$. Then*

$$\left|\left(K_n^*(f)\right)(x) - f(x)\right| \leq \left(\sum_{j=1}^{d}\left|\frac{\partial f(x)}{\partial x_j}\right|\right)$$

$$\left[\sum_{i=1}^{r} w_i \left[\left(\frac{\rho_i \|x\|_\infty + \lambda_i + 1}{n+\rho_i}\right) + \left(1 + \frac{\rho_i}{n+\rho_i}\right)\frac{T^*}{n^{1-\alpha}}\right]\right] + \tag{24.89}$$

$$d\sum_{i=1}^{r} w_i \left[\left(\frac{\rho_i \|x\|_\infty + \lambda_i + 1}{n+\rho_i}\right) + \left(1 + \frac{\rho_i}{n+\rho_i}\right)\frac{T^*}{n^{1-\alpha}}\right].$$

$$\max_{\widetilde{\alpha}:|\widetilde{\alpha}|=N} \omega_1\left(f_{\widetilde{\alpha}}, \left[\left(\frac{\rho_i \|x\|_\infty + \lambda_i + 1}{n+\rho_i}\right) + \left(1 + \frac{\rho_i}{n+\rho_i}\right)\frac{T^*}{n^{1-\alpha}}\right]\right).$$

Inequality (24.89) implies the pointwise convergence with rates of $\left(K_n^*(f)\right)(x)$ to $f(x)$, as $n \to \infty$, at speed $\frac{1}{n^{1-\alpha}}$.

Proof By (24.69). ∎

We also give

Theorem 24.18 *All here as in Theorem 24.10. Then*

$$\left|\left(M_n^*(f)\right)(x) - f(x)\right| \leq \sum_{l=1}^{N}\frac{1}{l!}\left(\left(\sum_{j=1}^{d}\left|\frac{\partial}{\partial x_j}\right|\right)^l f(x)\right).$$

$$\left(\frac{T^*}{n^{1-\alpha}} + \frac{1}{n}\right)^l + \frac{d^N}{N!}\left(\frac{T^*}{n^{1-\alpha}} + \frac{1}{n}\right)^N \max_{\widetilde{\alpha}:|\widetilde{\alpha}|=N} \omega_1\left(f_{\widetilde{\alpha}}, \frac{T^*}{n^{1-\alpha}} + \frac{1}{n}\right). \tag{24.90}$$

Inequality (24.90) implies the pointwise convergence with rates on $\left(M_n^*(f)\right)(x)$ to $f(x)$, as $n \to \infty$, at speed $\frac{1}{n^{1-\alpha}}$.

Proof Set

$$g_{\frac{k}{n}+\frac{i}{nr}}(t) = f\left(x + t\left(\frac{k}{n} + \frac{i}{nr} - x\right)\right), \quad 0 \leq t \leq 1. \tag{24.91}$$

Then we have

$$g^{(l)}_{\frac{k}{n}+\frac{i}{nr}}(t) = \left[\left(\sum_{j=1}^{d}\left(\frac{k_j}{n}+\frac{i_j}{nr_j}-x_j\right)\frac{\partial}{\partial x_j}\right)^l f\right]\left(x+t\left(\frac{k}{n}+\frac{i}{nr}-x\right)\right), \tag{24.92}$$

and

$$g_{\frac{k}{n}+\frac{i}{nr}}(0) = f(x). \tag{24.93}$$

By Taylor's formula, we get

$$f\left(\frac{k}{n}+\frac{i}{nr}\right) = g_{\frac{k}{n}+\frac{i}{nr}}(1) = \tag{24.94}$$

$$\sum_{l=0}^{N}\frac{g^{(l)}_{\frac{k}{n}+\frac{i}{nr}}(0)}{l!} + R_N\left(\frac{k}{n}+\frac{i}{nr},0\right),$$

where

$$R_N\left(\frac{k}{n}+\frac{i}{nr},0\right) =$$

$$\int_0^1\left(\int_0^{t_1}\cdots\left(\int_0^{t_{N-1}}\left(g^{(N)}_{\frac{k}{n}+\frac{i}{nr}}(t_N) - g^{(N)}_{\frac{k}{n}+\frac{i}{nr}}(0)\right)dt_N\right)\cdots\right)dt_1. \tag{24.95}$$

Here we denote by

$$f_{\widetilde{\alpha}} := \frac{\partial^{\widetilde{\alpha}} f}{\partial x^{\widetilde{\alpha}}},\ \widetilde{\alpha} := (\alpha_1,\ldots,\alpha_d),\ \alpha_j \in \mathbb{Z}^+,\ j=1,\ldots,d,$$

such that $|\widetilde{\alpha}| = \sum_{j=1}^{d}\alpha_j = N$. Thus

$$\left(M_n^*(f)\right)(x) = \frac{\sum_{k=\lceil nx-Tn^{\alpha}\rceil}^{[nx+Tn^{\alpha}]}\left(\sum_{i=1}^{r} w_i f\left(\frac{k}{n}+\frac{i}{nr}\right)\right) b\left(n^{1-\alpha}\left(x-\frac{k}{n}\right)\right)}{V(x)} =$$

$$\frac{\sum_{l=0}^{N}\frac{1}{l!}\sum_{k=\lceil nx-Tn^{\alpha}\rceil}^{[nx+Tn^{\alpha}]}\left(\sum_{i=1}^{r} w_i g^{(l)}_{\frac{k}{n}+\frac{i}{nr}}(0)\right) b\left(n^{1-\alpha}\left(x-\frac{k}{n}\right)\right)}{V(x)} + \tag{24.96}$$

$$\frac{\sum_{k=\lceil nx-Tn^{\alpha}\rceil}^{[nx+Tn^{\alpha}]}\left(\sum_{i=1}^{r} w_i R_N\left(\frac{k}{n}+\frac{i}{nr},0\right)\right) b\left(n^{1-\alpha}\left(x-\frac{k}{n}\right)\right)}{V(x)}.$$

Therefore it holds

$$\left(M_n^*(f)\right)(x)-f(x)= \tag{24.97}$$

$$\sum_{l=1}^{N}\frac{1}{l!}\frac{\sum_{k=\lceil nx-Tn^{\alpha}\rceil}^{[nx+Tn^{\alpha}]} b\left(n^{1-\alpha}\left(x-\frac{k}{n}\right)\right)}{V(x)}\left(\sum_{i=1}^{r} w_i g_{\frac{k}{n}+\frac{i}{nr}}^{(l)}(0)\right)+R^*,$$

where

$$R^*=\sum_{k=\lceil nx-Tn^{\alpha}\rceil}^{[nx+Tn^{\alpha}]}\frac{b\left(n^{1-\alpha}\left(x-\frac{k}{n}\right)\right)}{V(x)}\left(\sum_{i=1}^{r} w_i R_N\left(\frac{k}{n}+\frac{i}{nr},0\right)\right). \tag{24.98}$$

Consequently, we obtain

$$\left|\left(M_n^*(f)\right)(x)-f(x)\right|\le$$

$$\sum_{l=1}^{N}\frac{1}{l!}\frac{\sum_{k=\lceil nx-Tn^{\alpha}\rceil}^{[nx+Tn^{\alpha}]} b\left(n^{1-\alpha}\left(x-\frac{k}{n}\right)\right)}{V(x)}\left(\sum_{i=1}^{r} w_i \left|g_{\frac{k}{n}+\frac{i}{nr}}^{(l)}(0)\right|\right)+\left|R^*\right|. \tag{24.99}$$

Notice that

$$\left|g_{\frac{k}{n}+\frac{i}{nr}}^{(l)}(0)\right| \overset{(24.29)}{\le} \left(\left(\sum_{j=1}^{d}\left|\frac{\partial}{\partial x_j}\right|\right)^{l} f(x)\right)\left(\frac{T^*}{n^{1-\alpha}}+\frac{1}{n}\right)^{l}, \tag{24.100}$$

and so far we have

$$\left|\left(M_n^*(f)\right)(x)-f(x)\right|\le\sum_{l=1}^{N}\frac{1}{l!}\left(\left(\sum_{j=1}^{d}\left|\frac{\partial}{\partial x_j}\right|\right)^{l} f(x)\right)\left(\frac{T^*}{n^{1-\alpha}}+\frac{1}{n}\right)^{l}+\left|R^*\right|. \tag{24.101}$$

Next, we need to estimate $|R^*|$. For that, we observe ($0\le t_N\le 1$)

$$\left|g_{\frac{k}{n}+\frac{i}{nr}}^{(N)}(t_N)-g_{\frac{k}{n}+\frac{i}{nr}}^{(N)}(0)\right|=$$

$$|\left[\left(\sum_{j=1}^{d}\left(\frac{k_j}{n}+\frac{i_j}{nr_j}-x_j\right)\frac{\partial}{\partial x_j}\right)^N f\right]\left(x+t_N\left(\frac{k}{n}+\frac{i}{nr}-x\right)\right)- \tag{24.102}$$

$$\left[\left(\sum_{j=1}^{d}\left(\frac{k_j}{n}+\frac{i_j}{nr_j}-x_j\right)\frac{\partial}{\partial x_j}\right)^N f\right](x)| \overset{\text{(by (24.29), (24.30))}}{\leq}$$

$$d^N\left(\frac{T^*}{n^{1-\alpha}}+\frac{1}{n}\right)^N \max_{\widetilde{\alpha}:|\widetilde{\alpha}|=N}\omega_1\left(f_{\widetilde{\alpha}},\frac{T^*}{n^{1-\alpha}}+\frac{1}{n}\right). \tag{24.103}$$

Thus we find

$$\left|R_N\left(\frac{k}{n}+\frac{i}{nr},0\right)\right| \leq$$

$$\int_0^1\left(\int_0^{t_1}\cdots\left(\int_0^{t_{N-1}}\left|g^{(N)}_{\frac{k}{n}+\frac{i}{nr}}(t_N)-g^{(N)}_{\frac{k}{n}+\frac{i}{nr}}(0)\right|dt_N\right)\cdots\right)dt_1 \leq$$

$$\frac{d^N}{N!}\left(\frac{T^*}{n^{1-\alpha}}+\frac{1}{n}\right)^N \max_{\widetilde{\alpha}:|\widetilde{\alpha}|=N}\omega_1\left(f_{\widetilde{\alpha}},\frac{T^*}{n^{1-\alpha}}+\frac{1}{n}\right). \tag{24.104}$$

Hence we get

$$|R^*| \leq \sum_{k=\lceil nx-Tn^{\alpha}\rceil}^{[nx+Tn^{\alpha}]}\frac{b\left(n^{1-\alpha}\left(x-\frac{k}{n}\right)\right)}{V(x)}\left(\sum_{i=1}^{r}w_i\left|R_N\left(\frac{k}{n}+\frac{i}{nr},0\right)\right|\right) \leq \tag{24.105}$$

$$\frac{d^N}{N!}\left(\frac{T^*}{n^{1-\alpha}}+\frac{1}{n}\right)^N \max_{\widetilde{\alpha}:|\widetilde{\alpha}|=N}\omega_1\left(f_{\widetilde{\alpha}},\frac{T^*}{n^{1-\alpha}}+\frac{1}{n}\right), \tag{24.106}$$

proving the claim. ■

Corollary 24.19 (to Theorem 24.18) *Let all as in Corollary 24.11. Then*

$$\left|\left(M_n^*(f)\right)(x)-f(x)\right| \leq \frac{d^N}{N!}\left(\frac{T^*}{n^{1-\alpha}}+\frac{1}{n}\right)^N \max_{\widetilde{\alpha}:|\widetilde{\alpha}|=N}\omega_1\left(f_{\widetilde{\alpha}},\frac{T^*}{n^{1-\alpha}}+\frac{1}{n}\right). \tag{24.107}$$

Inequality (24.107) implies the pointwise convergence with rates of $\left(M_n^*(f)\right)(x)$ *to* $f(x)$, *as* $n\to\infty$, *at the high speed* $\frac{1}{n^{(1-\alpha)(N+1)}}$.

Proof By (24.90). ■

The uniform convergence with rates comes from

Corollary 24.20 (to Theorem 24.18) *Let all as in Corollary 24.12. Then*

$$\left\| M_n^* (f) - f \right\|_{\infty, \prod_{j=1}^{d} [-\gamma_j, \gamma_j]} \leq \sum_{l=1}^{N} \frac{1}{l!} \left(\left(\sum_{j=1}^{d} \left| \frac{\partial}{\partial x_j} \right| \right)^l \| f \|_{\infty, \prod_{j=1}^{d} [-\gamma_j, \gamma_j]} \right)$$

$$\left(\frac{T^*}{n^{1-\alpha}} + \frac{1}{n} \right)^l + \frac{d^N}{N!} \left(\frac{T^*}{n^{1-\alpha}} + \frac{1}{n} \right)^N \max_{\widetilde{\alpha}: |\widetilde{\alpha}| = N} \omega_1 \left(f_{\widetilde{\alpha}}, \frac{T^*}{n^{1-\alpha}} + \frac{1}{n} \right). \quad (24.108)$$

Inequality (24.108) implies the uniform convergence with rates of $M_n^* (f)$ to f, as $n \to \infty$, at the high speed $\frac{1}{n^{1-\alpha}}$.

Proof By (24.90). ■

Corollary 24.21 (to Theorem 24.18) *All as in Theorem 24.18 with $N = 1$. Then*

$$\left| \left(M_n^* (f) \right) (x) - f (x) \right| \leq \left[\sum_{j=1}^{d} \left| \frac{\partial f (x)}{\partial x_j} \right| \right.$$

$$\left. + d \max_{j \in \{1, \dots, d\}} \omega_1 \left(\frac{\partial f}{\partial x_j}, \frac{T^*}{n^{1-\alpha}} + \frac{1}{n} \right) \right] \left(\frac{T^*}{n^{1-\alpha}} + \frac{1}{n} \right). \quad (24.109)$$

Inequality (24.109) implies the pointwise convergence with rates of $\left(M_n^* (f) \right) (x)$ to $f (x)$, as $n \to \infty$, at speed $\frac{1}{n^{1-\alpha}}$.

Proof By (24.90). ■

Note 24.22 *We also observe that all the right hand sides of convergence inequalities (24.50), (24.66)–(24.69), (24.87)–(24.90), (24.107)–(24.109), are independent of b.*

Note 24.23 *We observe that*

$$H_n^* (1) = K_n^* (1) = M_n^* (1) = 1, \quad (24.110)$$

thus unitary operators.

Also, given that f is bounded, we obtain

$$\left\| H_n^* (f) \right\|_{\infty, \mathbb{R}^d} \leq \| f \|_{\infty, \mathbb{R}^d}, \quad (24.111)$$

$$\left\| K_n^* (f) \right\|_{\infty, \mathbb{R}^d} \leq \| f \|_{\infty, \mathbb{R}^d}, \quad (24.112)$$

and

$$\left\| M_n^*(f) \right\|_{\infty,\mathbb{R}^d} \leq \| f \|_{\infty,\mathbb{R}^d}. \tag{24.113}$$

Operators H_n^*, K_n^*, M_n^* are positive linear operators, and of course bounded operators directly by (24.111)–(24.113).

References

1. G.A. Anastassiou, Rate of convergence of some neural network operators to the unit-univariate case. J. Math. Anal. Appli. **212**, 237–262 (1997)
2. G.A. Anastassiou, Rate of convergence of some multivariate neural network operators to the unit. J. Comp. Math. Appl. **40**, 1–19 (2000)
3. G.A. Anastassiou, *Quantitative Approximations* (Chapman & Hall/CRC, Boca Raton, 2001)
4. G.A. Anastassiou, Rate of convergence of some neural network operators to the unit-univariate case, revisited. Vesnik **65**(4), 511–518 (2013)
5. G.A. Anastassiou, Rate of convergence of some multivariate neural network operators to the unit, revisited. J. Comput. Anal. Appl. **15**(7), 1300–1309 (2013)
6. G.A. Anastassiou, Approximation by Perturbed Neural Netwwork Operators. Submitted (2014)
7. G.A. Anastassiou, Approximations by Multivariate Perturbed Neural Network Operators. Submitted (2014)
8. A.R. Barron, Universal approximation bounds for superpositions of a sigmoidal function. IEEE Trans. Inf. Theory **39**, 930–945 (1993)
9. F.L. Cao, T.F. Xie, Z.B. Xu, The estimate for approximation error of neural networks: a constructive approach. Neurocomputing **71**, 626–630 (2008)
10. P. Cardaliaguet, G. Euvrard, Approximation of a function and its derivative with a Neural Network. Neural Netw. **5**, 207–220 (1992)
11. Z. Chen, F. Cao, The approximation operators with sigmoidal functions. Comput. Math. Appl. **58**, 758–765 (2009)
12. T.P. Chen, H. Chen, Universal approximation to nonlinear operators by neural networks with arbitrary activation functions and its applications to a dynamic system. IEEE Trans. Neural Netw. **6**, 911–917 (1995)
13. C.K. Chui, X. Li, Approximation by ridge functions and neural networks with one hidden layer. J. Approx. Theory **70**, 131–141 (1992)
14. D. Costarelli, R. Spigler, Approximation results for neural network operators activated by sigmoidal functions. Neural Netw. **44**, 101–106 (2013)
15. D. Costarelli, R. Spigler, Multivariate neural network operators with sigmoidal activation functions. Neural Netw. **48**, 72–77 (2013)
16. G. Cybenko, Approximation by superpositions of sigmoidal function. Math. Control Signals Syst. **2**, 303–314 (1989)
17. S. Ferrari, R.F. Stengel, Smooth function approximation using neural networks. IEEE Trans. Neural Netw. **16**, 24–38 (2005)
18. K.I. Funahashi, On the approximate realization of continuous mappings by neural networks. Neural Netw. **2**, 183–192 (1989)
19. N. Hahm, B.I. Hong, An approximation by neural networks with a fixed weight. Comput. Math. Appl. **47**, 1897–1903 (2004)
20. S. Haykin, *Neural Networks: A Comprehensive Foundation*, 2nd edn. (Prentice Hall, New York, 1998)
21. K. Hornik, M. Stinchombe, H. White, Multilayer feedforward networks are universal approximations. Neural Netw. **2**, 359–366 (1989)

22. K. Hornik, M. Stinchombe, H. White, Universal approximation of an unknown mapping and its derivatives using multilayer feedforward networks. Neural Netw. **3**, 551–560 (1990)
23. M. Leshno, V.Y. Lin, A. Pinks, S. Schocken, Multilayer feedforward networks with a nonpolynomial activation function can approximate any function. Neural Netw. **6**, 861–867 (1993)
24. V. Maiorov, R.S. Meir, Approximation bounds for smooth functions in $C\left(R^d\right)$ by neural and mixture networks. IEEE Trans. Neural Netw. **9**, 969–978 (1998)
25. Y. Makovoz, Uniform approximation by neural networks. J. Approx. Theory **95**, 215–228 (1998)
26. W. McCulloch, W. Pitts, A logical calculus of the ideas immanent in nervous activity. Bull. Math. Biophys. **7**, 115–133 (1943)
27. H.N. Mhaskar, C.A. Micchelli, Approximation by superposition of a sigmoidal function. Adv. Appl. Math. **13**, 350–373 (1992)
28. H.N. Mhaskar, C.A. Micchelli, Degree of approximation by neural networks with a single hidden layer. Adv. Appl. Math. **16**, 151–183 (1995)
29. T.M. Mitchell, *Machine Learning* (WCB-McGraw-Hill, New York, 1997)
30. D.D. Stancu, On a generalization of the Bernstein polynomials Studia Universitatis Babeş-Bolyai. Ser. Math. Phys. **14**, 31–45 (1969)
31. S. Suzuki, Constructive function approximation by three-layer artificial neural networks. Neural Netw. **11**, 1049–1058 (1998)
32. Z.B. Xu, F.L. Cao, The essential order of approximation for neural networks, Sci. China (Ser. F) **47**, 97–112 (2004)

Chapter 25
Voronovskaya Type Asymptotic Expansions for Perturbed Neural Networks

Here we study further perturbed normalized neural network operators of Cardaliaguet-Euvrard type. We derive univariate and multivariate Voronovskaya type asymptotic expansions for the error of approximation of these operators to the unit operator. It follows [9].

25.1 Introduction

Cardaliaguet and Euvrard were the first, see [12], to describe precisely and study neural network approximation operators to the unit operator. Namely they proved: be given $f : \mathbb{R} \to \mathbb{R}$ a continuous bounded function and b a centered bell-shaped function, then the functions

$$F_n(x) = \sum_{k=-n^2}^{n^2} \frac{f\left(\frac{k}{n}\right)}{In^{\alpha}} b\left(n^{1-\alpha}\left(x - \frac{k}{n}\right)\right),$$

where $I := \int_{-\infty}^{\infty} b(t)\,dt$, $0 < \alpha < 1$, converge uniformly on compacta to f.

You see above that the weights $\frac{f\left(\frac{k}{n}\right)}{In^{\alpha}}$ are explicitly known, for the first time shown in [12].

Furthermore the authors in [12] proved that: let $f : \mathbb{R}^p \to \mathbb{R}$, $p \in \mathbb{N}$, be a continuous bounded function and b a p-dimensional bell-shaped function. Then the functions

$$G_n(x) =$$

$$\sum_{k_1=-n^2}^{n^2} \cdots \sum_{k_p=-n^2}^{n^2} \frac{f\left(\frac{k_1}{n}, \ldots, \frac{k_p}{n}\right)}{In^{\alpha p}} b\left(n^{1-\alpha}\left(x_1 - \frac{k_1}{n}\right), \ldots, n^{1-\alpha}\left(x_p - \frac{k_p}{n}\right)\right),$$

G.A. Anastassiou, *Intelligent Systems II: Complete Approximation by Neural Network Operators*, Studies in Computational Intelligence 608,
DOI 10.1007/978-3-319-20505-2_25

where I is the integral of b on $\mathbb{R}^p$ and $0 < \alpha < 1$, converge uniformly on compacta to f.

Still the work [12] is qualitative and not quantitative.

The author in [1–3], see Chaps. 2–5, was the first to establish neural network approximations to continuous functions with rates, that is quantitative works, by very specifically defined neural network operators of Cardaliagnet-Euvrard and "Squashing" types, by employing the modulus of continuity of the engaged function or its high order derivative or partial derivatives, and producing very tight Jackson type inequalities. He treats there both the univariate and multivariate cases. The defining these operators "bell-shaped" and "squashing" function are assumed to be of compact support. Also in [3] he gives the Nth order asymptotic expansion for the error of weak approximation of these two operators to a special natural class of smooth functions, see Chaps. 4 and 5 there.

Though the work in [1–3], was quantitative, the rate of convergence was not precisely determined.

Finally the author in [6, 8], by normalizing his operators he achieved to determine the exact rates of convergence.

Recently the author in [10, 11] studied the convergence of perturbed cases of the above neural network operators. These perturbations actually occur in what we perceive in our computations from neural network operations. We continue here this last study by giving Voronovskaya type asymptotic expansions for the pointwise approximation of these perturbed operators to the unit operator, see also the related [5, 7].

For more about neural networks in general we refer to [13–18].

The chapter is presented in two parts, the univariate and multivariate.

25.2 Univariate Theory—Univariate Basics

Here the univariate activation function $b : \mathbb{R} \to \mathbb{R}_+$ is of compact support $[-T, T]$, $T > 0$. That is $b(x) > 0$ for any $x \in [-T, T]$, and clearly b may have jump discontinuities. Also the shape of the graph of b could be anything. Typically in neural networks approximation we take b as a sigmoidal function or bell-shaped function, of course here of compact support $[-T, T]$, $T > 0$.

Example 25.1 (i) b can be the characteristic function on $[-1, 1]$,
(ii) b can be that hat function over $[-1, 1]$, i.e.,

$$b(x) = \begin{cases} 1 + x, & -1 \leq x \leq 0, \\ 1 - x, & 0 < x \leq 1, \\ 0, & elsewhere, \end{cases}$$

(iii) the truncated sigmoidals

$$b(x) = \begin{cases} \frac{1}{1+e^{-x}} \text{ or } \tanh x \text{ or } \operatorname{erf}(x), \text{ for} x \in [-T, T], \text{ with large } T > 0, \\ 0, x \in \mathbb{R} - [-T, T], \end{cases}$$

(iv) the truncated Gompertz function

$$b(x) = \begin{cases} e^{-\alpha e^{-\beta x}}, x \in [-T, T]; \ \alpha, \beta > 0; \text{ large } T > 0, \\ 0, \ x \in \mathbb{R} - [-T, T], \end{cases}$$

So the general function b we will be using here covers all kinds of activation functions in univariate neural network approximations.

Typically we consider functions $f : \mathbb{R} \to \mathbb{R}$ that are either continuous and bounded, or uniformly continuous.

Let here the parameters $\mu, \nu \geq 0$; $\mu_i, \nu_i \geq 0$, $i = 1, \ldots, r \in \mathbb{N}$; $w_i \geq 0$: $\sum_{i=1}^{r} w_i = 1$; $0 < \alpha < 1$, $x \in \mathbb{R}$, $n \in \mathbb{N}$.

In this first part we study the asymptotic expansions of Voronovskaya type of the following one hidden layer univariate normalized neural network perturbed operators,

(i) the Stancu type (see [19])

$$(H_n(f))(x) = \frac{\sum_{k=-n^2}^{n^2} \left(\sum_{i=1}^{r} w_i f\left(\frac{k+\mu_i}{n+\nu_i}\right) \right) b\left(n^{1-\alpha}\left(x - \frac{k}{n}\right)\right)}{\sum_{k=-n^2}^{n^2} b\left(n^{1-\alpha}\left(x - \frac{k}{n}\right)\right)}, \quad (25.1)$$

(ii) the Kantorovich type

$$(K_n(f))(x) = \frac{\sum_{k=-n^2}^{n^2} \left(\sum_{i=1}^{r} w_i (n+\nu_i) \int_{\frac{k+\mu_i}{n+\nu_i}}^{\frac{k+\mu_i+1}{n+\nu_i}} f(t)\, dt \right) b\left(n^{1-\alpha}\left(x - \frac{k}{n}\right)\right)}{\sum_{k=-n^2}^{n^2} b\left(n^{1-\alpha}\left(x - \frac{k}{n}\right)\right)}, \quad (25.2)$$

and

(iii) the quadrature type

$$(M_n(f))(x) = \frac{\sum_{k=-n^2}^{n^2} \left(\sum_{i=1}^{r} w_i f\left(\frac{k}{n} + \frac{i}{nr}\right) \right) b\left(n^{1-\alpha}\left(x - \frac{k}{n}\right)\right)}{\sum_{k=-n^2}^{n^2} b\left(n^{1-\alpha}\left(x - \frac{k}{n}\right)\right)}. \quad (25.3)$$

Similar operators defined for bell-shaped functions and sample coefficients $f\left(\frac{k}{n}\right)$ were studied initially in [1–3, 5–8, 12].

Operator K_n in the corresponding Signal Processing context, represents the natural called "time-jitter" error, where the sample information is calculated in a perturbed neighborhood of $\frac{k+\mu}{n+\nu}$ rather than exactly at the node $\frac{k}{n}$.

The perturbed sample coefficients $f\left(\frac{k+\mu}{n+\nu}\right)$ with $0 \leq \mu \leq \nu$, were first used by Stancu [19], in a totally different context, generalizing Bernstein operators approximation on $C([0,1])$. For related approximation properties of these perturbed operators see [10, 11].

The terms in the ratio of sums (25.1)–(25.3) are nonzero, iff

$$\left|n^{1-\alpha}\left(x-\frac{k}{n}\right)\right| \leq T, \text{ i.e. } \left|x-\frac{k}{n}\right| \leq \frac{T}{n^{1-\alpha}} \tag{25.4}$$

iff

$$nx - Tn^{\alpha} \leq k \leq nx + Tn^{\alpha}. \tag{25.5}$$

In order to have the desired order of the numbers

$$-n^2 \leq nx - Tn^{\alpha} \leq nx + Tn^{\alpha} \leq n^2, \tag{25.6}$$

it is sufficiently enough to assume that

$$n \geq T + |x|. \tag{25.7}$$

When $x \in [-T, T]$ it is enough to assume $n \geq 2T$, which implies (25.6).

Proposition 25.2 ([1]) *Let $a \leq b$, $a, b \in \mathbb{R}$. Let $card(k)$ (≥ 0) be the maximum number of integers contained in $[a, b]$. Then*

$$\max(0, (b-a)-1) \leq card(k) \leq (b-a)+1. \tag{25.8}$$

Note 25.3 We would like to establish a lower bound on $card(k)$ over the interval $[nx - Tn^{\alpha}, nx + Tn^{\alpha}]$. From Proposition 25.2 we get that

$$card(k) \geq \max\left(2Tn^{\alpha} - 1, 0\right). \tag{25.9}$$

We obtain $card(k) \geq 1$, if

$$2Tn^{\alpha} - 1 \geq 1 \text{ iff } n \geq T^{-\frac{1}{\alpha}}. \tag{25.10}$$

So to have the desired order (25.6) and $card(k) \geq 1$ over $[nx - Tn^{\alpha}, nx + Tn^{\alpha}]$, we need to consider

$$n \geq \max\left(T + |x|, T^{-\frac{1}{\alpha}}\right). \tag{25.11}$$

Also notice that $card(k) \to +\infty$, as $n \to +\infty$.

Denote by $[\cdot]$ the integral part of a number and by $\lceil \cdot \rceil$ its ceiling.
So under assumption (25.11), the operators H_n, K_n, M_n, collapse to

(i)

$$(H_n(f))(x) = \frac{\sum_{k=\lceil nx-Tn^{\alpha}\rceil}^{[nx+Tn^{\alpha}]} \left(\sum_{i=1}^{r} w_i f\left(\frac{k+\mu_i}{n+\nu_i}\right)\right) b\left(n^{1-\alpha}\left(x-\frac{k}{n}\right)\right)}{\sum_{k=\lceil nx-Tn^{\alpha}\rceil}^{[nx+Tn^{\alpha}]} b\left(n^{1-\alpha}\left(x-\frac{k}{n}\right)\right)}, \tag{25.12}$$

(ii)

$$(K_n(f))(x) = \frac{\sum_{k=\lceil nx-Tn^{\alpha}\rceil}^{[nx+Tn^{\alpha}]} \left(\sum_{i=1}^{r} w_i (n+\nu_i) \int_{\frac{k+\mu_i}{n+\nu_i}}^{\frac{k+\mu_i+1}{n+\nu_i}} f(t)\,dt\right) b\left(n^{1-\alpha}\left(x-\frac{k}{n}\right)\right)}{\sum_{k=\lceil nx-Tn^{\alpha}\rceil}^{[nx+Tn^{\alpha}]} b\left(n^{1-\alpha}\left(x-\frac{k}{n}\right)\right)}, \tag{25.13}$$

and

(iii)

$$(M_n(f))(x) = \frac{\sum_{k=\lceil nx-Tn^{\alpha}\rceil}^{[nx+Tn^{\alpha}]} \left(\sum_{i=1}^{r} w_i f\left(\frac{k}{n}+\frac{i}{nr}\right)\right) b\left(n^{1-\alpha}\left(x-\frac{k}{n}\right)\right)}{\sum_{k=\lceil nx-Tn^{\alpha}\rceil}^{[nx+Tn^{\alpha}]} b\left(n^{1-\alpha}\left(x-\frac{k}{n}\right)\right)}. \tag{25.14}$$

We make

Remark 25.4 Let k as in (25.5). We observe that

$$\left|\frac{k+\mu}{n+\nu} - x\right| \leq \left|\frac{k}{n+\nu} - x\right| + \frac{\mu}{n+\nu}. \tag{25.15}$$

Next we see

$$\left|\frac{k}{n+\nu} - x\right| \leq \left|\frac{k}{n+\nu} - \frac{k}{n}\right| + \left|\frac{k}{n} - x\right| \overset{(25.4)}{\leq} \frac{\nu|k|}{n(n+\nu)} + \frac{T}{n^{1-\alpha}}$$

(by $|k| \leq \max\left(|nx - Tn^{\alpha}|, |nx + Tn^{\alpha}|\right) \leq n|x| + Tn^{\alpha}$)

$$\leq \left(\frac{\nu}{n+\nu}\right)\left(|x| + \frac{T}{n^{1-\alpha}}\right) + \frac{T}{n^{1-\alpha}}. \tag{25.16}$$

Consequently it holds the useful inequality

$$\left|\frac{k+\mu}{n+\nu}-x\right| \leq \left(\frac{\nu}{n+\nu}\right)\left(|x|+\frac{T}{n^{1-\alpha}}\right)+\frac{T}{n^{1-\alpha}}+\frac{\mu}{n+\nu}$$

$$=\left(\frac{\nu|x|+\mu}{n+\nu}\right)+\left(1+\frac{\nu}{n+\nu}\right)\frac{T}{n^{1-\alpha}}, \tag{25.17}$$

where $\mu, \nu \geq 0$, $0<\alpha<1$.

Also, by change of variable method, the operator K_n could be written conveniently as follows:
(ii)'

$$(K_n(f))(x) =$$

$$\frac{\sum_{k=\lceil nx-Tn^{\alpha}\rceil}^{[nx+Tn^{\alpha}]}\left(\sum_{i=1}^{r} w_i\,(n+\nu_i)\int_0^{\frac{1}{n+\nu_i}} f\left(t+\frac{k+\mu_i}{n+\nu_i}\right)dt\right) b\left(n^{1-\alpha}\left(x-\frac{k}{n}\right)\right)}{\sum_{k=\lceil nx-Tn^{\alpha}\rceil}^{[nx+Tn^{\alpha}]} b\left(n^{1-\alpha}\left(x-\frac{k}{n}\right)\right)}. \tag{25.18}$$

Let $N \in \mathbb{N}$, we denote by $AC^N(\mathbb{R})$ the space of functions f, such that $f^{(N-1)}$ is absolutely continuous function on compacta.

25.3 Univariate Results

We give our first univariate main result

Theorem 25.5 *Let $f \in AC^N(\mathbb{R})$, $N \in \mathbb{N}$, with $\|f^{(N)}\|_\infty := \|f^{(N)}\|_{\infty,\mathbb{R}} < \infty$, also $x \in \mathbb{R}$. Here $n \geq \max\left(T+|x|, T^{-\frac{1}{\alpha}}\right)$, $0<\alpha<1$, $n \in \mathbb{N}$, $T>0$. Then*

$$(H_n(f))(x) - f(x) = \sum_{j=1}^{N-1}\frac{f^{(j)}(x)}{j!}H_n\left((\cdot-x)^j, x\right) + o\left(\frac{1}{n^{(N-\varepsilon)(1-\alpha)}}\right), \tag{25.19}$$

where $0<\varepsilon\leq N$.

If $N=1$, the sum in (25.19) collapses.

The last (25.19) implies that

$$n^{(N-\varepsilon)(1-\alpha)}\left[(H_n(f))(x) - f(x) - \sum_{j=1}^{N-1}\frac{f^{(j)}(x)}{j!}H_n\left((\cdot-x)^j, x\right)\right] \to 0, \tag{25.20}$$

as $n\to\infty$, $0<\varepsilon\leq N$.

When $N=1$, or $f^{(j)}(x)=0$, for all $j = 1,\ldots,N-1$, then we derive

$$n^{(N-\varepsilon)(1-\alpha)}\left[(H_n(f))(x) - f(x)\right] \to 0, \tag{25.21}$$

as $n\to\infty$, $0<\varepsilon\leq N$.

Proof Let k as in (25.5). We observe that

$$w_i f\left(\frac{k+\mu_i}{n+\nu_i}\right) = \sum_{j=0}^{N-1} \frac{f^{(j)}(x)}{j!} w_i \left(\frac{k+\mu_i}{n+\nu_i} - x\right)^j + \tag{25.22}$$

$$w_i \int_x^{\frac{k+\mu_i}{n+\nu_i}} f^{(N)}(t) \frac{\left(\frac{k+\mu_i}{n+\nu_i} - t\right)^{N-1}}{(N-1)!} dt, \; i = 1, \ldots, r.$$

Call

$$V(x) = \sum_{k=\lceil nx - Tn^{\alpha} \rceil}^{[nx+Tn^{\alpha}]} b\left(n^{1-\alpha}\left(x - \frac{k}{n}\right)\right). \tag{25.23}$$

Hence

$$\frac{\left(\sum_{i=1}^{r} w_i f\left(\frac{k+\mu_i}{n+\nu_i}\right)\right) b\left(n^{1-\alpha}\left(x - \frac{k}{n}\right)\right)}{V(x)} =$$

$$\sum_{j=0}^{N-1} \frac{f^{(j)}(x)}{j!} \left(\sum_{i=1}^{r} w_i \left(\frac{k+\mu_i}{n+\nu_i} - x\right)^j\right) \frac{b\left(n^{1-\alpha}\left(x - \frac{k}{n}\right)\right)}{V(x)} + \tag{25.24}$$

$$\frac{b\left(n^{1-\alpha}\left(x - \frac{k}{n}\right)\right)}{V(x)} \left(\sum_{i=1}^{r} w_i \int_x^{\frac{k+\mu_i}{n+\nu_i}} f^{(N)}(t) \frac{\left(\frac{k+\mu_i}{n+\nu_i} - t\right)^{N-1}}{(N-1)!} dt\right).$$

Therefore it holds (see (25.12))

$$(H_n(f))(x) - f(x) - \tag{25.25}$$

$$\sum_{j=1}^{N-1} \frac{f^{(j)}(x)}{j!} \left(\sum_{k=\lceil nx - Tn^{\alpha} \rceil}^{[nx+Tn^{\alpha}]} \left\{\left(\sum_{i=1}^{r} w_i \left(\frac{k+\mu_i}{n+\nu_i} - x\right)^j\right) \frac{b\left(n^{1-\alpha}\left(x - \frac{k}{n}\right)\right)}{V(x)}\right\}\right)$$

$$= R(x),$$

where

$$R(x) = \frac{\sum_{k=\lceil nx - Tn^{\alpha} \rceil}^{[nx+Tn^{\alpha}]} b\left(n^{1-\alpha}\left(x - \frac{k}{n}\right)\right)}{V(x)}.$$

$$\left(\sum_{i=1}^{r} w_i \int_x^{\frac{k+\mu_i}{n+\nu_i}} f^{(N)}(t) \frac{\left(\frac{k+\mu_i}{n+\nu_i} - t\right)^{N-1}}{(N-1)!} dt\right). \tag{25.26}$$

Next we upper bound $R(x)$.

We notice the following:

(1) Case of $\frac{k+\mu_i}{n+\nu_i} \geq x$. Then

$$\left| \int_x^{\frac{k+\mu_i}{n+\nu_i}} f^{(N)}(t) \frac{\left(\frac{k+\mu_i}{n+\nu_i} - t\right)^{N-1}}{(N-1)!} dt \right| \leq \int_x^{\frac{k+\mu_i}{n+\nu_i}} \left| f^{(N)}(t) \right| \frac{\left(\frac{k+\mu_i}{n+\nu_i} - t\right)^{N-1}}{(N-1)!} dt \leq \tag{25.27}$$

$$\left\| f^{(N)} \right\|_\infty \frac{\left(\frac{k+\mu_i}{n+\nu_i} - x\right)^{N}}{N!} \overset{(25.17)}{\leq}$$

$$\left(\frac{\left\| f^{(N)} \right\|_\infty}{N!}\right) \left(\left(\frac{\nu_i |x| + \mu_i}{n + \nu_i}\right) + \left(1 + \frac{\nu_i}{n + \nu_i}\right) \frac{T}{n^{1-\alpha}}\right)^N. \tag{25.28}$$

(2) Case of $\frac{k+\mu_i}{n+\nu_i} \leq x$. Then

$$\left| \int_x^{\frac{k+\mu_i}{n+\nu_i}} f^{(N)}(t) \frac{\left(\frac{k+\mu_i}{n+\nu_i} - t\right)^{N-1}}{(N-1)!} dt \right| = \left| \int_{\frac{k+\mu_i}{n+\nu_i}}^{x} f^{(N)}(t) \frac{\left(t - \frac{k+\mu_i}{n+\nu_i}\right)^{N-1}}{(N-1)!} dt \right| \tag{25.29}$$

$$\leq \int_{\frac{k+\mu_i}{n+\nu_i}}^{x} \left| f^{(N)}(t) \right| \frac{\left(t - \frac{k+\mu_i}{n+\nu_i}\right)^{N-1}}{(N-1)!} dt \leq \left\| f^{(N)} \right\|_\infty \frac{\left(x - \frac{k+\mu_i}{n+\nu_i}\right)^{N}}{N!} \overset{(25.17)}{\leq}$$

$$\left(\frac{\left\| f^{(N)} \right\|_\infty}{N!}\right) \left(\left(\frac{\nu_i |x| + \mu_i}{n + \nu_i}\right) + \left(1 + \frac{\nu_i}{n + \nu_i}\right) \frac{T}{n^{1-\alpha}}\right)^N. \tag{25.30}$$

So in either case we get

$$\left| \int_x^{\frac{k+\mu_i}{n+\nu_i}} f^{(N)}(t) \frac{\left(\frac{k+\mu_i}{n+\nu_i} - t\right)^{N-1}}{(N-1)!} dt \right| \leq$$

$$\left(\frac{\left\| f^{(N)} \right\|_\infty}{N!}\right) \left(\left(\frac{\nu_i |x| + \mu_i}{n + \nu_i}\right) + \left(1 + \frac{\nu_i}{n + \nu_i}\right) \frac{T}{n^{1-\alpha}}\right)^N. \tag{25.31}$$

Consequently we obtain

$$|R(x)| \leq \left(\frac{\|f^{(N)}\|_\infty}{N!}\right)\sum_{i=1}^{r} w_i \left(\left(\frac{\nu_i |x| + \mu_i}{n+\nu_i}\right) + \left(1+\frac{\nu_i}{n+\nu_i}\right)\frac{T}{n^{1-\alpha}}\right)^N \leq \tag{25.32}$$

$$\left(\frac{\|f^{(N)}\|_\infty}{N!}\right)\max_{i\in\{1,\ldots,r\}}\left[\left(\left(\frac{\nu_i |x| + \mu_i}{n+\nu_i}\right) + \left(1+\frac{\nu_i}{n+\nu_i}\right)\frac{T}{n^{1-\alpha}}\right)^N\right] \leq$$

$$\left(\frac{\|f^{(N)}\|_\infty}{N!}\right)\max_{i\in\{1,\ldots,r\}}\left\{\left(\left(\frac{\nu_i |x| + \mu_i}{n^{1-\alpha}n^{\alpha}}\right) + \left(1+\frac{\nu_i}{n}\right)\frac{T}{n^{1-\alpha}}\right)^N\right\} =$$

$$\left(\frac{\|f^{(N)}\|_\infty}{N!}\right)\max_{i\in\{1,\ldots,r\}}\left\{\left(\left(\frac{\nu_i |x| + \mu_i}{n^{\alpha}}\right) + \left(1+\frac{\nu_i}{n}\right)T\right)^N\right\}\frac{1}{n^{(1-\alpha)N}} \leq \tag{25.33}$$

$$\left(\frac{\|f^{(N)}\|_\infty}{N!}\right)\max_{i\in\{1,\ldots,r\}}\left\{\left((\nu_i |x| + \mu_i) + (1+\nu_i)T\right)^N\right\}\frac{1}{n^{(1-\alpha)N}}.$$

We have proved that

$$|R(x)| \leq$$

$$\left(\frac{\|f^{(N)}\|_\infty}{N!}\right)\max_{i\in\{1,\ldots,r\}}\left\{\left((\nu_i |x| + \mu_i) + (1+\nu_i)T\right)^N\right\}\frac{1}{n^{(1-\alpha)N}} =: \frac{A}{n^{(1-\alpha)N}}. \tag{25.34}$$

That is we proved

$$|R(x)| = O\left(\frac{1}{n^{(1-\alpha)N}}\right), \tag{25.35}$$

and

$$|R(x)| = o(1). \tag{25.36}$$

And, letting $0 < \varepsilon \leq N$, we derive

$$\frac{|R(x)|}{\left(\frac{1}{n^{(N-\varepsilon)(1-\alpha)}}\right)} \leq \frac{A}{n^{(1-\alpha)N}} n^{(N-\varepsilon)(1-\alpha)} = \frac{A}{n^{\varepsilon(1-\alpha)}} \to 0, \tag{25.37}$$

as $n \to \infty$.

That is

$$|R(x)| = o\left(\frac{1}{n^{(N-\varepsilon)(1-\alpha)}}\right). \tag{25.38}$$

Clearly here we can rewrite (25.25), as

$$(H_n(f))(x) - f(x) - \sum_{j=1}^{N-1} \frac{f^{(j)}(x)}{j!} H_n\left((\cdot - x)^j, x\right) = R(x). \tag{25.39}$$

Based on (25.38) and (25.39) we derive (25.19). ■

We continue with

Theorem 25.6 *Let $f \in AC^N(\mathbb{R})$, $N \in \mathbb{N}$, with $\|f^{(N)}\|_\infty := \|f^{(N)}\|_{\infty,\mathbb{R}} < \infty$, also $x \in \mathbb{R}$. Here $n \geq \max\left(T + |x|, T^{-\frac{1}{\alpha}}\right)$, $0 < \alpha < 1$, $n \in \mathbb{N}$, $T > 0$. Then*

$$(K_n(f))(x) - f(x) = \sum_{j=1}^{N-1} \frac{f^{(j)}(x)}{j!} K_n\left((\cdot - x)^j, x\right) + o\left(\frac{1}{n^{(N-\varepsilon)(1-\alpha)}}\right), \tag{25.40}$$

where $0 < \varepsilon \leq N$.

If $N = 1$, the sum in (25.40) collapses.

The last (25.40) implies that

$$n^{(N-\varepsilon)(1-\alpha)}\left[(K_n(f))(x) - f(x) - \sum_{j=1}^{N-1} \frac{f^{(j)}(x)}{j!} K_n\left((\cdot - x)^j, x\right)\right] \to 0, \tag{25.41}$$

as $n \to \infty$, $0 < \varepsilon \leq N$.

When $N = 1$, or $f^{(j)}(x) = 0$, for all $j = 1, \ldots, N-1$, then we derive

$$n^{(N-\varepsilon)(1-\alpha)}\left[(K_n(f))(x) - f(x)\right] \to 0, \tag{25.42}$$

as $n \to \infty$, $0 < \varepsilon \leq N$.

Proof Let k as in (25.5). We observe that

$$\int_0^{\frac{1}{n+\nu_i}} f\left(t + \frac{k+\mu_i}{n+\nu_i}\right) dt = \sum_{j=0}^{N-1} \frac{f^{(j)}(x)}{j!} \int_0^{\frac{1}{n+\nu_i}} \left(t + \frac{k+\mu_i}{n+\nu_i} - x\right)^j dt +$$

$$\int_0^{\frac{1}{n+\nu_i}} \left(\int_x^{t+\frac{k+\mu_i}{n+\nu_i}} f^{(N)}(z) \frac{\left(t + \frac{k+\mu_i}{n+\nu_i} - z\right)^{N-1}}{(N-1)!} dz\right) dt, \tag{25.43}$$

$i = 1, \ldots, r$.

Hence it holds

$$\left(\sum_{i=1}^{r} w_i \left(n + \nu_i \right) \int_0^{\frac{1}{n+\nu_i}} f \left(t + \frac{k + \mu_i}{n + \nu_i} \right) dt \right) =$$

$$\sum_{j=0}^{N-1} \frac{f^{(j)}(x)}{j!} \left(\sum_{i=1}^{r} w_i \left(n + \nu_i \right) \int_0^{\frac{1}{n+\nu_i}} \left(t + \frac{k + \mu_i}{n + \nu_i} - x \right)^j dt \right) + \tag{25.44}$$

$$\left(\sum_{i=1}^{r} w_i \left(n + \nu_i \right) \int_0^{\frac{1}{n+\nu_i}} \left(\int_x^{t+\frac{k+\mu_i}{n+\nu_i}} f^{(N)}(z) \frac{\left(t + \frac{k+\mu_i}{n+\nu_i} - z \right)^{N-1}}{(N-1)!} dz \right) dt \right).$$

Call

$$V(x) = \sum_{k=\lceil nx - Tn^{\alpha} \rceil}^{[nx+Tn^{\alpha}]} b \left(n^{1-\alpha} \left(x - \frac{k}{n} \right) \right). \tag{25.45}$$

Therefore we obtain

$$\left(K_n(f) \right)(x) = \sum_{j=0}^{N-1} \frac{f^{(j)}(x)}{j!} \sum_{k=\lceil nx - Tn^{\alpha} \rceil}^{[nx+Tn^{\alpha}]} \frac{b \left(n^{1-\alpha} \left(x - \frac{k}{n} \right) \right)}{V(x)}. \tag{25.46}$$

$$\left(\sum_{i=1}^{r} w_i \left(n + \nu_i \right) \int_0^{\frac{1}{n+\nu_i}} \left(t + \frac{k + \mu_i}{n + \nu_i} - x \right)^j dt \right) + R(x),$$

where

$$R(x) = \sum_{k=\lceil nx - Tn^{\alpha} \rceil}^{[nx+Tn^{\alpha}]} \frac{b \left(n^{1-\alpha} \left(x - \frac{k}{n} \right) \right)}{V(x)}.$$

$$\left(\sum_{i=1}^{r} w_i \left(n + \nu_i \right) \int_0^{\frac{1}{n+\nu_i}} \left(\int_x^{t+\frac{k+\mu_i}{n+\nu_i}} f^{(N)}(z) \frac{\left(t + \frac{k+\mu_i}{n+\nu_i} - z \right)^{N-1}}{(N-1)!} dz \right) dt \right). \tag{25.47}$$

So far we have

$$\left(K_n(f) \right)(x) - f(x) - \sum_{j=1}^{N-1} \frac{f^{(j)}(x)}{j!} \sum_{k=\lceil nx - Tn^{\alpha} \rceil}^{[nx+Tn^{\alpha}]} \frac{b \left(n^{1-\alpha} \left(x - \frac{k}{n} \right) \right)}{V(x)}. \tag{25.48}$$

$$\left(\sum_{i=1}^{r} w_i \left(n+\nu_i\right) \int_0^{\frac{1}{n+\nu_i}} \left(t+\frac{k+\mu_i}{n+\nu_i}-x\right)^j dt\right) = R\left(x\right).$$

Next we upper bound $R\left(x\right)$.

As in the proof of Theorem 25.5 we find that

$$\left| \int_x^{t+\frac{k+\mu_i}{n+\nu_i}} f^{(N)}\left(z\right) \frac{\left(t+\frac{k+\mu_i}{n+\nu_i}-z\right)^{N-1}}{(N-1)!} dz \right| \leq$$

$$\left\| f^{(N)} \right\|_\infty \frac{\left| t+\frac{k+\mu_i}{n+\nu_i}-x \right|^N}{N!} \leq \left\| f^{(N)} \right\|_\infty \frac{\left(|t| + \left| \frac{k+\mu_i}{n+\nu_i}-x \right|\right)^N}{N!}. \tag{25.49}$$

Therefore we get

$$\left(n+\nu_i\right) \int_0^{\frac{1}{n+\nu_i}} \left| \int_x^{t+\frac{k+\mu_i}{n+\nu_i}} f^{(N)}\left(z\right) \frac{\left(t+\frac{k+\mu_i}{n+\nu_i}-z\right)^{N-1}}{(N-1)!} dz \right| dt \leq$$

$$\left(\frac{\left\| f^{(N)} \right\|_\infty}{N!}\right) \left(n+\nu_i\right) \int_0^{\frac{1}{n+\nu_i}} \left(|t| + \left| \frac{k+\mu_i}{n+\nu_i}-x \right|\right)^N dt \leq \tag{25.50}$$

$$\left(\frac{\left\| f^{(N)} \right\|_\infty}{N!}\right) \left(\frac{1}{n+\nu_i} + \left| \frac{k+\mu_i}{n+\nu_i}-x \right|\right)^N \overset{(25.17)}{\leq}$$

$$\left(\frac{\left\| f^{(N)} \right\|_\infty}{N!}\right) \left(\left(\frac{\nu_i |x| + \mu_i + 1}{n+\nu_i}\right) + \left(1+\frac{\nu_i}{n+\nu_i}\right) \frac{T}{n^{1-\alpha}} \right)^N. \tag{25.51}$$

Consequently we infer that

$$|R(x)| \leq \left(\frac{\left\| f^{(N)} \right\|_\infty}{N!}\right) \sum_{i=1}^{r} w_i \left[\left(\frac{\nu_i |x| + \mu_i + 1}{n+\nu_i}\right) + \left(1+\frac{\nu_i}{n+\nu_i}\right) \frac{T}{n^{1-\alpha}} \right]^N \leq \tag{25.52}$$

$$\left(\frac{\left\| f^{(N)} \right\|_\infty}{N!}\right) \max_{i\in\{1,\ldots,r\}} \left[\left(\frac{\nu_i |x| + \mu_i + 1}{n+\nu_i}\right) + \left(1+\frac{\nu_i}{n+\nu_i}\right) \frac{T}{n^{1-\alpha}} \right]^N \leq$$

$$\left(\frac{\left\| f^{(N)} \right\|_\infty}{N!}\right) \max_{i\in\{1,\ldots,r\}} \left[\left(\frac{\nu_i |x| + \mu_i + 1}{n^{1-\alpha} n^{\alpha}}\right) + \left(1+\frac{\nu_i}{n}\right) \frac{T}{n^{1-\alpha}} \right]^N =$$

$$\left(\frac{\|f^{(N)}\|_\infty}{N!}\right) \max_{i\in\{1,\ldots,r\}} \left[\left(\frac{\nu_i\,|x| + \mu_i + 1}{n^\alpha}\right) + \left(1 + \frac{\nu_i}{n}\right) T\right]^N \frac{1}{n^{(1-\alpha)N}} \leq \quad (25.53)$$

$$\left(\frac{\|f^{(N)}\|_\infty}{N!}\right) \max_{i\in\{1,\ldots,r\}} \left[(\nu_i\,|x| + \mu_i + 1) + (1 + \nu_i)\, T\right]^N \frac{1}{n^{(1-\alpha)N}}.$$

We have proved that

$$|R(x)| \leq$$

$$\left(\frac{\|f^{(N)}\|_\infty}{N!}\right) \max_{i\in\{1,\ldots,r\}} \left[(\nu_i\,|x| + \mu_i + 1) + (1 + \nu_i)\, T\right]^N \frac{1}{n^{(1-\alpha)N}} =: \frac{B}{n^{(1-\alpha)N}}. \quad (25.54)$$

That is we proved

$$|R(x)| = O\left(\frac{1}{n^{(1-\alpha)N}}\right), \quad (25.55)$$

and

$$|R(x)| = o(1). \quad (25.56)$$

And, letting $0 < \varepsilon \leq N$, we derive

$$\frac{|R(x)|}{\left(\frac{1}{n^{(N-\varepsilon)(1-\alpha)}}\right)} \leq \frac{B}{n^{(1-\alpha)N}} n^{(N-\varepsilon)(1-\alpha)} = \frac{B}{n^{\varepsilon(1-\alpha)}} \to 0, \quad (25.57)$$

as $n \to \infty$.

That is

$$|R(x)| = o\left(\frac{1}{n^{(N-\varepsilon)(1-\alpha)}}\right). \quad (25.58)$$

Clearly here we can rewrite (25.48), as

$$(K_n(f))(x) - f(x) - \sum_{j=1}^{N-1} \frac{f^{(j)}(x)}{j!} K_n\left((\cdot - x)^j, x\right) = R(x). \quad (25.59)$$

Based on (25.58) and (25.59) we derive (25.40). ■

We also give

Theorem 25.7 *Let* $f \in AC^N(\mathbb{R})$, $N \in \mathbb{N}$, *with* $\|f^{(N)}\|_\infty := \|f^{(N)}\|_{\infty,\mathbb{R}} < \infty$, *also* $x \in \mathbb{R}$. *Here* $n \geq \max\left(T + |x|, T^{-\frac{1}{\alpha}}\right)$, $0 < \alpha < 1$, $n \in \mathbb{N}$, $T > 0$. *Then*

$$(M_n(f))(x) - f(x) = \sum_{j=1}^{N-1} \frac{f^{(j)}(x)}{j!} M_n\left((\cdot - x)^j, x\right) + o\left(\frac{1}{n^{(N-\varepsilon)(1-\alpha)}}\right), \tag{25.60}$$

where $0 < \varepsilon \leq N$.

If $N = 1$, *the sum in (25.60) collapses.*

The last (25.60) implies that

$$n^{(N-\varepsilon)(1-\alpha)} \left[(M_n(f))(x) - f(x) - \sum_{j=1}^{N-1} \frac{f^{(j)}(x)}{j!} M_n\left((\cdot - x)^j, x\right) \right] \to 0, \tag{25.61}$$

as $n \to \infty$, $0 < \varepsilon \leq N$.

When $N = 1$, *or* $f^{(j)}(x) = 0$, *for all* $j = 1, \ldots, N-1$, *then we derive*

$$n^{(N-\varepsilon)(1-\alpha)} \left[(M_n(f))(x) - f(x) \right] \to 0, \tag{25.62}$$

as $n \to \infty$, $0 < \varepsilon \leq N$.

Proof Let k as in (25.5). Again by Taylor's formula we have that

$$\sum_{i=1}^{r} w_i f\left(\frac{k}{n} + \frac{i}{nr}\right) = \sum_{j=0}^{N-1} \frac{f^{(j)}(x)}{j!} \sum_{i=1}^{r} w_i \left(\frac{k}{n} + \frac{i}{nr} - x\right)^j + \tag{25.63}$$

$$\sum_{i=1}^{r} w_i \int_x^{\frac{k}{n}+\frac{i}{nr}} f^{(N)}(t) \frac{\left(\frac{k}{n} + \frac{i}{nr} - t\right)^{N-1}}{(N-1)!} dt.$$

Call

$$V(x) = \sum_{k=\lceil nx - Tn^{\alpha} \rceil}^{[nx + Tn^{\alpha}]} b\left(n^{1-\alpha}\left(x - \frac{k}{n}\right)\right). \tag{25.64}$$

Then

$$(M_n(f))(x) = \frac{\sum_{k=\lceil nx-Tn^{\alpha}\rceil}^{[nx+Tn^{\alpha}]} \left(\sum_{i=1}^{r} w_i f\left(\frac{k}{n} + \frac{i}{nr}\right)\right) b\left(n^{1-\alpha}\left(x - \frac{k}{n}\right)\right)}{V(x)}$$

$$= \sum_{j=0}^{N-1} \frac{f^{(j)}(x)}{j!} \frac{\sum_{k=\lceil nx-Tn^{\alpha}\rceil}^{[nx+Tn^{\alpha}]} b\left(n^{1-\alpha}\left(x - \frac{k}{n}\right)\right)}{V(x)}.$$

$$\sum_{i=1}^{r} w_i \left(\frac{k}{n} + \frac{i}{nr} - x\right)^j + \frac{\sum_{k=\lceil nx-Tn^{\alpha}\rceil}^{[nx+Tn^{\alpha}]} b\left(n^{1-\alpha}\left(x - \frac{k}{n}\right)\right)}{V(x)}. \tag{25.65}$$

$$\left(\sum_{i=1}^{r} w_i \int_x^{\frac{k}{n}+\frac{i}{nr}} f^{(N)}(t) \frac{\left(\frac{k}{n}+\frac{i}{nr}-t\right)^{N-1}}{(N-1)!} dt\right).$$

Therefore we get

$$(M_n(f))(x) - f(x) - \sum_{j=1}^{N-1} \frac{f^{(j)}(x)}{j!} \frac{\sum_{k=\lceil nx-Tn^{\alpha}\rceil}^{[nx+Tn^{\alpha}]} b\left(n^{1-\alpha}\left(x-\frac{k}{n}\right)\right)}{V(x)}.$$

$$\left(\sum_{i=1}^{r} w_i \left(\frac{k}{n}+\frac{i}{nr}-x\right)^j\right) = R(x), \tag{25.66}$$

where

$$R(x) = \frac{\sum_{k=\lceil nx-Tn^{\alpha}\rceil}^{[nx+Tn^{\alpha}]} b\left(n^{1-\alpha}\left(x-\frac{k}{n}\right)\right)}{V(x)}.$$

$$\left(\sum_{i=1}^{r} w_i \int_x^{\frac{k}{n}+\frac{i}{nr}} f^{(N)}(t) \frac{\left(\frac{k}{n}+\frac{i}{nr}-t\right)^{N-1}}{(N-1)!} dt\right). \tag{25.67}$$

As in the proof of Theorem 25.5 we obtain

$$\left|\int_x^{\frac{k}{n}+\frac{i}{nr}} f^{(N)}(t) \frac{\left(\frac{k}{n}+\frac{i}{nr}-t\right)^{N-1}}{(N-1)!} dt\right| \leq$$

$$\left(\frac{\|f^{(N)}\|_{\infty}}{N!}\right)\left|\frac{k}{n}+\frac{i}{nr}-x\right|^N \leq \left(\frac{\|f^{(N)}\|_{\infty}}{N!}\right)\left(\frac{1}{n}+\left|\frac{k}{n}-x\right|\right)^N \leq \tag{25.68}$$

$$\left(\frac{\|f^{(N)}\|_{\infty}}{N!}\right)\left(\frac{1}{n}+\frac{T}{n^{1-\alpha}}\right)^N. \tag{25.69}$$

Clearly then it holds

$$|R(x)| \leq \left(\frac{\|f^{(N)}\|_{\infty}}{N!}\right)\left(\frac{1}{n}+\frac{T}{n^{1-\alpha}}\right)^N =$$

$$\left(\frac{\|f^{(N)}\|_{\infty}}{N!}\right)\left(\frac{1}{n^{1-\alpha}n^{\alpha}}+\frac{T}{n^{1-\alpha}}\right)^N = \left(\frac{\|f^{(N)}\|_{\infty}}{N!}\right)\left(\frac{1}{n^{\alpha}}+T\right)^N \frac{1}{n^{(1-\alpha)N}} \leq \tag{25.70}$$

$$\left(\frac{\|f^{(N)}\|_{\infty}}{N!}\right)(1+T)^N \frac{1}{n^{(1-\alpha)N}}.$$

That is

$$|R(x)| \leq \left[\left(\frac{\|f^{(N)}\|_\infty}{N!}\right)(1+T)^N\right]\frac{1}{n^{(1-\alpha)N}} \tag{25.71}$$

$$=: \frac{C}{n^{(1-\alpha)N}}. \tag{25.72}$$

That is we proved

$$|R(x)| = O\left(\frac{1}{n^{(1-\alpha)N}}\right), \tag{25.73}$$

and

$$|R(x)| = o(1). \tag{25.74}$$

And, letting $0 < \varepsilon \leq N$, we derive

$$\frac{|R(x)|}{\left(\frac{1}{n^{(N-\varepsilon)(1-\alpha)}}\right)} \leq \frac{C}{n^{(1-\alpha)N}}n^{(N-\varepsilon)(1-\alpha)} = \frac{C}{n^{\varepsilon(1-\alpha)}} \to 0, \tag{25.75}$$

as $n \to \infty$.

That is

$$|R(x)| = o\left(\frac{1}{n^{(N-\varepsilon)(1-\alpha)}}\right). \tag{25.76}$$

Clearly here we can rewrite (25.66), as

$$(M_n(f))(x) - f(x) - \sum_{j=1}^{N-1}\frac{f^{(j)}(x)}{j!}M_n\left((\cdot - x)^j, x\right) = R(x). \tag{25.77}$$

Based on (25.76) and (25.77) we derive (25.60). ■

25.4 Multivariate Theory—Multivariate Basics

Here the activation function $b : \mathbb{R}^d \to \mathbb{R}_+$, $d \in \mathbb{N}$, is of compact support $B := \prod_{j=1}^{d}[-T_j, T_j]$, $T_j > 0$, $j = 1, \ldots, d$. That is $b(x) > 0$ for any $x \in B$, and clearly b may have jump discontinuities. Also the shape of the graph of b is immaterial.

Typically in neural networks approximation we take b to be a d-dimensional bell-shaped function (i.e. per coordinate is a centered bell-shaped function), or a product of univariate centered bell-shaped functions, or a product of sigmoid functions, in our case all them of compact support B.

Example 25.8 Take $b(x) = \beta(x_1)\beta(x_2)\ldots\beta(x_d)$, where β is any of the following functions, $j = 1, \ldots, d$:

(i) $\beta(x_j)$ is the characteristic function on $[-1, 1]$,
(ii) $\beta(x_j)$ is the hat function over $[-1, 1]$, that is,

$$\beta(x_j) = \begin{cases} 1 + x_j, & -1 \leq x_j \leq 0, \\ 1 - x_j, & 0 < x_j \leq 1, \\ 0, & elsewhere, \end{cases}$$

(iii) the truncated sigmoids

$$\beta(x_j) = \begin{cases} \frac{1}{1+e^{-x_j}} \text{ or } \tanh x_j \text{ or } \operatorname{erf}(x_j), \text{ for } x_j \in [-T_j, T_j], \text{ with large } T_j > 0, \\ 0, \; x_j \in \mathbb{R} - [-T_j, T_j], \end{cases}$$

(iv) the truncated Gompertz function

$$\beta(x_j) = \begin{cases} e^{-\alpha e^{-\beta x_j}}, x_j \in [-T_j, T_j]; \; \alpha, \beta > 0; \text{ large } T_j > 0, \\ 0, \; x_j \in \mathbb{R} - [-T_j, T_j], \end{cases}$$

The Gompertz functions are also sigmoid functions, with wide applications to many applied fields, e.g. demography and tumor growth modeling, etc.

Thus the general activation function b we will be using here includes all kinds of activation functions in neural network approximations.

Typically we consider functions $f : \mathbb{R}^d \to \mathbb{R}$ that either continuous and bounded, or uniformly continuous.

Let here the parameters: $0 < \alpha < 1$, $x = (x_1, \ldots, x_d) \in \mathbb{R}^d$, $n \in \mathbb{N}$; $r = (r_1, \ldots, r_d) \in \mathbb{N}^d$, $i = (i_1, \ldots, i_d) \in \mathbb{N}^d$, with $i_j = 1, 2, \ldots, r_j$, $j = 1, \ldots, d$; also let $w_i = w_{i_1,\ldots,i_d} \geq 0$, such that $\sum_{i_1=1}^{r_1}\sum_{i_2=1}^{r_2}\ldots\sum_{i_d=1}^{r_d} w_{i_1,\ldots,i_d} = 1$, in brief written as $\sum_{i=1}^{r} w_i = 1$. We further consider the parameters $k = (k_1, \ldots, k_d) \in \mathbb{Z}^d$; $\mu_i = (\mu_{i_1}, \ldots, \mu_{i_d}) \in \mathbb{R}_+^d$, $\nu_i = (\nu_{i_1}, \ldots, \nu_{i_d}) \in \mathbb{R}_+^d$; and $\lambda_i = \lambda_{i_1,\ldots,i_d}$, $\rho_i = \rho_{i_1,\ldots,i_d} \geq 0$. Call $\nu_i^{\min} = \min\{\nu_{i_1}, \ldots, \nu_{i_d}\}$ and $\|x\|_\infty = \max(|x_1|, \ldots, |x_d|)$.

In this second part we study the asymptotic expansions of Voronovskaya type of the following one hidden layer multivariate normalized neural network perturbed operators,

(i) the Stancu type (see [19])

$$\left(H_n^*(f)\right)(x) = \left(H_n^*(f)\right)(x_1, \ldots, x_d) = \tag{25.78}$$

$$\frac{\sum_{k=-n^2}^{n^2}\left(\sum_{i=1}^{r} w_i f\left(\frac{k+\mu_i}{n+\nu_i}\right)\right) b\left(n^{1-\alpha}\left(x-\frac{k}{n}\right)\right)}{\sum_{k=-n^2}^{n^2} b\left(n^{1-\alpha}\left(x-\frac{k}{n}\right)\right)} =$$

$$\frac{\sum_{k_1=-n^2}^{n^2}\cdots\sum_{k_d=-n^2}^{n^2}\left(\sum_{i_1=1}^{r}\cdots\sum_{i_d=1}^{r} w_{i_1,\ldots,i_d} f\left(\frac{k_1+\mu_{i_1}}{n+\nu_{i_1}},\ldots,\frac{k_d+\mu_{i_d}}{n+\nu_{i_d}}\right)\right)\cdot}{\sum_{k_1=-n^2}^{n^2}\cdots\sum_{k_d=-n^2}^{n^2} b\left(n^{1-\alpha}\left(x_1-\frac{k_1}{n}\right),\ldots,n^{1-\alpha}\left(x_d-\frac{k_d}{n}\right)\right)}$$

$$b\left(n^{1-\alpha}\left(x_1-\frac{k_1}{n}\right),\ldots,n^{1-\alpha}\left(x_d-\frac{k_d}{n}\right)\right),$$

(ii) the Kantorovich type

$$\left(K_n^*(f)\right)(x) = \tag{25.79}$$

$$\frac{\sum_{k=-n^2}^{n^2}\left(\sum_{i=1}^{r} w_i (n+\rho_i)^d \int_0^{\frac{1}{n+\rho_i}} f\left(t+\frac{k+\lambda_i}{n+\rho_i}\right)dt\right) b\left(n^{1-\alpha}\left(x-\frac{k}{n}\right)\right)}{\sum_{k=-n^2}^{n^2} b\left(n^{1-\alpha}\left(x-\frac{k}{n}\right)\right)} =$$

$$\sum_{k_1=-n^2}^{n^2}\cdots\sum_{k_d=-n^2}^{n^2}\left(\sum_{i_1=1}^{r}\cdots\sum_{i_d=1}^{r} w_{i_1,\ldots,i_d}\left(n+\rho_{i_1,\ldots,i_d}\right)^d\cdot\right.$$

$$\frac{\left.\int\cdots\int\cdots\int_0^{\frac{1}{n+\rho_{i_1,\ldots,i_d}}} f\left(t_1+\frac{k_1+\lambda_{i_1,\ldots,i_d}}{n+\rho_{i_1,\ldots,i_d}},\ldots,t_d+\frac{k_d+\lambda_{i_1,\ldots,i_d}}{n+\rho_{i_1,\ldots,i_d}}\right)dt_1\ldots dt_d\right)\cdot}{\sum_{k_1=-n^2}^{n^2}\cdots\sum_{k_d=-n^2}^{n^2} b\left(n^{1-\alpha}\left(x_1-\frac{k_1}{n}\right),\ldots,n^{1-\alpha}\left(x_d-\frac{k_d}{n}\right)\right)} \tag{25.80}$$

$$b\left(n^{1-\alpha}\left(x_1-\frac{k_1}{n}\right),\ldots,n^{1-\alpha}\left(x_d-\frac{k_d}{n}\right)\right),$$

and

(iii) the quadrature type

$$\left(M_n^*(f)\right)(x) = \frac{\sum_{k=-n^2}^{n^2}\left(\sum_{i=1}^{r} w_i f\left(\frac{k}{n}+\frac{i}{nr}\right)\right) b\left(n^{1-\alpha}\left(x-\frac{k}{n}\right)\right)}{\sum_{k=-n^2}^{n^2} b\left(n^{1-\alpha}\left(x-\frac{k}{n}\right)\right)} = \tag{25.81}$$

$$\frac{\sum_{k_1=-n^2}^{n^2}\cdots\sum_{k_d=-n^2}^{n^2}\left(\sum_{i_1=1}^{r_1}\cdots\sum_{i_d=1}^{r_d} w_{i_1,\ldots,i_d} f\left(\frac{k_1}{n}+\frac{i_1}{nr_1},\ldots,\frac{k_d}{n}+\frac{i_d}{nr_d}\right)\right)\cdot}{\sum_{k_1=-n^2}^{n^2}\cdots\sum_{k_d=-n^2}^{n^2} b\left(n^{1-\alpha}\left(x_1-\frac{k_1}{n}\right),\ldots,n^{1-\alpha}\left(x_d-\frac{k_d}{n}\right)\right)}$$

$$b\left(n^{1-\alpha}\left(x_1-\frac{k_1}{n}\right),\ldots,n^{1-\alpha}\left(x_d-\frac{k_d}{n}\right)\right).$$

Similar operators defined for d-dimensional bell-shaped activation functions and sample coefficients $f\left(\frac{k}{n}\right) = f\left(\frac{k_1}{n},\ldots,\frac{k_d}{n}\right)$ were studied initially in [1–3, 5–8, 12], etc. Also see the newer research in [10, 11].

The terms in the ratio of sums (25.78)–(25.81) can be nonzero, iff simultaneously

$$\left|n^{1-\alpha}\left(x_j-\frac{k_j}{n}\right)\right| \leq T_j, \text{ all } j=1,\ldots,d, \tag{25.82}$$

i.e. $\left|x_j-\frac{k_j}{n}\right| \leq \frac{T_j}{n^{1-\alpha}}$, all $j=1,\ldots,d$, iff

$$nx_j - T_j n^\alpha \leq k_j \leq nx_j + T_j n^\alpha, \text{ all } j=1,\ldots,d. \tag{25.83}$$

To have the order

$$-n^2 \leq nx_j - T_j n^\alpha \leq k_j \leq nx_j + T_j n^\alpha \leq n^2, \tag{25.84}$$

we need $n \geq T_j + |x_j|$, all $j=1,\ldots,d$. So (25.84) is true when we take

$$n \geq \max_{j\in\{1,\ldots,d\}} \left(T_j + |x_j|\right). \tag{25.85}$$

When $x \in B$ in order to have (25.84) it is enough to assume that $n \geq 2T^*$, where $T^* := \max\{T_1,\ldots,T_d\} > 0$. Consider

$$\widetilde{I}_j := \left[nx_j - T_j n^\alpha, nx_j + T_j n^\alpha\right], \; j=1,\ldots,d, \; n \in \mathbb{N}.$$

The length of $\widetilde{I}_j$ is $2T_j n^\alpha$. By Proposition 1 of [1], we get that the cardinality of $k_j \in \mathbb{Z}$ that belong to $\widetilde{I}_j := card\left(k_j\right) \geq \max\left(2T_j n^\alpha - 1, 0\right)$, any $j \in \{1,\ldots,d\}$. In order to have $card\left(k_j\right) \geq 1$, we need $2T_j n^\alpha - 1 \geq 1$ iff $n \geq T_j^{-\frac{1}{\alpha}}$, any $j \in \{1,\ldots,d\}$.

Therefore, a sufficient condition in order to obtain the order (25.84) along with the interval $\widetilde{I}_j$ to contain at least one integer for all $j=1,\ldots,d$ is that

$$n \geq \max_{j\in\{1,\ldots,d\}} \left\{T_j + |x_j|, T_j^{-\frac{1}{\alpha}}\right\}. \tag{25.86}$$

Clearly as $n \to +\infty$ we get that $card\left(k_j\right) \to +\infty$, all $j=1,\ldots,d$. Also notice that $card\left(k_j\right)$ equals to the cardinality of integers in $\left[\lceil nx_j - T_j n^\alpha\rceil, [nx_j + T_j n^\alpha]\right]$ for all $j=1,\ldots,d$.

From now on, in this part two we assume (25.86).

We denote by $T=(T_1,\ldots,T_d)$, $[nx+Tn^\alpha] = ([nx_1+T_1 n^\alpha],\ldots,[nx_d+T_d n^\alpha])$, and $\lceil nx - Tn^\alpha\rceil = (\lceil nx_1 - T_1 n^\alpha\rceil,\ldots,\lceil nx_d - T_d n^\alpha\rceil)$. Furthermore it holds

(i)

$$\left(H_n^*(f)\right)(x) = \left(H_n^*(f)\right)(x_1, \ldots, x_d) = \tag{25.87}$$

$$\frac{\sum_{k=\lceil nx-Tn^{\alpha}\rceil}^{[nx+Tn^{\alpha}]}\left(\sum_{i=1}^{r} w_i f\left(\frac{k+\mu_i}{n+\nu_i}\right)\right) b\left(n^{1-\alpha}\left(x-\frac{k}{n}\right)\right)}{\sum_{k=\lceil nx-Tn^{\alpha}\rceil}^{[nx+Tn^{\alpha}]} b\left(n^{1-\alpha}\left(x-\frac{k}{n}\right)\right)} =$$

$$\frac{\sum_{k_1=\lceil nx_1-T_1n^{\alpha}\rceil}^{[nx_1+T_1n^{\alpha}]}\cdots\sum_{k_d=\lceil nx_d-T_dn^{\alpha}\rceil}^{[nx_d+T_dn^{\alpha}]}\left(\sum_{i_1=1}^{r_1}\cdots\sum_{i_d=1}^{r_d} w_{i_1,\ldots,i_d} f\left(\frac{k_1+\mu_{i_1}}{n+\nu_{i_1}},\ldots,\frac{k_d+\mu_{i_d}}{n+\nu_{i_d}}\right)\right)\cdot}{\sum_{k_1=\lceil nx_1-T_1n^{\alpha}\rceil}^{[nx_1+T_1n^{\alpha}]}\cdots\sum_{k_d=\lceil nx_d-T_dn^{\alpha}\rceil}^{[nx_d+T_dn^{\alpha}]} b\left(n^{1-\alpha}\left(x_1-\frac{k_1}{n}\right),\ldots,n^{1-\alpha}\left(x_d-\frac{k_d}{n}\right)\right)}$$

$$b\left(n^{1-\alpha}\left(x_1-\frac{k_1}{n}\right),\ldots,n^{1-\alpha}\left(x_d-\frac{k_d}{n}\right)\right),$$

(ii)

$$\left(K_n^*(f)\right)(x) = \tag{25.88}$$

$$\frac{\sum_{k=\lceil nx-Tn^{\alpha}\rceil}^{[nx+Tn^{\alpha}]}\left(\sum_{i=1}^{r} w_i\left(n+\rho_i\right)\int_0^{\frac{1}{n+\rho_i}} f\left(t+\frac{k+\lambda_i}{n+\rho_i}\right)dt\right) b\left(n^{1-\alpha}\left(x-\frac{k}{n}\right)\right)}{\sum_{k=\lceil nx-Tn^{\alpha}\rceil}^{[nx+Tn^{\alpha}]} b\left(n^{1-\alpha}\left(x-\frac{k}{n}\right)\right)} =$$

$$\sum_{k_1=\lceil nx_1-T_1n^{\alpha}\rceil}^{[nx_1+T_1n^{\alpha}]}\cdots\sum_{k_d=\lceil nx_d-T_dn^{\alpha}\rceil}^{[nx_d+T_dn^{\alpha}]}\left(\sum_{i_1=1}^{r_1}\cdots\sum_{i_d=1}^{r_d} w_{i_1,\ldots,i_d}\left(n+\rho_{i_1,\ldots,i_d}\right)^d\cdot\right.$$

$$\frac{\left.\int\ldots\int\ldots\int_0^{\frac{1}{n+\rho_{i_1,\ldots,i_d}}} f\left(t_1+\frac{k_1+\lambda_{i_1,\ldots,i_d}}{n+\rho_{i_1,\ldots,i_d}},\ldots,t_d+\frac{k_d+\lambda_{i_1,\ldots,i_d}}{n+\rho_{i_1,\ldots,i_d}}\right)dt_1\ldots dt_d\right)\cdot}{\sum_{k_1=\lceil nx_1-T_1n^{\alpha}\rceil}^{[nx_1+T_1n^{\alpha}]}\cdots\sum_{k_d=\lceil nx_d-T_dn^{\alpha}\rceil}^{[nx_d+T_dn^{\alpha}]} b\left(n^{1-\alpha}\left(x_1-\frac{k_1}{n}\right),\ldots,n^{1-\alpha}\left(x_d-\frac{k_d}{n}\right)\right)} \tag{25.89}$$

$$b\left(n^{1-\alpha}\left(x_1-\frac{k_1}{n}\right),\ldots,n^{1-\alpha}\left(x_d-\frac{k_d}{n}\right)\right),$$

and

(iii)

$$\left(M_n^*(f)\right)(x) = \frac{\sum_{k=\lceil nx-Tn^{\alpha}\rceil}^{[nx+Tn^{\alpha}]}\left(\sum_{i=1}^{r} w_i f\left(\frac{k}{n}+\frac{i}{nr}\right)\right) b\left(n^{1-\alpha}\left(x-\frac{k}{n}\right)\right)}{\sum_{k=\lceil nx-Tn^{\alpha}\rceil}^{[nx+Tn^{\alpha}]} b\left(n^{1-\alpha}\left(x-\frac{k}{n}\right)\right)} = \tag{25.90}$$

$$\frac{\sum_{k_1=\lceil nx_1-T_1n^{\alpha}\rceil}^{[nx_1+T_1n^{\alpha}]}\cdots\sum_{k_d=\lceil nx_d-T_dn^{\alpha}\rceil}^{[nx_d+T_dn^{\alpha}]}\left(\sum_{i_1=1}^{r_1}\cdots\sum_{i_d=1}^{r_d}w_{i_1,\ldots,i_d}f\left(\frac{k_1}{n}+\frac{i_1}{nr_1},\ldots,\frac{k_d}{n}+\frac{i_d}{nr_d}\right)\right)\cdot}{\sum_{k_1=\lceil nx_1-T_1n^{\alpha}\rceil}^{[nx_1+T_1n^{\alpha}]}\cdots\sum_{k_d=\lceil nx_d-T_dn^{\alpha}\rceil}^{[nx_d+T_dn^{\alpha}]}b\left(n^{1-\alpha}\left(x_1-\frac{k_1}{n}\right),\ldots,n^{1-\alpha}\left(x_d-\frac{k_d}{n}\right)\right)}$$

$$b\left(n^{1-\alpha}\left(x_1-\frac{k_1}{n}\right),\ldots,n^{1-\alpha}\left(x_d-\frac{k_d}{n}\right)\right).$$

So if $\left|n^{1-\alpha}\left(x_j-\frac{k_j}{n}\right)\right|\leq T_j$, all $j=1,\ldots,d$, we get that

$$\left\|x-\frac{k}{n}\right\|_{\infty}\leq\frac{T^*}{n^{1-\alpha}}. \tag{25.91}$$

For convenience we call

$$V(x)=\sum_{k=\lceil nx-Tn^{\alpha}\rceil}^{[nx+Tn^{\alpha}]}b\left(n^{1-\alpha}\left(x-\frac{k}{n}\right)\right)=$$

$$\sum_{k_1=\lceil nx_1-T_1n^{\alpha}\rceil}^{[nx_1+T_1n^{\alpha}]}\cdots\sum_{k_d=\lceil nx_d-T_dn^{\alpha}\rceil}^{[nx_d+T_dn^{\alpha}]}b\left(n^{1-\alpha}\left(x_1-\frac{k_1}{n}\right),\ldots,n^{1-\alpha}\left(x_d-\frac{k_d}{n}\right)\right). \tag{25.92}$$

We make

Remark 25.9 Here always k is as in (25.84).

(I) We observe that

$$\left\|\frac{k+\mu_i}{n+\nu_i}-x\right\|_{\infty}\leq\left\|\frac{k}{n+\nu_i}-x\right\|_{\infty}+\left\|\frac{\mu_i}{n+\nu_i}\right\|_{\infty} \tag{25.93}$$

$$\leq\left\|\frac{k}{n+\nu_i}-x\right\|_{\infty}+\frac{\|\mu_i\|_{\infty}}{n+\nu_i^{\min}}.$$

Next we see

$$\left\|\frac{k}{n+\nu_i}-x\right\|_{\infty}\leq\left\|\frac{k}{n+\nu_i}-\frac{k}{n}\right\|_{\infty}+\left\|\frac{k}{n}-x\right\|_{\infty}\overset{(25.91)}{\leq}\left\|\frac{\nu_ik}{n(n+\nu_i)}\right\|_{\infty}+\frac{T^*}{n^{1-\alpha}} \tag{25.94}$$

$$\leq\|k\|_{\infty}\frac{\|\nu_i\|_{\infty}}{n\left(n+\nu_i^{\min}\right)}+\frac{T^*}{n^{1-\alpha}}=:(*).$$

We notice for $j=1,\ldots,d$ we get that

$$\left|k_j\right| \le n\left|x_j\right| + T_j n^{\alpha}.$$

Therefore

$$\|k\|_\infty \le \left\|n|x| + Tn^{\alpha}\right\|_\infty \le n\|x\|_\infty + T^* n^{\alpha}, \tag{25.95}$$

where $|x| = (|x_1|, \ldots, |x|_d)$.
Thus

$$(*) \le \left(n\|x\|_\infty + T^* n^{\alpha}\right)\frac{\|\nu_i\|_\infty}{n\left(n+\nu_i^{\min}\right)} + \frac{T^*}{n^{1-\alpha}} = \tag{25.96}$$

$$\left(\|x\|_\infty + T^* n^{\alpha-1}\right)\frac{\|\nu_i\|_\infty}{\left(n+\nu_i^{\min}\right)} + \frac{T^*}{n^{1-\alpha}}.$$

So we get

$$\left\|\frac{k}{n+\nu_i} - x\right\|_\infty \le \left(\|x\|_\infty + \frac{T^*}{n^{1-\alpha}}\right)\frac{\|\nu_i\|_\infty}{\left(n+\nu_i^{\min}\right)} + \frac{T^*}{n^{1-\alpha}}. \tag{25.97}$$

Consequently we obtain the useful

$$\left\|\frac{k+\mu_i}{n+\nu_i} - x\right\|_\infty \le \left(\frac{\|\nu_i\|_\infty\|x\|_\infty + \|\mu_i\|_\infty}{n+\nu_i^{\min}}\right) + \left(1 + \frac{\|\nu_i\|_\infty}{\left(n+\nu_i^{\min}\right)}\right)\frac{T^*}{n^{1-\alpha}}. \tag{25.98}$$

(II) We also have for

$$0 \le t_j \le \frac{1}{n+\rho_{i_1,\ldots,i_d}}, \; j = 1, \ldots, d, \tag{25.99}$$

that

$$\left\|t + \frac{k+\lambda_{i_1,\ldots,i_d}}{n+\rho_{i_1,\ldots,i_d}} - x\right\|_\infty \le \frac{1}{n+\rho_{i_1,\ldots,i_d}} + \left\|\frac{k+\lambda_{i_1,\ldots,i_d}}{n+\rho_{i_1,\ldots,i_d}} - x\right\|_\infty \le \tag{25.100}$$

$$\frac{1+\lambda_{i_1,\ldots,i_d}}{n+\rho_{i_1,\ldots,i_d}} + \left\|\frac{k}{n+\rho_{i_1,\ldots,i_d}} - x\right\|_\infty \le$$

$$\frac{1+\lambda_{i_1,\ldots,i_d}}{n+\rho_{i_1,\ldots,i_d}} + \left\|\frac{k}{n+\rho_{i_1,\ldots,i_d}} - \frac{k}{n}\right\|_\infty + \left\|\frac{k}{n} - x\right\|_\infty \le$$

$$\frac{1+\lambda_{i_1,\ldots,i_d}}{n+\rho_{i_1,\ldots,i_d}} + \frac{T^*}{n^{1-\alpha}} + \frac{\rho_{i_1,\ldots,i_d}\|k\|_\infty}{\left(n+\rho_{i_1,\ldots,i_d}\right)n} \le$$

$$\frac{1+\lambda_{i_1,\ldots,i_d}}{n+\rho_{i_1,\ldots,i_d}} + \frac{T^*}{n^{1-\alpha}} + \frac{\rho_{i_1,\ldots,i_d}}{n\left(n+\rho_{i_1,\ldots,i_d}\right)}\left(n\|x\|_\infty + T^* n^{\alpha}\right) = \tag{25.101}$$

$$\frac{1+\lambda_{i_1,\ldots,i_d}}{n+\rho_{i_1,\ldots,i_d}}+\frac{T^*}{n^{1-\alpha}}+\left(\frac{\rho_{i_1,\ldots,i_d}}{n+\rho_{i_1,\ldots,i_d}}\right)\left(\|x\|_\infty+\frac{T^*}{n^{1-\alpha}}\right). \tag{25.102}$$

We have found the useful

$$\left\|t+\frac{k+\lambda_{i_1,\ldots,i_d}}{n+\rho_{i_1,\ldots,i_d}}-x\right\|_\infty \leq \tag{25.103}$$

$$\left(\frac{\rho_{i_1,\ldots,i_d}\|x\|_\infty+\lambda_{i_1,\ldots,i_d}+1}{n+\rho_{i_1,\ldots,i_d}}\right)+\left(1+\frac{\rho_{i_1,\ldots,i_d}}{n+\rho_{i_1,\ldots,i_d}}\right)\frac{T^*}{n^{1-\alpha}}.$$

(III) We also observe that it holds

$$\left\|\frac{k}{n}+\frac{i}{nr}-x\right\|_\infty \leq \left\|\frac{k}{n}-x\right\|_\infty+\frac{1}{n}\left\|\frac{i}{r}\right\|_\infty \leq \frac{T^*}{n^{1-\alpha}}+\frac{1}{n}. \tag{25.104}$$

Next we follow [4, pp. 284–286].

About Multivariate Taylor Formula and Estimates

Let $\mathbb{R}^d$; $d \geq 2$; $z := (z_1, \ldots, z_d)$, $x_0 := (x_{01}, \ldots, x_{0d}) \in \mathbb{R}^d$. We consider the space of functions $AC^N(\mathbb{R}^d)$ with $f : \mathbb{R}^d \to \mathbb{R}$ be such that all partial derivatives of order $(N-1)$ are coordinatewise absolutely continuous functions on convex compacta, $N \in \mathbb{N}$. Also $f \in C^{N-1}(\mathbb{R}^d)$. Each Nth order partial derivative is denoted by $f_{\widetilde{\alpha}} := \frac{\partial^{\widetilde{\alpha}} f}{\partial x^{\widetilde{\alpha}}}$, where $\widetilde{\alpha} := (\alpha_1, \ldots, \alpha_N)$, $\alpha_j \in \mathbb{Z}^+$, $j = 1, \ldots, d$ and $|\widetilde{\alpha}| := \sum_{j=1}^d \alpha_j = N$. Consider $g_z(t) := f(x_0 + t(z - x_0))$, $t \geq 0$. Then

$$g_z^{(l)}(t) = \left[\left(\sum_{j=1}^d (z_j - x_{0j}) \frac{\partial}{\partial x_j}\right)^l f\right](x_{01} + t(z_1 - x_{01}), \ldots, x_{0N} + t(z_N - x_{0N})), \tag{25.105}$$

for all $l = 0, 1, 2, \ldots, N$.

Example 25.10 Let $d = N = 2$. Then

$$g_z(t) = f(x_{01} + t(z_1 - x_{01}), x_{02} + t(z_2 - x_{02})),\ t \in \mathbb{R},$$

and

$$g_z'(t) = (z_1 - x_{01}) \frac{\partial f}{\partial x_1}(x_0 + t(z - x_0)) + (z_2 - x_{02}) \frac{\partial f}{\partial x_2}(x_0 + t(z - x_0)). \tag{25.106}$$

Setting

$$(*) = (x_{01} + t(z_1 - x_{01}), x_{02} + t(z_2 - x_{02})) = (x_0 + t(z - x_0)),$$

we get

$$g_z''(t) = (z_1 - x_{01})^2 \frac{\partial f^2}{\partial x_1^2}(*) + (z_1 - x_{01})(z_2 - x_{02}) \frac{\partial f^2}{\partial x_2 \partial x_1}(*) +$$

$$(z_1 - x_{01})(z_2 - x_{02}) \frac{\partial f^2}{\partial x_1 \partial x_2}(*) + (z_2 - x_{02})^2 \frac{\partial f^2}{\partial x_2^2}(*). \tag{25.107}$$

Similarly, we have the general case of $d, N \in \mathbb{N}$ for $g_z^{(N)}(t)$.

We mention the following multivariate Taylor theorem.

Theorem 25.11 *Under the above assumptions we have*

$$f(z_1, \ldots, z_N) = g_z(1) = \sum_{l=0}^{N-1} \frac{g_z^{(l)}(0)}{l!} + R_N(z, 0), \tag{25.108}$$

where

$$R_N(z, 0) := \int_0^1 \left(\int_0^{t_1} \ldots \left(\int_0^{t_{N-1}} g_z^{(N)}(t_N)\, dt_N \right) \ldots \right) dt_1, \tag{25.109}$$

or

$$R_N(z, 0) = \frac{1}{(N-1)!} \int_0^1 (1-\theta)^{N-1} g_z^{(N)}(\theta)\, d\theta. \tag{25.110}$$

Notice that $g_z(0) = f(x_0)$.

We make

Remark 25.12 Assume here that

$$\| f_\alpha \|_{\infty, \mathbb{R}^d, N}^{\max} := \max_{|\alpha| = N} \| f_\alpha \|_{\infty, \mathbb{R}^d} < \infty. \tag{25.111}$$

Then

$$\left\| g_z^{(m)} \right\|_{\infty,[0,1]} = \left\| \left[\left(\sum_{j=1}^d (z_j - x_{0j}) \frac{\partial}{\partial x_j} \right)^N f \right] (x_0 + t(z - x_0)) \right\|_{\infty,[0,1]} \leq \tag{25.112}$$

$$\left(\sum_{j=1}^d |z_j - x_{0j}| \right)^N \| f_\alpha \|_{\infty, \mathbb{R}^d, N}^{\max},$$

that is

$$\left\| g_z^{(N)} \right\|_{\infty,[0,1]} \leq \left(\|z - x_0\|_{l_1} \right)^N \|f_\alpha\|_{\infty,\mathbb{R}^d,N}^{\max} < \infty. \quad (25.113)$$

Hence we get by (25.110) that

$$|R_N(z,0)| \leq \frac{\left\| g_z^{(N)} \right\|_{\infty,[0,1]}}{N!} < \infty. \quad (25.114)$$

And it holds

$$|R_N(z,0)| \leq \frac{\left(\|z - x_0\|_{l_1} \right)^N}{N!} \|f_\alpha\|_{\infty,\mathbb{R}^d,N}^{\max}, \quad (25.115)$$

$\forall\, z, x_0 \in \mathbb{R}^d$.

Inequality (25.115) will be an important tool in proving our main multivariate results.

25.5 Multivariate Results

We present our first multivariate main result

Theorem 25.13 *Let $f \in AC^N(\mathbb{R}^d) \cap C^{N-1}(\mathbb{R}^d)$, $d \in \mathbb{N} - \{1\}$, $N \in \mathbb{N}$, with $\|f_{\widetilde{\alpha}}\|_{\infty,\mathbb{R}^d,N}^{\max} < \infty$. Here $n \geq \max_{j \in \{1,\ldots,d\}} \left(T_j + |x_j|, T_j^{-\frac{1}{\alpha}} \right)$, where $x \in \mathbb{R}^d$, $0 < \alpha < 1$, $n \in \mathbb{N}$, $T_j > 0$. Then*

$$(H_n^*(f))(x) - f(x) =$$

$$\sum_{l=1}^{N-1} \left(\sum_{|\widetilde{\alpha}|=l} \left(\frac{f_{\widetilde{\alpha}}(x)}{\prod_{j=1}^{d} \alpha_j!} \right) H_n^* \left(\prod_{j=1}^{d} (\cdot - x_j)^{\alpha_j}, x \right) \right) + o\left(\frac{1}{n^{(N-\varepsilon)(1-\alpha)}} \right), \quad (25.116)$$

where $0 < \varepsilon \leq N$.

If $N = 1$, the sum in (25.116) collapses.

The last (25.116) implies that

$$n^{(N-\varepsilon)(1-\alpha)}.$$

$$\left[\left(H_n^*(f)\right)(x) - f(x) - \sum_{l=1}^{N-1} \left(\sum_{|\widetilde{\alpha}|=l} \left(\frac{f_{\widetilde{\alpha}}(x)}{\prod_{j=1}^{d} \alpha_j!} \right) H_n^* \left(\prod_{j=1}^{d} (\cdot - x_j)^{\alpha_j}, x \right) \right) \right] \to 0, \tag{25.117}$$

as $n \to \infty$, $0 < \varepsilon \leq N$.

When $N = 1$, *or* $f_{\widetilde{\alpha}}(x) = 0$, *for all* $\widetilde{\alpha} : |\widetilde{\alpha}| = l = 1, \ldots, N-1$, *then we derive*

$$n^{(N-\varepsilon)(1-\alpha)} \left[\left(H_n^*(f)\right)(x) - f(x) \right] \to 0, \tag{25.118}$$

as $n \to \infty$, $0 < \varepsilon \leq N$.

Proof Set

$$g_{\frac{k+\mu_i}{n+\nu_i}}(t) = f\left(x + t\left(\frac{k+\mu_i}{n+\nu_i} - x \right) \right), \quad 0 \leq t \leq 1. \tag{25.119}$$

Then we have

$$g^{(l)}_{\frac{k+\mu_i}{n+\nu_i}}(t) = \left[\left(\sum_{j=1}^{d} \left(\frac{k_j + \mu_{i_j}}{n + \nu_{i_j}} - x_j \right) \frac{\partial}{\partial x_j} \right)^l f \right] \tag{25.120}$$

$$\left(x_1 + t\left(\frac{k_1 + \mu_{i_1}}{n+\nu_{i_1}} - x_1 \right), \ldots, x_d + t\left(\frac{k_d + \mu_{i_d}}{n + \nu_{i_d}} - x_d \right) \right), \quad l = 0, \ldots, N,$$

and

$$g_{\frac{k+\mu_i}{n+\nu_i}}(0) = f(x). \tag{25.121}$$

By Taylor's formula, we get

$$f\left(\frac{k_1 + \mu_{i_1}}{n + \nu_{i_1}}, \ldots, \frac{k_d + \mu_{i_d}}{n + \nu_{i_d}} \right) = g_{\frac{k+\mu_i}{n+\nu_i}}(1) = \tag{25.122}$$

$$\sum_{l=0}^{N-1} \frac{g^{(l)}_{\frac{k+\mu_i}{n+\nu_i}}(0)}{l!} + R_N\left(\frac{k+\mu_i}{n+\nu_i}, 0 \right),$$

where

$$R_N\left(\frac{k+\mu_i}{n+\nu_i}, 0 \right) = \frac{1}{(N-1)!} \int_0^1 (1-\theta)^{N-1} g^{(N)}_{\frac{k+\mu_i}{n+\nu_i}}(\theta)\, d\theta. \tag{25.123}$$

Here we denote by

$$f_{\widetilde{\alpha}} := \frac{\partial^{\widetilde{\alpha}} f}{\partial x^{\widetilde{\alpha}}}, \widetilde{\alpha} := (\alpha_1, \ldots, \alpha_d), \alpha_j \in \mathbb{Z}^+, \ j = 1, \ldots, d, \tag{25.124}$$

such that $|\widetilde{\alpha}| = \sum_{j=1}^{d} \alpha_j = l, 0 \le l \le N$.

More precisely we can rewrite

$$f\left(\frac{k+\mu_i}{n+\nu_i}\right) - f(x) =$$

$$\sum_{l=1}^{N-1} \sum_{\left(\substack{\widetilde{\alpha}:=(\alpha_1,\ldots,\alpha_d),\ \alpha_j\in\mathbb{Z}^+, \\ j=1,\ldots,d,\ |\widetilde{\alpha}|=\sum_{j=1}^{d}\alpha_j=l}\right)} \left(\frac{1}{\prod_{j=1}^{d}\alpha_j!}\right) \left(\prod_{j=1}^{d}\left(\frac{k_j+\mu_{i_j}}{n+\nu_{i_j}} - x_j\right)^{\alpha_j}\right) f_{\widetilde{\alpha}}(x)$$

$$+ R_N\left(\frac{k+\mu_i}{n+\nu_i}, 0\right), \tag{25.125}$$

where

$$R_N\left(\frac{k+\mu_i}{n+\nu_i}, 0\right) = N \int_0^1 (1-\theta)^{N-1} \sum_{\left(\substack{\widetilde{\alpha}:=(\alpha_1,\ldots,\alpha_d),\ \alpha_j\in\mathbb{Z}^+, \\ j=1,\ldots,d,\ |\widetilde{\alpha}|=\sum_{j=1}^{d}\alpha_j=N}\right)} \left(\frac{1}{\prod_{j=1}^{d}\alpha_j!}\right) \cdot$$

$$\left(\prod_{j=1}^{d}\left(\frac{k_j+\mu_{i_j}}{n+\nu_{i_j}} - x_j\right)^{\alpha_j}\right) f_{\widetilde{\alpha}}\left(x + \theta\left(\frac{k+\mu_i}{n+\nu_i} - x\right)\right) d\theta. \tag{25.126}$$

Thus

$$\sum_{i=1}^{r} w_i f\left(\frac{k+\mu_i}{n+\nu_i}\right) = \sum_{l=0}^{N-1}\sum_{i=1}^{r} w_i \frac{g^{(l)}_{\frac{k+\mu_i}{n+\nu_i}}(0)}{l!} + \sum_{i=1}^{r} w_i R_N\left(\frac{k+\mu_i}{n+\nu_i}, 0\right), \tag{25.127}$$

and

$$\left(H_n^*(f)\right)(x) = \frac{\sum_{k=\lceil nx-Tn^{\alpha}\rceil}^{[nx+Tn^{\alpha}]}\left(\sum_{i=1}^{r} w_i f\left(\frac{k+\mu_i}{n+\nu_i}\right)\right) b\left(n^{1-\alpha}\left(x-\frac{k}{n}\right)\right)}{V(x)} =$$

$$\sum_{l=0}^{N-1} \frac{\sum_{k=\lceil nx-Tn^{\alpha}\rceil}^{[nx+Tn^{\alpha}]}\left(\sum_{i=1}^{r} w_i \frac{g^{(l)}_{\frac{k+\mu_i}{n+\nu_i}}(0)}{l!}\right) b\left(n^{1-\alpha}\left(x-\frac{k}{n}\right)\right)}{V(x)} + \tag{25.128}$$

$$\frac{\sum_{k=\lceil nx-Tn^{\alpha}\rceil}^{[nx+Tn^{\alpha}]} b\left(n^{1-\alpha}\left(x-\frac{k}{n}\right)\right)}{V(x)}\left(\sum_{i=1}^{r} w_i R_N\left(\frac{k+\mu_i}{n+\nu_i}, 0\right)\right).$$

Therefore it holds

$$\left(H_n^*(f)\right)(x) - f(x) -$$

$$\sum_{l=1}^{N-1} \frac{\sum_{k=\lceil nx-Tn^{\alpha}\rceil}^{[nx+Tn^{\alpha}]}\left(\sum_{i=1}^{r} w_i \frac{g^{(l)}_{\frac{k+\mu_i}{n+\nu_i}}(0)}{l!}\right) b\left(n^{1-\alpha}\left(x-\frac{k}{n}\right)\right)}{V(x)} = R^*(x), \tag{25.129}$$

where

$$R^*(x) = \sum_{k=\lceil nx-Tn^{\alpha}\rceil}^{[nx+Tn^{\alpha}]} \frac{b\left(n^{1-\alpha}\left(x-\frac{k}{n}\right)\right)}{V(x)}\left(\sum_{i=1}^{r} w_i R_N\left(\frac{k+\mu_i}{n+\nu_i}, 0\right)\right). \tag{25.130}$$

By (25.115) we get that

$$\left\| g^{(N)}_{\frac{k+\mu_i}{n+\nu_i}} \right\|_{\infty,[0,1]} \leq \left(\left\| \frac{k+\mu_i}{n+\nu_i} - x \right\|_{l_1}\right)^N \| f_{\widetilde{\alpha}} \|^{\max}_{\infty,\mathbb{R}^d,N} < \infty. \tag{25.131}$$

And furthermore it holds

$$\left| R_N\left(\frac{k+\mu_i}{n+\nu_i}, 0\right)\right| \leq \frac{\left(\left\| \frac{k+\mu_i}{n+\nu_i} - x \right\|_{l_1}\right)^N}{N!} \| f_{\widetilde{\alpha}} \|^{\max}_{\infty,\mathbb{R}^d,N}. \tag{25.132}$$

By (25.98) now we obtain

$$\left\| \frac{k+\mu_i}{n+\nu_i} - x \right\|_{l_1} \leq \tag{25.133}$$

$$d\left[\left(\frac{\|\nu_i\|_\infty \|x\|_\infty + \|\mu_i\|_\infty}{n+\nu_i^{\min}}\right) + \left(1+\frac{\|\nu_i\|_\infty}{n+\nu_i^{\min}}\right)\frac{T^*}{n^{1-\alpha}}\right],$$

and thus we have

$$\left|R_N\left(\frac{k+\mu_i}{n+\nu_i}, 0\right)\right| \leq$$

$$\frac{d^N}{N!}\left[\left(\frac{\|\nu_i\|_\infty \|x\|_\infty + \|\mu_i\|_\infty}{n+\nu_i^{\min}}\right) + \left(1+\frac{\|\nu_i\|_\infty}{n+\nu_i^{\min}}\right)\frac{T^*}{n^{1-\alpha}}\right]^N \|f_{\widetilde{\alpha}}\|_{\infty,\mathbb{R}^d,N}^{\max}. \tag{25.134}$$

Clearly now we can deduce

$$\left|R^*(x)\right| \leq \left(\frac{d^N}{N!}\|f_{\widetilde{\alpha}}\|_{\infty,\mathbb{R}^d,N}\right)\cdot$$

$$\sum_{i=1}^{r} w_i\left[\left(\frac{\|\nu_i\|_\infty \|x\|_\infty + \|\mu_i\|_\infty}{n+\nu_i^{\min}}\right) + \left(1+\frac{\|\nu_i\|_\infty}{n+\nu_i^{\min}}\right)\frac{T^*}{n^{1-\alpha}}\right]^N \leq \tag{25.135}$$

$$\left(\frac{d^N}{N!}\|f_{\widetilde{\alpha}}\|_{\infty,\mathbb{R}^d,N}\right)\cdot$$

$$\left(\max_{\substack{\text{all } i=(i_1,\ldots,i_d):\\ i_j=1,\ldots,r_j;\ j=1,\ldots,d}}\right)\left[\left(\frac{\|\nu_i\|_\infty \|x\|_\infty + \|\mu_i\|_\infty}{n+\nu_i^{\min}}\right) + \left(1+\frac{\|\nu_i\|_\infty}{n+\nu_i^{\min}}\right)\frac{T^*}{n^{1-\alpha}}\right]^N$$

$$\leq \left(\frac{d^N}{N!}\|f_{\widetilde{\alpha}}\|_{\infty,\mathbb{R}^d,N}\right)\cdot$$

$$\max_i\left[\left(\frac{\|\nu_i\|_\infty \|x\|_\infty + \|\mu_i\|_\infty}{n^{1-\alpha}n^\alpha}\right) + \left(1+\frac{\|\nu_i\|_\infty}{n}\right)\frac{T^*}{n^{1-\alpha}}\right]^N = \tag{25.136}$$

$$\left(\frac{d^N}{N!}\|f_{\widetilde{\alpha}}\|_{\infty,\mathbb{R}^d,N}\right)\cdot$$

$$\max_i\left[\left(\frac{\|\nu_i\|_\infty \|x\|_\infty + \|\mu_i\|_\infty}{n^\alpha}\right) + \left(1+\frac{\|\nu_i\|_\infty}{n}\right)T^*\right]^N \frac{1}{n^{(1-\alpha)N}} \leq$$

$$\left(\frac{d^N}{N!}\|f_{\widetilde{\alpha}}\|_{\infty,\mathbb{R}^d,N}\right)\cdot$$

$$\max_i\left[\left(\|\nu_i\|_\infty \|x\|_\infty + \|\mu_i\|_\infty\right) + \left(1+\|\nu_i\|_\infty\right)T^*\right]^N \frac{1}{n^{(1-\alpha)N}} =: \frac{\Lambda}{n^{(1-\alpha)N}}. \tag{25.137}$$

That is we proved

$$\left|R^*(x)\right| = O\left(\frac{1}{n^{(1-\alpha)N}}\right), \tag{25.138}$$

and

$$\left|R^*(x)\right| = o(1). \tag{25.139}$$

And, letting $0 < \varepsilon \leq N$, we derive

$$\frac{|R^*(x)|}{\left(\frac{1}{n^{(N-\varepsilon)(1-\alpha)}}\right)} \leq \frac{\Lambda}{n^{(1-\alpha)N}} n^{(N-\varepsilon)(1-\alpha)} = \frac{\Lambda}{n^{\varepsilon(1-\alpha)}} \to 0, \tag{25.140}$$

as $n \to \infty$.

I.e.

$$\left|R^*(x)\right| = o\left(\frac{1}{n^{(N-\varepsilon)(1-\alpha)}}\right). \tag{25.141}$$

Clearly here we can rewrite (25.129), as

$$\left(H_n^*(f)\right)(x) - f(x) - \sum_{l=1}^{N-1}\left(\sum_{|\widetilde{\alpha}|=l}\left(\frac{f_{\widetilde{\alpha}}(x)}{\prod_{j=1}^{d}\alpha_j!}\right) H_n^*\left(\prod_{j=1}^{d}(\cdot - x_j)^{\alpha_j}, x\right)\right) = R^*(x). \tag{25.142}$$

Based on (25.141) and (25.142) we derive (25.116). ■

We continue with

Theorem 25.14 *Let* $f \in AC^N(\mathbb{R}^d) \cap C^{N-1}(\mathbb{R}^d)$, $d \in \mathbb{N} - \{1\}$, $N \in \mathbb{N}$, *with* $\|f_{\widetilde{\alpha}}\|_{\infty,\mathbb{R}^d,N}^{\max} < \infty$. *Here* $n \geq \max_{j\in\{1,\ldots,d\}}\left(T_j + |x_j|, T_j^{-\frac{1}{\alpha}}\right)$, *where* $x \in \mathbb{R}^d$, $0 < \alpha < 1$, $n \in \mathbb{N}$, $T_j > 0$. *Then*

$$\left(K_n^*(f)\right)(x) - f(x) =$$

$$\sum_{l=1}^{N-1}\left(\sum_{|\widetilde{\alpha}|=l}\left(\frac{f_{\widetilde{\alpha}}(x)}{\prod_{j=1}^{d}\alpha_j!}\right) K_n^*\left(\prod_{j=1}^{d}(\cdot - x_j)^{\alpha_j}, x\right)\right) + o\left(\frac{1}{n^{(N-\varepsilon)(1-\alpha)}}\right), \tag{25.143}$$

where $0 < \varepsilon \leq N$.

If $N = 1$, *the sum in (25.143) collapses.*

The last (25.143) implies that

$$n^{(N-\varepsilon)(1-\alpha)}.$$

$$\left[\left(K_n^*(f)\right)(x) - f(x) - \sum_{l=1}^{N-1} \left(\sum_{|\widetilde{\alpha}|=l} \left(\frac{f_{\widetilde{\alpha}}(x)}{\prod_{j=1}^{d} \alpha_j!} \right) K_n^* \left(\prod_{j=1}^{d} (\cdot - x_j)^{\alpha_j}, x \right) \right) \right] \to 0, \tag{25.144}$$

as $n \to \infty$, $0 < \varepsilon \leq N$.

When $N = 1$, *or* $f_{\widetilde{\alpha}}(x) = 0$, *for all* $\widetilde{\alpha} : |\widetilde{\alpha}| = l = 1, \ldots, N-1$, *then we derive*

$$n^{(N-\varepsilon)(1-\alpha)} \left[\left(K_n^*(f)\right)(x) - f(x) \right] \to 0, \tag{25.145}$$

as $n \to \infty$, $0 < \varepsilon \leq N$.

Proof Set

$$g_{t+\frac{k+\lambda_i}{n+\rho_i}}(\lambda^*) = f\left(x + \lambda^* \left(t + \frac{k+\lambda_i}{n+\rho_i} - x\right)\right), \ 0 \leq \lambda^* \leq 1. \tag{25.146}$$

Then we have

$$g^{(l)}_{t+\frac{k+\lambda_i}{n+\rho_i}}(\lambda^*) =$$

$$\left[\left(\sum_{j=1}^{d} \left(t_j + \frac{k_j + \lambda_i}{n+\rho_i} - x_j \right) \frac{\partial}{\partial x_j} \right)^l f \right] \left(x + \lambda^* \left(t + \frac{k+\lambda_i}{n+\rho_i} - x \right) \right), \tag{25.147}$$

$l = 0, \ldots, N$, and

$$g_{t+\frac{k+\lambda_i}{n+\rho_i}}(0) = f(x). \tag{25.148}$$

By Taylor's formula, we get

$$f\left(t + \frac{k+\lambda_i}{n+\rho_i}\right) = g_{t+\frac{k+\lambda_i}{n+\rho_i}}(1) =$$

$$\sum_{l=0}^{N-1} \frac{g^{(l)}_{t+\frac{k+\lambda_i}{n+\rho_i}}(0)}{l!} + R_N\left(t + \frac{k+\lambda_i}{n+\rho_i}, 0\right), \tag{25.149}$$

where

$$R_N\left(t + \frac{k+\lambda_i}{n+\rho_i}, 0\right) = \frac{1}{(N-1)!} \int_0^1 (1-\theta)^{N-1} g^{(N)}_{t+\frac{k+\lambda_i}{n+\rho_i}}(\theta)\, d\theta. \tag{25.150}$$

Here we denote by

$$f_{\widetilde{\alpha}} := \frac{\partial^{\widetilde{\alpha}} f}{\partial x^{\widetilde{\alpha}}}, \widetilde{\alpha} := (\alpha_1, \ldots, \alpha_d), \alpha_j \in \mathbb{Z}^+, \ j = 1, \ldots, d, \tag{25.151}$$

such that $|\widetilde{\alpha}| = \sum_{j=1}^{d} \alpha_j = l, 0 \leq l \leq N.$

More precisely we can rewrite

$$f\left(t + \frac{k + \lambda_i}{n + \rho_i}\right) - f(x) =$$

$$\sum_{l=1}^{N-1} \sum_{\left(\substack{\widetilde{\alpha}:=(\alpha_1,\ldots,\alpha_d),\ \alpha_j \in \mathbb{Z}^+, \\ j=1,\ldots,d,\ |\widetilde{\alpha}|=\sum_{j=1}^{d} \alpha_j = l}\right)} \left(\frac{1}{\prod_{j=1}^{d} \alpha_j!}\right) \left(\prod_{j=1}^{d} \left(\left(t_j + \frac{k_j + \lambda_i}{n + \rho_i}\right) - x_j\right)^{\alpha_j}\right) f_{\widetilde{\alpha}}(x)$$

$$+ R_N\left(t + \frac{k + \lambda_i}{n + \rho_i}, 0\right), \tag{25.152}$$

where

$$R_N\left(t + \frac{k + \lambda_i}{n + \rho_i}, 0\right) = N \int_0^1 (1 - \theta)^{N-1} \cdot$$

$$\sum_{\left(\substack{\widetilde{\alpha}:=(\alpha_1,\ldots,\alpha_d),\ \alpha_j \in \mathbb{Z}^+, \\ j=1,\ldots,d,\ |\widetilde{\alpha}|=\sum_{j=1}^{d} \alpha_j = N}\right)} \left(\frac{1}{\prod_{j=1}^{d} \alpha_j!}\right) \left(\prod_{j=1}^{d} \left(\left(t_j + \frac{k_j + \lambda_i}{n + \rho_i}\right) - x_j\right)^{\alpha_j}\right) \cdot \tag{25.153}$$

$$f_{\widetilde{\alpha}}\left(x + \theta\left(\left(t + \frac{k + \lambda_i}{n + \rho_i}\right) - x\right)\right) d\theta.$$

Thus it holds

$$\sum_{i=1}^{r} w_i (n + \rho_i)^d \int_0^{\frac{1}{n+\rho_i}} f\left(t + \frac{k + \lambda_i}{n + \rho_i}\right) dt = \tag{25.154}$$

$$\sum_{l=0}^{N-1} \frac{\sum_{i=1}^{r} w_i (n+\rho_i)^d \int_0^{\frac{1}{n+\rho_i}} g^{(l)}_{t+\frac{k+\lambda_i}{n+\rho_i}} (0)\, dt}{l!} +$$

$$\sum_{i=1}^{r} w_i (n+\rho_i)^d \int_0^{\frac{1}{n+\rho_i}} R_N \left(t + \frac{k+\lambda_i}{n+\rho_i}, 0 \right) dt.$$

Hence it holds

$$\left(K_n^* (f) \right)(x) =$$

$$\frac{\sum_{k=\lceil nx-Tn^{\alpha} \rceil}^{[nx+Tn^{\alpha}]} \left(\sum_{i=1}^{r} w_i (n+\rho_i)^d \int_0^{\frac{1}{n+\rho_i}} f\left(t + \frac{k+\lambda_i}{n+\rho_i} \right) dt \right) b\left(n^{1-\alpha} \left(x - \frac{k}{n} \right) \right)}{V(x)} =$$

$$\sum_{l=0}^{N-1} \frac{1}{l!} \sum_{k=\lceil nx-Tn^{\alpha} \rceil}^{[nx+Tn^{\alpha}]} \frac{b\left(n^{1-\alpha} \left(x - \frac{k}{n} \right) \right)}{V(x)} \left(\sum_{i=1}^{r} w_i (n+\rho_i)^d \int_0^{\frac{1}{n+\rho_i}} g^{(l)}_{t+\frac{k+\lambda_i}{n+\rho_i}} (0)\, dt \right) + \tag{25.155}$$

$$\sum_{k=\lceil nx-Tn^{\alpha} \rceil}^{[nx+Tn^{\alpha}]} \frac{b\left(n^{1-\alpha} \left(x - \frac{k}{n} \right) \right)}{V(x)} \left(\sum_{i=1}^{r} w_i (n+\rho_i)^d \int_0^{\frac{1}{n+\rho_i}} R_N \left(t + \frac{k+\lambda_i}{n+\rho_i}, 0 \right) dt \right).$$

So we see that

$$\left(K_n^* (f) \right)(x) - f(x) -$$

$$\sum_{l=1}^{N-1} \frac{1}{l!} \sum_{k=\lceil nx-Tn^{\alpha} \rceil}^{[nx+Tn^{\alpha}]} \frac{b\left(n^{1-\alpha} \left(x - \frac{k}{n} \right) \right)}{V(x)} \left(\sum_{i=1}^{r} w_i (n+\rho_i)^d \int_0^{\frac{1}{n+\rho_i}} g^{(l)}_{t+\frac{k+\lambda_i}{n+\rho_i}} (0)\, dt \right) \tag{25.156}$$

$$= R^* (x),$$

where

$$R^* (x) =$$

$$\sum_{k=\lceil nx-Tn^{\alpha} \rceil}^{[nx+Tn^{\alpha}]} \frac{b\left(n^{1-\alpha} \left(x - \frac{k}{n} \right) \right)}{V(x)} \left(\sum_{i=1}^{r} w_i (n+\rho_i)^d \int_0^{\frac{1}{n+\rho_i}} R_N \left(t + \frac{k+\lambda_i}{n+\rho_i}, 0 \right) dt \right). \tag{25.157}$$

By (25.115) we get that

$$\left\| g^{(N)}_{t+\frac{k+\lambda_i}{n+\rho_i}} \right\|_{\infty,[0,1]} \leq \left(\left\| t + \frac{k+\lambda_i}{n+\rho_i} - x \right\|_{l_1} \right)^N \| f_{\widetilde{\alpha}} \|^{\max}_{\infty,\mathbb{R}^d,N} < \infty. \tag{25.158}$$

And furthermore it holds

$$\left|R_N\left(t+\frac{k+\lambda_i}{n+\rho_i},0\right)\right| \leq$$

$$\frac{\left(\left\|t+\frac{k+\lambda_i}{n+\rho_i}-x\right\|_{l_1}\right)^N}{N!}\|f_{\widetilde{\alpha}}\|_{\infty,\mathbb{R}^d,N}^{\max}. \tag{25.159}$$

By (25.103) now we obtain

$$\left\|t+\frac{k+\lambda_i}{n+\rho_i}-x\right\|_{l_1} \leq \tag{25.160}$$

$$d\left[\left(\frac{\rho_i\|x\|_\infty+\lambda_i+1}{n+\rho_i}\right)+\left(1+\frac{\rho_i}{n+\rho_i}\right)\frac{T^*}{n^{1-\alpha}}\right],$$

and thus we have

$$\left|R_N\left(t+\frac{k+\lambda_i}{n+\rho_i},0\right)\right| \leq \left(\frac{d^N}{N!}\|f_{\widetilde{\alpha}}\|_{\infty,\mathbb{R}^d,N}^{\max}\right)\cdot$$

$$\left[\left(\frac{\rho_i\|x\|_\infty+\lambda_i+1}{n+\rho_i}\right)+\left(1+\frac{\rho_i}{n+\rho_i}\right)\frac{T^*}{n^{1-\alpha}}\right]^N. \tag{25.161}$$

Clearly now we can infer

$$\left|R^*(x)\right| \leq \left(\frac{d^N}{N!}\|f_{\widetilde{\alpha}}\|_{\infty,\mathbb{R}^d,N}^{\max}\right)\cdot$$

$$\sum_{i=1}^{r} w_i\left[\left(\frac{\rho_i\|x\|_\infty+\lambda_i+1}{n+\rho_i}\right)+\left(1+\frac{\rho_i}{n+\rho_i}\right)\frac{T^*}{n^{1-\alpha}}\right]^N \leq \tag{25.162}$$

$$\left(\frac{d^N}{N!}\|f_{\widetilde{\alpha}}\|_{\infty,\mathbb{R}^d,N}^{\max}\right)\cdot$$

$$\max_{\left(\substack{\text{all } i=(i_1,\ldots,i_d):\\ i_j=1,\ldots,r_j;\ j=1,\ldots,d}\right)_i}\left[\left(\frac{\rho_i\|x\|_\infty+\lambda_i+1}{n+\rho_i}\right)+\left(1+\frac{\rho_i}{n+\rho_i}\right)\frac{T^*}{n^{1-\alpha}}\right]^N \leq$$

$$\left(\frac{d^N}{N!}\|f_{\widetilde{\alpha}}\|_{\infty,\mathbb{R}^d,N}^{\max}\right)\cdot$$

$$\max_i \left[\left(\frac{\rho_i \|x\|_\infty + \lambda_i + 1}{n^{1-\alpha} n^{\alpha}} \right) + \left(1 + \frac{\rho_i}{n}\right) \frac{T^*}{n^{1-\alpha}} \right]^N = \tag{25.163}$$

$$\left(\frac{d^N}{N!} \| f_{\widetilde{\alpha}} \|_{\infty, \mathbb{R}^d, N}^{\max} \right).$$

$$\max_i \left[\left(\frac{\rho_i \|x\|_\infty + \lambda_i + 1}{n^{\alpha}} \right) + \left(1 + \frac{\rho_i}{n}\right) T^* \right]^N \frac{1}{n^{(1-\alpha)N}} \leq$$

$$\left(\frac{d^N}{N!} \| f_{\widetilde{\alpha}} \|_{\infty, \mathbb{R}^d, N}^{\max} \right).$$

$$\max_i \left[(\rho_i \|x\|_\infty + \lambda_i + 1) + (1 + \rho_i) T^* \right]^N \frac{1}{n^{(1-\alpha)N}} =: \frac{D}{n^{(1-\alpha)N}}. \tag{25.164}$$

That is we proved

$$\left| R^* (x) \right| = O \left(\frac{1}{n^{(1-\alpha)N}} \right), \tag{25.165}$$

and

$$\left| R^* (x) \right| = o (1). \tag{25.166}$$

And, letting $0 < \varepsilon \leq N$, we derive

$$\frac{|R^* (x)|}{\left(\frac{1}{n^{(N-\varepsilon)(1-\alpha)}} \right)} \leq \frac{D}{n^{(1-\alpha)N}} n^{(N-\varepsilon)(1-\alpha)} = \frac{D}{n^{\varepsilon(1-\alpha)}} \to 0, \tag{25.167}$$

as $n \to \infty$.

I.e.

$$\left| R^* (x) \right| = o \left(\frac{1}{n^{(N-\varepsilon)(1-\alpha)}} \right). \tag{25.168}$$

Clearly here we can rewrite (25.156), as

$$\left(K_n^* (f) \right) (x) - f (x) - \sum_{l=1}^{N-1} \left(\sum_{|\widetilde{\alpha}|=l} \left(\frac{f_{\widetilde{\alpha}}(x)}{\prod_{j=1}^{d} \alpha_j!} \right) K_n^* \left(\prod_{j=1}^{d} (\cdot - x_j)^{\alpha_j}, x \right) \right) = R^* (x). \tag{25.169}$$

Based on (25.168) and (25.169) we derive (25.143). ■

We finish with

Theorem 25.15 *Let* $f \in AC^N(\mathbb{R}^d) \cap C^{N-1}(\mathbb{R}^d)$, $d \in \mathbb{N} - \{1\}$, $N \in \mathbb{N}$, *with* $\| f_{\widetilde{\alpha}} \|_{\infty,\mathbb{R}^d,N}^{\max} < \infty$. *Here* $n \geq \max_{j\in\{1,\ldots,d\}} \left(T_j + |x_j|, T_j^{-\frac{1}{\alpha}} \right)$, *where* $x \in \mathbb{R}^d$, $0 < \alpha < 1$, $n \in \mathbb{N}$, $T_j > 0$. *Then*

$$\left(M_n^*(f)\right)(x) - f(x) =$$

$$\sum_{l=1}^{N-1} \left(\sum_{|\widetilde{\alpha}|=l} \left(\frac{f_{\widetilde{\alpha}}(x)}{\prod_{j=1}^{d} \alpha_j!} \right) M_n^* \left(\prod_{j=1}^{d} (\cdot - x_j)^{\alpha_j}, x \right) \right) + o\left(\frac{1}{n^{(N-\varepsilon)(1-\alpha)}} \right), \tag{25.170}$$

where $0 < \varepsilon \leq N$.

If $N = 1$, *the sum in (25.170) collapses.*

The last (25.170) implies that

$$n^{(N-\varepsilon)(1-\alpha)}.$$

$$\left[\left(M_n^*(f)\right)(x) - f(x) - \sum_{l=1}^{N-1} \left(\sum_{|\widetilde{\alpha}|=l} \left(\frac{f_{\widetilde{\alpha}}(x)}{\prod_{j=1}^{d} \alpha_j!} \right) M_n^* \left(\prod_{j=1}^{d} (\cdot - x_j)^{\alpha_j}, x \right) \right) \right] \to 0, \tag{25.171}$$

as $n \to \infty$, $0 < \varepsilon \leq N$.

When $N = 1$, *or* $f_{\widetilde{\alpha}}(x) = 0$, *for all* $\widetilde{\alpha} : |\widetilde{\alpha}| = l = 1, \ldots, N-1$, *then we derive*

$$n^{(N-\varepsilon)(1-\alpha)} \left[\left(M_n^*(f)\right)(x) - f(x) \right] \to 0, \tag{25.172}$$

as $n \to \infty$, $0 < \varepsilon \leq N$.

Proof Set

$$g_{\frac{k}{n}+\frac{i}{nr}}(t) = f\left(x + t\left(\frac{k}{n} + \frac{i}{nr} - x \right) \right), \quad 0 \leq t \leq 1. \tag{25.173}$$

Then we have

$$g_{\frac{k}{n}+\frac{i}{nr}}^{(l)}(t) = \left[\left(\sum_{j=1}^{d} \left(\frac{k_j}{n} + \frac{i_j}{nr_j} - x_j \right) \frac{\partial}{\partial x_j} \right)^l f \right] \left(x + t\left(\frac{k}{n} + \frac{i}{nr} - x \right) \right), \tag{25.174}$$

$l = 0, \ldots, N$, and

$$g_{\frac{k}{n}+\frac{i}{nr}}(0) = f(x). \tag{25.175}$$

By Taylor's formula, we get

$$f\left(\frac{k}{n}+\frac{i}{nr}\right)=g_{\frac{k}{n}+\frac{i}{nr}}(1)=$$

$$\sum_{l=0}^{N-1}\frac{g^{(l)}_{\frac{k}{n}+\frac{i}{nr}}(0)}{l!}+R_N\left(\frac{k}{n}+\frac{i}{nr},0\right), \tag{25.176}$$

where

$$R_N\left(\frac{k}{n}+\frac{i}{nr},0\right)=\frac{1}{(N-1)!}\int_0^1(1-\theta)^N g^{(N)}_{\frac{k}{n}+\frac{i}{nr}}(\theta)\,d\theta. \tag{25.177}$$

Here we denote by

$$f_{\tilde{\alpha}}:=\frac{\partial^{\tilde{\alpha}}f}{\partial x^{\tilde{\alpha}}},\ \tilde{\alpha}:=(\alpha_1,\ldots,\alpha_d),\ \alpha_j\in\mathbb{Z}^+,\ \ j=1,\ldots,d, \tag{25.178}$$

such that $|\tilde{\alpha}|=\sum_{j=1}^{d}\alpha_j=l, 0\le l\le N.$

More precisely we can rewrite

$$f\left(\frac{k}{n}+\frac{i}{nr}\right)-f(x)=$$

$$\sum_{l=1}^{N-1}\sum_{\tilde{\alpha}:|\tilde{\alpha}|=l}\left(\frac{f_{\tilde{\alpha}}(x)}{\prod_{j=1}^{d}\alpha_j!}\right)\left(\prod_{j=1}^{d}\left(\frac{k_j}{n}+\frac{i_j}{nr_j}-x_j\right)^{\alpha_j}\right)$$

$$+R_N\left(\frac{k}{n}+\frac{i}{nr},0\right), \tag{25.179}$$

where

$$R_N\left(\frac{k}{n}+\frac{i}{nr},0\right)=N\int_0^1(1-\theta)^{N-1}\cdot$$

$$\sum_{\tilde{\alpha}:|\tilde{\alpha}|=N}\left(\frac{1}{\prod_{j=1}^{d}\alpha_j!}\right)\left(\prod_{j=1}^{d}\left(\frac{k_j}{n}+\frac{i_j}{nr_j}-x_j\right)^{\alpha_j}\right)\cdot \tag{25.180}$$

$$f_{\widetilde{\alpha}}\left(x+\theta\left(\frac{k}{n}+\frac{i}{nr}-x\right)\right)d\theta.$$

Thus

$$\left(M_n^*(f)\right)(x)=\frac{\sum_{k=\lceil nx-Tn^{\alpha}\rceil}^{[nx+Tn^{\alpha}]}\left(\sum_{i=1}^{r}w_i f\left(\frac{k}{n}+\frac{i}{nr}\right)\right)b\left(n^{1-\alpha}\left(x-\frac{k}{n}\right)\right)}{V(x)}=$$

$$\frac{\sum_{l=0}^{N-1}\frac{1}{l!}\sum_{k=\lceil nx-Tn^{\alpha}\rceil}^{[nx+Tn^{\alpha}]}\left(\sum_{i=1}^{r}w_i g_{\frac{k}{n}+\frac{i}{nr}}^{(l)}(0)\right)b\left(n^{1-\alpha}\left(x-\frac{k}{n}\right)\right)}{V(x)}+ \tag{25.181}$$

$$\frac{\sum_{k=\lceil nx-Tn^{\alpha}\rceil}^{[nx+Tn^{\alpha}]}\left(\sum_{i=1}^{r}w_i R_N\left(\frac{k}{n}+\frac{i}{nr},0\right)\right)b\left(n^{1-\alpha}\left(x-\frac{k}{n}\right)\right)}{V(x)}.$$

Therefore it holds

$$\left(M_n^*(f)\right)(x)-f(x)-$$

$$\sum_{l=1}^{N-1}\frac{1}{l!}\frac{\sum_{k=\lceil nx-Tn^{\alpha}\rceil}^{[nx+Tn^{\alpha}]}b\left(n^{1-\alpha}\left(x-\frac{k}{n}\right)\right)}{V(x)}\left(\sum_{i=1}^{r}w_i g_{\frac{k}{n}+\frac{i}{nr}}^{(l)}(0)\right)=R^*(x), \tag{25.182}$$

where

$$R^*(x)=\sum_{k=\lceil nx-Tn^{\alpha}\rceil}^{[nx+Tn^{\alpha}]}\frac{b\left(n^{1-\alpha}\left(x-\frac{k}{n}\right)\right)}{V(x)}\left(\sum_{i=1}^{r}w_i R_N\left(\frac{k}{n}+\frac{i}{nr},0\right)\right). \tag{25.183}$$

By (25.115) we get that

$$\left\|g_{\frac{k}{n}+\frac{i}{nr}}^{(N)}\right\|_{\infty,[0,1]}\leq\left(\left\|\frac{k}{n}+\frac{i}{nr}-x\right\|_{l_1}\right)^N\|f_{\widetilde{\alpha}}\|_{\infty,\mathbb{R}^d,N}^{\max}<\infty. \tag{25.184}$$

And furthermore it holds

$$\left|R_N\left(\frac{k}{n}+\frac{i}{nr},0\right)\right|\leq$$

$$\frac{\left(\left\|\frac{k}{n}+\frac{i}{nr}-x\right\|_{l_1}\right)^N}{N!}\|f_{\widetilde{\alpha}}\|_{\infty,\mathbb{R}^d,N}^{\max}. \tag{25.185}$$

By (25.104) we obtain

$$\left\| \frac{k}{n} + \frac{i}{nr} - x \right\|_{l_1} \leq d \left(\frac{T^*}{n^{1-\alpha}} + \frac{1}{n} \right), \tag{25.186}$$

and thus we have

$$\left| R_N \left(\frac{k}{n} + \frac{i}{nr}, 0 \right) \right| \leq \left(\frac{d^N}{N!} \left\| f_{\widetilde{\alpha}} \right\|_{\infty, \mathbb{R}^d, N}^{\max} \right) \left(\frac{T^*}{n^{1-\alpha}} + \frac{1}{n} \right)^N. \tag{25.187}$$

Clearly we get also that

$$\left| R^* (x) \right| \leq \left(\frac{d^N}{N!} \left\| f_{\widetilde{\alpha}} \right\|_{\infty, \mathbb{R}^d, N}^{\max} \right) \left(\frac{T^*}{n^{1-\alpha}} + \frac{1}{n^{1-\alpha} n^{\alpha}} \right)^N = \tag{25.188}$$

$$\left(\frac{d^N}{N!} \left\| f_{\widetilde{\alpha}} \right\|_{\infty, \mathbb{R}^d, N}^{\max} \right) \left(T^* + \frac{1}{n^{\alpha}} \right)^N \frac{1}{n^{(1-\alpha)N}} \leq$$

$$\left[\left(\frac{d^N}{N!} \left\| f_{\widetilde{\alpha}} \right\|_{\infty, \mathbb{R}^d, N}^{\max} \right) \left(T^* + 1 \right)^N \right] \frac{1}{n^{(1-\alpha)N}}. \tag{25.189}$$

That is we find

$$\left| R^* (x) \right| \leq \left[\left(\frac{d^N}{N!} \left\| f_{\widetilde{\alpha}} \right\|_{\infty, \mathbb{R}^d, N}^{\max} \right) \left(1 + T^* \right)^N \right] \frac{1}{n^{(1-\alpha)N}} =: \frac{E}{n^{(1-\alpha)N}}. \tag{25.190}$$

That is we proved

$$\left| R^* (x) \right| = O \left(\frac{1}{n^{(1-\alpha)N}} \right), \tag{25.191}$$

and

$$\left| R^* (x) \right| = o (1). \tag{25.192}$$

And, letting $0 < \varepsilon \leq N$, we derive

$$\frac{\left| R^* (x) \right|}{\left(\frac{1}{n^{(N-\varepsilon)(1-\alpha)}} \right)} \leq \frac{E}{n^{(1-\alpha)N}} n^{(N-\varepsilon)(1-\alpha)} = \frac{E}{n^{\varepsilon(1-\alpha)}} \to 0, \tag{25.193}$$

as $n \to \infty$.

I.e.

$$\left| R^* (x) \right| = o \left(\frac{1}{n^{(N-\varepsilon)(1-\alpha)}} \right). \tag{25.194}$$

Clearly here we can rewrite (25.182), as

$$\left(M_n^*(f)\right)(x) - f(x) - \sum_{l=1}^{N-1} \left(\sum_{\widetilde{\alpha}:|\widetilde{\alpha}|=l} \left(\frac{f_{\widetilde{\alpha}}(x)}{\prod_{j=1}^{d} \alpha_j!} \right) M_n^* \left(\prod_{j=1}^{d} (\cdot - x_j)^{\alpha_j}, x \right) \right) = R^*(x). \tag{25.195}$$

Based on (25.194) and (25.195) we derive (25.170). ■

References

1. G.A. Anastassiou, Rate of convergence of some neural network operators to the unit-univariate case. J. Math. Anal. Appl. **212**, 237–262 (1997)
2. G.A. Anastassiou, Rate of convergence of some multivariate neural network operators to the unit. J. Comp. Math. Appl. **40**, 1–19 (2000)
3. G.A. Anastassiou, *Quantitative Approximations* (Chapman & Hall/CRC, Boca Raton, New York, 2001)
4. G.A. Anastassiou, *Advanced Inequalities* (World Scientific Publishing Company, Singapore, New Jersey, 2011)
5. G.A. Anastassiou, Multivariate Voronovskaya type asymptotic expansions for normalized bell and squashing type neural network operators. Neural Parallel Sci. Comput. **20**, 1–10 (2012)
6. G.A. Anastassiou, Rate of convergence of some neural network operators to the unit-univariate case, revisited. Vesnik **65**(4), 511–518 (2013)
7. G.A. Anastassiou, Fractional Voronovskaya type asymptotic expansions for bell and squashing type neural network operators. J. Comput. Anal. Appl. **15**(7), 1231–1239 (2013)
8. G.A. Anastassiou, Rate of convergence of some multivariate neural network operators to the unit, revisited. J. Comput. Anal. Appl. **15**(7), 1300–1309 (2013)
9. G.A. Anastassiou, Voronovskaya type asymptotic expansions for perturbed neural network operators. New Math. Nat. Comput. Accepted 2014
10. G.A. Anastassiou, Approximation by perturbed neural network operators. Submitted 2014
11. G.A. Anastassiou, Approximations by multivariate perturbed neural network operators. Submitted 2014
12. P. Cardaliaguet, G. Euvrard, Approximation of a function and its derivative with a neural network. Neural Netw. **5**, 207–220 (1992)
13. Z. Chen, F. Cao, The approximation operators with sigmoidal functions. Comput. Math. Appl. **58**, 758–765 (2009)
14. D. Costarelli, R. Spigler, Approximation results for neural network operators activated by sigmoidal functions. Neural Netw. **44**, 101–106 (2013)
15. D. Costarelli, R. Spigler, Multivariate neural network operators with sigmoidal activation functions. Neural Netw. **48**, 72–77 (2013)
16. S. Haykin, *Neural Networks: A Comprehensive Foundation*, 2nd edn. (Prentice Hall, New York, 1998)
17. W. McCulloch, W. Pitts, A logical calculus of the ideas immanent in nervous activity. Bull. Math. Biophys. **7**, 115–133 (1943)
18. T.M. Mitchell, *Machine Learning* (WCB-McGraw-Hill, New York, 1997)
19. D.D. Stancu, On a generalization of the Bernstein polynomials. Studia Universitatis Babeş-Bolyai, Series Mathematica-Physica **14**, 31–45 (1969)

Chapter 26
Approximation Using Fuzzy Perturbed Neural Networks

This chapter deals with the determination of the rate of convergence to the unit of each of three newly introduced here fuzzy perturbed normalized neural network operators of one hidden layer. These are given through the fuzzy modulus of continuity of the involved fuzzy number valued function or its high order fuzzy derivative and that appears in the right-hand side of the associated fuzzy Jackson type inequalities. The activation function is very general, especially it can derive from any sigmoid or bell-shaped function. The right hand sides of our fuzzy convergence inequalities do not depend on the activation function. The sample fuzzy functionals are of Stancu, Kantorovich and Quadrature types. We give applications for the first fuzzy derivative of the involved function. It follows [14].

26.1 Introduction

The Cardaliaguet-Euvrard real neural network operators were studied extensively in [15], where the authors among many other things proved that these operators converge uniformly on compacta, to the unit over continuous and bounded functions. Our fuzzy "perturbed normalized neural network operators" are motivated and inspired by the "bell" and "squashing functions" of [15]. The work in [15] is qualitative where the used bell-shaped function is general. However, our work, though greatly motivated by [15], is quantitative and the used activation functions are of compact support. Here we extend to fuzzy level our initial real work, see [12]. We derive a series of fuzzy Jackson type inequalities giving close upper bounds to the errors in approximating the unit operator by the above fuzzy perturbed neural network induced operators. All involved constants there are well determined. These are pointwise and uniform estimates involving the first fuzzy modulus of continuity of the engaged fuzzy function or the fuzzy derivative of the function under approximation. We give all necessary background of the fuzzy calculus needed.

G.A. Anastassiou, *Intelligent Systems II: Complete Approximation by Neural Network Operators*, Studies in Computational Intelligence 608,
DOI 10.1007/978-3-319-20505-2_26

Initial work of the subject was done in [10, 11]. These works motivated the current chapter.

26.2 Fuzzy Mathematical Analysis Background

We need the following basic background

Definition 26.1 (*see* [21]) Let $\mu : \mathbb{R} \to [0, 1]$ with the following properties:

(i) is normal, i.e., $\exists\, x_0 \in \mathbb{R} : \mu(x_0) = 1$.
(ii) $\mu(\lambda x + (1 - \lambda) y) \geq \min\{\mu(x), \mu(y)\}$, $\forall\, x, y \in \mathbb{R}$, $\forall\, \lambda \in [0, 1]$ (μ is called a convex fuzzy subset).
(iii) μ is upper semicontinuous on $\mathbb{R}$, i.e., $\forall\, x_0 \in \mathbb{R}$ and $\forall\, \varepsilon > 0$, $\exists$ neighborhood $V(x_0) : \mu(x) \leq \mu(x_0) + \varepsilon$, $\forall\, x \in V(x_0)$.
(iv) the set $\overline{\text{supp}(\mu)}$ is compact in $\mathbb{R}$ (where $\text{supp}(\mu) := \{x \in \mathbb{R} : \mu(x) > 0\}$).
We call μ a fuzzy real number. Denote the set of all μ with $\mathbb{R}_{\mathcal{F}}$.
E.g., $\chi_{\{x_0\}} \in \mathbb{R}_{\mathcal{F}}$, for any $x_0 \in \mathbb{R}$, where $\chi_{\{x_0\}}$ is the characteristic function at x_0.

For $0 < r \leq 1$ and $\mu \in \mathbb{R}_{\mathcal{F}}$ define $[\mu]^r := \{x \in \mathbb{R} : \mu(x) \geq r\}$ and $[\mu]^0 := \overline{\{x \in \mathbb{R} : \mu(x) \geq 0\}}$.

Then it is well known that for each $r \in [0, 1]$, $[\mu]^r$ is a closed and bounded interval of $\mathbb{R}$ [17].

For $u, v \in \mathbb{R}_{\mathcal{F}}$ and $\lambda \in \mathbb{R}$, we define uniquely the sum $u \oplus v$ and the product $\lambda \odot u$ by

$$[u \oplus v]^r = [u]^r + [v]^r, \quad [\lambda \odot u]^r = \lambda [u]^r, \ \forall r \in [0, 1],$$

where

$[u]^r + [v]^r$ means the usual addition of two intervals (as subsets of $\mathbb{R}$) and $\lambda [u]^r$ means the usual product between a scalar and a subset of $\mathbb{R}$ (see, e.g., [21]).

Notice $1 \odot u = u$ and it holds

$$u \oplus v = v \oplus u, \ \lambda \odot u = u \odot \lambda.$$

If $0 \leq r_1 \leq r_2 \leq 1$ then $[u]^{r_2} \subseteq [u]^{r_1}$. Actually $[u]^r = \left[u_-^{(r)}, u_+^{(r)}\right]$, where $u_-^{(r)} \leq u_+^{(r)}$, $u_-^{(r)}, u_+^{(r)} \in \mathbb{R}$, $\forall\, r \in [0, 1]$.

For $\lambda > 0$ one has $\lambda u_{\pm}^{(r)} = (\lambda \odot u)_{\pm}^{(r)}$, respectively.

Define

$$D : \mathbb{R}_{\mathcal{F}} \times \mathbb{R}_{\mathcal{F}} \to \mathbb{R}_+$$

by

$$D(u, v) := \sup_{r \in [0,1]} \max \left\{ \left| u_-^{(r)} - v_-^{(r)} \right|, \left| u_+^{(r)} - v_+^{(r)} \right| \right\},$$

where

$$[v]^r = \left[v_-^{(r)}, v_+^{(r)}\right]; \ u, v \in \mathbb{R}_{\mathcal{F}}.$$

We have that D is a metric on $\mathbb{R}_{\mathcal{F}}$.

Then $(\mathbb{R}_{\mathcal{F}}, D)$ is a complete metric space, see [21, 22].

Here $\sum^{*}$ stands for fuzzy summation and $\widetilde{0} := \chi_{\{0\}} \in \mathbb{R}_{\mathcal{F}}$ is the neural element with respect to $\oplus$, i.e.,

$$u \oplus \widetilde{0} = \widetilde{0} \oplus u = u, \forall u \in \mathbb{R}_{\mathcal{F}}.$$

Denote

$$D^* (f, g) := \sup_{x \in X \subseteq \mathbb{R}} D(f, g),$$

where $f, g : X \to \mathbb{R}_{\mathcal{F}}$.

We mention

Definition 26.2 Let $f : X \subseteq \mathbb{R} \to \mathbb{R}_{\mathcal{F}}$, X interval, we define the (first) fuzzy modulus of continuity of f by

$$\omega_1^{(\mathcal{F})} (f, \delta)_X = \sup_{x,y \in X, |x-y| \le \delta} D(f(x), f(y)), \delta > 0.$$

We define by $C_{\mathcal{F}}^{U}(\mathbb{R})$ the space of fuzzy uniformly continuous functions from $\mathbb{R} \to \mathbb{R}_{\mathcal{F}}$, also $C_{\mathcal{F}}(\mathbb{R})$ is the space of fuzzy continuous functions on $\mathbb{R}$, and $C_B(\mathbb{R}, \mathbb{R}_{\mathcal{F}})$ is the fuzzy continuous and bounded functions.

We mention

Proposition 26.3 ([6]) *Let $f \in C_{\mathcal{F}}^{U}(X)$. Then $\omega_1^{(\mathcal{F})}(f, \delta)_X < \infty$, for any $\delta > 0$.*

Proposition 26.4 ([6]) *It holds*

$$\lim_{\delta \to 0} \omega_1^{(\mathcal{F})} (f, \delta)_X = \omega_1^{(\mathcal{F})} (f, 0)_X = 0,$$

iff $f \in C_{\mathcal{F}}^{U}(X)$.

Proposition 26.5 ([6]) *Here $[f]^r = \left[f_-^{(r)}, f_+^{(r)}\right]$, $r \in [0, 1]$. Let $f \in C_{\mathcal{F}}(\mathbb{R})$. Then $f_\pm^{(r)}$ are equicontinuous with respect to $r \in [0, 1]$ over $\mathbb{R}$, respectively in $\pm$.*

Note 26.6 It is clear by Propositions 26.4 and 26.5, that if $f \in C_{\mathcal{F}}^{U}(\mathbb{R})$, then $f_\pm^{(r)} \in C_U(\mathbb{R})$ (uniformly continuous on $\mathbb{R}$). Clearly also if $f \in C_B(\mathbb{R}, \mathbb{R}_{\mathcal{F}})$, then $f_\pm^{(r)} \in C_B(\mathbb{R})$ (continuous and bounded functions on $\mathbb{R}$).

Proposition 26.7 *Let $f : \mathbb{R} \to \mathbb{R}_{\mathcal{F}}$. Assume that $\omega_1^{\mathcal{F}}(f,\delta)_X$, $\omega_1\left(f_-^{(r)},\delta\right)_X$, $\omega_1\left(f_+^{(r)},\delta\right)_X$ are finite for any $\delta > 0$, $r \in [0,1]$, where X any interval of $\mathbb{R}$.*
Then

$$\omega_1^{(\mathcal{F})}(f,\delta)_X = \sup_{r\in[0,1]} \max\left\{\omega_1\left(f_-^{(r)},\delta\right)_X, \omega_1\left(f_+^{(r)},\delta\right)_X\right\}.$$

Proof Similar to Proposition 14.15, p. 246 of [7]. ∎

We need

Remark 26.8 ([4]) Here $r \in [0,1]$, $x_i^{(r)}, y_i^{(r)} \in \mathbb{R}$, $i = 1,\ldots,m \in \mathbb{N}$. Suppose that

$$\sup_{r\in[0,1]} \max\left(x_i^{(r)}, y_i^{(r)}\right) \in \mathbb{R}, \text{for } i = 1,\ldots,m.$$

Then one sees easily that

$$\sup_{r\in[0,1]} \max\left(\sum_{i=1}^m x_i^{(r)}, \sum_{i=1}^m y_i^{(r)}\right) \le \sum_{i=1}^m \sup_{r\in[0,1]} \max\left(x_i^{(r)}, y_i^{(r)}\right). \tag{26.1}$$

We need

Definition 26.9 Let $x, y \in \mathbb{R}_{\mathcal{F}}$. If there exists $z \in \mathbb{R}_{\mathcal{F}} : x = y \oplus z$, then we call z the H-difference on x and y, denoted $x - y$.

Definition 26.10 ([19]) Let $T := [x_0, x_0 + \beta] \subset \mathbb{R}$, with $\beta > 0$. A function $f : T \to \mathbb{R}_{\mathcal{F}}$ is H-difference at $x \in T$ if there exists an $f'(x) \in \mathbb{R}_{\mathcal{F}}$ such that the limits (with respect to D)

$$\lim_{h\to 0+} \frac{f(x+h) - f(x)}{h}, \lim_{h\to 0+} \frac{f(x) - f(x-h)}{h} \tag{26.2}$$

exist and are equal to $f'(x)$.

We call f' the H-derivative or fuzzy derivative of f at x.

Above is assumed that the H-differences $f(x+h) - f(x)$, $f(x) - f(x-h)$ exist in $\mathbb{R}_{\mathcal{F}}$ in a neighborhood of x.

Higher order H-fuzzy derivatives are defined the obvious way, like in the real case.

We denote by $C_{\mathcal{F}}^N(\mathbb{R})$, $N \ge 1$, the space of all N-times continuously H-fuzzy differentiable functions from $\mathbb{R}$ into $\mathbb{R}_{\mathcal{F}}$.

We mention

Theorem 26.11 ([18]) *Let $f : \mathbb{R} \to \mathbb{R}_{\mathcal{F}}$ be H-fuzzy differentiable. Let $t \in \mathbb{R}$, $0 \le r \le 1$. Clearly*

$$[f(t)]^r = \left[f(t)_-^{(r)}, f(t)_+^{(r)}\right] \subseteq \mathbb{R}.$$

Then $(f(t))_\pm^{(r)}$ *are differentiable and*

$$[f'(t)]^r = \left[\left(f(t)_-^{(r)}\right)', \left(f(t)_+^{(r)}\right)'\right].$$

I.e.

$$(f')_\pm^{(r)} = \left(f_\pm^{(r)}\right)', \ \forall r \in [0, 1].$$

Remark 26.12 ([5]) Let $f \in C_{\mathcal{F}}^N(\mathbb{R})$, $N \geq 1$. Then by Theorem 26.11 we obtain

$$\left[f^{(i)}(t)\right]^r = \left[\left(f(t)_-^{(r)}\right)^{(i)}, \left(f(t)_+^{(r)}\right)^{(i)}\right],$$

for $i = 0, 1, 2, \ldots, N$, and in particular we have that

$$\left(f^{(i)}\right)_\pm^{(r)} = \left(f_\pm^{(r)}\right)^{(i)},$$

for any $r \in [0, 1]$, all $i = 0, 1, 2, \ldots, N$.

Note 26.13 ([5]) Let $f \in C_{\mathcal{F}}^N(\mathbb{R})$, $N \geq 1$. Then by Theorem 26.11 we have $f_\pm^{(r)} \in C^N(\mathbb{R})$, for any $r \in [0, 1]$.

We need also a particular case of the Fuzzy Henstock integral ($\delta(x) = \frac{\delta}{2}$), see [21].

Definition 26.14 ([16], *p.* 644) Let $f : [a, b] \to \mathbb{R}_{\mathcal{F}}$. We say that f is Fuzzy-Riemann integrable to $I \in \mathbb{R}_{\mathcal{F}}$ if for any $\varepsilon > 0$, there exists $\delta > 0$ such that for any division $P = \{[u, v]; \xi\}$ of $[a, b]$ with the norms $\Delta(P) < \delta$, we have

$$D\left(\sum_P^* (v-u) \odot f(\xi), I\right) < \varepsilon.$$

We write

$$I := (FR)\int_a^b f(x)\,dx. \tag{26.3}$$

We mention

Theorem 26.15 ([17]) *Let* $f : [a, b] \to \mathbb{R}_{\mathcal{F}}$ *be fuzzy continuous. Then*

$$(FR)\int_a^b f(x)\,dx$$

exists and belongs to $\mathbb{R}_{\mathcal{F}}$*, furthermore it holds*

$$\left[(FR)\int_a^b f(x)\,dx\right]^r = \left[\int_a^b (f)_-^{(r)}(x)\,dx,\ \int_a^b (f)_+^{(r)}(x)\,dx\right], \tag{26.4}$$

$\forall\, r \in [0, 1]$.

26.3 Real Neural Networks Approximation Basics (see [12])

Here the activation function $b : \mathbb{R} \to \mathbb{R}_+$ is of compact support $[-T, T]$, $T > 0$. That is $b(x) > 0$ for any $x \in [-T, T]$, and clearly b may have jump discontinuities. Also the shape of the graph of b could be anything. Typically in neural networks approximation we take b as a sigmoidal function or bell-shaped function, of course here of compact support $[-T, T]$, $T > 0$.

Example 26.16 (i) b can be the characteristic function on $[-1, 1]$,

(ii) b can be that hat function over $[-1, 1]$, i.e.,

$$b(x) = \begin{cases} 1 + x, & -1 \le x \le 0, \\ 1 - x, & 0 < x \le 1, \\ 0, \text{elsewhere}, \end{cases}$$

(iii) the truncated sigmoidals

$$b(x) = \begin{cases} \frac{1}{1+e^{-x}} \text{ or } \tanh x \text{ or } \operatorname{erf}(x), \text{for } x \in [-T, T], \text{ with large } T > 0, \\ 0, x \in \mathbb{R} - [-T, T], \end{cases}$$

(iv) the truncated Gompertz function

$$b(x) = \begin{cases} e^{-\gamma e^{-\beta x}}, x \in [-T, T]; \gamma, \beta > 0;\ \text{large } T > 0, \\ 0,\ x \in \mathbb{R} - [-T, T], \end{cases}$$

The Gompertz functions are also sigmoidal functions, with wide applications to many applied fields, e.g. demography and tumor growth modeling, etc.

So the general function b we will be using here covers all kinds of activation functions in neural network approximations.

In the real case we consider functions $f : \mathbb{R} \to \mathbb{R}$ that are either continuous and bounded, or uniformly continuous.

Let here the parameters $\mu, \nu \ge 0$; $\mu_i, \nu_i \ge 0$, $i = 1, \ldots, \overline{r} \in \mathbb{N}$; $w_i \ge 0$: $\sum_{i=1}^{\overline{r}} w_i = 1$; $0 < \alpha < 1$, $x \in \mathbb{R}$, $n \in \mathbb{N}$.

We use here the first modulus of continuity

$$\omega_1 (f, \delta) := \sup_{\substack{x, y \in \mathbb{R} \\ |x - y| \leq \delta}} |f (x) - f (y)| ,$$

and given that f is uniformly continuous we get $\lim_{\delta \to 0} \omega_1 (f, \delta) = 0$.

Here we mention from [12] about the pointwise convergence with rates over $\mathbb{R}$, to the unit operator, of the following one hidden layer normalized neural network perturbed operators,

(i) the Stancu type [19]

$$(H_n (f)) (x) = \frac{\sum_{k=-n^2}^{n^2} \left(\sum_{i=1}^{\overline{r}} w_i f \left(\frac{k+\mu_i}{n+\nu_i} \right) \right) b \left(n^{1-\alpha} \left(x - \frac{k}{n} \right) \right)}{\sum_{k=-n^2}^{n^2} b \left(n^{1-\alpha} \left(x - \frac{k}{n} \right) \right)}, \tag{26.5}$$

(ii) the Kantorovich type

$$(K_n (f)) (x) = \frac{\sum_{k=-n^2}^{n^2} \left(\sum_{i=1}^{\overline{r}} w_i (n + \nu_i) \int_{\frac{k+\mu_i}{n+\nu_i}}^{\frac{k+\mu_i+1}{n+\nu_i}} f (t)\, dt \right) b \left(n^{1-\alpha} \left(x - \frac{k}{n} \right) \right)}{\sum_{k=-n^2}^{n^2} b \left(n^{1-\alpha} \left(x - \frac{k}{n} \right) \right)}, \tag{26.6}$$

and
(iii) the quadrature type

$$(M_n (f)) (x) = \frac{\sum_{k=-n^2}^{n^2} \left(\sum_{i=1}^{\overline{r}} w_i f \left(\frac{k}{n} + \frac{i}{n\overline{r}} \right) \right) b \left(n^{1-\alpha} \left(x - \frac{k}{n} \right) \right)}{\sum_{k=-n^2}^{n^2} b \left(n^{1-\alpha} \left(x - \frac{k}{n} \right) \right)}. \tag{26.7}$$

Similar operators defined for bell-shaped functions and sample coefficients $f \left(\frac{k}{n} \right)$ were studied initially in [1–3, 8, 9, 15] etc.

Here we care about the generalized perturbed cases of these operators (see [12, 13]).

Operator K_n in the corresponding Signal Processing context, represents the natural called "time-jitter" error, where the sample information is calculated in a perturbed neighborhood of $\frac{k+\mu}{n+\nu}$ rather than exactly at the node $\frac{k}{n}$.

The perturbed sample coefficients $f \left(\frac{k+\mu}{n+\nu} \right)$ with $0 \leq \mu \leq \nu$, were first used by Stancu [19], in a totally different context, generalizing Bernstein operators approximation on $C ([0, 1])$.

The terms in the ratio of sums (26.5)–(26.7) are nonzero, iff

$$\left|n^{1-\alpha}\left(x-\frac{k}{n}\right)\right| \leq T, \text{ i.e. } \left|x-\frac{k}{n}\right| \leq \frac{T}{n^{1-\alpha}} \tag{26.8}$$

iff

$$nx - Tn^{\alpha} \leq k \leq nx + Tn^{\alpha}. \tag{26.9}$$

In order to have the desired order of the numbers

$$-n^2 \leq nx - Tn^{\alpha} \leq nx + Tn^{\alpha} \leq n^2, \tag{26.10}$$

it is sufficiently enough to assume that

$$n \geq T + |x|. \tag{26.11}$$

When $x \in [-T, T]$ it is enough to assume $n \geq 2T$, which implies (26.10).

Proposition 26.17 ([1]) *Let $a \leq b$, $a, b \in \mathbb{R}$. Let $card(k)$ (≥ 0) be the maximum number of integers contained in $[a, b]$. Then*

$$\max(0, (b-a)-1) \leq card(k) \leq (b-a)+1. \tag{26.12}$$

Note 26.18 We would like to establish a lower bound on $card(k)$ over the interval $[nx - Tn^{\alpha}, nx + Tn^{\alpha}]$. From Proposition 26.17 we get that

$$card(k) \geq \max\left(2Tn^{\alpha} - 1, 0\right). \tag{26.13}$$

We obtain $card(k) \geq 1$, if

$$2Tn^{\alpha} - 1 \geq 1 \text{ iff } n \geq T^{-\frac{1}{\alpha}}. \tag{26.14}$$

So to have the desired order (26.10) and $card(k) \geq 1$ over $[nx - Tn^{\alpha}, nx + Tn^{\alpha}]$, we need to consider

$$n \geq \max\left(T + |x|, T^{-\frac{1}{\alpha}}\right). \tag{26.15}$$

Also notice that $card(k) \to +\infty$, as $n \to +\infty$.

Denote by $[\cdot]$ the integral part of a number and by $\lceil\cdot\rceil$ its ceiling.

So under assumption (26.15), the operators H_n, K_n, M_n, collapse to

(i)

$$(H_n(f))(x) = \frac{\sum_{k=\lceil nx-Tn^{\alpha}\rceil}^{[nx+Tn^{\alpha}]}\left(\sum_{i=1}^{\overline{r}} w_i f\left(\frac{k+\mu_i}{n+\nu_i}\right)\right) b\left(n^{1-\alpha}\left(x-\frac{k}{n}\right)\right)}{\sum_{k=\lceil nx-Tn^{\alpha}\rceil}^{[nx+Tn^{\alpha}]} b\left(n^{1-\alpha}\left(x-\frac{k}{n}\right)\right)}, \tag{26.16}$$

(ii)

$$(K_n(f))(x) = \frac{\sum_{k=\lceil nx-Tn^{\alpha}\rceil}^{[nx+Tn^{\alpha}]} \left(\sum_{i=1}^{\bar{r}} w_i (n+\nu_i) \int_{\frac{k+\mu_i}{n+\nu_i}}^{\frac{k+\mu_i+1}{n+\nu_i}} f(t)\, dt \right) b\left(n^{1-\alpha}\left(x-\frac{k}{n}\right)\right)}{\sum_{k=\lceil nx-Tn^{\alpha}\rceil}^{[nx+Tn^{\alpha}]} b\left(n^{1-\alpha}\left(x-\frac{k}{n}\right)\right)}, \tag{26.17}$$

and
(iii)

$$(M_n(f))(x) = \frac{\sum_{k=\lceil nx-Tn^{\alpha}\rceil}^{[nx+Tn^{\alpha}]} \left(\sum_{i=1}^{\bar{r}} w_i f\left(\frac{k}{n}+\frac{i}{n\bar{r}}\right) \right) b\left(n^{1-\alpha}\left(x-\frac{k}{n}\right)\right)}{\sum_{k=\lceil nx-Tn^{\alpha}\rceil}^{[nx+Tn^{\alpha}]} b\left(n^{1-\alpha}\left(x-\frac{k}{n}\right)\right)}. \tag{26.18}$$

From [12] we will mention and use the following results.

Theorem 26.19 *Let $x \in \mathbb{R}$, $T > 0$ and $n \in \mathbb{N}$ such that $n \geq \max\left(T + |x|, T^{-\frac{1}{\alpha}}\right)$. Then*

$$|(H_n(f))(x) - f(x)| \leq \sum_{i=1}^{\bar{r}} w_i \omega_1 \left(f, \left(\frac{\nu_i |x| + \mu_i}{n+\nu_i}\right) + \left(1 + \frac{\nu_i}{n+\nu_i}\right)\frac{T}{n^{1-\alpha}}\right) \tag{26.19}$$
$$\leq \max_{i\in\{1,\dots,\bar{r}\}} \left\{ \omega_1 \left(f, \left(\frac{\nu_i |x| + \mu_i}{n+\nu_i}\right) + \left(1 + \frac{\nu_i}{n+\nu_i}\right)\frac{T}{n^{1-\alpha}}\right)\right\}.$$

Corollary 26.20 *Let $x \in [-T^*, T^*]$, $T^* > 0$, $n \in \mathbb{N} : n \geq \max\left(T + T^*, T^{-\frac{1}{\alpha}}\right)$, $T > 0$. Then*

$$\|H_n(f) - f\|_{\infty,[-T^*,T^*]} \leq \sum_{i=1}^{\bar{r}} w_i \omega_1 \left(f, \left(\frac{\nu_i T^* + \mu_i}{n+\nu_i}\right) + \left(1 + \frac{\nu_i}{n+\nu_i}\right)\frac{T}{n^{1-\alpha}}\right) \tag{26.20}$$
$$\leq \max_{i\in\{1,\dots,\bar{r}\}} \left\{ \omega_1 \left(f, \left(\frac{\nu_i T^* + \mu_i}{n+\nu_i}\right) + \left(1 + \frac{\nu_i}{n+\nu_i}\right)\frac{T}{n^{1-\alpha}}\right)\right\}.$$

Theorem 26.21 *Let $x \in \mathbb{R}$, $T > 0$ and $n \in \mathbb{N}$ such that $n \geq \max\left(T + |x|, T^{-\frac{1}{\alpha}}\right)$. Then*

$$|(K_n(f))(x) - f(x)| \leq \tag{26.21}$$
$$\max_{i\in\{1,\dots,\bar{r}\}} \left\{ \omega_1 \left(f, \left(\frac{\nu_i |x| + \mu_i + 1}{n+\nu_i}\right) + \left(1 + \frac{\nu_i}{n+\nu_i}\right)\frac{T}{n^{1-\alpha}}\right)\right\}.$$

Corollary 26.22 *Let $x \in \left[-T^*, T^*\right]$, $T^* > 0$, $n \in \mathbb{N} : n \geq \max\left(T + T^*, T^{-\frac{1}{\alpha}}\right)$, $T > 0$. Then*

$$\|K_n(f) - f\|_{\infty,[-T^*,T^*]} \leq \tag{26.22}$$

$$\max_{i\in\{1,\ldots,\overline{r}\}} \left\{\omega_1\left(f, \left(\frac{\nu_i T^* + \mu_i + 1}{n + \nu_i}\right) + \left(1 + \frac{\nu_i}{n+\nu_i}\right)\frac{T}{n^{1-\alpha}}\right)\right\}.$$

Theorem 26.23 *Let $x \in \mathbb{R}$, $T > 0$ and $n \in \mathbb{N}$ such thatn $\geq \max\left(T + |x|, T^{-\frac{1}{\alpha}}\right)$. Then*

$$|M_n(f)(x) - f(x)| \leq \omega_1\left(f, \frac{1}{n} + \frac{T}{n^{1-\alpha}}\right). \tag{26.23}$$

Corollary 26.24 *Let $x \in \left[-T^*, T^*\right]$, $T^* > 0$, $n \in \mathbb{N} : n \geq \max\left(T + T^*, T^{-\frac{1}{\alpha}}\right)$, $T > 0$. Then*

$$\|M_n(f) - f\|_{\infty,[-T^*,T^*]} \leq \omega_1\left(f, \frac{1}{n} + \frac{T}{n^{1-\alpha}}\right). \tag{26.24}$$

In (26.19)–(26.24) all converges are at the speed $\frac{1}{n^{1-\alpha}}$.

Again from [12], taking into account the differentiation order of f we obtain:

Theorem 26.25 *Let $x \in \mathbb{R}$, $T > 0$ and $n \in \mathbb{N}$ such that $n \geq \max\left(T + |x|, T^{-\frac{1}{\alpha}}\right)$, $0 < \alpha < 1$. Let $f \in C^N(\mathbb{R})$, $N \in \mathbb{N}$, such that $f^{(N)}$ is uniformly continuous or is continuous and bounded. Then*

$$|(H_n(f))(x) - f(x)| \leq$$

$$\sum_{j=1}^{N} \frac{\left|f^{(j)}(x)\right|}{j!} \left\{\sum_{i=1}^{\overline{r}} w_i \left[\left(\frac{\nu_i |x| + \mu_i}{n+\nu_i}\right) + \left(1 + \frac{\nu_i}{n+\nu_i}\right)\frac{T}{n^{1-\alpha}}\right]^j\right\} + \tag{26.25}$$

$$\sum_{i=1}^{\overline{r}} w_i \omega_1\left(f^{(N)}, \left(\frac{\nu_i |x| + \mu_i}{n+\nu_i}\right) + \left(1 + \frac{\nu_i}{n+\nu_i}\right)\frac{T}{n^{1-\alpha}}\right) \cdot$$

$$\frac{\left(\left(\frac{\nu_i|x|+\mu_i}{n+\nu_i}\right) + \left(1 + \frac{\nu_i}{n+\nu_i}\right)\frac{T}{n^{1-\alpha}}\right)^N}{N!}.$$

Inequality (26.25) implies the pointwise convergence with rates on $(H_n(f))(x)$ to $f(x)$, as $n \to \infty$, at speed $\frac{1}{n^{1-\alpha}}$.

Corollary 26.26 *All as in Theorem 26.25, plus* $f^{(j)}(x) = 0$, $j = 1, \ldots, N$. *Then*

$$|(H_n(f))(x) - f(x)| \leq$$

$$\sum_{i=1}^{\overline{r}} w_i \omega_1 \left(f^{(N)}, \left(\frac{\nu_i |x| + \mu_i}{n + \nu_i} \right) + \left(1 + \frac{\nu_i}{n + \nu_i} \right) \frac{T}{n^{1-\alpha}} \right) \cdot$$

$$\frac{\left(\left(\frac{\nu_i |x| + \mu_i}{n + \nu_i} \right) + \left(1 + \frac{\nu_i}{n + \nu_i} \right) \frac{T}{n^{1-\alpha}} \right)^N}{N!}, \tag{26.26}$$

a convergence at the high speed of $\frac{1}{n^{(1-\alpha)(N+1)}}$.

Corollary 26.27 *Let* $x \in [-T^*, T^*]$, $T^* > 0$; $T > 0$ *and* $n \in \mathbb{N}$ *such that* $n \geq \max\left(T + T^*, T^{-\frac{1}{\alpha}}\right)$. *Let* $f \in C^N(\mathbb{R})$, $N \in \mathbb{N}$, *such that* $f^{(N)}$ *is uniformly continuous or is continuous and bounded. Then*

$$\|H_n(f) - f\|_{\infty,[-T^*,T^*]} \leq \tag{26.27}$$

$$\sum_{j=1}^{N} \frac{\|f^{(j)}\|_{\infty,[-T^*,T^*]}}{j!} \left\{ \sum_{i=1}^{\overline{r}} w_i \left[\left(\frac{\nu_i T^* + \mu_i}{n + \nu_i} \right) + \left(1 + \frac{\nu_i}{n + \nu_i} \right) \frac{T}{n^{1-\alpha}} \right]^j \right\} +$$

$$\sum_{i=1}^{\overline{r}} w_i \omega_1 \left(f^{(N)}, \left(\frac{\nu_i T^* + \mu_i}{n + \nu_i} \right) + \left(1 + \frac{\nu_i}{n + \nu_i} \right) \frac{T}{n^{1-\alpha}} \right) \cdot$$

$$\frac{\left(\left(\frac{\nu_i T^* + \mu_i}{n + \nu_i} \right) + \left(1 + \frac{\nu_i}{n + \nu_i} \right) \frac{T}{n^{1-\alpha}} \right)^N}{N!}.$$

Corollary 26.28 *All as in Theorem 26.25, case of* $N = 1$. *It holds*

$$|(H_n(f))(x) - f(x)| \leq$$

$$|f'(x)| \left\{ \sum_{i=1}^{\overline{r}} w_i \left[\left(\frac{\nu_i |x| + \mu_i}{n + \nu_i} \right) + \left(1 + \frac{\nu_i}{n + \nu_i} \right) \frac{T}{n^{1-\alpha}} \right] \right\} + \tag{26.28}$$

$$\sum_{i=1}^{\overline{r}} w_i \omega_1 \left(f', \left(\frac{\nu_i |x| + \mu_i}{n + \nu_i} \right) + \left(1 + \frac{\nu_i}{n + \nu_i} \right) \frac{T}{n^{1-\alpha}} \right) \cdot$$

$$\left(\left(\frac{\nu_i |x| + \mu_i}{n + \nu_i} \right) + \left(1 + \frac{\nu_i}{n + \nu_i} \right) \frac{T}{n^{1-\alpha}} \right).$$

Theorem 26.29 *Same assumptions as in Theorem 26.25, with* $0 < \alpha < 1$. *Then*

$$|(K_n(f))(x) - f(x)| \leq \sum_{j=1}^{N} \frac{\left|f^{(j)}(x)\right|}{j!} \cdot \tag{26.29}$$

$$\left(\sum_{i=1}^{\overline{r}} w_i \left(\left(\frac{\nu_i |x| + \mu_i + 1}{n + \nu_i}\right) + \left(1 + \frac{\nu_i}{n + \nu_i}\right) \frac{T}{n^{1-\alpha}}\right)^j\right) +$$

$$\sum_{i=1}^{\overline{r}} w_i \left\{ \omega_1 \left(f^{(N)}, \left(\frac{\nu_i |x| + \mu_i + 1}{n + \nu_i}\right) + \left(1 + \frac{\nu_i}{n + \nu_i}\right) \frac{T}{n^{1-\alpha}} \right) \cdot \right.$$

$$\left. \frac{\left(\left(\frac{\nu_i |x| + \mu_i + 1}{n + \nu_i}\right) + \left(1 + \frac{\nu_i}{n + \nu_i}\right) \frac{T}{n^{1-\alpha}}\right)^N}{N!} \right\}.$$

Inequality (26.29) implies the pointwise convergence with rates of $(K_n(f))(x)$ *to* $f(x)$, *as* $n \to \infty$, *at the speed* $\frac{1}{n^{1-\alpha}}$.

Corollary 26.30 *All as in Theorem 26.25, plus* $f^{(j)}(x) = 0$, $j = 1, \ldots, N$; $0 < \alpha < 1$. *Then*

$$|(K_n(f))(x) - f(x)| \leq$$

$$\sum_{i=1}^{\overline{r}} w_i \left\{ \omega_1 \left(f^{(N)}, \left(\frac{\nu_i |x| + \mu_i + 1}{n + \nu_i}\right) + \left(1 + \frac{\nu_i}{n + \nu_i}\right) \frac{T}{n^{1-\alpha}} \right) \cdot \right. \tag{26.30}$$

$$\left. \frac{\left(\left(\frac{\nu_i |x| + \mu_i + 1}{n + \nu_i}\right) + \left(1 + \frac{\nu_i}{n + \nu_i}\right) \frac{T}{n^{1-\alpha}}\right)^N}{N!} \right\},$$

a convergence at speed $\frac{1}{n^{(1-\alpha)(N+1)}}$.

Corollary 26.31 *Let* $x \in \left[-T^*, T^*\right]$, $T^* > 0$; $T > 0$ *and* $n \in \mathbb{N}$ *such that* $n \geq \max\left(T + T^*, T^{-\frac{1}{\alpha}}\right)$, $0 < \alpha < 1$. *Let* $f \in C^N(\mathbb{R})$, $N \in \mathbb{N}$, *such that* $f^{(N)}$ *is uniformly continuous or is continuous and bounded. Then*

$$\|K_n(f) - f\|_{\infty,[-T^*,T^*]} \leq \sum_{j=1}^{N} \frac{\left\|f^{(j)}\right\|_{\infty,[-T^*,T^*]}}{j!} \cdot \tag{26.31}$$

$$\left(\sum_{i=1}^{\overline{r}} w_i \left[\left(\frac{\nu_i T^* + \mu_i + 1}{n + \nu_i}\right) + \left(1 + \frac{\nu_i}{n + \nu_i}\right)\frac{T}{n^{1-\alpha}}\right]^j\right) +$$

$$\sum_{i=1}^{\overline{r}} w_i \left\{\omega_1\left(f^{(N)}, \left(\frac{\nu_i T^* + \mu_i + 1}{n + \nu_i}\right) + \left(1 + \frac{\nu_i}{n + \nu_i}\right)\frac{T}{n^{1-\alpha}}\right) \cdot \right.$$

$$\left.\frac{\left(\left(\frac{\nu_i T^* + \mu_i + 1}{n + \nu_i}\right) + \left(1 + \frac{\nu_i}{n + \nu_i}\right)\frac{T}{n^{1-\alpha}}\right)^N}{N!}\right\}.$$

Corollary 26.32 *All as in Theorem 26.25, case of $N = 1$. It holds*

$$|(K_n(f))(x) - f(x)| \leq$$

$$\left|f'(x)\right| \left(\sum_{i=1}^{\overline{r}} w_i \left(\left(\frac{\nu_i |x| + \mu_i + 1}{n + \nu_i}\right) + \left(1 + \frac{\nu_i}{n + \nu_i}\right)\frac{T}{n^{1-\alpha}}\right)\right) + \quad (26.32)$$

$$\sum_{i=1}^{\overline{r}} w_i \left\{\omega_1\left(f', \left(\frac{\nu_i |x| + \mu_i + 1}{n + \nu_i}\right) + \left(1 + \frac{\nu_i}{n + \nu_i}\right)\frac{T}{n^{1-\alpha}}\right) \cdot \right.$$

$$\left.\left(\left(\frac{\nu_i |x| + \mu_i + 1}{n + \nu_i}\right) + \left(1 + \frac{\nu_i}{n + \nu_i}\right)\frac{T}{n^{1-\alpha}}\right)\right\}.$$

Theorem 26.33 *Let all as in Theorem 26.25. Then*

$$|(M_n(f))(x) - f(x)| \leq \sum_{j=1}^{N} \frac{\left|f^{(j)}(x)\right|}{j!}\left[\frac{T}{n^{1-\alpha}} + \frac{1}{n}\right]^j + \quad (26.33)$$

$$\omega_1\left(f^{(N)}, \frac{T}{n^{1-\alpha}} + \frac{1}{n}\right)\frac{\left(\frac{T}{n^{1-\alpha}} + \frac{1}{n}\right)^N}{N!}.$$

Inequality (26.33) implies the pointwise convergence with rates of $(M_n(f))(x)$ to $f(x)$, as $n \to \infty$, at the speed $\frac{1}{n^{1-\alpha}}$.

Corollary 26.34 *All as in Theorem 26.25, plus $f^{(j)}(x) = 0$, $j = 1, \ldots, N$. Then*

$$|(M_n(f))(x) - f(x)| \leq \omega_1\left(f^{(N)}, \frac{T}{n^{1-\alpha}} + \frac{1}{n}\right)\frac{\left(\frac{T}{n^{1-\alpha}} + \frac{1}{n}\right)^N}{N!}, \quad (26.34)$$

a convergence at speed $\frac{1}{n^{(1-\alpha)(N+1)}}$.

Corollary 26.35 *All here as in Corollary 26.27. Then*

$$\|M_n(f)-f\|_{\infty,[-T^*,T^*]} \le \sum_{j=1}^{N} \frac{\left\|f^{(j)}\right\|_{\infty,[-T^*,T^*]}}{j!}\left(\frac{T}{n^{1-\alpha}}+\frac{1}{n}\right)^j \tag{26.35}$$

$$+\,\omega_1\left(f^{(N)},\frac{T}{n^{1-\alpha}}+\frac{1}{n}\right)\frac{\left(\frac{T}{n^{1-\alpha}}+\frac{1}{n}\right)^N}{N!}.$$

Finally we mention also from [12].

Corollary 26.36 *All as in Theorem 26.25, $N=1$ case. It holds*

$$|(M_n(f))(x)-f(x)| \le \left[\left|f'(x)\right|+\omega_1\left(f',\frac{T}{n^{1-\alpha}}+\frac{1}{n}\right)\right]\left(\frac{T}{n^{1-\alpha}}+\frac{1}{n}\right). \tag{26.36}$$

Remark 26.37 (1) By change of variable method, the operator K_n could be written conveniently as follows:

(ii)'

$$(K_n(f))(x)=$$

$$\frac{\sum_{k=\lceil nx-Tn^{\alpha}\rceil}^{[nx+Tn^{\alpha}]}\left(\sum_{i=1}^{\overline{r}} w_i\,(n+\nu_i)\int_0^{\frac{1}{n+\nu_i}} f\left(t+\frac{k+\mu_i}{n+\nu_i}\right)dt\right)b\left(n^{1-\alpha}\left(x-\frac{k}{n}\right)\right)}{\sum_{k=\lceil nx-Tn^{\alpha}\rceil}^{[nx+Tn^{\alpha}]} b\left(n^{1-\alpha}\left(x-\frac{k}{n}\right)\right)}. \tag{26.37}$$

(2) Next we apply the principle of iterated suprema. Let $W\subseteq\mathbb{R}$ and $f:W\to\mathbb{R}_{\mathcal{F}}$ with $D^*(f,\widetilde{o})<\infty$, that is f is a fuzzy bounded function. Then easily we derive that

$$D^*(f,\widetilde{o})=\sup_{r\in[0,1]}\max\left\{\left\|f_-^{(r)}\right\|_{\infty},\left\|f_+^{(r)}\right\|_{\infty}\right\}, \tag{26.38}$$

where $\|\cdot\|_{\infty}$ is the supremum norm of the function over W, and

$$D^*(f,\widetilde{o})=\sup_{x\in W} D(f(x),\widetilde{o}).$$

26.4 Fuzzy Neural Network Approximations

Here we consider $f\in C_{\mathcal{F}}^{U}(\mathbb{R})$ or $f\in C_B(\mathbb{R},\mathbb{R}_{\mathcal{F}})$, b is as in Sect. 26.3, $0<\alpha<1$, also the rest of parameters, are as in Sect. 26.3. For $x\in\mathbb{R}$, we take always that $n\ge\max\left(T+|x|,T^{-\frac{1}{\alpha}}\right)$. The fuzzy analog of operators H_n, K_n, M_n, see (26.16)–(26.18) and (26.37) follows, $n\in\mathbb{N}$.

We define the corresponding fuzzy operators next

(i)

$$\left(H_n^{\mathcal{F}}(f)\right)(x) = \frac{\sum_{k=\lceil nx-Tn^{\alpha}\rceil}^{[nx+Tn^{\alpha}]*} \left(\sum_{i=1}^{\overline{r}*} w_i \odot f\left(\frac{k+\mu_i}{n+\nu_i}\right)\right) \odot b\left(n^{1-\alpha}\left(x-\frac{k}{n}\right)\right)}{\sum_{k=\lceil nx-Tn^{\alpha}\rceil}^{[nx+Tn^{\alpha}]} b\left(n^{1-\alpha}\left(x-\frac{k}{n}\right)\right)}, \tag{26.39}$$

(ii)

$$\left(K_n^{\mathcal{F}}(f)\right)(x) =$$

$$\frac{\sum_{k=\lceil nx-Tn^{\alpha}\rceil}^{[nx+Tn^{\alpha}]*} \left(\sum_{i=1}^{\overline{r}*} w_i (n+\nu_i) \odot (FR)\int_0^{\frac{1}{n+\nu_i}} f\left(t+\frac{k+\mu_i}{n+\nu_i}\right) dt\right) \odot b\left(n^{1-\alpha}\left(x-\frac{k}{n}\right)\right)}{\sum_{k=\lceil nx-Tn^{\alpha}\rceil}^{[nx+Tn^{\alpha}]} b\left(n^{1-\alpha}\left(x-\frac{k}{n}\right)\right)}, \tag{26.40}$$

and

(iii)

$$\left(M_n^{\mathcal{F}}(f)\right)(x) = \frac{\sum_{k=\lceil nx-Tn^{\alpha}\rceil}^{[nx+Tn^{\alpha}]*} \left(\sum_{i=1}^{\overline{r}*} w_i \odot f\left(\frac{k}{n}+\frac{i}{n\overline{r}}\right)\right) \odot b\left(n^{1-\alpha}\left(x-\frac{k}{n}\right)\right)}{\sum_{k=\lceil nx-Tn^{\alpha}\rceil}^{[nx+Tn^{\alpha}]} b\left(n^{1-\alpha}\left(x-\frac{k}{n}\right)\right)}. \tag{26.41}$$

Call

$$V(x) = \sum_{k=\lceil nx-Tn^{\alpha}\rceil}^{[nx+Tn^{\alpha}]} b\left(n^{1-\alpha}\left(x-\frac{k}{n}\right)\right). \tag{26.42}$$

We notice that ($r \in [0,1]$)

$$\left[\left(H_n^{\mathcal{F}}(f)\right)(x)\right]^r = \sum_{k=\lceil nx-Tn^{\alpha}\rceil}^{[nx+Tn^{\alpha}]} \left(\sum_{i=1}^{\overline{r}} w_i \left[f\left(\frac{k+\mu_i}{n+\nu_i}\right)\right]^r\right) \frac{b\left(n^{1-\alpha}\left(x-\frac{k}{n}\right)\right)}{V(x)}$$

$$= \sum_{k=\lceil nx-Tn^{\alpha}\rceil}^{[nx+Tn^{\alpha}]} \left(\sum_{i=1}^{\overline{r}} w_i \left[f_-^{(r)}\left(\frac{k+\mu_i}{n+\nu_i}\right), f_+^{(r)}\left(\frac{k+\mu_i}{n+\nu_i}\right)\right]\right) \frac{b\left(n^{1-\alpha}\left(x-\frac{k}{n}\right)\right)}{V(x)}$$

$$= \left[\sum_{k=\lceil nx-Tn^{\alpha}\rceil}^{[nx+Tn^{\alpha}]} \left(\sum_{i=1}^{\overline{r}} w_i f_-^{(r)}\left(\frac{k+\mu_i}{n+\nu_i}\right)\right) \frac{b\left(n^{1-\alpha}\left(x-\frac{k}{n}\right)\right)}{V(x)},\right. \tag{26.43}$$

$$\left.\sum_{k=\lceil nx-Tn^{\alpha}\rceil}^{[nx+Tn^{\alpha}]}\left(\sum_{i=1}^{\overline{r}} w_i f_{+}^{(r)}\left(\frac{k+\mu_i}{n+\nu_i}\right)\right)\frac{b\left(n^{1-\alpha}\left(x-\frac{k}{n}\right)\right)}{V(x)}\right]$$

$$=\left[\left(H_n\left(f_{-}^{(r)}\right)\right)(x),\left(H_n\left(f_{+}^{(r)}\right)\right)(x)\right].$$

We have proved that

$$\left(H_n^{\mathcal{F}}(f)\right)_{\pm}^{(r)}=H_n\left(f_{\pm}^{(r)}\right),\ \forall\, r\in[0,1], \tag{26.44}$$

respectively.

For convenience also we call

$$A(x)=\frac{b\left(n^{1-\alpha}\left(x-\frac{k}{n}\right)\right)}{V(x)}. \tag{26.45}$$

We observe that

$$\left[\left(K_n^{\mathcal{F}}(f)\right)(x)\right]^r=$$

$$\sum_{k=\lceil nx-Tn^{\alpha}\rceil}^{[nx+Tn^{\alpha}]}\left(\sum_{i=1}^{\overline{r}} w_i(n+\nu_i)\left[\int_0^{\frac{1}{n+\nu_i}} f\left(t+\frac{k+\mu_i}{n+\nu_i}\right)dt\right]^r\right)A(x)\overset{(26.4r)}{=}$$

$$\sum_{k=\lceil nx-Tn^{\alpha}\rceil}^{[nx+Tn^{\alpha}]}\left(\sum_{i=1}^{\overline{r}} w_i(n+\nu_i)\left[\int_0^{\frac{1}{n+\nu_i}} f_{-}^{(r)}\left(t+\frac{k+\mu_i}{n+\nu_i}\right)dt,\right.\right.$$

$$\left.\left.\int_0^{\frac{1}{n+\nu_i}} f_{+}^{(r)}\left(t+\frac{k+\mu_i}{n+\nu_i}\right)dt\right]\right)A(x)= \tag{26.46}$$

$$\left[\sum_{k=\lceil nx-Tn^{\alpha}\rceil}^{[nx+Tn^{\alpha}]}\left(\sum_{i=1}^{\overline{r}} w_i(n+\nu_i)\int_0^{\frac{1}{n+\nu_i}} f_{-}^{(r)}\left(t+\frac{k+\mu_i}{n+\nu_i}\right)dt\right)A(x),\right.$$

$$\left.\sum_{k=\lceil nx-Tn^{\alpha}\rceil}^{[nx+Tn^{\alpha}]}\left(\sum_{i=1}^{\overline{r}} w_i(n+\nu_i)\int_0^{\frac{1}{n+\nu_i}} f_{+}^{(r)}\left(t+\frac{k+\mu_i}{n+\nu_i}\right)dt\right)A(x)\right]$$

$$=\left[\left(K_n\left(f_{-}^{(r)}\right)\right)(x),\left(K_n\left(f_{+}^{(r)}\right)\right)(x)\right].$$

We have proved that

$$\left(K_n^{\mathcal{F}}(f)\right)_{\pm}^{(r)} = K_n\left(f_{\pm}^{(r)}\right), \quad \forall r \in [0,1], \tag{26.47}$$

respectively.

By linear change of variable of fuzzy Riemann integrals, see [7, pp. 242–243], we easily obtain that

$$\left(K_n^{\mathcal{F}}(f)\right)(x) =$$

$$\frac{\sum_{k=\lceil nx-Tn^{\alpha}\rceil}^{[nx+Tn^{\alpha}]^*}\left(\sum_{i=1}^{\overline{r}*} w_i\,(n+\nu_i)\odot(FR)\int_{\frac{k+\mu_i}{n+\nu_i}}^{\frac{k+\mu_i+1}{n+\nu_i}} f(t)\,dt\right)\odot b\left(n^{1-\alpha}\left(x-\frac{k}{n}\right)\right)}{\sum_{k=\lceil nx-Tn^{\alpha}\rceil}^{[nx+Tn^{\alpha}]} b\left(n^{1-\alpha}\left(x-\frac{k}{n}\right)\right)}, \tag{26.48}$$

which is the fuzzy analog of $(K_n(f))(x)$ as in (26.17).

Next we observe that

$$\left[\left(M_n^{\mathcal{F}}(f)\right)(x)\right]^r = \sum_{k=\lceil nx-Tn^{\alpha}\rceil}^{[nx+Tn^{\alpha}]}\left(\sum_{i=1}^{\overline{r}} w_i\left[f\left(\frac{k}{n}+\frac{i}{n\overline{r}}\right)\right]^r\right)A(x) =$$

$$\sum_{k=\lceil nx-Tn^{\alpha}\rceil}^{[nx+Tn^{\alpha}]}\left(\sum_{i=1}^{\overline{r}} w_i\left[f_-^{(r)}\left(\frac{k}{n}+\frac{i}{n\overline{r}}\right), f_+^{(r)}\left(\frac{k}{n}+\frac{i}{n\overline{r}}\right)\right]\right)A(x) = \tag{26.49}$$

$$\left[\sum_{k=\lceil nx-Tn^{\alpha}\rceil}^{[nx+Tn^{\alpha}]}\left(\sum_{i=1}^{\overline{r}} w_i f_-^{(r)}\left(\frac{k}{n}+\frac{i}{n\overline{r}}\right)\right)A(x),\right.$$

$$\left.\sum_{k=\lceil nx-Tn^{\alpha}\rceil}^{[nx+Tn^{\alpha}]}\left(\sum_{i=1}^{\overline{r}} w_i f_+^{(r)}\left(\frac{k}{n}+\frac{i}{n\overline{r}}\right)\right)A(x)\right]$$

$$=\left[\left(M_n\left(f_-^{(r)}\right)\right)(x), \left(M_n\left(f_+^{(r)}\right)\right)(x)\right].$$

That is proving

$$\left(M_n^{\mathcal{F}}(f)\right)_{\pm}^{(r)} = M_n\left(f_{\pm}^{(r)}\right), \quad \forall r \in [0,1], \tag{26.50}$$

respectively.

We present our first main result

Theorem 26.38 *It holds*

$$D\left(\left(H_n^{\mathcal{F}}(f)\right)(x), f(x)\right) \leq$$

$$\omega_1^{(\mathcal{F})}\left(f, \max_{i\in\{1,\ldots,\overline{r}\}}\left[\left(\frac{\nu_i|x|+\mu_i}{n+\nu_i}\right)+\left(1+\frac{\nu_i}{n+\nu_i}\right)\frac{T}{n^{1-\alpha}}\right]\right), \tag{26.51}$$

where $\omega_1^{(\mathcal{F})}(f,\cdot) = \omega_1^{(\mathcal{F})}(f,\cdot)_{\mathbb{R}}$.

Notice that (26.51) implies $H_n^{\mathcal{F}}(f) \xrightarrow{D} f$, *pointwise, as* $n \to \infty$ *when* $f \in C_{\mathcal{F}}^U(\mathbb{R})$, *at speed* $\frac{1}{n^{1-\alpha}}$.

Proof We observe that

$$D\left(\left(H_n^{\mathcal{F}}(f)\right)(x), f(x)\right) =$$

$$\sup_{r\in[0,1]}\max\left\{\left|\left(H_n^{\mathcal{F}}(f)\right)_-^{(r)}(x) - f_-^{(r)}(x)\right|, \left|\left(H_n^{\mathcal{F}}(f)\right)_+^{(r)}(x) - f_+^{(r)}(x)\right|\right\} \overset{(26.44)}{=}$$

$$\tag{26.52}$$

$$\sup_{r\in[0,1]}\max\left\{\left|\left(H_n\left(f_-^{(r)}\right)\right)(x) - f_-^{(r)}(x)\right|, \left|\left(H_n\left(f_+^{(r)}\right)\right)(x) - f_+^{(r)}(x)\right|\right\} \overset{(26.19)}{\leq}$$

$$\sup_{r\in[0,1]}\max\left\{\max_{i\in\{1,\ldots,\overline{r}\}}\left\{\omega_1\left(f_-^{(r)}, \left(\frac{\nu_i|x|+\mu_i}{n+\nu_i}\right)+\left(1+\frac{\nu_i}{n+\nu_i}\right)\frac{T}{n^{1-\alpha}}\right)\right\},\right.$$

$$\left.\max_{i\in\{1,\ldots,\overline{r}\}}\left\{\omega_1\left(f_+^{(r)}, \left(\frac{\nu_i|x|+\mu_i}{n+\nu_i}\right)+\left(1+\frac{\nu_i}{n+\nu_i}\right)\frac{T}{n^{1-\alpha}}\right)\right\}\right\} =$$

$$\sup_{r\in[0,1]}\max\left\{\omega_1\left(f_-^{(r)}, \max_{i\in\{1,\ldots,\overline{r}\}}\left[\left(\frac{\nu_i|x|+\mu_i}{n+\nu_i}\right)+\left(1+\frac{\nu_i}{n+\nu_i}\right)\frac{T}{n^{1-\alpha}}\right]\right),\right.$$

$$\left.\omega_1\left(f_+^{(r)}, \max_{i\in\{1,\ldots,\overline{r}\}}\left[\left(\frac{\nu_i|x|+\mu_i}{n+\nu_i}\right)+\left(1+\frac{\nu_i}{n+\nu_i}\right)\frac{T}{n^{1-\alpha}}\right]\right)\right\} =$$

(by Proposition 26.7)

$$\omega_1^{(\mathcal{F})}\left(f, \max_{i\in\{1,\ldots,\overline{r}\}}\left[\left(\frac{\nu_i|x|+\mu_i}{n+\nu_i}\right)+\left(1+\frac{\nu_i}{n+\nu_i}\right)\frac{T}{n^{1-\alpha}}\right]\right), \tag{26.53}$$

proving the claim. ■

Corollary 26.39 *Let* $x \in [-T^*, T^*]$, $T^* > 0$, $n \in \mathbb{N} : n \geq \max\left(T+T^*, T^{-\frac{1}{\alpha}}\right)$, $T > 0$. *Then*

$$D^*\left(H_n^{\mathcal{F}}(f), f\right)_{[-T^*,T^*]} \leq$$

$$\omega_1^{(\mathcal{F})}\left(f, \max_{i\in\{1,\ldots,\overline{r}\}}\left[\left(\frac{\nu_i T^* + \mu_i}{n+\nu_i}\right) + \left(1 + \frac{\nu_i}{n+\nu_i}\right)\frac{T}{n^{1-\alpha}}\right]\right). \tag{26.54}$$

Notice that (26.54) implies $H_n^{\mathcal{F}}(f) \overset{D^*}{\to} f$, uniformly, as $n \to \infty$ when $f \in C_{\mathcal{F}}^U(\mathbb{R})$, at speed $\frac{1}{n^{1-\alpha}}$.

Proof By (26.51). ■

We continue with

Theorem 26.40 *It holds*

$$D\left(\left(K_n^{\mathcal{F}}(f)\right)(x), f(x)\right) \leq$$

$$\omega_1^{(\mathcal{F})}\left(f, \max_{i\in\{1,\ldots,\overline{r}\}}\left[\left(\frac{\nu_i |x| + \mu_i + 1}{n+\nu_i}\right) + \left(1 + \frac{\nu_i}{n+\nu_i}\right)\frac{T}{n^{1-\alpha}}\right]\right). \tag{26.55}$$

Notice that (26.55) implies $K_n^{\mathcal{F}}(f) \overset{D}{\to} f$, pointwise, as $n \to \infty$ when $f \in C_{\mathcal{F}}^U(\mathbb{R})$, at speed $\frac{1}{n^{1-\alpha}}$.

Proof Similar to Theorem 26.38, we use (26.47) and (26.21) into $D\left(\left(K_n^{\mathcal{F}}(f)\right)(x), f(x)\right)$. Finally we use Proposition 26.7. ■

Corollary 26.41 *All as in Corollary 26.39. Then*

$$D^*\left(\left(K_n^{\mathcal{F}}(f), f\right)\right)_{[-T^*,T^*]} \leq \tag{26.56}$$

$$\omega_1^{(\mathcal{F})}\left(f, \max_{i\in\{1,\ldots,\overline{r}\}}\left[\left(\frac{\nu_i T^* + \mu_i + 1}{n+\nu_i}\right) + \left(1 + \frac{\nu_i}{n+\nu_i}\right)\frac{T}{n^{1-\alpha}}\right]\right).$$

Notice that (26.56) implies $K_n^{\mathcal{F}}(f) \overset{D^*}{\to} f$, uniformly, as $n \to \infty$ when $f \in C_{\mathcal{F}}^U(\mathbb{R})$, at speed $\frac{1}{n^{1-\alpha}}$.

Proof By (26.55). ■

We also have

Theorem 26.42 *It holds*

$$D\left(\left(M_n^{\mathcal{F}}(f)\right)(x), f(x)\right) \leq \omega_1^{(\mathcal{F})}\left(f, \frac{1}{n} + \frac{T}{n^{1-\alpha}}\right). \tag{26.57}$$

Notice that (26.57) implies $M_n^{\mathcal{F}}(f) \overset{D}{\to} f$, pointwise, as $n \to \infty$ when $f \in C_{\mathcal{F}}^U(\mathbb{R})$, at speed $\frac{1}{n^{1-\alpha}}$.

Proof Similar to Theorem 26.38, we use (26.50) and (26.23) into $D\left(\left(M_n^{\mathcal{F}}(f)\right)(x), f(x)\right)$. Finally we use Proposition 26.7. ■

Corollary 26.43 *All as in Corollary 26.39. Then*

$$D^*\left(\left(M_n^{\mathcal{F}}(f), f\right)\right)_{[-T^*,T^*]} \leq \omega_1^{(\mathcal{F})}\left(f, \frac{1}{n} + \frac{T}{n^{1-\alpha}}\right). \tag{26.58}$$

Notice that (26.58) implies $M_n^{\mathcal{F}}(f) \overset{D^*}{\to} f$, uniformly, as $n \to \infty$ when $f \in C_{\mathcal{F}}^U(\mathbb{R})$, at speed $\frac{1}{n^{1-\alpha}}$.

Proof By (26.57). ■

We proceed to the following results where we use the smoothness of a fuzzy derivative of f.

Theorem 26.44 *Let $f \in C_{\mathcal{F}}^N(\mathbb{R})$, $N \in \mathbb{N}$, with H-fuzzy $f^{(N)}$ either fuzzy continuous and bounded or fuzzy uniformly continuous.*

Here $x \in \mathbb{R}$, $T > 0$, $n \in \mathbb{N} : n \geq \max\left(T + |x|, T^{-\frac{1}{\alpha}}\right)$, $0 < \alpha < 1$, the rest of parameters as in Sect. 26.3. Then

$$D\left(\left(H_n^{\mathcal{F}}(f)\right)(x), f(x)\right) \leq$$

$$\sum_{j=1}^{N} \frac{D\left(f^{(j)}(x), \widetilde{o}\right)}{j!}\left\{\sum_{i=1}^{\overline{r}} w_i \left[\left(\frac{\nu_i |x| + \mu_i}{n + \nu_i}\right) + \left(1 + \frac{\nu_i}{n + \nu_i}\right)\frac{T}{n^{1-\alpha}}\right]^j\right\} +$$

$$\sum_{i=1}^{\overline{r}} w_i \frac{\left(\left(\frac{\nu_i |x| + \mu_i}{n + \nu_i}\right) + \left(1 + \frac{\nu_i}{n + \nu_i}\right)\frac{T}{n^{1-\alpha}}\right)^N}{N!}. \tag{26.59}$$

$$\omega_1^{(\mathcal{F})}\left(f^{(N)}, \left(\frac{\nu_i |x| + \mu_i}{n + \nu_i}\right) + \left(1 + \frac{\nu_i}{n + \nu_i}\right)\frac{T}{n^{1-\alpha}}\right).$$

Inequality (26.59) implies the pointwise convergence with rates on $\left(H_n^{\mathcal{F}}(f)\right)(x) \overset{D}{\to} f(x)$, as $n \to \infty$, at speed $\frac{1}{n^{1-\alpha}}$.

Proof We observe that

$$D\left(\left(H_n^{\mathcal{F}}(f)\right)(x), f(x)\right) =$$

$$\sup_{r\in[0,1]} \max\left\{\left|\left(H_n^{\mathcal{F}}(f)\right)_-^{(r)}(x) - f_-^{(r)}(x)\right|, \left|\left(H_n^{\mathcal{F}}(f)\right)_+^{(r)}(x) - f_+^{(r)}(x)\right|\right\} \overset{(26.44)}{=}$$

$$\sup_{r\in[0,1]} \max\left\{\left|\left(H_n\left(f_-^{(r)}\right)\right)(x) - f_-^{(r)}(x)\right|, \left|\left(H_n\left(f_+^{(r)}\right)\right)(x) - f_+^{(r)}(x)\right|\right\} \overset{(26.25)}{\leq} \tag{26.60}$$

$$\sup_{r\in[0,1]} \max\left\{\sum_{j=1}^{N} \frac{\left|\left(f_-^{(r)}\right)^{(j)}(x)\right|}{j!} \left\{\sum_{i=1}^{\bar{r}} w_i \left[\left(\frac{\nu_i|x|+\mu_i}{n+\nu_i}\right) + \left(1+\frac{\nu_i}{n+\nu_i}\right)\frac{T}{n^{1-\alpha}}\right]^j\right\}\right.$$

$$+\sum_{i=1}^{\bar{r}} w_i \omega_1\left(\left(f_-^{(r)}\right)^{(N)}, \left(\frac{\nu_i|x|+\mu_i}{n+\nu_i}\right) + \left(1+\frac{\nu_i}{n+\nu_i}\right)\frac{T}{n^{1-\alpha}}\right) \cdot$$

$$\frac{\left(\left(\frac{\nu_i|x|+\mu_i}{n+\nu_i}\right) + \left(1+\frac{\nu_i}{n+\nu_i}\right)\frac{T}{n^{1-\alpha}}\right)^N}{N!},$$

$$\sum_{j=1}^{N} \frac{\left|\left(f_+^{(r)}\right)^{(j)}(x)\right|}{j!} \left\{\sum_{i=1}^{\bar{r}} w_i \left[\left(\frac{\nu_i|x|+\mu_i}{n+\nu_i}\right) + \left(1+\frac{\nu_i}{n+\nu_i}\right)\frac{T}{n^{1-\alpha}}\right]^j\right\} +$$

$$\sum_{i=1}^{\bar{r}} w_i \omega_1\left(\left(f_+^{(r)}\right)^{(N)}, \left(\frac{\nu_i|x|+\mu_i}{n+\nu_i}\right) + \left(1+\frac{\nu_i}{n+\nu_i}\right)\frac{T}{n^{1-\alpha}}\right) \cdot$$

$$\left.\frac{\left(\left(\frac{\nu_i|x|+\mu_i}{n+\nu_i}\right) + \left(1+\frac{\nu_i}{n+\nu_i}\right)\frac{T}{n^{1-\alpha}}\right)^N}{N!}\right\} = \tag{26.61}$$

(by Remark 26.12)

$$\sup_{r\in[0,1]} \max\left\{\sum_{j=1}^{N} \frac{\left|\left(f^{(j)}\right)_-^{(r)}(x)\right|}{j!} \left\{\sum_{i=1}^{\bar{r}} w_i \left[\left(\frac{\nu_i|x|+\mu_i}{n+\nu_i}\right) + \left(1+\frac{\nu_i}{n+\nu_i}\right)\frac{T}{n^{1-\alpha}}\right]^j\right\}\right.$$

$$+\sum_{i=1}^{\bar{r}} w_i \omega_1\left(\left(f^{(N)}\right)_-^{(r)}, \left(\frac{\nu_i|x|+\mu_i}{n+\nu_i}\right) + \left(1+\frac{\nu_i}{n+\nu_i}\right)\frac{T}{n^{1-\alpha}}\right) \cdot$$

$$\frac{\left(\left(\frac{\nu_i|x|+\mu_i}{n+\nu_i}\right) + \left(1+\frac{\nu_i}{n+\nu_i}\right)\frac{T}{n^{1-\alpha}}\right)^N}{N!},$$

$$\sum_{j=1}^{N} \frac{\left|\left(f^{(j)}\right)_{+}^{(r)}(x)\right|}{j!} \left\{ \sum_{i=1}^{\overline{r}} w_i \left[\left(\frac{\nu_i |x| + \mu_i}{n + \nu_i} \right) + \left(1 + \frac{\nu_i}{n + \nu_i} \right) \frac{T}{n^{1-\alpha}} \right]^j \right\} + \tag{26.62}$$

$$\sum_{i=1}^{\overline{r}} w_i \omega_1 \left(\left(f^{(N)}\right)_{+}^{(r)}, \left(\frac{\nu_i |x| + \mu_i}{n + \nu_i} \right) + \left(1 + \frac{\nu_i}{n + \nu_i} \right) \frac{T}{n^{1-\alpha}} \right) \cdot$$

$$\left. \frac{\left(\left(\frac{\nu_i |x| + \mu_i}{n + \nu_i} \right) + \left(1 + \frac{\nu_i}{n + \nu_i} \right) \frac{T}{n^{1-\alpha}} \right)^N}{N!} \right\} \overset{\text{(by (26.1))}}{\leq}$$

$$\sum_{j=1}^{N} \frac{1}{j!} \left\{ \sum_{i=1}^{\overline{r}} w_i \left[\left(\frac{\nu_i |x| + \mu_i}{n + \nu_i} \right) + \left(1 + \frac{\nu_i}{n + \nu_i} \right) \frac{T}{n^{1-\alpha}} \right]^j \right\} \cdot$$

$$\sup_{r \in [0,1]} \max \left\{ \left| \left(f^{(j)}\right)_{-}^{(r)}(x) \right|, \left| \left(f^{(j)}\right)_{+}^{(r)}(x) \right| \right\} + \tag{26.63}$$

$$\sum_{i=1}^{\overline{r}} w_i \frac{\left(\left(\frac{\nu_i |x| + \mu_i}{n + \nu_i} \right) + \left(1 + \frac{\nu_i}{n + \nu_i} \right) \frac{T}{n^{1-\alpha}} \right)^N}{N!}$$

$$\sup_{r \in [0,1]} \max \left\{ \omega_1 \left(\left(f^{(N)}\right)_{-}^{(r)}, \left(\frac{\nu_i |x| + \mu_i}{n + \nu_i} \right) + \left(1 + \frac{\nu_i}{n + \nu_i} \right) \frac{T}{n^{1-\alpha}} \right), \right.$$

$$\left. \omega_1 \left(\left(f^{(N)}\right)_{+}^{(r)}, \left(\frac{\nu_i |x| + \mu_i}{n + \nu_i} \right) + \left(1 + \frac{\nu_i}{n + \nu_i} \right) \frac{T}{n^{1-\alpha}} \right) \right\} =$$

(by definition of D and Proposition 26.7)

$$\sum_{j=1}^{N} \frac{1}{j!} \left\{ \sum_{i=1}^{\overline{r}} w_i \left[\left(\frac{\nu_i |x| + \mu_i}{n + \nu_i} \right) + \left(1 + \frac{\nu_i}{n + \nu_i} \right) \frac{T}{n^{1-\alpha}} \right]^j \right\} D\left(f^{(j)}(x), \widetilde{o} \right) +$$

$$\sum_{i=1}^{\overline{r}} w_i \frac{\left(\left(\frac{\nu_i |x| + \mu_i}{n + \nu_i} \right) + \left(1 + \frac{\nu_i}{n + \nu_i} \right) \frac{T}{n^{1-\alpha}} \right)^N}{N!} \cdot$$

$$\omega_1^{(F)} \left(f^{(N)}, \left(\frac{\nu_i |x| + \mu_i}{n + \nu_i} \right) + \left(1 + \frac{\nu_i}{n + \nu_i} \right) \frac{T}{n^{1-\alpha}} \right), \tag{26.64}$$

proving the claim. ■

Corollary 26.45 (to Theorem 26.44) *Additionally assume that* $D\left(f^{(j)}(x), \widetilde{o}\right) = 0$, $j = 1, \ldots, N$. *Then*

$$D\left(\left(H_n^{\mathcal{F}}(f)\right)(x), f(x)\right) \leq$$

$$\sum_{i=1}^{\overline{r}} w_i \frac{\left(\left(\frac{\nu_i |x| + \mu_i}{n+\nu_i}\right) + \left(1 + \frac{\nu_i}{n+\nu_i}\right)\frac{T}{n^{1-\alpha}}\right)^N}{N!}.$$

$$\omega_1^{(\mathcal{F})}\left(f^{(N)}, \left(\frac{\nu_i |x| + \mu_i}{n+\nu_i}\right) + \left(1 + \frac{\nu_i}{n+\nu_i}\right)\frac{T}{n^{1-\alpha}}\right). \tag{26.65}$$

Inequality (26.65) implies the pointwise convergence with rates of $\left(H_n^{\mathcal{F}}(f)\right)(x) \xrightarrow{D} f(x)$, *as* $n \to \infty$, *at high speed of* $\frac{1}{n^{(1-\alpha)(N+1)}}$.

Proof By (26.59). ■

Corollary 26.46 (to Theorem 26.44) *Here we take* $x \in \left[-T^*, T^*\right]$, $T^* > 0$; $T > 0$ *and* $n \in \mathbb{N} : n \geq \max\left(T + T^*, T^{-\frac{1}{\alpha}}\right)$. *Let* $f \in C_{\mathcal{F}}^N(\mathbb{R})$, $N \in \mathbb{N}$, $f^{(N)}$ *is either fuzzy continuous and bounded or fuzzy uniformly continuous;* $0 < \alpha < 1$. *Then*

$$D^*\left(H_n^{\mathcal{F}}(f), f\right)_{[-T^*,T^*]} \leq$$

$$\sum_{j=1}^{N} \frac{D^*\left(f^{(j)}, \widetilde{o}\right)_{[-T^*,T^*]}}{j!}\left\{\sum_{i=1}^{\overline{r}} w_i \left[\left(\frac{\nu_i T^* + \mu_i}{n+\nu_i}\right) + \left(1 + \frac{\nu_i}{n+\nu_i}\right)\frac{T}{n^{1-\alpha}}\right]^j\right\} +$$

$$\sum_{i=1}^{\overline{r}} w_i \frac{\left(\left(\frac{\nu_i T^* + \mu_i}{n+\nu_i}\right) + \left(1 + \frac{\nu_i}{n+\nu_i}\right)\frac{T}{n^{1-\alpha}}\right)^N}{N!}. \tag{26.66}$$

$$\omega_1^{(\mathcal{F})}\left(f^{(N)}, \left(\frac{\nu_i T^* + \mu_i}{n+\nu_i}\right) + \left(1 + \frac{\nu_i}{n+\nu_i}\right)\frac{T}{n^{1-\alpha}}\right).$$

Inequality (26.66) implies the uniform convergence with rates of $H_n^{\mathcal{F}}(f) \xrightarrow{D^*} f$, *as* $n \to \infty$, *at speed* $\frac{1}{n^{1-\alpha}}$.

Proof By (26.59). ■

Corollary 26.47 (to Theorem 26.44) *Case* $N = 1$. *It holds*

$$D\left(\left(H_n^{\mathcal{F}}(f)\right)(x), f(x)\right) \leq$$

$$\left[D\left(f'(x), \widetilde{o}\right) + \omega_1^{(\mathcal{F})}\left(f', \left(\frac{\nu_i |x| + \mu_i}{n + \nu_i}\right) + \left(1 + \frac{\nu_i}{n + \nu_i}\right)\frac{T}{n^{1-\alpha}}\right)\right] \cdot$$

$$\left\{\sum_{i=1}^{\overline{r}} w_i \left[\left(\frac{\nu_i |x| + \mu_i}{n + \nu_i}\right) + \left(1 + \frac{\nu_i}{n + \nu_i}\right)\frac{T}{n^{1-\alpha}}\right]\right\}. \tag{26.67}$$

Inequality (26.67) implies the pointwise convergence with rates of $\left(H_n^{\mathcal{F}}(f)\right)(x) \overset{D}{\to} f(x)$*, as* $n \to \infty$*, at speed* $\frac{1}{n^{1-\alpha}}$.

Proof By (26.59). ■

We continue with

Theorem 26.48 *Let* $f \in C_{\mathcal{F}}^N(\mathbb{R})$*,* $N \in \mathbb{N}$*, with H-fuzzy* $f^{(N)}$ *either fuzzy continuous and bounded or fuzzy uniformly continuous.*

Here $x \in \mathbb{R}$*,* $T > 0$*,* $n \in \mathbb{N} : n \geq \max\left(T + |x|, T^{-\frac{1}{\alpha}}\right)$*,* $0 < \alpha < 1$*, the rest of parameters as in Sect. 26.3. Then*

$$D\left(\left(K_n^{\mathcal{F}}(f)\right)(x), f(x)\right) \leq$$

$$\sum_{j=1}^{N} \frac{D\left(f^{(j)}(x), \widetilde{o}\right)}{j!}\left\{\sum_{i=1}^{\overline{r}} w_i \left[\left(\frac{\nu_i |x| + \mu_i + 1}{n + \nu_i}\right) + \left(1 + \frac{\nu_i}{n + \nu_i}\right)\frac{T}{n^{1-\alpha}}\right]^j\right\} +$$

$$\sum_{i=1}^{\overline{r}} w_i \frac{\left(\left(\frac{\nu_i |x| + \mu_i + 1}{n + \nu_i}\right) + \left(1 + \frac{\nu_i}{n + \nu_i}\right)\frac{T}{n^{1-\alpha}}\right)^N}{N!}. \tag{26.68}$$

$$\omega_1^{(\mathcal{F})}\left(f^{(N)}, \left(\frac{\nu_i |x| + \mu_i + 1}{n + \nu_i}\right) + \left(1 + \frac{\nu_i}{n + \nu_i}\right)\frac{T}{n^{1-\alpha}}\right).$$

Inequality (26.68) implies the pointwise convergence with rates on $\left(K_n^{\mathcal{F}}(f)\right)(x) \overset{D}{\to} f(x)$*, as* $n \to \infty$*, at speed* $\frac{1}{n^{1-\alpha}}$.

Proof Similar to Theorem 26.44. Here we use (26.47), Theorem 26.29, Remark 26.12 along with (26.1), the definition of D and Proposition 26.7. ■

Corollary 26.49 (to Theorem 26.48) *Additionally assume that* $D\left(f^{(j)}(x), \widetilde{o}\right) = 0$*,* $j = 1, \ldots, N$*. Then*

$$D\left(\left(K_n^{\mathcal{F}}(f)\right)(x), f(x)\right) \leq$$

$$\sum_{i=1}^{\overline{r}} w_i \frac{\left(\left(\frac{\nu_i |x| + \mu_i + 1}{n + \nu_i}\right) + \left(1 + \frac{\nu_i}{n + \nu_i}\right)\frac{T}{n^{1-\alpha}}\right)^N}{N!}. \tag{26.69}$$

$$\omega_1^{(\mathcal{F})}\left(f^{(N)}, \left(\frac{\nu_i |x| + \mu_i + 1}{n + \nu_i}\right) + \left(1 + \frac{\nu_i}{n + \nu_i}\right)\frac{T}{n^{1-\alpha}}\right).$$

Inequality (26.69) implies the pointwise convergence with rates of $\left(K_n^{\mathcal{F}}(f)\right)(x) \overset{D}{\rightarrow} f(x)$, *as* $n \rightarrow \infty$, *at high speed of* $\frac{1}{n^{(1-\alpha)(N+1)}}$.

Proof By (26.68). ■

Corollary 26.50 (to Theorem 26.48) *Here we take* $x \in \left[-T^*, T^*\right]$, $T^* > 0$; $T > 0$ *and* $n \in \mathbb{N} : n \geq \max\left(T + T^*, T^{-\frac{1}{\alpha}}\right)$. *Let* $f \in C_{\mathcal{F}}^N(\mathbb{R})$, $N \in \mathbb{N}$, $f^{(N)}$ *is either fuzzy continuous and bounded or fuzzy uniformly continuous;* $0 < \alpha < 1$. *Then*

$$D^*\left(K_n^{\mathcal{F}}(f), f\right)_{[-T^*,T^*]} \leq$$

$$\sum_{j=1}^{N} \frac{D^*\left(f^{(j)}, \widetilde{o}\right)_{[-T^*,T^*]}}{j!} \left\{\sum_{i=1}^{\overline{r}} w_i \left[\left(\frac{\nu_i T^* + \mu_i + 1}{n + \nu_i}\right) + \left(1 + \frac{\nu_i}{n + \nu_i}\right)\frac{T}{n^{1-\alpha}}\right]^j\right\} +$$

$$\sum_{i=1}^{\overline{r}} w_i \frac{\left(\left(\frac{\nu_i T^* + \mu_i + 1}{n + \nu_i}\right) + \left(1 + \frac{\nu_i}{n + \nu_i}\right)\frac{T}{n^{1-\alpha}}\right)^N}{N!}. \tag{26.70}$$

$$\omega_1^{(\mathcal{F})}\left(f^{(N)}, \left(\frac{\nu_i T^* + \mu_i + 1}{n + \nu_i}\right) + \left(1 + \frac{\nu_i}{n + \nu_i}\right)\frac{T}{n^{1-\alpha}}\right).$$

Inequality (26.70) implies the uniform convergence with rates of $K_n^{\mathcal{F}}(f) \overset{D^*}{\rightarrow} f$, *as* $n \rightarrow \infty$, *at speed* $\frac{1}{n^{1-\alpha}}$.

Proof By (26.68). ■

Corollary 26.51 (to Theorem 26.48) *Case* $N = 1$. *It holds*

$$D\left(\left(K_n^{\mathcal{F}}(f)\right)(x), f(x)\right) \leq$$

$$\left[D\left(f'(x), \widetilde{o}\right) + \omega_1^{(\mathcal{F})}\left(f', \left(\frac{\nu_i |x| + \mu_i + 1}{n + \nu_i}\right) + \left(1 + \frac{\nu_i}{n + \nu_i}\right)\frac{T}{n^{1-\alpha}}\right)\right].$$

$$\left\{\sum_{i=1}^{\overline{r}} w_i \left[\left(\frac{\nu_i |x| + \mu_i + 1}{n + \nu_i}\right) + \left(1 + \frac{\nu_i}{n + \nu_i}\right)\frac{T}{n^{1-\alpha}}\right]\right\}. \tag{26.71}$$

Inequality (26.71) implies the pointwise convergence with rates of $\left(K_n^{\mathcal{F}}(f)\right)(x) \overset{D}{\rightarrow} f(x)$, *as* $n \rightarrow \infty$, *at speed* $\frac{1}{n^{1-\alpha}}$.

Proof By (26.68). ∎

We also have

Theorem 26.52 *All here as in Theorem 26.44. Then*

$$D\left(\left(M_n^{\mathcal{F}}(f)\right)(x), f(x)\right) \leq \sum_{j=1}^{N} \frac{D\left(f^{(j)}(x), \widetilde{o}\right)}{j!}\left[\frac{T}{n^{1-\alpha}}+\frac{1}{n}\right]^j +$$

$$\omega_1^{(\mathcal{F})}\left(f^{(N)}, \frac{T}{n^{1-\alpha}}+\frac{1}{n}\right)\frac{\left(\frac{T}{n^{1-\alpha}}+\frac{1}{n}\right)^N}{N!}. \tag{26.72}$$

Inequality (26.72) implies the pointwise convergence with rates on $\left(M_n^{\mathcal{F}}(f)\right)(x) \overset{D}{\to} f(x)$*, as* $n \to \infty$*, at the speed* $\frac{1}{n^{1-\alpha}}$.

Proof As in Theorem 26.44. We use here (26.50), Theorem 26.33, Remark 26.12, (26.1), definition of D and Proposition 26.7. ∎

Corollary 26.53 (to Theorem 26.52) *Additionally assume that* $D\left(f^{(j)}(x), \widetilde{o}\right) = 0$, $j = 1, \ldots, N$. *Then*

$$D\left(\left(M_n^{\mathcal{F}}(f)\right)(x), f(x)\right) \leq$$

$$\omega_1^{(\mathcal{F})}\left(f^{(N)}, \frac{T}{n^{1-\alpha}}+\frac{1}{n}\right)\frac{\left(\frac{T}{n^{1-\alpha}}+\frac{1}{n}\right)^N}{N!}. \tag{26.73}$$

Inequality (26.73) implies the pointwise convergence with rates of $\left(M_n^{\mathcal{F}}(f)\right)(x) \overset{D}{\to} f(x)$*, as* $n \to \infty$*, at high speed of* $\frac{1}{n^{(1-\alpha)(N+1)}}$.

Proof By (26.72). ∎

Corollary 26.54 (to Theorem 26.52) *Here all as in Corollary 26.46. Then*

$$D^*\left(M_n^{\mathcal{F}}(f), f\right)_{[-T^*,T^*]} \leq$$

$$\sum_{j=1}^{N} \frac{D^*\left(f^{(j)}, \widetilde{o}\right)_{[-T^*,T^*]}}{j!}\left[\frac{T}{n^{1-\alpha}}+\frac{1}{n}\right]^j + \omega_1^{(\mathcal{F})}\left(f^{(N)}, \frac{T}{n^{1-\alpha}}+\frac{1}{n}\right)\frac{\left(\frac{T}{n^{1-\alpha}}+\frac{1}{n}\right)^N}{N!}. \tag{26.74}$$

Inequality (26.74) implies the uniform convergence with rates of $M_n^{\mathcal{F}}(f) \overset{D^*}{\to} f$*, as* $n \to \infty$*, at speed* $\frac{1}{n^{1-\alpha}}$.

Proof By (26.72). ∎

Corollary 26.55 (to Theorem 26.52) *Case $N = 1$. It holds*

$$D\left(\left(M_n^{\mathcal{F}}(f)\right)(x), f(x)\right) \leq$$

$$\left[D\left(f'(x), \tilde{o}\right) + \omega_1^{(\mathcal{F})}\left(f', \frac{T}{n^{1-\alpha}} + \frac{1}{n}\right)\right]\left(\frac{T}{n^{1-\alpha}} + \frac{1}{n}\right). \quad (26.75)$$

Inequality (26.75) implies the pointwise convergence with rates of $\left(M_n^{\mathcal{F}}(f)\right)(x) \overset{D}{\rightarrow} f(x)$, as $n \rightarrow \infty$, at the speed $\frac{1}{n^{1-\alpha}}$.

Proof By (26.72). ■

Remark 26.56 All real neural network approximation results listed here were transferred to the fuzzy setting.

References

1. G.A. Anastassiou, Rate of convergence of some neural network operators to the unit-univariate case. J. Math. Anal. Appl. **212**, 237–262 (1997)
2. G.A. Anastassiou, Rate of convergence of some multivariate neural network operators to the unit. J. Comp. Math. Appl. **40**, 1–19 (2000)
3. G.A. Anastassiou, *Quantitative Approximations* (Chapman & Hall/CRC, Boca Raton, New York, 2001)
4. G.A. Anastassiou, Fuzzy approximation by fuzzy convolution type operators. Comput. Math. **48**, 1369–1386 (2004)
5. G.A. Anastassiou, Higher order fuzzy Korovkin theory via inequalities. Commun. Appl. Anal. **10**(2), 359–392 (2006)
6. G.A. Anastassiou, Fuzzy Korovkin theorems and inequalities. J. Fuzzy Math. **15**(1), 169–205 (2007)
7. G.A. Anastassiou, *Fuzzy Mathematics: Approximation Theory* (Springer, Heidelberg, New York, 2010)
8. G.A. Anastassiou, Rate of convergence of some neural network operators to the unit-univariate case, revisited. Vesnik **65**(4), 511–518 (2013)
9. G.A. Anastassiou, Rate of convergence of some multivariate neural network operators to the unit, revisited. J. Comput. Anal. Appl. **15**(7), 1300–1309 (2013)
10. G.A. Anastassiou, Fuzzy fractional approximations by fuzzy normalized bell and squashing type neural network operators. J. Fuzzy Math. **22**(1), 139–156 (2014)
11. G.A. Anastassiou, Higher order multivariate fuzzy approximation by basic neural network operators. Cubo **16**(03), 21–35 (2014)
12. G.A. Anastassiou, Approximation by perturbed neural network operators. Submitted 2014
13. G.A. Anastassiou, Approximations by multivariate perturbed neural network operators. Submitted 2014
14. G.A. Anastassiou, Approximation by fuzzy perturbed neural network operators. Submitted 2014
15. P. Cardaliaguet, G. Euvrard, Approximation of a function and its derivative with a neural network. Neural Netw. **5**, 207–220 (1992)
16. S. Gal, in *Approximation Theory in Fuzzy Setting*, ed. by G. Anastassiou. Handbook ofAnalytic-Computational Methods in Applied Mathematics, Chap. 13 (Chapman & Hall/CRC, Boca Raton, 2000), pp. 617–666

17. R. Goetschel Jr, W. Voxman, Elementary fuzzy calculus. Fuzzy Sets Syst. **18**, 31–43 (1986)
18. O. Kaleva, Fuzzy differential equations. Fuzzy Sets Syst. **24**, 301–317 (1987)
19. D.D. Stancu, On a generalization of the Bernstein polynomials. Studia Universitatis Babeş-Bolyai, Series Mathematica-Physica **14**, 31–45 (1969)
20. C. Wu, Z. Gong, On Henstock integrals of interval-valued functions and fuzzy-valued functions. Fuzzy Sets Syst. **115**(3), 377–391 (2000)
21. C. Wu, Z. Gong, On Henstock integral of fuzzy number valued functions (I). Fuzzy Sets Syst. **120**(3), 523–532 (2001)
22. C. Wu, M. Ma, On embedding problem of fuzzy number spaces: Part 1. Fuzzy Sets Syst. **44**, 33–38 (1991)

Chapter 27
Multivariate Fuzzy Perturbed Neural Network Approximations

This chapter studies the determination of the rate of convergence to the unit of each of three newly introduced here multivariate fuzzy perturbed normalized neural network operators of one hidden layer. These are given through the multivariate fuzzy modulus of continuity of the involved multivariate fuzzy number valued function or its high order fuzzy partial derivatives and that appears in the right-hand side of the associated fuzzy multivariate Jackson type inequalities. The multivariate activation function is very general, especially it can derive from any sigmoid or bell-shaped function. The right hand sides of our multivariate fuzzy convergence inequalities do not depend on the activation function. The sample multivariate fuzzy functionals are of Stancu, Kantorovich and Quadrature types. We give applications for the first fuzzy partial derivatives of the involved function. It follows [13].

27.1 Introduction

The Cardaliaguet-Euvrard real neural network operators were studied extensively in [16], where the authors among many other things proved that these operators converge uniformly on compacta, to the unit over continuous and bounded functions. Our fuzzy "multivariate perturbed normalized neural network operators" are motivated and inspired by the "multivariate bell" and "multivariate squashing functions" of [16]. The work in [16] is qualitative where the used multivariate bell-shaped function is general. However, our work, though greatly motivated by [16], is quantitative and the used multivariate activation functions are of compact support. Here we extend to the fuzzy environment our initial real work, see [15]. We derive a series of multivariate fuzzy Jackson type inequalities giving close upper bounds to the errors in approximating the unit operator by the above multivariate fuzzy perturbed neural network induced operators. All involved constants there are well determined. These are pointwise and uniform estimates involving the multivariate first fuzzy modulus of continuity of the engaged multivariate fuzzy function or the fuzzy partial

G.A. Anastassiou, *Intelligent Systems II: Complete Approximation by Neural Network Operators*, Studies in Computational Intelligence 608,
DOI 10.1007/978-3-319-20505-2_27

derivatives of the function under approximation. We give all necessary background of the multivariate fuzzy calculus needed.

Initial work of the subject was done in [11, 12]. These works motivate the current work.

27.2 Fuzzy Multivariate Real Analysis Background

We need the following basic background

Definition 27.1 (*see* [23]) Let $\mu : \mathbb{R} \to [0, 1]$ with the following properties:

(i) is normal, i.e., $\exists\, x_0 \in \mathbb{R} : \mu(x_0) = 1$.
(ii) $\mu(\lambda x + (1-\lambda) y) \geq \min\{\mu(x), \mu(y)\}, \forall\, x, y \in \mathbb{R}, \forall\, \lambda \in [0, 1]$ (μ is called a convex fuzzy subset).
(iii) μ is upper semicontinuous on $\mathbb{R}$, i.e., $\forall\, x_0 \in \mathbb{R}$ and $\forall\, \varepsilon > 0$, $\exists$ neighborhood $V(x_0) : \mu(x) \leq \mu(x_0) + \varepsilon, \forall\, x \in V(x_0)$.
(iv) the set $\overline{\operatorname{supp}(\mu)}$ is compact in $\mathbb{R}$ (where $\operatorname{supp}(\mu) := \{x \in \mathbb{R} : \mu(x) > 0\}$).

We call μ a fuzzy real number. Denote the set of all μ with $\mathbb{R}_{\mathcal{F}}$.

E.g., $\chi_{\{x_0\}} \in \mathbb{R}_{\mathcal{F}}$, for any $x_0 \in \mathbb{R}$, where $\chi_{\{x_0\}}$ is the characteristic function at x_0.

For $0 < r \leq 1$ and $\mu \in \mathbb{R}_{\mathcal{F}}$ define

$$[\mu]^r := \{x \in \mathbb{R} : \mu(x) \geq r\} \tag{27.1}$$

and

$$[\mu]^0 := \overline{\{x \in \mathbb{R} : \mu(x) \geq 0\}}.$$

Then it is well known that for each $r \in [0, 1]$, $[\mu]^r$ is a closed and bounded interval of $\mathbb{R}$ [18].

For $u, v \in \mathbb{R}_{\mathcal{F}}$ and $\lambda \in \mathbb{R}$, we define uniquely the sum $u \oplus v$ and the product $\lambda \odot u$ by

$$[u \oplus v]^r = [u]^r + [v]^r, \quad [\lambda \odot u]^r = \lambda [u]^r, \; \forall\, r \in [0, 1],$$

where $[u]^r + [v]^r$ means the usual addition of two intervals (as subsets of $\mathbb{R}$) and $\lambda [u]^r$ means the usual product between a scalar and a subset of $\mathbb{R}$ (see, e.g., [23]).

Notice $1 \odot u = u$ and it holds

$$u \oplus v = v \oplus u, \; \lambda \odot u = u \odot \lambda.$$

If $0 \leq r_1 \leq r_2 \leq 1$ then

$$[u]^{r_2} \subseteq [u]^{r_1}.$$

Actually $[u]^r = \left[u_-^{(r)}, u_+^{(r)}\right]$, where $u_-^{(r)} \leq u_+^{(r)}$, $u_-^{(r)}, u_+^{(r)} \in \mathbb{R}, \forall\, r \in [0, 1]$.

For $\lambda > 0$ one has $\lambda u_{\pm}^{(r)} = (\lambda \odot u)_{\pm}^{(r)}$, respectively.

Define $D : \mathbb{R}_{\mathcal{F}} \times \mathbb{R}_{\mathcal{F}} \to \mathbb{R}_{\mathcal{F}}$ by

$$D(u, v) := \sup_{r \in [0,1]} \max \left\{ \left| u_{-}^{(r)} - v_{-}^{(r)} \right|, \left| u_{+}^{(r)} - v_{+}^{(r)} \right| \right\}, \tag{27.2}$$

where

$$[v]^r = \left[v_{-}^{(r)}, v_{+}^{(r)} \right]; \; u, v \in \mathbb{R}_{\mathcal{F}}.$$

We have that D is a metric on $\mathbb{R}_{\mathcal{F}}$.

Then $(\mathbb{R}_{\mathcal{F}}, D)$ is a complete metric space, see [23, 24].

Let $f, g : \mathbb{R}^m \to \mathbb{R}_{\mathcal{F}}$. We define the distance

$$D^*(f, g) := \sup_{x \in \mathbb{R}^m} D(f(x), g(x)).$$

Here $\sum^{*}$ stands for fuzzy summation and $\widetilde{0} := \chi_{\{0\}} \in \mathbb{R}_{\mathcal{F}}$ is the neutral element with respect to $\oplus$, i.e.,

$$u \oplus \widetilde{0} = \widetilde{0} \oplus u = u, \; \forall \, u \in \mathbb{R}_{\mathcal{F}}.$$

We need

Remark 27.2 ([4]). Here $r \in [0, 1]$, $x_i^{(r)}, y_i^{(r)} \in \mathbb{R}$, $i = 1, \ldots, m \in \mathbb{N}$. Suppose that

$$\sup_{r \in [0,1]} \max \left(x_i^{(r)}, y_i^{(r)} \right) \in \mathbb{R}, \text{ for } i = 1, \ldots, m.$$

Then one sees easily that

$$\sup_{r \in [0,1]} \max \left(\sum_{i=1}^{m} x_i^{(r)}, \sum_{i=1}^{m} y_i^{(r)} \right) \leq \sum_{i=1}^{m} \sup_{r \in [0,1]} \max \left(x_i^{(r)}, y_i^{(r)} \right). \tag{27.3}$$

Definition 27.3 Let $f \in C(\mathbb{R}^m)$, $m \in \mathbb{N}$, which is bounded or uniformly continuous, we define $(h > 0)$

$$\omega_1(f, h) := \sup_{\text{all } x_i, x_i' \in \mathbb{R}, \, |x_i - x_i'| \leq h, \text{ for } i=1,\ldots,m} \left| f(x_1, \ldots, x_m) - f(x_1', \ldots, x_m') \right|. \tag{27.4}$$

Definition 27.4 Let $f : \mathbb{R}^m \to \mathbb{R}_{\mathcal{F}}$, we define the fuzzy modulus of continuity of f by

$$\omega_1^{(\mathcal{F})}(f, \delta) = \sup_{x, y \in \mathbb{R}^m, \, |x_i - y_i| \leq \delta, \text{ for } i=1,\ldots,m} D(f(x), f(y)), \; \delta > 0, \tag{27.5}$$

where $x = (x_1, \ldots, x_m)$, $y = (y_1, \ldots, y_m)$.

For $f : \mathbb{R}^m \to \mathbb{R}_{\mathcal{F}}$, we use

$$[f]^r = \left[f_-^{(r)}, f_+^{(r)} \right], \tag{27.6}$$

where $f_\pm^{(r)} : \mathbb{R}^m \to \mathbb{R}$, $\forall\, r \in [0, 1]$.

We need

Proposition 27.5 *Let $f : \mathbb{R}^m \to \mathbb{R}_{\mathcal{F}}$. Assume that $\omega_1^{\mathcal{F}}(f, \delta)$, $\omega_1\left(f_-^{(r)}, \delta\right)$, $\omega_1\left(f_+^{(r)}, \delta\right)$ are finite for any $\delta > 0$, $r \in [0, 1]$.*

Then

$$\omega_1^{(\mathcal{F})}(f, \delta) = \sup_{r \in [0,1]} \max\left\{ \omega_1\left(f_-^{(r)}, \delta\right), \omega_1\left(f_+^{(r)}, \delta\right) \right\}. \tag{27.7}$$

Proof By Proposition 1 of [8]. ■

We define by $C_{\mathcal{F}}^U(\mathbb{R}^m)$ the space of fuzzy uniformly continuous functions from $\mathbb{R}^m \to \mathbb{R}_{\mathcal{F}}$, also $C_{\mathcal{F}}(\mathbb{R}^m)$ is the space of fuzzy continuous functions on $\mathbb{R}^m$, and $C_B(\mathbb{R}^m, \mathbb{R}_{\mathcal{F}})$ is the fuzzy continuous and bounded functions.

We mention

Proposition 27.6 ([6]) *Let $f \in C_{\mathcal{F}}^U(\mathbb{R}^m)$. Then $\omega_1^{(\mathcal{F})}(f, \delta) < \infty$, for any $\delta > 0$.*

Proposition 27.7 ([6]) *It holds*

$$\lim_{\delta \to 0} \omega_1^{(\mathcal{F})}(f, \delta) = \omega_1^{(\mathcal{F})}(f, 0) = 0, \tag{27.8}$$

iff $f \in C_{\mathcal{F}}^U(\mathbb{R}^m)$.

Proposition 27.8 ([6]) *Let $f \in C_{\mathcal{F}}(\mathbb{R}^m)$. Then $f_\pm^{(r)}$ are equicontinuous with respect to $r \in [0, 1]$ over $\mathbb{R}^m$, respectively in $\pm$.*

Note 27.9 *It is clear by Propositions 27.5, 27.7, that if $f \in C_{\mathcal{F}}^U(\mathbb{R}^m)$, then $f_\pm^{(r)} \in C_U(\mathbb{R}^m)$ (uniformly continuous on $\mathbb{R}^m$). Clearly also if $f \in C_B(\mathbb{R}^m, \mathbb{R}_{\mathcal{F}})$, then $f_\pm^{(r)} \in C_B(\mathbb{R}^m)$ (continuous and bounded functions on $\mathbb{R}^m$).*

We need

Definition 27.10 Let $x, y \in \mathbb{R}_{\mathcal{F}}$. If there exists $z \in \mathbb{R}_{\mathcal{F}} : x = y \oplus z$, then we call z the H-difference on x and y, denoted $x - y$.

Definition 27.11 ([23]) Let $T := [x_0, x_0 + \beta] \subset \mathbb{R}$, with $\beta > 0$. A function $f : T \to \mathbb{R}_{\mathcal{F}}$ is H-difference at $x \in T$ if there exists an $f'(x) \in \mathbb{R}_{\mathcal{F}}$ such that the limits (with respect to D)

$$\lim_{h\to 0+}\frac{f(x+h)-f(x)}{h},\quad \lim_{h\to 0+}\frac{f(x)-f(x-h)}{h} \tag{27.9}$$

exist and are equal to $f'(x)$.

We call f' the H-derivative or fuzzy derivative of f at x.

Above is assumed that the H-differences $f(x+h)-f(x)$, $f(x)-f(x-h)$ exist in $\mathbb{R}_{\mathcal{F}}$ in a neighborhood of x.

Definition 27.12 We denote by $C_{\mathcal{F}}^{N}(\mathbb{R}^m)$, $N\in\mathbb{N}$, the space of all N-times fuzzy continuously differentiable functions from $\mathbb{R}^m$ into $\mathbb{R}_{\mathcal{F}}$.

Here fuzzy partial derivatives are defined via Definition 27.11 in the obvious way as in the ordinary real case.

We mention

Theorem 27.13 ([19]) *Let $f:[a,b]\subseteq\mathbb{R}\to\mathbb{R}_{\mathcal{F}}$ be H-fuzzy differentiable. Let $t\in[a,b]$, $0\le r\le 1$. Clearly*

$$[f(t)]^r=\left[f(t)_-^{(r)},f(t)_+^{(r)}\right]\subseteq\mathbb{R}.$$

Then $(f(t))_{\pm}^{(r)}$ are differentiable and

$$[f'(t)]^r=\left[\left(f(t)_-^{(r)}\right)',\left(f(t)_+^{(r)}\right)'\right].$$

I.e.

$$(f')_{\pm}^{(r)}=\left(f_{\pm}^{(r)}\right)',\quad\forall\, r\in[0,1]. \tag{27.10}$$

Remark 27.14 (*see also* [5]) Let $f\in C_{\mathcal{F}}^{N}(\mathbb{R})$, $N\ge 1$. Then by Theorem 27.13 we obtain $f_{\pm}^{(r)}\in C^N(\mathbb{R})$ and

$$\left[f^{(i)}(t)\right]^r=\left[\left(f(t)_-^{(r)}\right)^{(i)},\left(f(t)_+^{(r)}\right)^{(i)}\right],$$

for $i=0,1,2,\ldots,N$, and in particular we have that

$$\left(f^{(i)}\right)_{\pm}^{(r)}=\left(f_{\pm}^{(r)}\right)^{(i)}, \tag{27.11}$$

for any $r\in[0,1]$.

Let $f\in C_{\mathcal{F}}^{N}(\mathbb{R}^m)$, denote $f_{\widetilde{\alpha}}:=\frac{\partial^{\widetilde{\alpha}}f}{\partial x^{\widetilde{\alpha}}}$, where $\widetilde{\alpha}:=(\widetilde{\alpha_1},\ldots,\widetilde{\alpha_m})$, $\widetilde{\alpha_i}\in\mathbb{Z}^+$, $i=1,\ldots,m$ and

$$0<|\widetilde{\alpha}|:=\sum_{i=1}^{m}\widetilde{\alpha_i}\le N,\quad N>1.$$

Then by Theorem 27.13 we get that

$$\left(f_{\pm}^{(r)}\right)_{\widetilde{\alpha}} = (f_{\widetilde{\alpha}})_{\pm}^{(r)}, \ \forall\, r \in [0,1], \tag{27.12}$$

and any $\widetilde{\alpha} : |\widetilde{\alpha}| \leq N$. Here $f_{\pm}^{(r)} \in C^N(\mathbb{R}^m)$.

For the definition of general fuzzy integral we follow [19] next.

Definition 27.15 Let (Ω, Σ, μ) be a complete σ-finite measure space. We call $F : \Omega \to \mathbb{R}_{\mathcal{F}}$ measurable iff $\forall$ closed $B \subseteq \mathbb{R}$ the function $F^{-1}(B) : \Omega \to [0,1]$ defined by

$$F^{-1}(B)(w) := \sup_{x \in B} F(w)(x), \text{ all } w \in \Omega$$

is measurable, see [20].

Theorem 27.16 ([20]) *For* $F : \Omega \to \mathbb{R}_{\mathcal{F}}$,

$$F(w) = \{(F_{-}^{(r)}(w), F_{+}^{(r)}(w)) | 0 \leq r \leq 1\},$$

the following are equivalent

(1) F is measurable,

(2) $\forall\, r \in [0,1]$, $F_{-}^{(r)}$, $F_{+}^{(r)}$ *are measurable.*

Following [20], given that for each $r \in [0,1]$, $F_{-}^{(r)}$, $F_{+}^{(r)}$ are integrable we have that the parametrized representation

$$\left\{\left(\int_A F_{-}^{(r)} d\mu, \int_A F_{+}^{(r)} d\mu\right) | 0 \leq r \leq 1\right\}$$

is a fuzzy real number for each $A \in \Sigma$.

The last fact leads to

Definition 27.17 ([20]) A measurable function $F : \Omega \to \mathbb{R}_{\mathcal{F}}$,

$$F(w) = \{(F_{-}^{(r)}(w), F_{+}^{(r)}(w)) | 0 \leq r \leq 1\}$$

is called integrable if for each $r \in [0,1]$, $F_{\pm}^{(r)}$ are integrable, or equivalently, if $F_{\pm}^{(0)}$ are integrable.

In this case, the fuzzy integral of F over $A \in \Sigma$ is defined by

$$\int_A F d\mu := \left\{\left(\int_A F_{-}^{(r)} d\mu, \int_A F_{+}^{(r)} d\mu\right) | 0 \leq r \leq 1\right\}. \tag{27.13}$$

By [20] F is integrable iff $w \to \|F(w)\|_{\mathcal{F}}$ is real-valued integrable. Here

$$\|u\|_{\mathcal{F}} := D\left(u, \widetilde{0}\right), \ \forall\, u \in \mathbb{R}_{\mathcal{F}}.$$

We need also

Theorem 27.18 ([20]) *Let $F, G : \Omega \to \mathbb{R}_{\mathcal{F}}$ be integrable. Then*
(1) Let $a, b \in \mathbb{R}$, then $aF + bG$ is integrable and for each $A \in \Sigma$,

$$\int_A (aF + bG)\, d\mu = a \int_A F d\mu + b \int_A G d\mu;$$

(2) $D(F, G)$ is a real-valued integrable function and for each $A \in \Sigma$,

$$D\left(\int_A F d\mu, \int_A G d\mu\right) \leq \int_A D(F, G)\, d\mu. \tag{27.14}$$

In particular,

$$\left\| \int_A F d\mu \right\|_{\mathcal{F}} \leq \int_A \|F\|_{\mathcal{F}}\, d\mu.$$

Above μ could be the Lebesgue measure, with all the basic properties valid here too.

Basically here we have

$$\left[\int_A F d\mu\right]^r := \left[\int_A F_-^{(r)} d\mu, \int_A F_+^{(r)} d\mu\right], \tag{27.15}$$

i.e.

$$\left(\int_A F d\mu\right)_{\pm}^{(r)} = \int_A F_{\pm}^{(r)} d\mu, \tag{27.16}$$

$\forall\, r \in [0, 1]$, respectively.

In this chapter we use the fuzzy integral with respect to Lebesgue measure on $\mathbb{R}^m$. See also Fubini's theorem from [20].

We also need

Notation 27.19 *We denote*

$$\left(\sum_{i=1}^{2} D\left(\frac{\partial}{\partial x_i}, \widetilde{o}\right)\right)^2 f(x) := \tag{27.17}$$

$$D\left(\frac{\partial^2 f(x_1, x_2)}{\partial x_1^2}, \widetilde{o}\right) + D\left(\frac{\partial^2 f(x_1, x_2)}{\partial x_2^2}, \widetilde{o}\right) + 2D\left(\frac{\partial^2 f(x_1, x_2)}{\partial x_1 \partial x_2}, \widetilde{o}\right).$$

In general we denote ($j = 1, \ldots, N$)

$$\left(\sum_{i=1}^{m} D\left(\frac{\partial}{\partial x_i}, \widetilde{0}\right)\right)^{j} f(x) := \tag{27.18}$$

$$\sum_{(j_1,\ldots,j_m)\in\mathbb{Z}_+^m : \sum_{i=1}^m j_i = j} \frac{j!}{j_1! j_2! \ldots j_m!} D\left(\frac{\partial^j f(x_1,\ldots,x_m)}{\partial x_1^{j_1} \partial x_2^{j_2} \ldots \partial x_m^{j_m}}, \widetilde{0}\right).$$

27.3 Real Neural Networks Multivariate Approximation Basics (see [15])

Here the activation function $b : \mathbb{R}^d \to \mathbb{R}_+$, $d \in \mathbb{N}$, is of compact support $\overline{B} := \prod_{j=1}^{d} [-T_j, T_j]$, $T_j > 0$, $j = 1, \ldots, d$. That is $b(x) > 0$ for any $x \in \overline{B}$, and clearly b may have jump discontinuities. Also the shape of the graph of b is immaterial.

Typically in neural networks approximation we take b to be a d-dimensional bell-shaped function (i.e. per coordinate is a centered bell-shaped function), or a product of univariate centered bell-shaped functions, or a product of sigmoid functions, in our case all of them are of compact support $\overline{B}$.

Example 27.20 Take $b(x) = \beta(x_1)\beta(x_2)\ldots\beta(x_d)$, where β is any of the following functions, $j = 1, \ldots, d$:

(i) $\beta(x_j)$ is the characteristic function on $[-1, 1]$,
(ii) $\beta(x_j)$ is the hat function over $[-1, 1]$, that is,

$$\beta(x_j) = \begin{cases} 1 + x_j, & -1 \le x_j \le 0, \\ 1 - x_j, & 0 < x_j \le 1, \\ 0, & \textit{elsewhere}, \end{cases}$$

(iii) the truncated sigmoids

$$\beta(x_j) = \begin{cases} \frac{1}{1+e^{-x_j}} \text{ or } \tanh x_j \text{ or } \operatorname{erf}(x_j), \text{ for } x_j \in [-T_j, T_j], \text{ with large } T_j > 0, \\ 0, \; x_j \in \mathbb{R} - [-T_j, T_j], \end{cases}$$

(iv) the truncated Gompertz function

$$\beta(x_j) = \begin{cases} e^{-\alpha e^{-\beta x_j}}, \; x_j \in [-T_j, T_j]; \; \alpha, \beta > 0; \text{ large } T_j > 0, \\ 0, \; x_j \in \mathbb{R} - [-T_j, T_j], \end{cases}$$

The Gompertz functions are also sigmoid functions, with wide applications to many applied fields, e.g. demography and tumor growth modeling, etc.

Thus the general activation function b we will be using here includes all kinds of activation functions in neural network approximations.

Here we consider functions $f : \mathbb{R}^d \to \mathbb{R}$ that either continuous and bounded, or uniformly continuous.

Let here the parameters: $0 < \alpha < 1$, $x = (x_1, \ldots, x_d) \in \mathbb{R}^d$, $n \in \mathbb{N}$; $\overline{r} = (r_1, \ldots, r_d) \in \mathbb{N}^d$, $i = (i_1, \ldots, i_d) \in \mathbb{N}^d$, with $i_j = 1, 2, \ldots, r_j$, $j = 1, \ldots, d$; also let $w_i = w_{i_1,\ldots,i_d} \geq 0$, such that $\sum_{i_1=1}^{r_1} \sum_{i_2=1}^{r_2} \cdots \sum_{i_d=1}^{r_d} w_{i_1,\ldots,i_d} = 1$, in brief written as $\sum_{i=1}^{\overline{r}} w_i = 1$. We further consider the parameters $k = (k_1, \ldots, k_d) \in \mathbb{Z}^d$; $\mu_i = (\mu_{i_1}, \ldots, \mu_{i_d}) \in \mathbb{R}_+^d$, $\nu_i = (\nu_{i_1}, \ldots, \nu_{i_d}) \in \mathbb{R}_+^d$; and $\lambda_i = \lambda_{i_1,\ldots,i_d}$, $\rho_i = \rho_{i_1,\ldots,i_d} \geq 0$; $\mu, \nu \geq 0$. Call $\nu_i^{\min} = \min\{\nu_{i_1}, \ldots, \nu_{i_d}\}$.

We use here the first modulus of continuity, with $\delta > 0$,

$$\omega_1(f, \delta) := \sup_{\substack{x, y \in \mathbb{R}^d \\ \|x - y\|_\infty \leq \delta}} |f(x) - f(y)|, \tag{27.19}$$

where $\|x\|_\infty = \max(|x_1|, \ldots, |x_d|)$. Notice (27.19) is equivalent to (27.4).

Given that f is uniformly continuous we get $\lim_{\delta \to 0} \omega_1(f, \delta) = 0$.

Here we mention from [15] about the pointwise convergence with rates over $\mathbb{R}^d$, to the unit operator, of the following one hidden layer multivariate normalized neural network perturbed operators,

(i) the Stancu type (see [21])

$$(H_n(f))(x) = (H_n(f))(x_1, \ldots, x_d) = \tag{27.20}$$

$$\frac{\sum_{k=-n^2}^{n^2} \left(\sum_{i=1}^{\overline{r}} w_i f\left(\frac{k+\mu_i}{n+\nu_i}\right) \right) b\left(n^{1-\alpha}\left(x - \frac{k}{n}\right)\right)}{\sum_{k=-n^2}^{n^2} b\left(n^{1-\alpha}\left(x - \frac{k}{n}\right)\right)} =$$

$$\frac{\sum_{k_1=-n^2}^{n^2} \cdots \sum_{k_d=-n^2}^{n^2} \left(\sum_{i_1=1}^{r_1} \cdots \sum_{i_d=1}^{r_d} w_{i_1,\ldots,i_d} f\left(\frac{k_1+\mu_{i_1}}{n+\nu_{i_1}}, \ldots, \frac{k_d+\mu_{i_d}}{n+\nu_{i_d}}\right)\right) \cdot}{\sum_{k_1=-n^2}^{n^2} \cdots \sum_{k_d=-n^2}^{n^2} b\left(n^{1-\alpha}\left(x_1 - \frac{k_1}{n}\right), \ldots, n^{1-\alpha}\left(x_d - \frac{k_d}{n}\right)\right)}$$

$$b\left(n^{1-\alpha}\left(x_1 - \frac{k_1}{n}\right), \ldots, n^{1-\alpha}\left(x_d - \frac{k_d}{n}\right)\right),$$

(ii) the Kantorovich type

$$(K_n(f))(x) = \tag{27.21}$$

$$\frac{\sum_{k=-n^2}^{n^2}\left(\sum_{i=1}^{\overline{r}} w_i (n+\rho_i)^d \int_0^{\frac{1}{n+\rho_i}} f\left(t+\frac{k+\lambda_i}{n+\rho_i}\right)dt\right) b\left(n^{1-\alpha}\left(x-\frac{k}{n}\right)\right)}{\sum_{k=-n^2}^{n^2} b\left(n^{1-\alpha}\left(x-\frac{k}{n}\right)\right)} =$$

$$\sum_{k_1=-n^2}^{n^2} \cdots \sum_{k_d=-n^2}^{n^2} \left(\sum_{i_1=1}^{r_1} \cdots \sum_{i_d=1}^{r_d} w_{i_1,\ldots,i_d} \left(n+\rho_{i_1,\ldots,i_d}\right)^d \cdot\right.$$

$$\frac{\left.\int \cdots \int \int_0^{\frac{1}{n+\rho_{i_1,\ldots,i_d}}} \cdots \int f\left(t_1+\frac{k_1+\lambda_{i_1,\ldots,i_d}}{n+\rho_{i_1,\ldots,i_d}}, \ldots, t_d+\frac{k_d+\lambda_{i_1,\ldots,i_d}}{n+\rho_{i_1,\ldots,i_d}}\right) dt_1 \ldots dt_d\right) \cdot}{\sum_{k_1=-n^2}^{n^2} \cdots \sum_{k_d=-n^2}^{n^2} b\left(n^{1-\alpha}\left(x_1-\frac{k_1}{n}\right), \ldots, n^{1-\alpha}\left(x_d-\frac{k_d}{n}\right)\right)} \tag{27.22}$$

$$b\left(n^{1-\alpha}\left(x_1-\frac{k_1}{n}\right), \ldots, n^{1-\alpha}\left(x_d-\frac{k_d}{n}\right)\right),$$

and

(iii) the quadrature type

$$(M_n(f))(x) = \frac{\sum_{k=-n^2}^{n^2}\left(\sum_{i=1}^{\overline{r}} w_i f\left(\frac{k}{n}+\frac{i}{n\overline{r}}\right)\right) b\left(n^{1-\alpha}\left(x-\frac{k}{n}\right)\right)}{\sum_{k=-n^2}^{n^2} b\left(n^{1-\alpha}\left(x-\frac{k}{n}\right)\right)} = \tag{27.23}$$

$$\frac{\sum_{k_1=-n^2}^{n^2} \cdots \sum_{k_d=-n^2}^{n^2}\left(\sum_{i_1=1}^{r_1} \cdots \sum_{i_d=1}^{r_d} w_{i_1,\ldots,i_d} f\left(\frac{k_1}{n}+\frac{i_1}{nr_1}, \ldots, \frac{k_d}{n}+\frac{i_d}{nr_d}\right)\right) \cdot}{\sum_{k_1=-n^2}^{n^2} \cdots \sum_{k_d=-n^2}^{n^2} b\left(n^{1-\alpha}\left(x_1-\frac{k_1}{n}\right), \ldots, n^{1-\alpha}\left(x_d-\frac{k_d}{n}\right)\right)}$$

$$b\left(n^{1-\alpha}\left(x_1-\frac{k_1}{n}\right), \ldots, n^{1-\alpha}\left(x_d-\frac{k_d}{n}\right)\right).$$

Similar operators defined for d-dimensional bell-shaped activation functions and sample coefficients $f\left(\frac{k}{n}\right) = f\left(\frac{k_1}{n}, \ldots, \frac{k_d}{n}\right)$ were studied initially in [1–3, 9, 10, 16], etc.

Here we care about the multivariate generalized perturbed cases of these operators (see [14, 15]).

Operator K_n in the corresponding Signal Processing context, represents the natural so called "time-jitter" error, where the sample information is calculated in a perturbed neighborhood of $\frac{k+\mu}{n+\nu}$ rather than exactly at the node $\frac{k}{n}$.

The perturbed sample coefficients $f\left(\frac{k+\mu}{n+\nu}\right)$ with $0 \leq \mu \leq \nu$, were first used by Stancu [21], in a totally different context, generalizing Bernstein operators approximation on $C([0,1])$.

The terms in the ratio of sums (27.20)–(27.23) can be nonzero, iff simultaneously

$$\left| n^{1-\alpha}\left(x_j - \frac{k_j}{n}\right)\right| \leq T_j, \text{ all } j = 1, \ldots, d, \tag{27.24}$$

i.e. $\left|x_j - \frac{k_j}{n}\right| \leq \frac{T_j}{n^{1-\alpha}}$, all $j = 1, \ldots, d$, iff

$$nx_j - T_j n^\alpha \leq k_j \leq nx_j + T_j n^\alpha, \text{ all } j = 1, \ldots, d. \tag{27.25}$$

To have the order

$$-n^2 \leq nx_j - T_j n^\alpha \leq k_j \leq nx_j + T_j n^\alpha \leq n^2, \tag{27.26}$$

we need $n \geq T_j + |x_j|$, all $j = 1, \ldots, d$. So (27.26) is true when we take

$$n \geq \max_{j \in \{1,\ldots,d\}} \left(T_j + |x_j|\right). \tag{27.27}$$

When $x \in \overline{B}$ in order to have (27.26) it is enough to assume that $n \geq 2T^*$, where $T^* := \max\{T_1, \ldots, T_d\} > 0$. Consider

$$\widetilde{I}_j := \left[nx_j - T_j n^\alpha, nx_j + T_j n^\alpha\right], \ j = 1, \ldots, d, \ n \in \mathbb{N}.$$

The length of $\widetilde{I}_j$ is $2T_j n^\alpha$. By Proposition 1 of [1], we get that the cardinality of $k_j \in \mathbb{Z}$ that belong to $\widetilde{I}_j := card\left(k_j\right) \geq \max\left(2T_j n^\alpha - 1, 0\right)$, any $j \in \{1, \ldots, d\}$. In order to have $card\left(k_j\right) \geq 1$, we need $2T_j n^\alpha - 1 \geq 1$ iff $n \geq T_j^{-\frac{1}{\alpha}}$, any $j \in \{1, \ldots, d\}$.

Therefore, a sufficient condition in order to obtain the order (27.26) along with the interval $\widetilde{I}_j$ to contain at least one integer for all $j = 1, \ldots, d$ is that

$$n \geq \max_{j \in \{1,\ldots,d\}} \left\{T_j + |x_j|, T_j^{-\frac{1}{\alpha}}\right\}. \tag{27.28}$$

Clearly as $n \to +\infty$ we get that $card\left(k_j\right) \to +\infty$, all $j = 1, \ldots, d$. Also notice that $card\left(k_j\right)$ equals to the cardinality of integers in $\left[\lceil nx_j - T_j n^\alpha \rceil, \left[nx_j + T_j n^\alpha\right]\right]$ for all $j = 1, \ldots, d$. Here $[\cdot]$ denotes the integral part of the number while $\lceil \cdot \rceil$ denotes its ceiling.

From now on, in this chapter we assume (27.28).

We denote by $T = (T_1, \ldots, T_d)$, $[nx + Tn^\alpha] = ([nx_1 + T_1 n^\alpha], \ldots, [nx_d + T_d n^\alpha])$, and $\lceil nx - Tn^\alpha \rceil = (\lceil nx_1 - T_1 n^\alpha \rceil, \ldots, \lceil nx_d - T_d n^\alpha \rceil)$. Furthermore it holds

(i)

$$(H_n(f))(x) = (H_n(f))(x_1, \ldots, x_d) = \tag{27.29}$$

$$\frac{\sum_{k=\lceil nx-Tn^{\alpha}\rceil}^{[nx+Tn^{\alpha}]} \left(\sum_{i=1}^{\bar{r}} w_i f\left(\frac{k+\mu_i}{n+\nu_i}\right) \right) b\left(n^{1-\alpha}\left(x-\frac{k}{n}\right)\right)}{\sum_{k=\lceil nx-Tn^{\alpha}\rceil}^{[nx+Tn^{\alpha}]} b\left(n^{1-\alpha}\left(x-\frac{k}{n}\right)\right)} =$$

$$\frac{\sum_{k_1=\lceil nx_1-T_1n^{\alpha}\rceil}^{[nx_1+T_1n^{\alpha}]} \cdots \sum_{k_d=\lceil nx_d-T_dn^{\alpha}\rceil}^{[nx_d+T_dn^{\alpha}]} \left(\sum_{i_1=1}^{r_1} \cdots \sum_{i_d=1}^{r_d} w_{i_1,\ldots,i_d} f\left(\frac{k_1+\mu_{i_1}}{n+\nu_{i_1}}, \ldots, \frac{k_d+\mu_{i_d}}{n+\nu_{i_d}}\right) \right) \cdot}{\sum_{k_1=\lceil nx_1-T_1n^{\alpha}\rceil}^{[nx_1+T_1n^{\alpha}]} \cdots \sum_{k_d=\lceil nx_d-T_dn^{\alpha}\rceil}^{[nx_d+T_dn^{\alpha}]} b\left(n^{1-\alpha}\left(x_1-\frac{k_1}{n}\right), \ldots, n^{1-\alpha}\left(x_d-\frac{k_d}{n}\right)\right)}$$

$$b\left(n^{1-\alpha}\left(x_1-\frac{k_1}{n}\right), \ldots, n^{1-\alpha}\left(x_d-\frac{k_d}{n}\right)\right),$$

(ii)

$$(K_n(f))(x) = \tag{27.30}$$

$$\frac{\sum_{k=\lceil nx-Tn^{\alpha}\rceil}^{[nx+Tn^{\alpha}]} \left(\sum_{i=1}^{\bar{r}} w_i (n+\rho_i)^d \int_0^{\frac{1}{n+\rho_i}} f\left(t+\frac{k+\lambda_i}{n+\rho_i}\right) dt \right) b\left(n^{1-\alpha}\left(x-\frac{k}{n}\right)\right)}{\sum_{k=\lceil nx-Tn^{\alpha}\rceil}^{[nx+Tn^{\alpha}]} b\left(n^{1-\alpha}\left(x-\frac{k}{n}\right)\right)} =$$

$$\sum_{k_1=\lceil nx_1-T_1n^{\alpha}\rceil}^{[nx_1+T_1n^{\alpha}]} \cdots \sum_{k_d=\lceil nx_d-T_dn^{\alpha}\rceil}^{[nx_d+T_dn^{\alpha}]} \left(\sum_{i_1=1}^{r_1} \cdots \sum_{i_d=1}^{r_d} w_{i_1,\ldots,i_d} \left(n+\rho_{i_1,\ldots,i_d}\right)^d \cdot \right.$$

$$\frac{\left. \int \cdots \int_0^{\frac{1}{n+\rho_{i_1,\ldots,i_d}}} \cdots \int f\left(t_1+\frac{k_1+\lambda_{i_1,\ldots,i_d}}{n+\rho_{i_1,\ldots,i_d}}, \ldots, t_d+\frac{k_d+\lambda_{i_1,\ldots,i_d}}{n+\rho_{i_1,\ldots,i_d}}\right) dt_1 \ldots dt_d \right) \cdot}{\sum_{k_1=\lceil nx_1-T_1n^{\alpha}\rceil}^{[nx_1+T_1n^{\alpha}]} \cdots \sum_{k_d=\lceil nx_d-T_dn^{\alpha}\rceil}^{[nx_d+T_dn^{\alpha}]} b\left(n^{1-\alpha}\left(x_1-\frac{k_1}{n}\right), \ldots, n^{1-\alpha}\left(x_d-\frac{k_d}{n}\right)\right)} \tag{27.31}$$

$$b\left(n^{1-\alpha}\left(x_1-\frac{k_1}{n}\right), \ldots, n^{1-\alpha}\left(x_d-\frac{k_d}{n}\right)\right),$$

and

(iii)

$$(M_n(f))(x) = \frac{\sum_{k=\lceil nx-Tn^{\alpha}\rceil}^{[nx+Tn^{\alpha}]} \left(\sum_{i=1}^{\bar{r}} w_i f\left(\frac{k}{n}+\frac{i}{n\bar{r}}\right) \right) b\left(n^{1-\alpha}\left(x-\frac{k}{n}\right)\right)}{\sum_{k=\lceil nx-Tn^{\alpha}\rceil}^{[nx+Tn^{\alpha}]} b\left(n^{1-\alpha}\left(x-\frac{k}{n}\right)\right)} = \tag{27.32}$$

$$\frac{\sum_{k_1=\lceil nx_1-T_1n^{\alpha}\rceil}^{[nx_1+T_1n^{\alpha}]}\cdots\sum_{k_d=\lceil nx_d-T_dn^{\alpha}\rceil}^{[nx_d+T_dn^{\alpha}]}\left(\sum_{i_1=1}^{r_1}\cdots\sum_{i_d=1}^{r_d}w_{i_1,\ldots,i_d}f\left(\frac{k_1}{n}+\frac{i_1}{nr_1},\ldots,\frac{k_d}{n}+\frac{i_d}{nr_d}\right)\right)\cdot}{\sum_{k_1=\lceil nx_1-T_1n^{\alpha}\rceil}^{[nx_1+T_1n^{\alpha}]}\cdots\sum_{k_d=\lceil nx_d-T_dn^{\alpha}\rceil}^{[nx_d+T_dn^{\alpha}]}b\left(n^{1-\alpha}\left(x_1-\frac{k_1}{n}\right),\ldots,n^{1-\alpha}\left(x_d-\frac{k_d}{n}\right)\right)}$$

$$b\left(n^{1-\alpha}\left(x_1-\frac{k_1}{n}\right),\ldots,n^{1-\alpha}\left(x_d-\frac{k_d}{n}\right)\right).$$

So if $\left|n^{1-\alpha}\left(x_j-\frac{k_j}{n}\right)\right|\leq T_j$, all $j=1,\ldots,d$, we get that

$$\left\|x-\frac{k}{n}\right\|_{\infty}\leq\frac{T^*}{n^{1-\alpha}}. \tag{27.33}$$

For convinience we call

$$V(x)=\sum_{k=\lceil nx-Tn^{\alpha}\rceil}^{[nx+Tn^{\alpha}]}b\left(n^{1-\alpha}\left(x-\frac{k}{n}\right)\right)=$$

$$\sum_{k_1=\lceil nx_1-T_1n^{\alpha}\rceil}^{[nx_1+T_1n^{\alpha}]}\cdots\sum_{k_d=\lceil nx_d-T_dn^{\alpha}\rceil}^{[nx_d+T_dn^{\alpha}]}b\left(n^{1-\alpha}\left(x_1-\frac{k_1}{n}\right),\ldots,n^{1-\alpha}\left(x_d-\frac{k_d}{n}\right)\right). \tag{27.34}$$

We make

Remark 27.21 (*see* [15]) Here always k is as in (27.26).
(I) We have that

$$\left\|\frac{k+\mu_i}{n+\nu_i}-x\right\|_{\infty}\leq\left(\frac{\|\nu_i\|_{\infty}\|x\|_{\infty}+\|\mu_i\|_{\infty}}{n+\nu_i^{\min}}\right)+\left(1+\frac{\|\nu_i\|_{\infty}}{\left(n+\nu_i^{\min}\right)}\right)\frac{T^*}{n^{1-\alpha}}. \tag{27.35}$$

Hence we derive

$$\omega_1\left(f,\left\|\frac{k+\mu_i}{n+\nu_i}-x\right\|_{\infty}\right)\leq$$

$$\omega_1\left(f,\left(\frac{\|\nu_i\|_{\infty}\|x\|_{\infty}+\|\mu_i\|_{\infty}}{n+\nu_i^{\min}}\right)+\left(1+\frac{\|\nu_i\|_{\infty}}{\left(n+\nu_i^{\min}\right)}\right)\frac{T^*}{n^{1-\alpha}}\right), \tag{27.36}$$

with dominant speed of convergence $\frac{1}{n^{1-\alpha}}$.
(II) We also have for

$$0\leq t_j\leq\frac{1}{n+\rho_{i_1,\ldots,i_d}},\ j=1,\ldots,d, \tag{27.37}$$

that it holds

$$\left\| t+\frac{k+\lambda_{i_1,\ldots,i_d}}{n+\rho_{i_1,\ldots,i_d}}-x\right\|_\infty \leq$$

$$\left(\frac{\rho_{i_1,\ldots,i_d}\|x\|_\infty+\lambda_{i_1,\ldots,i_d}+1}{n+\rho_{i_1,\ldots,i_d}}\right)+\left(1+\frac{\rho_{i_1,\ldots,i_d}}{n+\rho_{i_1,\ldots,i_d}}\right)\frac{T^*}{n^{1-\alpha}}, \tag{27.38}$$

and

$$\omega_1\left(f,\left\| t+\frac{k+\lambda_{i_1,\ldots,i_d}}{n+\rho_{i_1,\ldots,i_d}}-x\right\|_\infty\right) \leq$$

$$\omega_1\left(f,\left(\frac{\rho_{i_1,\ldots,i_d}\|x\|_\infty+\lambda_{i_1,\ldots,i_d}+1}{n+\rho_{i_1,\ldots,i_d}}\right)+\left(1+\frac{\rho_{i_1,\ldots,i_d}}{n+\rho_{i_1,\ldots,i_d}}\right)\frac{T^*}{n^{1-\alpha}}\right), \tag{27.39}$$

with dominant speed $\frac{1}{n^{1-\alpha}}$.
(III) We also have

$$\left\| \frac{k}{n}+\frac{i}{n\overline{r}}-x\right\|_\infty \leq \frac{T^*}{n^{1-\alpha}}+\frac{1}{n}, \tag{27.40}$$

and

$$\omega_1\left(f,\left\| \frac{k}{n}+\frac{i}{n\overline{r}}-x\right\|_\infty\right) \leq \omega_1\left(f,\frac{T^*}{n^{1-\alpha}}+\frac{1}{n}\right), \tag{27.41}$$

with dominant speed $\frac{1}{n^{1-\alpha}}$.

Inequalities (27.35)–(27.41) were essential in the proofs of the next Theorems 27.22–27.27, proved and presented in [15].

So we mention from [15] to use in this chapter the following results.

Theorem 27.22 *Let $x \in \mathbb{R}^d$ and $n \in \mathbb{N}$ such that $n \geq \max_{j\in\{1,\ldots,d\}}\left(T_j+|x_j|, T_j^{-\frac{1}{\alpha}}\right)$, $T_j > 0$, $0<\alpha<1$; $f \in C_B\left(\mathbb{R}^d\right)$ or $f \in C_U\left(\mathbb{R}^d\right)$. Then*

$$|(H_n(f))(x)-f(x)| \leq$$

$$\sum_{i=1}^{\overline{r}} w_i\omega_1\left(f,\left(\frac{\|\nu_i\|_\infty\|x\|_\infty+\|\mu_i\|_\infty}{n+\nu_i^{\min}}\right)+\left(1+\frac{\|\nu_i\|_\infty}{n+\nu_i^{\min}}\right)\frac{T^*}{n^{1-\alpha}}\right) = \tag{27.42}$$

$$\sum_{i_1=1}^{r_1}\cdots\sum_{i_d=1}^{r_d} w_{i_1,\ldots,i_d}\omega_1\left(f,\left(\frac{\|\nu_i\|_\infty\|x\|_\infty+\|\mu_i\|_\infty}{n+\nu_i^{\min}}\right)+\left(1+\frac{\|\nu_i\|_\infty}{n+\nu_i^{\min}}\right)\frac{T^*}{n^{1-\alpha}}\right),$$

where $i=(i_1,\ldots,i_d)$.

Theorem 27.23 *All assumptions as in Theorem 27.22. Then*

$$|(K_n(f))(x) - f(x)| \leq$$

$$\sum_{i=1}^{\bar{r}} w_i \omega_1\left(f, \left(\frac{\rho_i \|x\|_\infty + \lambda_i + 1}{n + \rho_i}\right) + \left(1 + \frac{\rho_i}{n + \rho_i}\right)\frac{T^*}{n^{1-\alpha}}\right) = \tag{27.43}$$

$$\sum_{i_1=1}^{r_1} \cdots \sum_{i_d=1}^{r_d} w_{i_1,\ldots,i_d}\omega_1\left(f, \left(\frac{\rho_{i_1,\ldots,i_d}\|x\|_\infty + \lambda_{i_1,\ldots,i_d} + 1}{n + \rho_{i_1,\ldots,i_d}}\right) + \left(1 + \frac{\rho_{i_1,\ldots,i_d}}{n + \rho_{i_1,\ldots,i_d}}\right)\frac{T^*}{n^{1-\alpha}}\right).$$

Theorem 27.24 *All here as in Theorem 27.22. Then*

$$|M_n(f)(x) - f(x)| \leq \omega_1\left(f, \frac{T^*}{n^{1-\alpha}} + \frac{1}{n}\right). \tag{27.44}$$

All convergences in (27.42)–(27.44) are at the rate of $\frac{1}{n^{1-\alpha}}$, when $f \in C_U(\mathbb{R}^d)$.

Theorem 27.25 *Let $x \in \mathbb{R}^d$ and $n \in \mathbb{N}$ such that $n \geq \max_{j\in\{1,\ldots,d\}}\left(T_j + |x_j|, T_j^{-\frac{1}{\alpha}}\right)$, $T_j > 0$, $0 < \alpha < 1$. Let also $f \in C^N(\mathbb{R}^d)$, $N \in \mathbb{N}$, such that all of its partial derivatives $f_{\widetilde{\alpha}}$ of order N, $\widetilde{\alpha} : |\widetilde{\alpha}| = \sum_{j=1}^{d}\alpha_j = N$, are uniformly continuous or continuous and bounded. Then*

$$|(H_n(f))(x) - f(x)| \leq \sum_{l=1}^{N}\frac{1}{l!}\left(\left(\sum_{j=1}^{d}\left|\frac{\partial}{\partial x_j}\right|\right)^l f(x)\right).$$

$$\left[\sum_{i=1}^{\bar{r}} w_i\left[\left(\frac{\|v_i\|_\infty\|x\|_\infty + \|\mu_i\|_\infty}{n + v_i^{\min}}\right) + \left(1 + \frac{\|v_i\|_\infty}{n + v_i^{\min}}\right)\frac{T^*}{n^{1-\alpha}}\right]^l\right] +$$

$$\frac{d^N}{N!}\sum_{i=1}^{\bar{r}} w_i\left[\left(\frac{\|v_i\|_\infty\|x\|_\infty + \|\mu_i\|_\infty}{n + v_i^{\min}}\right) + \left(1 + \frac{\|v_i\|_\infty}{n + v_i^{\min}}\right)\frac{T^*}{n^{1-\alpha}}\right]^N.$$

$$\max_{\widetilde{\alpha}:|\widetilde{\alpha}|=N}\omega_1\left(f_{\widetilde{\alpha}}, \left[\left(\frac{\|v_i\|_\infty\|x\|_\infty + \|\mu_i\|_\infty}{n + v_i^{\min}}\right) + \left(1 + \frac{\|v_i\|_\infty}{n + v_i^{\min}}\right)\frac{T^*}{n^{1-\alpha}}\right]\right). \tag{27.45}$$

Inequality (27.45) implies the pointwise convergence with rates on $(H_n(f))(x)$ to $f(x)$, as $n \to \infty$, at speed $\frac{1}{n^{1-\alpha}}$.

Theorem 27.26 *All here as in Theorem 27.25. Then*

$$|(K_n(f))(x) - f(x)| \leq$$

$$\sum_{l=1}^{N} \frac{1}{l!} \left(\sum_{i=1}^{\overline{r}} w_i \left[\left(\frac{\rho_i \|x\|_\infty + \lambda_i + 1}{n + \rho_i} \right) + \left(1 + \frac{\rho_i}{n + \rho_i} \right) \frac{T^*}{n^{1-\alpha}} \right]^l \right) \cdot$$

$$\left(\left(\sum_{j=1}^{d} \left| \frac{\partial}{\partial x_j} \right| \right)^l f(x) \right) + \tag{27.46}$$

$$\frac{d^N}{N!} \sum_{i=1}^{\overline{r}} w_i \left[\left(\frac{\rho_i \|x\|_\infty + \lambda_i + 1}{n + \rho_i} \right) + \left(1 + \frac{\rho_i}{n + \rho_i} \right) \frac{T^*}{n^{1-\alpha}} \right]^N \cdot$$

$$\max_{\widetilde{\alpha}:|\widetilde{\alpha}|=N} \omega_1 \left(f_{\widetilde{\alpha}}, \left[\left(\frac{\rho_i \|x\|_\infty + \lambda_i + 1}{n + \rho_i} \right) + \left(1 + \frac{\rho_i}{n + \rho_i} \right) \frac{T^*}{n^{1-\alpha}} \right] \right).$$

Inequality (27.46) implies the pointwise convergence with rates of $(K_n(f))(x)$ *to* $f(x)$, *as* $n \to \infty$, *at speed* $\frac{1}{n^{1-\alpha}}$.

Theorem 27.27 *All here as in Theorem 27.25. Then*

$$|(M_n(f))(x) - f(x)| \leq \sum_{l=1}^{N} \frac{1}{l!} \left(\left(\sum_{j=1}^{d} \left| \frac{\partial}{\partial x_j} \right| \right)^l f(x) \right) \cdot$$

$$\left(\frac{T^*}{n^{1-\alpha}} + \frac{1}{n} \right)^l + \frac{d^N}{N!} \left(\frac{T^*}{n^{1-\alpha}} + \frac{1}{n} \right)^N \max_{\widetilde{\alpha}:|\widetilde{\alpha}|=N} \omega_1 \left(f_{\widetilde{\alpha}}, \frac{T^*}{n^{1-\alpha}} + \frac{1}{n} \right). \tag{27.47}$$

Inequality (27.47) implies the pointwise convergence with rates on $(M_n(f))(x)$ *to* $f(x)$, *as* $n \to \infty$, *at speed* $\frac{1}{n^{1-\alpha}}$.

In this chapter we extend Theorems 27.22–27.27 to the fuzzy setting and environment. We give also important special cases applications of these fuzzy main results.

27.4 Fuzzy Multivariate Neural Network Approximations

Here we consider $f \in C_{\mathcal{F}}^{U}(\mathbb{R}^d)$ or $f \in C_B(\mathbb{R}^d, \mathbb{R}_{\mathcal{F}})$, b is as in Sect. 27.3, $0 < \alpha < 1$, also the rest of the parameters are as in Sect. 27.3. For $x \in \mathbb{R}^d$, we take always that $n \geq \max_{j \in \{1,\ldots,d\}} \left(T_j + |x_j|, T_j^{-\frac{1}{\alpha}} \right)$, see (27.28). The fuzzy analogs of the operators

H_n, K_n, M_n, see (27.29), (27.30) together with (27.31), and (27.32), respectivelly, follow, $n \in \mathbb{N}$.

We define the corresponding fuzzy multivariate neural network operators next:

(i)

$$\left(H_n^{\mathcal{F}}(f)\right)(x) = \left(H_n^{\mathcal{F}}(f)\right)(x_1, \ldots, x_d) =$$

$$\frac{\sum_{k=\lceil nx-Tn^{\alpha}\rceil}^{[nx+Tn^{\alpha}]^*} \left(\sum_{i=1}^{\overline{r}*} w_i \odot f\left(\frac{k+\mu_i}{n+\nu_i}\right)\right) \odot b\left(n^{1-\alpha}\left(x-\frac{k}{n}\right)\right)}{\sum_{k=\lceil nx-Tn^{\alpha}\rceil}^{[nx+Tn^{\alpha}]} b\left(n^{1-\alpha}\left(x-\frac{k}{n}\right)\right)} = \tag{27.48}$$

$$\frac{\sum_{k_1=\lceil nx_1-T_1n^{\alpha}\rceil}^{[nx_1+T_1n^{\alpha}]} \cdots \sum_{k_d=\lceil nx_d-T_dn^{\alpha}\rceil}^{[nx_d+T_dn^{\alpha}]^*} \left(\sum_{i_1=1}^{r_1} \cdots \sum_{i_d=1}^{r_d*} w_{i_1,\ldots,i_d} \odot f\left(\frac{k_1+\mu_{i_1}}{n+\nu_{i_1}}, \ldots, \frac{k_d+\mu_{i_d}}{n+\nu_{i_d}}\right)\right) \odot}{\sum_{k_1=\lceil nx_1-T_1n^{\alpha}\rceil}^{[nx_1+T_1n^{\alpha}]} \cdots \sum_{k_d=\lceil nx_d-T_dn^{\alpha}\rceil}^{[nx_d+T_dn^{\alpha}]} b\left(n^{1-\alpha}\left(x_1-\frac{k_1}{n}\right), \ldots, n^{1-\alpha}\left(x_d-\frac{k_d}{n}\right)\right)}$$

$$b\left(n^{1-\alpha}\left(x_1-\frac{k_1}{n}\right), \ldots, n^{1-\alpha}\left(x_d-\frac{k_d}{n}\right)\right),$$

(ii)

$$\left(K_n^{\mathcal{F}}(f)\right)(x) =$$

$$\frac{\sum_{k=\lceil nx-Tn^{\alpha}\rceil}^{[nx+Tn^{\alpha}]^*} \left(\sum_{i=1}^{\overline{r}*} w_i (n+\rho_i)^d \odot \int_0^{\frac{1}{n+\rho_i}} f\left(t+\frac{k+\lambda_i}{n+\rho_i}\right) dt\right) \odot b\left(n^{1-\alpha}\left(x-\frac{k}{n}\right)\right)}{\sum_{k=\lceil nx-Tn^{\alpha}\rceil}^{[nx+Tn^{\alpha}]} b\left(n^{1-\alpha}\left(x-\frac{k}{n}\right)\right)} =$$

$$\sum_{k_1=\lceil nx_1-T_1n^{\alpha}\rceil}^{[nx_1+T_1n^{\alpha}]} \cdots \sum_{k_d=\lceil nx_d-T_dn^{\alpha}\rceil}^{[nx_d+T_dn^{\alpha}]*} \left(\sum_{i_1=1}^{r_1} \cdots \sum_{i_d=1}^{r_d*} w_{i_1,\ldots,i_d} \left(n+\rho_{i_1,\ldots,i_d}\right)^d \odot \right. \tag{27.49}$$

$$\frac{\left. \int \cdots \int_0^{\frac{1}{n+\rho_{i_1,\ldots,i_d}}} \cdots \int f\left(t_1+\frac{k_1+\lambda_{i_1,\ldots,i_d}}{n+\rho_{i_1,\ldots,i_d}}, \ldots, t_d+\frac{k_d+\lambda_{i_1,\ldots,i_d}}{n+\rho_{i_1,\ldots,i_d}}\right) dt_1 \ldots dt_d \right) \odot}{\sum_{k_1=\lceil nx_1-T_1n^{\alpha}\rceil}^{[nx_1+T_1n^{\alpha}]} \cdots \sum_{k_d=\lceil nx_d-T_dn^{\alpha}\rceil}^{[nx_d+T_dn^{\alpha}]} b\left(n^{1-\alpha}\left(x_1-\frac{k_1}{n}\right), \ldots, n^{1-\alpha}\left(x_d-\frac{k_d}{n}\right)\right)}$$

$$b\left(n^{1-\alpha}\left(x_1-\frac{k_1}{n}\right), \ldots, n^{1-\alpha}\left(x_d-\frac{k_d}{n}\right)\right),$$

and

(iii)

$$\left(M_n^{\mathcal{F}}(f)\right)(x) = \frac{\sum_{k=\lceil nx-Tn^{\alpha}\rceil}^{[nx+Tn^{\alpha}]^*}\left(\sum_{i=1}^{\overline{r}*} w_i \odot f\left(\frac{k}{n}+\frac{i}{n\overline{r}}\right)\right)\odot b\left(n^{1-\alpha}\left(x-\frac{k}{n}\right)\right)}{\sum_{k=\lceil nx-Tn^{\alpha}\rceil}^{[nx+Tn^{\alpha}]} b\left(n^{1-\alpha}\left(x-\frac{k}{n}\right)\right)} = \tag{27.50}$$

$$\frac{\sum_{k_1=\lceil nx_1-T_1n^{\alpha}\rceil}^{[nx_1+T_1n^{\alpha}]}\cdots\sum_{k_d=\lceil nx_d-T_dn^{\alpha}\rceil}^{[nx_d+T_dn^{\alpha}]^*}\left(\sum_{i_1=1}^{r_1}\cdots\sum_{i_d=1}^{r_d*} w_{i_1,\dots,i_d}\odot f\left(\frac{k_1}{n}+\frac{i_1}{nr_1},\dots,\frac{k_d}{n}+\frac{i_d}{nr_d}\right)\right)\odot}{\sum_{k_1=\lceil nx_1-T_1n^{\alpha}\rceil}^{[nx_1+T_1n^{\alpha}]}\cdots\sum_{k_d=\lceil nx_d-T_dn^{\alpha}\rceil}^{[nx_d+T_dn^{\alpha}]} b\left(n^{1-\alpha}\left(x_1-\frac{k_1}{n}\right),\dots,n^{1-\alpha}\left(x_d-\frac{k_d}{n}\right)\right)}$$

$$b\left(n^{1-\alpha}\left(x_1-\frac{k_1}{n}\right),\dots,n^{1-\alpha}\left(x_d-\frac{k_d}{n}\right)\right).$$

We notice that ($r \in [0,1]$)

$$\left[\left(H_n^{\mathcal{F}}(f)\right)(x)\right]^r = \sum_{k=\lceil nx-Tn^{\alpha}\rceil}^{[nx+Tn^{\alpha}]}\left(\sum_{i=1}^{\overline{r}} w_i\left[f\left(\frac{k+\mu_i}{n+\nu_i}\right)\right]^r\right)\frac{b\left(n^{1-\alpha}\left(x-\frac{k}{n}\right)\right)}{V(x)}$$

$$= \sum_{k=\lceil nx-Tn^{\alpha}\rceil}^{[nx+Tn^{\alpha}]}\left(\sum_{i=1}^{\overline{r}} w_i\left[f_-^{(r)}\left(\frac{k+\mu_i}{n+\nu_i}\right), f_+^{(r)}\left(\frac{k+\mu_i}{n+\nu_i}\right)\right]\right)\frac{b\left(n^{1-\alpha}\left(x-\frac{k}{n}\right)\right)}{V(x)}$$

$$= \left[\sum_{k=\lceil nx-Tn^{\alpha}\rceil}^{[nx+Tn^{\alpha}]}\left(\sum_{i=1}^{\overline{r}} w_i f_-^{(r)}\left(\frac{k+\mu_i}{n+\nu_i}\right)\right)\frac{b\left(n^{1-\alpha}\left(x-\frac{k}{n}\right)\right)}{V(x)},\right. \tag{27.51}$$

$$\left.\sum_{k=\lceil nx-Tn^{\alpha}\rceil}^{[nx+Tn^{\alpha}]}\left(\sum_{i=1}^{\overline{r}} w_i f_+^{(r)}\left(\frac{k+\mu_i}{n+\nu_i}\right)\right)\frac{b\left(n^{1-\alpha}\left(x-\frac{k}{n}\right)\right)}{V(x)}\right]$$

$$= \left[\left(H_n\left(f_-^{(r)}\right)\right)(x), \left(H_n\left(f_+^{(r)}\right)\right)(x)\right].$$

We have proved that

$$\left(H_n^{\mathcal{F}}(f)\right)_{\pm}^{(r)} = H_n\left(f_{\pm}^{(r)}\right), \quad \forall\, r \in [0,1], \tag{27.52}$$

respectively.

For convinience also we call

$$A(x) = \frac{b\left(n^{1-\alpha}\left(x-\frac{k}{n}\right)\right)}{V(x)}. \tag{27.53}$$

We observe that

$$\left[\left(K_n^{\mathcal{F}}(f)\right)(x)\right]^r =$$

$$\sum_{k=\lceil nx-Tn^{\alpha}\rceil}^{[nx+Tn^{\alpha}]}\left(\sum_{i=1}^{\bar{r}} w_i (n+\rho_i)^d \left[\int_0^{\frac{1}{n+\rho_i}} f\left(t+\frac{k+\lambda_i}{n+\rho_i}\right)dt\right]^r\right)A(x) \overset{(27.15)}{=}$$

$$\sum_{k=\lceil nx-Tn^{\alpha}\rceil}^{[nx+Tn^{\alpha}]}\left(\sum_{i=1}^{\bar{r}} w_i (n+\rho_i)^d \left[\int_0^{\frac{1}{n+\rho_i}} f_-^{(r)}\left(t+\frac{k+\lambda_i}{n+\rho_i}\right)dt,\right.\right.$$

$$\left.\left.\int_0^{\frac{1}{n+\rho_i}} f_+^{(r)}\left(t+\frac{k+\lambda_i}{n+\rho_i}\right)dt\right]\right)A(x) = \tag{27.54}$$

$$\left[\sum_{k=\lceil nx-Tn^{\alpha}\rceil}^{[nx+Tn^{\alpha}]}\left(\sum_{i=1}^{\bar{r}} w_i (n+\rho_i)^d \int_0^{\frac{1}{n+\rho_i}} f_-^{(r)}\left(t+\frac{k+\lambda_i}{n+\rho_i}\right)dt\right)A(x),\right.$$

$$\left.\sum_{k=\lceil nx-Tn^{\alpha}\rceil}^{[nx+Tn^{\alpha}]}\left(\sum_{i=1}^{\bar{r}} w_i (n+\rho_i)^d \int_0^{\frac{1}{n+\rho_i}} f_+^{(r)}\left(t+\frac{k+\lambda_i}{n+\rho_i}\right)dt\right)A(x)\right]$$

$$= \left[\left(K_n\left(f_-^{(r)}\right)\right)(x), \left(K_n\left(f_+^{(r)}\right)\right)(x)\right].$$

We have proved that

$$\left(K_n^{\mathcal{F}}(f)\right)_{\pm}^{(r)} = K_n\left(f_{\pm}^{(r)}\right), \quad \forall\, r \in [0,1], \tag{27.55}$$

respectively.

Next we observe that

$$\left[\left(M_n^{\mathcal{F}}(f)\right)(x)\right]^r = \sum_{k=\lceil nx-Tn^{\alpha}\rceil}^{[nx+Tn^{\alpha}]}\left(\sum_{i=1}^{\bar{r}} w_i \left[f\left(\frac{k}{n}+\frac{i}{n\bar{r}}\right)\right]^r\right)A(x) =$$

$$\sum_{k=\lceil nx-Tn^{\alpha}\rceil}^{[nx+Tn^{\alpha}]}\left(\sum_{i=1}^{\bar{r}} w_i \left[f_-^{(r)}\left(\frac{k}{n}+\frac{i}{n\bar{r}}\right), f_+^{(r)}\left(\frac{k}{n}+\frac{i}{n\bar{r}}\right)\right]\right)A(x) = \tag{27.56}$$

$$\left[\sum_{k=\lceil nx-Tn^{\alpha}\rceil}^{[nx+Tn^{\alpha}]}\left(\sum_{i=1}^{\bar{r}} w_i f_-^{(r)}\left(\frac{k}{n}+\frac{i}{n\bar{r}}\right)\right)A(x),\right.$$

$$\sum_{k=\lceil nx-Tn^{\alpha}\rceil}^{[nx+Tn^{\alpha}]}\left(\sum_{i=1}^{\overline{r}} w_i f_+^{(r)}\left(\frac{k}{n}+\frac{i}{n\overline{r}}\right)\right) A(x)\Bigg]$$

$$=\left[\left(M_n\left(f_-^{(r)}\right)\right)(x),\left(M_n\left(f_+^{(r)}\right)\right)(x)\right].$$

That is proving

$$\left(M_n^{\mathcal{F}}(f)\right)_{\pm}^{(r)}=M_n\left(f_{\pm}^{(r)}\right),\quad \forall\, r\in[0,1], \tag{27.57}$$

respectively.

It follows are main results.

We present

Theorem 27.28 *All as at the begining of Sect. 27.4. It holds*

$$D\left(\left(H_n^{\mathcal{F}}(f)\right)(x), f(x)\right)\leq$$

$$\sum_{i=1}^{\overline{r}} w_i \omega_1^{(\mathcal{F})}\left(f,\left(\frac{\|\nu_i\|_\infty\|x\|_\infty+\|\mu_i\|_\infty}{n+\nu_i^{\min}}\right)+\left(1+\frac{\|\nu_i\|_\infty}{n+\nu_i^{\min}}\right)\frac{T^*}{n^{1-\alpha}}\right)= \tag{27.58}$$

$$\sum_{i_1=1}^{r_1}\cdots\sum_{i_d=1}^{r_d} w_{i_1,\ldots,i_d}\omega_1^{(\mathcal{F})}\left(f,\left(\frac{\|\nu_i\|_\infty\|x\|_\infty+\|\mu_i\|_\infty}{n+\nu_i^{\min}}\right)+\left(1+\frac{\|\nu_i\|_\infty}{n+\nu_i^{\min}}\right)\frac{T^*}{n^{1-\alpha}}\right).$$

Proof We observe that

$$D\left(\left(H_n^{\mathcal{F}}(f)\right)(x), f(x)\right)=$$

$$\sup_{r\in[0,1]}\max\left\{\left|\left(H_n^{\mathcal{F}}(f)\right)_-^{(r)}(x)-f_-^{(r)}(x)\right|,\left|\left(H_n^{\mathcal{F}}(f)\right)_+^{(r)}(x)-f_+^{(r)}(x)\right|\right\}\overset{(27.52)}{=}$$

$$\sup_{r\in[0,1]}\max\left\{\left|\left(H_n\left(f_-^{(r)}\right)\right)(x)-f_-^{(r)}(x)\right|,\left|\left(H_n\left(f_+^{(r)}\right)\right)(x)-f_+^{(r)}(x)\right|\right\}\overset{(27.42)}{\leq} \tag{27.59}$$

$$\sup_{r\in[0,1]}\max\left\{\sum_{i=1}^{\overline{r}} w_i\omega_1\left(f_-^{(r)},\left(\frac{\|\nu_i\|_\infty\|x\|_\infty+\|\mu_i\|_\infty}{n+\nu_i^{\min}}\right)+\left(1+\frac{\|\nu_i\|_\infty}{n+\nu_i^{\min}}\right)\frac{T^*}{n^{1-\alpha}}\right),\right.$$

$$\left.\sum_{i=1}^{\overline{r}} w_i\omega_1\left(f_+^{(r)},\left(\frac{\|\nu_i\|_\infty\|x\|_\infty+\|\mu_i\|_\infty}{n+\nu_i^{\min}}\right)+\left(1+\frac{\|\nu_i\|_\infty}{n+\nu_i^{\min}}\right)\frac{T^*}{n^{1-\alpha}}\right)\right\}\overset{(27.3)}{\leq}$$

$$\sum_{i=1}^{\bar{r}} w_i \sup_{r\in[0,1]} \max\left\{\omega_1\left(f_-^{(r)}, \left(\frac{\|\nu_i\|_\infty \|x\|_\infty + \|\mu_i\|_\infty}{n+\nu_i^{\min}}\right) + \left(1+\frac{\|\nu_i\|_\infty}{n+\nu_i^{\min}}\right)\frac{T^*}{n^{1-\alpha}}\right),\right. \tag{27.60}$$

$$\left.\omega_1\left(f_+^{(r)}, \left(\frac{\|\nu_i\|_\infty \|x\|_\infty + \|\mu_i\|_\infty}{n+\nu_i^{\min}}\right) + \left(1+\frac{\|\nu_i\|_\infty}{n+\nu_i^{\min}}\right)\frac{T^*}{n^{1-\alpha}}\right)\right\} \overset{(27.7)}{=}$$

$$\sum_{i=1}^{\bar{r}} w_i \omega_1^{(\mathcal{F})}\left(f, \left(\frac{\|\nu_i\|_\infty \|x\|_\infty + \|\mu_i\|_\infty}{n+\nu_i^{\min}}\right) + \left(1+\frac{\|\nu_i\|_\infty}{n+\nu_i^{\min}}\right)\frac{T^*}{n^{1-\alpha}}\right), \tag{27.61}$$

proving the claim. ■

Theorem 27.29 *All as at the beginning of Sect. 27.4. It holds*

$$D\left(\left(K_n^{\mathcal{F}}(f)\right)(x), f(x)\right) \leq$$

$$\sum_{i=1}^{\bar{r}} w_i \omega_1^{(\mathcal{F})}\left(f, \left(\frac{\rho_i \|x\|_\infty + \lambda_i + 1}{n+\rho_i}\right) + \left(1+\frac{\rho_i}{n+\rho_i}\right)\frac{T^*}{n^{1-\alpha}}\right) = \tag{27.62}$$

$$\sum_{i_1=1}^{r_1} \cdots \sum_{i_d=1}^{r_d} w_{i_1,\ldots,i_d} \omega_1^{(\mathcal{F})}\left(f, \left(\frac{\rho_{i_1,\ldots,i_d}\|x\|_\infty + \lambda_{i_1,\ldots,i_d}+1}{n+\rho_{i_1,\ldots,i_d}}\right) + \left(1+\frac{\rho_{i_1,\ldots,i_d}}{n+\rho_{i_1,\ldots,i_d}}\right)\frac{T^*}{n^{1-\alpha}}\right).$$

Proof Using (27.43), (27.55), similar to Theorem 27.28. ■

Theorem 27.30 *All as in Theorem 27.28. It holds*

$$D\left(\left(M_n^{\mathcal{F}}(f)\right)(x), f(x)\right) \leq \omega_1^{(\mathcal{F})}\left(f, \frac{T^*}{n^{1-\alpha}} + \frac{1}{n}\right). \tag{27.63}$$

Proof Using (27.44), (27.57), similar to Theorem 27.28. ■

Remark 27.31 When $f \in C_{\mathcal{F}}^U(\mathbb{R}^d)$, as $n \to \infty$, from Theorems 27.28–27.30 we obtain the pointwise convergence with rates of $\left(H_n^{\mathcal{F}}(f)\right)(x) \overset{D}{\to} f(x)$, $\left(K_n^{\mathcal{F}}(f)\right)(x) \overset{D}{\to} f(x)$ and $\left(M_n^{\mathcal{F}}(f)\right)(x) \overset{D}{\to} f(x)$, at the speed of $\frac{1}{n^{1-\alpha}}$.

In the next three corollaries we take $x \in \prod_{j=1}^{d}[-\gamma_j, \gamma_j] \subset \mathbb{R}^d$, $\gamma_j > 0$, $\gamma^* = \max\{\gamma_1, \ldots, \gamma_d\}$ and $n \in \mathbb{N}$ such that $n \geq \max_{j\in\{1,\ldots,d\}}\left\{T_j + \gamma_j, T_j^{-\frac{1}{\alpha}}\right\}$.

We derive the following

Corollary 27.32 (to Theorem 27.28) *It holds*

$$D^*\left(H_n^{\mathcal{F}}(f), f\right)_{\prod_{j=1}^{d}[-\gamma_j,\gamma_j]} \leq$$

$$\sum_{i=1}^{\overline{r}} w_i \omega_1^{(\mathcal{F})}\left(f, \left(\frac{\|\nu_i\|_\infty \gamma^* + \|\mu_i\|_\infty}{n+\nu_i^{\min}}\right) + \left(1 + \frac{\|\nu_i\|_\infty}{n+\nu_i^{\min}}\right)\frac{T^*}{n^{1-\alpha}}\right) = \quad (27.64)$$

$$\sum_{i_1=1}^{r_1} \cdots \sum_{i_d=1}^{r_d} w_{i_1,\dots,i_d} \omega_1^{(\mathcal{F})}\left(f, \left(\frac{\|\nu_i\|_\infty \gamma^* + \|\mu_i\|_\infty}{n+\nu_i^{\min}}\right) + \left(1 + \frac{\|\nu_i\|_\infty}{n+\nu_i^{\min}}\right)\frac{T^*}{n^{1-\alpha}}\right).$$

Proof By (27.58). ∎

Corollary 27.33 (to Theorem 27.29) *It holds*

$$D^*\left(K_n^{\mathcal{F}}(f), f\right)_{\prod_{j=1}^{d}[-\gamma_j,\gamma_j]} \leq$$

$$\sum_{i=1}^{\overline{r}} w_i \omega_1^{(\mathcal{F})}\left(f, \left(\frac{\rho_i \gamma^* + \lambda_i + 1}{n+\rho_i}\right) + \left(1 + \frac{\rho_i}{n+\rho_i}\right)\frac{T^*}{n^{1-\alpha}}\right) = \quad (27.65)$$

$$\sum_{i_1=1}^{r_1} \cdots \sum_{i_d=1}^{r_d} w_{i_1,\dots,i_d} \omega_1^{(\mathcal{F})}\left(f, \left(\frac{\rho_{i_1,\dots,i_d} \gamma^* + \lambda_{i_1,\dots,i_d} + 1}{n+\rho_{i_1,\dots,i_d}}\right) + \left(1 + \frac{\rho_{i_1,\dots,i_d}}{n+\rho_{i_1,\dots,i_d}}\right)\frac{T^*}{n^{1-\alpha}}\right).$$

Proof By (27.62). ∎

Corollary 27.34 (to Theorem 27.30) *It holds*

$$D^*\left(M_n^{\mathcal{F}}(f), f\right)_{\prod_{j=1}^{d}[-\gamma_j,\gamma_j]} \leq \omega_1^{(\mathcal{F})}\left(f, \frac{T^*}{n^{1-\alpha}} + \frac{1}{n}\right). \quad (27.66)$$

Proof By (27.63). ∎

Remark 27.35 When $f \in C_{\mathcal{F}}^{U}(\mathbb{R}^d)$, as $n \to \infty$, from Corollaries 27.32–27.34 we obtain the uniform convergence with rates of $H_n^{\mathcal{F}}(f) \overset{D^*}{\to} f$, $K_n^{\mathcal{F}}(f) \overset{D^*}{\to} f$ and $M_n^{\mathcal{F}}(f) \overset{D^*}{\to} f$, at the speed of $\frac{1}{n^{1-\alpha}}$.

Next we present higher order of fuzzy approximation results based on the high order fuzzy differentiability of the approximated function.

Theorem 27.36 *Let $x \in \mathbb{R}^d$ and $n \in \mathbb{N}$ such that $n \geq \max_{j\in\{1,\dots,d\}}\left(T_j + |x_j|, T_j^{-\frac{1}{\alpha}}\right)$, $T_j > 0$, $0 < \alpha < 1$. Let also $f \in C_{\mathcal{F}}^{N}(\mathbb{R}^d)$, $N \in \mathbb{N}$, such that all of its fuzzy partial* *derivatives* $f_{\widetilde{\alpha}}$ *of order* N, $\widetilde{\alpha} : |\widetilde{\alpha}| = \sum_{j=1}^{d} \alpha_j = N$, *are fuzzy uniformly continuous or fuzzy continuous and bounded. Then*

$$D\left(\left(H_n^{\mathcal{F}}(f)\right)(x), f(x)\right) \leq$$

$$\sum_{i=1}^{N} \frac{1}{l!} \left[\sum_{i=1}^{\bar{r}} w_i \left[\left(\frac{\|v_i\|_\infty \|x\|_\infty + \|\mu_i\|_\infty}{n + v_i^{\min}} \right) + \left(1 + \frac{\|v_i\|_\infty}{n + v_i^{\min}} \right) \frac{T^*}{n^{1-\alpha}} \right]^l \right]$$

$$\cdot \left[\left(\sum_{j=1}^{d} D\left(\frac{\partial}{\partial x_j}, \widetilde{o} \right) \right)^l f(x) \right] +$$

$$\frac{d^N}{N!} \sum_{i=1}^{\bar{r}} w_i \left[\left(\frac{\|v_i\|_\infty \|x\|_\infty + \|\mu_i\|_\infty}{n + v_i^{\min}} \right) + \left(1 + \frac{\|v_i\|_\infty}{n + v_i^{\min}} \right) \frac{T^*}{n^{1-\alpha}} \right]^N \cdot \quad (27.67)$$

$$\max_{\widetilde{\alpha}:|\widetilde{\alpha}|=N} \omega_1^{(\mathcal{F})} \left(f_{\widetilde{\alpha}}, \left[\left(\frac{\|v_i\|_\infty \|x\|_\infty + \|\mu_i\|_\infty}{n + v_i^{\min}} \right) + \left(1 + \frac{\|v_i\|_\infty}{n + v_i^{\min}} \right) \frac{T^*}{n^{1-\alpha}} \right] \right).$$

Proof We observe that

$$D\left(\left(H_n^{\mathcal{F}}(f)\right)(x), f(x)\right) =$$

$$\sup_{r\in[0,1]} \max \left\{ \left| \left(H_n^{\mathcal{F}}(f)\right)_-^{(r)}(x) - f_-^{(r)}(x) \right|, \left| \left(H_n^{\mathcal{F}}(f)\right)_+^{(r)}(x) - f_+^{(r)}(x) \right| \right\} \overset{(27.52)}{=}$$

$$\sup_{r\in[0,1]} \max \left\{ \left| \left(H_n\left(f_-^{(r)}\right)\right)(x) - f_-^{(r)}(x) \right|, \left| \left(H_n\left(f_+^{(r)}\right)\right)(x) - f_+^{(r)}(x) \right| \right\} \overset{(27.45)}{\leq}$$

$$\sup_{r\in[0,1]} \max \left\{ \sum_{l=1}^{N} \frac{1}{l!} \left(\left(\sum_{j=1}^{d} \left| \frac{\partial}{\partial x_j} \right| \right)^l f_-^{(r)}(x) \right) \cdot \right. \quad (27.68)$$

$$\left[\sum_{i=1}^{\bar{r}} w_i \left[\left(\frac{\|v_i\|_\infty \|x\|_\infty + \|\mu_i\|_\infty}{n + v_i^{\min}} \right) + \left(1 + \frac{\|v_i\|_\infty}{n + v_i^{\min}} \right) \frac{T^*}{n^{1-\alpha}} \right]^l \right] +$$

$$\frac{d^N}{N!} \sum_{i=1}^{\bar{r}} w_i \left[\left(\frac{\|v_i\|_\infty \|x\|_\infty + \|\mu_i\|_\infty}{n + v_i^{\min}} \right) + \left(1 + \frac{\|v_i\|_\infty}{n + v_i^{\min}} \right) \frac{T^*}{n^{1-\alpha}} \right]^N \cdot$$

$$\max_{\widetilde{\alpha}:|\widetilde{\alpha}|=N} \omega_1 \left(\left(f_-^{(r)} \right)_{\widetilde{\alpha}}, \left[\left(\frac{\|\nu_i\|_\infty \|x\|_\infty + \|\mu_i\|_\infty}{n + \nu_i^{\min}} \right) + \left(1 + \frac{\|\nu_i\|_\infty}{n + \nu_i^{\min}} \right) \frac{T^*}{n^{1-\alpha}} \right] \right),$$

$$\sum_{l=1}^{N} \frac{1}{l!} \left(\left(\sum_{j=1}^{d} \left| \frac{\partial}{\partial x_j} \right| \right)^l f_+^{(r)}(x) \right) \cdot$$

$$\left[\sum_{i=1}^{\overline{r}} w_i \left[\left(\frac{\|\nu_i\|_\infty \|x\|_\infty + \|\mu_i\|_\infty}{n + \nu_i^{\min}} \right) + \left(1 + \frac{\|\nu_i\|_\infty}{n + \nu_i^{\min}} \right) \frac{T^*}{n^{1-\alpha}} \right]^l \right] +$$

$$\frac{d^N}{N!} \sum_{i=1}^{\overline{r}} w_i \left[\left(\frac{\|\nu_i\|_\infty \|x\|_\infty + \|\mu_i\|_\infty}{n + \nu_i^{\min}} \right) + \left(1 + \frac{\|\nu_i\|_\infty}{n + \nu_i^{\min}} \right) \frac{T^*}{n^{1-\alpha}} \right]^N \cdot$$

$$\max_{\widetilde{\alpha}:|\widetilde{\alpha}|=N} \omega_1 \left(\left(f_+^{(r)} \right)_{\widetilde{\alpha}}, \left[\left(\frac{\|\nu_i\|_\infty \|x\|_\infty + \|\mu_i\|_\infty}{n + \nu_i^{\min}} \right) + \left(1 + \frac{\|\nu_i\|_\infty}{n + \nu_i^{\min}} \right) \frac{T^*}{n^{1-\alpha}} \right] \right) \Bigg\}$$

$$\overset{(27.12)}{=} \sup_{r \in [0,1]} \max \left\{ \sum_{l=1}^{N} \frac{1}{l!} \left(\left(\sum_{j=1}^{d} \left| \frac{\partial}{\partial x_j} \right| \right)^l f(x) \right)_-^{(r)} \cdot \right.$$

$$\left[\sum_{i=1}^{\overline{r}} w_i \left[\left(\frac{\|\nu_i\|_\infty \|x\|_\infty + \|\mu_i\|_\infty}{n + \nu_i^{\min}} \right) + \left(1 + \frac{\|\nu_i\|_\infty}{n + \nu_i^{\min}} \right) \frac{T^*}{n^{1-\alpha}} \right]^l \right] +$$

$$\frac{d^N}{N!} \sum_{i=1}^{\overline{r}} w_i \left[\left(\frac{\|\nu_i\|_\infty \|x\|_\infty + \|\mu_i\|_\infty}{n + \nu_i^{\min}} \right) + \left(1 + \frac{\|\nu_i\|_\infty}{n + \nu_i^{\min}} \right) \frac{T^*}{n^{1-\alpha}} \right]^N \cdot$$

$$\max_{\widetilde{\alpha}:|\widetilde{\alpha}|=N} \omega_1 \left(\left(f_{\widetilde{\alpha}} \right)_-^{(r)}, \left[\left(\frac{\|\nu_i\|_\infty \|x\|_\infty + \|\mu_i\|_\infty}{n + \nu_i^{\min}} \right) + \left(1 + \frac{\|\nu_i\|_\infty}{n + \nu_i^{\min}} \right) \frac{T^*}{n^{1-\alpha}} \right] \right),$$

$$\sum_{l=1}^{N} \frac{1}{l!} \left(\left(\sum_{j=1}^{d} \left| \frac{\partial}{\partial x_j} \right| \right)^l f(x) \right)_+^{(r)} \cdot \tag{27.69}$$

$$\left[\sum_{i=1}^{\overline{r}} w_i \left[\left(\frac{\|\nu_i\|_\infty \|x\|_\infty + \|\mu_i\|_\infty}{n + \nu_i^{\min}} \right) + \left(1 + \frac{\|\nu_i\|_\infty}{n + \nu_i^{\min}} \right) \frac{T^*}{n^{1-\alpha}} \right]^l \right] +$$

$$\frac{d^N}{N!}\sum_{i=1}^{\overline{r}} w_i \left[\left(\frac{\|\nu_i\|_\infty \|x\|_\infty + \|\mu_i\|_\infty}{n+\nu_i^{\min}}\right) + \left(1+\frac{\|\nu_i\|_\infty}{n+\nu_i^{\min}}\right)\frac{T^*}{n^{1-\alpha}}\right]^N \cdot$$

$$\left.\max_{\widetilde{\alpha}:|\widetilde{\alpha}|=N} \omega_1\left(\left(f_{\widetilde{\alpha}}\right)_+^{(r)}, \left[\left(\frac{\|\nu_i\|_\infty \|x\|_\infty + \|\mu_i\|_\infty}{n+\nu_i^{\min}}\right) + \left(1+\frac{\|\nu_i\|_\infty}{n+\nu_i^{\min}}\right)\frac{T^*}{n^{1-\alpha}}\right]\right)\right\}$$

$$\overset{(27.3)}{\leq} \sum_{l=1}^{N} \frac{1}{l!}\left[\sum_{i=1}^{\overline{r}} w_i \left[\left(\frac{\|\nu_i\|_\infty \|x\|_\infty + \|\mu_i\|_\infty}{n+\nu_i^{\min}}\right) + \left(1+\frac{\|\nu_i\|_\infty}{n+\nu_i^{\min}}\right)\frac{T^*}{n^{1-\alpha}}\right]^l\right]$$

$$\cdot \sup_{r\in[0,1]} \max\left\{\left(\left(\sum_{j=1}^{d}\left|\frac{\partial}{\partial x_j}\right|\right)^l f(x)\right)_-^{(r)}, \left(\left(\sum_{j=1}^{d}\left|\frac{\partial}{\partial x_j}\right|\right)^l f(x)\right)_+^{(r)}\right\} +$$

$$\frac{d^N}{N!}\sum_{i=1}^{\overline{r}} w_i \left[\left(\frac{\|\nu_i\|_\infty \|x\|_\infty + \|\mu_i\|_\infty}{n+\nu_i^{\min}}\right) + \left(1+\frac{\|\nu_i\|_\infty}{n+\nu_i^{\min}}\right)\frac{T^*}{n^{1-\alpha}}\right]^N \cdot \quad (27.70)$$

$$\max_{\widetilde{\alpha}:|\widetilde{\alpha}|=N} \sup_{r\in[0,1]} \max\left\{\omega_1\left(\left(f_{\widetilde{\alpha}}\right)_-^{(r)}, \left[\left(\frac{\|\nu_i\|_\infty \|x\|_\infty + \|\mu_i\|_\infty}{n+\nu_i^{\min}}\right) + \left(1+\frac{\|\nu_i\|_\infty}{n+\nu_i^{\min}}\right)\frac{T^*}{n^{1-\alpha}}\right]\right),\right.$$

$$\left.\omega_1\left(\left(f_{\widetilde{\alpha}}\right)_+^{(r)}, \left[\left(\frac{\|\nu_i\|_\infty \|x\|_\infty + \|\mu_i\|_\infty}{n+\nu_i^{\min}}\right) + \left(1+\frac{\|\nu_i\|_\infty}{n+\nu_i^{\min}}\right)\frac{T^*}{n^{1-\alpha}}\right]\right)\right\}$$

$$\overset{(\text{by}(27.3),(27.7),(27.18))}{\leq} \sum_{l=1}^{N} \frac{1}{l!}\left[\sum_{i=1}^{\overline{r}} w_i \left[\left(\frac{\|\nu_i\|_\infty \|x\|_\infty + \|\mu_i\|_\infty}{n+\nu_i^{\min}}\right) + \left(1+\frac{\|\nu_i\|_\infty}{n+\nu_i^{\min}}\right)\frac{T^*}{n^{1-\alpha}}\right]^l\right]$$

$$\cdot\left[\left(\sum_{j=1}^{d} D\left(\frac{\partial}{\partial x_j}, \widetilde{o}\right)\right)^l f(x)\right] +$$

$$\frac{d^N}{N!}\sum_{i=1}^{\overline{r}} w_i \left[\left(\frac{\|\nu_i\|_\infty \|x\|_\infty + \|\mu_i\|_\infty}{n+\nu_i^{\min}}\right) + \left(1+\frac{\|\nu_i\|_\infty}{n+\nu_i^{\min}}\right)\frac{T^*}{n^{1-\alpha}}\right]^N \cdot \quad (27.71)$$

$$\max_{\widetilde{\alpha}:|\widetilde{\alpha}|=N} \omega_1^{(\mathcal{F})}\left(f_{\widetilde{\alpha}}, \left[\left(\frac{\|\nu_i\|_\infty \|x\|_\infty + \|\mu_i\|_\infty}{n+\nu_i^{\min}}\right) + \left(1+\frac{\|\nu_i\|_\infty}{n+\nu_i^{\min}}\right)\frac{T^*}{n^{1-\alpha}}\right]\right).$$

The theorem is proved. ■

Similarly we find

Theorem 27.37 *All as in Theorem 27.36. Then*

$$D\left(\left(K_n^{\mathcal{F}}(f)\right)(x), f(x)\right) \leq$$

$$\sum_{i=1}^{N}\frac{1}{l!}\left[\sum_{i=1}^{\overline{r}} w_i\left[\left(\frac{\rho_i\|x\|_\infty+\lambda_i+1}{n+\rho_i}\right)+\left(1+\frac{\rho_i}{n+\rho_i}\right)\frac{T^*}{n^{1-\alpha}}\right]^l\right]$$

$$\cdot\left[\left(\sum_{j=1}^{d} D\left(\frac{\partial}{\partial x_j},\widetilde{o}\right)\right)^l f(x)\right]+$$

$$\frac{d^N}{N!}\sum_{i=1}^{\overline{r}} w_i\left[\left(\frac{\rho_i\|x\|_\infty+\lambda_i+1}{n+\rho_i}\right)+\left(1+\frac{\rho_i}{n+\rho_i}\right)\frac{T^*}{n^{1-\alpha}}\right]^N\cdot \tag{27.72}$$

$$\max_{\widetilde{\alpha}:|\widetilde{\alpha}|=N}\omega_1^{(\mathcal{F})}\left(f_{\widetilde{\alpha}},\left[\left(\frac{\rho_i\|x\|_\infty+\lambda_i+1}{n+\rho_i}\right)+\left(1+\frac{\rho_i}{n+\rho_i}\right)\frac{T^*}{n^{1-\alpha}}\right]\right).$$

Proof Similar to Theorem 27.36, using (27.46), (27.55), etc. ■

We continue with

Theorem 27.38 *All as in Theorem 27.36. Then*

$$D\left(\left(M_n^{\mathcal{F}}(f)\right)(x), f(x)\right) \leq$$

$$\sum_{i=1}^{N}\frac{1}{l!}\left(\left(\sum_{j=1}^{d} D\left(\frac{\partial}{\partial x_j},\widetilde{o}\right)\right)^l f(x)\right)\left(\frac{T^*}{n^{1-\alpha}}+\frac{1}{n}\right)^l+$$

$$\frac{d^N}{N!}\left(\frac{T^*}{n^{1-\alpha}}+\frac{1}{n}\right)^N\max_{\widetilde{\alpha}:|\widetilde{\alpha}|=N}\omega_1^{(\mathcal{F})}\left(f_{\widetilde{\alpha}},\frac{T^*}{n^{1-\alpha}}+\frac{1}{n}\right). \tag{27.73}$$

Proof Similar to Theorem 27.36, using (27.47), (27.57), etc. ■

Note 27.39 *Inequalities (27.67), (27.72) and (27.73) imply the pointwise convergence with rates of* $\left(H_n^{\mathcal{F}}(f)\right)(x) \overset{D}{\rightarrow} f(x)$, $\left(K_n^{\mathcal{F}}(f)\right)(x) \overset{D}{\rightarrow} f(x)$ *and* $\left(M_n^{\mathcal{F}}(f)\right)(x) \overset{D}{\rightarrow} f(x)$, *as* $n\rightarrow\infty$, *at speed* $\frac{1}{n^{1-\alpha}}$.

Note 27.40 *In the next three corolaries additionally we assume that*

$$f_{\widetilde{\alpha}}(x)=\widetilde{o}, \text{for all } \widetilde{\alpha}:|\widetilde{\alpha}|=1,\ldots,N, \tag{27.74}$$

for $x \in \mathbb{R}^d$ *fixed.*

The last implies $D(f_{\widetilde{\alpha}}(x), \widetilde{o}) = 0$, *and by (27.18) we obtain*

$$\left[\left(\sum_{j=1}^{d} D\left(\frac{\partial}{\partial x_j}, \widetilde{o}\right)\right)^l f(x)\right] = 0, \tag{27.75}$$

for $l = 1, \ldots, N$.

So we derive the following special results.

Corollary 27.41 *Let* $x \in \mathbb{R}^d$ *and* $n \in \mathbb{N}$ *such that* $n \geq \max_{j \in \{1,\ldots,d\}} \left(T_j + |x_j|, T_j^{-\frac{1}{\alpha}}\right)$, $T_j > 0$, $0 < \alpha < 1$. *Let also* $f \in C_{\mathcal{F}}^N(\mathbb{R}^d)$, $N \in \mathbb{N}$, *such that all of its fuzzy partial derivatives* $f_{\widetilde{\alpha}}$ *of order* N, $\widetilde{\alpha} : |\widetilde{\alpha}| = \sum_{j=1}^{d} \alpha_j = N$, *are fuzzy uniformly continuous or fuzzy continuous and bounded. Assume also (27.74). Then*

$$D\left(\left(H_n^{\mathcal{F}}(f)\right)(x), f(x)\right) \leq$$

$$\frac{d^N}{N!} \sum_{i=1}^{\overline{r}} w_i \left[\left(\frac{\|\nu_i\|_\infty \|x\|_\infty + \|\mu_i\|_\infty}{n + \nu_i^{\min}}\right) + \left(1 + \frac{\|\nu_i\|_\infty}{n + \nu_i^{\min}}\right) \frac{T^*}{n^{1-\alpha}}\right]^N \cdot \tag{27.76}$$

$$\max_{\widetilde{\alpha}:|\widetilde{\alpha}|=N} \omega_1^{(\mathcal{F})}\left(f_{\widetilde{\alpha}}, \left[\left(\frac{\|\nu_i\|_\infty \|x\|_\infty + \|\mu_i\|_\infty}{n + \nu_i^{\min}}\right) + \left(1 + \frac{\|\nu_i\|_\infty}{n + \nu_i^{\min}}\right) \frac{T^*}{n^{1-\alpha}}\right]\right).$$

Proof By (27.67) and (27.75). ■

Corollary 27.42 *All as in Corollary 27.41. Then*

$$D\left(\left(K_n^{\mathcal{F}}(f)\right)(x), f(x)\right) \leq$$

$$\frac{d^N}{N!} \sum_{i=1}^{\overline{r}} w_i \left[\left(\frac{\rho_i \|x\|_\infty + \lambda_i + 1}{n + \rho_i}\right) + \left(1 + \frac{\rho_i}{n + \rho_i}\right) \frac{T^*}{n^{1-\alpha}}\right]^N \cdot \tag{27.77}$$

$$\max_{\widetilde{\alpha}:|\widetilde{\alpha}|=N} \omega_1^{(\mathcal{F})}\left(f_{\widetilde{\alpha}}, \left[\left(\frac{\rho_i \|x\|_\infty + \lambda_i + 1}{n + \rho_i}\right) + \left(1 + \frac{\rho_i}{n + \rho_i}\right) \frac{T^*}{n^{1-\alpha}}\right]\right).$$

Proof By (27.72) and (27.75). ■

Corollary 27.43 *All as in Corollary 27.41. Then*

$$D\left(\left(M_n^{\mathcal{F}}(f)\right)(x), f(x)\right) \leq$$

$$\frac{d^N}{N!}\left(\frac{T^*}{n^{1-\alpha}}+\frac{1}{n}\right)^N \max_{\widetilde{\alpha}:|\widetilde{\alpha}|=N} \omega_1^{(\mathcal{F})}\left(f_{\widetilde{\alpha}}, \frac{T^*}{n^{1-\alpha}}+\frac{1}{n}\right). \tag{27.78}$$

Proof By (27.73) and (27.75). ■

By (27.76)–(27.78) we get fuzzy pointwise convergence with rates at high speed $\frac{1}{n^{(1-\alpha)(N+1)}}$.

We need

Lemma 27.44 ([7], p. 131) *Let K be a compact subset of the real normed vector space $(V, \|\cdot\|)$ and $f \in C_{\mathcal{F}}(K)$ (space of continuous fuzzy real number valued functions on K). Then f is a fuzzy bounded function.*

Assumption 27.45 In the next three corollaries we consider $x \in G = \prod_{j=1}^{d}\left[-\gamma_j, \gamma_j\right] \subset \mathbb{R}^d$, $\gamma_j > 0$, $\gamma^* = \max\{\gamma_1, \ldots, \gamma_d\}$ and $n \in \mathbb{N} : n \geq \max_{j\in\{1,\ldots,d\}}\left(T_j + \gamma_j, T_j^{-\frac{1}{\alpha}}\right)$, $T_j > 0$, $0 < \alpha < 1$. Let also $f \in C_{\mathcal{F}}^N\left(\mathbb{R}^d\right)$, $N \in \mathbb{N}$, such that all of its fuzzy partials $f_{\widetilde{\alpha}}$ of order N, $\widetilde{\alpha} : |\widetilde{\alpha}| = N$, are fuzzy uniformly continuous or fuzzy continuous and bounded.

Using Lemma 27.44, Assumption 27.45 along with (27.18) and subadditivity of $\|\cdot\|_\infty$, we clearly obtain that

$$\left\|\left(\sum_{j=1}^{d} D\left(\frac{\partial}{\partial x_j}, \widetilde{o}\right)\right)^l f(x)\right\|_{\infty,G} < \infty, \tag{27.79}$$

for all $l = 1, \ldots, N$.

We define

$$D^*(f, g)|_G = \sup_{x\in G} D(f(x), g(x)),$$

where $f, g : G \to \mathbb{R}_{\mathcal{F}}$.

We give

Corollary 27.46 *We suppose Assumption 27.45. Then*

$$D^{*}\left(H_{n}^{\mathcal{F}}(f), f\right)|_{G} \leq$$

$$\sum_{i=1}^{N}\frac{1}{l!}\left[\sum_{i=1}^{\overline{r}}w_{i}\left[\left(\frac{\|\nu_{i}\|_{\infty}\gamma^{*}+\|\mu_{i}\|_{\infty}}{n+\nu_{i}^{\min}}\right)+\left(1+\frac{\|\nu_{i}\|_{\infty}}{n+\nu_{i}^{\min}}\right)\frac{T^{*}}{n^{1-\alpha}}\right]^{l}\right]$$

$$\cdot\left\|\left(\sum_{j=1}^{d}D\left(\frac{\partial}{\partial x_{j}},\widetilde{o}\right)\right)^{l}f(x)\right\|_{\infty,G}+ \qquad (27.80)$$

$$\frac{d^{N}}{N!}\sum_{i=1}^{\overline{r}}w_{i}\left[\left(\frac{\|\nu_{i}\|_{\infty}\gamma^{*}+\|\mu_{i}\|_{\infty}}{n+\nu_{i}^{\min}}\right)+\left(1+\frac{\|\nu_{i}\|_{\infty}}{n+\nu_{i}^{\min}}\right)\frac{T^{*}}{n^{1-\alpha}}\right]^{N}\cdot$$

$$\max_{\widetilde{\alpha}:|\widetilde{\alpha}|=N}\omega_{1}^{(\mathcal{F})}\left(f_{\widetilde{\alpha}},\left[\left(\frac{\|\nu_{i}\|_{\infty}\gamma^{*}+\|\mu_{i}\|_{\infty}}{n+\nu_{i}^{\min}}\right)+\left(1+\frac{\|\nu_{i}\|_{\infty}}{n+\nu_{i}^{\min}}\right)\frac{T^{*}}{n^{1-\alpha}}\right]\right).$$

Proof By (27.67) and (27.79). ∎

Corollary 27.47 *We suppose Assumption 27.45. Then*

$$D^{*}\left(K_{n}^{\mathcal{F}}(f), f\right)|_{G} \leq$$

$$\sum_{i=1}^{N}\frac{1}{l!}\left[\sum_{i=1}^{\overline{r}}w_{i}\left[\left(\frac{\rho_{i}\gamma^{*}+\lambda_{i}+1}{n+\rho_{i}}\right)+\left(1+\frac{\rho_{i}}{n+\rho_{i}}\right)\frac{T^{*}}{n^{1-\alpha}}\right]^{l}\right]$$

$$\cdot\left\|\left(\sum_{j=1}^{d}D\left(\frac{\partial}{\partial x_{j}},\widetilde{o}\right)\right)^{l}f(x)\right\|_{\infty,G}+ \qquad (27.81)$$

$$\frac{d^{N}}{N!}\sum_{i=1}^{\overline{r}}w_{i}\left[\left(\frac{\rho_{i}\gamma^{*}+\lambda_{i}+1}{n+\rho_{i}}\right)+\left(1+\frac{\rho_{i}}{n+\rho_{i}}\right)\frac{T^{*}}{n^{1-\alpha}}\right]^{N}\cdot$$

$$\max_{\widetilde{\alpha}:|\widetilde{\alpha}|=N}\omega_{1}^{(\mathcal{F})}\left(f_{\widetilde{\alpha}},\left[\left(\frac{\rho_{i}\gamma^{*}+\lambda_{i}+1}{n+\rho_{i}}\right)+\left(1+\frac{\rho_{i}}{n+\rho_{i}}\right)\frac{T^{*}}{n^{1-\alpha}}\right]\right).$$

Proof By (27.72) and (27.79). ∎

Corollary 27.48 *We suppose Assumption 27.45. Then*

$$D^*\left(M_n^{\mathcal{F}}(f), f\right)|_G \leq$$

$$\sum_{i=1}^{N}\frac{1}{l!}\left\|\left(\sum_{j=1}^{d} D\left(\frac{\partial}{\partial x_j}, \widetilde{o}\right)\right)^l f(x)\right\|_{\infty,G}\left(\frac{T^*}{n^{1-\alpha}}+\frac{1}{n}\right)^l +$$

$$\frac{d^N}{N!}\left(\frac{T^*}{n^{1-\alpha}}+\frac{1}{n}\right)^N \max_{\widetilde{\alpha}:|\widetilde{\alpha}|=N}\omega_1^{(\mathcal{F})}\left(f_{\widetilde{\alpha}}, \frac{T^*}{n^{1-\alpha}}+\frac{1}{n}\right). \qquad (27.82)$$

Proof By (27.73) and (27.79). ■

Note 27.49 *Inequalities (27.80)–(27.82) imply the uniform convergence on G with rates of $H_n^{\mathcal{F}}(f) \overset{D^*}{\to} f$, $K_n^{\mathcal{F}}(f) \overset{D^*}{\to} f$ and $M_n^{\mathcal{F}}(f) \overset{D^*}{\to} f$, as $n \to \infty$, at speed $\frac{1}{n^{1-\alpha}}$.*

We continue with

Corollary 27.50 (to Theorem 27.36) *Case of $N = 1$. It holds*

$$D\left(\left(H_n^{\mathcal{F}}(f)\right)(x), f(x)\right) \leq$$

$$\left[\sum_{i=1}^{\overline{r}} w_i\left[\left(\frac{\|\nu_i\|_\infty\|x\|_\infty+\|\mu_i\|_\infty}{n+\nu_i^{\min}}\right)+\left(1+\frac{\|\nu_i\|_\infty}{n+\nu_i^{\min}}\right)\frac{T^*}{n^{1-\alpha}}\right]\right] \qquad (27.83)$$

$$\cdot\left(\sum_{j=1}^{d} D\left(\frac{\partial f}{\partial x_j}, \widetilde{o}\right)\right)+$$

$$d\left\{\sum_{i=1}^{\overline{r}} w_i\left[\left(\frac{\|\nu_i\|_\infty\|x\|_\infty+\|\mu_i\|_\infty}{n+\nu_i^{\min}}\right)+\left(1+\frac{\|\nu_i\|_\infty}{n+\nu_i^{\min}}\right)\frac{T^*}{n^{1-\alpha}}\right]\cdot\right.$$

$$\left.\max_{j\in\{1,\ldots,d\}}\omega_1^{(\mathcal{F})}\left(\frac{\partial f}{\partial x_j}, \left[\left(\frac{\|\nu_i\|_\infty\|x\|_\infty+\|\mu_i\|_\infty}{n+\nu_i^{\min}}\right)+\left(1+\frac{\|\nu_i\|_\infty}{n+\nu_i^{\min}}\right)\frac{T^*}{n^{1-\alpha}}\right]\right)\right\}.$$

Proof By (27.67). ■

Corollary 27.51 (to Theorem 27.37) *Case of* $N = 1$. *Then*

$$D\left(\left(K_n^{\mathcal{F}}(f)\right)(x), f(x)\right) \leq$$

$$\left(\sum_{i=1}^{\overline{r}} w_i \left[\left(\frac{\rho_i \|x\|_\infty + \lambda_i + 1}{n+\rho_i}\right) + \left(1 + \frac{\rho_i}{n+\rho_i}\right)\frac{T^*}{n^{1-\alpha}}\right]\right) \tag{27.84}$$

$$\cdot\left(\sum_{j=1}^{d} D\left(\frac{\partial f}{\partial x_j}, \widetilde{o}\right)\right) +$$

$$d\left\{\sum_{i=1}^{\overline{r}} w_i \left[\left(\frac{\rho_i \|x\|_\infty + \lambda_i + 1}{n+\rho_i}\right) + \left(1 + \frac{\rho_i}{n+\rho_i}\right)\frac{T^*}{n^{1-\alpha}}\right]\cdot\right.$$

$$\left.\max_{j\in\{1,\ldots,d\}} \omega_1^{(\mathcal{F})}\left(\frac{\partial f}{\partial x_j}, \left[\left(\frac{\rho_i \|x\|_\infty + \lambda_i + 1}{n+\rho_i}\right) + \left(1 + \frac{\rho_i}{n+\rho_i}\right)\frac{T^*}{n^{1-\alpha}}\right]\right)\right\}.$$

Proof By (27.72). ■

We finish with

Corollary 27.52 (to Theorem 27.38) *Case* $N = 1$. *Then*

$$D\left(\left(M_n^{\mathcal{F}}(f)\right)(x), f(x)\right) \leq \tag{27.85}$$

$$\left[\sum_{j=1}^{d} D\left(\frac{\partial f}{\partial x_j}, \widetilde{o}\right) + d \max_{j\in\{1,\ldots,d\}} \omega_1^{(\mathcal{F})}\left(\frac{\partial f}{\partial x_j}, \frac{T^*}{n^{1-\alpha}} + \frac{1}{n}\right)\right]\left(\frac{T^*}{n^{1-\alpha}} + \frac{1}{n}\right).$$

Proof By (27.73). ■

Note 27.53 *Inequalities (27.83)–(27.85) imply the pointwise convergence with rates of* $H\left(\mathcal{F}_n(f)\right)(x) \overset{D}{\to} f(x)$, $\left(K_n^{\mathcal{F}}(f)\right)(x) \overset{D}{\to} f(x)$ *and* $\left(M_n^{\mathcal{F}}(f)\right)(x) \overset{D}{\to} f(x)$, *as* $n \to \infty$, *at speed* $\frac{1}{n^{1-\alpha}}$.

References

1. G.A. Anastassiou, Rate of convergence of some neural network operators to the unit-univariate case. J. Math. Anal. Appl. **212**, 237–262 (1997)
2. G.A. Anastassiou, Rate of convergence of some multivariate neural network operators to the unit. J. Comp. Math. Appl. **40**, 1–19 (2000)

3. G.A. Anastassiou, *Quantitative Approximations* (Chapman&Hall/CRC, Boca Raton, 2001)
4. G.A. Anastassiou, Fuzzy approximation by fuzzy convolution type operators. Comput. Math. **48**, 1369–1386 (2004)
5. G.A. Anastassiou, Higher order Fuzzy Korovkin Theory via inequalities. Commun. Appl. Anal. **10**(2), 359–392 (2006)
6. G.A. Anastassiou, Fuzzy Korovkin Theorems and inequalities. J. Fuzzy Math. **15**(1), 169–205 (2007)
7. G.A. Anastassiou, *Fuzzy Mathematics: Approximation Theory* (Springer, New York, 2010)
8. G.A. Anastassiou, Higher order multivariate fuzzy approximation by multivariate fuzzy wavelet type and neural network operators. J. Fuzzy Math. **19**(3), 601–618 (2011)
9. G.A. Anastassiou, Rate of convergence of some neural network operators to the unit-univariate case, revisited. Vesnik **65**(4), 511–518 (2013)
10. G.A. Anastassiou, Rate of convergence of some multivariate neural network operators to the unit, revisited. J. Comput. Anal. Appl. **15**(7), 1300–1309 (2013)
11. G.A. Anastassiou, Fuzzy fractional approximations by fuzzy normalized bell and squashing type neural network operators. J. Fuzzy Math. **22**(1), 139–156 (2014)
12. G.A. Anastassiou, Higher order multivariate fuzzy approximation by basic neural network operators. Cubo **16**(03), 21–35 (2014). www.tjnsa.com (electronic)
13. G.A. Anastassiou, Multivariate fuzzy perturbed neural network operators approximation. J. Nonlinear Sci. Appl. **7**, 383–406 (2014)
14. G.A. Anastassiou, Approximation by Perturbed Neural Network Operators, Submitted 2014
15. G.A. Anastassiou, Approximations by Multivariate Perturbed Neural Network Operators, Submitted 2014
16. P. Cardaliaguet, G. Euvrard, Approximation of a function and its derivative with a neural network. Neural Netw. **5**, 207–220 (1992)
17. S. Gal, in *Approximation Theory in Fuzzy Setting, Chapter 13*, Handbook of Analytic-Computational Methods in Applied Mathematics, ed. by G. Anastassiou (Chapman & Hall/CRC, Boca Raton, 2000), pp. 617–666
18. R. Goetschel Jr, W. Voxman, Elementary fuzzy calculus. Fuzzy Sets Syst. **18**, 31–43 (1986)
19. O. Kaleva, Fuzzy differential equations. Fuzzy Sets Syst. **24**, 301–317 (1987)
20. Y.K. Kim, B.M. Ghil, Integrals of fuzzy-number-valeud functions. Fuzzy Sets and Systems **86**, 213–222 (1997)
21. D.D. Stancu, On a generalization of the Bernstein polynomials. Studia Universitatis Babeş-Bolyai, Series Mathematica-Physica **14**, 31–45 (1969)
22. Wu Congxin, Gong Zengtai, On Henstock integrals of interval-valued functions and fuzzy-valued functions. Fuzzy Sets Syst. **115**(3), 377–391 (2000)
23. C. Wu, Z. Gong, On Henstock integral of fuzzy number valued functions (I). Fuzzy Sets Syst. **120**(3), 523–532 (2001)
24. C. Wu, M. Ma, On embedding problem of fuzzy number spaces: Part 1. Fuzzy Sets Syst. **44**, 33–38 (1991)

Chapter 28
Multivariate Fuzzy-Random Perturbed Neural Network Approximations

In this chapter we study the rate of multivariate pointwise and uniform convergences in the q-mean to the Fuzzy-Random unit operator of perturbed multivariate normalized Fuzzy-Random neural network operators of Stancu, Kantorovich and Quadrature types. These multivariate Fuzzy-Random operators arise in a natural way among multivariate Fuzzy-Random neural networks. The rates are given via multivariate Probabilistic-Jackson type inequalities using the multivariate Fuzzy-Random modulus of continuity of the involved multivariate Fuzzy-Random function. Also some interesting results in multivariate Fuzzy-Random Analysis are given of independent merit. It follows [10].

28.1 Fuzzy Random Analysis Basics

We begin with

Definition 28.1 (*see* [17]) Let $\mu : \mathbb{R} \to [0, 1]$ with the following properties:

(i) is normal, i.e., $\exists\, x_0 \in \mathbb{R} : \mu(x_0) = 1$.
(ii) $\mu(\lambda x + (1-\lambda) y) \geq \min\{\mu(x), \mu(y)\}, \forall\, x, y \in \mathbb{R}, \forall\, \lambda \in [0, 1]$ (μ is called a convex fuzzy subset).
(iii) μ is upper semicontinuous on $\mathbb{R}$, i.e., $\forall\, x_0 \in \mathbb{R}$ and $\forall\, \varepsilon > 0$, $\exists$ neighborhood $V(x_0) : \mu(x) \leq \mu(x_0) + \varepsilon, \forall\, x \in V(x_0)$.
(iv) the set $\overline{\text{supp}(\mu)}$ is compact in $\mathbb{R}$ (where $\text{supp}(\mu) := \{x \in \mathbb{R}; \mu(x) > 0\}$).

We call μ a fuzzy real number. Denote the set of all μ with $\mathbb{R}_{\mathcal{F}}$.

E.g., $\chi_{\{x_0\}} \in \mathbb{R}_{\mathcal{F}}$, for any $x_0 \in \mathbb{R}$, where $\chi_{\{x_0\}}$ is the characteristic function at x_0.

For $0 < r \leq 1$ and $\mu \in \mathbb{R}_{\mathcal{F}}$ define $[\mu]^r := \{x \in \mathbb{R} : \mu(x) \geq r\}$ and $[\mu]^0 := \overline{\{x \in \mathbb{R} : \mu(x) > 0\}}$.

G.A. Anastassiou, *Intelligent Systems II: Complete Approximation by Neural Network Operators*, Studies in Computational Intelligence 608,
DOI 10.1007/978-3-319-20505-2_28

Then it is well known that for each $r \in [0, 1]$, $[\mu]^r$ is a closed and bounded interval of $\mathbb{R}$. For $u, v \in \mathbb{R}_{\mathcal{F}}$ and $\lambda \in \mathbb{R}$, we define uniquely the sum $u \oplus v$ and the product $\lambda \odot u$ by

$$[u \oplus v]^r = [u]^r + [v]^r, \quad [\lambda \odot u]^r = \lambda [u]^r, \quad \forall\, r \in [0, 1],$$

where $[u]^r + [v]^r$ means the usual addition of two intervals (as subsets of $\mathbb{R}$) and $\lambda [u]^r$ means the usual product between a scalar and a subset of $\mathbb{R}$ (see, e.g., [17]). Notice $1 \odot u = u$ and it holds $u \oplus v = v \oplus u$, $\lambda \odot u = u \odot \lambda$. If $0 \le r_1 \le r_2 \le 1$ then $[u]^{r_2} \subseteq [u]^{r_1}$. Actually $[u]^r = \left[u_-^{(r)}, u_+^{(r)}\right]$, where $u_-^{(r)} < u_+^{(r)}$, $u_-^{(r)}, u_+^{(r)} \in \mathbb{R}$, $\forall$ $r \in [0, 1]$.

Define

$$D : \mathbb{R}_{\mathcal{F}} \times \mathbb{R}_{\mathcal{F}} \to \mathbb{R}_+ \cup \{0\}$$

by

$$D(u, v) := \sup_{r \in [0,1]} \max \left\{ \left| u_-^{(r)} - v_-^{(r)} \right|, \left| u_+^{(r)} - v_+^{(r)} \right| \right\},$$

where $[v]^r = \left[v_-^{(r)}, v_+^{(r)}\right]$; $u, v \in \mathbb{R}_{\mathcal{F}}$. We have that D is a metric on $\mathbb{R}_{\mathcal{F}}$. Then $(\mathbb{R}_{\mathcal{F}}, D)$ is a complete metric space, see [17], with the properties

$$\begin{aligned} D(u \oplus w, v \oplus w) &= D(u, v), \quad \forall\, u, v, w \in \mathbb{R}_{\mathcal{F}}, \\ D(k \odot u, k \odot v) &= |k|\, D(u, v), \quad \forall\, u, v \in \mathbb{R}_{\mathcal{F}}, \forall\, k \in \mathbb{R}, \\ D(u \oplus v, w \oplus e) &\le D(u, w) + D(v, e), \quad \forall\, u, v, w, e \in \mathbb{R}_{\mathcal{F}}. \end{aligned} \tag{28.1}$$

Let $f, g : \mathbb{R} \to \mathbb{R}_{\mathcal{F}}$ be fuzzy real number valued functions. The distance between f, g is defined by

$$D^*(f, g) := \sup_{x \in \mathbb{R}} D(f(x), g(x)).$$

On $\mathbb{R}_{\mathcal{F}}$ we define a partial order by "$\le$": $u, v \in \mathbb{R}_{\mathcal{F}}$, $u \le v$ iff $u_-^{(r)} \le v_-^{(r)}$ and $u_+^{(r)} \le v_+^{(r)}$, $\forall\, r \in [0, 1]$.

We need

Lemma 28.2 ([5]) *For any $a, b \in \mathbb{R} : a \cdot b \ge 0$ and any $u \in \mathbb{R}_{\mathcal{F}}$ we have*

$$D(a \odot u, b \odot u) \le |a - b| \cdot D(u, \widetilde{o}), \tag{28.2}$$

where $\widetilde{o} \in \mathbb{R}_{\mathcal{F}}$ is defined by $\widetilde{o} := \chi_{\{0\}}$.

Lemma 28.3 ([5])

(i) *If we denote* $\widetilde{o} := \chi_{\{0\}}$, *then* $\widetilde{o} \in \mathbb{R}_{\mathcal{F}}$ *is the neutral element with respect to* $\oplus$, *i.e.*, $u \oplus \widetilde{o} = \widetilde{o} \oplus u = u$, $\forall\, u \in \mathbb{R}_{\mathcal{F}}$.

(ii) *With respect to* $\widetilde{o}$, *none of* $u \in \mathbb{R}_{\mathcal{F}}$, $u \neq \widetilde{o}$ *has opposite in* $\mathbb{R}_{\mathcal{F}}$.

(iii) *Let* $a, b \in \mathbb{R} : a \cdot b \geq 0$, *and any* $u \in \mathbb{R}_{\mathcal{F}}$, *we have* $(a + b) \odot u = a \odot u \oplus b \odot u$. *For general* $a, b \in \mathbb{R}$, *the above property is false.*

(iv) *For any* $\lambda \in \mathbb{R}$ *and any* $u, v \in \mathbb{R}_{\mathcal{F}}$, *we have* $\lambda \odot (u \oplus v) = \lambda \odot u \oplus \lambda \odot v$.

(v) *For any* $\lambda, \mu \in \mathbb{R}$ *and* $u \in \mathbb{R}_{\mathcal{F}}$, *we have* $\lambda \odot (\mu \odot u) = (\lambda \cdot \mu) \odot u$.

(vi) *If we denote* $\|u\|_{\mathcal{F}} := D(u, \widetilde{o})$, $\forall\, u \in \mathbb{R}_{\mathcal{F}}$, *then* $\|\cdot\|_{\mathcal{F}}$ *has the properties of a usual norm on* $\mathbb{R}_{\mathcal{F}}$, *i.e.*,

$$\begin{gathered} \|u\|_{\mathcal{F}} = 0 \text{ iff } u = \widetilde{o}, \ \|\lambda \odot u\|_{\mathcal{F}} = |\lambda| \cdot \|u\|_{\mathcal{F}}, \\ \|u \oplus v\|_{\mathcal{F}} \leq \|u\|_{\mathcal{F}} + \|v\|_{\mathcal{F}}, \ \|u\|_{\mathcal{F}} - \|v\|_{\mathcal{F}} \leq D(u, v). \end{gathered} \tag{28.3}$$

Notice that $(\mathbb{R}_{\mathcal{F}}, \oplus, \odot)$ is not a linear space over $\mathbb{R}$; and consequently $(\mathbb{R}_{\mathcal{F}}, \|\cdot\|_{\mathcal{F}})$ is not a normed space.

As in Remark 4.4 ([5]) one can show easily that a sequence of operators of the form

$$L_n(f)(x) := \sum_{k=0}^{n*} f\left(x_{k_n}\right) \odot w_{n,k}(x), \quad n \in \mathbb{N}, \tag{28.4}$$

($\sum^*$ denotes the fuzzy summation) where $f : \mathbb{R}^d \to \mathbb{R}_{\mathcal{F}}$, $x_{k_n} \in \mathbb{R}^d$, $d \in \mathbb{N}$, $w_{n,k}(x)$ real valued weights, are linear over $\mathbb{R}^d$, i.e.,

$$L_n(\lambda \odot f \oplus \mu \odot g)(x) = \lambda \odot L_n(f)(x) \oplus \mu \odot L_n(g)(x), \tag{28.5}$$

$\forall \lambda, \mu \in \mathbb{R}$, any $x \in \mathbb{R}^d$; $f, g : \mathbb{R}^d \to \mathbb{R}_{\mathcal{F}}$. (Proof based on Lemma 28.3 (iv).)

We further need

Definition 28.4 (*see also* [15], *Definition 13.16, p. 654*) Let $(X, \mathcal{B}, P)$ be a probability space. A fuzzy-random variable is a $\mathcal{B}$-measurable mapping $g : X \to \mathbb{R}_{\mathcal{F}}$ (i.e., for any open set $U \subseteq \mathbb{R}_{\mathcal{F}}$, in the topology of $\mathbb{R}_{\mathcal{F}}$ generated by the metric D, we have

$$g^{-1}(U) = \{s \in X; g(s) \in U\} \in \mathcal{B}). \tag{28.6}$$

The set of all fuzzy-random variables is denoted by $\mathcal{L}_{\mathcal{F}}(X, \mathcal{B}, P)$. Let $g_n, g \in \mathcal{L}_{\mathcal{F}}(X, \mathcal{B}, P)$, $n \in \mathbb{N}$ and $0 < q < +\infty$. We say $g_n(s) \overset{\text{“}q\text{-mean”}}{\underset{n \to +\infty}{\rightarrow}} g(s)$ if

$$\lim_{n \to +\infty} \int_X D(g_n(s), g(s))^q \, P(ds) = 0. \tag{28.7}$$

Remark 28.5 (see [15], p. 654) If $f, g \in \mathcal{L}_{\mathcal{F}}(X, \mathcal{B}, P)$, let us denote $F : X \to \mathbb{R}_+ \cup \{0\}$ by $F(s) = D(f(s), g(s))$, $s \in X$. Here, F is $\mathcal{B}$-measurable, because $F = G \circ H$, where $G(u, v) = D(u, v)$ is continuous on $\mathbb{R}_{\mathcal{F}} \times \mathbb{R}_{\mathcal{F}}$, and $H : X \to \mathbb{R}_{\mathcal{F}} \times \mathbb{R}_{\mathcal{F}}$, $H(s) = (f(s), g(s))$, $s \in X$, is $\mathcal{B}$-measurable. This shows that the above convergence in q-mean makes sense.

Definition 28.6 (*see* [15], *p. 654, Definition 13.17*) Let $(T, \mathcal{T})$ be a topological space. A mapping $f : T \to \mathcal{L}_{\mathcal{F}}(X, \mathcal{B}, P)$ will be called fuzzy-random function (or fuzzy-stochastic process) on T. We denote $f(t)(s) = f(t, s)$, $t \in T$, $s \in X$.

Remark 28.7 (see [15], p. 655) Any usual fuzzy real function $f : T \to \mathbb{R}_{\mathcal{F}}$ can be identified with the degenerate fuzzy-random function $f(t, s) = f(t)$, $\forall\, t \in T$, $s \in X$.

Remark 28.8 (see [15], p. 655) Fuzzy-random functions that coincide with probability one for each $t \in T$ will be consider equivalent.

Remark 28.9 (see [15], p. 655) Let $f, g : T \to \mathcal{L}_{\mathcal{F}}(X, \mathcal{B}, P)$. Then $f \oplus g$ and $k \odot f$ are defined pointwise, i.e.,

$$(f \oplus g)(t, s) = f(t, s) \oplus g(t, s),$$
$$(k \odot f)(t, s) = k \odot f(t, s), \; t \in T, s \in X.$$

Definition 28.10 (*see also Definition 13.18, pp. 655–656,* [15]) For a fuzzy-random function $f : \mathbb{R}^d \to \mathcal{L}_{\mathcal{F}}(X, \mathcal{B}, P)$, $d \in \mathbb{N}$, we define the (first) fuzzy-random modulus of continuity

$$\Omega_1^{(\mathcal{F})}(f, \delta)_{L^q} =$$

$$\sup \left\{ \left(\int_X D^q(f(x, s), f(y, s)) P(ds) \right)^{\frac{1}{q}} : x, y \in \mathbb{R}^d, \; \|x - y\|_{l_1} \le \delta \right\}, \tag{28.8}$$

$0 < \delta$, $1 \le q < \infty$.

Definition 28.11 ([9]) Here $1 \le q < +\infty$. Let $f : \mathbb{R}^d \to \mathcal{L}_{\mathcal{F}}(X, \mathcal{B}, P)$, $d \in \mathbb{N}$, be a fuzzy random function. We call f a (q-mean) uniformly continuous fuzzy random function over $\mathbb{R}^d$, iff $\forall\, \varepsilon > 0\; \exists\, \delta > 0$: whenever $\|x - y\|_{l_1} \le \delta$, $x, y \in \mathbb{R}^d$, implies that

$$\int_X (D(f(x, s), f(y, s)))^q P(ds) \le \varepsilon. \tag{28.9}$$

We denote it as $f \in C_{FR}^{U_q}(\mathbb{R}^d)$.

Proposition 28.12 ([9]) *Let* $f \in C_{FR}^{U_q}(\mathbb{R}^d)$. *Then* $\Omega_1^{(\mathcal{F})}(f,\delta)_{L^q} < \infty$, *any* $\delta > 0$.

Proposition 28.13 ([9]) *Let* $f, g : \mathbb{R}^d \to \mathcal{L}_{\mathcal{F}}(X, \mathcal{B}, P)$, $d \in \mathbb{N}$, *be fuzzy random functions. It holds*

(i) $\Omega_1^{(\mathcal{F})}(f,\delta)_{L^q}$ *is nonnegative and nondecreasing in* $\delta > 0$.

(ii) $\lim_{\delta \downarrow 0} \Omega_1^{(\mathcal{F})}(f,\delta)_{L^q} = \Omega_1^{(\mathcal{F})}(f,0)_{L^q} = 0$, *iff* $f \in C_{FR}^{U_q}(\mathbb{R}^d)$.

(iii) $\Omega_1^{(\mathcal{F})}(f,\delta_1+\delta_2)_{L^q} \le \Omega_1^{(\mathcal{F})}(f,\delta_1)_{L^q} + \Omega_1^{(\mathcal{F})}(f,\delta_2)_{L^q}$, $\delta_1, \delta_2 > 0$.

(iv) $\Omega_1^{(\mathcal{F})}(f,n\delta)_{L^q} \le n\Omega_1^{(\mathcal{F})}(f,\delta)_{L^q}$, $\delta > 0$, $n \in \mathbb{N}$.

(v) $\Omega_1^{(\mathcal{F})}(f,\lambda\delta)_{L^q} \le \lceil\lambda\rceil\, \Omega_1^{(\mathcal{F})}(f,\delta)_{L^q} \le (\lambda+1)\, \Omega_1^{(\mathcal{F})}(f,\delta)_{L^q}$, $\lambda > 0$, $\delta > 0$, *where* $\lceil\cdot\rceil$ *is the ceiling of the number.*

(vi) $\Omega_1^{(\mathcal{F})}(f \oplus g,\delta)_{L^q} \le \Omega_1^{(\mathcal{F})}(f,\delta)_{L^q} + \Omega_1^{(\mathcal{F})}(g,\delta)_{L^q}$, $\delta > 0$. *Here* $f \oplus g$ *is a fuzzy random function.*

(vii) $\Omega_1^{(\mathcal{F})}(f,\cdot)_{L^q}$ *is continuous on* $\mathbb{R}_+$, *for* $f \in C_{FR}^{U_q}(\mathbb{R}^d)$.

We give

Definition 28.14 ([7]) Let $f(t,s)$ be a stochastic process from $\mathbb{R}^d \times (X, \mathcal{B}, P)$ into $\mathbb{R}$, $d \in \mathbb{N}$, where $(X, \mathcal{B}, P)$ is a probability space. We define the q-mean multivariate first moduli of continuity of f by

$$\Omega_1(f,\delta)_{L^q} :=$$

$$\sup\left\{\left(\int_X |f(x,s) - f(y,s)|^q P(ds)\right)^{\frac{1}{q}} : x, y \in \mathbb{R}^d,\ \|x-y\|_{l_1} \le \delta\right\}, \tag{28.10}$$

$\delta > 0$, $1 \le q < \infty$.

For more see [7].

We mention

Proposition 28.15 ([9]) *Assume that* $\Omega_1^{(\mathcal{F})}(f,\delta)_{L^q}$ *is finite,* $\delta > 0$, $1 \le q < \infty$. *Then*

$$\Omega_1^{(\mathcal{F})}(f,\delta)_{L^q} \ge \sup_{r\in[0,1]} \max\left\{\Omega_1\left(f_-^{(r)},\delta\right)_{L^q}, \Omega_1\left(f_+^{(r)},\delta\right)_{L^q}\right\}. \tag{28.11}$$

The reverse direction "$\le$" "is not possible".

Remark 28.16 ([9]) For each $s \in X$ we define the usual first modulus of continuity of $f(\cdot,s)$ by

$$\omega_1^{(\mathcal{F})}(f(\cdot,s),\delta) := \sup_{\substack{x,y\in\mathbb{R}^d \\ \|x-y\|_{l_1}\le\delta}} D(f(x,s), f(y,s)),\ \delta > 0. \tag{28.12}$$

Therefore

$$D^q\left(f(x,s), f(y,s)\right) \leq \left(\omega_1^{(\mathcal{F})}\left(f(\cdot,s),\delta\right)\right)^q, \tag{28.13}$$

$\forall\, s \in X$ and $x, y \in \mathbb{R}^d : \|x-y\|_{l_1} \leq \delta$, $\delta > 0$.

Hence it holds

$$\left(\int_X D^q\left(f(x,s), f(y,s)\right) P(ds)\right)^{\frac{1}{q}} \leq \left(\int_X \left(\omega_1^{(\mathcal{F})}\left(f(\cdot,s),\delta\right)\right)^q P(ds)\right)^{\frac{1}{q}}, \tag{28.14}$$

$\forall\, x, y \in \mathbb{R}^d : \|x-y\|_{l_1} \leq \delta$.

We have that

$$\Omega_1^{(\mathcal{F})}(f,\delta)_{L^q} \leq \left(\int_X \left(\omega_1^{(\mathcal{F})}\left(f(\cdot,s),\delta\right)\right)^q P(ds)\right)^{\frac{1}{q}}, \tag{28.15}$$

under the assumption that the right hand side of (28.15) is finite.

The reverse "$\geq$" of the last (28.15) is not true.

Also we have

Proposition 28.17 ([6]) *(i) Let $Y(t,\omega)$ be a real valued stochastic process such that Y is continuous in $t \in [a,b]$. Then Y is jointly measurable in (t,ω).*

(ii) Further assume that the expectation $(E|Y|)(t) \in C([a,b])$, or more generally $\int_a^b (E|Y|)(t)\,dt$ makes sense and is finite. Then

$$E\left(\int_a^b Y(t,\omega)\,dt\right) = \int_a^b (EY)(t)\,dt. \tag{28.16}$$

According to [14], p. 94 we have the following

Definition 28.18 Let $(Y, \mathcal{T})$ be a topological space, with its σ-algebra of Borel sets $\mathcal{B} := \mathcal{B}(Y,\mathcal{T})$ generated by $\mathcal{T}$. If $(X, \mathcal{S})$ is a measurable space, a function $f : X \to Y$ is called measurable iff $f^{-1}(B) \in \mathcal{S}$ for all $B \in \mathcal{B}$.

By Theorem 4.1.6 of [14], p. 89 f as above is measurable iff

$$f^{-1}(C) \in \mathcal{S} \text{ for all } C \in \mathcal{T}.$$

We would need

Theorem 28.19 (see [14], p. 95) *Let $(X, \mathcal{S})$ be a measurable space and (Y, d) be a metric space. Let f_n be measurable functions from X into Y such that for all $x \in X$, $f_n(x) \to f(x)$ in Y. Then f is measurable. I.e., $\lim_{n\to\infty} f_n = f$ is measurable.*

We need also

Proposition 28.20 *Let f, g be fuzzy random variables from $\mathcal{S}$ into $\mathbb{R}_{\mathcal{F}}$. Then*
(i) Let $c \in \mathbb{R}$, then $c \odot f$ is a fuzzy random variable.
(ii) $f \oplus g$ is a fuzzy random variable.

For the definition of general fuzzy integral we follow [16] next.

Definition 28.21 Let (Ω, Σ, μ) be a complete σ-finite measure space. We call $F : \Omega \to R_{\mathcal{F}}$ measurable iff $\forall$ closed $B \subseteq \mathbb{R}$ the function $F^{-1}(B) : \Omega \to [0, 1]$ defined by

$$F^{-1}(B)(w) := \sup_{x \in B} F(w)(x), \text{ all } w \in \Omega \tag{28.17}$$

is measurable, see [16].

Theorem 28.22 ([16]) *For $F : \Omega \to \mathbb{R}_{\mathcal{F}}$,*

$$F(w) = \left\{ \left(F_-^{(r)}(w), F_+^{(r)}(w) \right) | 0 \leq r \leq 1 \right\}, \tag{28.18}$$

the following are equivalent

(1) F is measurable,
(3) $\forall r \in [0, 1]$, $F_-^{(r)}$, $F_+^{(r)}$ are measurable.

Following [16], given that for each $r \in [0, 1]$, $F_-^{(r)}$, $F_+^{(r)}$ are integrable we have that the parametrized representation

$$\left\{ \left(\int_A F_-^{(r)} d\mu, \int_A F_+^{(r)} d\mu \right) | 0 \leq r \leq 1 \right\} \tag{28.19}$$

is a fuzzy real number for each $A \in \Sigma$.

The last fact leads to

Definition 28.23 ([16]) A measurable function $F : \Omega \to \mathbb{R}_{\mathcal{F}}$,

$$F(w) = \left\{ \left(F_-^{(r)}(w), F_+^{(r)}(w) \right) | 0 \leq r \leq 1 \right\}$$

is called integrable if for each $r \in [0, 1]$, $F_{\pm}^{(r)}$ are integrable, or equivalently, if $F_{\pm}^{(0)}$ are integrable.

In this case, the fuzzy integral of F over $A \in \Sigma$ is defined by

$$\int_A F d\mu := \left\{ \left(\int_A F_-^{(r)} d\mu, \int_A F_+^{(r)} d\mu \right) | 0 \leq r \leq 1 \right\}.$$

By [16], F is integrable iff $w \to \|F(w)\|_{\mathcal{F}}$ is real-valued integrable.

Here denote

$$\|u\|_{\mathcal{F}} := D\left(u, \widetilde{0}\right), \ \forall\, u \in \mathbb{R}_{\mathcal{F}}.$$

We need also

Theorem 28.24 ([16]) *Let $F, G : \Omega \to \mathbb{R}_{\mathcal{F}}$ be integrable. Then*

(1) Let $a, b \in \mathbb{R}$, then $aF + bG$ is integrable and for each $A \in \Sigma$,

$$\int_A (aF + bG)\, d\mu = a \int_A F d\mu + b \int_A G d\mu; \tag{28.20}$$

(2) $D(F, G)$ is a real—valued integrable function and for each $A \in \Sigma$,

$$D\left(\int_A F d\mu, \int_A G d\mu\right) \leq \int_A D(F, G)\, d\mu. \tag{28.21}$$

In particular,

$$\left\| \int_A F d\mu \right\|_{\mathcal{F}} \leq \int_A \|F\|_{\mathcal{F}}\, d\mu. \tag{28.22}$$

Above μ could be the Lebesgue measure, in this chapter the multivariate Lebesgue measure, with all the basic properties valid here too.

Basically here we have

$$\left[\int_A F d\mu\right]^r = \left[\int_A F_-^{(r)} d\mu, \int_A F_+^{(r)} d\mu\right], \tag{28.23}$$

i.e.

$$\left(\int_A F d\mu\right)_{\pm}^{(r)} = \int_A F_{\pm}^{(r)} d\mu, \ \forall\, r \in [0, 1]. \tag{28.24}$$

Next we state Fubini's theorem for fuzzy number-valued functions and fuzzy number-valued integrals, see [16].

Theorem 28.25 *Let $(\Omega_1, \Sigma_1, \mu_1)$ and $(\Omega_2, \Sigma_2, \mu_2)$ be two complete σ-finite measure spaces, and let $(\Omega_1 \times \Omega_2, \Sigma_1 \times \Sigma_2, \mu_1 \times \mu_2)$ be their product measure space. If a fuzzy number-valued function $F : \Omega_1 \times \Omega_2 \to \mathbb{R}_{\mathcal{F}}$ is $\mu_1 \times \mu_2$-integrable, then*

(1) the fuzzy-number-valued function $F(\cdot, \omega_2) : \omega_1 \to F(\omega_1, \omega_2)$ is μ_1-integrable for $\omega_2 \in \Omega_2$, μ_2-a.e.,
(2) the fuzzy-number-valued function $\omega_2 \to \int_{\Omega_1} F(\omega_1, \omega_2)\, d\mu_1(\omega_1)$ is μ_2-integrable,

and

(3)

$$\int_{\Omega_1\times\Omega_2} Fd(\mu_1\times\mu_2) = \int_{\Omega_2}\left(\int_{\Omega_1} F(\omega_1,\omega_2)\,d\mu_1(\omega_1)\right)d\mu_2(\omega_2) = \tag{28.25}$$

$$\int_{\Omega_1}\left(\int_{\Omega_2} F(\omega_1,\omega_2)\,d\mu_2(\omega_2)\right)d\mu_1(\omega_1).$$

We further mention

Theorem 28.26 *Let* $f : \prod_{i=1}^{N}[a_i,b_i] \to L_{\mathcal{F}}(X,\mathcal{B},P)$, *be a fuzzy-random function. We assume that* $f\left(\overrightarrow{t},s\right)$ *is fuzzy continuous in* $\overrightarrow{t} \in \prod_{i=1}^{N}[a_i,b_i]$, *for any* $s \in X$. *Then* $\int_{\prod_{i=1}^{N}[a_i,b_i]} f\left(\overrightarrow{t},s\right)d\overrightarrow{t}$ *exists in* $\mathbb{R}_{\mathcal{F}}$ *and it is a fuzzy-random variable in* $s \in X$.

Proof By definition of fuzzy integral we notice here that

$$\int_{\prod_{i=1}^{N}[a_i,b_i]} f\left(\overrightarrow{t},s\right)d\overrightarrow{t} =$$

$$\left\{\left(\int_{\prod_{i=1}^{N}[a_i,b_i]} f_-^{(r)}\left(\overrightarrow{t},s\right)d\overrightarrow{t},\int_{\prod_{i=1}^{N}[a_i,b_i]} f_+^{(r)}\left(\overrightarrow{t},s\right)d\overrightarrow{t}\right)|0\le r\le 1\right\}, \tag{28.26}$$

where $d\overrightarrow{t}$ is the multivariate Lebesgue measure on $\prod_{i=1}^{N}[a_i,b_i]$. Because $f\left(\overrightarrow{t},s\right)$ is fuzzy continuous in $\overrightarrow{t}$, we get that $f_{\pm}^{(r)}\left(\overrightarrow{t},s\right)$, $0\le r\le 1$, are real valued continuous in $\overrightarrow{t}$, for each $s \in X$. Hence the real integrals $\int_{\prod_{i=1}^{N}[a_i,b_i]} f_{\pm}^{(r)}\left(\overrightarrow{t},s\right)d\overrightarrow{t}$ are multivariate Riemann integrals that exist, for each $s \in X$.

Thus $\int_{\prod_{i=1}^{N}[a_i,b_i]} f\left(\overrightarrow{t},s\right)d\overrightarrow{t} \in \mathbb{R}_{\mathcal{F}}$, i.e. it exists.

By Theorem 28.19 and the definition of multivariate Riemann integral we get that $\int_{\prod_{i=1}^{N}[a_i,b_i]} f_{\pm}^{(r)}\left(\overrightarrow{t},s\right)d\overrightarrow{t}$ are P-measurable functions in $s \in X$.

Taking into account (28.24) and Theorem 28.22 we derive that $\int_{\prod_{i=1}^{N}[a_i,b_i]} f\left(\overrightarrow{t},s\right)d\overrightarrow{t}$ is a fuzzy-random variable in $s \in X$. ■

28.2 Main Results

We are motivated by [9, 12, 13].

Here the activation function $b : \mathbb{R}^d \to \mathbb{R}_+$, $d \in \mathbb{N}$, is of compact support $B := \prod_{j=1}^{d} [-T_j, T_j]$, $T_j > 0$, $j = 1, \ldots, d$. That is $b(x) > 0$ for any $x \in B$, and clearly b may have jump discontinuities. Also the shape of the graph of b is immaterial.

Typically in neural networks approximation we take b to be a d-dimensional bell-shaped function (i.e. per coordinate is a centered bell-shaped function), or a product of univariate centered bell-shaped functions, or a product of sigmoid functions, in our case all them of compact support B.

Example 28.27 Take $b(x) = \beta(x_1)\beta(x_2)\ldots\beta(x_d)$, where β is any of the following functions, $i = 1, \ldots, d$:

(i) $\beta(x_j)$ is the characteristic function on $[-1, 1]$,

(ii) $\beta(x_j)$ is the hat function over $[-1, 1]$, that is,

$$\beta(x_j) = \begin{cases} 1 + x_j, & -1 \le x_j \le 0, \\ 1 - x_j, & 0 < x_j \le 1, \\ 0, & \text{elsewhere}, \end{cases} \tag{28.27}$$

(iii) the truncated sigmoids

$$\beta(x_j) = \begin{cases} \frac{1}{1+e^{-x_j}} \text{ or } \tanh x_j \text{ or } \operatorname{erf}(x_j), \text{ for } x_j \in [-T_j, T_j], \text{ with large } T_j > 0, \\ 0, \; x_j \in \mathbb{R} - [-T_j, T_j], \end{cases}$$

(iv) the truncated Gompertz function

$$\beta(x_j) = \begin{cases} e^{-\alpha e^{-\beta x_j}}, x_j \in [-T_j, T_j]; \; \alpha, \beta > 0; \text{ large } T_j > 0, \\ 0, \; x_j \in \mathbb{R} - [-T_j, T_j], \end{cases}$$

The Gompertz functions are also sigmoid functions, with wide applications to many applied fields, e.g. demography and tumor growth modeling, etc.

Thus the general activation function b we will be using here includes all kinds of activation functions in neural network approximations.

Here we consider functions $f \in C_{FR}^{U_q}(\mathbb{R}^d)$.

Let here the parameters: $0 < \alpha < 1$, $x = (x_1, \ldots, x_d) \in \mathbb{R}^d$, $n \in \mathbb{N}$; $\overline{r} = (r_1, \ldots, r_d) \in \mathbb{N}^d$, $i = (i_1, \ldots, i_d) \in \mathbb{N}^d$, with $i_j = 1, 2, \ldots, r_j$, $j = 1, \ldots, d$; also let $w_i = w_{i_1,\ldots,i_d} \ge 0$, such that $\sum_{i_1=1}^{r_1} \sum_{i_2=1}^{r_2} \cdots \sum_{i_d=1}^{r_d} w_{i_1,\ldots,i_d} = 1$, in brief written as $\sum_{i=1}^{\overline{r}} w_i = 1$. We further consider the parameters $k = (k_1, \ldots, k_d) \in \mathbb{Z}^d$; $\mu_i =$

$\left(\mu_{i_1},\ldots,\mu_{i_d}\right)\in\mathbb{R}_+^d$, $\nu_i=\left(\nu_{i_1},\ldots,\nu_{i_d}\right)\in\mathbb{R}_+^d$; and $\lambda_i=\lambda_{i_1,\ldots,i_d}$, $\rho_i=\rho_{i_1,\ldots,i_d}\geq 0$. Call $\nu_i^{\min}=\min\{\nu_{i_1},\ldots,\nu_{i_d}\}$.

In this chapter we study in q-mean ($1\leq q<\infty$) the pointwise and uniform convergences with rates over $\mathbb{R}^d$, to the fuzzy-random unit operator, of the following fuzzy-random normalized one hidden layer multivariate perturbed neural network operators, where $s\in X$, $(X,\mathcal{B},P)$ a probability space, $n\in\mathbb{N}$,

(i) the Stancu type

$$\left(H_n^{FR}(f)\right)(x,s)=\left(H_n^{FR}(f)\right)(x_1,\ldots,x_d,s)= \tag{28.28}$$

$$\frac{\sum_{k=-n^2}^{n^2*}\left(\sum_{i=1}^{\overline{r}*}w_i\odot f\left(\frac{k+\mu_i}{n+\nu_i},s\right)\right)\odot b\left(n^{1-\alpha}\left(x-\frac{k}{n}\right)\right)}{\sum_{k=-n^2}^{n^2}b\left(n^{1-\alpha}\left(x-\frac{k}{n}\right)\right)}=$$

$$\frac{\sum_{k_1=-n^2}^{n^2*}\cdots\sum_{k_d=-n^2}^{n^2*}\left(\sum_{i_1=1}^{r_1*}\cdots\sum_{i_d=1}^{r_d*}w_{i_1,\ldots,i_d}\odot f\left(\frac{k_1+\mu_{i_1}}{n+\nu_{i_1}},\ldots,\frac{k_d+\mu_{i_d}}{n+\nu_{i_d}},s\right)\right)\odot}{\sum_{k_1=-n^2}^{n^2}\cdots\sum_{k_d=-n^2}^{n^2}b\left(n^{1-\alpha}\left(x_1-\frac{k_1}{n}\right),\ldots,n^{1-\alpha}\left(x_d-\frac{k_d}{n}\right)\right)}$$

$$b\left(n^{1-\alpha}\left(x_1-\frac{k_1}{n}\right),\ldots,n^{1-\alpha}\left(x_d-\frac{k_d}{n}\right)\right),$$

(ii) the Kantorovich type

$$\left(K_n^{FR}(f)\right)(x,s)= \tag{28.29}$$

$$\frac{\sum_{k=-n^2}^{n^2*}\left(\sum_{i=1}^{\overline{r}*}w_i\left(n+\rho_i\right)^d\odot\int_0^{\frac{1}{n+\rho_i}}f\left(t+\frac{k+\lambda_i}{n+\rho_i},s\right)dt\right)\odot b\left(n^{1-\alpha}\left(x-\frac{k}{n}\right)\right)}{\sum_{k=-n^2}^{n^2}b\left(n^{1-\alpha}\left(x-\frac{k}{n}\right)\right)}=$$

$$\sum_{k_1=-n^2}^{n^2*}\cdots\sum_{k_d=-n^2}^{n^2*}\left(\sum_{i_1=1}^{r_1*}\cdots\sum_{i_d=1}^{r_d*}w_{i_1,\ldots,i_d}\left(n+\rho_{i_1,\ldots,i_d}\right)^d\odot\right.$$

$$\frac{\left.\int\cdots\int_0^{\frac{1}{n+\rho_{i_1,\ldots,i_d}}}\cdots\int f\left(t_1+\frac{k_1+\lambda_{i_1,\ldots,i_d}}{n+\rho_{i_1,\ldots,i_d}},\ldots,t_d+\frac{k_d+\lambda_{i_1,\ldots,i_d}}{n+\rho_{i_1,\ldots,i_d}},s\right)dt_1\ldots dt_d\right)\odot}{\sum_{k_1=-n^2}^{n^2}\cdots\sum_{k_d=-n^2}^{n^2}b\left(n^{1-\alpha}\left(x_1-\frac{k_1}{n}\right),\ldots,n^{1-\alpha}\left(x_d-\frac{k_d}{n}\right)\right)} \tag{28.30}$$

$$b\left(n^{1-\alpha}\left(x_1-\frac{k_1}{n}\right),\ldots,n^{1-\alpha}\left(x_d-\frac{k_d}{n}\right)\right),$$

and

(iii) the quadrature type

$$\left(M_n^{FR}(f)\right)(x,s) = \frac{\sum_{k=-n^2}^{n^2 *}\left(\sum_{i=1}^{\overline{r}*} w_i \odot f\left(\frac{k}{n}+\frac{i}{n\overline{r}}, s\right)\right) \odot b\left(n^{1-\alpha}\left(x-\frac{k}{n}\right)\right)}{\sum_{k=-n^2}^{n^2} b\left(n^{1-\alpha}\left(x-\frac{k}{n}\right)\right)} = \tag{28.31}$$

$$\frac{\sum_{k_1=-n^2}^{n^2 *} \cdots \sum_{k_d=-n^2}^{n^2 *}\left(\sum_{i_1=1}^{r_1 *} \cdots \sum_{i_d=1}^{r_d *} w_{i_1,\ldots,i_d} \odot f\left(\frac{k_1}{n}+\frac{i_1}{nr_1}, \ldots, \frac{k_d}{n}+\frac{i_d}{nr_d}, s\right)\right) \odot}{\sum_{k_1=-n^2}^{n^2} \cdots \sum_{k_d=-n^2}^{n^2} b\left(n^{1-\alpha}\left(x_1-\frac{k_1}{n}\right), \ldots, n^{1-\alpha}\left(x_d-\frac{k_d}{n}\right)\right)}$$

$$b\left(n^{1-\alpha}\left(x_1-\frac{k_1}{n}\right), \ldots, n^{1-\alpha}\left(x_d-\frac{k_d}{n}\right)\right).$$

Similar operators defined for d-dimensional bell-shaped activation functions and sample coefficients $f\left(\frac{k}{n}\right) = f\left(\frac{k_1}{n}, \ldots, \frac{k_d}{n}\right)$ were studied initially in [9].

In this chapter we assume that

$$n \geq \max_{j\in\{1,\ldots,d\}}\left\{T_j + |x_j|, T_j^{-\frac{1}{\alpha}}\right\}, \tag{28.32}$$

see also [3], p. 91.

So, by (28.32) we can rewrite the operators as follows ($[\cdot]$ is the integral part and $\lceil\cdot\rceil$ the ceiling of a number).

We denote by $T = (T_1, \ldots, T_d)$, $[nx + Tn^{\alpha}] = ([nx_1 + T_1 n^{\alpha}], \ldots, [nx_d + T_d n^{\alpha}])$, $\lceil nx - Tn^{\alpha}\rceil = (\lceil nx_1 - T_1 n^{\alpha}\rceil, \ldots, \lceil nx_d - T_d n^{\alpha}\rceil)$. It holds

(i)

$$\left(H_n^{FR}(f)\right)(x,s) = \left(H_n^{FR}(f)\right)(x_1, \ldots, x_d, s) = \tag{28.33}$$

$$\frac{\sum_{k=\lceil nx - Tn^{\alpha}\rceil}^{[nx+Tn^{\alpha}]*}\left(\sum_{i=1}^{\overline{r}*} w_i \odot f\left(\frac{k+\mu_i}{n+\nu_i}, s\right)\right) \odot b\left(n^{1-\alpha}\left(x-\frac{k}{n}\right)\right)}{\sum_{k=\lceil nx - Tn^{\alpha}\rceil}^{[nx+Tn^{\alpha}]} b\left(n^{1-\alpha}\left(x-\frac{k}{n}\right)\right)} =$$

$$\frac{\sum_{k_1=\lceil nx_1 - T_1 n^{\alpha}\rceil}^{[nx_1+T_1 n^{\alpha}]*} \cdots \sum_{k_d=\lceil nx_d - T_d n^{\alpha}\rceil}^{[nx_d+T_d n^{\alpha}]*}\left(\sum_{i_1=1}^{r_1 *} \cdots \sum_{i_d=1}^{r_d *} w_{i_1,\ldots,i_d} \odot f\left(\frac{k_1+\mu_{i_1}}{n+\nu_{i_1}}, \ldots, \frac{k_d+\mu_{i_d}}{n+\nu_{i_d}}, s\right)\right) \odot}{\sum_{k_1=\lceil nx_1 - T_1 n^{\alpha}\rceil}^{[nx_1+T_1 n^{\alpha}]} \cdots \sum_{k_d=\lceil nx_d - T_d n^{\alpha}\rceil}^{[nx_d+T_d n^{\alpha}]} b\left(n^{1-\alpha}\left(x_1-\frac{k_1}{n}\right), \ldots, n^{1-\alpha}\left(x_d-\frac{k_d}{n}\right)\right)}$$

$$b\left(n^{1-\alpha}\left(x_1-\frac{k_1}{n}\right), \ldots, n^{1-\alpha}\left(x_d-\frac{k_d}{n}\right)\right),$$

(ii)

$$\left(K_n^{FR}(f)\right)(x,s) =$$

$$\frac{\sum_{k=\lceil nx-Tn^{\alpha}\rceil}^{[nx+Tn^{\alpha}]*}\left(\sum_{i=1}^{\overline{r}*} w_i\,(n+\rho_i)^d \odot \int_0^{\frac{1}{n+\rho_i}} f\left(t+\frac{k+\lambda_i}{n+\rho_i}, s\right) dt\right)\odot b\left(n^{1-\alpha}\left(x-\frac{k}{n}\right)\right)}{\sum_{k=\lceil nx-Tn^{\alpha}\rceil}^{[nx+Tn^{\alpha}]} b\left(n^{1-\alpha}\left(x-\frac{k}{n}\right)\right)} = \quad (28.34)$$

$$\sum_{k_1=\lceil nx_1-T_1n^{\alpha}\rceil}^{[nx_1+T_1n^{\alpha}]*} \cdots \sum_{k_d=\lceil nx_d-T_dn^{\alpha}\rceil}^{[nx_d+T_dn^{\alpha}]*} \left(\sum_{i_1=1}^{r_1*} \cdots \sum_{i_d=1}^{r_d*} w_{i_1,\ldots,i_d}\left(n+\rho_{i_1,\ldots,i_d}\right)^d \odot\right.$$

$$\frac{\left.\int\cdots\int_0^{\frac{1}{n+\rho_{i_1,\ldots,i_d}}}\cdots\int f\left(t_1+\frac{k_1+\lambda_{i_1,\ldots,i_d}}{n+\rho_{i_1,\ldots,i_d}},\ldots,t_d+\frac{k_d+\lambda_{i_1,\ldots,i_d}}{n+\rho_{i_1,\ldots,i_d}}, s\right) dt_1\ldots dt_d\right)\odot}{\sum_{k_1=\lceil nx_1-T_1n^{\alpha}\rceil}^{[nx_1+T_1n^{\alpha}]}\cdots\sum_{k_d=\lceil nx_d-T_dn^{\alpha}\rceil}^{[nx_d+T_dn^{\alpha}]} b\left(n^{1-\alpha}\left(x_1-\frac{k_1}{n}\right),\ldots,n^{1-\alpha}\left(x_d-\frac{k_d}{n}\right)\right)} \quad (28.35)$$

$$b\left(n^{1-\alpha}\left(x_1-\frac{k_1}{n}\right),\ldots,n^{1-\alpha}\left(x_d-\frac{k_d}{n}\right)\right),$$

and

(iii)

$$\left(M_n^{FR}(f)\right)(x,s) = \frac{\sum_{k=\lceil nx-Tn^{\alpha}\rceil}^{[nx+Tn^{\alpha}]*}\left(\sum_{i=1}^{\overline{r}*} w_i \odot f\left(\frac{k}{n}+\frac{i}{n\overline{r}}, s\right)\right)\odot b\left(n^{1-\alpha}\left(x-\frac{k}{n}\right)\right)}{\sum_{k=\lceil nx-Tn^{\alpha}\rceil}^{[nx+Tn^{\alpha}]} b\left(n^{1-\alpha}\left(x-\frac{k}{n}\right)\right)} = \quad (28.36)$$

$$\frac{\sum_{k_1=\lceil nx_1-T_1n^{\alpha}\rceil}^{[nx_1+T_1n^{\alpha}]*}\cdots\sum_{k_d=\lceil nx_d-T_dn^{\alpha}\rceil}^{[nx_d+T_dn^{\alpha}]*}\left(\sum_{i_1=1}^{r_1*}\cdots\sum_{i_d=1}^{r_d*} w_{i_1,\ldots,i_d}\odot f\left(\frac{k_1}{n}+\frac{i_1}{nr_1},\ldots,\frac{k_d}{n}+\frac{i_d}{nr_d}, s\right)\right)\odot}{\sum_{k_1=\lceil nx_1-T_1n^{\alpha}\rceil}^{[nx_1+T_1n^{\alpha}]}\cdots\sum_{k_d=\lceil nx_d-T_dn^{\alpha}\rceil}^{[nx_d+T_dn^{\alpha}]} b\left(n^{1-\alpha}\left(x_1-\frac{k_1}{n}\right),\ldots,n^{1-\alpha}\left(x_d-\frac{k_d}{n}\right)\right)}$$

$$b\left(n^{1-\alpha}\left(x_1-\frac{k_1}{n}\right),\ldots,n^{1-\alpha}\left(x_d-\frac{k_d}{n}\right)\right).$$

So if $\left|n^{1-\alpha}\left(x_j-\frac{k_j}{n}\right)\right| \leq T_j$, all $j=1,\ldots,d$, we get that

$$\left\|x-\frac{k}{n}\right\|_{l_1} \leq \frac{\|T\|_{l_1}}{n^{1-\alpha}}. \quad (28.37)$$

For convenience we call

$$V(x) = \sum_{k=\lceil nx-Tn^{\alpha}\rceil}^{[nx+Tn^{\alpha}]} b\left(n^{1-\alpha}\left(x-\frac{k}{n}\right)\right) =$$

$$\sum_{k_1=\lceil nx_1-T_1n^{\alpha}\rceil}^{[nx_1+T_1n^{\alpha}]} \cdots \sum_{k_d=\lceil nx_d-T_dn^{\alpha}\rceil}^{[nx_d+T_dn^{\alpha}]} b\left(n^{1-\alpha}\left(x_1-\frac{k_1}{n}\right),\ldots,n^{1-\alpha}\left(x_d-\frac{k_d}{n}\right)\right). \tag{28.38}$$

We make

Remark 28.28 Here always k is as in

$$-n^2 \leq nx_j - T_jn^{\alpha} \leq k_j \leq nx_j + T_jn^{\alpha} \leq n^2, \tag{28.39}$$

(I) We observe that

$$\left\|\frac{k+\mu_i}{n+\nu_i}-x\right\|_{l_1} \leq \left\|\frac{k}{n+\nu_i}-x\right\|_{l_1} + \left\|\frac{\mu_i}{n+\nu_i}\right\|_{l_1}$$

$$\leq \left\|\frac{k}{n+\nu_i}-x\right\|_{l_1} + \frac{\|\mu_i\|_{l_1}}{n+\nu_i^{\min}}. \tag{28.40}$$

Next see that

$$\left\|\frac{k}{n+\nu_i}-x\right\|_{l_1} \leq \left\|\frac{k}{n+\nu_i}-\frac{k}{n}\right\|_{l_1} + \left\|\frac{k}{n}-x\right\|_{l_1} \leq \frac{\|k\nu_i\|_{l_1}}{n\left(n+\nu_i^{\min}\right)} + \frac{\|T\|_{l_1}}{n^{1-\alpha}} \tag{28.41}$$

$$\leq \frac{\|k\|_{l_2}\|\nu_i\|_{l_2}}{n\left(n+\nu_i^{\min}\right)} + \frac{\|T\|_{l_1}}{n^{1-\alpha}} =: (*). \tag{28.42}$$

We notice for $j = 1, \ldots, d$ we get

$$|k_j| \leq n|x_j| + T_jn^{\alpha}.$$

Hence

$$\|k\|_{l_2} \leq \left\|n|x| + Tn^{\alpha}\right\|_{l_2} \leq n\|x\|_{l_2} + \|T\|_{l_2} n^{\alpha}, \tag{28.43}$$

where $|x| = (|x_1|, \ldots, |x_d|)$.

Thus

$$(*) \leq \frac{\left(n\|x\|_{l_2} + \|T\|_{l_2} n^{\alpha}\right)\|\nu_i\|_{l_2}}{n\left(n+\nu_i^{\min}\right)} + \frac{\|T\|_{l_1}}{n^{1-\alpha}}. \tag{28.44}$$

So we get

$$\left\|\frac{k}{n+\nu_i}-x\right\|_{l_1} \leq \frac{\left(n\|x\|_{l_2} + \|T\|_{l_2} n^{\alpha}\right)\|\nu_i\|_{l_2}}{n\left(n+\nu_i^{\min}\right)} + \frac{\|T\|_{l_1}}{n^{1-\alpha}} = \tag{28.45}$$

$$\left(\frac{\|\nu_i\|_{l_2} \|x\|_{l_2}}{n + \nu_i^{\min}} \right) + \frac{\|\nu_i\|_{l_2} \|T\|_{l_2}}{n^{1-\alpha} \left(n + \nu_i^{\min}\right)} + \frac{\|T\|_{l_1}}{n^{1-\alpha}}.$$

Hence it holds

$$\left\| \frac{k + \mu_i}{n + \nu_i} - x \right\|_{l_1} \leq$$

$$\left(\frac{\|\nu_i\|_{l_2} \|x\|_{l_2} + \|\mu_i\|_{l_1}}{n + \nu_i^{\min}} \right) + \frac{1}{n^{1-\alpha}} \left(\|T\|_{l_1} + \frac{\|\nu_i\|_{l_2} \|T\|_{l_2}}{\left(n + \nu_i^{\min}\right)} \right). \tag{28.46}$$

(II) We also have for

$$0 \leq t_j \leq \frac{1}{n + \rho_i}, \; j = 1, \ldots, d, \tag{28.47}$$

that

$$\left\| t + \frac{k + \lambda_i}{n + \rho_i} - x \right\|_{l_1} \leq \|t\|_{l_1} + \left\| \frac{k + \lambda_i}{n + \rho_i} - x \right\|_{l_1} \leq$$

$$\frac{d}{n + \rho_i} + \left\| \frac{k + \lambda_i}{n + \rho_i} - x \right\|_{l_1} \leq \tag{28.48}$$

$$\frac{d}{n + \rho_i} + \left\| \frac{\lambda_i}{n + \rho_i} \right\|_{l_1} + \left\| \frac{k}{n + \rho_i} - x \right\|_{l_1} =$$

$$\frac{d}{n + \rho_i} + \frac{d\lambda_i}{n + \rho_i} + \left\| \frac{k}{n + \rho_i} - x \right\|_{l_1} =$$

$$\frac{d\left(1 + \lambda_i\right)}{n + \rho_i} + \left\| \frac{k}{n + \rho_i} - x \right\|_{l_1} \leq \tag{28.49}$$

$$\frac{d\left(1 + \lambda_i\right)}{n + \rho_i} + \left\| \frac{k}{n + \rho_i} - \frac{k}{n} \right\|_{l_1} + \left\| \frac{k}{n} - x \right\|_{l_1} \leq$$

$$\frac{d\left(1 + \lambda_i\right)}{n + \rho_i} + \frac{\rho_i}{n\left(n + \rho_i\right)} \|k\|_{l_1} + \frac{\|T\|_{l_1}}{n^{1-\alpha}} \leq \tag{28.50}$$

$$\frac{d\left(1 + \lambda_i\right)}{n + \rho_i} + \frac{\rho_i}{n\left(n + \rho_i\right)} \left(n\|x\|_{l_1} + \|T\|_{l_1} n^{\alpha}\right) + \frac{\|T\|_{l_1}}{n^{1-\alpha}} =$$

$$\frac{d(1+\lambda_i)}{n+\rho_i} + \frac{\rho_i \|x\|_{l_1}}{(n+\rho_i)} + \frac{\rho_i \|T\|_{l_1}}{n^{1-\alpha}(n+\rho_i)} + \frac{\|T\|_{l_1}}{n^{1-\alpha}} = \tag{28.51}$$

$$\left(\frac{\rho_i \|x\|_{l_1} + d(\lambda_i + 1)}{n+\rho_i}\right) + \frac{\|T\|_{l_1}}{n^{1-\alpha}}\left(1 + \frac{\rho_i}{n+\rho_i}\right). \tag{28.52}$$

We have found that

$$\left\| t + \frac{k+\lambda_i}{n+\rho_i} - x \right\|_{l_1} \leq$$

$$\left(\frac{\rho_i \|x\|_{l_1} + d(\lambda_i + 1)}{n+\rho_i}\right) + \frac{\|T\|_{l_1}}{n^{1-\alpha}}\left(1 + \frac{\rho_i}{n+\rho_i}\right). \tag{28.53}$$

(III) We observe that

$$\left\| \frac{k}{n} + \frac{i}{n\overline{r}} - x \right\|_{l_1} \leq \left\| \frac{k}{n} - x \right\|_{l_1} + \frac{1}{n}\left\| \frac{i}{\overline{r}} \right\|_{l_1} \leq \frac{\|T\|_{l_1}}{n^{1-\alpha}} + \frac{d}{n}. \tag{28.54}$$

Convergence results follow.

Theorem 28.29 *Let $f \in C_{FR}^{U_q}(\mathbb{R}^d)$, $1 \leq q < +\infty$, $x \in \mathbb{R}^d$ and $n \in \mathbb{N}$ such that $n \geq \max_{j \in \{1,\dots,d\}} \left(T_j + |x_j|, T_j^{-\frac{1}{\alpha}}\right)$, $0 < \alpha < 1$, $T_j > 0$. Then*

$$\left(\int_X D^q\left(\left(H_n^{\mathcal{F}R}(f)\right)(x,s), f(x,s)\right) P(ds)\right)^{\frac{1}{q}} \leq$$

$$\sum_{i=1}^{\overline{r}} w_i \Omega_1^{(\mathcal{F})}\left(f, \left(\frac{\|\nu_i\|_{l_2}\|x\|_{l_2} + \|\mu_i\|_{l_1}}{n+\nu_i^{\min}}\right) + \frac{1}{n^{1-\alpha}}\left(\|T\|_{l_1} + \frac{\|\nu_i\|_{l_2}\|T\|_{l_2}}{(n+\nu_i^{\min})}\right)\right)_{L^q} = \tag{28.55}$$

$$\sum_{i_1=1}^{r_1} \cdots \sum_{i_d=1}^{r_d} w_{i_1,\dots,i_d} \cdot$$

$$\Omega_1^{(\mathcal{F})}\left(f, \left(\frac{\|\nu_i\|_{l_2}\|x\|_{l_2} + \|\mu_i\|_{l_1}}{n+\nu_i^{\min}}\right) + \frac{1}{n^{1-\alpha}}\left(\|T\|_{l_1} + \frac{\|\nu_i\|_{l_2}\|T\|_{l_2}}{(n+\nu_i^{\min})}\right)\right)_{L^q}, \tag{28.56}$$

where $i = (i_1, \dots, i_d)$.

As $n \to \infty$, we get that

$$\left(H_n^{\mathcal{F}R}(f)\right)(x,s) \overset{\text{“}q\text{-mean”}}{\rightarrow} f(x,s)$$

with rates at the speed $\frac{1}{n^{1-\alpha}}$.

Proof We may write

$$\left(H_n^{\mathcal{F}R}(f)\right)(x,s) = \sum_{k=\lceil nx-Tn^{\alpha}\rceil}^{[nx+Tn^{\alpha}]*} \left(\sum_{i=1}^{\overline{r}*} w_i \odot f\left(\frac{k+\mu_i}{n+\nu_i}, s\right)\right) \odot \frac{b\left(n^{1-\alpha}\left(x-\frac{k}{n}\right)\right)}{V(x)}. \tag{28.57}$$

We observe that

$$D\left(\left(H_n^{\mathcal{F}R}(f)\right)(x,s), f(x,s)\right) =$$

$$D\left(\sum_{k=\lceil nx-Tn^{\alpha}\rceil}^{[nx+Tn^{\alpha}]*} \left(\sum_{i=1}^{\overline{r}*} w_i \odot f\left(\frac{k+\mu_i}{n+\nu_i}, s\right)\right) \odot \frac{b\left(n^{1-\alpha}\left(x-\frac{k}{n}\right)\right)}{V(x)}, f(x,s)\odot 1\right) = \tag{28.58}$$

$$D\left(\sum_{k=\lceil nx-Tn^{\alpha}\rceil}^{[nx+Tn^{\alpha}]*} \left(\sum_{i=1}^{\overline{r}*} w_i \odot f\left(\frac{k+\mu_i}{n+\nu_i}, s\right)\right) \odot \frac{b\left(n^{1-\alpha}\left(x-\frac{k}{n}\right)\right)}{V(x)},\right.$$

$$\left.\left(\sum_{i=1}^{\overline{r}*} w_i \odot f(x,s)\right) \odot \frac{V(x)}{V(x)}\right) = \tag{28.59}$$

$$D\left(\sum_{k=\lceil nx-Tn^{\alpha}\rceil}^{[nx+Tn^{\alpha}]*} \left(\sum_{i=1}^{\overline{r}*} w_i \odot f\left(\frac{k+\mu_i}{n+\nu_i}, s\right)\right) \odot \frac{b\left(n^{1-\alpha}\left(x-\frac{k}{n}\right)\right)}{V(x)},\right.$$

$$\left.\sum_{k=\lceil nx-Tn^{\alpha}\rceil}^{[nx+Tn^{\alpha}]*} \left(\sum_{i=1}^{\overline{r}*} w_i \odot f(x,s)\right) \odot \frac{b\left(n^{1-\alpha}\left(x-\frac{k}{n}\right)\right)}{V(x)}\right) \overset{(28.1)}{\leq}$$

$$\sum_{k=\lceil nx-Tn^{\alpha}\rceil}^{[nx+Tn^{\alpha}]} \frac{b\left(n^{1-\alpha}\left(x-\frac{k}{n}\right)\right)}{V(x)} \left(\sum_{i=1}^{\overline{r}} w_i D\left(f\left(\frac{k+\mu_i}{n+\nu_i}, s\right), f(x,s)\right)\right). \tag{28.60}$$

Thus it holds so far that

$$D\left(\left(H_n^{\mathcal{F}R}(f)\right)(x,s), f(x,s)\right) \leq$$

$$\sum_{k=\lceil nx-Tn^{\alpha}\rceil}^{[nx+Tn^{\alpha}]} \frac{b\left(n^{1-\alpha}\left(x-\frac{k}{n}\right)\right)}{V(x)} \left(\sum_{i=1}^{\overline{r}} w_i D\left(f\left(\frac{k+\mu_i}{n+\nu_i}, s\right), f(x,s)\right)\right). \tag{28.61}$$

Hence

$$\left(\int_X D^q\left(\left(H_n^{\mathcal{F}R}(f)\right)(x,s), f(x,s)\right)P(ds)\right)^{\frac{1}{q}} \leq$$

$$\sum_{k=\lceil nx-Tn^{\alpha}\rceil}^{[nx+Tn^{\alpha}]} \frac{b\left(n^{1-\alpha}\left(x-\frac{k}{n}\right)\right)}{V(x)}.$$

$$\left(\sum_{i=1}^{\overline{r}} w_i \left(\int_X D^q\left(f\left(\frac{k+\mu_i}{n+\nu_i}, s\right), f(x,s)\right)P(ds)\right)^{\frac{1}{q}}\right) \leq \tag{28.62}$$

$$\sum_{k=\lceil nx-Tn^{\alpha}\rceil}^{[nx+Tn^{\alpha}]} \frac{b\left(n^{1-\alpha}\left(x-\frac{k}{n}\right)\right)}{V(x)} \left(\sum_{i=1}^{\overline{r}} w_i \Omega_1^{(\mathcal{F})}\left(f, \left\|\frac{k+\mu_i}{n+\nu_i} - x\right\|_{l_1}\right)_{L^q}\right) \overset{(28.46)}{\leq}$$

(28.63)

$$\sum_{i=1}^{\overline{r}} w_i \Omega_1^{(\mathcal{F})}\left(f, \left(\frac{\|\nu_i\|_{l_2}\|x\|_{l_2} + \|\mu_i\|_{l_1}}{n+\nu_i^{\min}}\right) + \frac{1}{n^{1-\alpha}}\left(\|T\|_{l_1} + \frac{\|\nu_i\|_{l_2}\|T\|_{l_2}}{\left(n+\nu_i^{\min}\right)}\right)\right)_{L^q},$$

proving the claim. ■

Corollary 28.30 (to Theorem 28.29) *Let* $x \in \prod_{j=1}^{d}[-\gamma_j, \gamma_j] \subset \mathbb{R}^d$, $\gamma_j > 0$, $\gamma := (\gamma_1, \ldots, \gamma_d)$ *and* $n \in \mathbb{N}$ *such that* $n \geq \max_{j\in\{1,\ldots,d\}}\{T_j + \gamma_j, T_j^{-\frac{1}{\alpha}}\}$. *Then*

$$\left\|\left(\int_X D^q\left(\left(H_n^{\mathcal{F}R}(f)\right)(x,s), f(x,s)\right)P(ds)\right)^{\frac{1}{q}}\right\|_{\infty, \prod_{j=1}^{d}[-\gamma_j,\gamma_j]} \leq$$

$$\sum_{i=1}^{\overline{r}} w_i \Omega_1^{(\mathcal{F})}\left(f, \left(\frac{\|\nu_i\|_{l_2}\|\gamma\|_{l_2} + \|\mu_i\|_{l_1}}{n+\nu_i^{\min}}\right) + \frac{1}{n^{1-\alpha}}\left(\|T\|_{l_1} + \frac{\|\nu_i\|_{l_2}\|T\|_{l_2}}{\left(n+\nu_i^{\min}\right)}\right)\right)_{L^q}. \tag{28.64}$$

We continue with

Theorem 28.31 *All assumptions as in Theorem 28.29. Additionally assume that* $f(t,s)$ *is fuzzy continuous in* $t \in \mathbb{R}^d$, *for each* $s \in X$. *Then*

$$\left(\int_X D^q\left(\left(K_n^{\mathcal{F}R}(f)\right)(x,s), f(x,s)\right)P(ds)\right)^{\frac{1}{q}} \leq$$

$$\sum_{i=1}^{\overline{r}} w_i \Omega_1^{(\mathcal{F})}\left(f, \left(\frac{\rho_i \|x\|_{l_1} + d(\lambda_i+1)}{n+\rho_i}\right) + \frac{\|T\|_{l_1}}{n^{1-\alpha}}\left(1+\frac{\rho_i}{n+\rho_i}\right)\right)_{L^q} = \tag{28.65}$$

$$\sum_{i_1=1}^{r_1} \cdots \sum_{i_d=1}^{r_d} w_{i_1,\ldots,i_d} \Omega_1^{(\mathcal{F})}\left(f, \left(\frac{\rho_i \|x\|_{l_1} + d(\lambda_i+1)}{n+\rho_i}\right) + \frac{\|T\|_{l_1}}{n^{1-\alpha}}\left(1+\frac{\rho_i}{n+\rho_i}\right)\right)_{L^q}.$$

As $n \to \infty$, we get

$$\left(K_n^{\mathcal{F}R}(f)\right)(x,s) \overset{\text{"q-mean"}}{\to} f(x,s)$$

with rates at the speed $\frac{1}{n^{1-\alpha}}$.

Proof By Remark 28.5 the function $F(t,s) := D\left(f\left(t+\frac{k+\lambda_i}{n+\rho_i}, s\right), f(x,s)\right)$ is $\mathcal{B}$-measurable over X with respect to s. Also F is continuous in $t \in \mathbb{R}^d$, similar reasoning as in explanation of Remark 28.5, for each $s \in X$.

Thus for $F(t,s)$, by joint measurability theorem of Caratheodory, see [1], p. 156, we get that $F(t,s)$ is jointly measurable over $\mathbb{R}^d \times X$. Hence in what follow we can apply Fubini's theorem.

We may rewrite

$$\left(K_n^{\mathcal{F}R}(f)\right)(x,s) = \sum_{k=\lceil nx-Tn^{\alpha}\rceil}^{[nx+Tn^{\alpha}]*}\left(\sum_{i=1}^{\overline{r}*} w_i (n+\rho_i)^d \odot \int_0^{\frac{1}{n+\rho_i}} f\left(t+\frac{k+\lambda_i}{n+\rho_i}, s\right)dt\right)$$

$$\odot \frac{b\left(n^{1-\alpha}\left(x-\frac{k}{n}\right)\right)}{V(x)}. \tag{28.66}$$

We observe that

$$D\left(\left(K_n^{\mathcal{F}R}(f)\right)(x,s), f(x,s)\right) =$$

$$D\left(\sum_{k=\lceil nx-Tn^{\alpha}\rceil}^{[nx+Tn^{\alpha}]*}\left(\sum_{i=1}^{\overline{r}*} w_i (n+\rho_i)^d \odot \int_0^{\frac{1}{n+\rho_i}} f\left(t+\frac{k+\lambda_i}{n+\rho_i}, s\right)dt\right)\right.$$

$$\left.\odot \frac{b\left(n^{1-\alpha}\left(x-\frac{k}{n}\right)\right)}{V(x)},\right.$$

$$\sum_{k=\lceil nx-Tn^{\alpha}\rceil}^{[nx+Tn^{\alpha}]*}\left(\left(\sum_{i=1}^{\overline{r}*} w_i\,(n+\rho_i)^d \odot \int_0^{\frac{1}{n+\rho_i}} f(x,s)\,dt\right)\odot \frac{b\left(n^{1-\alpha}\left(x-\frac{k}{n}\right)\right)}{V(x)}\right) \overset{(28.1)}{\leq} \tag{28.67}$$

$$\sum_{k=\lceil nx-Tn^{\alpha}\rceil}^{[nx+Tn^{\alpha}]} \frac{b\left(n^{1-\alpha}\left(x-\frac{k}{n}\right)\right)}{V(x)}\cdot$$

$$\left(\sum_{i=1}^{\overline{r}*} w_i\,(n+\rho_i)^d\, D\left(\int_0^{\frac{1}{n+\rho_i}} f\left(t+\frac{k+\lambda_i}{n+\rho_i},s\right)dt, \int_0^{\frac{1}{n+\rho_i}} f(x,s)\,dt\right)\right) \overset{(28.21)}{\leq}$$

$$\sum_{k=\lceil nx-Tn^{\alpha}\rceil}^{[nx+Tn^{\alpha}]} \frac{b\left(n^{1-\alpha}\left(x-\frac{k}{n}\right)\right)}{V(x)}\cdot$$

$$\left(\sum_{i=1}^{\overline{r}} w_i\,(n+\rho_i)^d \int_0^{\frac{1}{n+\rho_i}} D\left(f\left(t+\frac{k+\lambda_i}{n+\rho_i},s\right), f(x,s)\right)dt\right). \tag{28.68}$$

That is it holds

$$D\left(\left(K_n^{\mathcal{F}R}(f)\right)(x,s), f(x,s)\right) \leq \sum_{k=\lceil nx-Tn^{\alpha}\rceil}^{[nx+Tn^{\alpha}]} \frac{b\left(n^{1-\alpha}\left(x-\frac{k}{n}\right)\right)}{V(x)}\cdot$$

$$\left(\sum_{i=1}^{\overline{r}} w_i\,(n+\rho_i)^d \int_0^{\frac{1}{n+\rho_i}} D\left(f\left(t+\frac{k+\lambda_i}{n+\rho_i},s\right), f(x,s)\right)dt\right). \tag{28.69}$$

Hence (let $p, q > 1 : \frac{1}{p}+\frac{1}{q}=1$)

$$\left(\int_X D^q\left(\left(K_n^{\mathcal{F}R}(f)\right)(x,s), f(x,s)\right)P(ds)\right)^{\frac{1}{q}} \leq \sum_{k=\lceil nx-Tn^{\alpha}\rceil}^{[nx+Tn^{\alpha}]} \frac{b\left(n^{1-\alpha}\left(x-\frac{k}{n}\right)\right)}{V(x)}\cdot$$

$$\left(\sum_{i=1}^{\overline{r}} w_i\,(n+\rho_i)^d \left(\int_X\left(\int_0^{\frac{1}{n+\rho_i}} D\left(f\left(t+\frac{k+\lambda_i}{n+\rho_i},s\right), f(x,s)\right)dt\right)^q P(ds)\right)^{\frac{1}{q}}\right) \tag{28.70}$$

(by Hölder's inequality)

$$\leq \sum_{k=\lceil nx-Tn^{\alpha}\rceil}^{[nx+Tn^{\alpha}]} \frac{b\left(n^{1-\alpha}\left(x-\frac{k}{n}\right)\right)}{V(x)}\left(\sum_{i=1}^{\overline{r}} w_i\,(n+\rho_i)^d\right.$$

$$\left(\int_X \frac{1}{(n+\rho_i)^{\frac{dq}{p}}} \left(\int_0^{\frac{1}{n+\rho_i}} D^q \left(f\left(t+\frac{k+\lambda_i}{n+\rho_i}, s\right), f(x,s)\right) dt\right) P(ds)\right)^{\frac{1}{q}}\Bigg)$$

$$\overset{\text{(by Fubini's theorem)}}{=} \sum_{k=\lceil nx-Tn^{\alpha}\rceil}^{[nx+Tn^{\alpha}]} \frac{b\left(n^{1-\alpha}\left(x-\frac{k}{n}\right)\right)}{V(x)} \left(\sum_{i=1}^{\overline{r}} w_i \frac{(n+\rho_i)^d}{(n+\rho_i)^{\frac{d}{p}}}\right.$$

$$\left.\left(\int_0^{\frac{1}{n+\rho_i}} \left(\int_X D^q \left(f\left(t+\frac{k+\lambda_i}{n+\rho_i}, s\right), f(x,s)\right) P(ds)\right) dt\right)^{\frac{1}{q}}\right) \overset{(28.8)}{\leq}$$

$$\sum_{k=\lceil nx-Tn^{\alpha}\rceil}^{[nx+Tn^{\alpha}]} \frac{b\left(n^{1-\alpha}\left(x-\frac{k}{n}\right)\right)}{V(x)} \left(\sum_{i=1}^{\overline{r}} w_i \frac{(n+\rho_i)^d}{(n+\rho_i)^{\frac{d}{p}}}\right.$$

$$\left.\left(\int_0^{\frac{1}{n+\rho_i}} \Omega_1^{(\mathcal{F})}\left(f, \left\|t+\frac{k+\lambda_i}{n+\rho_i}-x\right\|_{l_1}\right)_{L^q}^q dt\right)^{\frac{1}{q}}\right) \overset{(28.53)}{\leq} \qquad (28.71)$$

$$\sum_{k=\lceil nx-Tn^{\alpha}\rceil}^{[nx+Tn^{\alpha}]} \frac{b\left(n^{1-\alpha}\left(x-\frac{k}{n}\right)\right)}{V(x)} \left(\sum_{i=1}^{\overline{r}} w_i \frac{(n+\rho_i)^d}{(n+\rho_i)^{\frac{d}{p}}}\right.$$

$$\left.\frac{1}{(n+\rho_i)^{\frac{d}{p}}} \Omega_1^{(\mathcal{F})}\left(f, \left(\frac{\rho_i \|x\|_{l_1} + d(\lambda_i+1)}{n+\rho_i}\right) + \frac{\|T\|_{l_1}}{n^{1-\alpha}}\left(1+\frac{\rho_i}{n+\rho_i}\right)\right)_{L^q}\right) =$$

$$\sum_{i=1}^{\overline{r}} w_i \Omega_1^{(\mathcal{F})}\left(f, \left(\frac{\rho_i \|x\|_{l_1} + d(\lambda_i+1)}{n+\rho_i}\right) + \frac{\|T\|_{l_1}}{n^{1-\alpha}}\left(1+\frac{\rho_i}{n+\rho_i}\right)\right)_{L^q}, \qquad (28.72)$$

proving the claim. ■

The case of $q = 1$ goes through in a similar and simpler way.

Corollary 28.32 (to Theorem 28.31) *Let* $x \in \prod_{j=1}^{d} [-\gamma_j, \gamma_j] \subset \mathbb{R}^d$, $\gamma_j > 0$, $\gamma := (\gamma_1, \ldots, \gamma_d)$ *and* $n \in \mathbb{N}$ *such that* $n \geq \max_{j\in\{1,\ldots,d\}} \{T_j + \gamma_j, T_j^{-\frac{1}{\alpha}}\}$. *Then*

$$\left\| \left(\int_X D^q \left(\left(K_n^{\mathcal{F}R}(f)\right)(x,s), f(x,s)\right) P(ds)\right)^{\frac{1}{q}} \right\|_{\infty, \prod_{j=1}^{d}[-\gamma_j,\gamma_j]} \leq \qquad (28.73)$$

$$\sum_{i=1}^{\overline{r}} w_i \Omega_1^{(\mathcal{F})} \left(f, \left(\frac{\rho_i \|\gamma\|_{l_1} + d(\lambda_i + 1)}{n + \rho_i} \right) + \frac{\|T\|_{l_1}}{n^{1-\alpha}} \left(1 + \frac{\rho_i}{n + \rho_i} \right) \right)_{L^q}.$$

We finish with

Theorem 28.33 *All as in Theorem 28.29. Then*

$$\left(\int_X D^q \left(\left(M_n^{\mathcal{F}R}(f) \right)(x,s), f(x,s) \right) P(ds) \right)^{\frac{1}{q}} \leq$$

$$\Omega_1^{(\mathcal{F})} \left(f, \frac{\|T\|_{l_1}}{n^{1-\alpha}} + \frac{d}{n} \right)_{L^q}. \tag{28.74}$$

As $n \to \infty$, we get that

$$\left(M_n^{\mathcal{F}R}(f) \right)(x,s) \overset{\text{"q-mean"}}{\to} f(x,s)$$

with rates at the speed $\frac{1}{n^{1-\alpha}}$.

Proof We may rewrite

$$\left(M_n^{\mathcal{F}R}(f) \right)(x,s) = \sum_{k=\lceil nx - Tn^{\alpha} \rceil}^{[nx+Tn^{\alpha}]*} \left(\sum_{i=1}^{\overline{r}*} w_i \odot f\left(\frac{k}{n} + \frac{i}{n\overline{r}}, s \right) \right) \odot \frac{b\left(n^{1-\alpha}\left(x - \frac{k}{n}\right)\right)}{V(x)}. \tag{28.75}$$

We observe that

$$D\left(\left(M_n^{\mathcal{F}R}(f) \right)(x,s), f(x,s) \right) =$$

$$D\left(\sum_{k=\lceil nx - Tn^{\alpha} \rceil}^{[nx+Tn^{\alpha}]*} \left(\sum_{i=1}^{\overline{r}*} w_i \odot f\left(\frac{k}{n} + \frac{i}{n\overline{r}}, s \right) \right) \odot \frac{b\left(n^{1-\alpha}\left(x - \frac{k}{n}\right)\right)}{V(x)}, \right.$$

$$\left. \sum_{k=\lceil nx - Tn^{\alpha} \rceil}^{[nx+Tn^{\alpha}]*} \left(\sum_{i=1}^{\overline{r}*} w_i \odot f(x,s) \right) \odot \frac{b\left(n^{1-\alpha}\left(x - \frac{k}{n}\right)\right)}{V(x)} \right) \overset{(28.1)}{\leq}$$

$$\sum_{k=\lceil nx - Tn^{\alpha} \rceil}^{[nx+Tn^{\alpha}]} \frac{b\left(n^{1-\alpha}\left(x - \frac{k}{n}\right)\right)}{V(x)} \left(\sum_{i=1}^{\overline{r}} w_i D\left(f\left(\frac{k}{n} + \frac{i}{n\overline{r}}, s \right), f(x,s) \right) \right). \tag{28.76}$$

Hence it holds

$$\left(\int_X D^q\left(\left(M_n^{\mathcal{F}R}(f)\right)(x,s), f(x,s)\right) P(ds)\right)^{\frac{1}{q}} \leq$$

$$\sum_{k=\lceil nx-Tn^{\alpha}\rceil}^{[nx+Tn^{\alpha}]} \frac{b\left(n^{1-\alpha}\left(x-\frac{k}{n}\right)\right)}{V(x)}$$

$$\left(\sum_{i=1}^{\overline{r}} w_i \left(\int_X D^q\left(f\left(\frac{k}{n}+\frac{i}{n\overline{r}}, s\right), f(x,s)\right) P(ds)\right)^{\frac{1}{q}}\right) \leq \tag{28.77}$$

$$\sum_{k=\lceil nx-Tn^{\alpha}\rceil}^{[nx+Tn^{\alpha}]} \frac{b\left(n^{1-\alpha}\left(x-\frac{k}{n}\right)\right)}{V(x)} \left(\sum_{i=1}^{\overline{r}} w_i \Omega_1^{(\mathcal{F})}\left(f, \left\|\frac{k}{n}+\frac{i}{n\overline{r}}-x\right\|_{l_1}\right)_{L^q}\right) \overset{(28.54)}{\leq}$$

$$\Omega_1^{(\mathcal{F})}\left(f, \frac{\|T\|_{l_1}}{n^{1-\alpha}}+\frac{d}{n}\right)_{L^q}, \tag{28.78}$$

proving the claim. ■

Corollary 28.34 (to Theorem 28.33) *Here all as in Corollary 28.30. Then*

$$\left\|\left(\int_X D^q\left(\left(M_n^{\mathcal{F}R}(f)\right)(x,s), f(x,s)\right) P(ds)\right)^{\frac{1}{q}}\right\|_{\infty, \prod_{j=1}^{d}[-\gamma_j,\gamma_j]} \leq$$

$$\Omega_1^{(\mathcal{F})}\left(f, \frac{\|T\|_{l_1}}{n^{1-\alpha}}+\frac{d}{n}\right)_{L^q}. \tag{28.79}$$

Comment 28.35 *All the convergence fuzzy random results of this chapter can have real analogous results for real valued random functions (stochastic processes) using* $\Omega_1(f,\delta)_{L^q}$, *see (28.10).*

References

1. C.D. Aliprantis, O. Burkinshaw, *Principles of Real Analysis*, 3rd edn. (Academic Press Inc, San Diego, 1998)
2. G.A. Anastassiou, Rate of convergence of some neural network operators to the unit-univariate case. J. Math. Anal. Appl. **212**, 237–262 (1997)
3. G.A. Anastassiou, *Quantitative Approximation* (Chapman & Hall/CRC, Boca Raton, 2001)
4. G.A. Anastassiou, Rate of convergence of fuzzy neural network operators, univariate case. J. Fuzzy Math. **10**(3), 755–780 (2002)

5. G.A. Anastassiou, S. Gal, On a fuzzy trigonometric approximation theorem of Weierstrass-type. J. Fuzzy Math. **9**(3), 701–708 (2001)
6. G.A. Anastassiou, in *Univariate Fuzzy-Random Neural Network Approximation Operators*, ed. by G. Anastassiou of special session Computational Methods in Analysis. Computers and Mathematics with Applications, Special issue/Proceedings, vol. 48 (AMS meeting in Orlando, Florida, 2004), pp. 1263–1283, Nov 2002
7. G.A. Anastassiou, Multivariate stochastic Korovkin theory given quantitatively. Math. Comput. Model. **48**, 558–580 (2008)
8. G.A. Anastassiou, *Fuzzy Mathematics: Approximation Theory* (Springer, New York, 2010)
9. G.A. Anastassiou, Multivariate fuzzy-random normalized neural network approximation operators. Ann. Fuzzy Math. Inform. **6**(1), 191–212 (2013)
10. G.A. Anastassiou, Multivariate fuzzy-random perturbed neural network approximation. J. Concr. Appl. Math. **13**(3–4), 279–300 (2015)
11. G.A. Anastassiou, Higher order multivariate fuzzy approximation by basic neural network operators. Cubo **16**(03), 21–35 (2014)
12. G.A. Anastassiou, *Approximation by Multivariate Perturbed Neural Network Operators*. Submitted (2014)
13. P. Cardaliaguet, G. Euvrard, Approximation of a function and its derivative with a neural network. Neural Netw. **5**, 207–220 (1992)
14. R.M. Dudley, *Real Analysis and Probability* (Wadsworth & Brooks/Cole Mathematics Series, Pacific Grove, 1989)
15. S. Gal, in *Approximation Theory in Fuzzy Setting*, ed. by G. Anastassiou. Handbook of Analytic-Computational Methods in Applied Mathematics, Chap. 13, pp. 617–666 (Chapman & Hall/CRC, Boca Raton, 2000)
16. Y.K. Kim, B.M. Ghil, Integrals of fuzzy-number-valued functions. Fuzzy Sets Syst. **86**, 213–222 (1997)
17. W. Congxin, G. Zengtai, On Henstock integrals of fuzzy number valued functions (I). Fuzzy Sets Syst. **115**(3), 377–391 (2000)

Index

G.A. Anastassiou, *Intelligent Systems II: Complete Approximation by Neural Network Operators*, Studies in Computational Intelligence 608,
DOI 10.1007/978-3-319-20505-2

Printed in the United States
By Bookmasters

Printed in the United States
By Bookmasters